AF376238

E. Nieschlag H.M. Behre (Eds.) **Testosterone**

Springer

Berlin
Heidelberg
New York
Barcelona
Budapest
Hong Kong
London
Milan
Paris
Santa Clara
Singapore
Tokyo

E. Nieschlag H. M. Behre (Eds.)

Testosterone

Action – Deficiency – Substitution

Second Edition

Editorial Assistant: S. Nieschlag

With Contributions by
L. E. Atkinson H. M. Behre S. Bhasin R. Bross
R. Casaburi Y.-L. Chang K. Christiansen
B. Couzinet A. von Eckardstein J. S. Finkelstein
J. Frick L. J. Gooren D. J. Handelsman A. Jungwirth
J. M. Kaufman P. Y. Liu W. A. Meikle E. Nieschlag
C. A. Quigley V. A. Randall F. F. G. Rommerts
E. Rovan G. Schaison W. Schänzer P. J. Snyder
T. W. Storer A. Vermeulen G. F. Weinbauer

With 95 Figures and 28 Tables

Springer

Prof. Dr. med. Eberhard Nieschlag
Priv.-Doz. Dr. med. Hermann M. Behre
Institute of Reproductive Medicine, University of Münster
Domagkstrasse 11, D-48129 Münster, Germany

Editorial Assistant: *Susan Nieschlag, MA*

ISBN-13:978-3-642-72187-8   e-ISBN-13:978-3-642-72185-4
DOI: 10.1007/978-3-642-72185-4

Library of Congress Cataloging-in Publication Data
Testosterone: action, deficiency, substitution / E. Nieschlag, H.M. Behre (eds.); editorial
assistant, S. Nieschlag; with contributions by L.E. Atkinson ... [et al.]. – 2nd ed. Includes
bibliographical references and index. ISBN-13:978-3-642-72187-8
1.Testosterone – Therapeutic use. 2. Testosterone – Physiological effect. I. Nieschlag. E.
II. Behre, H.M. (Hermann M.), 1961– . RM296.5.T47T47 1998 615'.366–dc21

Production: PRO EDIT GmbH, Heidelberg
Illustrations: G. Hippmann, Nürnberg
Typesetting (Data conversion): K+V Fotosatz GmbH, Beerfelden
Cover-Design: E. Kirchner, Heidelberg

SPIN 10566757   13/3135-5 4 3 2 1 0 – Printed on acid-free paper

Preface

The first edition of "Testosterone: Action, Deficiency, Substitution" was published in 1990. Since then our understanding of the hormone that turns males into men has tremendously increased. Therefore, the editors felt that a second extended edition of the book is warranted in order to summarize established and recent findings in the field and to present the reader with an up-to-date textbook. The increased mass of knowledge is reflected by the growth of the volume from 14 to 20 chapters.

In the updated edition the biochemistry and metabolism of androgens have been complemented by extensive information on the molecular biology of the androgen receptor and its disorders. The key role of testosterone in spermatogenesis is now better defined. We have a more complete understanding of the psychotropic effects of testosterone and know so much about the different target organs and functions that individual chapters deal with testosterone and the prostate, lipids and the cardiovascular system, hair, bones and muscles. The general chapter on pharmacology and clinical uses of testosterone, in particular in male hypogonadism, is extended by pharmacokinetic studies on testosterone preparations and individual substitution modalities using testosterone esters as well as implants and advanced transdermal applications. The physiologic basis and possible clinical applications of testosterone in non-gonadal diseases, in male senescence, in hormonal male contraception and in transsexuals are discussed. The last chapter describes the role of "investigative" steroid biochemistry applied to tracking anabolic steroid abuse.

In order to synchronize the writing of the various chapters the authors met on January 22–25, 1998 at Castle Elmau, Bavaria, for final editing of their previously submitted chapters. This guaranteed that all chapters were concluded simultaneously and reflect precisely the state of current knowledge. The contributors deserve our appreciation for their prompt and reliable cooperation in this venture. We are grateful for a generous grant from SmithKline Beecham Pharma (Munich) facilitating the meeting of the authors in combination with a Symposion open to professionals interested in the topic. The splendid isolation of Castle Elmau high up in the Bavarian Alps provided a congenial atmosphere for a productive meeting.

Much of the editorial work was performed by Susan Nieschlag M.A. whose untiring professional help is gratefully acknowledged. Bärbel Bahnes, Kerstin Neuhaus and Angelika Schick helped with the word processing of the manuscripts. Finally we would like to thank Dr. Carol Bacchus and the Springer-Verlag for supporting the project and producing the book.

Münster/Schloß Elmau, April 1998

Eberhard Nieschlag
Hermann M. Behre

Table of Contents

List of Contributors

Atkinson, L. E.
ALZA Corporation,
950 Page Mill Road,
Palo Alto, CA 94394-0802, USA

Behre, H. M.
Institute of Reproductive Medicine,
University of Münster,
Domagkstr. 11, 48129 Münster,
Germany

Bhasin, S.
Division of Endocrinology,
Metabolism and Molecular Medicine,
Charles R. Drew University
of Medicine and Science,
1621 E 120th Street, MP-02,
Los Angeles, CA 90059, USA

Bross, R.
Division of Endocrinology,
Metabolism and Molecular Medicine,
Charles R. Drew University
of Medicine and Science,
1621 E 120th Street, MP-02,
Los Angeles, CA 90059, USA

Casaburi, R.
Division of Respiratory Medicine,
Pulmonary Physiology, and Critical
Care Medicine,
Harbor-UCLA Medical Center,
Torrance CA 90502, USA

Chang, Y.-L.
ALZA Corporation,
950 Page Mill Road,
Palo Alto, CA 94304-0802, USA

Christiansen, K.
Institute for Human Biology
of the University,
Allende-Platz 2,
20146 Hamburg, Germany

Couzinet, B.
Department of Endocrinology
and Reproductive Medicine,
Centre Hospitalier de Bicetre,
78, rue du General Leclerc,
94275 Kremlin-Bicetre, France

von Eckardstein, A.
Institute of Clinical Chemistry
and Laboratory Medicine, Central
Laboratory, Albert-Schweitzer-Str. 33,
48129 Münster, Germany

Finkelstein, J. S.
Endocrinology Unit, Massachusetts
General Hospital, Bulfinch 327 –
Fruit Street, Boston, MA 02114, USA

Frick, J.
Department of Urology,
Hospital Salzburg, General Hospital,
Müllner Hauptstr. 48,
5020 Salzburg, Austria

Gooren, L. J.
Dept. of Endocrinology,
Free University Hospital,
P. O. Box 7057,
1007 MB Amsterdam,
The Netherlands

Handelsman, D. J.
Andrology Unit, Department of Medicine and Obstetrics/Gynecology
(D 02), University of Sydney,
Sydney 2006, NSW, Australia

Jungwirth, A.
Department of Urology,
Salzburg General Hospital,
Müllner Hauptstr. 48,
5020 Salzburg, Austria

Kaufman, J. M.
Department of Endocrinology,
Academish Ziekenhuis,
185 de Pintelaan,
9000 Gent, Belgium

Liu, P. Y.
Andrology Unit, Department of
Medicine and Obstetrics/Gynecology
(D 02), University of Sydney,
Sydney 2006, NSW, Australia

Meikle, W. A.
Division of Endocrinology,
University of Utah Medical School,
50 North Medical Drive,
Salt Lake City, UT 84132-0001, USA

Nieschlag, E.
Institute of Reproductive Medicine,
University of Münster,
Domagkstr. 11, 48129 Münster,
Germany

Quigley, C. A.
Department of Pediatrics,
Indiana University Medical Center,
702 Barnhill Drive,
Indianapolis, Indiana 46202-5225,
USA

Randall, V. A.
Department of Biomedical Sciences,
Richmond Building, University of
Bradford, Richmond Road,
Bradford BD7 1DP, United Kingdom

Rommerts, F. F. G.
Department of Endocrinology
and Reproduction,
Erasmus University Rotterdam,
P. O. Box 1738, 3000 DR Rotterdam,
The Netherlands

Rovan, E.
Department Urology,
Salzburg General Hospital,
Müllner Hauptstr. 48,
5020 Salzburg, Austria

Schaison, G.
Department of Endocrinology
and Reproductive Medicine,
Centre Hospitalier de Bicetre,
78, rue du General Leclerc,
94275 Kremlin-Bicetre, France

Schänzer, W.
Institute of Biochemistry,
German Sports University Köln,
Carl-Diem-Weg 6,
50933 Köln, Germany

Snyder, P. J.
Department of Medicine,
University of Pennsylvania,
520 Stemmler Hall,
3450 Hamilton Walk,
Philadelphia, PA 19104-6149, USA

Storer, T. W.
Laboratory of Exercise Science,
El Camino College,
Torrance, CA 90502, USA

Vermeulen, A.
Maaltemeers 33,
9052 Ghent-Zwijnaarde, Belgium

Weinbauer, G. F.
Institute of Reproductive Medicine,
University of Münster,
Domagkstr. 11, 48129 Münster,
Germany

1 Testosterone: An overview of biosynthesis, transport, metabolism and nongenomic actions

Focko F. G. Rommerts

Contents

1.1 Introduction

In the male androgens are essential for the development and maintenance of specific reproductive tissues such as testis, prostate, epididymis, seminal vesicle and penis, as well as for other characteristic male properties such as increased muscle strength, hair growth, etc. (Mooradian et al. 1987). In order to maintain the androgen concentration at appropriate levels, the production rates of androgens must be balanced against the rates of metabolic clearance and excretion. The action of androgens in target cells depends on the amount of steroid which can penetrate into the cells, the extent of metabolic conversion within the cells, the interactions with the receptor proteins and finally, upon the action of the androgen receptors at the genomic level.

The biochemical aspects of production, metabolism, transport and action of androgens will be discussed in separate paragraphs. Where possible, data obtained from human tissues will be emphasized. This chapter will deal with only the major and general aspects. A more extensive description of these topics and more can be found in the book "The Leydig Cell" edited by AH Payne, MP Hardy and LD Russell (1996).

1.2 Biosynthetic pathways

1.2.1 General

In the human male, testosterone is the major circulating androgen. More than 95% is secreted by the testis, which produces approximately 6–7 mg per day (Coffey 1988). The metabolic steps required for the conversion of cholesterol into androgens take place in approximately 500 million Leydig cells which constitute only a few percent of the total testicular volume. Although Leydig cells are of major importance for the generation of the circulating androgenic hormones, the adrenal cortex also contributes to the production of androgens. The production of steroids is not limited to classical endocrine glands but small amounts can also be produced in brain cells (Baulieu 1997). Although the contribution of cells in the nervous system to the circulating hormone levels is very small, local production of steroids can be physiologically very important for local action. High local levels can be generated when the transport and clearance are low.

Since Leydig cells are most important for the production of androgens, the steroidogenic pathways in these cells will be described in more detail; the steroidogenic reactions which occur in other cell types show many similarities. The enzymes and intermediates involved in this reaction cascade are depicted in Figure 1.1. The pathways for biosynthesis of androgens and the regulation thereof have been reviewed extensively and the reader is referred to these reviews for detailed information (Ewing and Zirkin 1983; Hall 1988; Payne and O'Shaughnessy 1996; Rommerts and Brinkmann 1981; Rommerts and Cooke 1988; Rommerts and van der Molen 1989; Saez 1989, 1994; Stocco and Clark 1996). Additional references will be given when specific data are discussed.

The precursor in the synthesis of steroids is cholesterol. This substrate may be synthesized *de novo* from acetate but it may also be taken up from plasma lipoproteins. For human Leydig cells the low density lipoprotein fraction seems to be the predominant extracellular store of cholesterol (Freeman and Rommerts 1996). In addition, intracellular lipid droplets which contain cholesterol esters may function as intracellular stores of cholesterol. The relative contributions of synthesis and cholesterol supply from lipoproteins or lipid droplets depend on the species and the extent of stimulation of steroid

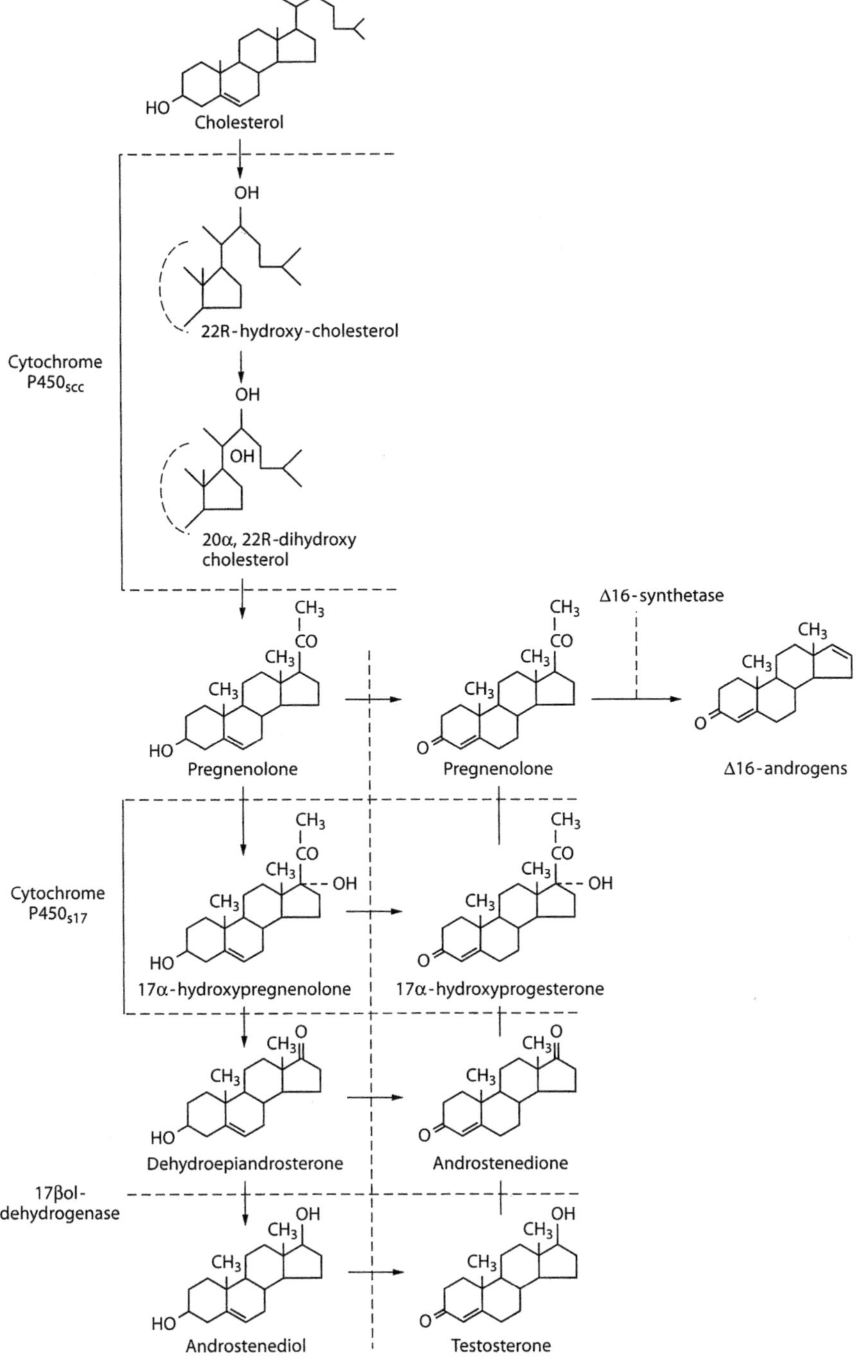

Fig. 1.1. Steroidogenic pathways in the human testis

production. In general, Leydig cells possess a large capacity for endogenous cholesterol synthesis and the capacity for storage and uptake of cholesterol via lipid droplets seems to be limited. It appears that the cholesterol in the plasma membranes acts as the main and most readily available pool of cholesterol. A vesicle-mediated transport system involving an endosomal/lysosomal network seems to act as the conveyer belt for intracellular cholesterol transport to the mitochondria. The supply of cholesterol to the outer membrane of the mitochondria also requires transfer proteins and in this process sterol carrier protein$_2$ (SCP$_2$) could play an important role (van Noort et al. 1988). This protein could facilitate cholesterol trafficking inside the cell in conjunction with the cytoskeleton and the vesicular system, but although many suggestions have been made in this direction, there is still no definite proof for this model. An important question in this respect is whether changes in intracellular cholesterol trafficking under the influence of LH (luteinizing hormone) are a consequence of utilization of cholesterol at the mitochondrial level followed by a re-equilibration process or whether LH actively directs cholesterol movement to the mitochondria.

Whatever mechanisms operate, the ultimate result of the coupled intracellular transport mechanisms is regulation of the availability of cholesterol at the level of the mitochondria for production of pregnenolone (C$_{21}$) from cholesterol (C$_{27}$). The cleavage of the side chain of cholesterol inside the mitochondria which results in the formation of pregnenolone is the start of the steroidogenic cascade. Subsequently, pregnenolone is converted to a variety of C$_{19}$-steroids by enzymes in the endoplasmic reticulum. The biosynthesis of the biologically active androgens is thus the result of a stepwise degradation of the biologically inactive pregnenolone. This process is catalyzed by oxidative enzymes, many of which are members of a group of heme-containing proteins called cytochromes P450. As can be seen in Figure 1.1, the specific steroidogenic P450 enzymes can catalyze different although related reactions. The precise pathways which are utilized for the formation of testosterone most probably depend on the properties and amounts of the various enzymes as well as on the composition of the membrane into which these steroid-converting enzymes are integrated. Under normal conditions the total capacity of the pregnenolone-converting enzyme system in humans is insufficient to convert all available pregnenolone into testosterone. As a result many intermediates in the form of progesterone derivatives leak out of the Leydig cells. This illustrates that the rate-limiting step for the production of testosterone is localized at the level of the endoplasmic reticulum, whereas the rate-determining step for steroidogenesis, regulated by luteinizing hormone, is at the level of the cholesterol side chain cleavage activity in the mitochondria (van Haren et al. 1989).

1.2.2 Steroids different from testosterone

Some specific intermediates of the steroidogenic cascade are worth mentioning when testosterone substitution is practised. In the human, as well as in the porcine testis, pregnenolone and progesterone can also be oxidized to 16-androgens which can be further metabolized to androstenone (5α-androst-16-en-3-one) and androstenol (5α-androst-16-en-3β- or 3α-ol) in sweat glands (Weusten et al. 1987 a). Although these steroids are not recognized as biologically active steroids in a classical sense, they clearly act as pheromones in pigs. Humans can also perceive these pheromones but there are less convincing data about the ultimate responses (Comfort 1971). Another specific testicular metabolite derived from progesterone is 3α-hydroxy-4-pregnen-20-one, produced by immature Sertoli cells from rats. This steroid was reported to suppress FSH secretion by the pituitary cells in a specific way (Wiebe 1997). The biological effects of these metabolites of progesterone are of interest since receptors for these metabolites have not been reported. The formation and function of other testosterone metabolites such as 17β-estradiol and 5α-dihydrotestosterone will be discussed later. Before describing the cascade of events regulated by luteinizing hormone, more details about the steroidogenic reactions in the mitochondria and endoplasmic reticulum will be given.

1.2.3 Regulation of cholesterol side chain cleavage activity

The cholesterol side chain cleavage enzyme (P450$_{scc}$) responsible for the initiation of the steroidogenic process is located in the inner membrane of the mitochondria. This inner mitochondrial membrane contains small amounts of cholesterol. The availability of cholesterol in the inner membrane is thus at least one of the rate-limiting factors for the generation of pregnenolone from cholesterol. Other factors which are of importance are the amount and activity of the P450$_{scc}$ enzyme and the capacity for delivering reducing equivalents from NADPH to the P450$_{scc}$ via flavoproteins and iron-containing proteins (see Figure 1.2).

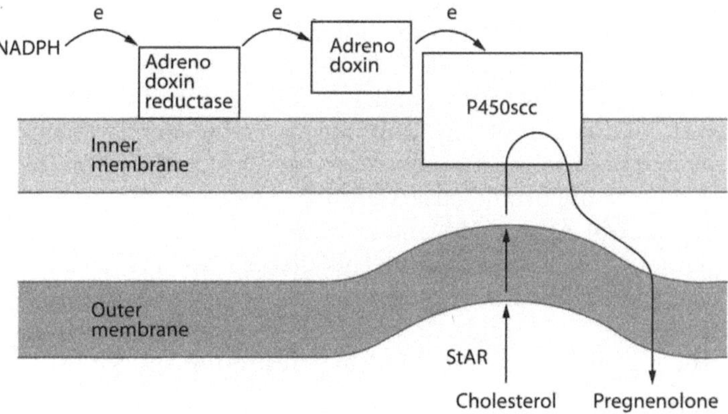

Fig. 1.2. The cholesterol side-chain cleavage system in mitochondria

The capacity of the electron transport system appears not to be rate-limiting for steroid production since more than 10-fold higher rates of pregnenolone production can be obtained in Leydig cells when, instead of cholesterol, the more soluble intermediate 22R-hydroxycholesterol is provided. The various steps in the side chain cleavage process i.e. the hydroxylations at C_{22} and C_{20} followed by cleavage of the bond between C_{20} and C_{22} are all catalyzed by P450$_{scc}$. The affinity of the enzyme for the intermediates and the conversion rates are high and no significant amounts of intermediates can be measured in mitochondria. So far there have been no convincing indications that the P450$_{scc}$ protein can directly be regulated by phosphorylation. The mechanism by which cholesterol is transferred from the outer mitochondrial membrane to the inner mitochondrial membrane has recently been elucidated, although many questions about details still exist. For a long time it was known that one or more unknown labile regulatory proteins are involved in regulating the hormone dependent intra-mitochondrial cholesterol trafficking. Recently, it has been shown that the steroidogenesis activator protein (StAR) fulfils the criteria for this labile protein. This does not only hold for Leydig cells but applies to all cells in the adrenal gland and ovary that are active in hormone-dependent synthesis of steroids (Stocco 1997).

The transcription of the StAR gene during embryonal development is regulated by the transcription regulator SF-1, an orphan receptor, which also regulates the expression of the genes for the P450 enzymes (Clark et al. 1995; Parker and Schimmer 1997; Rice et al. 1991). It thus appears that all essential elements for hormone-dependent steroidogenesis are regulated in a coordinated fashion. The rate of transcription of the StAR gene is controlled by hormones but under normal conditions there is always enough messenger RNA to sustain a steady state production of a 37 kDa protein precursor. This 37kDa protein can be transported to the mitochondria where it interacts with proteins on the outer mitochondrial membrane. As a result of this interaction the StAR protein is imported into the matrix of the mitochondrion and during this transport process "contact sites" are formed between the outer and inner membrane of the mitochondrion. It is thought that cholesterol can be transferred from the outer to the inner membrane via these transient "contact sites". Since StAR is continuously processed and ultimately inactivated inside the mitochondrion during this insertion process, continuous synthesis and probably also hormone-dependent phosphorylation of StAR are required to maintain hormone activated steroid production.

A very important observation in favour of the important role of StAR in the control of steroidogenesis came from studies on the disease lipoid congenital adrenal hyperplasia. This disease is characterised by an accumulation of cholesterol within Leydig and adrenal cells and an inability of the patients to synthesize steroids. It could be shown that mutations in the StAR gene which caused truncation and inactivation of the StAR protein were the cause of this disease (Lin et al. 1995). These clinical data further support the physiological importance of StAR for activated steroid production. Yet it is known that steroids can also be produced in tissues without StAR such as the placenta. This indicates that other mitochondrial proteins such as ligands for the mitochondrial benzodiazepine receptor could also facilitate formation of "contact sites" and intra-mitochondrial transfer of cholesterol (Papadopoulos 1993). However, StAR is very likely the long sought protein that is involved in the rapid regulation of steroid production.

1.2.4 Regulation of pregnenolone metabolism

The first product of the cholesterol side chain cleavage process, the biologically inactive steroid pregnenolone, is further metabolized by enzymes present in the endoplasmic reticulum. Much has been learned about the primary structure and the biosynthesis of various P450 enzymes after application of new techniques such as protein chemistry and molecular biology (reviewed by Miller 1988). This can be illustrated by enzyme activities which convert C_{21}-pregnenolone to C_{19}-steroids. It was previously thought that 17α-hydroxylase and $C_{17,20}$-lyase activity reside in separate enzymes which could be differentially regulated by hormones (Rommerts and Brinkmann 1981; Smals et al. 1980). When it was suggested that the 17α-hydroxylase and $C_{17,20}$-lyase activities may reside in only one protein, the question emerged how the two activities could be regulated differentially in the adrenal and in the testis. In the adrenal 17α-hydroxylation is predominant, whereas in the testis cleavage of the C_{17-20} bond is the major enzyme activity. This question became more important when results of transfection studies showed that both enzyme activities were indeed present in a single protein, $P450_{C17}$, coded by one gene, CYP 17 (Zuber et al. 1986). It appears that the differential expression of enzyme activities in the testis and the adrenal depends on the micro-environment of the enzyme in the endoplasmic reticulum; the relatively high levels of P450 reductase and cytochrome b_5 in the testis have been proposed to generate more reduction power and thereby promote the formation of androgens (Hall 1991). Results from a study by Zhang et al. (1995), however, show that differential expression of hydroxylase or lyase activity by $P450_{C17}$ also depends on the degree of phosphorylation of the protein. If natural stimuli can be identified that can regulate the phosphorylation of $P450_{C17}$, this could be the first demonstration of physiological regulation of a steroidogenic enzyme through phosphorylation. This possibility for regulation of the microsomal cytochrome $P450_{C17}$ activity appears to be of less importance for the testis, since in this organ the lyase activity is predominant.

The synthesis of $P450_{C17}$ in the testis is under the control of LH, via cAMP stimulation of CYP17 gene expression. Although CYP17 gene expression is clearly regulated by cAMP dependent mechanisms, cAMP responsive elements have not been found in the CYP17 promoter (Payne and O'Shaughnessy 1996). Deficiency of $P450_{C17}$ is rare but a few cases of XY individuals with female phenotypes have been reported (Monno et al. 1993). The degradation of the enzyme can be enhanced in the presence of elevated

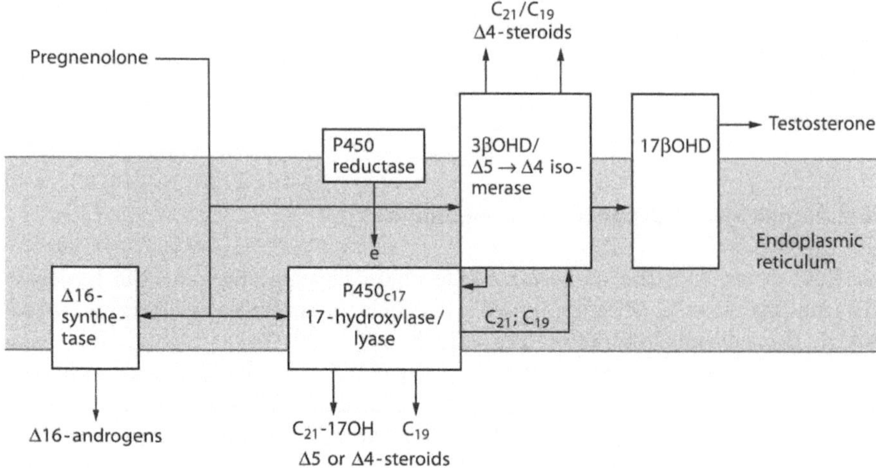

Fig. 1.3. Metabolism of pregnenolone in endoplasmic reticulum

levels of steroids by oxygen-mediated damage. Although this steroid-mediated inactivation has been shown to occur *in vitro*, it is unknown whether this process plays a role in the regulation of enzyme activity *in vivo* at low oxygen tension (Payne et al. 1985).

The presence of many steroid-converting enzymes allows many different possible pathways for the conversion of pregnenolone into testosterone (see Fig. 1.1). Depending on whether pregnenolone is converted initially by the 3β-hydroxysteroid dehydrogenase $\Delta5$-$\Delta4$-isomerase complex or the $P450_{C17}$ enzyme, a $\Delta4$- or $\Delta5$-pathway predominates. In the human testis most of the steroids are formed via the $\Delta5$-pathway with dehydroepiandrosterone (DHEA) as the first C_{19} intermediate (Weusten et al. 1987b) (see Fig. 1.3). The enzyme 3β-hydroxysteroid dehydrogenase $\Delta5$-$\Delta4$-isomerase (Fig 1.3: 3βOHD) catalyzes the conversion of $\Delta5$-3β-hydroxysteroids to $\Delta4$-3-ketosteroids, an essential step in the biosynthetic pathway. The dehydrogenase and isomerase activities are catalyzed by one protein coded by one gene (Lachane et al.1990). Although the two enzyme activities are carried out by one single protein, it appears that separate sites on the molecule mediate the specific enzyme activities (Luu-The et al. 1991). Different isoforms of this enzyme are expressed in steroidogenic but also in non-steroidogenic tissues. In the human testis the type II isoenzyme with almost equal affinity for dehydroepiandrosterone and pregnenolone is expressed. Several point mutations of the gene which affect intracellular location and the affinity for the substrate have been identified (Rhéaume et al. 1995).

The final step in the biosynthetic pathway of testosterone is the reduction of the 17-keto group by the 17β-hydroxysteroid dehydrogenase (Fig 1.3 17βHSD). This enzyme activity is represented by five different isoforms which are present in many tissues (Andersson and Moghrabi 1997). An interesting feature of 17βHSD type 2 is that the enzyme also possesses 20αHSD activity. In the testis the type 3 isoform is present, mainly in the Leydig cells. Although generally present in the body, deficiency of the testicular activity of 17βHSD accounts for most defects in testosterone biosynthesis in the human (Geissler et al. 1994; Labrie et al. 1997).

When all steroid-converting activities are taken together, the pregnenolone-converting enzymes present in the smooth endoplasmic reticulum function in close cooperation and act as a metabolic trap for pregnenolone, released by the mitochondria.

1.3 Regulation of androgen synthesis by LH

1.3.1 General

Luteinizing hormone and follicle stimulating hormone (FSH) are required for the development and maintenance of testicular functions. LH is the most important hormone for control of Leydig cell functions, but other hormones and locally produced factors also play a role. These hormones regulate steroid production by controlling the metabolic activities in existing cells, but they also control the size of the Leydig cell population via control of proliferation and differentiation (Chemes 1996). In the human fetal period around 14 weeks of gestation there is a sharp increase in the number and activity of Leydig cells. In this developmental period the maternal hCG plays an important role for the regulation of the Leydig cell activities. It is less clear what controls Leydig cell development before this period and there is good evidence that early Leydig cell development and onset of steroid production take place without gonadotropin stimulation. During postnatal development of the human testis major changes occur in the Leydig cell population. In the early neonatal period gonadotropins stimulate the development and activity of fetal Leydig cells to such an extent that during the first three months of life peripheral and testicular levels of testosterone are similar to those found during puberty. In the following period during the first year of life these fetal Leydig cells regress via ill-defined mechanisms and a dormant phase persists until puberty. During puberty a second wave of proliferation and differentiation occurs under the influence of rising plasma LH levels. This ultimately leads to the adult population of Leydig cells (Chemes 1996; Saez 1994).

Many studies on the short-term effects of LH have been carried out with isolated cells from rats and other animals and much has been learned about the mechanisms of hormone action involved in the rapid stimulation of steroidogenesis. For investigations on long-term or tropic effects of hormones, these *in vitro* systems are less suitable since isolated cells change their phenotypic properties after prolonged culture periods due to the absence of their natural environment. Tumour cell lines are constant in their functional properties and are thus a better choice for these investigations of long-term

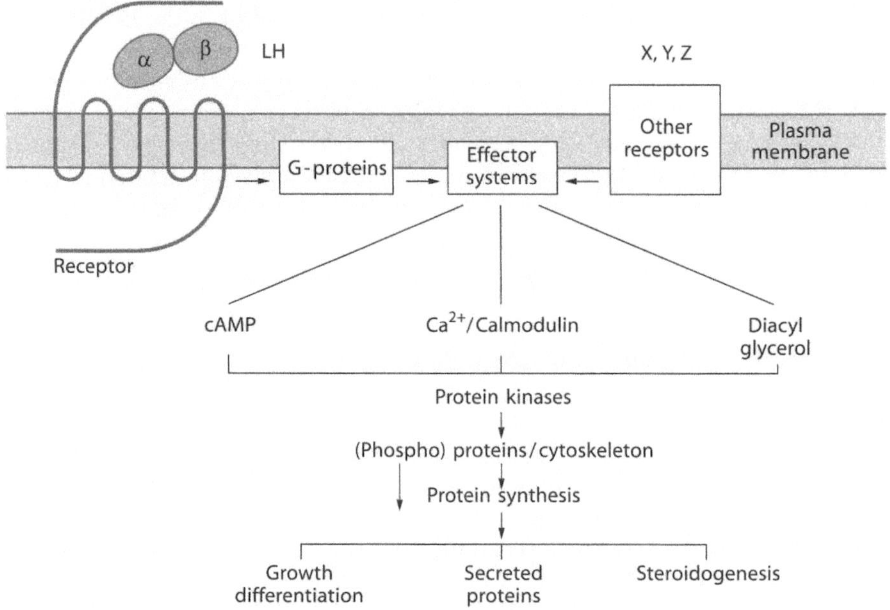

Fig. 1.4. Model for pleiotropic regulation of Leydig cell functions by LH and other local signalling molecules (X,Y,Z)

tropic effects. However, the regulatory systems in tumour cells and normal cells are often different and there is limited agreement on the detailed mechanisms for regulation of the steroidogenic enzymes.

The principles for regulation of Leydig cell function are similar to those for other somatic cells. Via their receptors, protein hormones and growth factors control phosphorylation of important cellular proteins, either via direct activation of kinases or indirectly via elevations in the level of intracellular second messengers. The covalently modified or newly synthesized proteins can then affect a variety of activities in discrete subcellular structures (plasma membrane, mitochondria, cytoskeleton, nuclei, etc.). As a result of the complex interplay of these intracellular activities, different physiological responses are generated (protein secretion, steroid production, growth, energy production, etc.). Since one hormone often stimulates multiple reactions (pleiotropic response) which may show different response kinetics, it is difficult to link specific transducing systems to particular responses (see Fig. 1.4).

1.3.2 Stimulatory actions of LH

LH is the most important hormone for regulation of Leydig cell number and functions. LH acts on Leydig cells via LH receptors of which the structure has been known since 1989 (Loosfelt et al. 1989; McFarland et al. 1989).

Much about the importance of the LH receptor and LH-dependent receptor activation has been learned from studies dealing with receptor mutations (Themmen et al. 1997) Activating mutations cause precocious puberty as a consequence of increased androgen production during the fetal and postnatal period, even in the absence of gonadotropic stimulation. The clinical manifestations depend on the severity of the LH receptor mutations. As can be expected, inactivating mutations of the LH receptor give rise to male pseudohermaphroditism caused by Leydig cell hypoplasia. Again, depending on the type of the mutation, receptors can be completely resistant or can still show a diminished response.

These observations indicate that the functional properties of the Leydig cells in these patients mainly depend on the gonadotropic stimulation of the LH receptor and that paracrine systems within the testis cannot or can only partly compensate for the lack of LH receptor stimulation.

Activated LH receptors stimulate adenylate cyclase via GTP binding proteins and this results in increased production of cyclic AMP, but also other products may be formed as a consequence of LH receptor activation (Cooke 1996; Rommerts and Cooke 1988; Saez 1994). Although cAMP can increase steroid production, there has been doubt as to whether cAMP is the only second messenger of the action of LH. Low concentrations of LH stimulate steroidogenesis without detectable changes in intracellular cAMP levels. Specific intracellular pools of cAMP have been postulated to explain these observations, but the results may also indicate that at low levels of LH other messenger systems operate. Since rapid changes in intracellular calcium ion levels have been detected after administration of hormones (sometimes oscillations occur; Berridge and Galione 1988) calcium may also play an important role in signal-transduction. For the Leydig cells calcium ions and calmodulin are also essential for full steroidogenic activities but it is less clear at which level calcium plays a role. Phospholipids, specific phospholipases and products of phospholipid metabolism such as leukotrienes are of paramount importance in signal transduction of many types. These compounds have also been detected in rat Leydig cells but it remains to be demonstrated to what extent they are essential for the effects of LH on steroidogenesis, especially in the human testis. If phospholipid metabolism is not or less involved in the action of LH, it still could be essential for the expression of biological effects of locally produced factors on Leydig cell function.

The activation of various signal transduction pathways in (rat) Leydig cells causes activation of different classes of protein kinases and kinase linked pathways. A major part of these kinase–linked pathways project to the nucleus where kinases or nuclear localized phosphoproteins could mediate the tropic effects of LH by regulation of gene expression. Transcription regulation of the steroidogenic enzymes has been investigated in great detail. The results of these studies show that the protein kinase A pathway is of predominant importance for controlling the promoter regions of most of these genes. However, only a limited number of cAMP responsive elements has been detected and regulation of transcription of steroidogenic enzymes clearly depends on a complex interaction between different transcription factors. Other phosphoproteins may, in connection with the cytoskeleton and StAR, play a role in the intracellular transfer of cholesterol. This is considered the rapid control of steroidogenesis. The tropic control of steroidogenic activities in mitochondria and the smooth endoplasmic reticulum is mainly exerted by regulation of the biosynthesis of the steroidogenic enzymes via increased levels of mRNAs. No activation of these enzymes by phosphorylation has been shown so far, although the differential regulation of the 17α-hydroxylase and $C_{17,20}$-lyase activities earlier mentioned may be the exception. Regulation of the amount of these proteins is a relatively slow process and it can take several hours before enzymatic activities change after stimulation with LH. The steroid-transforming activities of the enzymes in the endoplasmic reticulum are also affected by the levels of endogenous steroid precursors and endproducts and product inhibition has been shown. Since the pattern of accumulated intermediates often depends on the external conditions it is difficult to make general conclusions on this regulatory aspect of steroidogenesis and for specific information the reader is referred to a review by Gower and Cooke (1983). At this stage it is important to stress again that the capacity of the enzymes in the endoplasmic reticulum to convert pregnenolone into biologically active steroids is often smaller than the steroidogenic capacity, the generation of pregnenolone. At a low rate of steroidogenesis the output of androgens depends most likely on the production of pregnenolone, whereas at a high steroidogenic flux the pregnenolone converting capacity may become rate limiting for the output of androgens.

Although the most important Leydig cell products are steroids, Leydig cells also produce protein products like IGF-1 and other growth factors. These products are mainly important for paracrine or autocrine regulatory events within the testis.

1.3.3 Adaptation of Leydig cells

Most of the initial effects of LH or hCG on steroid production are stimulatory, as can be seen from the rapid increase of the testosterone concentration in plasma. In rats this response is much faster and more pronounced than in men (Huhtaniemi et al. 1983; Saez 1989). Although studies with isolated cells from human testes have shown that a small proportion of the human cells can respond more or less similarly to those in rats, it is unknown why the magnitude of the steroidogenic response of the intact human testis *in vivo* is small (Simpson et al. 1987). In both species the period of stimulated steroid production after hCG administration *in vivo* is followed by a period of diminished output of testosterone. This transient steroidogenic desensitization is the result of four different phenomena:
1) the coupling between the LH receptors and the adenylate cyclase is diminished, probably as a result of receptor phosphorylation;

2) the number of LH receptors is also decreased due to an increased rate of receptor internalisation;
3) at the same time the mRNA level for the LH receptors has also decreased due to a higher turnover of this mRNA pool;
4) the activities of the steroidogenic activities in the endoplasmic reticulum are lower.

In different species these LH induced adaptive responses do not always occur to the same extent. This drop in androgen production, which occurs in rats 24–36 h after hCG administration, is accompanied by a rise in the secretion of 17α-hydroxyprogesterone, indicating a partial block at the level of the C_{17-20}-lyase activity. In the same period the production of 17β-estradiol increases. Estrogens have therefore been implicated as a causal factor in the development of steroidogenic lesions (Saez 1989). These changes in the steroidogenic properties of Leydig cells, however, can also be mediated by other mechanisms (Brinkmann et al. 1982). Following a period of diminished androgen production, plasma testosterone levels rise over the next period of 3–4 days whereas 17α-hydroxyprogesterone levels decrease.

This biphasic response of steroid production depends on the stage of the development of the Leydig cells. In prepubertal children and in hypogonadal adult men, one injection of hCG induces a sustained rise in testosterone without significant changes in 17α-hydroxyprogesterone and 17β-estradiol plasma levels. After-long term treatment with hCG, an adult pattern of response is observed in both groups (reviewed by Forest 1989). It therefore appears that the long-term steroidogenic response of Leydig cells depends on previous exposure to gonadotropins. After exposure of rat Leydig cells to high doses of gonadotropins, the degree of stimulation of adenylate cyclase is greatly diminished within several hours and after 24 h the number of LH receptors is diminished (Saez 1989). These phenomena have been described as desensitization and receptor down regulation respectively and they have often been used to explain the decreased production of androgens which develops after the initial stimulation. However, as mentioned earlier, Leydig cells are still active in the production of steroids other than androgens. Moreover, the regulation of steroidogenic activities is complex and not all changes in cellular activities are synchronized in time. For instance, in Leydig cells from mature rats isolated 10 days after injections of hCG (administered at day 0 and day 7), LH receptors were down regulated and adenylate cyclase desensitized. However, LH-dependent androgen production was increased (Calvo et al. 1984). Similarly, in the period of low receptor number and reduced cyclic AMP response, LH-dependent prostaglandin production in the testis is very high (Haour et al. 1979) and Leydig cells show hypertrophy (Hugson and de Kretser 1984). High doses of hCG also stimulate cell proliferation of Leydig cells (Teerds et al. 1988). It seems therefore that Leydig cells can adapt their activities to changes in the environment such as exposure to hCG. Depending on the species, the function of interest and the time interval after exposure to hCG, this adaptation process can include stimulatory and inhibitory changes.

1.4 Regulation of androgen synthesis by locally produced factors

Leydig cells in the testis are surrounded by other cells localised either in the seminiferous tubules, such as Sertoli cells, or in the interstitial tissue, such as macrophages. Many observations indicate that these neighbouring cells can influence the function of Leydig cells in a paracrine fashion. On the other hand, Leydig cells themselves can also produce growth factors, which can stimulate or inhibit Leydig cell steroidogenesis in an autocrine fashion (see Fig 1.5). FSH stimulates development of Leydig cells, probably via Sertoli cell products. Disturbances in the spermatogenic epithelium also affect Leydig cells. Moreover, conditioned media from Sertoli cells or seminiferous tubules can modify the steroidogenic activities of Leydig cells, suggesting the presence of many stimulatory and inhibitory components (reviewed by Gnessi et al. 1997; Saez 1989, 1994; Sharpe 1983 and 1996). Although the existence of paracrine regulating systems can be inferred from these data, there is hardly any consistency in the various reports. In most studies the results appear to depend chiefly on the species used, the techniques applied for the isolation of cells or secretion products, cell culture conditions, etc. Even when one batch of secreted Sertoli cell products was used to regulate steroid production in one standardised Leydig cell preparation, it was found that the short-term effects of the Sertoli cell products were stimulatory, whereas the long-term effects were strongly inhibitory (van Haren et al. 1995). These long-term inhibitory effects of Sertoli cell products *in vitro*, are in sharp contrast with the long-term stimulatory effects that Sertoli cells exert on Leydig cells *in vivo*. In a limited number of studies specific (recombinant) growth factors have been used, but this has not resolved the existing confusion. Another question is whether specific products, which are active when added to isolated Leydig cells in a chemically defined medium, are also effective *in vivo* when they act together with many other local products. *In vitro* experiments

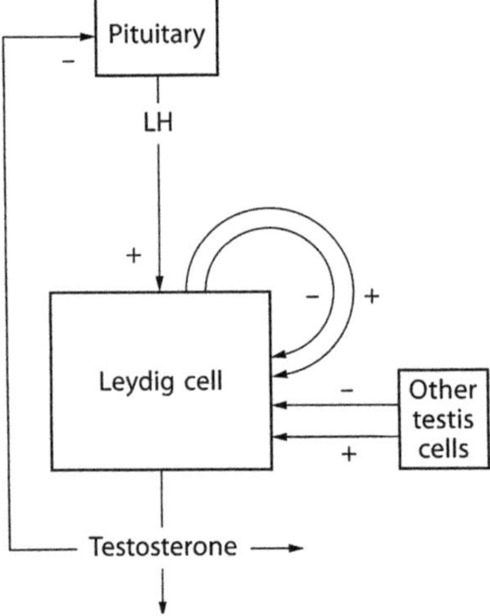

Fig. 1.5. Regulation of Leydig cell steroidogenesis by LH and locally produced factors

have already demonstrated that the activity of a certain compound can be inhibitory or stimulatory, depending on the previous short-term or permanent exposure to other signal molecules. The action of growth factors is thus context dependent (Sporn and Roberts 1988) or, in other words, it can be a part of a cellular signalling language with the individual growth factors functioning as the letters of the alphabet. Just as letters in the alphabet can form words with a particular meaning, a certain combination of growth factors may give a (more) useful message to the cell than the isolated growth factors. It will not be difficult to understand that this cellular language can become complex when many signal molecules are involved (see also Fig. 1.5).

There is now agreement that individual local secretion products can only be considered as potential paracrine or autocrine factors if four criteria are fulfilled:

1) the molecule should regulate at least one biological activity of the target cell;
2) the molecule must be secreted in adequate quantities to guarantee a physiological response;
3) regulation of secretion of the molecule must be possible;
4) changes in the local concentration of the molecule should influence the properties of the target cell *in vivo*.

Since it is almost impossible to fulfill all these criteria for one particular compound, this section on the physiological relevance of local regulation could be very short. However, since local regulatory systems are in general very important for specific cell function and because so many research efforts have been made to understand these systems, a brief summary of the past and the present of paracrine regulation will be given before some conclusions will be made.

The appreciation of local regulation and the shift from endocrine research to paracrine/autocrine research started approximately 25 years ago when it was shown that FSH could stimulate steroidogenesis in rat Leydig cells (Odell et al. 1973). A second wave occurred when LHRH or LHRH-like molecules became available and when direct effects on Leydig cells could be shown (Hsueh and Jones 1981). This period of initial great excitement was followed by a period of disappointment when it was not possible to show significant secretion of endogenous "LHRH-like" compounds by testicular cells.

In the following years results became available from many studies using growth factors which are known to be produced in the testis such as IGF-1, TGFβ, EGF/TGFα, FGF, PDGF, inhibin/activin, interleukin, TNFα, etc. (reviewed by Gnessi et al. 1997; Saez 1994; Schlatt et al. 1997). In hundreds of publications many inhibitory and stimulatory effects of these compounds on steroidogenesis *in vitro* could be shown. Again many speculations were made on the biological relevance of these findings for the regulation of Leydig cells in the testis. However, when the results of all these investigations are taken together, it is still not clear how important these molecules, alone or together, are for the regulation of Leydig cells *in vivo*. When using information obtained from *in vitro* experiments to explain Leydig cell functioning *in situ*, two typical aspects of Leydig cells *in vivo* should not be forgotten. Firstly, Leydig cells *in vivo* are surrounded by an interstitial fluid which will average out specific paracrine influences of neighbouring cells. Secondly, the Leydig cells are a part of a closed negative feedback system with the brain and the pituitary as the regulator of LH secretion. Local testicular influences which result in activation or inhibition of steroid production will therefore be compensated by alterations in the LH secretion (see Fig. 1.5).

The problems connected with an understanding of the physiological role of paracrine factors in general will be illustrated by discussing the possible paracrine role of IGF-1 for Leydig cell steroidogenesis (for a review of the experimental findings related to this subject, see Lin 1996), because all data on production and effects make this molecule a most promising paracrine regulator. IGF-1 is produced locally in many tissues and in combination with the six different IGF binding proteins, the IFG/IGF-BP system has been proposed as a super-system for fine tuning of local hormone action. Since IGF-1, IGF binding proteins and specific proteases can be produced by the target cells themselves as well as by neighbouring cells, autocrine and paracrine regulatory systems can be integrated (Collett-Solberg and Cohen 1996). Many *in vitro* studies have shown that this IGF-1 system can also influence Leydig cells. IGF-1 can enhance the stimulatory effects of LH/hCG on Leydig cells through specific IGF-1 receptors and IGF-1 and also some IGF binding proteins can be produced by the Leydig cell itself. One could therefore postulate that part of the stimulatory action of LH on steroid production is mediated by an external and obligatory IGF-1 loop. Other testicular cells that can secrete compounds into the interstitial fluid can amplify or inhibit this external loop by either increasing IGF-1 concentration or by decreasing IGF-1 action through increased production of the specific binding proteins. Thus in theory the Leydig cell appears to be surrounded by a network of regulatory molecules with IGF-1 as co-stimulator and fine tuner of LH and the binding proteins as fine tuners of IGF-1 action. In contrast with this model, IGF-1 did not affect LH-induced cholesterol side chain cleavage enzyme activity in cultured Leydig cells (van Haren et al. 1992).

There is general agreement that *in vitro* investigations cannot answer the question on the physiological importance of the IGF system for the Leydig cell and that transgenic animals with specific gene knockouts could shed new light on this problem. In this connection Baker et al. 1996 showed that mutant male mice with an inactive IGF-1 gene were reduced in size and were infertile. However, a close inspection showed that, although the size of the testis was approximately 40% of the normal size and the total (not the free levels) peripheral testosterone levels were approximately 20% of the normal values, sperm cells of these mutant mice were able to fertilize normal oocytes. Moreover, the infertility of the males was caused by the absence of mating behaviour. These results show that in the complete absence of the IGF-1 system indeed the normal growth of the animal is disturbed, but the data also show that the Leydig cells are still functional and that the testis, although reduced in size, can still produce active spermatozoa.

It appears from this study, as well as from many other studies using knockout animals, that questions about the importance of a particular gene product (such as IGF-1) cannot be answered with a firm yes or no. It is very likely, as experience from many other studies with knockout animals for growth factors and regulating molecules has shown, that the function of one defective gene can be compensated by other components of the external cellular regulatory network. It becomes increasingly apparent that physiological regulation of cell function is more than regulation of a simple linear process. Cell regulation involves many regulatory systems; adaptive epigenetic networks and redundancy are now well-known features of cellular regulatory systems (Strohman 1993). An answer to the question of the importance of the IGF-1 system for the regulation of steroid production in a normal organism can only be obtained when the amount of IGF-1 can be increased or decreased near the Leydig cells within the testis after normal development of the animal. This can be accomplished with animals in which genes in specific cell types can be knocked out conditionally. However, these animals are not yet available for this purpose.

In summary, regulation of steroid production by LH not only involves complex interactions of transcription factors for regulating gene expression in the nucleus and other regulatory pathways in the extranuclear compartment of the cell, but a similar level of complexity of local regulating molecules exists on the outside of the cell. Although this extracellular local regulatory system is complex, it seems to be very flexible and adaptive, as shown by gene knockout experiments. In contrast to the flexibility of local regulatory systems, LH receptor activation is always required for proper cell function. This has clearly been illustrated by the abnormalities due to activating and inactivating mutations in the LH receptor gene, as discussed previously. LH thus appears to be high in the hierarchy of regulating molecules. This is in accordance with its role as an endocrine regulator. Another argument for a subordinate role of the paracrine system is that for regulation of steroidogenesis in testicular Leydig cells there are no reports that the local regulatory network can modify the properties of the Leydig cell to such an extent that for maintenance of peripheral testosterone levels abnormally high or low levels of LH are required. A disbalance between LH and testosterone levels can occur, however, when genetic defects in steroidogenic enzymes or in the LH receptor cause insufficient production of testosterone. Under these conditions the feedback system tries to compensate this deficiency with high levels of LH. In light of the many problems that still exist in our understanding of the paracrine control of rodent Leydig cells, the mechanisms involved in paracrine regulation of human Leydig cells remain totally obscure. When all the above discussed information is taken together it is not difficult to conclude that the main regulator for the steroid production of the human Leydig cells is still LH.

1.5 Secretion of steroids

1.5.1 Trafficking inside cells

The formation of androgens from pregnenolone in the smooth endoplasmic reticulum of Leydig cells is the result of interactions between different membrane-bound enzymes with steroids as mobile elements. Within the smooth endoplasmic reticulum, the trafficking of the steroids is to a great extent influenced by the affinities of the enzymes and binding properties of other membrane components. Apart from these binding entities the movements of steroids in tissues appear to depend mostly on diffusion (reviewed by Mendel 1989). There are no reports that steroids are like secreted proteins which are released from vesicles after fusion with the cell membrane. Conjugated steroids, however, cannot easily pass cell membranes by diffusion and for this class of steroids specific transport systems are required (Mulder et al. 1973). Binding proteins for androgens or other steroids can play an important role in decreasing the concentration of unbound steroids outside the cell and in this way enhance the diffusion process, but there are no indications that such proteins are also important inside the cell. Another argument against an important role of specific transport systems for steroids in Leydig cells is the observation that secretion of pregnenolone, which is normally very low, can become as high as secretion of testosterone when metabolism of pregnenolone to testosterone is inhibited (van Haren et al. 1989).

Although diffusion is probably the main driving force for steroid-trafficking within the cell, it has been shown that multidrug resistance proteins in the plasma membrane can increase the rate of transport of steroids over the plasma membrane (Ueda et al. 1992). If these transport proteins can also transport androgens and if they are localised in the Leydig cell, they could aid in the secretion of androgens. Alternatively these steroid transport proteins could alter the steady state concentrations of androgens in target cells. However, since the transport activities of these proteins have mainly been investigated with corticosteroids (Gruol and Bourgeois 1997) the relevance of this transport system for androgens is unknown.

1.5.2 Trafficking between the testicular compartments

Steroids such as pregnenolone, progesterone and testosterone not only rapidly pass the Leydig cell membranes but they can also equilibrate rapidly between different testicular compartments (van Doorn et al. 1974). As a result of this continuous supply of testosterone by the Leydig cells, the surrounding testicular cells and binding proteins are saturated and all tissue compartments are exposed to a high level of unbound testosterone (Rommerts et al. 1976). The secretion pattern in the testis is thus most likely to be determined by the amount of steroids that is produced inside the tissue, the permeability characteristics of the membranes and the binding proteins in various testicular fluids. The Leydig cells in the testis are surrounded by an interstitial fluid which is rich in plasma proteins and the cells are also in close contact with blood vessels. The preferential direction of secretion in the testis is mainly determined by the concentration gradient and flow rates of the various fluids. Since the blood flow is much higher than the flow of the interstitial fluid, most of the unconjugated steroids diffuse from the interstitial space to the blood and leave the testis via the venous blood (Maddocks and Sharpe 1989). The boar testis is an exception which supports this view. In this species steroid sulphates, which cannot readily diffuse through the walls which surround the interstitial space, were 13 to 35-fold more concentrated in interstitial fluid than in venous blood (Setchell et al. 1983). During passage of the venous blood through the pampiniform plexus the primary venous blood is diluted approximately 2-fold by incoming arterial blood. This occurs through anastomoses present in this network of interacting blood vessels (Noordhuizen-Stassen et al. 1985).

The presence of relatively high levels of dihydrotestosterone in human spermatic venous blood has been taken as evidence that dihydrotestosterone is produced in the testis (Hammond et al. 1977). However, other studies have shown virtually no 5α-reductase in human testis (Miautani et al. 1977). Although these data are contradictory, dihydrotestosterone is most likely an epididymal steroid and it is conceivable that dihydrotestosterone produced in the epididymis is transported to the testicular venous blood during the dilution process that occurs in the pampiniform plexus. Although dihydrotestosterone is not produced in the testis from testosterone, another derivative from testosterone, 17β-estradiol, is produced by Leydig cells. The testicular contribution to the total estrogen production, however, is small (in the order of 20%) as compared to the peripheral aromatization. The local production of 17β-estradiol may be of great importance for regulating Leydig cell functions, for instance in the previously mentioned development of steroidogenic lesions.

1.6 Transport of androgens in the body

In the periphery steroids equilibrate rapidly between various organs and blood, as becomes clear from identical levels of free testosterone in saliva and blood (Wang et al. 1981). Almost all cells in the periphery are thus exposed to the same concentration of unbound steroids, except when local metabolism is more active than the rate of supply of testosterone. The total concentration of steroids in target tissues and body fluids is mainly dependent on the presence of binding proteins such as sex hormone binding globulin (SHBG), and albumin (see Fig. 1.6). Binding proteins in body fluids can act as a storage for steroids which have a high rate of metabolism during passage of blood through the liver (Mendel 1989). In this way, extensive metabolism of active steroids can be inhibited. However, the presence of SHBG and

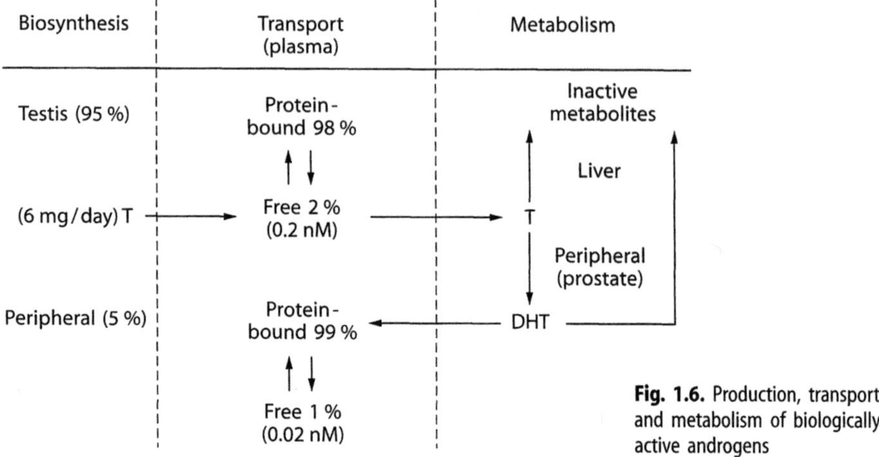

Fig. 1.6. Production, transport and metabolism of biologically active androgens

albumin in body fluids is not essential for steroid homeostasis. This can be inferred from analbuminemic rats which possess neither SHBG nor albumin. In these rats, which are fertile, the total plasma concentration of testosterone is close to the free concentration in normal rats (Mendel et al. 1989). This free testosterone concentration is within the range of the affinity constant of the androgen receptor K_d 10^{-9}–10^{-10}M. Changes in the peripheral free testosterone concentration can therefore be directly sensed by the androgen receptor if there is equilibration between the exterior and interior of androgen target cells.

It is not easy to envisage how steroids such as dihydrotestosterone and 17β-estradiol, which can be present in picomolar concentrations (free), can be biologically active when the affinity constants for the receptors are in the (sub)nanomolar range. However, it has become clear that steroid receptors can also be (partly) activated, independent of steroids, by phosphorylation. It is not impossible that such "predisposed receptors" can be activated further by binding very small amounts of steroid (O'Malley et al. 1995). In some cases local production can furnish these active steroids, as for instance, dihydrotestosterone in the prostate (Coffey 1988) and 17β-estradiol in the brain (Michael et al. 1986). If the capacity of such local production is sufficiently high, this could explain the apparent discrepancy between receptor affinity and plasma availability. Alternatively, specific transport systems for steroids into target cells are a possibility, as discussed previously.

1.7 Metabolism of testosterone

The steady-state level of biologically active steroids in the body as a whole is determined by the rate of synthesis and the rate of degradation. To maintain a steady-state concentration of active steroids in a target cell a similar balance between supply and removal must be maintained. Different factors contribute to the control of the balance. Outside the target cell these factors are: flow rate of biological fluid (blood or lymph), release from binding proteins, transport through membranes, connective tissue cell layers and sometimes inactivation of steroid during transport. Inside the target cell local activation or inactivation reactions and outward transport play a role. Alterations in the rate of degradation of androgens induced by disease, ageing, treatment with drugs etc. are therefore as important as changes in the rate of synthesis. As an example it could be possible that an altered androgen metabolism during ageing may be responsible for a local hyperandrogenic state in the prostate, leading to benign prostatic hyperplasia (Ishimaru et al. 1977).

There are several possibilities for the metabolism of testosterone (see Fig. 1.7). Aromatization or reduction of the $\Delta 4$ bond of testosterone give rise to 17β-estradiol and 5α-dihydrotestosterone, respectively. These steroids have completely different biological activities since they interact with discrete receptors in the cell. Actions of testosterone on target tissues are therefore significantly modulated by metabolic reactions (see Fig. 1.7). When a target cell is estrogen–dependent, its aromatase activity and the supply of androgen substrate are of major importance for determining the rate of synthesis of estrogens. In the human the aromatase cytochrome P450 enzyme (p450arom) is encoded by a single gene (CYP 19). This gene is expressed in many tissues

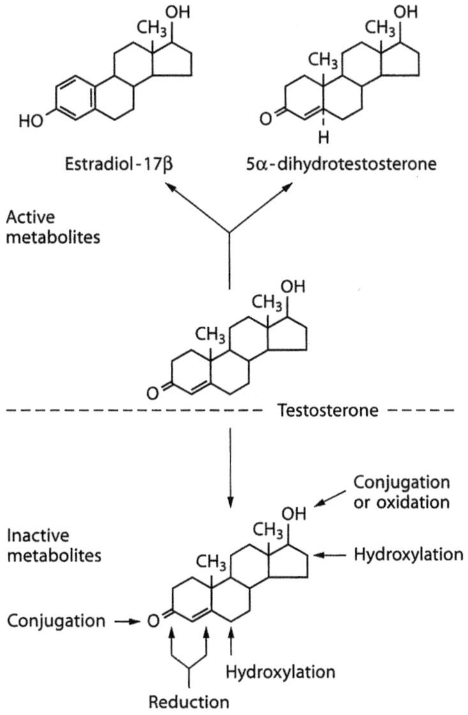

Fig. 1.7. Various possibilities for metabolism of testosterone

including the placenta, testis, fat tissue, liver, brain, hair, follicles and the brain. A very few cases of complete aromatase deficiency due to a gene defect have been noted (Bulun 1996; Morishima et al. 1995). The activity of 17βHSD, especially the type 2 isoform that favours oxidative reactions, determines how much of the active estradiol is converted to biologically inactive estrone (Andersson and Moghrabi 1997).

For proper action of androgens it is often necessary to convert testosterone into 5α-dihydrotestosterone before it can activate the receptor. Two isoforms of 5α-reductase exist and isoform 2 is most important because deficiencies of this reductase are correlated with abnormal clinical manifestations such as some cases of human male pseudohermaphroditism (Wilson et al. 1993). To establish a critical steady-state concentration of DHT, not only the activity of the 5α-reductase must be high enough, but also the metabolism of DHT must be low. In the prostate of the dog the activity of the reductive 3α/3β- steroid dehydrogenase activities are low and this favours the formation of DHT. The low rate of metabolism through the 3α/3β- dehydrogenase pathway may be the consequence of a low expression of one or both of these enzymes in the prostate, but it may also be possible that within one cell there is a balance between oxidative and reductive actions of two different isoenzymes. A recent report showing that rat and human prostate contain an oxidative 3α-hydroxysteroid dehydrogenase that can convert 5α-androstane-diol back to dihydrotestosterone (Biswas and Russell 1997) supports this hypothesis. Such a "back conversion" could also explain an old observation that 5α-androstane-3α,17β-diol is a more potent androgen for maintaining epididymis function than dihydrotestosterone or testosterone (Lubicz-Nawrocki 1973).

Prostate tissue also contains the type 2 17β-hydroxysteroid dehydrogenase that is primarily oxidative in nature. This enzyme does not metabolize DHT but it does convert testosterone to androstenedione, especially when 5α-reductase is inhibited (George 1997). In muscle there is high activity of 3αHSD and low 5α-reductase (Luke and Coffey 1994). This combination of enzymes seems to operate to optimize the amount of testosterone for testosterone-dependent receptor stimulation in this cell type.

In other target cells such as the skin and the hair follicle, the level of DHT as the most active ligand, can depend on the supply of testosterone and conversion to DHT on one hand, balanced by the catabolism of DHT via reducing 3α/3β-steroid dehydrogenases and glucuronidation on the other hand (Rittmaster 1994). Oxidizing activities of 3α-steroid dehydrogenases may, however, offset this inactivation of DHT (Penning 1997). A network of steroid metabolizing enzymes thus operates to establish a pattern of active and inactive androgen metabolites. Not only is the balance of enzyme activities within one cell type important for regulating the level of the active ligands in the target cell, but also enzyme activities in surrounding cells can be of great importance. It has been shown that in the kidney the 11β-hydroxysteroid dehydrogenase in cell layers that surround the target cells for mineralocorticosteroids can inactivate glucocorticoids and so prevent these glucocorticoids from inappropriately competing with aldosterone for the mineralocorticoid receptors (White et al.

1997). Although a similar "metabolic shield" has not been shown to protect target cells against certain androgen actions, a similar system could also operate to increase the selectivity of androgen action in particular cell types.

Although the balance of specific "activating" and "inactivating" steroid conversions is of great importance for the manipulation of the androgen response of the target cells, they are less important for overall degradation and clearance of androgens. The pathway for degradation of androgens in various tissues is determined by the profile of enzymes involved in the inactivation process. Enzymes that are active in degrading androgens are 5α- and 5β-steroid reductases, 17β-hydroxysteroid dehydrogenase and 3α- and 3β-hydroxysteroid dehydrogenases. In addition to these enzymes which convert existing functional groups, androgens can also be hydroxylated at the 6, 7, 15 or 16 position (Träger 1977). Most of these androgen metabolites are intrinsically inactive. The 5β-androgenic metabolites are a special group of compounds which stimulate the production of heme in bone marrow and liver (Besa and Bullock 1981). These biological effects are not mediated by the classical androgen receptor. Thus steroid metabolites that cannot bind to nuclear steroid receptors can still express biological activity (see also next section). Metabolism should therefore not always be considered as an inactivating pathway preparing a steroid for excretion. Androsterone (3α-hydroxy-5α-androstane-17-one) and etiocholanolone (3α-hydroxy-5β-androstane-17-one) are the most abundant urinary androgen metabolites.

Some androgen metabolites are excreted as free steroids, whereas others are conjugated. These conjugated steroids carry a charged group such as a sulphate or a glucuronide group on the 3- or 17-position. Dehydroepiandrosterone-sulphate is a well-known example of a conjugated steroid which is produced by the adrenal cortex and which is present in the circulation at micromolar levels without a clear physiological function, although certain abnormalities may have connections with alterations in the plasma levels of DHEA-sulphate (Ebeling and Koivisto 1994). In adult men, glucuronides of the 5β-androstane compounds are most abundant. The majority of the catabolic reactions take place in the liver, but the prostate and the skin also contribute significantly to the metabolism of androgens.

All steroid-metabolizing enzymes together constitute a network for transforming androgens in excretion products that finally leave the body via the urine or the skin. The flux through this network is great because the overall half-life of testosterone in men is only 12 minutes. It is clear that for maintenance of a constant level of testosterone in the body, this breakdown must be balanced by a continuous supply from the testis.

1.8 Non-genomic effects of androgens

The major action of androgenic hormones is through direct activation of DNA transcription via high affinity interactions with the androgen receptor. Information on the physiology and pathophysiology of these androgen receptor actions will be given in the next chapter. In this section, complementary non-genomic effects of androgens will be discussed.

In recent years, a variety of rapid non-genomic effects of steroids, often involving ion fluxes, have been documented for ligands such as progesterone (Baldi et al. 1995; Brucker and Lipford 1995), estrogens (Aronica et al. 1994; Mermelstein et al. 1996), corticosterone (Ibarrola et al. 1991), aldosterone (Christ et al. 1995; Wehling et al. 1994), and vitamin D (Sergeev and Rhoten 1995). Androgens have also been shown to induce rapid calcium fluxes in a variety of classical androgen-dependent cell types. Calcium fluxes following addition of picomolar concentrations of testosterone have been reported in rat heart myocytes (Koenig et al. 1989) and male rat osteoblasts (Lieberherr and Grosse 1994). Steinsapir et al. (1991) have shown increases in intracellular calcium in human prostate cancer cells after addition of 5α-dihydrotestosterone and testosterone within two minutes. In most of these studies it could be shown that the calcium fluxes in the cell were caused by a transmembrane influx of extracellular calcium through the plasma membrane (see Fig. 1.8). Although steroid-mediated calcium signalling has been demonstrated in many cell types, the structure of the receptive unit and the biological significance of these membrane effects are yet unknown. Recently the cDNA of a possible membrane receptor for plant steroids (brassinosteroids) was cloned (Li and Chory 1997).

Most of these alternative, non-genomic effects of steroids could add complementary signals to the genomic effects of these steroids, but there are a few examples where the nongenomic steroid effects are linked to biological responses which cannot be mediated by the classical steroid response pathway. It is known that testosterone can modify the susceptibility of T-cells to infectious diseases. Recently, effects of testosterone and testosterone covalently coupled to albumin on the calcium flux through the plasma membrane of T-cells were reported (Benten et al. 1997). Since T-cells do not possess

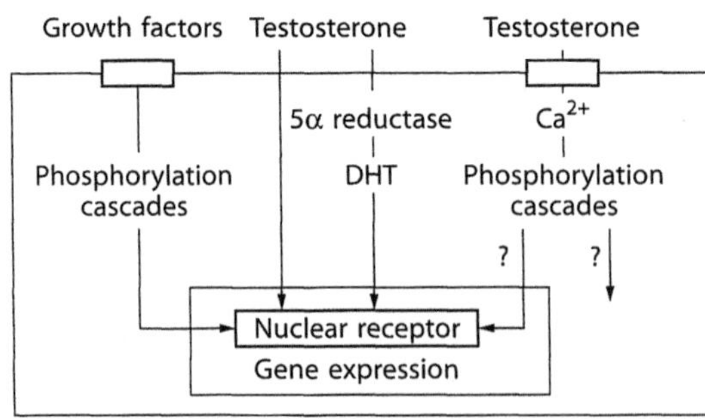

Fig. 1.8. Genomic and nongenomic actions of testosterone

classical androgen receptors, this biological response indicates the involvement of plasma membrane receptors for the expression of the androgen effects. Another example of unconventional androgen action is the rapid activation of membrane receptors in the olfactory sensory neurons by nanomolar concentrations of androstenon (Δ16-5α-androsten-3-one) (Snyder et al. 1988).

The dependency of spermatogenesis on high levels of testosterone is a good example of a local intratesticular action of testosterone that cannot easily be explained on the basis of the properties of the classical nuclear receptor. Since the levels of testosterone required for maintaining spermatogenesis are much higher than the saturation level of the high-affinity androgen receptor, an alternative sensing system, different from the classical receptor, might operate (Rommerts 1988 and 1992). In eels it has been shown that nanomolar concentrations of 11-keto testosterone for which no nuclear receptor has been found, are essential for maintaining spermatogenesis *in vitro*, whereas high concentrations of testosterone or dihydrotestosterone were inactive (Miura et al. 1991). For rodents Gorczynska and Handelsman (1995) presented evidence for an alternative response system in rats when they showed that freshly isolated immature Sertoli cells could respond to testosterone, testosterone covalently bound to albumin and 5α-dihydrotestosterone with increases in calcium levels within 40 seconds. Although rather unphysiological, high doses of 300 nM – 3 μM testosterone and 5α-dihydrotestosterone were necessary; these data support the view that alternative low-affinity interactions of steroids with yet ill-defined receptive structures in the plasma membrane may be essential for maintenance of spermatogenesis and coordination of this process with steroidogenesis, especially in seasonal breeders.

So far the non-genomic effects of steroids have received much less attention than the genomic effects. This may be due to the fact that hardly anything is known about the structure and only to some extent about the functional properties of this membranous responsive system and that most (but not all, as discussed previously) of the physiological effects of normal and defective androgen action can be explained by the properties of the classical nuclear receptor alone. Although simple explanations may be satisfactory and attractive, we know that biological systems in general are fine-regulated by networks with other molecules and new signalling pathways and connections are reported continuously. The importance of external regulatory networks for the outcome of steroid hormone action in the nucleus was stressed by O'Malley et al. (1995), when he proposed that membrane transduction pathways, activated by growth factors which interact with the nuclear receptor via intracellular phosphorylation cascades, may "set the nuclear receptor thermostat" for responses to steroids. In a similar fashion the signal transduction pathways activated by the membrane effects of steroids could also influence its genomic actions, indicating that nuclear and membrane effects of steroids are probably more closely linked than previously thought (see Fig. 1.8).

1.9 Key messages

- Steroidogenesis is the cleavage of the carbon chain of cholesterol inside the mitochondria with formation of the biologically inactive steroid, pregnenolone, as endproduct. Specific enzymes in the endoplasmic reticulum of the Leydig cells form a metabolic network to catalyse the transformation of pregnenolone into the biologically active testosterone.
- LH regulates the transport of cholesterol to the inside of the mitochondria (short-term regulation of steroidogenesis) as well as the profile and activities of the pregnenolone metabolising enzymes (long-term regulation of steroidogenesis).
- The physiological role of specific paracrine factors for regulation of Leydig cell steroidogenesis is less clear than the role of LH.
- For extracellular and intracellular transport of cholesterol, specific transport sytems are required. In contrast, steroids diffuse through tissues without specific transport systems and, as a result, all cells in the body "see" roughly the same concentration of unbound testosterone.
- In target cells transformations of testosterone into more active ligands can take place when the rate of inactivation is low.
- The reponse of target cells depends on the occupancy of the receptors which in turn depends on the intracellular concentration of unbound (androgen) ligands.
- In addition to genomic effects of androgens through the nuclear receptor, androgens can also stimulate non-genomic effects through interactions with the cell membrane.
- Disturbances in the biosynthesis of androgens through enzyme defects or in the actions of androgen are often the cause of abnormal sexual differentiation.

1.10 References

Andersson S, Moghrabi N (1997) Physiology and molecular genetics of 17β-hydroxysteroid dehydrogenases. Steroids 62:143–147

Aronica SM, Kraus WL, Katzenellenbogen BS (1994) Estrogen action via the cAMP signalling pathway – stimulation of adenylate cyclase and cAMP regulated gene transcription. Proc Nat Acad Sci 91:8517–8521

Baker J, Hardy MP, Zhou J, Bondy C, Lupu F, Bellve AR, Efstratiadis A (1996) Effects of an IGF-1 gene null mutation on mouse reproduction. Molec Endocr 10:903–918

Baldi E, Krausz C, Luconi M, Bonaccorsi L, Maggi M, Forti G (1995) Actions of progesterone on human sperm – a model of non-genomic effects of steroids. J Steroid Biochem Mol Biol 53:199–203

Baulieu EE (1997) Neurosteroids: of the nervous system, by the nervous system, for the nervous system In Conn PM (ed) Rec Prog Horm Res vol 52, pp 1–32

Benten WPM, Lieberherr M, Sekeris CE, Wunderlich F (1997) Testosterone induces Ca^{2+} influx via non-genomic surface receptors in activated T cells. FEBS Letters 407:211–214

Berridge MJ, Galione A (1988) Cytosolic calcium oscillators. Fed Proc Am Soc Exp Biol 2:3074–3082

Besa CE, Bullock LP (1981) The role of the androgen receptor in erythropoiesis. Endocrinology 109:1983–1989

Biswas MG, Russell DW (1997) Expression cloning and characterization of oxidative 17β- and 3α-hydroxysteroid dehydrogenases from rat and human prostate. J Biol Chem 272:15959–15966

Brinkmann AO, Leemborg FG, Rommerts FFG, van der Molen HJ (1982) Translocation of the testicular oestradiol receptor is not an obligatory step in the gonadotropin-induced inhibition of C17-20-lyase. Endocrinology 110:1834–1836

Brucker C, Lipford GB (1995) The human sperm acrosome reaction – physiology and regulatory mechanisms – an update. Hum Reprod Update 1:51–62

Bulun SE (1996) Aromatase deficiency in women and men: would you have predicted the phenotypes? (clinical review). J Clin Endocr Metab 81:867–871

Calvo JC, Radicella JP, Pignataro OP, Charreau EH (1984) Effect of a second injection of human chorionic gonadotropin on the desensitized Leydig cells. Mol Cell Endocr 34:31–38

Chemes HE (1996) Leydig cell development in humans. In Payne AH, Hardy MP, Russell LD (eds) The Leydig cell. Cache River Press, Vienna, Il, pp 176–202

Christ M, Douwes K, Eisen C, Bechtner G, Theisen K, Wehling M (1995) Rapid non-genomic effects of aldosterone on sodium transport in rat vascular smooth muscle cells: involvement of the Na$^+$/ H$^+$ – antiport. Hypertension 25:117–123

Clark BJ, Soo SC, Caron KM, Ikeda Y, Parker KL, Stocco DM (1995) Hormonal and developmental regulation of the steroidogenic acute regulatory (StAR) protein. Mol Endocrinol 9:1346–1355

Coffey DS (1988) Androgen action and the sex accessory tissues. In Knobil E, Neill J (eds) The physiology of reproduction. Raven Press, New York, pp 1081–1119

Collett-Solberg PF, Cohen P (1996) The role of the insulin-like growth factor binding proteins and the IGFBP proteases in modulation IGF action. Endocrinol Metab Clin North America 25:591–614

Comfort A (1971) Likelihood of human pheromones. Nature 230:432–433

Cooke BA (1996) Transduction of the luteinizing hormone signal within the Leydig cell. In Payne AH, Hardy MP, Russell LD (eds) The Leydig cell. Cache River Press, Vienna, Il, pp 352–366

Ebeling P, Koivisto VA (1994) Physiological importance of dehydroepiadrosterone. Lancet 343:1479–1481

Ewing LL, Zirkin B (1983) Leydig cell structure and steroidogenic function. Rec Progr Horm Res 39:599–635

Forest MG (1989) Physiological changes in circulating androgens. Pediatr Adolesc Endocrinol 19:104–129

Freeman DA, Rommerts FFG (1996) Regulation of Leydig cell cholesterol transport. In Payne AH, Hardy MP, Russell LD (eds) The Leydig cell. Cache River Press, Vienna, Il, pp 232–239

Geissler WM, Davis DL, Wu l, Bradshaw KD, Patel S, Mendonca BB, Elliston KO, Wilson JD, Russell DW, Andersson S (1994) Male pseudohermaphroditism caused by mutations of testicular 17β-hydroxysteroid dehydrogenase 3. Nature Genet 7:34–39

George FW (1997) Androgen metabolism in the prostate of the finasteride treated adult rat: a possible explanation for the differential action of testosterone and 5α-dihydrotestosterone during development of the male urogenital tract. Endocrinology 138:871–877

Gnessi L, Fabri A, Spera G (1997) Gonadal peptides as mediators of development and functional control of the testis: An integrated system with hormones and local environment. Endocr Rev 18:541–609

Gorczynska E, Handelsman DJ (1995) Androgens rapidly increase the cytosolic calcium concentration in Sertoli cells. Endocrinology 136:2052–2059

Gower DB, Cooke GM (1983) Regulation of steroid-transforming enzymes by endogenous steroids. J Steroid Biochem 19:1527–1556

Gruol DJ, Bourgeois S (1997) Chemosensitizing steroids: glucocorticoid receptor agonists capable of inhibiting P-glycoprotein function. Cancer Res 54:720–727

Hall PF (1988) Testicular steroid synthesis: organization and regulation. In Knobil E, Neill J (eds) The physiology of reproduction. Raven Press, New York, pp 975–998

Hall PF (1991) Cytochrome P450 C21$_{scc}$: one enzyme with two actions:hydroxylase and lyase. J Steroid Biochem Mol Biol 40:527–532

Hammond GL, Ruokonen A, Kontturi M, Koskela E, Vihko R (1977) The simultaneous radioimmunoassay of seven steroids in human spermatic and peripheral venous blood. J Clin Endocrinol Metab 45:16–24

Haour F, Kovanetzova B, Dray F, Saez JM (1979) hCG induced prostaglandin E2 and F2a release in adult rat testis: role in Leydig cell desensitization to hCG. Life Sci 24:2151–2158

Hsueh AJW, Jones PBC (1981) Extrapituitary actions of gonadotropin-releasing hormone Endocr Rev 2:437–461

Hugson YM, de Kretser DM (1984) Acute responses of Leydig cells to hCG: evidence for early hypertrophy of Leydig cells. Mol Cell Endocr 35:75–82

Huhtaniemi I, Bolton NJ, Martikainen H, Vihko R (1983) Comparison of serum steroid responses to a single injection of hCG in man and rat. J Steroid Biochem 19:1147–1151

Ibarrola I, Ogiza K, Marino A, Macarulla JM, Trueba M (1991) Steroid hormone specifically binds to rat kidney plasma membrane. J Bioenerget Biomemb 23:919–926

Ishimaru T, Pages L, Horton R (1977) Altered metabolism of androgens in elderly men with benign prostatic hyperplasia. J Clin Endocrinol Metab 45:695–701

Koenig H, Fan CC, Goldstone AD, Lu CY, Trout JJ (1989) Polyamines mediate androgenic stimulation of calcium fluxes and membrane transport in rat heart myocytes. Circ Res 64 415–426

Labrie F, Luu-The V, Lin SX, Labrie C, Simard J, Breton R, Belanger A (1997) The key role of 17β-hydroxysteroid dehydrogenases in sex steroid biology. Steroids 62:148–158

Lachane Y, Luu-The V, Labrie C, Simard C, Dumont M, de Launou Y, Guerin S, Leblanc G, Labrie F (1990) Characterization of the structure-activity relationship of rat types I and II 3β hydroxysteroid dehydrogenase Δ5-Δ4-isomerase gene and its expression in mammalian cells. J Biol Chem 265:20469–20475

Li J, Chory J (1997) A putative leucine-rich repeat receptor kinase involved in brassinosteroid signal transduction. Cell 90:929–938

Lieberherr M, Grosse B (1994) Androgens increase intracellular calcium concentrations and inositol 1,4,5-trisphosphate and diacylglycerol formation via a pertussis toxin sensitive G protein. J Biol Chem 269:7219–7223

Lin D, Sugawara T, Strauss JF, Clark BJ, Stocco DM, Saenger P, Rogol A, Millar WL (1995) Indispensable role of steroidogenic acute regulatory protein in adrenal and gonadal steroidogenesis. Science 267:1828–1831

Lin T (1996) Insulin-like growth factor-I regulation of the Leydig cell In: Payne AH, Hardy MP, Russell LD (eds) The Leydig cell. Cache River Press, Vienna, Il, pp 478–491

Loosfelt H, Misrahi M, Atger M, Salesse R, Thi MTVH-L, Jolivet A, Guiochon-Mantel A, Sar S, Jallal B, Garnier J, Milgrom E (1989) Cloning and sequencing of porcine LH-hCG receptor cDNA: variants lacking transmembrane domain. Science 245:525–528

Lubicz-Nawrocki CM (1973) The effect of metabolites of testosterone on the viability of hamster epididymal spermatozoa. J Endocrinol 58:193–198

Luke MC, Coffey DS (1994) The male sex accessory tissues: structure, androgen action and physiology In Knobil E, Neill JD (eds) The physiology of reproduction, Raven Press, New York pp 1435–1480

Luu-The V, Takahashi Y, de Lanoit M, Dumont M, Lachane Y, Labrie F (1991) Evidence for distinct dehydrogenase and isomerase sites within a single 3β-hydroxysteroid dehydrogenase Δ5-ene-Δ4-ene isomerase protein. Biochemistry 30:8861–8865

Maddocks S, Sharpe RM (1989) Dynamics of testosterone secretion by the rat testis: implications for measurement of the intratesticular levels of testosterone. J Endocr 122:323–329

McFarland KC, Sprengel R, Phillips HS, Koher M, Rosemblit N, Nikolics K, Segaloff DL, Seeburg PH (1989) Lutropin-choriogonadotropin receptor: an unusual member of the G protein-coupled receptor family. Science 245:494–499

Mendel CM (1989) The free hormone hypothesis: a physiologically based mathematical model. Endocr Rev 10:232–274

Mendel CM, Murai JT, Siiteri PK, Monroe SE, Inove M (1989) Conservation of free but not total or non-sex-hormone binding-globulin-bound testosterone in serum from nagase analbuminemic rats. Endocrinology 124:3128–2130

Mermelstein PG, Becker JB, Surmeier DJ (1996) Estradiol reduces calcium currents in rat neostriatal neurons via a membrane receptor. J Neurosci 16:595–604

Michael RP, Bonsall RW, Rees HD (1986) Accumulation of 3H testosterone and 3H estradiol in the brain of the female primate: evidence for the aromatization hypothesis. Endocrinology 118:1935–1944

Miller WL (1988) Molecular biology of steroid hormone synthesis. Endocr Rev 9:295–318

Miura T, Yamauchi K, Takahashi H, Nagahama Y (1991) Hormonal induction of all stages of spermatogenesis in vitro in the male Japanese eel *(Anguilla japonica)*. Proc Natl Acad Sci USA 88:5774–5778

Miautani S, Tsujimura T, Akashi S, Matsumoto K (1977) Lack of metabolism of progesterone, testosterone and pregnenolone to 5α-products in monkey and human testes compared with rodent testes. J Clin Endocrinol Metab 44:1023–1031

Monno S, Ogawa H, Date T, Fujioka WL, Millar WL, Kobayashi M (1993) Mutation of histidine -373 to leucine in cytochrome-P450c17 causes 17α-hydroxylase deficiency. J Biol Chem 268:25811–25817

Mooradian AD, Morley JE, Korenman SG (1987) Biological actions of androgens. Endocr Rev 8:1–27

Morishima A, Grumbach MM, Simpson ER, Fisher C, Qin K (1995) Aromatase deficiency in male and female siblings caused by a novel mutation and the physiological role of oestrogens. J Clin Endocr Metab 80:3689–3698

Mulder E, Lamers-Stahlhofen GJM, van der Molen HJ (1973) Interaction between steroids and membranes. Uptake of steroids and steroid sulphates by resealed erythrocyte ghosts. J Steroid Biochem 4:369–379

Noordhuizen-Stassen EW, Charbon GA, de Jong FH, Wensing CJG (1985) Functional arterio-venous anastomoses between the testicular artery and the pampiniform plexus in the spermatic cord of rams. J Reprod Fert 75:193–201

Odell WD, Swerdloff RS, Jacobs HS and Hescox MA (1973) FSH induction of sensitivity to LH; one cause of sexual maturation in the male rat. Endocrinology 92:160–165

O'Malley BW, Schrader WT, Mani S, Smith C, Weigel NL, Conneely OM, Clark JH (1995) An alternative ligand-independent pathway for activation of steroid receptors. Rec Prog Horm Res 50:333–353

Papadopoulos V (1993) Peripheral-type benzo-diazepine/diazepam binding inhibitor receptor: biological role in steroidogenic cell function. Endocr Rev 14:22–240

Parker KL, Schimmer BP (1997) Steroidogenic factor 1: A key determinant of endocrine development and function. Endocr Rev 18:361–377

Payne AH, Quinn PG, Rani CSS (1985) Regulation of microsomal cytochrome P-450 enzymes and testosterone production in Leydig cells. Rec Progr Horm Res 41:153–185

Payne AH, Hardy MP, Russell LD (eds) (1996) The Leydig cell. Cache River Press, Vienna, Il.

Payne AH, O'Shaughnessy (1996) Structure, function, and regulation of steroidogenic enzymes in the Leydig cell. In Payne AH, Hardy MP, Russell LD (eds) The Leydig cell. Cache River Press, Vienna, Il, pp 260–285

Penning TM (1997) Molecular endocrinology of hydroxysteroid dehydrogenases. Endocr Reviews 18:281–305

Rice DA, Mouw AR, Bogerd AM, Parker KL (1991) A shared promoter element regulates the expression of three steroidogenic enzymes. Mol Endocrinol 5:1552–1556

Rittmaster RS (1994) Finasteride. New Eng J Med 330:120–125

Rhéaume E, Sanchez F, Mebarki F, Gagnon E, Carel JC, Chaussain JL, Morel Y, Labrie F, Simard JN (1995) Identification and characterization of the G15D mutation found in a male patient with 3β-hydroxysteroid dehydrogenase (3β-HSD) deficiency – alteration of the putative NAD- binding domain of type-II 3β-HSD Biochem. 34:2893–2990

Rommerts FFG (1988) How much androgen is required for maintenance of spermatogenesis? J Endocr 116:7–9

Rommerts FFG (1992) Cell surface actions of steroids: A complementary mechanism for regulation of spermatogenesis? In Spermatogenesis, Fertilization and Contraception (Molecular, Cellular and Endocrine Events in Male Reproduction), Schering Foundation Workshop 4, Nieschlag E and Habenicht UF (eds), Springer Verlag, Berlin pp 1–19

Rommerts FFG, Brinkmann AO (1981) Modulation of steroidogenic activities in testis Leydig cells. Mol Cell Endocr 21:15–28

Rommerts FFG, Cooke BA (1988) The mechanisms of action of luteinizing hormone. II. Transducing systems and biological effects. In: Cooke BA, King RJB, van der Molen HJ (eds) Hormones and their Actions, part II, Elsevier Science Publishers BV (Biomedical Division), Amsterdam, pp 163–180

Rommerts FFG, Grootegoed JA, van der Molen HJ (1976) Physiological role for androgen binding protein-steroid complex in testis? Steroids 22:43–49

Rommerts FFG, van der Molen HJ (1989) Testicular steroidogenesis. In: Burger H, de Kretser D (eds) The testis, 2nd ed. Raven Press, New York, pp 303–328

Saez JM (1989) Endocrine and paracrine regulation of testicular functions. Pediatr Adolesc Endocrinol 19:37–55

Saez JM (1994) Leydig cells: endocrine, paracrine and autocrine regulation. Endocr Rev 16:574–626

Schlatt S, Meinhardt A, Nieschlag E (1997) Paracrine regulation of cellular interactions in the testis: factors in search of a function. Eur J Endocr 137:107–117

Sergeev IN, Rhoten WB (1995), 1, 25-dihydroxyvitamin D-3 evokes oscillations of intracellular calcium in a pancreatic beta cell line. Endocrinology 136:2852–2861

Setchell BP, Laurie MS, Flint APF, Heap RB (1983) Transport of free and conjugated steroids from the boar testis in lymph, venous blood and rete testis fluid. J Endocr 96:127–136

Sharpe RM (1983) Local control of testicular function. Quart J Exp Physiol 68:265–282

Sharpe RM (1993) Experimental evidence for Sertoli-germ cell and Sertoli-Leydig cell interactions. In Russell LD, Griswold MD (eds) The Sertoli cell. Cache River Press, Vienna, Il, pp 391–418

Simpson BJB, Wu FCW, Sharpe RM (1987) Isolation of human Leydig cells which are highly responsive to human chorionic gonadotropin. J Clin Endocrinol Metab 65:415–422

Smals AGH, Pieters GFFM, Lozekoot DC, Benraad TJ, Kloppenborg PWC (1980) Dissociated responses of plasma testosterone and 17-hydroxyprogesterone to single or repeated human chorionic gonadotropin administration in normal men. J Clin Endocrinol Metab 50:190–193

Snyder SH, Sklar PB, Pevsner J (1988) Molecular mechanisms of olfaction. J Biol Chem 263:13971–13974

Sporn MB, Roberts AB (1988) Peptide growth factors are multifunctional. Nature 132:217–219

Steinsapir J, Socci R, Reinach P (1991) Effects of androgen on intercellular calcium of LNCaP cells. Biochem Biophys Res Comm 179:90–96

Stocco DM (1997) A StAR search: Implications in controlling steroidogenesis. Biol Reprod 56:328–336

Stocco DM, Clark BJ (1996) Regulation of the acute production of steroids in steroidogenic cells. Endocr Rev 17:221–244

Strohman RC (1993) Ancient genomes, wise bodies, unhealthy people: limits of genetic thinking in biology and medicine. Perspec Biol Med 37:112–145

Teerds KJ, de Rooij DG, Rommerts FFG, Wensing CJG (1988) The regulation of the proliferation and differentiation of Leydig cell precursors after EDS administration of daily hCG treatment. J Androl 9:343–351

Themmen APN, Martens JWM, Brunner HG (1997) Gonadotropin receptor mutations (review). J Endocrinol 153:179–183

Träger L (1977) Steroidhormone: Biosynthese, Stoffwechsel, Wirkung. Springer Verlag, Berlin, pp 164–197.

Ueda K, Okamura N, Hirai M, Tanigawara Y, Saeki T, Kioka N, Komano T, Hori R (1992) Human P-glycoprotein transports cortisol, aldosterone and dexamethasone but not progesterone. J Biol Chem 267:24248–24252

van Doorn LG, de Bruijn HWA, Galjaard H, van der Molen HJ (1974) Intercellular transport of steroids in the infused rabbit testis. Biol Reprod 10:47–53

van Haren L, Cailleau J, Rommerts FFG (1989) Measurement of steroidogenesis in rodent Leydig cells: a comparison between pregnenolone and testosterone production. Mol Cell Endocr 65:157–164

van Haren L, Flinterman JF, Orly J, Rommerts FFG (1992) Luteinizing hormone induction of the cholesterol side-chain cleavage enzyme in cultured immature rat Leydig cells: no role for insulin-like growth factor-I? Molec Cell Endocr 87:57–67

van Haren L, Flinterman JF, Rommerts FFG (1995) Inhibition of luteinizing hormone-dependent induction of cholesterol side-chain cleavage enzyme in immature rat Leydig cells by Sertoli cell products. Eur J Endocrinol 132:627–634

van Noort M, Rommerts FFG, van Amerongen A, Wirtz KWA (1988) Intracellular redistribution of SCP2 in Leydig cells after hormonal stimulation may contribute to increased pregnenolone production. Biochem Biophys Res Commun 154:60–65

Wang C, Plymate E, Nieschlag E, Paulsen CA (1981) Salivary testosterone in men. Further evidence of a direct correlation with free serum testosterone. J Clin Endocrinol Metab 53:1021–1024

Wehling M, Ulsenheimer A, Schneider M, Neylon C, Christ M (1994) Rapid effects of aldosterone on free intracellular calcium in vascular smooth muscle and endothelial cells: subcellular localization of calcium release by single cell imaging. Biochem Biophys Res Comm 204:475–481

Weusten JJAM, Smals AGH, Hofman JA, Kloppenborg PWC, Benraad ThJ (1987a) The sex pheromone precursor androsta-5,16 dien-3αol is a major early metabolite in in vitro pregnenolone metabolism in human testicular homogenates. J Clin Endocrinol Metab 65:753–756

Weusten JJAM, Smals AGH, Hofman JA, Kloppenborg PWC, Benraad TJ (1987b) Early time sequence in pregnenolone metabolism to testosterone in homogenates of human and rat testis. Endocrinology 120:1909–1913

White PC, Mune T, Agarwal AR (1997) 11β-hydroxysteroid dehydrogenase and the syndrome of apparent mineralocorticoid excess. Endocr Rev 18:135–156

Wiebe JP (1997) Nongenomic actions of steroids on gonadotropin release. In Conn PM (ed) Rec Prog Horm Res vol 52, pp 71–102

Wilson JD (1988) Androgen abuse by athletes. Endocr Rev 9:181–199

Wilson JD, Griffin JE, Russell DW (1993) Steroid 5α-reductase 2 deficiency. Endocr Rev 14:577–593

Zhang LH, Rodriguez H, Ohno S, Miller WL (1995) Serine phosphorylation of human P450c17 increases 17, 20-lyase activity: Implications for adrenarche and the polycystic ovary syndrome. Proc Natl Acad Sci USA 92:10619–10623

Zuber MX, Simpson ER, Waterman MR (1986) Expression of bovine 17α-hydroxylase cytochrome P450cDNA in non-steroidogenic (COS1) cells. Science 234:1258–1260

2 The androgen receptor: Physiology and pathophysiology

Charmian A. Quigley

Contents

2.1 Introduction

The androgen receptor (AR[1]) is a member of the steroid receptor family of nuclear transcription factors, whose action requires the binding of ligand. Upon binding of androgens the AR undergoes a conformational change that converts it to the active DNA-binding state. Interaction of the active ligand-bound AR with the hormone response element enhancer sequence in the promoter region of a target gene, in the presence of appropriate co-factors, alters the rate of transcription of the target gene. These events underlie the mechanism of androgen action. AR is found ubiquitously throughout mammalian tissues, implying that androgens have diverse roles not only in development and maintenance of male sexual function, but also in many other sexually dimorphic processes, from modulation of immune function, to development of neural tissues.

Defective function of the AR results in a state of androgen resistance – the androgen insensitivity syndrome (AIS). AIS represents an archetypal example of a hormone resistance disorder: androgens are secreted by the testes in normal or increased amounts; however, due to defective AR function there is loss of target organ response to the hormone and the effects of androgens are diminished or absent. The molecular cloning of the AR made it possible to dissect the various structural domains and functional roles of the AR in great detail *in vitro*, and to determine the molecular basis of the defective AR function that characterizes AIS. More than 200 mutations have now been reported in the AR gene, and the specific effects of these diverse mutations have provided insights into structure-function relationships of particular domains, and even individual amino acid residues of the AR.

Molecular abnormalities of the AR gene are found in two other important clinical disorders – spinal and bulbar muscular atrophy (SBMA; Kennedy disease) and prostate cancer. Patients with SBMA suffer progressive deterioration of spinal and bulbar motor neurons, in addition to having a number of features of androgen resistance. The AR of affected individuals contains an expanded amino-terminal polyglutamine tract, similar to that found in pro-

---

[1] Throughout this chapter, unless otherwise stated, the abbreviation AR refers to the human androgen receptor, hAR.

teins associated with other neurodegenerative diseases such as Huntington Chorea. In carcinoma of the prostate gland (CaP), particularly in metastatic or recurrent disease, the AR undergoes somatic mutation at the level of the prostate itself, however, it is normal in the patients' genomic DNA. In many cases, the development of somatic AR gene mutations coincides with the loss of hormone responsiveness of the tumor. These three diverse clinical disorders – AIS, SBMA and CaP – serve to highlight the importance of the AR as a transcription factor with diverse roles and diverse defects. This chapter reviews the physiology and pathophysiology of the AR, with particular attention to its role in the pathophysiology of AIS.

2.2 Androgen receptor gene

2.2.1 Genomic localization; structure of AR gene and mRNA

The complementary DNA (cDNA) encoding the human AR was cloned in 1988 (Chang et al. 1988a, 1988b; Lubahn et al. 1988a, 1988b; Tilley et al. 1989; Trapman et al. 1988) and subsequently localized to the long arm of the X chromosome at Xq11–q12 (Brown et al. 1989; Mahtani et al. 1991). The sequence of the intron-exon boundaries of the human AR gene was reported in 1989 (Lubahn et al. 1989) allowing rapid evolution in the understanding of the molecular basis of AIS by the use of polymerase chain reaction (PCR) and related techniques of molecular biology (for review see Quigley et al. 1995) (Fig. 2.1).

The AR gene is a single-copy gene that spans 75–90 kilobases (kb) of genomic DNA (Brinkmann et al. 1989; Kuiper et al. 1989; Lubahn et al. 1989). Its protein coding region (approximately 2800 base pair [bp] open reading frame) comprises 8 exons, designated 1–8 or A–H (Lubahn et al. 1989; Marcelli et al. 1990a). The exons are separated by introns up to 26 kb in size (Kuiper et al. 1989). Towards the 5' end of exon 1 is a CAG triplet repeat region that contains an average of 21±2 repeats (La Spada et al. 1991), with a range of 11–31 in the normal mixed-sex population (Edwards et al. 1991, 1992) and 14–35 in males (Macke et al. 1993). This repeat region is highly polymorphic, as there are 20 different allele sizes, 80% of females are heterozygous for the size of the repeat at this locus and there is variation in average repeat size between racial groups, the most frequent allele size in the black population being 18 repeats, compared with 21 in whites (Edwards et al. 1992). This region serves as a useful polymorphic genetic marker for inheritance patterns of X-chromosomal conditions including, but not limited to, AR defects. As discussed in section 2.7.2.3., expansion of this CAG repeat is the molecular defect identified in patients with SBMA (La Spada et al. 1991).

Northern blot analysis of poly(A+) RNA isolated from a number of human tissues reveals the presence of a major species of approximately 10-kb,

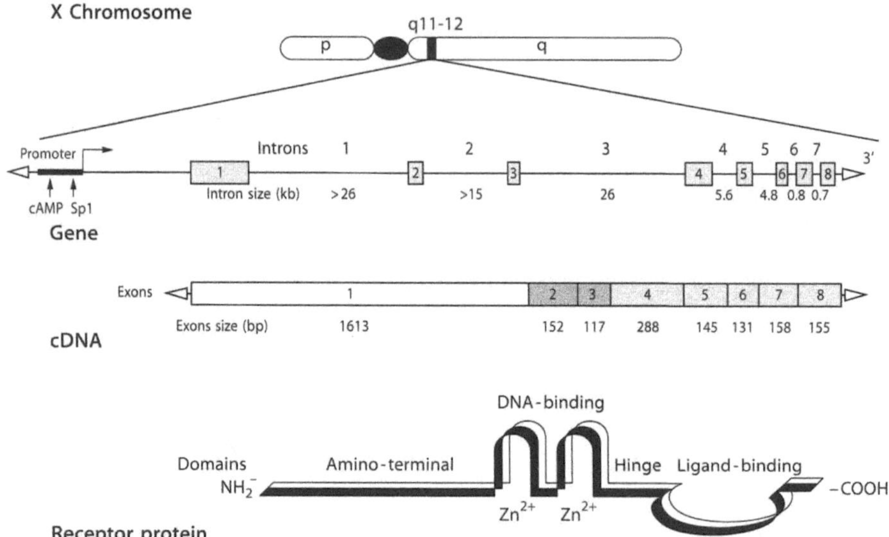

Fig. 2.1. The X-chromosomal locus of the AR gene and structural organization of the gene and protein. *Top:* The AR gene is located in the pericentromeric region of the long arm of the X chromosome at Xq11–12 and spans 75–90 kb of genomic DNA. The promoter region does not contain a TATA box but does contain binding sites for cAMP and SP1. The 8 coding exons (numbered within boxes) are separated by introns ranging from 0.7 kb to over 26 kb in size (numbered above the diagram; intron size below the diagram) (Kuiper et al. 1989). *Middle:* The androgen receptor cDNA comprises a coding region of 2757 bp. The sizes of the 8 exons are given below the diagram. Exon 1 encodes the amino-terminal domain; exons 2 and 3 encode the DNA-binding domain; the 5' region of exon 4 encodes the hinge region including the nuclear targeting signal; the 3' portion of exon 4 and exons 5–8 encode the LBD (Lubahn et al. 1989). *Lower:* The main functional domains of the androgen receptor are, from left to right: the amino-terminal domain, which contains the polymorphic polyglutamine region; the DNA-binding domain, comprising a pair of zinc fingers; the hinge region, containing the nuclear-targeting sequence; the ligand-binding domain. This diagram is not representative of actual 3-dimensional structure

and a less abundant ~7-kb species that results from differential processing of precursor mRNA (Faber et al. 1991; Lubahn et al. 1988 a; Wolf et al. 1993). The difference in size between the mRNA species and the translational open reading frame is due to the presence of untranslated 5' and 3' sequences.

2.2.2 Regulation of AR expression

Studies of the promoter for the human AR gene have determined that the principal site of transcription initiation (designated TIS I) is an adenosine (A) residue located approximately 1100 bp 5' of the translation-initiating methionine codon (Faber et al. 1991, 1993; Tilley et al. 1990). The minimal region of the promoter necessary for activity comprises nucleotides –74 to +87 surrounding the transcription initiation site (Takane and McPhaul 1996). The promoter of the AR gene does not contain a typical TATA or

CAAT box but does contain G-C rich elements, including a binding site for the ubiquitous mammalian transcription factor SP1, a homopurine stretch comprised of alternating A and G residues a cyclic AMP response element, binding sites for a variety of other transcription factors and sequences similar to binding sites for helix-loop-helix proteins (Baarends et al. 1990; Lindzey et al. 1993; Mizokami et al. 1994; Song et al. 1993; Tilley et al. 1990). A number of half-palindromic potential hormone response elements (HREs) for AR, GR and PR have been found within the promoter of the rat AR gene and one HRE half-site for estrogen receptor (ER) also has been reported (Baarends et al. 1990; Song et al. 1993). Such sequences may be important for regulation of AR mRNA expression by steroid hormones.

Androgenic regulation of AR expression – both at the transcriptional and at the translational level – is complex, being age, time, and cell- or tissue-type dependent. Furthermore, there are contrasting effects of androgens at the protein versus the mRNA level even within the same cell type (Evans and Hughes 1985; Gad et al. 1988; Krongrad et al. 1991; Wolf et al. 1993). At the protein level, androgens stabilize the receptor in transfected cells, reducing protein degradation and resulting in an increase in receptor level. This mechanism likely accounts for the apparent up-regulation of androgen binding after pre-incubation of fibroblast monolayers with androgen (Evans and Hughes 1985; Gad et al. 1988; Kemppainen et al. 1992; Krongrad et al. 1991; Pinsky et al. 1983; Syms et al. 1985), and perhaps for the higher receptor levels in male than in female fetal rats, and in rats exposed to high levels of androgen prenatally (Bentvelsen et al. 1993). In contrast, the overall effect of androgens at the mRNA level is down-regulation (Burnstein et al. 1995; Krongrad et al. 1991; Shan et al. 1990; Wolf et al. 1993). For example, blockade of androgen action by treatment with anti-androgens results in more than 4-fold increase in testicular AR mRNA content (Dankbar et al. 1995). Both androgen and dexamethasone down-regulate AR mRNA in COS 1 cells (Burnstein et al. 1995). However, in some cells, such as PC3 prostate cancer cells and osteoblasts, AR mRNA is up-regulated by androgens (Dai and Burnstein 1996; Wiren et al. 1997). AR autoregulation is mediated by sequences within the promoter and coding region of the AR (Burnstein et al. 1995; Dai and Burnstein 1996; Wiren et al. 1997). Hormones other than steroids also regulate AR in various cell or tissue types. For example, FSH up-regulates AR mRNA and protein in Sertoli cells, probably via cyclic AMP, while epidermal growth factor down-regulates AR mRNA in LNCaP prostate cancer cells (Blok et al. 1989; Henttu and Vihko 1993; Lindzey et al. 1993; Sanborn et al. 1991).

2.3 Androgen receptor protein

The AR is a member of the group of four closely-related steroid receptors that also includes the progesterone receptor (PR), glucocorticoid receptor

(GR) and mineralocorticoid receptor (MR) – sometimes referred to as the "GR-like" receptors. These four receptors, related by their sequence homology and also by their ability to activate target gene transcription via the same HRE, comprise a sub-group of the larger and more diverse family of nuclear transcription factors that includes the estrogen receptor (ER), thyroid hormone receptors (TRα and TRβ), vitamin D receptor (VDR), retinoic acid receptor (RAR), retinoid X receptor (RXR), and a number of related receptors for which ligands have not yet been identified in many cases. Members of this family interact directly with their target genes to up- or down-regulate gene transcription (Adler et al. 1991; Claessens et al. 1989; De Vos et al. 1991; Ho et al. 1993; Rennie et al. 1993; Tan et al. 1992).

2.3.1 Functional domains and structural organization of the AR

The predominant form of the AR is a 110–114 kDa protein of 910–919 amino acids (a.a.) (Jenster et al. 1991; Quarmby et al. 1990a; Wilson et al. 1992). The variation in reported length is due to the variable lengths of the triplet repeats in exon 1 of the AR coding sequence (Chang et al. 1988a, 1988b; Lubahn et al. 1988a, 1989; Tilley et al. 1989; Trapman et al. 1988)[2]. A smaller 87 kDa form of the AR, lacking approximately 190 amino acids from the amino terminus, is produced from an alternative translation-initiation methionine codon located within exon 1 (Wilson and McPhaul 1994). This truncated receptor (designated AR-A) comprises approximately 4–26% of detectable AR, depending on the specific tissue (Wilson and McPhaul 1996). The *in vivo* relevance of the shorter form of the AR is at present unknown. The full-length AR (the only form discussed hereafter) is a single polypeptide comprised of relatively discrete functional domains: an amino-terminal domain; a DNA-binding domain (DBD); a hinge region; and a ligand-binding domain (LBD) (Fig. 2.1). Although the domains are distinct, they interact to alter each other's behavior: for example, until the LBD is occupied by androgen it blocks transactivation by the amino-terminal domain and the amino-terminal domain blocks DNA binding. The three-dimensional structure of the AR is as yet unknown, but it is likely to be similar to that of other nuclear receptors (NRs), for which the three-dimensional structure of the DBD and LBD have been established by nuclear resonance spectroscopy and crystallographic analysis (Härd et al. 1990; Luisi et al. 1991; Wurtz et al. 1996).

---

[2] The nucleotide and a.a. numbering used in this chapter is according to the report of Lubahn et al. (1989). In correlating reports from various laboratories the difference in reported AR length of up to nine a.a. should be accounted for using the following example: arginine 774 in the sequence of Lubahn et al. (1989) corresponds to arginine 773 in the sequence published by Chang et al. (1988b), to arginine 772 in the sequence reported by Tilley et al. (1989) and to arginine 765 in the sequence published by Trapman et al. (1988).

2.3.1.1 The amino-terminal domain

Encoded by exon 1, this is the largest domain of the AR, comprising over half of the AR protein (residues 1–~537). The amino-terminal domain is the most variable in size and least homologous in sequence between members of the steroid receptor family. Within this domain is a transcription activation region – activation function 1 (AF-1) – located between amino acids 141 and 338 (Simental et al. 1991). There are also other sub-regions within the amino terminus involved in regulation of gene transcription (Jenster et al. 1991, 1995; Palvimo et al. 1993; Rundlett et al. 1990; Simental et al. 1991). The amino-terminal domain of human, rat and mouse AR contains a number of homopolymeric amino acid stretches. The most amino-terminal of these, encoded by the polymorphic CAG triplet repeat, is a stretch of 21 ± 2 glutamine residues (~a.a. 58–79) (La Spada et al. 1991). Further downstream is a stretch of nine proline residues (~a.a. 372–379) that does not vary in size, and closer to the DNA-binding domain is a polymorphic region of about 24 glycine residues (~a.a. 449–472). These regions, particularly the polyglutamine and polyproline regions, may be important in transcriptional regulation via protein-protein interactions with other transcription factors (Gerber et al. 1994; McEwan and Gustafsson 1997; Palvimo et al. 1996). In addition to its roles in transcription regulation, the amino terminus probably contributes to the three-dimensional structure and conformation of the receptor molecule through interaction with other regions of the protein (Evans 1988; Langley et al. 1995). As an example, relative shortening of the amino-terminal polyglutamine tract found in one PAIS kindred (Batch et al. 1993a; McPhaul et al. 1991a) was associated with increased thermolability of androgen binding, suggesting that amino-terminal/LBD interactions are required for the maintenance of stable androgen binding.

2.3.1.2 The DNA-binding domain (DBD)

The central region of the AR encoded by exons 2 and 3 contains the DNA-binding domain (a.a. 559–624). Within this region four cysteine residues, invariably present in all steroid receptors, bind a zinc ion in each of two loop structures known as *"zinc fingers"*. The first zinc finger is encoded by exon 2 and the second by exon 3. The amino acid sequence of this domain is the most highly-conserved region among members of the steroid receptor family, the AR zinc fingers having approximately 80% amino acid identity with those of MR, PR and GR (Freedman 1992). The DBD determines the specificity of receptor interaction with DNA (Berg 1989; Danielsen et al. 1989; Freedman 1992; Green et al. 1988; Truss and Beato 1993), with each of the zinc fingers subserving distinct functions. The first zinc finger is responsible for recognition of the target DNA sequence, three amino acids at its base (glycine 577, serine 578, valine 581) being particularly critical (Berg 1989; Danielsen et al. 1989; Freedman 1992; Green et al. 1988). This trio, also present in GR, PR and MR, appears to make direct contact with HREs of target genes (Marschke et al. 1993). The second zinc finger, which is highly basic, stabilizes DNA-receptor interaction by contact with the DNA phosphate backbone (Berg 1989). The three-dimensional structure of the DBD of GR and ER has been determined by crystallographic and nuclear magnetic resonance analysis to comprise a number of structural elements including reverse turns, a B-sheet and the two fingers themselves, each followed by an alpha helix which are oriented perpendicular to each other forming a compact hydrophobic core (Freedman 1992; Luisi et al. 1991; Zhang et al. 1997). The first finger folds over the hydrophobic core and the second projects out from the core (Freedman 1992).

2.3.1.3 The nuclear targeting signal and hinge region

The nuclear targeting signal is a bipartite sequence consisting of two clusters of basic residues separated by ten amino acids located at a.a. 617–633[3] in the DBD and hinge region (Jenster et al. 1993; Zhou et al. 1994). This 17 amino acid sequence mediates the transfer of AR from the cytoplasm to the nucleus by as yet unknown mechanisms that require interaction between the amino-terminal domain and LBD (Jenster et al. 1991, 1993; Simental et al. 1991; Zhou et al. 1994). The hinge, located between the DBD and the LBD, encoded by the 5′ region of exon 4, is a region of low sequence homology between AR and other steroid receptors. The hinge appears to be involved in conformational changes of the AR induced by binding of androgens and antiandrogens and may serve as an interface for interactions with other proteins such as c-Jun (part of the AP-1 transcription complex)(Bubulya et al. 1996; Kuil and Mulder 1994; Moilanen et al. 1997). In addition, the hinge contains one of the AR phosphorylation sites (see Section 2.6.2) required for optimal AR transcriptional activity (Zhou et al. 1995).

2.3.1.4 The ligand-binding domain (LBD)

The ligand-binding domain (LBD), encoded by the 3′ portion of exon 4 and exons 5–8, encompasses approximately the carboxy-terminal one-third of the protein, a.a. 670–919. The a.a. of this region displays about 50% identity with the corresponding residues in GR, MR and PR. A principal function of the LBD is the specific, high-affinity binding of androgens. This region is also likely the binding site for inhibitory proteins such as the 90-kDa heat shock protein and has a role in other receptor functions including dimerization and transcriptional regulation, via a region referred to as activation function 2 (AF-2) (Housley et al. 1990; Jenster et al. 1991; Moilanen et al. 1997; Nemoto et al. 1994; Simental et al. 1991; Smith and Toft 1993; Wong et al. 1993). Crystallographic studies of the three-dimensional structure of the LBD of a

---

[3] The location of the nuclear targeting sequence reported by Jenster et al. (1993) is essentially equivalent to that reported by Zhou et al. (1994) when the difference in a.a. numbering scheme between the two laboratories is accounted for.

number of steroid receptors predicts a consensus structure comprising 12α-helices arranged in a three-layer sandwich-like configuration with minor regions of variation (Renaud et al. 1995; Wurtz et al. 1996). The actual site at which the ligand binds is buried in a hydrophobic pocket formed by interaction of 24 residues in helices 1, 3, 5, 11 and 12, and loops 6, 7, 11 and 12 (Wurtz et al. 1996). A 'lid' for the binding cavity is formed by helices 11 and 12. Renaud et al. (1995) suggest that ligand binding alters the structure of the LBD to a more compact one, less vulnerable to the actions of proteases. These conformational changes, particularly in the 11th and 12th helices, create the surface required for interaction with transcriptional cofactors, contributing to the transcriptional activity of nuclear receptors.

2.4 Tissue distribution of AR

In keeping with the diverse biological roles of androgens, androgen binding is present almost ubiquitously throughout the tissues of the body. AR protein and/or mRNA expression have been detected in a wide array of genital and non-genital tissues and cell lines, in a variety of species, from the human (Kimura et al. 1993; Sar et al. 1990; Tilley et al. 1990) to the frog (Fischer et al. 1993). In the human, AR mRNA is detectable by Northern blot analysis in testis, prostate and genital skin fibroblasts, liver, the prostate cancer cell line LNCaP (Lymph Node Carcinoma of the Prostate) and the breast cancer cell lines T47D and MCF-7 (Nakagama et al. 1991; Quarmby et al. 1990b; Tilley et al. 1990; Wolf et al. 1993). The human AR protein is detectable by immunoblot analysis (Western blot) in extracts of genital skin fibroblasts (Beitel et al. 1994b; Wilson et al. 1992) and by immunohistochemistry in the nuclei of glandular epithelial cells of the prostate, male and female fetal external genitalia, cultured foreskin fibroblasts of normal and AIS-affected human subjects, ovary, prepubertal testis, fetal cartilaginous tissue, sebaceous glands and hair follicles (Ben-Hur et al. 1997; Blauer et al. 1991; De Bellis et al. 1992; Horie et al. 1992; Kalloo et al. 1993; Quigley et al. 1992a) and (C.A. Quigley, F.S. French and M. Sar, unpublished). Immunohistochemical study of paraffin-embedded human tissues has demonstrated AR in the nuclei of a wide variety of tissues, including sweat glands, hair follicles, cardiac muscle, vascular and gastrointestinal smooth muscle cells, thyroid follicular cells, adrenal cortical cells, prostate, liver, pineal gland and various cell types in the temporal cortex (Hinchliffe et al. 1996; Kimura et al. 1993; Loda et al. 1994; Luboshitzky et al. 1997; Puy et al. 1995). Androgen binding studies confirm the wide distribution of AR, generally indicating greater AR expression in genital than in non-genital tissues (Griffin et al. 1976). In general, there is a similar amount and distribution of AR in male and female genital skin (Griffin et al. 1976; Mowszowicz et al. 1981).

2.5 Ontogeny of AR expression in fetal and postnatal life

Expression of AR in the human fetus is present even prior to the onset of testicular androgen secretion and androgen binding is greater in genital than non-genital tissues during weeks 8–22 of gestation (Sultan et al. 1980, 1982). There is a marked similarity in distribution and intensity of AR staining in the external genitalia of male and female fetuses at 18–22 weeks gestation (Kalloo et al. 1993), a finding that explains the propensity for female fetuses to virilize when exposed to supranormal androgen concentrations *in utero*.

The few studies that have systematically examined the expression of AR with age have demonstrated somewhat conflicting results, likely due to methodological differences. Sultan et al. (1982) found no change in AR level in the neonatal period during the time of the postnatal testosterone surge, and Mowsowicz et al. (1981) demonstrated no age-related change in cytosolic AR in adult men. Similarly, Fichman et al. (1981) found that the total amount of AR was fairly constant with age, however, a greater proportion of the total receptor was located in cell nuclei during times of higher androgen exposure such as infancy and puberty. In contrast, in an analysis of 49 males from infancy to 59 years of age, Roehrborn et al. (1987), found a marked age-related change in foreskin total AR level. AR level was high in the newborn, dropping to a nadir at one year of age, then rising slowly again throughout childhood to peak at 17–18 years of age. The fact that the amount of AR increases before the teenage surge in testicular androgen production suggests that factors beyond androgens are involved in stimulation of AR expression during this time. However, the dramatic increase in foreskin AR content during puberty, paralleling the increase in serum testosterone, supports a stimulatory effect of testosterone on AR expression or stability during the time of most marked penile growth. Roehrborn et al. (1987) noted that after cessation of penile growth in the late teenage years, foreskin AR content declined steadily to a nadir at about 30–40 years of age, and remained at this low level thereafter.

2.6 Molecular mechanisms of AR action

2.6.1 Ligand binding

The unliganded AR binds the chaperone protein Heat Shock Protein 90 (HSP90) and perhaps related HSPs (Carson-Jurica et al. 1990; Fang et al. 1996; Housley et al. 1990; Smith and Toft 1993). This interaction retains the AR in an inactive state and is important for maintenance of a conformation optimal for high affinity ligand binding. Thus the binding of HSP90 contributes to regulation of AR hormone binding affinity (Fang et al. 1996). Androgens are transported to target tissues largely in a protein-bound state, where they dissociate from carrier proteins, diffuse into target cells and bind to the

AR. Wurtz et al. (1996) have proposed a consensus structure and mechanism of action for the LBD of nuclear receptors, including AR. They suggest that the ligand is attracted into the ligand-binding pocket by electrostatic forces, inducing a conformational change that involves repositioning of helices and loops and results in enclosure of the ligand within the pocket by a 'lid' formed by helix 12. Once the ligand is bound, the structure of the LBD becomes more compact, presumably facilitating subsequent steps in the pathway of receptor action. While these events have not been proven for AR, it is likely that the general principles will hold true. It is known that androgen binding induces a conformational change in the AR – perhaps as a result of removal of proteins such as HSP90, unmasking certain functional domains. These changes facilitate receptor dimerization, nuclear transport and interaction with target DNA (Colvard and Wilson 1987; Grino et al. 1987), and culminate in regulation of target gene transcription (Fig. 2.2). In addition, androgen binding stabilizes the receptor, presumably preventing proteolysis, resulting in higher levels of immunoreactive receptor and higher levels of bind-

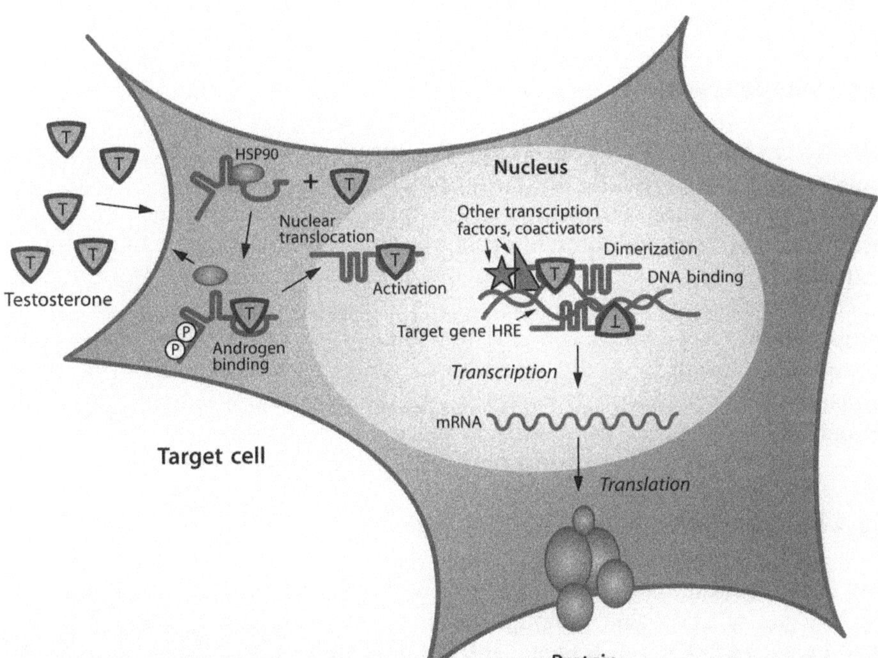

Fig. 2.2. The molecular events involved in induction of an androgen effect within the target cell. The initial step is the binding of androgen (T) to the cytosolic AR. This results removal of inhibitory proteins such as HSP90 (oval), increase in receptor phosphorylation (P), and activation of the receptor to its DNA-binding state. After translocation to the cell nucleus, the androgen-receptor complex undergoes dimerization and binding to a hormone response element (HRE), regulatory DNA sequence in or near a target gene, in concert with other transcription-regulating proteins, including the AR-specific co-activator, ARA70. This interaction of the receptor dimer with its target gene regulates gene transcription, either by enhancing or repressing it

ing (Kemppainen et al. 1992; Krongrad et al. 1991). Binding of androgen is necessary for subsequent DNA binding and transcriptional activity of the wild-type receptor. However, a mutant AR with *in vitro* deletion of its LBD induces transcription in the absence of hormone (Jenster et al. 1991; Simental et al. 1991), suggesting that in the absence of androgens, the LBD acts as a repressor of AR transactivation (transcription activation) regions, perhaps due to an inhibitory conformation adopted by the unliganded protein.

A variety of androgens and other steroids are bound by AR: the AR has highest affinity for DHT, followed by testosterone, primarily because dissociation of the hormone-receptor complex occurs more slowly with DHT than with testosterone (Wilson and French 1976). Wild-type AR has low affinity for adrenal androgens such as dehydroepiandrosterone and androstenedione (Culig et al. 1993) and for non-androgenic steroids such as progesterone and estradiol (Wilson and French 1976); however, some mutant ARs display increased affinity for non-androgenic steroids and antiandrogens. The higher the affinity of ligand for receptor, the greater the transcriptional activity of the receptor, indicating that receptor activation is ligand specific (Wong et al. 1993).

2.6.2 Phosphorylation

The AR is phosphorylated at serine and threonine residues throughout the protein, particularly in the amino-terminal domain (serine 81 and 94) and hinge region (serine 650), inducing an increase in the apparent molecular mass of the protein. Receptor phosphorylation is a post-translational process common to steroid receptors and other transcriptionally-active proteins, mediated by protein kinases A and C. AR phosphorylation is increased by androgens and appears to enhance the transcriptional activity of the receptor, perhaps by altering the interaction of ligand-activated AR with other proteins of the transcription complex (Ikonen et al. 1994; Jenster et al. 1994; Kuiper and Brinkmann 1995; Zhou et al. 1995).

2.6.3 Nuclear localization

The AR is believed to be located in the cytoplasmic compartment prior to the binding of androgen, although its precise intracellular location is uncertain. Once bound by androgens, the AR is a nuclear protein. Immunocytochemical studies have demonstrated that nuclear uptake of the AR is an androgen-dependent process, mediated via the nuclear targeting signal located in the C-terminal segment of the second zinc finger and the hinge region. The mechanism by which this signal mediates nuclear transport of the receptor is unknown but may involve hormone-mediated interactions between the amino- and carboxy-terminal regions of the protein (Jenster et al. 1993; Simental et al. 1991; Zhou et al. 1994).

2.6.4 Dimerization, DNA binding and transactivation

Wild-type AR forms homodimers which appear to result from interactions between the amino terminus of one AR molecule and the LBD of another, producing an anti-parallel orientation of receptors upon dimerization (Doesburg et al. 1997; Langley et al. 1995). Receptor dimerization is a hormone-dependent process, presumably because a specific hormone-induced conformation is required to allow the protein molecules to interact in a stable fashion. The exact sequence of events in dimerization/DNA binding is unclear. In some studies the interaction of receptor with DNA appears to be the primary event, and receptor monomers subsequently assemble on the DNA to form dimers (Berg 1989; Green et al. 1988; Luisi et al. 1991; Wong et al. 1993). However, other data for GR suggest that receptor dimerization occurs in the absence of DNA (Wrange et al. 1989), presumably followed by binding of the receptor dimer to the DNA target site. Whichever the case for AR, it is known that AR homodimers bind via their DBDs to HREs of target genes and their flanking DNA (Marschke et al. 1993; Palvimo et al. 1993). This interaction results in regulation of the rate of transcription of these genes (either activation or repression), probably through interactions with other components of the transcription complex near the transcription start site of the gene (Archer et al. 1991; Beato 1989; Cullen et al. 1993; Ptashne 1988). A number of activator regions including AF-1 in the amino-terminus, and AF-2 in the LBD, may be required for optimal AR transactivation function at specific promoters and within specific cell types.

2.6.5 Androgen response elements; AR target genes

The AR binds to a simple HRE consensus sequence (a 15 bp partial palindrome-5'-AGNACAnnnTGTNCT-3'), which is also bound by GR and PR (Cato et al. 1987; Claessens et al. 1989; Denison et al. 1989; Ham et al. 1988; Lieberman et al. 1993; Marschke et al. 1993; Nordeen et al. 1990; Rundlett et al. 1990; Tan et al. 1992; Truss and Beato 1993). This overlap in activity between AR, PR and GR at certain HREs may allow for synergistic actions of androgens and glucocorticoids (Ho et al. 1993; Marschke et al. 1994). In addition, recent studies suggest that the AR may heterodimerize with GR at the common HRE and influence each other's transcriptional activity, perhaps underlying the differential physiological effects of androgens and glucocorticoids (Chen et al. 1997). Specificity of AR action at the common HRE is likely conferred by interactions with AR-specific cofactors. The most potent androgen response element (ARE) discovered to date is an 11-bp sequence consisting of a pair of overlapping direct repeats of the consensus GRE (Zhou et al. 1997). In addition, AR may act via selective complex AREs comprised of a number of interacting elements, including the HRE itself, and recognition sequences for other transcription control factors (Adler et al. 1991; Ho et al. 1993; Marschke et al. 1994; Rennie et al. 1993). AREs have been discovered and AR regulation demonstrated for a number of genes including the probasin gene (Rennie et al. 1993), the gene encoding prostate specific antigen (PSA) (Zhang et al. 1997), the factor IX gene (Morgan et al. 1997) and genes for prostatic binding protein (Claessens et al. 1989, 1993) and cystatin-related proteins in the rat (DeVos et al. 1997). In addition, a number of androgen-regulated genes of unknown function have been identified by reverse-transcriptase PCR of human foreskin fibroblasts (Nitsche et al. 1996), including genes whose expression is repressed by androgens. One interesting testosterone-induced gene identified by this strategy was that encoding testican, a testicular proteoglycan that may be involved in cell adhesion, migration and proliferation. Such a protein would be an obvious candidate for involvement in the processes of genital morphogenesis.

2.6.6 Protein-protein interactions

AR regulation of transcription is enhanced by interaction with a 70-Kd, 614 amino acid protein known as ARA70 (Yeh and Chang 1996). The 10-fold enhancement of AR transcriptional activity in prostate cancer cells is ligand dependent and specific: it is present with testosterone and DHT but not with hydroxyflutamide (Yeh and Chang 1996). AR has direct interactions with the basal transcription factors TFIIF and TBP (TATA-box-binding protein) via the AR amino terminus and this interaction appears to stimulate recruitment of the transcriptional machinery to the promoter region of the target gene (McEwan and Gustafsson 1997). The c-Jun component of AP-1 also has a direct interaction with AR, probably via interaction of the leucine zipper region of c-Jun and the DBD of AR (Sato et al. 1997). However, this interaction represses AR transactivation. Similarly, interaction between an amino-terminal region of AR and RelA, a member of the NF-kappaB family, represses AR-mediated transactivation (Palvimo et al. 1996). It is likely that protein-protein interactions between AR and other transcription factors and cofactors are numerous and may have significant influences on the transcriptional function of AR.

2.6.7 Ligand-independent AR activation

Interestingly, there is evidence for ligand-independent activation of AR transcriptional activity by the peptide growth factors particularly insulin-like growth factor I (IGF-I) and to a lesser extent, keratinocyte growth factor (KGF) and epidermal growth factor (EGF) (Culig et al. 1995). In addition, the protein kinase A activator, forskolin, can activate AR to its transcriptionally-competent state in the absence of androgens (Nazareth and Weigel 1996), likely via an effect on its phosphorylation state; this activity requires the DBD of the receptor and may be important in the androgen-independent growth of prostatic tumors.

2.7 Role of AR in fetal sexual differentiation

Normal human sexual development begins with the establishment of genetic sex at fertilization, and is followed by sex *determination* (gonad development) and sex *differentiation* (genital development). Human embryos of both genetic sexes develop identically for the first six weeks of gestation, being endowed with bipotential gonadal tissue, two sets of internal genital ducts – Wolffian (mesonephric) and Müllerian (paramesonephric) – and undifferentiated external genitalia. Differentiation of the testes from the indifferent gonadal ridge begins at around 6 weeks' gestation, directed by SRY, a DNA-binding transcription factor, which is the product of the *SRY* gene (*Sex*-determining *Region* of the *Y*) located on the short arm of the Y chromosome (Sinclair et al. 1990). There is likely interaction from other autosomal or X-chromosomal genes in the process of testis development (Bardoni et al. 1994; Tommerup et al. 1993). Since the testes of individuals with AIS develop normally – at least initially – it can be deduced that testicular differentiation is not an androgen-dependent process.

Following development of the testis, the events of male sex differentiation take two paths, one inhibitory (Müllerian duct regression), the other stimulatory (Wolffian duct development) (Fig. 2.3). Regression of the Müllerian ducts at six to eight weeks' gestation precludes development of female internal genital structures – Fallopian tubes, uterus and upper vagina. Müllerian duct regression is stimulated by the Sertoli cell glycoprotein anti-Müllerian hormone (AMH) (Josso et al. 1991), also known as Müllerian inhibiting substance (MIS) (Lee and Donahoe 1993). AMH is a member of the transforming growth factor-β family of peptides and acts via a serine-threonine kinase receptor, the AMH-R. The stimulatory events of male sex differentiation – stabilization of the Wolffian duct system and subsequent differentiation into the epididymides, vasa deferentia and seminal vesicles – requires the presence of androgens and a functional AR. AR can be detected in fetal skin prior to testosterone secretion by fetal Leydig cells, which begins around eight weeks, under the control of unknown factors, and peaks between 11 and 18 weeks' gestation, under the control of maternal chorionic gonadotropin. Development of the male internal genital ducts occurs between 9 and 13 weeks' gestation, induced by the action of testosterone itself (rather than DHT, since the enzyme 5-α reductase 2, required for production of DHT from testosterone, is not expressed in these tissues until about 13 weeks' gestation). In this context, testosterone is probably acting via a paracrine effect due to its high local concentrations in the vicinity of the Wolffian ducts, which are in close proximity to the testes. Development of the prostate and prostatic urethra from the urogenital sinus, and masculinization of the external genital primordia – the genital tubercle, urethral folds and labioscrotal swellings – into the penis, penile urethra and scrotum, also occurs between 9 and 13 weeks' gestation (Fig. 2.3). These latter events require the more potent androgen DHT, and 5-α reductase 2 is expressed in these tissues at the appropriate time. Mediation of the effects of both testosterone and DHT re-

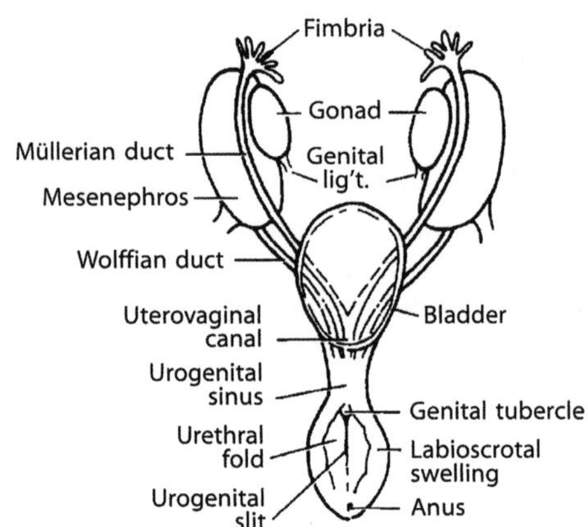

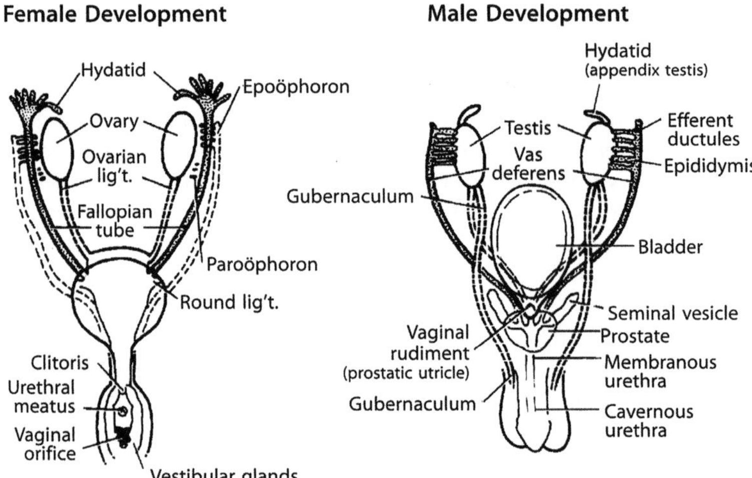

Fig. 2.3. a Differentiation of fetal internal genitalia. Top figure shows internal genitalia at approximately 6 weeks' gestation, before differentiation begins. The fetus has two complete sets of internal ducts – mesonephric (Wolffian) and paramesonephric (Müllerian). Lower left figure shows female internal genital development, with regression of Wolffian duct system (dotted) and development of Müllerian ducts into fallopian tubes, uterus and upper vagina. Lower right figure shows male internal development, with regression of Mullerian ducts (dotted) and development of Wolffian ducts into epididymis, vas deferens and seminal vesicles. **b** Differentiation of fetal external genitalia. Top – indifferent external genitalia at approximately 6 weeks' gestation. Left – female phenotypic development in the absence of androgens or a functional AR; Right – male phenotypic development in the presence of normal concentration of androgens and functional AR

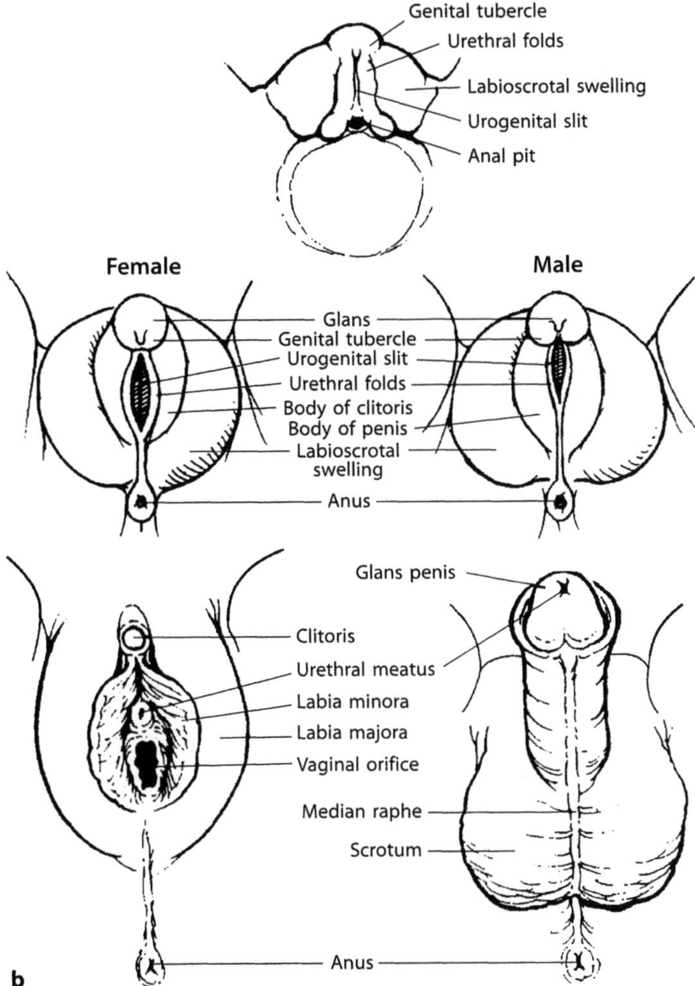

Fig. 2.3 b   **b**

quires a functional AR to induce expression of androgen-dependent genes necessary for internal and external genital masculinization (Imperato-McGinley et al. 1992; Reyes et al. 1989; Sultan et al. 1982; Wilson et al. 1981, 1993; Word et al. 1989).

2.8. Defects of the androgen receptor

2.8.1 Androgen insensitivity syndromes

It has been suggested that the syndromes of androgen insensitivity represent the single most common identifiable cause of male pseudohermaphroditism

Fig. 2.4. Individuals with various clinical forms of AIS. **a.** Complete AIS (grade 7) – no pubic hair in adulthood; **b.** AIS grade 6 – pubic hair present in adulthood in individuals with female phenotype; **c.** AIS grade 5 – mild posterior labial fusion and clitoral enlargement (child); **d.** AIS grade 4 – genital ambiguity with a clitoral-like phallus, a urethral orifice on the perineum, and smooth labioscrotal folds which contain testes. The androgen receptor gene of this child contains a single base mutation that converts arginine 840 to histidine in the LBD of the receptor (De Bellis et al. 1994). Reprinted with permission. The Endocrine Society; **e.** AIS grade 4 – small penis with perineal hypospadias; some rugosity of unfused labioscrotal folds present; **f.** AIS grade 2–3 – somewhat small penis, hypospadias with urethral orifice at the base of the penis, fused scrotum, bilateral cryptorchidism. Photograph kindly provided by Dr. Martin Ritzén, The Karolinska Institute, Stockholm

(Savage et al. 1978). Affected individuals have a 46,XY karyotype and normally-developed testes. The external genital phenotype varies greatly (Fig. 2.4) and the various AIS subtypes were previously designated by a variety of syndromic titles, suggesting the existence of discrete disorders (testicular feminization, partial AIS, Reifenstein syndrome, Gilbert-Dreyfus syndrome, Lubs syndrome, Goldwater syndrome, infertile male syndrome, etc.). In 1965 the etiology of AIS was suggested as "an absence or alteration of the target site, making the peripheral organs refractory to the effects of androgens" (French et al. 1965). The terms *androgen insensitivity* or *androgen resistance* were subsequently adopted by most clinicians, in place of the original term *testicular feminization*, both because they more accurately reflect the nature of the disorder, and because they are considered to be psychosocially preferable. The androgen insensitivity syndrome is the term preferred by the worldwide support group for individuals and families with AIS – the Androgen Insensitivity Syndrome Support Group (AIS-SG). Molecular analysis of the AR gene in individuals with AIS has clarified the fact that, rather than being discrete disorders, the various clinical forms of AIS represent a continuum, resulting from mutations of variable type and severity in the AR gene. In order to facilitate evaluation of the relationship between clinical phenotype and structural and functional defects of the AR, the following phenotypic classification of individuals with AIS was proposed, but is not intended to replace detailed clinical description of the physical findings (Quigley et al. 1995). The grading system for genital phenotype in AIS is modeled on the Prader classification for congenital adrenal hyperplasia (Prader 1954): genital phenotype of an individual with AIS is graded 1–7 in order of increasing severity of androgen resistance and thus increasingly more female phenotype (Fig. 2.5)[4].

---

[4] This classification system is employed only in cases where there is adequate clinical information provided in the report, on which to base a judgment. Where insufficient clinical data are available, the simpler designation of *complete AIS* or *partial AIS* has been used. If there is any indication of androgen action, such as clitoromegaly or pubic hair, the designation of *partial AIS* has been given.

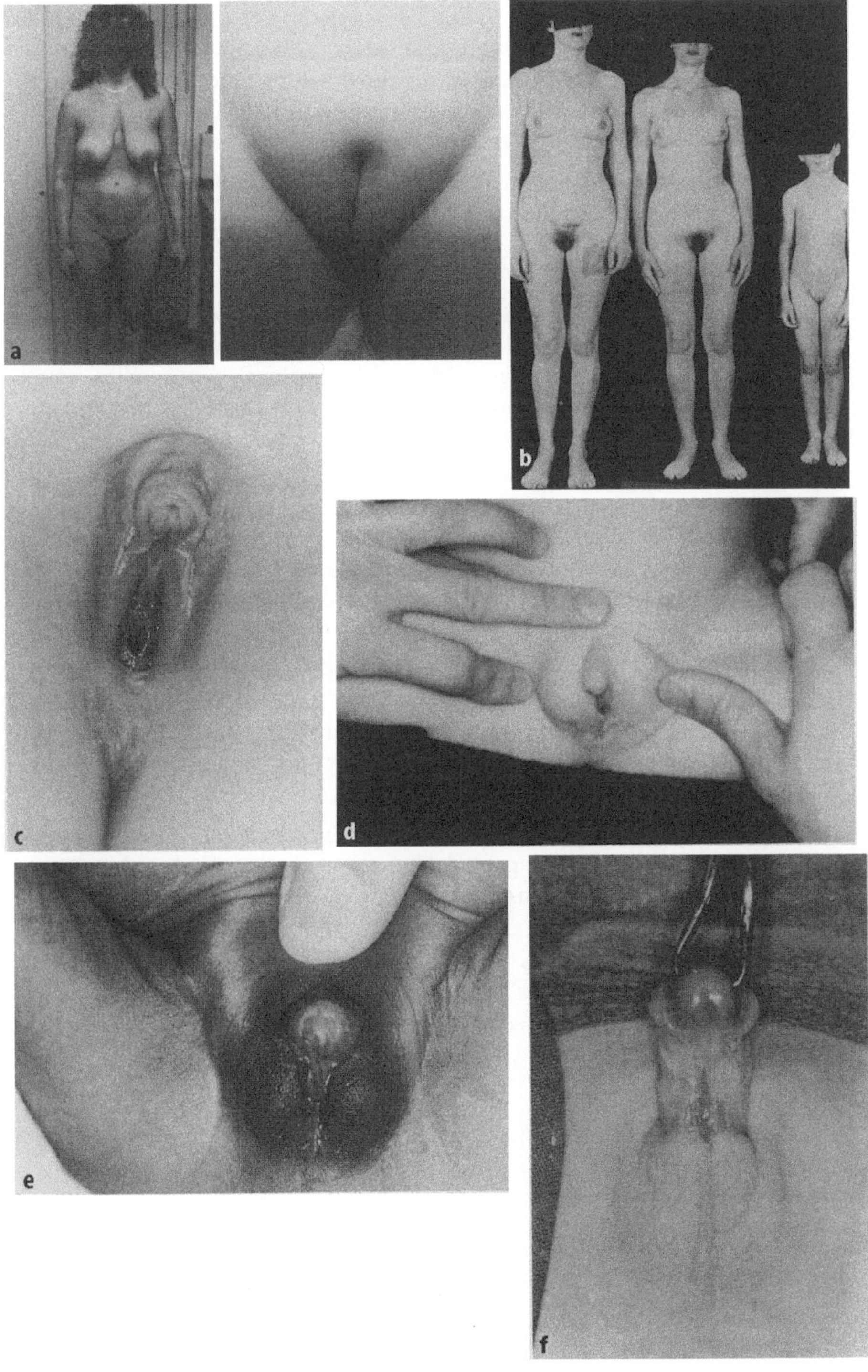

Androgen insensitivity grade 1 (male phenotype): individuals with normal male external genitalia such as infertile males with azoospermia and hormonal features of androgen resistance, those with reduced virilization at puberty (so-called "minimal" androgen resistance) or those with SBMA, who have normal androgen responsiveness in fetal life.

Androgen insensitivity grade 2 (partial AIS with male phenotype): individuals who have an unequivocally male phenotype, but who have mildly defective fetal masculinization, manifest by defects such as isolated hypospadias and/or smaller than average penile size.

Androgen insensitivity grade 3 (partial AIS with male phenotype): individuals with predominantly male phenotype but with more severely defective masculinization *in utero,* as evidenced by perineal hypospadias, small penis, with cryptorchidism and/or bifid scrotum. Individuals with Reifenstein syndrome exemplify this grade.

Androgen insensitivity grade 4 (partial AIS with ambiguous phenotype): individuals with ambiguous phenotype, with severely limited masculinization evidenced by a phallic structure that is intermediate between a clitoris and a penis, generally accompanied by a urogenital sinus with perineal orifice, and labioscrotal folds with or without rugation and posterior fusion.

Androgen insensitivity grade 5 (partial AIS with female phenotype): individuals with essentially female phenotype (i.e. minimal fetal androgen action), including separate urethral and vaginal orifices, with minimal androgenization evidenced by mild clitoromegaly or a small degree of posterior labial fusion.

Androgen insensitivity grade 6 (partial AIS with female phenotype): individuals with a normal female genital phenotype (i.e. no fetal androgen action), who develop androgen-dependent pubic and/or axillary hair at puberty.

Androgen insensitivity grade 7 (complete AIS; CAIS): individuals with female phenotype and absence of pubic or axillary sexual hair after puberty. AIS grades 6 and 7 are indistinguishable before puberty and such individuals are classified as grade 6/7 until completion of puberty.

Breast development, ranging from mild gynecomastia to Tanner stage V female breasts, can occur with all grades of AIS, tending to be more pronounced with the more severe grades.

2.8.1.1 Clinical features

Complete AIS (CAIS; AIS Grade 7). Complete AIS is relatively rare. The most accurate prevalence figure currently available is approximately 1:20,400 male

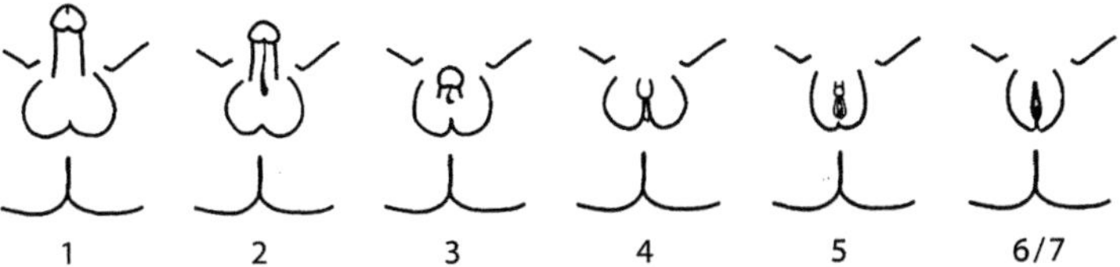

Fig. 2.5. Schematic representation of grading scheme for clinical classification of AIS. Grades are numbered 1–7 in order of increasing severity (more defective masculinization). Grade 1: normal masculinization in utero; Grade 2: male phenotype with mild defect in masculinization – isolated hypospadias; Grade 3: male phenotype with severe defect in masculinization – small penis, perineoscrotal hypospadias, bifid scrotum &/ or cryptorchidism; Grade 4: severe genital ambiguity – clitoral like phallus, labioscrotal folds, single perineal orifice; Grade 5: female phenotype with posterior labial fusion and clitoromegaly; Grade 6/7: female phenotype (grade 6 if pubic hair present in adulthood)

births, estimated on the basis of a large Danish nationwide prospective patient registry (Bangsbøll et al. 1992). Individuals with classic CAIS have completely female external genitalia, sometimes with underdevelopment of the clitoris and labia minora or majora (a feature more obvious after puberty), and have absence of sexual hair in the pubic region and axillae. Older reports of "pubic hair" in individuals with CAIS probably describe patients who either have an extreme form of partial AIS (grade 6 – see table above), or are mistakenly describing estrogen-induced vellus hair as "pubic" hair. The presence of true pubic hair, even in an individual with an entirely female phenotype, is evidence for some degree of androgen responsiveness, indicative of a severe form of partial AIS, rather than CAIS. Individuals with CAIS have a blind-ending vagina of variable depth, sometimes as diminutive as 1–2 cm. There is no uterus. Although it is generally stated that Müllerian structures are present only rarely in CAIS (Dodge et al. 1985; Heller et al. 1992; Oka et al. 1984; Swanson and Coronel. 1993; Ulloa-Aguirre et al. 1990), detailed and more comprehensive histopathological studies indicate that residual Müllerian structures, such as small fallopian tubes, may be present in up to one third of cases (Bale et al. 1992; Rutgers and Scully 1991). Remnants of Wolffian structures such as vestigial vas deferens or epididymis, may also be found (Morris 1953; Pettersson and Bonnier 1937; Wilkins 1950). Diagnosis in infancy is often prompted by the presence of one or more inguinal herniae in a phenotypically female infant (which, on exploration, are found to contain testes).

Individuals with CAIS whose testes are *in situ* feminize spontaneously at puberty, with development of adult breasts under the influence of testicular estrogens unopposed by the effects of androgens. A number of women with complete AIS report small nipples (personal communication, AIS-SG). The age at onset of puberty has not been studied systematically, but tends to be delayed relative to that of 46,XX females, being more consistent with that of males, suggesting a direct role of androgens in induction of pubertal activation of the hypothalamic-pituitary-gonadal axis. Skeletal maturation occurs at a rate similar to that of normal males (i.e. slower than in girls). However, in six 46,XY teenagers with complete AIS the pubertal growth spurt was found to be similar in timing and magnitude to that of normal girls. Women with AIS generally are taller than average for females, their mean adult height falling between that of normal males and females. In addition, women with AIS are sometimes reported to have a eunuchoid body habitus (Hauser 1963; Morris 1953; Morris and Mahesh 1963; Ritzen 1992; Smith et al. 1985; van Gelderen 1986; Varrela et al. 1984; Wilkins 1950; Zachmann et al. 1986).

Partial AIS (PAIS; AIS Grades 1–6). Because of the variability of the clinical manifestations and the existence of subtle or atypical forms of androgen resistance, the prevalence of partial forms of AIS is unknown. Nevertheless, as a group, these disorders are probably at least as common as CAIS (Aiman and Griffin 1982; Aiman et al. 1979; Morrow et al. 1987).

The partial or incomplete forms of AIS comprise a wide spectrum of clinical phenotypes: individuals with the most severe form of PAIS have a completely female phenotype, manifesting evidence of limited androgen responsiveness only at puberty with the development of pubic hair (AIS grade 6) (De Bellis et al. 1994). Somewhat greater androgen responsiveness is seen in individuals who have a female phenotype with mild clitoromegaly and/or slight labial fusion (AIS grade 5). Individuals with more significant androgen responsiveness, associated with scrotalized labia and a phallic structure intermediate between clitoris and penis, are those who present the greatest clinical challenge and are typically described as having "ambiguous" genitalia. Such individuals would generally represent AIS grade 4. In some cases, particularly those referred to as Reifenstein Syndrome, there is more extensive masculinization, the affected individuals having an essentially male, but undermasculinized, phenotype with small penis, perineal hypospadias and variable cryptorchidism (AIS grade 3). A small number of studies also suggest that, in its mildest forms of expression, PAIS may be manifest simply by uncomplicated hypospadias (Batch et al. 1993b; Keenan et al. 1984), by infertility in a phenotypically normal male (Aiman and Griffin 1982; Aiman et al. 1979; Akin et al. 1991; Morrow et al. 1987) or even by gynecomastia and androgen binding abnormalities in a fertile male (Grino et al. 1988; Pinsky et al. 1984). Notably, individuals within a single PAIS kindred may have quite different genital phenotypes, such that sex of rearing may vary from one affected member of the family to another (Batch et al. 1993a; Brown et al. 1982; Evans et al. 1997; Forest et al. 1990; Grino et al. 1989; Rodien et al. 1996; Wilson et al. 1974). Wolffian structures may develop to a variable extent in PAIS, depending upon the degree of sensitivity/resistance to androgen. Thus the epididymides, vasa deferentia and seminal vesicles may vary from rudimentary to fully-formed.

At puberty, virilization and/or feminization may occur depending upon the hormonal status of the individual. For example, if the testes remain in situ, the normal or increased secretion of testosterone will probably induce a degree of virilization proportional to the degree of masculinization that occurred *in utero*. As in CAIS, feminization of breasts and body contours occurs as a result of relatively high estrogen concentrations in the presence of androgen resistance.

2.8.1.2 Developmental pathophysiology

Despite the ubiquitous distribution of the AR, most events of embryogenesis occur normally in the 46,XY androgen insensitive fetus, indicating lack of androgen requirement for most developmental processes. However, certain events including the differentiation and development of the internal and external genitalia and to a greater or lesser degree, descent of the testes, are disturbed due to their requirement for androgen action.

Development of the testes. Testicular development *in utero* occurs normally in the fetus with AIS. Immature spermatogonia (germ cells) are present in the testes at birth and during childhood. However, there is progressive loss of spermatogonia with increasing age: testes of teenage AIS patients contain only occasional spermatogonia, and no germ cells are present in adult AIS testes. Spermatocytes and more mature germ cells are absent at all ages (Bale et al. 1992; Müller 1984; Rutgers and Scully 1991). These findings are likely to reflect both the requirement of testicular androgen responsiveness for spermatogenesis and the prolonged exposure of the testes to the elevated temperature of non-scrotal locations. Leydig and Sertoli cell development and function occur normally and therefore appear to be androgen-independent processes.

Genital morphogenesis. Both internal and external genital morphogenesis are disturbed, reflecting the dependence of these processes on AR function. There is reduced or absent development of the epididymis, vas deferens and seminal vesicles from the Wolffian duct anlagen, the labioscrotal folds fail to enlarge and fuse to form the scrotum, there is lack of phallic growth and the phallic urethra is absent or incompletely formed. External masculinization is believed to depend mainly on the presence of DHT and there is evidence that the activity of the enzyme 5α-reductase 2, required for conversion of testosterone to DHT, is androgen-dependent (Mowszowicz et al. 1983). Activity of 5α-reductase 2 is reduced in some individuals with androgen insensitivity presumably secondary to the AR defect. In such individuals, the consequent reduced availability of DHT in the external genital target cells during embryogenesis may contribute to the defective masculinization of individuals with partial forms of AIS. In one PAIS kindred the difference in phenotype between affected siblings was ascribed to the secondary deficiency of 5α-reductase 2 in the more severely affected individual (Boehmer et al. 1997; Imperato-McGinley et al. 1982; Jukier et al. 1984; Kuttenn et al. 1979; Mauvais-Jarvis et al. 1970).

Sertoli cell function *in utero* appears to be normal in most individuals with AIS, since AMH concentrations are normal and there is partial or complete regression of the Müllerian ducts (Harbison et al. 1991; Josso 1994). However, in about one third of subjects Müllerian duct regression is incomplete, perhaps due to reduced expression or defective function of the AMH receptor, since there is evidence that AMH-R expression is androgen regulated (Baarends et al. 1994; Rutgers and Scully 1991).

Testicular descent. Substantial debate exists regarding the determinants of testicular descent and the role of androgens in this process. Testicular descent is believed to be driven in large part by the gubernaculum, a ligamentous structure that connects the inferior pole of the testis to the base of the scrotum. Gubernacular migration appears to be under the influence of a neuropeptide – calcitonin gene-regulated peptide (CGRP) – secreted by the genitofemoral nerve, probably under some degree of androgenic control. AR is detectable in gubernacular tissue (George and Peterson 1988) and testicular descent is inhibited by anti-androgens in a time-specific fashion (Spencer JR et al. 1991; Spencer et al. 1993; van der Schoot 1992): these compounds inhibit outgrowth of the gubernaculum, however, if administered after gubernacular outgrowth has begun they do not inhibit testicular descent (Husmann and McPhaul 1991; 1992).

Few studies have systematically examined the question of testicular descent in AIS. However, Hutson (1986) found 35 of 36 testes of 16 children with AIS and two with androgen deficiency to be at or beyond the level of the inguinal ring. Partial testicular descent in individuals with CAIS suggests that androgen action is certainly not the sole, and may not be the primary, determinant of the transabdominal phase of testicular descent. Furthermore, the cryptorchidism of individuals with defects of AMH secretion or action, suggests a role for this hormone particularly in the transinguinal phase of testicular descent (Josso et al. 1991; Knebelmann et al. 1991).

Testicular neoplasia. Testicular tumors of germ cell and non-germ cell (Sertoli cell and interstitial or Leydig cell) precursors occur with increased frequency in individuals with AIS, however series are small and it is unclear whether the incidence is any greater than that seen in simple cryptorchidism. Germ cell neoplasia is usually low grade, such as carcinoma-*in-situ* (CIS; intratubular germ cell neoplasia, IGCN), however seminomas (malignant germ cell tumors, sometimes called germinomas or dysgerminomas) have been reported occasionally. A number of small studies suggest that CIS may be more common in individuals with PAIS than in those with CAIS. CIS was found in the seminiferous tubules of three of eight children with PAIS (the youngest of whom was only two months old), but in none of the four children with CAIS (Müller and Skakkebaek 1984). In another study, intratubular germ cell neoplasia was found in the testes of five of eight prepubertal children with PAIS (Cassio et al. 1990). In contrast, two studies found no cases of intratubular germ cell neoplasia in a total of 36 children and adolescents with complete or partial AIS, although it was unclear how the diagnosis of AIS was made in one study (Bale et al. 1992; Ramani et al. 1993). Seminomas have been reported in post-pubertal patients, generally over 30 years of age; the youngest such patient was a 14-year old who, notably, had metastatic disease. The risk of malignant germ cell tumors increases with age climbing from about 3% at 20 years, to 30% at 50 years. Tumors of non-germ cell origin also tend to be low grade, the most frequent being adenomas of Sertoli cells or Leydig cells. Sertoli cell adenomas are common, occurring in 10% of patients reported by Rutgers and Scully. Leydig cell adenomas are less common, however Leydig cell hyperplasia is well-recognized and in one study of six patients whose testes were removed at 12–18 years of age, many of the Leydig cells were abnormal, with pleomorphic nuclei and bizarre cellular shapes, resembling immature fetal Leydig cells, perhaps induced by exposure to chronically increased concentrations of LH. Mixed tumors of a benign type, such as hamartomatous nodules, containing germ cell, Leydig cell and Sertoli cell elements, are common. Scully reported one or more testicular hamartomas in 25% of women with AIS in one series, and as many as 63% in another. Malignant mixed tumors, such as teratoma, gonadoblastoma, sex cord tumor or embryonal carcinoma, are rare (Collins et al. 1993; Hurt et al. 1989; Jockenhövel et al. 1993; Lecca et al. 1988; Manuel et al. 1976; Morris 1953; Morris and Mahesh 1963; O'Connel et al. 1973; Pelliniemi et al. 1980; Ramaswamy et al. 1985; Rutgers and Scully 1991; Scully 1981).

The overall frequency of gonadal neoplasia in individuals with AIS is difficult to judge, due to the small size of most series, and a variety of sampling biases. The overall risk for gonadal tumor in AIS has been estimated to be 6–9% (Rutgers and Scully 1991; Scully 1981). Manuel et al. (1976) found no tumors in 23 patients of their own and only seven tumors in 82 AIS cases in the literature (8.5%); in a retrospective survey utilizing a national patient registry in Denmark, Bangsbøll et al. (1992) reported non-malignant gonadal tumors in 4 of 21 patients ranging in age from newborn to 68 years at the time of AIS diagnosis – the youngest patient in this series with a tumor was a 14-year-old with a Sertoli cell nodule. No gonadal tumors were found in another series of 14 individuals with AIS (Lukusa et al. 1991). Apart from the children with carcinoma-in-situ noted above, the youngest age at which a gonadal tumor has been reported is 14 years (Bangsbøll et al. 1992; Hurt et al. 1989; Manuel et al. 1976). All patients with gonadal tumors have been post-pubertal.

2.8.1.3 Endocrine findings

Studies during the 1960s helped to clarify the biology of AIS, establishing that affected individuals had normal testicular biosynthesis of testosterone, normal androgen metabolism and estradiol production. Generalized tissue resistance to androgens in CAIS was established when an affected individual was found to have no metabolic response to testosterone or DHT in nitrogen balance studies. In addition, studies of an individual with PAIS demonstrated a limited anabolic response to high doses of administered androgens indicating that the partial form of AIS was the result of reduced responsiveness to androgens (French et al. 1965, 1966; Morris and Mahesh 1963; Rosenfield et al. 1971; Strickland and French 1969).

Adults with AIS have reduced sensitivity of the hypothalamus and pituitary gland to negative feedback regulation of gonadotropin secretion by sex steroids, primarily due to loss of sensitivity to androgens. Thus LH secretion is increased, stimulating Leydig cell testosterone production. Classically (although not universally), post-pubertal individuals with CAIS or PAIS, whose testes are *in situ*, have increased serum concentrations of LH, normal (or sometimes increased) follicle stimulating hormone (FSH) and testosterone, relative to those of normal males (Campo et al. 1979; Cicognani et al. 1989; French et al. 1965, 1966; Imperato-McGinley et al. 1982; Judd et al. 1972; Morris and Mahesh 1963; Southren et al. 1965; van Look et al. 1977). Studies of 24-hour LH secretion pattern reveal an increase in total LH production, due to an increase in LH pulse frequency and amplitude (Boyar et al. 1978). In contrast to the findings in adults, prepubertal children with CAIS generally have testosterone and LH concentrations in the normal range for age (Cicognani et al. 1989; Faiman and Winter 1974). Limited studies in newborns and infants with CAIS indicate that LH and testosterone are not increased in this age group (J.-L. Chaussain, personal communication; C.A. Quigley and F.S. French, unpublished observations). Furthermore, the transient surge of LH and testosterone that occurs in the normal male infant at around 6 weeks of age ("mini-puberty"), does not appear to occur in infants with complete AIS (J.-L. Chaussain, personal communication; C.A. Quigley and F.S. French, unpublished observations), suggesting that the neonatal LH rise in normal male infants requires androgen-mediated hypothalamic-pituitary imprinting. In contrast to infants with CAIS, those with PAIS (the few whose perinatal hormone status has been reported) do appear to have increased LH and testosterone secretion, presumably due to the occurrence of some androgenic priming of the hypothalamic-pituitary axis during fetal life (Lee et al. 1986; Nagel et al. 1986), (J.-L. Chaussain, personal communication). In the period of hypothalamic-pituitary-gonadal "quiescence" between infancy and puberty, LH and testosterone are in the normal range for age (Cicognani et al. 1989). Some patients have a significantly increased ratio of testosterone to DHT (Imperato-McGinley et al. 1982). Although not as high as seen in patients with 5α-reductase deficiency, this disturbance suggests a functional or secondary form of 5α-reductase deficiency.

Estrogen production – mainly by the testes due to increased LH stimulation of Leydig cells, and to a lesser extent by aromatization of androstenedione and testosterone in peripheral tissues – is increased in individuals with AIS to about twice that of normal males (Griffin and Wilson 1989; MacDonald et al. 1979). Plasma concentrations or urinary excretion of estrogens are at or above the upper limit of normal in individuals with AIS compared with those of normal males (Boyar et al. 1978; Deshpande et al. 1965; French et al. 1965, 1966; Morris and Mahesh 1963). Although serum estradiol concentrations in individuals with AIS are usually equivalent to those seen in the follicular phase of the normal female menstrual cycle, they are inadequate to fully suppress LH *in vivo*, suggesting either that there is a distinct requirement for AR-regulated negative feedback, or that the estradiol concentrations required to suppress male hypothalamic GnRH secretion are greater than can be achieved by testicular estradiol production. Gonadectomy removes the existing negative feedback upon hypothalamic GnRH secretion and results in a marked increase in the already raised LH and FSH concentrations. The raised gonadotropins of gonadectomized individuals with AIS can be suppressed by administration of ethinyl estradiol (80–200 mcg/day), indicating that the control of male gonadotropin secretion is at least in part estrogen-mediated (Imperato-McGinley et al. 1982; van Look et al. 1977; Zarate et al. 1974).

Due to the effects of estradiol stimulation, sex hormone binding globulin (SHBG) concentrations are substantially higher in normal adult women than in men. Serum SHBG concentrations of individuals with CAIS whose testes are *in situ* are similar to those of normal females, while those of individuals with PAIS are intermediate between those of normal males and normal females, indicating a continuum in regulation of an androgen-responsive gene (Mauvais-Jarvis et al. 1970; Rosenfield et al. 1971). Castrated individuals with CAIS who are not receiving estrogen replacement have SHBG concentrations close to those of normal males. The incomplete suppression of SHBG in response to administration of the anabolic steroid stanozolol, has been used as an aid in the differential diagnosis between PAIS and other forms of male pseudohermaphroditism and as a functional assessment of androgen responsiveness in various forms of AIS (Sinnecker et al. 1997; Sinnecker and Kohler 1989). A modified version of this test evaluates the SHBG response to the increase in serum testosterone induced by a standard hCG stimulation test (Bertelloni et al. 1997).

2.8.1.4 Androgen binding studies

Determination of the etiology of androgen insensitivity at the cellular level in humans became approachable in the 1970's with the establishment of assays of androgen binding in cultured genital skin fibroblasts (Keenan et al. 1975). In these studies, Keenan and co-workers (1974) established the existence of a specific intracellular receptor for androgens, and subsequently implicated defective function of this receptor in the pathogenesis of AIS in humans, confirming the hypothesis established earlier in rodent models (Gehring et al. 1971). Numerous subsequent studies of androgen binding in genital skin fibroblasts revealed that the androgen binding disorders in this syndrome are as heterogeneous as the clinical picture: defects range from complete absence of androgen binding, to binding of reduced affinity, binding of normal affinity with reduced capacity, or binding with qualitative abnormalities such as thermolability (that is, reduced androgen binding capacity at temperatures above 37 °C), increased ligand dissociation rate or altered ligand specificity (Amrhein et al. 1976; 1977; Brown and Migeon 1981; Brown et al. 1982; Evans et al. 1984; Griffin 1979; Griffin and Wilson 1989; Griffin et al. 1976;

Hughes and Evans 1987, 1988; Kaufman et al. 1976, 1979; Pinsky et al. 1981, 1984, 1985; Warne et al. 1984).

Studies of androgen binding in cultured genital skin fibroblasts were the gold standard for diagnosis of AIS until the advent of molecular techniques, and have elucidated the cause of AIS at the cellular level in many cases. The parallel development of androgen binding assays in different laboratories led to a variety of descriptive terms including: *receptor-negative* (usually referring to unmeasurable binding, but occasionally including individuals with very low but detectable binding); *receptor-deficient or receptor-reduced* (generally referring to binding of reduced capacity, but occasionally describing binding of reduced affinity); *receptor-positive* (usually referring to quantitatively normal, qualitatively abnormal androgen binding). This latter term is sometimes also used to describe the situation in which no abnormality of binding can be detected, implying a defect in the androgen response pathway beyond the AR itself. At other times the term *post-receptor* defects is used. Individuals who are described as "receptor-negative" have the phenotype of complete AIS. However, apart from this, there is no consistent correlation between the quantity or quality of androgen binding and the external genital phenotype of the affected individual.

2.8.1.5 Genetic basis

The inherited nature of AIS was initially observed by Pettersson and Bonnier (1937) who determined that the disorder was transmitted only by women. Morris and Mahesh (1963) subsequently noted that the partial and complete forms of the syndrome did not occur in the same pedigree. Studies of the *Tfm* mouse established the X-chromosomal localization of the *Tfm* locus (Lyon and Hawkes 1970; Ohno and Lyon 1970). The locus for human AIS was subsequently found to be homologous to the *Tfm* locus in the mouse and by analogy, human AIS was assigned an X-chromosomal locus (Migeon et al. 1981). Later studies demonstrated that the AR gene is located at Xq11–q12 in a region of the X chromosome that has high evolutionary conservation in all mammals including marsupials and monotremes, dating it to a common ancestor present about 150 million years ago (Brown et al. 1989; Mahtani et al. 1991; Spencer et al. 1991).

2.8.1.6 Molecular pathophysiology

With the advent of molecular biologic techniques for analysis of the AR gene (for details see Quigley et al. 1995), a wide variety of defects in the AR gene have been identified and characterized in individuals with various forms of AIS. To date, approximately 250 mutations have been reported, making the AR the most frequently mutant human transcription factor known. For complete and up-to-date information regarding AR gene mutations, the reader is advised to consult the Androgen Receptor Gene Mutations Database on the Internet at http://www.mcgill.ca/androgendb.

In summary, the AR defects reported to date in patients with AIS include:
- Complete or partial AR gene deletions;
- Deletions or insertions of small numbers of nucleotides;
- Splice junction mutations;
- Single base mutations introducing nonsense codons;
- Missense mutations introducing amino acid substitutions.

CAIS is associated with numerous molecular defects including complete and partial AR gene deletions, single base mutations that introduce premature termination codons into the AR gene or disrupt the splicing of the mRNA, and missense mutations that cause single amino acid substitutions in the protein. PAIS is most commonly associated with amino acid substitutions in the LBD and DBD, only rarely being associated with small-scale deletions. AR mutations are found rarely in patients with isolated hypospadias.

At the functional level, two primary types of AR defect are associated with AIS: abnormalities of androgen binding and abnormalities of DNA binding, either of which may disturb transcriptional activity of the AR. Complete absence of androgen binding (generally referred to as receptor negative), may be caused by various molecular defects including deletions of all or part of the AR gene, mutations that introduce a premature termination codon into the AR gene or disturb mRNA splicing (such mutations often abolish AR protein expression) and missense mutations that cause amino acid substitutions in the LBD. Retention of androgen binding associated with qualitative or quantitative binding defects (referred to as receptor-positive or receptor-deficient), occurs in association with a variety of amino acid substitutions in the LBD. Amino acid substitutions or deletions in the DBD have their major impact on the DNA-binding function of the receptor-androgen binding studies in skin fibroblasts generally yield normal results.

AR Gene Deletions. A variety of deletions of all or part of the AR gene have been described in individuals with CAIS (Fig. 2.6a), however deletions are uncommon, representing less than 10% of reported AR gene mutations.

Complete deletion of the AR gene. Complete deletion of the AR gene has been reported in three apparently unrelated families. In two unrelated patients there was mental retardation in addition to CAIS, while neurocognitive function was entirely normal in the third family. Subsequent studies of genomic DNA of all three complete deletion pedigrees suggested the presence of a locus for non-specific mental retardation within the X-chromosomal segment bounded by markers DXS1 and DXS905, which includes the AR gene. This locus is not the AR gene itself, but likely a nearby gene (Davies et al. 1997; Quigley et al. 1992b; Trifiro et al. 1991a).

Complete deletion of the AR gene represents the "null" phenotype of AIS, since there is no AR present whatsoever. Important clinical features of the patients reported by Quigley et al. (1992b) include: normal physical, neurological and intellectual development and health of the affected individuals,

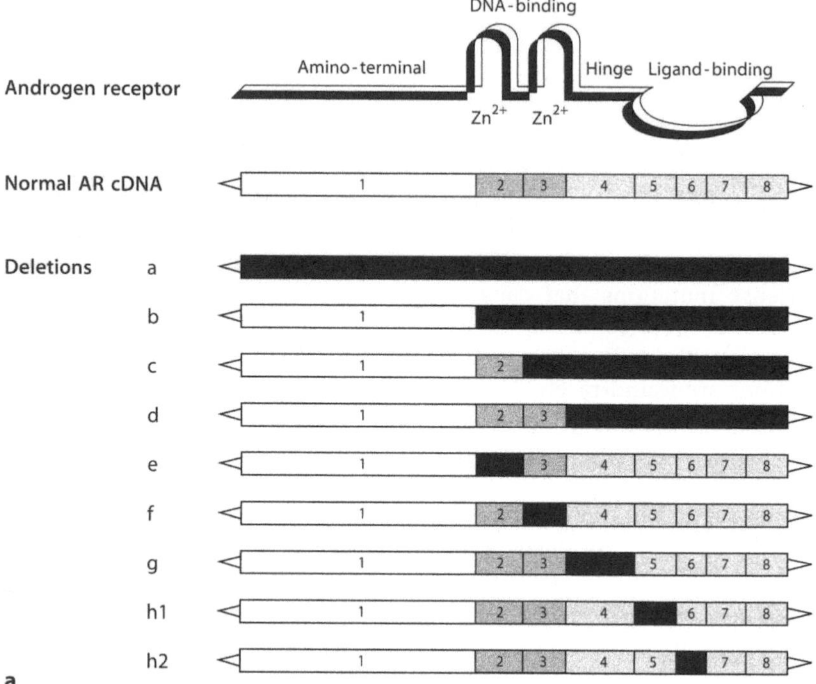

Fig. 2.6. a Summary of deletions reported in the androgen receptor gene. A schematic of the AR protein and normal AR cDNA are shown at the top. Deletions, indicated by dark shading, have been reported as follows: a. Complete deletion of the AR gene (Davies et al. 1997; Quigley et al. 1992b; Trifiro et al. 1991a); b. Deletion of exons 2–8 (Jakubiczka et al. 1997); c. Deletion of exons 3–8 (Brown et al. 1993); d. Deletion of exons 4–8 (Brown et al. 1988); e. Deletion of exon 2 (Quigley et al. 1992c); f. Deletion of exon 3, producing an internally-deleted protein (Quigley et al. 1992a); g. Deletion of exon 4, in an infertile man (Akin et al. 1991); h1 and h2. Deletion of exon 5 or exons 6 and 7 respectively, in members of the same family (MacLean et al. 1993). **b** Diagram of the AR cDNA with locations of premature termination codons (upper diagram). The asterisks indicate cryptic termination codons generated downstream of a frameshift mutation [deletion of one nucleotide (introduces stop codon at Leu172); insertion of four nucleotides (introduces stop codon at Asp232) and deletion of one nucleotide (generates stop codon at Cys619)]. Splicing mutations (lower diagram) – see section 2.8.1.6, page [65] for details of individual mutations. The mutation marked with 'x' represents an intronic deletion that includes the pre-mRNA splicing branch site (Ris-Stalpers et al. 1994b)

apart from the anomalies of the reproductive system; hypoplastic labia majora of the affected women (suggesting that androgens may have a role in normal development of the female external genitalia); delayed puberty of the obligate carriers (suggesting a role for androgens in priming the hypothalamic-pituitary-gonadal axis of for pubertal development in the female).

Partial deletions of the AR gene. A variety of partial AR gene deletions has been reported in association with various clinical manifestations of AIS. Deletions of exons 4–8 (Brown et al. 1988; Meyer, III. et al. 1975) and 3–8 (Brown et al. 1993) have been reported in the AR genes of kindreds with absent androgen

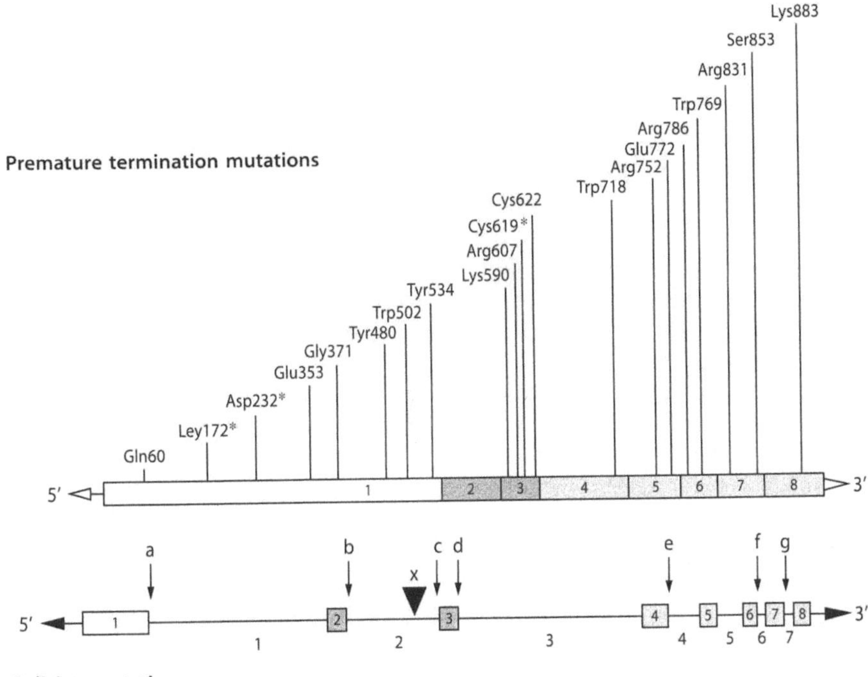

b

Fig. 2.6 b

binding in genital skin fibroblasts, and AIS grades 6 and 7 respectively. A unique CAIS family with a complex set of AR mutations was reported by MacLean et al. (1993). Two 46,XY siblings, their mother and maternal grandmother had deletion of exon 5, encoding amino acids 725–772 at the heart of the LBD. The AR gene of the affected 46,XY maternal aunt retained exon 5; however there was deletion of exons 6 and 7, encoding amino acids 773–868 of the LBD. These deletions had one break point in common in intron 4, however, the exon 5 deletion included about 5 kb upstream of the breakpoint, while the exon 6/7 deletion extended 5 kb downstream of the common breakpoint. Possible mechanisms underlying these unusual molecular defects include "illegitimate" recombination (crossing over between non-homologous sequences) or the insertion into intron 5 of a transposon-like element (a moveable DNA sequence capable of excision from or insertion into the human genome).

A number of deletions of different single exons of the AR gene also have been reported. The effect of such a deletion upon receptor expression and function depends primarily upon two factors: first, the state of the translational reading frame in the absence of the particular exon, and second, the region of the AR encoded by the deleted exon. Because of the sizes of the exons and the nature of the intron/exon boundaries of the AR gene, only exons 3 and 4 can be deleted in frame, and thereby theoretically allow production of an AR protein. Isolated deletion of exon 2 was discovered in two siblings with CAIS (AIS grade 7). This deletion disturbs the translational reading frame of the AR mRNA so is functionally equivalent to deletion of the whole AR gene (Quigley et al. 1992 c). In another family with CAIS, exon 3 of the AR gene was deleted. Because this deletion does not disturb the translational reading frame of the AR mRNA, an internally-deleted mutant AR is produced, lacking the second zinc finger of the DBD, but retaining the LBD. The mutant AR had severely reduced DNA-binding activity and lacked transcriptional activity *in vitro*, in keeping with the CAIS phenotype of the affected siblings. Notably, this mutant AR displayed increased high-affinity androgen binding activity, perhaps reflecting increased AR expression and protein level due to failure of AR auto-regulation (Quigley et al. 1992 a).

Deletion of approximately 6 kb from intron 2 of the AR gene, including the putative pre-mRNA splicing branch site was reported in a family with grade 3 AIS, manifest by hypospadias, small penis, lack of male body hair and gynecomastia, whose androgen binding affinity and capacity in cultured genital skin fibroblasts was within the normal range (Ris-Stalpers et al. 1994 b). This intronic deletion resulted in aberrant splicing of about 90% of the subjects' AR mRNA, such that exon 3 was spliced out, predicting loss of the second zinc finger. The remaining 10% of the transcribed AR mRNA was normal and this low level of expression of normal AR was believed to be responsible for the retained masculinization of the affected individuals. This study and another by McPhaul et al. (1991 a) suggests that as little as 10% of normal AR function may be adequate to induce the activity of certain androgen-responsive genes involved in male sex differentiation.

Nonsense Mutations. Single base changes in the AR gene that convert an amino acid codon into a translation termination (stop) codon (nonsense or amber mutations) have been reported in exons 1 and 3–8 (Brown et al. 1993; De Bellis et al. 1992; Marcelli et al. 1990 a, 1990 b; McPhaul et al. 1991 b; Pinsky et al. 1992; Ris-Stalpers et al. 1992; Sai et al. 1990; Tincello et al. 1992; Trifiro et al. 1991 b; Vasiliou et al. 1994; Wilson et al. 1992; Zoppi et al. 1993) (Fig. 2.6b). At the molecular level, two major disturbances may contribute to the functional defect associated with such mutations: first, their effects on AR protein structure and second, their effects upon mRNA stability and thereby on receptor protein expression.

Immunoblotting and immunocytochemical studies indicate that AR protein is expressed in genital skin fibroblasts of some individuals whose AR gene contains a premature termination codon (De Bellis et al. 1992; Marcelli et al. 1990 a, 1990 b; Ris-Stalpers et al. 1992). However, such proteins are likely to be truncated at either the carboxy-terminal or the amino-terminal end, depending on the location of the nonsense codon. Premature termination codons most commonly cause carboxy-terminal truncation associated with absent androgen binding and CAIS. Amino-terminally truncated receptors may occur if the mutant AR gene contains an alternative translation start site downstream of the premature termination codon, as was found in a pair of siblings with CAIS whose AR gene contained a single base mutation in exon 1 that introduced a premature termination codon at position 60 (Zoppi et al. 1993), with re-initiation of protein translation from a cryptic initiation codon at position 188. When this mutant AR gene was expressed *in vitro* the smaller 87-kDa AR species (AR-A) was observed [the predominant AR species (AR-B) is 110–114 kDa]. This amino-terminally truncated protein retained the LBD; genital skin fibroblasts retained about one-quarter of normal androgen binding capacity and the rate of dissociation of bound androgen was increased. Since the receptor lacked the amino-terminal transcription-regulating region it was transcriptionally inactive.

The second factor likely to contribute to the severity of the functional defect of an AR gene containing a premature termination codon is a reduction in steady state levels of AR mRNA, probably due to an increase in the rate of mRNA degradation. This phenomenon has been demonstrated in the *Tfm* mouse, in which the presence of a premature termination codon in exon 1 destabilizes the AR mRNA, resulting in a low steady state mRNA level, perhaps due to loss of mRNA protection that occurs when mRNA is released prematurely from the translation apparatus (Charest et al. 1991; He et al. 1991). The location of the premature stop codon seems to influence its effect upon mRNA level. For example, the presence of a premature termination codon in exon 3 was associated with a low AR mRNA level in genital skin fibroblasts (Marcelli et al. 1990a), while a premature termination codon in exon 6 (Marcelli et al. 1990b) or exon 8 (Trifiro et al. 1991b) had no effect on mRNA level. These studies suggest that premature stop codons occurring relatively 5′ in the mRNA may have a more marked effect on mRNA level than those occurring more 3′ (Fig 2.6b).

Frameshift Mutations

Splice-junction mutations. The intron-exon junctions of the AR gene have the typical splice consensus sequences found in essentially all eucaryotic genes, including those encoding the other members of the steroid receptor family. These sequences (GT-splice donor, at the 5′ end of the intron; AG-splice acceptor, at the 3′ end of the intron) are absolutely required for proper mRNA splicing. In addition, it has recently been found that the adenine at the +3 position of the splice donor site (GT*A*) is also critical for proper AR gene splicing, and is present in the majority of eucaryotic genes (Trifiro et al. 1997). A number of single base mutations affecting the consensus splice donor/acceptor nucleotides in the AR gene have been reported. The loss of a normal splice site generally results in use of a cryptic splice site elsewhere, leading to splicing out of important coding regions of the mRNA. Translation of the disturbed mRNA would produce an internally truncated and/or structurally disorganized protein. Examples include the loss of 41 amino acids from the LBD associated with absence of androgen binding and CAIS, a receptor truncation further downstream in the LBD also associated with loss of androgen binding and CAIS and an adenine to thymine transversion at position +3 of the intron 6 splice-donor site resulting in skipping of exon 6, associated with negligible fibroblast androgen binding and CAIS. In contrast, a splicing abnormality at the exon 3/intron 3 junction in a subject with CAIS was associated with normal androgen binding, presumably because the missplicing affects the DBD, rather than the LBD (Evans et al. 1991; Gottlieb et al. 1997; Pinsky et al. 1992; Ris-Stalpers et al. 1990; Trifiro et al. 1997; Yong et al. 1994).

Nucleotide Insertions and Deletions. Deletions or insertions of small numbers of nucleotides other than three (i.e. a complete codon) disturb the reading frame of the AR gene, introducing a premature termination codon downstream of the mutation, and thereby preventing expression of a functional AR protein. Examples include deletion of one nucleotide from exon 1, or insertion of four nucleotides into exon 1 (Batch et al. 1992). Although reinitiation of translation could theoretically occur further 3', the absence of androgen binding in genital skin fibroblasts, and the CAIS phenotype of the subjects, indicate that no functional AR protein is produced.

Single Codon Deletions

Single codon (3 bp) do not disturb the translational reading frame of the AR gene and are thus ostensibly less disruptive than large scale deletions and frameshifts. Nevertheless, these mutations are associated with major disturbance of AR function, due to loss of a critical a.a. residue or to alteration in AR protein tertiary structure or folding. This is reflected by the CAIS phenotype of affected individuals. For example, deletion of codon 582 or 583 from exon 2 removes one of a pair of strictly-conserved phenylalanine residues from a region between the first and second zinc fingers believed to contribute to a hydrophobic core that stabilizes the three dimensional structure of the DBD. Deletion of the arginine 615 (immediately carboxy terminal to the second zinc finger), removes a residue that may contribute to the tetrahedral arrangement of the zinc-coordinating cysteine residues at the base of the second zinc finger. These mutations are associated with marked reduction of DNA binding and transcriptional activity of the mutant AR. Deletion of asparagine 692 near the amino-terminal end of the LBD was associated with qualitative androgen binding abnormalities, including low affinity and increased thermolability of binding in an individual with CAIS (Batch et al. 1992; Beitel et al. 1994b; El-Awady et al. 1993; Freedman 1992; Pinsky et al. 1992).

Missense Mutations

Missense mutations – single base mutations that cause amino acid substitutions – are the most common type of mutation found in the AR gene. By the end of 1997, more than 120 distinct missense mutations had been identified at over 100 different codons in the AR gene in more than 200 individuals or families (Fig. 2.7). About half of the reported mutations have been found in at least two apparently unrelated individuals, the remainder being apparently unique to a single individual.

Approximately 80% of the reported substitutions are in the LBD, all but two of the remainder being in the DBD. Certain types of amino acid residue appear to be more common sites of substitution than others. About 75% of AR amino acid substitutions affect either highly-charged residues, such as arginine (positively charged) or aspartic or glutamic acids (negatively charged) or hydrophobic residues, such as leucine, valine and methionine. Highly charged, hydrophilic residues are generally found on the outer surfaces of proteins and hydrophobic residues within the interior, so that substitution of such a.a. likely causes significant conformational change. For simplicity, missense mutations in each of the three major functional domains of the receptor are summarized separately below.

The amino terminus: Although the amino-terminal domain encompasses approximately half of the AR's 919 residues and is required for regulation of target gene transcription, reports of mutations in this region are scarce; however, a silent polymorphism replacing guanine with adenine at the third base of codon 211 (CAG→CAA – both encode glutamine) is present in the AR genes of 14% of controls (Hiort et al. 1994). The infrequency of reported mutations in exon 1 is also striking in light of the fact that this exon contains numerous CpG codons which are recognize as mutational "hot spots" (Youssoufian et al. 1986). Two functionally relevant mutations have been described: In a family with PAIS a mutation that changed the second residue of the amino-terminal domain from glutamic acid to lysine was found (Choong et al. 1996 b). The mutant AR had an increased rate of dissociation of bound androgen and reduced transcriptional activity. In addition, the efficiency of translation of a cDNA containing this mutation was reduced, likely compounding the functional defect by reducing AR protein level. Further downstream was reported a mutation that converts proline 390 to arginine likely causing a substantial alteration of protein structure since proline residues are associated with sharp turns in DNA. It is of interest that although the substitution was located in the amino terminus, there was absent androgen binding in genital skin fibroblasts and the clinical phenotype of CAIS (Vasiliou et al. 1994 and Pinsky, L. personal communication), highlighting the importance of interactions between the amino-terminal and ligand-binding domains for normal AR function.

A number of factors potentially contribute to the paucity of reported exon 1 missense mutations in AIS. It is possible that mutations occur in this region with a frequency equal to those elsewhere but have gone undetected for lack of investigation (due to the relatively more difficult technical aspects of examining this region because of its size and high G-C content), or by producing neutral substitutions (amino acid changes that do not affect protein activity). In support of this latter hypothesis, it is notable that certain stretches of homopolymeric amino acids in the amino-terminus are highly

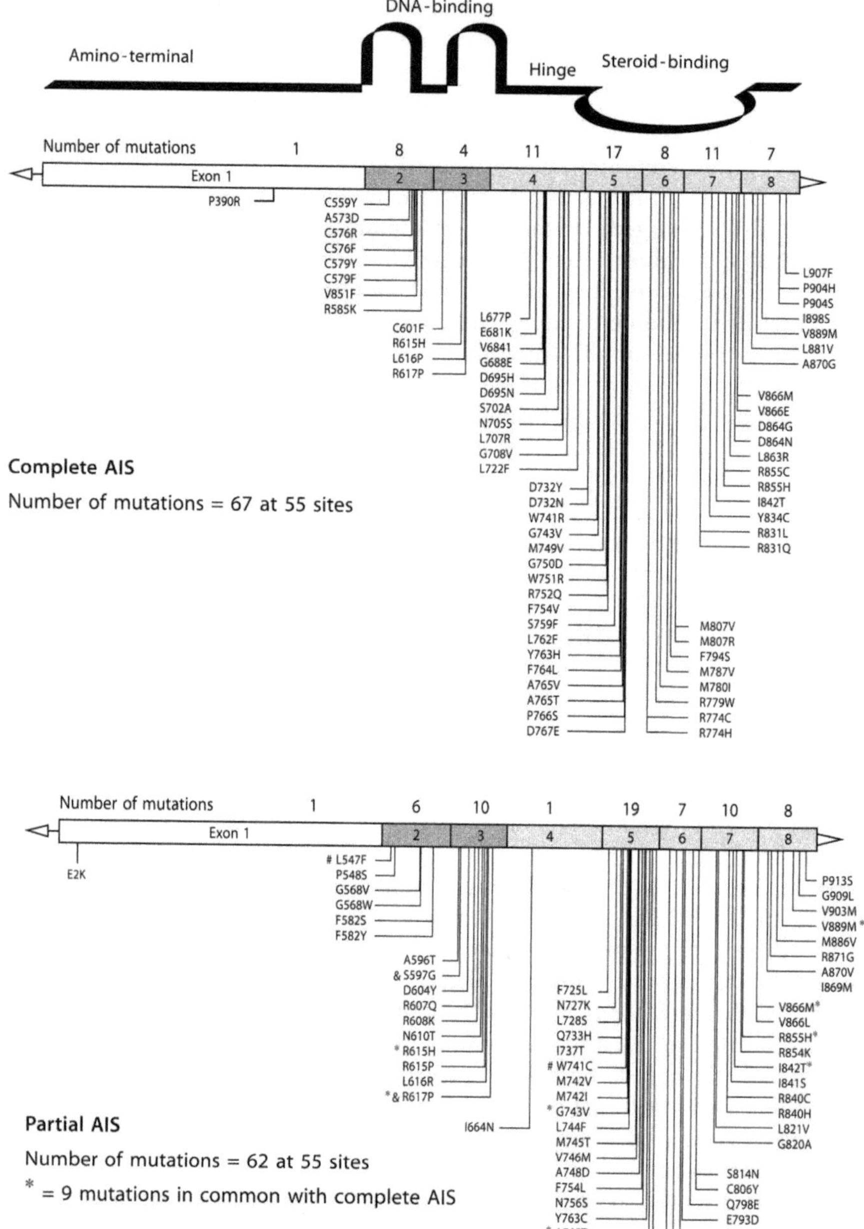
DNA-binding
Amino-terminal
Hinge
Steroid-binding
Number of mutations
1   8   4   11   17   8   11   7
Exon 1
2   3   4   5   6   7   8
P390R
C559Y
A573D
C576R
C576F
C579Y
C579F
V851F
R585K
C601F
R615H
L616P
R617P
L677P
E681K
V684I
G688E
D695H
D695N
S702A
N705S
L707R
G708V
L722F
L907F
P904H
P904S
I898S
V889M
L881V
A870G
V866M
V866E
D864G
D864N
L863R
R855C
R855H
I842T
Y834C
R831L
R831Q
D732Y
D732N
W741R
G743V
M749V
G750D
W751R
R752Q
F754V
S759F
L762F
Y763H
F764L
A765V
A765T
P766S
D767E
M807V
M807R
F794S
M787V
M780I
R779W
R774C
R774H
Complete AIS
Number of mutations = 67 at 55 sites

Number of mutations
1   6   10   1   19   7   10   8
Exon 1
2   3   4   5   6   7   8
E2K
L547F
P548S
G568V
G568W
F582S
F582Y
A596T
& S597G
D604Y
R607Q
R608K
N610T
* R615H
R615P
L616R
* & R617P
F725L
N727K
L728S
Q733H
I737T
W741C
M742V
M742I
* G743V
L744F
M745T
V746M
A748D
F754L
N756S
Y763C
* A765T
E772A
E772G
I664N
P913S
G909L
V903M
V889M *
M886V
R871G
A870V
I869M
V866M*
V866L
R855H*
R854K
I842T*
I841S
R840C
R840H
L821V
G820A
S814N
C806Y
Q798E
E793D
L790F
M780I
R774H*
Partial AIS
Number of mutations = 62 at 55 sites
* = 9 mutations in common with complete AIS

◀──

Fig. 2.7. Single amino acid substitutions in the androgen receptor in individuals with complete AIS (upper) and partial AIS (lower), located with respect to the androgen receptor cDNA. Only distinct mutations are shown, not multiple instances of the same mutation. A line diagram of the receptor protein is shown above the androgen receptor cDNA, for orientation with Fig. 1. Substitutions are identified by the single letter amino acid code for the native and substituting amino acids, flanking the codon number (according to the sequence of Lubahn et al. (1989)). The single letter codes for amino acids are: A, alanine: C, cysteine; D, aspartic, acid; E, glutamic acid; F, phenylalanine; G, glycine; H, histidine; I, isoleucine; K, lysine; L, leucine; M, methionine; N, asparagine; P, proline; Q, glutamine; R, arginine; S, serine; T, threonine; V, valine; W, tryptophan; Y, tyrosine. The two substitutions marked with (&) were found in one individual (S597G and R617P). The two substitutions marked with (#) were found in one individual (L547F and W741C). Mutations marked with an asterisk (*) have been reported in both CAIS and PAIS

polymorphic (Edwards et al. 1992; Lubahn et al. 1989; Marcelli et al. 1990 a). In addition, the amino-terminal region is the least well-conserved among members of the steroid receptor family, suggesting that some variability in protein structure is tolerated within this region. McPhaul et al. (1993 and personal communication) found no exon 1 mutations in the fully-sequenced AR genes of over 30 individuals with AIS. In addition, screening studies of exon 1 using denaturing gradient gel electrophoresis (DGGE) analysis or single strand conformational polymorphism (SSCP) analysis, revealed no evidence of missence mutations in a total of 21 patients (Batch et al. 1992; De Bellis et al. 1992).

Although amino-terminal mutations appear to be infrequent in AIS, it is interesting to note that exon 1 mutations have been reported in a number of prostate cancer specimens. In addition, the amino-terminal domain has been implicated in the pathogenesis of SBMA, due to variation in the CAG triplet repeat length, as discussed in detail below.

The DNA-binding domain (DBD). In as many as one third of patients with clinical and endocrine features typical of AIS the androgen binding characteristics of their genital skin fibroblasts are quantitatively and qualitatively normal (Amrhein et al. 1976; Hughes and Evans 1987; Kaufman et al. 1979). These findings previously lead to the conclusion that there was a defect in the pathway of androgen action beyond the AR itself (termed "post-receptor" defects). However, with the advent of molecular biologic analysis of the AR gene in the 1990s it became clear that amino acid substitutions in the DBD accounted for the defective AR action of many individuals with AIS whose androgen binding is normal (Fig. 2.8). *In vitro* studies of DBD-mutant ARs reveal that complete AIS is associated with absence of receptor-DNA binding, while PAIS is associated with retention of some DNA binding (De Bellis et al. 1994; Quigley et al. 1992 a; Zoppi et al. 1992). Other functions such as dimerization of receptor and protein-protein interactions may also be affected by mutations in the DBD.

Four cysteine residues, invariantly present in the homologous locations in all steroid receptors, co-ordinately bind a zinc ion in each of the two fingers. Alteration of any of these cysteines would disrupt the entire structure of the

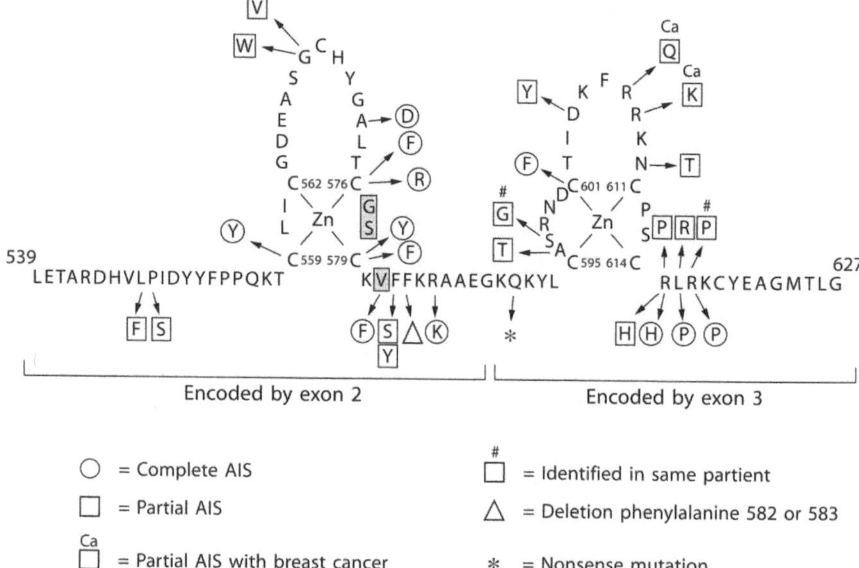

Encoded by exon 2   Encoded by exon 3

○ = Complete AIS   □# = Identified in same partient

□ = Partial AIS   △ = Deletion phenylalanine 582 or 583

□Ca = Partial AIS with breast cancer   * = Nonsense mutation

Fig. 2.8. Amino acid substitutions in the AR DNA-binding domain, caused by mutations in exons 2 and 3 of the AR gene. The DNA-binding domain of steroid receptors is arranged as two zinc fingers. The shaded, boxed amino acids at the base of the first zinc finger have been shown in GR and ER to confer DNA response element specificity. Substitutions associated with complete AIS are shown in circles. Those found in individuals with partial AIS are shown in squares. The two substitutions associated with male breast cancer are indicated by Ca. Two substitutions found in a single individual with partial AIS are indicated with #. Deletion of phenylalanine residue 582 or 583 is indicated with a triangle. The introduction of a premature termination codon in place of lysine 590, is indicated with a large asterisk

finger and thereby the DNA-binding function of the receptor. Substitutions of one of the cysteine residues in the first zinc finger have been found in a number of individuals with CAIS (Chang et al. 1991; Sultan et al. 1993; Zoppi et al. 1992). *In vitro* studies revealed essentially normal androgen binding but defective DNA binding and marked impairment of receptor transcription activation function.

Due to their highly basic nature, substitutions involving arginine residues within the DBD also have profound effects upon receptor function. Examples are provided by the substitution of leucine 616 with arginine, and the replacement of arginine 617 by proline, both of which were found in patients with CAIS. Arginine 617, located just carboxy-terminal to the second zinc finger, lies in a region of the second zinc finger believed to be critical for formation of the alpha-helical conformation necessary for DNA binding (Freedman 1992; Härd et al. 1990; Luisi et al. 1991). The presence of a proline residue would tend to introduce a bend in the alpha helix, severely disturbing structure and function of the DBD of the mutant AR. The presence of arginine in place of leucine at position 616 (2nd zinc finger) likely disturbs the hydrophobic core of the DNA-binding domain (Freedman 1992). Not surprisingly, these mutations disrupt DNA binding and transcriptional activity (De Bellis et al. 1994; Marcelli et al. 1991b; Zoppi et al. 1992).

Two mutations affecting the second zinc finger of the AR are particularly interesting because of their association with male breast cancer in individuals with PAIS – a situation not previously reported in studies of AIS-affected individuals. The AR genes of both families contained mutations that replaced one of a pair of arginine residues at the tip of the second zinc finger. In the first family, in which breast cancer developed in two affected males at 55 and 75 years of age, arginine 607 (conserved in AR, ER, GR and MR) was replaced by glutamine (Wooster et al. 1992). In the second family, in which breast cancer was detected in a young man only 38 years old, the highly-conserved residue arginine 608 was replaced by lysine (Lobaccaro et al. 1993a, 1993d). Subsequent studies revealed reduced transactivation function of the mutant ARs, with no evidence of aberrant transcription regulation via an ERE (Poujol et al. 1997). Although the specific molecular mechanisms are unknown, the occurrence of mutations at adjacent loci in two unrelated families with partial AIS and male breast cancer suggests a more than chance association between the mutant AR and oncogenesis. However, it is also noteworthy that breast cancer has not been reported in association with CAIS, nor in patients with PAIS due to mutations in the LBD, suggesting that unopposed estrogen effect *per se* is not of prime importance, but rather, that the mutant DBD itself may play a specific role in this particular pathology. It could be suggested that, in contrast with mutations that simply impair AR function, these DNA-binding domain mutant ARs may have a specific functional disturbance, perhaps of their protein-protein interactions, that impairs repression of certain oncogenes.

The ligand-binding domain (LBD). The LBD of the AR is encoded by the 3' portion of exon 4 and exons 5–8 of the AR gene. Single base mutations that cause amino acid substitutions and alter androgen binding have been reported in the AR genes of individuals with CAIS and PAIS in each of these exons (Fig. 2.7), causing a wide range of androgen binding abnormalities, including complete absence of androgen binding, reduced affinity or capacity of binding, and qualitative disturbances, such as altered steroid specificity, increased rate of ligand dissociation, and increased thermolability of binding. Attempts to correlate abnormalities of androgen binding with the molecular defect have proven difficult. However, some of the specific features of LBD mutations that affect AR function include the location of the substitution, the nature of the amino acid substituted, the nature of the amino acid that replaced it, and probably other factors such as the effect of the mutation on protein expression and stability.

Distribution and nature of mutations in the LBD. Alignment of AR, PR, GR and MR amino acid sequences reveals regions of high conservation within this receptor subgroup, and the overall amino acid identity between the LBDs of AR and at least 2 other members of its subfamily is about 50%. Highly-conserved residues are the most common sites of amino acid substitution in the AR, about 75% of the amino acid substitutions in the AR LBD occurring at residues identical in at least two other members of the subfamily. These observations, while not surprising, serve to confirm the assumption that such highly-conserved residues have critical roles in receptor structure and function. Overall there is no difference in location of mutations that totally abrogate androgen binding, versus those that induce subtle qualitative changes, however, mutations in exon 4 appear to cause severe disruption or AR function, since 11 of 12 such mutations are associated with CAIS and only one with PAIS. Notably, there is a distinct clustering of mutations in exon 5 in both CAIS and PAIS, accounting for about 35% of the amino acid substitutions in the LBD, although exon 5 represents only 20% of the AR coding sequence. Furthermore, amino acid substitutions have occurred at approximately 60% of the residues encoded by exon 5, compared with approximately 20% of amino acid residues encoded by other exons. These figures imply a particular importance of this cluster of amino acids, and it is noted that this region of the LBD is the most highly-conserved between members of the steroid receptor subfamily to which the AR belongs. The AR has 63–70% sequence identity with PR, MR and GR in this region (Fig. 2.9), compared with 38–55% through the remainder of the LBD. This region represents helices 4/5 of the consensus nuclear receptor LBD structure and is not directly involved in formation of the ligand-binding pocket (Wurtz et al. 1996). Nevertheless, it must play other critical roles, such as stabilizing the region around the pocket, and binding of the 90-kDa heatshock protein as determined for GR (Danielsen et al. 1986; Housley et al. 1990; Pratt et al. 1988). McPhaul et al. noted that this mutational cluster between codons 728 and 774 of the AR is similar to a mutational cluster in the thyroid hormone receptor, further highlighting the functional importance of this region (McPhaul et al. 1992; Weiss and Refetoff 1992).

Mutational hot spots. Four sites in the LBD – arginine residues 774, 840 and 855 and valine 866 – together account for almost one quarter of the missense

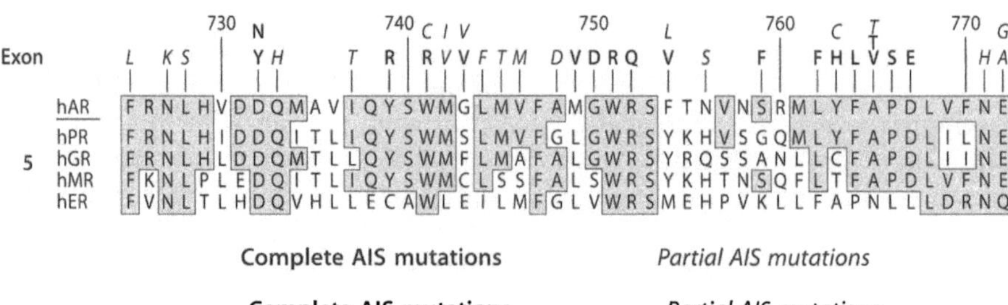

Fig. 2.9. Amino acid sequence of the LBD of human AR, PR, GR, MR and ER encoded by exons 4–8 in human AR. Areas of sequence identity between hAR and other receptors are boxed. Within the central region of the LBD about 80% of mutations affect amino acid residues that are conserved in at least 3 members of this receptor sub-family. Substitutions found associated with complete AIS are shown in bold. Substitutions associated with partial AIS are shown in italics

mutations reported to date. At these codons alternative amino acid substitutions are produced depending upon the base of the triplet that undergoes mutation and the nature of that mutation. These mutations highlight the complexity of the effects of amino acid substitutions upon AR function.

Mutations at codon 774 in exon 6 have been found in 11 apparently unrelated individuals with CAIS or PAIS, replacing the native arginine with either cysteine or histidine (Gottlieb et al. 1997). Both substitutions disrupt androgen binding; in general the cysteine substitution (Brown et al. 1990; Marcelli et al. 1991 a; Mebarki et al. 1993; Prior et al. 1992) being more deleterious than the histidine (Batch et al. 1992; De Bellis et al. 1992; Prior et al. 1992). Binding disturbances associated with these mutations include complete absence of androgen binding, reduced affinity and capacity, and increased thermolability. Disturbances of transcription regulation are associated with these binding defects. Arginine 774, located within helix 6 of the LBD, is modestly conserved being present in PR and GR, suggesting that it does not fulfill a role unique to AR. However, it may be important for maintaining the LBD in a stable conformation (Marcelli et al. 1991 a).

Arginine 840, which lies within the putative helix 9 of the LBD and is unique to AR (most other steroid receptors have lysine at this site), is also the site of substitution by either cysteine or histidine in 3 and 11 cases respectively (Beitel et al. 1994 a; De Bellis et al. 1994; Ghirri and Brown 1993; Gottlieb et al. 1997; Hiort et al. 1993; Imai et al. 1994; Imasaki et al. 1994; Lumbroso et al. 1994; McPhaul et al. 1992). However, in contrast to the arginine 774 substitution which causes CAIS in some cases, individuals with either cysteine or histidine at position 840 have PAIS phenotypes. In keeping with this, these mutant ARs retain some androgen binding and transcriptional activity at supraphysiological androgen concentrations (De Bellis et al. 1994). These findings probably reflect the fact that arginine 840 lies outside the ligand-binding pocket itself and the structure of the ligand-binding pocket is not grossly disturbed by this substitution.

Multiple mutations (18 cases) also have been reported at arginine 855 (Gottlieb et al. 1997), an extremely highly-conserved residue, present in GR, MR, PR, ER, ERR (estrogen-related receptor) and COUP (chicken ovalbumin upstream promoter). This residue is located in the putative helix 10 of the LBD and lies within a hydrophobic sub-region shown in mouse ER to be involved in receptor dimerization (Fawell et al. 1990). Thus disturbance of dimerization, in addition to ligand binding, could compound the functional defect of receptors with substitutions in this region. ARs that contain cysteine at position 855 have absent androgen binding and are associated with CAIS in all cases reported to date (Brown et al. 1993; De Bellis et al. 1992; Lobaccaro et al. 1993 c; McPhaul et al. 1992; Tincello et al. 1992); replacement of arginine 855 with histidine generally has a less deleterious effect upon receptor function, evidenced by partial masculinization in some affected individuals (Batch et al. 1993 b; Chang et al. 1991; McPhaul et al. 1992). Notably, this mutant AR is associated with phenotypic variation between affected individuals, most individuals being reported as having PAIS, with a few reported as having CAIS (McPhaul et al. 1992).

The apparently high mutation rate at these sites in part reflects the fact that each of the four 'hot-spots' contains a CpG dinucleotide (cytosine/guanine pair); these doublets are subject to a high frequency of cytosine to thymine (C–T) transitions by a process involving methylation of cytosine to 5-methylcytosine, followed by spontaneous deamination to thymine (Youssoufian et al. 1986). The three hot spots described above involve arginine residues, and approximately 20% of all AR mutations have occurred at arginine codons which may in part reflecting a detection bias, since arginine is a highly basic amino acid and its substitution by an acidic or neutral amino acid will alter the charge of the region and thus the function of the receptor more noticeably than substitution of a less highly-charged amino acid.

Effect of mutations on qualitative aspects of androgen binding. Certain amino acid substitutions in the AR LBD have little effect on equilibrium binding affinity, but instead alter qualitative aspects of binding such as thermostability, androgen dissociation rate or ligand specificity. Thermolability of binding – that is, a lower androgen binding capacity at 41°C than at 37°C – has been reported in association with a variety of amino acid substitutions in the LBD (Batch et al. 1992, 1993a, 1993b; Beitel et al. 1994a; Gottlieb et al. 1997; Imasaki et al. 1994; Kasumi et al. 1993; Marcelli et al. 1991a; McPhaul et al. 1991a; McPhaul et al. 1992; Nakao et al. 1993; Prior et al. 1992; Ris-Stalpers et al. 1994a). However there is no predictability to the type of mutation associated with this binding defect: examples of mutations causing thermolability include glycine 820 (conserved in GR, MR and PR (Kasumi et al. 1993)) to alanine, glycine 743 (non-conserved) to valine (Nakao et al. 1993) and arginine 840 (non-conserved) to cysteine (Beitel et al. 1994a; McPhaul et al. 1992). Substitutions that cause thermolability of androgen binding appear to be most commonly found in individuals with PAIS, but how this defect relates to dysfunction of the AR *in vivo* is unknown.

A variety of LBD substitutions may alter AR ligand-binding specificity. Replacement of serine 814, located just beyond the eighth helix of the ligand-binding pocket and conserved in GR, MR, VDR and TR, with asparagine (asparagine is present in PR and ER) (Pinsky et al. 1992) produces an AR with "androgen-selective" binding – affinity of binding is reduced and dissociation rate increased with DHT, but is normal with methyltrieneolone. This mutation is associated with a variable phenotype in affected individuals, ranging from AIS grade 1 in one family (normal male genitalia with infertility and gynecomastia) (Pinsky et al. 1984) to AIS grade 4 (severe undermasculinization with labioscrotal folds, and a clitoral phallus) in another (Pinsky et al. 1985). Replacement of valine 866 (helix 10) with methionine (Lubahn et al. 1989) is associated with reduced receptor affinity for testosterone and DHT, and a relative increase in affinity for progesterone (PR has methionine at this site). Presence of leucine at position 866 results in a ligand-specific alteration in dissociation rate – an increased rate of dissociation of methyltrieneolone and mibolerone (synthetic, non-metabolizable androgens), with normal dissociation rate of DHT (Brown et al. 1982; Kazemi-Esfarjani et al. 1993). Another site at which substitution alters ligand specificity is arginine 871 (helix 10), which, when replaced by glycine as found in a patient with grade 1 AIS, demonstrates variable rates of androgen dissociation and thermolability in the presence of different androgens (Kaufman et al. 1990; Pinsky et al. 1992). Altered ligand specificity is also seen with the mutant AR in the prostate cancer cell line, LNCaP (Harris et al. 1990; Veldscholte et al. 1990).

Increased rate of dissociation of bound androgen despite normal equilibrium binding affinity has been reported for a number of AR mutations, including aspartic acid 695 to histidine or asparagine, tyrosine 763 to cysteine and valine 889 to methionine (Batch et al. 1993a; De Bellis et al. 1994; Kazemi-Esfarjani et al. 1993; McPhaul et al. 1991a; McPhaul et al. 1992; Ris-Stalpers et al. 1991). Apart from the functional disturbance that the rapid dissociation of ligand imposes upon the receptor, this defect may also predispose the receptor to proteolytic degradation, as discussed below.

Effect of mutations on receptor level. In addition to altering receptor function directly, certain amino acid substitutions appear to reduce receptor protein level presumably by affecting post-translational processing or degradation of the receptor. Reduced levels of immunoreactive AR have been demonstrated in genital skin fibroblasts of a number of individuals whose ARs have amino acid substitutions in the LBD of the receptor (Wilson et al. 1992). Correct folding of the polypeptide may be disturbed by the amino acid substitution, predisposing the mutant receptor to proteolytic degradation. Furthermore, there is evidence that androgen binding itself is important in maintenance of receptor level by stabilizing the receptor and protecting it from degradation (Kemppainen et al. 1992). Thus a mutation that increases the rate of androgen dissociation from the receptor may allowing degradation to occur more readily; the reduced quantity of receptor would be reflected by diminished androgen binding capacity. The reduced levels of AR protein, in concert with the intrinsic dysfunction of the receptor, could exacerbate the functional degree of androgen resistance. An example is provided by a mutant AR containing methionine in place of valine 889, associated with grade 6 AIS. Despite an equilibrium binding affinity close to normal, this receptor has reduced binding capacity and increased androgen dissociation rate. This mutant retains some transcriptional activity, but only at supraphysiological androgen concentrations, which presumably compensate for the rapid dissociation and stabilize the receptor, restoring binding capacity and transcriptional function (De Bellis et al. 1994).

2.8.1.7 Is there a genotype/phenotype correlation for AR mutations?

It has long been noted that the relationship between the genital phenotype of individuals with AIS and the androgen binding characteristics of their cultured genital skin fibroblasts is inconsistent. With the advent of molecular analysis of the AR gene, it had been hoped that it would be possible to establish correlations between the molecular defects, and either the androgen binding characteristics or the clinical phenotype associated with different mutant receptors. However, the large number and wide variety of AR gene mutations has precluded this. Individuals with CAIS have AR defects ranging from complete deletion of the gene, to a wide array of a.a. substitutions in the DBD or LBD that disrupt DNA binding or androgen binding to a greater or lesser degree. Individuals with PAIS have an almost equally varied collection of AR defects, although even small deletions are extremely rare. AR mutations are found only rarely in patients with isolated hypospadias (Allera et al. 1995; Hiort et al. 1994; Sutherland et al. 1996). In general, a.a. substitutions associated with CAIS tend to be more radical in nature and affect more highly-conserved residues than those associated with PAIS. However, there are many exceptions to this generalization, and a number of instances in which the same mutation has been found in kindreds with CAIS on the one hand and PAIS on the other. Complete and partial forms of AIS do not coexist within a family (Morris and Mahesh 1963) however, phenotypic variation between and within PAIS kindreds is well described (Batch et al. 1993a; Brown et al. 1982; Forest et al. 1990; Grino et al. 1989; Wilson et al. 1974). The finding that the same substitution may be associated with CAIS in one kindred and PAIS in another, or with different phenotypes within the same PAIS kindred, indicates that additional factors (for example variation in testosterone level at a critical period) influence the effects of the mutant AR upon the development of the external genital phenotype (Batch et al. 1993a; Boehmer et al. 1997; Evans et al. 1997; Lobaccaro et al. 1993e; McPhaul et al. 1992; Nakao et al. 1993; Rodien et al. 1996). Furthermore, while *in vitro* analyses of the transcription-activating function of mutant receptors may provide some correlation with phenotype (Brown et al. 1990; Kazemi-Esfarjani et al. 1993; Klocker et al. 1992; Marcelli et al. 1990a, 1990b; Marcelli et al. 1991a, 1991b; McPhaul et al. 1991a; Prior et al. 1992; Ris-Stalpers et al. 1991; Zoppi et al. 1992), these studies provide no insight into the phenotypic variation between affected individuals with the same mutation. For these reasons it has not been and likely will not be possible to establish any clear correlation between genotype and phenotype.

Finally, in a number of subjects with clinical and endocrine features entirely consistent with AIS, it has not been possible to detect a mutation anywhere within the coding region of the AR gene (Batch et al. 1992; Morel et al. 1994) and (M.J. McPhaul, personal communication). While in some cases this may be due to diagnostic or technical failures, particularly where rapid screening methods have been utilized, the apparent absence of mutation in the AR coding region in such cases raises the possibility of defects in regions of the AR gene not yet examined such as the promoter, in genes upstream or downstream of the AR, or in factors required for full AR activity *in vivo*.

2.8.1.8 Carrier detection and prenatal diagnosis

Since the AR is encoded by an X-chromosomal gene, the syndromes of androgen insensitivity follow an X-linked recessive pattern of inheritance. Therefore any obligate carrier female has a one in two chance that her 46,XY offspring will be affected, and a one in two chance that her 46,XX offspring will be a carrier. Delayed puberty (Kaufman et al. 1976; Quigley et al. 1992b; Sai et al. 1990) and/or reduced or asymmetric development of pubic and axillary hair (McKusick 1990; Morris and Mahesh 1963; Pinsky et al. 1985; Pion et al. 1965; Puck et al. 1960) are reported by some AIS carriers, however these findings are inconsistent. In the absence of reliable clinical signs, carrier detection in female members of an AIS-affected family is most accurately performed by molecular analysis of the AR gene, using one of a variety of molecular techniques. Restriction fragment length polymorphism (RFLP) analysis is hampered by the fact that only 18% of women are heterozygous for the *Hind*III polymorphism within the AR gene (Brown et al. 1989; Quigley et al. 1992b). Other carrier determination strategies have utilized PCR/DGGE or PCR/SSCP screening and allele-specific PCR. Carrier status determination is most straightforward in families in which a specific single base mutation has been identified. Mutation detection in potential carriers (or subsequent affected individuals) may then be performed simply and rapidly by screening or sequencing only the region of interest (Batch et al. 1993a; De Bellis et al. 1994; Hiort et al. 1993; Lobaccaro et al. 1993e; Lumbroso et al. 1993; Yong et al. 1994). If the mutation alters a restriction enzyme cleavage site, mutation detection is further simplified by digestion of PCR-amplified DNA with the appropriate restriction enzyme (Imasaki et al. 1994; Jakubicza and Werder 1992; Lobaccaro et al. 1993b; Ris-Stalpers et al. 1991; Sai et al. 1990; Trifiro et al. 1991b). The simple and sensitive technique of allele-specific PCR, in which one of the pair of oligonucleotide primers contains the mutant sequence, further streamlines the approach to carrier detection, since only DNA containing the mutant sequence will amplify in the presence of the mutant primer (Klocker et al. 1992). Analysis of the polymorphic exon 1 CAG repeat for variation in repeat number may also be useful in carrier status determination, provided that the mother is heterozygous for the polymorphic repeat (that she carries two alleles of different CAG repeat lengths), and that one allele segregates with the mutant AR gene in a previous affected child (McPhaul et al. 1991a; Mebarki et al. 1991, 1993; Ris-Stalpers et al. 1994a).

2.8.1.9 Clinical management

In the early days of the molecular analysis of AR defects, it was hoped that such studies would provide objective management strategies to aid assignment of sex-of-rearing and optimize long-term psychosexual outcome. Unfortunately, although molecular studies of the AR gene in AIS have provided significant information regarding AR structure and function, they have not lived up to expectations regarding their value in clinical management. This is most striking in light of the phenotypic variation noted between and within kindreds with the same mutation as discussed above. Thus, at the present time, clinical factors remain of primary importance in assignment of sex-of-rearing for an infant with AIS.

The appropriate sex-of-rearing of AIS-affected individuals is reasonably straightforward for those with the mildest (grade 1, 2) and most severe (grades 5–7) degrees, where there is little doubt about the predominant genital phenotype. However, the optimal sex-of-rearing and management for individuals with grades 3 and 4 AIS is a subject of great concern and debate (Diamond and Sigmundson 1997b; Grant 1995; Reiner 1997). While it has for the last 20 years been fairly standard practice to recommend female sex-of-rearing for individuals whose penile size or structure was considered "inadequate", this practice is being re-examined in the light of long-term outcome of some affected individuals. For example, individuals with PAIS and ambiguous genitalia reared as females generally have significant vaginal hypoplasia. Creation of a vagina capacious enough for intercourse requires complex, major vaginal reconstructive surgery. These surgeries have a high complication rate with problems such as strictures, inadequate vaginal size, painful intercourse, intravaginal hair growth, vaginal dryness or excessive mucus production, depending on the surgical procedure used; approximately 25% of patients require repeated surgery and sexual function may be poor (Newman et al. 1992). In addition, the clitoral cosmetic surgery (clitoroplasty or clitoral recession) commonly utilized in patients reared as females has the potential to impair clitoral sensation. Significant psychosocial impact is likely for a young woman undergoing such major procedures to the sexual organs. In light of these facts it should be noted that there is no objective evidence to indicate that the psychosexual outcome of a woman who has undergone such major genital surgery is in any way superior to that of a man who has a small or incompletely-formed penis. In addition, the effect of prenatal androgen exposure to the brain must be considered in terms of its potential impact on gender identity, even in those whose tissues are to some extent resistant to androgens. The long-held tenet that gender identity is a plastic characteristic, determined by environmental factors in the first two years of life, is unsupported by concrete data, and has been challenged by studies of individuals with various genital anomalies who undergo spontaneous gender transversion from their female sex-of-rearing to male (their prenatal hormonal sex) (Diamond and Sigmundson 1997a; Money 1974). In addition, the psychosexual outcome of women with congenital adrenal hyperplasia, prenatally exposed to higher than normal androgen concentrations, provides further evidence of the effect of prenatal androgens on gender identity and bonding behavior (deVere White et al. 1994). Comprehensive long-term outcome data for individuals with PAIS are lacking and further studies are required. Nevertheless, these factors suggest that the traditional approach of assigning female sex-of-rearing in PAIS cases with genital ambiguity should be reevaluated.

Notably, there has been some success with high dose testosterone treatment of patients with PAIS during early childhood or the teenage years, despite definite molecular defects in the AR gene and abnormal androgen-binding (Grino et al. 1989; Hiort et al. 1993; McPhaul et al. 1991 a). Two teenage brothers with AIS grade 3 had androgen binding that was selectively abnormal, the dissociation of methyltrieneolone and mibolerone from the receptor being abnormally rapid, while the dissociation rate of DHT was normal. In the presence of a high concentration of DHT, transcriptional activation by the mutant receptor containing a valine to leucine substitution at position 866 approached that of the normal receptor (Kazemi-Esfarjani et al. 1993). These young men had substantial virilization in response to 500 mg testosterone enanthate administered every two weeks (Hiort et al. 1993). Such data provide evidence that in PAIS cases associated with mutations in the LBD, it may be possible to partially compensate for the AR dysfunction by administration of supraphysiological amounts of the androgens that bind the mutant AR most stably or with greatest affinity. The use of tests such as the SHBG response to stanozolol or hCG stimulation testing in some cases correlates with response to administered testosterone. However, these tests may not be completely reliable – in one reported case there was substantial SHBG suppression without change in penile size in response to high dose non-aromatizable androgens (Keely et al. 1993).

One further aspect of management of AIS that warrants discussion is the timing of gonadectomy. In individuals with retained androgen responsiveness reared as females, early gonadectomy is necessary to prevent virilization at puberty. However, this is not the case for those with AIS grade 6/7. The approach to the timing of gonadectomy has fluctuated over the years, and also seems to vary geographically. In the USA the trend over the last decade or so has been to recommend gonadectomy during infancy or early childhood, on the premise that this would be psychologically preferable for the affected individual, since discussion of the diagnosis and details of the karyotype and gonadal tissue with the patient would be avoided. However, there are no studies addressing the optimal timing of gonadectomy from the psychological standpoint. In fact, there may be both physiological and psychological advantages to delayed gonadectomy. There is evidence that women with AIS have reduced bone mineralization (Munoz-Torres et al. 1995; Soule et al. 1995 and Finkelstein J. personal communication); however, systematic studies addressing the role of the timing of gonadectomy in the etiology of this problem have not been undertaken. Nevertheless, it would seem logical to suggest that endogenous estrogenization would be superior to that available by exogenous means, particularly when issues such as compliance with therapy are considered. Furthermore, there is a growing drive from support groups for individuals with AIS and other disorders of sexual differentiation for physicians to provide full disclosure to patients and to involve them in the decision-making process regarding issues such as genital and gonadal surgery. These factors, in addition to the educational and psychosocial background of the family, should be incorporated into the decision-making process. Families facing management of these disorders will require psychological support and counseling, and AIS affected individuals themselves may require such support lifelong.

2.8.2 Spinal and bulbar muscular atrophy
(SBMA; Kennedy Disease; bulbospinal neuronopathy)

2.8.2.1 Clinical features

The X-linked motor neuron disorder known as spinal and bulbar muscular atrophy (SBMA), Kennedy disease, or bulbospinal neuronopathy has features of AR dysfunction both at the clinical and at the molecular level. The major disturbance in affected individuals is mid-adult onset (usually 30–50 years of age) of slowly-progressive atrophy of the spinal and bulbar muscles manifest by cramps, tremor and weakness, mainly of the tongue, facial muscles and proximal limb girdle muscles. Degeneration of spinal motor neurons and muscle wasting is associated with elevated serum creatine kinase and signs of muscle degeneration in muscle biopsy specimens (Harding et al. 1982). In one case, post-mortem examination revealed atrophy and loss of the anterior horn cells of the spinal cord (Kennedy et al. 1968). Sensory neuropathy or abnormal sensory nerve conduction velocity is also reported in some cases (Guidetti et al. 1996). In addition to its features of neurologic disease SBMA is also a non-classical form of mild, late-onset androgen resistance. Affected individuals have normal male external genitalia; signs of androgen resistance, including development of gynecomastia in more than half the patients, generally appear after the third decade of life. Notably, this is often the earliest manifestation of the condition (Belsham et al. 1992a; Harding et al. 1982). Serum testosterone concentrations are generally normal, while estradiol and LH are variably increased (Arbizu et al. 1983; Ertekin and Sirin 1993; Harding et al. 1982). Azoospermia, impotence and testicular atrophy may also develop in these men, at least half of whom were previously fertile. Germinal failure with normal Leydig cells was reported in one affected man (Arbizu et al. 1983), while marked Leydig cell involution was found in another (Hausmanowa-Petrusewicz et al. 1983). Phenotypic expression of SBMA is variable with respect to age at onset and severity of neurological symptoms, both between and within families. Interestingly, there is an increased incidence of non insulin-dependent diabetes mellitus in affected individuals (Barkhaus et al. 1982; Ertekin and Sirin 1993; Harding et al. 1982; Kennedy et al. 1968; Schoenen et al. 1979), raising a question of the relationship between androgen and insulin action. Recently it has been determined that a founder effect is contributes significantly to the occurrence of SBMA in Japan (Tanaka et al. 1996).

2.8.2.2 AR expression and androgen binding

Studies of AR expression and androgen binding in SBMA are limited and conflicting. In one study, no AR was detectable by immunostaining in scrotal skin of three patients (Matsuura et al. 1992). Androgen binding capacity in genital skin fibroblasts is reduced in some affected individuals, but normal in others (Danek et al. 1994; Warner et al. 1992). Reduction of androgen binding affinity was observed in one study and the severity of the patients' gynecomastia and testicular atrophy showed a strong correlation with affinity of androgen binding in genital skin fibroblasts (MacLean et al. 1994). Furthermore, a modest correlation existed between the androgen binding affinity and the number of CAG triplet repeats in exon 1 of the AR gene (described below). Examination of AR expression in neural tissues also provides conflicting data. Normal intense AR immunostaining of motor neurons was found in an autopsy study of 2 patients (Ogata et al. 1994), while reduced AR mRNA expression in motor neurons was found in another study (Nakamura et al. 1997).

2.8.2.3 The molecular defect – expansion of a trinucleotide repeat in exon 1 of the AR gene

The normal AR gene contains a polymorphic stretch of CAG triplets in exon 1 of average length 21±2 repeats, [range 11–31 or 14–35]. In patients with SBMA this repeat segment is expanded to 40–62 repeats (Amato et al. 1993; Belsham et al. 1992b; Biancalana et al. 1992; Choi et al. 1993; Edwards et al. 1992; Igarashi et al. 1992; Kaspar et al. 1994; La Spada et al. 1991, 1992; Macke et al. 1993). This type of mutation, known as dynamic mutation (Richards and Sutherland 1992), has been reported in at least 10 other neurodegenerative disorders, including myotonic dystrophy (MD) (Lavedan et al. 1993), the fragile X syndrome (FraX) (Caskey et al. 1992), spinocerebellar ataxia type 1 (SCA 1) (Orr et al. 1993), Huntington disease (HD) (The Huntington's Disease Collaborative Research Group. 1993) and dentato-rubro-pallido-luysian atrophy (DRPLA) (Koide et al. 1994; Nagafuchi et al. 1994). In this group of disorders the repeated DNA segment displays meiotic instability: the size of the repeat increases (or occasionally contracts) with succeeding generations, perhaps due to slippage of DNA polymerase during DNA replication. Two studies have also reported this phenomenon in SBMA, noting that the instability of the repeat is greater in male than in female meioses (Biancalana et al. 1992; La Spada et al. 1992). In addition, variation in CAG repeat length between different sperm from the same SBMA-affected individual has been described (Zhang et al. 1995).

There is a correlation between the size of the expanded segment and the severity of the clinical disease in these triplet repeat-associated neurodegenerative conditions (Richards and Sutherland 1992). Most studies have demonstrated a similar correlation in kindreds with SBMA (Biancalana et al. 1992; Doyu et al. 1992; La Spada et al. 1992; MacLean et al. 1994; Shimada et al.

1995). In general, the greater the number of CAG repeats, the lower the age of onset and greater the severity of muscle weakness. Surprisingly, La Spada et al. (1992) found that this correlation applied only to the neurologic features of the disease, not to the features of androgen insensitivity and noted that there was intrafamilial variation in disease severity in individuals with identical repeat size. Thus, factors other than repeat size are likely to contribute to the phenotypic variability in this disorder, as in more typical forms of androgen resistance.

2.8.2.4 Molecular pathophysiology

Many lines of evidence point to a role for androgens/AR in neural function. Radiolabelled androgens have been shown to concentrate in the nuclei of motor neurons of rat spinal cord (Sar and Stumpf 1977), indicating the presence of AR in these cells, and AR is also present in rat cranial nerve nuclei (Yu and McGinnis 1986). Androgens regulate the development of a sexually dimorphic spinal motor nucleus (Goldstein and Sengelaub 1992), the length of motorneuron dendrites (Kurz et al. 1986) and the branching of neurites (Lustig et al. 1994). Further, the anterior horn cell degeneration that is the hallmark of SBMA suggests that the AR plays a role in maintenance of normal motor neuron function (or, alternatively, that a mutant AR directly induces anterior horn cell dysfunction). One study showed that expansion of the CAG repeat caused a reduction of AR mRNA and protein expression in a neuronal cell line *in vitro* (Choong et al. 1996a). Although the animal and *in vitro* data confirm the presence of, and implicate a role for, AR in neural tissue, the absence of neurologic dysfunction in numerous reported cases of AIS, particularly those caused by the null mutation (Quigley et al. 1992b), raises a question regarding the absolute requirement for AR in normal neural function. The development of neurodegenerative changes in patients with SBMA, in contrast with the normal neuromuscular status of individuals with AIS, suggests that the pathogenesis of the neurodegenerative features of SBMA is not related directly to androgen insensitivity, but involves cellular interactions unique to the receptor containing the expanded glutamine segment.

The mechanism by which the expanded CAG repeat causes disease in this and the other triplet-expansion disorders remains unclear, but is generally postulated to reflect a gain-of-function mutation. The codon CAG encodes the amino acid glutamine and the expanded triplet repeat in SBMA, SCA 1, Huntington disease and DRPLA encodes an enlarged polyglutamine tract in the respective protein in each case. Polyglutamine stretches occur in other transcription-regulating proteins (Gerber et al. 1994; Orr et al. 1993) and are believed to facilitate protein-protein interaction (Perutz et al. 1994). One hypothesis to explain the pathophysiologic effect of the expanded polyglutamine segment of proteins like the AR suggests that the polyglutamine region acts as a polar "zipper" joining transcription factors bound to separate DNA fragments. The expanded segment could induce abnormal interactions between transcription factors or these large protein complexes might gradually precipitate within affected neurons (Perutz et al. 1994). In addition, there is evidence for a role of polyglutamine segments in transcriptional regulation. Gerber et al. reported that maximal transactivation by transcription factors containing homopolymeric regions, was present when the protein contained 10–30 glutamine residues and found a reduced level of transcriptional activity of factors containing glutamine tracts of more than 30 residues (Gerber et al. 1994). The polyglutamine segment in the AR is also involved in transcription regulation. However, the effect of expansion of the segment upon AR transcriptional activity varies between studies, some demonstrating reduced transactivation in the presence of the expansion and others showing wild-type activity (Chamberlain et al. 1994; Choong et al. 1996a; Gao et al. 1996; Mhatre et al. 1993; Nakajima et al. 1996; Neuschmid-Kaspar et al. 1996). Although the gain-of-function theory seems most appropriate to explain the basis of disease in SBMA, the recessive nature of the condition, evidenced by lack of clinical manifestations in the great majority of heterozygous females, argues somewhat against this, since it would be expected that the presence of one abnormal AR allele would result in production of sufficient abnormal protein to cause disease. Perhaps the abnormal AR may lose function with respect to interaction with certain genes and gain function in its interaction with others (Mhatre et al. 1993).

2.8.2.5 Clinical management

High dose oral testosterone therapy has been utilized in patients with SBMA without convincing efficacy (Danek et al. 1994; Goldenberg and Bradley 1996; Neuschmid-Kaspar et al. 1996). Prenatal diagnosis of SBMA in the fetus of an obligate carrier can be undertaken on the basis of analysis of the size of the AR gene CAG repeat (Yapijakis et al. 1996), and this technique may also be used to aid in differential diagnosis between SBMA and other motor neuron diseases (Ferlini et al. 1995).

2.8.3 Prostate cancer

Prostate cancer (CaP) is the second most frequent cause of cancer deaths (after lung cancer) among males in the U.S. (Schoenberg et al. 1994b). Because of the pivotal role of androgens and the AR in development of the prostate, much attention has been focused on their potential role in prostate cancer. One well-documented feature of prostate carcinoma is that although medical or surgical castration initially induces remission of cancer growth, most CaPs subsequently become resistant to androgen-ablative therapy – that is, they become androgen independent. As Chapter 9 presents a detailed review of prostate cancer only a brief summary is provided here.

Mutations in the AR gene have been reported in DNA extracted from a small proportion of CaP specimens (Culig et al. 1993; Gaddipati et al. 1994; Hakimi et al. 1996; Newmark et al. 1992; Schoenberg et al. 1994b). However these are somatic mutations, present in the cancerous tissues only, not in the patients' germ line (genomic) DNA; furthermore, mutations are much more frequent in metastatic or recurrent prostate cancer than in primary CaP. There are a number of interesting contrasts between the patterns of mutations seen in CaP and AIS. First, there is an apparent clustering of mutations in exon 4 of the AR gene, encoding amino acids 628–724 which comprise the hinge region and the amino-terminal portion of the LBD. A number of these amino acid substitutions are located in putative helices 1 and 3 of the LBD which contribute to the ligand binding pocket; in AIS the predominant cluster of mutations occurs in exon 5, encoding amino acids 725–772 which form helices 5 and 6. Second, a significant number of mutations have been found in exon 1 in CaP, contrasting with the paucity of exon 1 mutations in AIS. Third, no mutations have been found in exons 2 or 3, encoding the AR zinc fingers, a frequent site of mutations in AIS. Because of the role of polyglutamine segments in transcription factors, and the polymorphic nature of the CAG triplet repeat in exon 1 of the AR gene, this region has been examined as a potential site of abnormality in CaP. One study found no change in triplet repeat length (Ruizeveld de Winter et al. 1994), however in another study somatic contraction of the region from 24 repeats in genomic DNA to 18 repeats in tumor tissue was found in one patient (Schoenberg et al. 1994b). Further suggestion of a relationship between CaP and CAG repeat length is suggested by the finding that individuals whose AR gene had <20 CAG repeats were at increased risk for prostate cancer (Ingles et al. 1997); in another study, 56% of patients with localized CaP had ≤18 CAG repeats in exon 1, compared with only 18% of the general population (Schoenberg et al. 1994a). It is also of interest and possible relevance to note that the rate of prostate cancer is higher, and the CAG repeat region shorter in African Americans than in the Caucasian population (Ingles et al. 1997). In addition, there may be a correlation between CAG repeat length and age at onset of prostate cancer, with shorter CAG repeats being associated with younger age at onset (Hardy et al. 1996).

Mutant ARs in CaP may show altered ligand specificity, suggesting that the mutant AR may aberrantly mediate responses to steroids other than its cognate ligands testosterone and DHT, such as adrenal steroids or testosterone metabolites (Culig et al. 1993, 1996). This may be an important factor in tumor progression following androgen ablation therapy. The AR of the human prostate cancer cell line LNCaP, which contains the same threonine to alanine substitution at position 877 as reported in certain patients, also displays this altered specificity of ligand binding (Harris et al. 1990, 1991; Ris-Stalpers et al. 1993; Veldscholte et al. 1990). The mutant AR binds and is transcriptionally active in the presence of androgens, the antiandrogen hydroxyflutamide, progesterone and estradiol; reflecting this, LNCaP cells can be stimulated to grow by androgens, hydroxyflutamide, progesterone and estradiol. The altered ligand specificity of mutant ARs in advanced CaP indicates a possible role for AR in tumor progression. Although androgens generally down-regulate AR mRNA, they up-regulate AR mRNA expression in certain prostate cancer cell lines (Dai et al. 1996). This phenomenon may contribute to progression or recurrence of prostate cancer, since increased expression of the normal AR gene in carcinomatous cells would enable these aberrant cells to optimize the available low androgen concentrations for cell growth and clonal expansion (Koivisto et al. 1996, 1997; Visakorpi et al. 1995). While AR mutations are found fairly often in advanced CaP, mutations appear to be uncommon in early prostate cancer and their role in oncogenesis is uncertain. Finally it should be noted that the finding of mutations in other genes (such as the tumor suppressor genes, retinoblastoma (*RB*) (Bookstein et al. 1990) and *p53* (Isaacs et al. 1991)) in CaP tissues and cell lines, raises the possibility that the mutations detected in the AR gene represent an epiphenomenon, reflecting the genetic instability of neoplastic tissues, rather than being specific to AR.

2.9 Key messages

- The AR is a ligand-activated nuclear transcription factor encoded by a single copy gene located at Xq11–q12 and the AR is expressed ubiquitously throughout the body.
- Normal function of the AR is required for male sex differentiation *in utero*, and for normal post-natal virilization and sexual function.
- The AR is structurally homologous to other steroid hormone receptors, comprising three major functional domains – the amino-terminal domain, the DNA-binding domain, containing two zinc fingers, and the ligand-binding domain.
- By analogy with the crystal structure of other receptors, the ligand-binding domain is likely to be arranged as a series of 12 alpha helices.
- A variety of steroids can bind to the AR. The steroids for which the AR has highest affinity are dihydrotestosterone and testosterone. The AR must be able to bind androgen to be transcriptionally active.
- Upon ligand binding the AR forms homodimers via interactions between the amino and carboxy termini of receptor molecules.
- The liganded homodimer binds to hormone response element DNA sequences in the promoter regions of target genes.
- AR/DNA binding involves interactions with other transcriptionally active proteins. These interactions regulate the rate of transcription of target genes.
- The androgen insensitivity syndromes (AIS) are a continuum of clinical disorders resulting from varying degrees of target tissue resistance to androgens.
- A wide variety of mutations in the AR gene, ranging from complete deletion of the gene, to single base mutations that result in amino acid substitutions, have been reported in AIS.

- Missence mutations in AIS cluster in exon 5 of the AR gene, with a second less prominent cluster in exon 7.
- There is no clear genotype/phenotype correlation for AR mutations in AIS.
- Apart from the effects on development of the reproductive tract, AR mutations associated with AIS appear to cause no obvious disorders in other systems.
- The neurodegenerative disease spinal and bulbar muscular atrophy (SBMA) is caused by expansion of the exon 1 CAG repeat, encoding the amino-terminal polyglutamine tract of the AR.
- Somatic mutations of the AR gene are found in a minority of prostate cancer specimens, particularly in cases of metastatic or recurrent cancer. The mutations are distributed differently from those seen in AIS.
- AR mutations associated with prostate cancer commonly alter ligand binding specificity of the receptor.
- There may be an association of a shorter CAG repeat length in exon 1, with an increased risk of prostate cancer.

2.10 References

Adler AJ, Scheller A, Hoffman Y, Robins DM (1991) Multiple components of a complex androgen dependent enhancer. Mol Endocrinol 5:1587–1596

Aiman J, Griffin JE, Gazak JM, Wilson JD, MacDonald PC (1979) Androgen insensitivity as a cause of infertility in otherwise normal men. N Engl J Med 300:223–227

Aiman J, Griffin JE (1982) The frequency of androgen receptor deficiency in infertile men. J Clin Endocrinol Metab 54:725–732

Akin JW, Behzadian A, Tho SPT, McDonough PG (1991) Evidence for a partial deletion in the androgen receptor gene in a phenotypic male with azoospermia. Am J Obstet Gynecol 165:1891–1894

Allera A, Herbst MA, Griffin JE, Wilson JD, Schweikert HU, McPhaul MJ (1995) Mutations of the androgen receptor coding sequence are infrequent in patients with isolated hypospadias. J Clin Endocrinol Metab 80(9):2697–2699

Amato AA, Prior TW, Barohn RJ, Snyder P, Papp A, Mendell JR (1993) Kennedy's disease: a clinicopathologic correlation with mutations in the androgen receptor gene. Neurology 43:791–794

Amrhein JA, Meyer WJ, III, Jones HW, Jr, Migeon CJ (1976) Androgen insensitivity in man: evidence for genetic heterogeneity. Proc Natl Acad Sci USA 73:891–894

Amrhein JA, Jones Klingensmith G, Walsh PC, McKusick VA, Migeon CJ (1977) Partial androgen insensitivity: the Reifenstein syndrome revisited. N Engl J Med 297:350–356

Arbizu T, Santamaría J, Gomez JM, Quílez A, Serra JM (1983) A family with adult spinal and bulbar muscular atrophy, X-linked inheritance and associated testicular failure. J Neurol Sci 59:371–382

Archer TK, Cordingley MG, Wolford RG, Hager GL (1991) Transcription factor access is mediated by accurately positioned nucleosomes on the mouse mammary tumor virus promoter. Mol Cell Biol 11:688–698

Baarends WM, Themmen APN, Blok LJ, Mackenbach P, Brinkmann AO, Meijer D, Faber PW, Trapman J, Grootegoed JA (1990) The rat androgen receptor gene promoter. Mol Cell Endocrinol 74:75–84

Baarends WM, van Helmond MJL, Post M, van der Schoot PJCM, Hoogerbrugge JW, de Winter JP, Uilenbroek JTJ, Karels B, Wilming LG, Meijers JHC, et al. (1994) A novel member of the transmembrane serine/threonine kinase receptor family is specifically expressed in the gonads and in mesenchymal cells adjacent to the müllerian duct. Development 120:189–197

Bale PM, Howard NJ, Wright JE (1992) Male pseudohermaphroditism in XY children with female phenotype. Pediatr Pathol 12:29–49

Bangsbøll S, Qvist I, Lebech PE, Lewinsky M (1992) Testicular feminization syndrome and associated gonadal tumors in Denmark. Acta Obstet Gynecol Scand 71:63–66

Bardoni B, Zanaria E, Guioli S, Floridia G, Worley KC, Tonini G, Ferrante E, Chiumello G, McCabe ERB, Fraccaro M, et al. (1994) A dosage sensitive locus at chromosome Xp21 is involved in male to female sex reversal. Nature Genet 7:497–501

Barkhaus PE, Kennedy WR, Stern LZ, Harrington RB (1982) Hereditary proximal and bulbar motor neuron disease of late onset. A report of six cases. Arch Neurol 39:112–116

Batch JA, Williams DM, Davies HR, Brown BD, Evans BAJ, Hughes IA, Patterson MN (1992) Androgen receptor gene mutations identified by SSCP in fourteen subjects with androgen insensitivity syndrome. Hum Mol Genet 1:497–503

Batch JA, Davies HR, Evans BAJ, Hughes IA, Patterson MN (1993a) Phenotypic variation and detection of carrier status in the partial androgen insensitivity syndrome. Arch Dis Child 68:453–457

Batch JA, Evans BAJ, Hughes IA, Patterson MN (1993b) Mutations in the androgen receptor gene identified in perineal hypospadias. J Med Genet 30:198–201

Beato M (1989) Gene regulation by steroid hormones. Cell 56:335–344

Beitel LK, Kazemi-Esfarjani P, Kaufman M, Lumbroso R, DiGeorge AM, Killinger DW, Trifiro MA, Pinsky L (1994a) Substitution of arginine-839 by cysteine or histidine in the androgen receptor causes different receptor phenotypes in cultured cells and coordinate degrees of clinical androgen resistance. J Clin Invest 94:546–554

Beitel LK, Prior L, Vasiliou DM, Gottlieb B, Kaufman M, Lumbroso R, Alvarado C, McGillivray B, Trifiro M, Pinsky L (1994b) Complete androgen insensitivity due to mutations in the probable α-helical segments of the DNA-binding domain in the human androgen receptor. Hum Mol Genet 3:21–27

Belsham DD, Greenberg CR, Faiman C, Yee WC, Wrogemann K (1992a) Clinical applications of mutation analysis in families with androgen receptor defects: complete and partial androgen insensitivity syndrome and spinal and bulbar muscular atrophy [Abstract]. 74th Annual Meeting of the Endocrine Society:142-Abstract #361

Belsham DD, Yee W-C, Greenberg CR, Wrogemann K (1992b) Analysis of the CAG repeat region of the androgen receptor gene in a kindred with X-linked spinal and bulbar muscular atrophy. J Neurol Sci 112:133–138

Ben-Hur H, Thole HH, Mashiah A, Insler V, Berman V, Shezen E, Elias D, Zuckerman A, Ornoy A (1997) Estrogen, progesterone and testosterone receptors in fetal cartilaginous tissue. Calcified Tissue Int 60(6):520–526

Bentvelsen FM, McPhaul MJ, Wilson JD, George FW (1993) Developmental regulation of the androgen receptor of the fetal rat [Abstract]. 75th Annual Meeting of the Endocrine Society Abstract #600

Berg J (1989) DNA-binding specificity of steroid receptors. Cell 57:1065–1068

Bertelloni S, Federico G, Baroncelli GI, Cavallo L, Corsello G, Liotta A, Rigon F, Saggese G (1997) Biochemical selection of prepubertal patients with androgen insensitivity syndrome by sex hormone-binding globulin response to the human chorionic gonadotropin test. Pediatr Res 41(2):266–271

Biancalana V, Serville F, Pommier J, Julien J, Hanauer A, Mandel JL (1992) Moderate instability of the trinuleotide repeat in spino bulbar muscular atrophy. Hum Mol Genet 1:255–258

Blauer M, Vaalasti A, Pauli SL, Ylikomi T, Joensuu T, Tuohimaa P (1991) Location of androgen receptor in human skin. J Invest Dermatol 97(2):264–268

Blok LJ, Mackenbach P, Trapman J, Themmen APN, Brinkmann AO, Grootegoed JA (1989) Follicle stimulating hormone regulates androgen receptor mRNA in Sertoli cells. Mol Cell Endocrinol 63:267–271

Boehmer ALM, Brinkmann AO, Verleun-Mooijman MCT, Niermeijer MF, Nijman JM, Drop SLS (1997) Extreme phenotypic variation within one family with androgen insensitivty syndrome (AIS): an explanation [Abstract]. 79th Annual Meeting of the Endocrine Society:Abstract #P2-502

Bookstein R, Rio P, Madreperla SA, Hong F, Allred C, Grizzle WE, Lee W (1990) Promoter deletion and loss of retinoblastoma gene expression in human prostate carcinoma. Proc Natl Acad Sci USA 87:7762–7766

Boyar RM, Moore RJ, Rosner W, Aiman J, Chipman J, Madden JD, Marks JF, Griffin JE (1978) Studies of gonadotropin-gonadal dynamics in patients with androgen insensitivity. J Clin Endocrinol Metab 47:1116–1122

Brinkmann AO, Faber PW, van Rooij HCJ, Kuiper GGJM, Ris C, Klaassen P, van der Korput JAGM, Voorhorst MM, van Laar JH, Mulder E, et al. (1989) The human androgen receptor: domain structure, genomic organization and regulation of expression. J Steroid Biochem 34:307–310

Brown CJ, Goss SJ, Lubahn DB, Joseph DR, Wilson EM, French FS, Willard HF (1989) Androgen receptor locus on the human X chromosome: regional localization to Xq11-12 and description of a DNA polymorphism. Am J Hum Genet 44:264–269

Brown TR, Maes M, Rothwell SW, Migeon CJ (1982) Human complete androgen insensitivity with normal dihydrotestosterone receptor binding capacity in cultured genital skin fibroblasts: evidence for a qualitative abnormality of the receptor. J Clin Endocrinol Metab 55:61–69

Brown TR, Lubahn DB, Wilson EM, Joseph DR, French FS, Migeon CJ (1988) Deletion of the steroid binding domain of the human androgen receptor gene in one family with complete androgen insensitivity syndrome: evidence for further genetic heterogeneity in this syndrome. Proc Natl Acad Sci USA 85:8151–8155

Brown TR, Lubahn DB, Wilson EM, French FS, Migeon CJ, Corden JL (1990) Functional characterization of naturally occurring mutant androgen receptors from subjects with complete androgen insensitivity. Mol Endocrinol 4:1759–1772

Brown TR, Scherer PA, Chang Y, Migeon CJ, Ghirri P, Murono K, Zhou Z (1993) Molecular genetics of human androgen insensitivity. Eur J Pediatr 152 (Suppl 2):S62–S69

Brown TR, Migeon CJ (1981) Cultured human skin fibroblasts: a model for the study of androgen action. Mol Cell Biochem 36:3–22

Bubulya A, Wise SC, Shen XQ, Burmeister LA, Shemshedini L (1996) c-Jun can mediate androgen receptor-induced transactivation. J Biol Chem 271(40):24583–24589

Burnstein KL, Maiorino CA, Dai JL, Cameron DJ (1995) Androgen and glucocorticoid regulation of androgen receptor cDNA expression. Mol Cell Endocrinol 115(2):177–186

Campo S, Stivel M, Nicolau G, Monteagudo C, Rivarola M (1979) Testicular function in post pubertal male pseudohermaphroditism. Clin Endocrinol 11:481–490

Carson-Jurica MA, Schrader WT, O'Malley BW (1990) Steroid receptor family: structure and functions. Endocr Rev 11:201–220

Caskey CT, Pizzuti A, Fu Y, Fenwick RG, Jr, Nelson DL (1992) Triplet repeat mutations in human disease. Science 256:784–789

Cassio A, Cacciari E, D'Errico A, Balsamo A, Grigioni FW, Pascucci MG, Bacci F, Tacconi M, Mancini AM (1990) Incidence of intratubular germ cell neoplasia in androgen insensitivity syndrome. Acta Endocrinol (Copenh) 123:416–422

Cato ACB, Henderson D, Ponta H (1987) The hormone response element of the mouse mammary tumor virus DNA mediates the progestin and androgen induction of transcription in the proviral long terminal repeat region. EMBO J 6:363–368

Chamberlain NL, Driver ED, Miesfeld RL (1994) The length and location of CAG trinucleotide repeats in the androgen receptor N-terminal domian affect transactivation function. Nucleic Acids Res 22(15):3181–3186

Chang C, Kokontis J, Liao S (1988a) Structural analysis of complementary DNA and amino acid sequences of human and rat androgen receptors. Proc Natl Acad Sci USA 85:7211–7215

Chang C, Kokontis J, Liao S (1988b) Molecular cloning of human and rat complementary DNA encoding androgen receptors. Science 240:324–326

Chang YT, Migeon CJ, Brown TR (1991) Human androgen insensitivity due to androgen receptor gene point mutations in subjects with normal androgen receptor levels but impaired biological activity [Abstract]. 73rd Annual Meeting of the Endocrine Society:37-Abs #28

Charest NJ, Zhou Z, Lubahn DB, Olsen KL, Wilson EM, French FS (1991) A frameshift mutation destabilizes androgen receptor messenger RNA in the *Tfm* mouse. Mol Endocrinol 5:573–581

Chen S, Wang J, Yu G, Liu W, Pearce D (1997) Androgen and glucocorticoid receptor heterodimer formation. A possible mechanism for mutual inhibition of transcriptional activity. J Biol Chem 272(22):14087–14092

Choi W-T, MacLean HE, Chu S, Warne GL, Zajac JD (1993) Kennedy's disease: genetic diagnosis of an inherited form of motor neuron disease. Aust N Z J Med 23:187–192

Choong CS, Kemppainen JA, Zhou Z-X, Wilson EM (1996a) Reduced androgen receptor gene expression with first exon CAG repeat expansion. Mol Endocrinol 10(12):1527–1535

Choong CS, Quigley CA, French FS, Wilson EM (1996b) A novel missense mutation in the amino terminal domain of the human androgen receptor gene in a family with partial androgen insensitivity syndrome causes reduced efficiency of protein translation. J Clin Invest 98:1423–1431

Cicognani A, Cacciari E, Tacconi M, Pascucci MG, Tonioli S, Pirazzoli P, Balsamo A (1989) Effect of gonadectomy on growth hormone, IGF-I and sex steroids in children with complete and incomplete androgen insensitivity. Acta Endocrinol (Copenh) 121:777–783

Claessens F, Celis L, Peeters B, Heyns W, Verhoeven G, Rombauts W (1989) Functional characterization of an androgen response element in the first intron of the C3 (1) gene of prostatic binding protein. Biochem Biophys Res Commun 164:833–840

Claessens F, Celis L, De Vos P, Peeters B, Heyns W, Verhoeven G, Rombauts W (1993) Intronic androgen response elements of prostatic binding protein genes. Biochem Biophys Res Commun 191:688–694

Collins GM, Kim DU, Logrono R, Rickert RR, Zablow A, Breen JL (1993) Pure seminoma arising in androgen insensitivity syndrome (testicular feminization syndrome). Mod Pathol 6:89–93

Colvard DS, Wilson EM (1987) Transformation of the 10S androgen receptor in a hamster ductus deferens tumor cell line. Endocrinology 121:931–940

Culig Z, Hobisch A, Cronauer MV, Cato ACB, Hittmair A, Radmayr C, Eberle J, Bartsch G, Klocker H (1993) Mutant androgen receptor detected in an advanced-stage prostatic carcinoma is activated by adrenal androgens and progesterone. Mol Endocrinol 7:1541–1550

Culig Z, Hobisch A, Cronauer MV, Hittmair A, Radmayr C, Bartsch G, Klocker H (1995) Activation of androgen receptor by polypeptide growth factors and cellular regulators. World J Urol 13:285–289

Culig Z, Stober J, Gast A, Peterziel H, Hobisch A, Radmayr C, Hittmair A, Bartsch G, Cato ACB, Klocker H (1996) Activation of two mutant androgen receptors from human prostatic carcinoma by adrenal androgens and metabolic derivatives of testosterone. Cancer Detect Prev 20(1):68–75

Cullen KE, Kladde MP, Seyfred MA (1993) Interaction between transcription regulatory regions of prolactin chromatin. Science 261:203–206

Dai JL, Maiorino CA, Gkonos PJ, Burnstein KL (1996) Androgenic up-regulation of androgen receptor cDNA expression in androgen-independent prostate cancer cells. Steroids 61(9):531–539

Dai JL, Burnstein KL (1996) Two androgen response elements in the androgen receptor coding region are required for cell-specific up-regulation of receptor messenger RNA. Mol Endocrinol 10(12):1582–1594

Danek A, Witt TN, Mann K, Schweikert HU, Romalo G, La Spada AR, Fischbeck KH (1994) Decrease in androgen binding and effect of androgen treatment in a case of X-linked bulbospinal neuropathy. Clin Investigator 72(11):892–897

Danielsen M, Northrop JP, Ringold GM (1986) The mouse glucocorticoid receptor: mapping of functional domains by cloning, sequencing and expression of wild-type and mutant receptor proteins. EMBO J 5:2513–2522

Danielsen M, Hinck L, Ringold GM (1989) Two amino acids within the knuckle of the first zinc finger specify DNA response element activation by the glucocorticoid receptor. Cell 57:1131–1138

Dankbar B, Sohn M, Nieschlag E, Gromoll J (1995) Quantification of androgen receptor and follicle stimulating hormone receptor mRNA levels in human and monkey testes by a ribonuclease-protection assay. Int J Androl 18(2):88–96

Davies HR, Hughes IA, Savage MO, Quigley CA, Trifiro M, Pinsky L, Brown TR, Patterson MN (1997) Androgen insensitivity with mental retardation: a contiguous gene syndrome? J Med Gen 34(2):158–160

De Bellis A, Quigley CA, Cariello NF, El-Awady MK, Sar M, Lane MV, Wilson EM, French FS (1992) Single base mutations in the androgen receptor gene causing complete androgen insensitivity: rapid detection by a modified denaturing gradient gel electrophoresis technique. Mol Endocrinol 6:1909–1920

De Bellis A, Quigley CA, Marschke KB, El-Awady MK, Lane MV, Smith EP, Sar M, Wilson EM, French FS (1994) Characterization of mutant androgen receptors causing partial androgen insensitivity syndrome. J Clin Endocrinol Metab 78:513–522

De Vos P, Claessens F, Winderickx J, Van Dijck P, Celis L, Peeters B, Rombauts W, Heyns W, Verhoeven G (1991) Interaction of androgen response elements with the DNA-binding domain of the rat androgen receptor expressed in *Escherichia coli*. J Biol Chem 266:3439–3443

Denison SH, Sands A, Tindall DJ (1989) A tyrosine aminotransferase glucocorticoid response element also mediates androgen enhancement of gene expression. Endocrinology 124:1091–1093

Deshpande N, Wang DY, Bulbrook RD, McMillian M (1965) Hormone studies in cases of testicular feminization. Steroids 6:437–449

deVere White RW, Gumerlock PH, Chamberlain SR, Lee F, Siders DB (1994) Human androgen receptor analysis in prostate cancer [Abstract]. J Urol 151 (5):468A-Abst #962

Devos A, Claessens F, Alen P, Winderickx J, Heyns W, Rombauts W, Peeters B (1997) Identification of a functional androgen-response element in the exon 1-coding sequence of the cystatin-related protein gene crp2. Mol Endocrinol 11(8):1033–1043

Diamond M, Sigmundson HK (1997a) Management of intersexuality: guidelines for dealing with individuals with ambiguous genitalia. Arch Pediat Adolescent Med 151:1046–1050

Diamond M, Sigmundson HK (1997b) Sex reassignment at birth. Long term review and clinical implications. Arch Pediat Adol Med 151:298–304

Dodge ST, Finkelston MS, Miyazawa K (1985) Testicular feminization with incomplete Müllerian regression. Fertil Steril 43:937–938

Doesburg P, Kuil CW, Berrevoets CA, Steketee K, Faber PW, Mulder E, Brinkmann AO, Trapman J (1997) Functional in vivo interaction between the amino-terminal, transactivation domain and the ligand binding domain of the androgen receptor. Biochemistry 36(5):1052–1064

Doyu M, Sobue G, Mukai E, Kachi T, Yasuda T, Mitsuma T, Takahashi A (1992) Severity of X-linked recessive bulbospinal neuronopathy correlates with size of the tandem CAG repeat in the androgen receptor gene. Ann Neurol 32:707–710

Edwards A, Civitello A, Hammond HA, Caskey CT (1991) DNA typing and genetic mapping with trimeric and tetrameric tandem repeats. Am J Hum Genet 49:746–756

Edwards A, Hammond HA, Jin L, Caskey CT, Chakraborty R (1992) Genetic variation at five trimeric and tetrameric tandem repeat loci in four human population groups. Genomics 12:241–253

El-Awady MK, Marschke KB, De Bellis A, Quigley CA (1993) Natural and site directed mutations in the zinc finger region of the human androgen receptor alter transcriptional activity [Abstract]. 75th Annual Meeting of the Endocrine Society:203-Abs #610

Ertekin C, Sirin H (1993) X-linked bulbospinal muscular atrophy (Kennedy's syndrome): a report of three cases. Acta Neurol Scand 87:56–61

Evans BAJ, Jones TR, Hughes IA (1984) Studies of the androgen receptor in dispersed fibroblasts: investigation of patients with androgen insensitivity. Clin Endocrinol 30:93–105

Evans BAJ, Ismail RA, France T, Hughes IA (1991) Analysis of the androgen receptor gene structure in a patient with complete androgen insensitivity syndrome [Abstract]. J Endocrinol Suppl 129:Abstract #65

Evans BAJ, Hughes IA, Bevan CL, Patterson MN, Gregory JW (1997) Phenotypic diversity in siblings with partial androgen insensitivity syndrome. Arch Dis Child 76(6):529–531

Evans BAJ, Hughes IA (1985) Augmentation of androgen-receptor binding in vitro: studies in normals and patients with androgen insensitivity. Clin Endocrinol 23:567–577

Evans RM (1988) The steroid and thyroid hormone receptor superfamily. Science 240:889–895

Faber PW, van Rooij HCJ, van der Korput HAGM, Baarends WM, Brinkmann AO, Grootegoed JA, Trapman J (1991) Characterization of the human androgen receptor transcription unit. J Biol Chem 266:10743–10749

Faber PW, van Rooij HCJ, Schipper HJ, Brinkmann AO, Trapman J (1993) Two different, overlapping pathways of transcription initiation are active on the TATA-less human androgen receptor promoter. The role of Sp1. J Biol Chem 268:9296–9301

Faiman C, Winter JSD (1974) The control of gonadotropin secretion in complete testicular feminization. J Clin Endocrinol Metab 39:631–638

Fang Y, Fliss AE, Robins DM, Caplan AJ (1996) Hsp90 regulates androgen receptor hormone binding affinity in vivo. J Biol Chem 271(45):28697–28702

Fawell SE, Lees JA, White R, Parker MG (1990) Characterization and colocalization of steroid binding and dimerization activities in the mouse estrogen receptor. Cell 60:953–962

Ferlini A, Patrosso MC, Guidetti D, Merlini L, Uncini A, Ragno M, Plasmati R, Fini S, Repetto M, Vezzoni P, et al. (1995) Androgen receptor gene (CAG)n repeat analysis in the differential diagnosis between Kennedy disease and other motoneuron disorders. Am J Med Genet 55(1):105–111

Fichman KR, Nyberg LM, Bujnovsky P, Brown TR, Walsh PC (1981) The ontogeny of the androgen receptor in human foreskin. J Clin Endocrinol Metab 52:919–923

Fischer L, Catz D, Kelley D (1993) An androgen receptor mRNA isoform associated with hormone induced cell proliferation. Proc Natl Acad Sci USA 90:8254–8258

Forest MG, Mollard P, David M, Morel Y, Bertrand J (1990) Syndrome d'insensibilite incomplete aux androgenes. Difficultes du diagnostic et de la conduit a tenir. Arch Fr Pediatr 47:107–113

Freedman LP (1992) Anatomy of the steroid receptor zinc finger region. Endocr Rev 13:129–145

French FS, Baggett B, Van Wyk JJ, Talbert LM, Hubbard WR, Johnston FR, Weaver RP, Forchielli E, Rao GS, Sarda IR (1965) Testicular feminization: clinical, morphological and biochemical studies. J Clin Endocrinol Metab 25:661–677

French FS, Van Wyk JJ, Baggett B, Easterling WE, Talbert LM, Johnston FR, Forchielli E, Dey AC (1966) Further evidence of a target organ defect in the syndrome of testicular feminization. J Clin Endocrinol Metab 26:493–503

Gad YZ, Berkovitz GD, Migeon CJ, Brown TR (1988) Studies of up-regulation of androgen receptors in genital skin fibroblasts. Mol Cell Endocrinol 57:205–213

Gaddipati JP, McLeod DG, Heidenberg HB, Sesterhenn IS, Finger MJ, Moul JW, Srivastava S (1994) Frequent detection of codon 877 mutation in the androgen receptor gene in advanced prostate cancers. Cancer Res 54:2861–2864

Gao T, Marcelli M, McPhaul MJ (1996) Transcriptional activation and transient expression of the human androgen receptor. J Steroid Biochem Mol Biol 59(1):9–20

Gehring U, Tomkins GM, Ohno S (1971) Effect of the androgen-insensitivity mutation on a cytoplasmic receptor for dihydrotestosterone. Nature New Biol 232:106–107

George FW, Peterson KG (1988) Partial characterization of the androgen receptor of the newborn rat gubernaculum. Biol Reprod 39:536–539

Gerber H, Seipel K, Georgiev O, Höfferer M, Hug M, Rusconi S, Schaffner W (1994) Transcriptional activation modulated by homopolymeric glutamine and proline stretches. Science 263:808–811

Ghirri P, Brown TR (1993) Improved detection of point mutations in the human androgen receptor gene by denaturing gradient gel electrophoresis of DNA heteroduplexes under stringent denaturing conditions [Abstract]. Pediatr Res 33(5):Abstract #95

Goldenberg JN, Bradley WG (1996) Testosterone therapy and the pathogenesis of Kennedy's disease (X linked bulbospinal muscular atrophy). J Neurol Sci 135(2):158–161

Goldstein LA, Sengelaub DR (1992) Timing and duration of dihydrotestosterone treatment affect the development of motoneuron number and morphology in a sexually dimorphic rat spinal nucleus. J Comp Neurol 326:147–157

Gottlieb B, Trifiro M, Lumbroso R, Pinsky L [Anonymous] (1997) The androgen receptor gene mutations database. 〈None Specified〉 http://www.mcgill.ca/androgendb/: World Wide Web. Anonymous

Grant DB (1995) Ethical issues in children with genital ambiguity. Brit J Urol 76 (Suppl. 2.):75–78

Green S, Kumar V, Theulaz I, Wahli W, Chambon P (1988) The N-terminal DNA-binding 'zinc finger' of the oestrogen and glucocorticoid receptors determines target gene specificity. EMBO J 7:3037–3044

Griffin JE, Punyashthiti K, Wilson JD (1976) Dihydrotestosterone binding by cultured human fibroblasts. Comparison of cells from control subjects and from patients with hereditary male pseudohermaphroditism due to androgen resistance. J Clin Invest 57:1342–1351

Griffin JE (1979) Testicular feminization associated with a thermolabile androgen receptor in cultured human fibroblasts. J Clin Invest 64:1624–1631

Griffin JE, Wilson JD (1989) The androgen resistance syndromes: 5α-reductase deficiency, testicular feminization, and related disorders. In: Scriver CR, Beaudet AL, Sly WS, Valle D (eds). The metabolic basis of inherited disease. (6th ed) New York:McGraw-Hill pp 1919–1944

Grino PB, Griffin JE, Wilson JD (1987) Transformation of the androgen receptor to the deoxyribonucleic acid-binding state: studies in homogenates and intact cells. Endocrinology 120:1914–1920

Grino PB, Griffin JE, Cushard WG, Jr, Wilson JD (1988) A mutation of the androgen receptor associated with partial androgen resistance, familial gynecomastia and fertility. J Clin Endocrinol Metab 66:754–761

Grino PB, Isidro-Gutierrez RF, Griffin JE, Wilson JD (1989) Androgen resistance associated with a qualitative abnormality of the androgen receptor and responsive to high dose androgen therapy. J Clin Endocrinol Metab 68:578–584

Guidetti D, Vescovini E, Motti L, Ghidoni E, Gemignani F, Marbini A, Patrosso MC, Ferlini A, Solime F (1996) X-linked bulbar and spinal muscular atrophy, or Kennedy disease: clinical, neurophysiological, neuropathological, neuropsychological and molecular study of a large family. J Neurol Sci 135(2):140–148

Hakimi JM, Rondinelli RH, Schoenberg MP, Barrack ER (1996) Androgen receptor gene structure and function in prostate cancer. World J Urol 14(5):329–337

Ham J, Thomson A, Needham M, Webb P, Parker M (1988) Characterization of response elements for androgens, glucocorticoids and progestins in mouse mammary tumour virus. Nucleic Acids Res 16:5263–5276

Harbison MD, Magid ML, Josso N, Mininberg DT, New MI (1991) Anti-Müllerian hormone in three intersex conditions. Ann Genet (Paris) 34:226–232

Harding AE, Thomas PK, Baraitser M, Bradbury PG, Morgan-Hughes JA, Ponsford JR (1982) X-linked recessive bulbospinal neuronopathy: a report of ten cases. J Neurol Neurosurg Psychiatry 45:1012–1019

Hardy DO, Scher HI, Bogenreider T, Sabbatini P, Zhang ZF, Nanus DM, Catterall JF (1996) Androgen receptor CAG repeat lengths in prostate cancer: correlation ith age of onset. J Clin Endocrin Metab 81(12):4400–4405

Harris SE, Rong Z, Harris MA, Lubahn DB (1990) Androgen receptor in human prostate carcinoma LNCaP/ADEP cells contains a mutation which alters the specificity of the steroid-dependent transcriptional activation region [Abstract]. 72nd Annual Meeting of the Endocrine Society Abstract #275

Harris SE, Harris MA, Rong Z, Hall J, Judge S, French FS, Joseph DR, Lubahn DB, Simental JA, Wilson EM (1991) Androgen regulation of HBGF-1 (αFGF) mRNA and characterization of the androgen-receptor mRNA in the human prostate carcinoma cell line – LNCaP/A-DEP. In: Karr JP, Coffey DS, Smith RG, Tindall DJ (eds). Molecular and cellular biology of prostate cancer. New York: Plenum Press pp 315–330

Hauser GA (1963) Testicular feminization. In: Overzier C, editor. Intersexuality. New York:Academic Press pp 255–276

Hausmanowa-Petrusewicz I, Borkowska J, Janczewski Z (1983) X-linked adult form of spinal muscular atrophy. J Neurol 229:175–188

Härd T, Kellenbach E, Boelens R, Maler BA, Dahlman K, Freedman LP, Carlstedt-Duke J, Yamamoto KR, Gustafsson J-A, Kaptein R (1990) Solution structure of the glucocorticoid receptor DNA-binding domain. Science 249:157–160

He WW, Kumar MV, Tindall DJ (1991) A frame-shift mutation in the androgen receptor gene causes complete androgen insensitivity in the testicular-feminized mouse. Nucleic Acids Res 19:2373–2378

Heller DS, Ranzini A, Futterweit W, Dottino P, Deligdisch L (1992) Müllerian remnants in complete androgen insensitivity syndrome. Int J Fertil 37:283–285

Henttu P, Vihko P (1993) Growth factor regulation of gene expression in the human prostatic carcinoma cell line LNCaP. Cancer Res 53:1051–1058

Hinchliffe SA, Woods S, Gray S, Burt AD (1996) Cellular distribution of androgen receptors in the liver. J Clin Pathol 49(5):418–420

Hiort O, Huang Q, Sinnecker GHG, Sadeghi-Nejad A, Kruse K, Wolfe HJ, Yandell DW (1993) Single strand conformational polymorphism analysis of androgen receptor gene mutations in patients with androgen insensitivity syndromes: application for diagnosis, genetic counseling and therapy. J Clin Endocrinol Metab 77:262–266

Hiort O, Klauber G, Cendron M, Sinnecker GHG, Keim L, Schwinger E, Wolfe HJ, Yandell DW (1994) Molecular characterization of the androgen receptor gene in boys with hypospadias. Eur J Pediatr 153:317–321

Ho K, Marschke KB, Tan J-A, Power SGA, Wilson EM, French FS (1993) A complex response element in intron 1 of the androgen-regulated 20-kDa protein gene displays cell type-dependent androgen receptor specificity. J Biol Chem 268:27226–27235

Horie K, Takakura K, Fujiwara H, Suginami H, Liao S, Mori T (1992) Immunohistochemical localization of androgen receptor in the human ovary throughout the menstrual cycle in relation to oestrogen and progesterone receptor expression. Hum Reprod 7:184–190

Housley PR, Sanchez ER, Danielsen M, Ringold GM, Pratt WB (1990) Evidence that the conserved region in the steroid binding domain of the glucocorticoid receptor is required for both optimal binding of hsp90 and protection from proteolytic cleavage. A two site model for hsp90 binding to the steroid-binding domain. J Biol Chem 265:12778–12781

Hughes IA, Evans BAJ (1987) Androgen insensitvity in forty-nine patients: classification based on clinical and androgen receptor phenotypes. Horm Res 28:25–29

Hughes IA, Evans BAJ (1988) The fibroblast as a model for androgen resistant states. Clin Endocrinol 28:565–579

Hurt WG, Bodurtha JN, McCall JB, Ali MM (1989) Seminoma in pubertal patient with androgen insensitivity syndrome. Am J Obstet Gynecol 161:530–531

Husmann DA, McPhaul MJ (1991) Time-specific androgen blockade with flutamide inhibits testicular descent in the rat. Endocrinology 129:1409–1416

Husmann DA, McPhaul MJ (1992) Reversal of flutamide-induced cryptorchidism by prenatal time specific androgens. Endocrinology 131:1711–1715

Hutson JM (1986) Testicular feminization: a model for testicular descent in mice and men. J Pediatr Surg 21:195–198

Igarashi S, Tanno Y, Onodera O, Yamazaki M, Sato S, Ishikawa A, Miyatani N, Nagashima M, Ishikawa Y, Sahashi K, et al. (1992) Strong correlation between the number of CAG repeats in androgen receptor genes and the clinical onset of features of spinal and bulbar muscular atrophy. Neurology 42:2300–2302

Ikonen T, Palvimo JJ, Kallio PJ, Reinikainen P, Janne OA (1994) Stimulation of androgen-regulated transactivation by modulators of protein phosphorylation. Endocrinology 135:1359–1366

Imai A, Ohno T, Furui T, Tamaya T (1994) A frame-shift and point mutations of the androgen receptor gene cause androgen resistance [Abstract]. Program of the 76th Annual Meeting of the Endocrine Society:634-Abs #1733

Imasaki K, Hasegawa T, Okabe T, Sakai Y, Haji M, Takayanagi R, Nawata H (1994) Single amino acid substitution (840Arg→His) in the hormone-binding domain of the androgen receptor leads to incomplete androgen insensitivity syndrome associated with a thermolabile androgen receptor. Eur J Endocrinol 130:569–574

Imperato-McGinley J, Peterson RE, Gautier T, Cooper G, Danner R, Arthur A, Morris PL, Sweeney WJ, Shackleton C (1982) Hormonal evaluation of a large kindred with complete androgen insensitivity: evidence for secondary 5α-reductase deficiency. J Clin Endocrinol Metab 54:931–941

Imperato-McGinley J, Sanchez RS, Spencer JR, Yee B, Vaughn ED (1992) Comparison of the effects of the 5α-reductase inhibitor finasteride and the antiandrogen flutamide on prostate and genital differentiation: dose-response studies. Endocrinology 131:1149–1156

Ingles SA, Ross RK, Yu MC, Irvine RA, La Pera G, Haile RW, Coetzee GA (1997) Association of prostate cancer risk with genetic polymorphisms in vitamin D receptor and androgen receptor. Journal of the National Cancer Insititute 89(2):166–170

Isaacs WB, Carter BS, Ewing CM (1991) Wild-type p53 suppresses growth of human prostate cancer cells containing mutant p53 alleles. Cancer Res 51:4716–4720

Jakubicza S, Werder EA (1992) Point mutation in the steroid-binding domain of the androgen receptor gene in a family with complete androgen insensitivity syndrome (CAIS). Hum Genet 90:311–312

Jakubiczka S, Nedel S, Werder EA, Schleiermacher E, Theile U, Wolff G, Wieacker P (1997) Mutations of the androgen receptor gene in patients with complete androgen insensitivity. Hum Mutat 9(1):57–61

Jenster G, van der Korput HAGM, van Vroonhoven C, van der Kwast TH, Trapman J, Brinkmann AO (1991) Domains of the human androgen receptor involved in steroid binding, transcriptional activation and subcellular localization. Mol Endocrinol 5:1396–1404

Jenster G, Trapman J, Brinkmann AO (1993) Nuclear import of the humn androgen receptor. Biochem J 293:761–768

Jenster G, de Ruiter PE, van der Korput HAGM, Kuiper GGJM, Trapman J, Brinkmann AO (1994) Changes in the abundance of androgen receptor isotypes: effects of ligand treatment, glutamine-stretch variation, and mutation of putative phosphorylation sites. Biochemistry 33(47):14064–14072

Jenster G, van der Korput HAGM, Trapman J, Brinkmann AO (1995) Identifaction of two transcription activation units in the N-terminal domain or the human androgen receptor. J Biol Chem 270(13):7341–7346

Jockenhövel F, Rutgers JKL, Mason JS, Griffin JE, Swerdloff RS (1993) Leydig cell neoplasia in a patient with Reifenstein syndrome. Exp Clin Endocrinol 101:365–370

Josso N, Boussin L, Knebelmann B, Nihoul-Fékété C, Picard J (1991) Anti-Müllerian hormone and intersex states. Trends Endocrinol Metab 2:227–233

Josso N, Rey R, Lordereau-Richard I, Cate RL (1994) Anti-Müllerian hormone and testicular development: Clinical and hormonal aspects. In: Dufau ML, Fabbri A, (eds). Cell and molecular biology of the testis. Rome, Ares-Serono Symposia pp 7–15

Judd HL, Hamilton CR, Barlow JJ, Yen SSC, Kliman B (1972) Androgen and gonadotropin dynamics in testicular feminization syndrome. J Clin Endocrinol 34:229–234

Jukier L, Kaufman M, Pinsky L, Peterson RE (1984) Partial androgen resistance associated with secondary 5α-reductase deficiency: identification of a novel qualitative androgen receptor defect and clinical implications. J Clin Endocrinol Metab 59:679–688

Kalloo NB, Gearhart JP, Barrack ER (1993) Sexually dimorphic expression of estrogen receptors, but not of androgen receptors in human fetal external genitalia. J Clin Endocrinol Metab 77(3):692–698

Kaspar F, Muigg A, Ransmayr G, Bartsch G, Klocker H (1994) Androgen receptor gene mutation in two brothers with X-linked spinomuscular atrophy associated with androgen insensitivity [Abstract]. J Urol 151 (5):467A-Abst #959

Kasumi H, Komori S, Yamasaki N, Shima H, Isojima S (1993) Single nucleotide substitution of the androgen receptor gene in a case with receptor-positive androgen insensitivity (complete form). Acta Endocrinol (Copenh) 128:355–360

Kaufman M, Straisfeld C, Pinsky L (1976) Male pseudohermaphroditism presumably due to target organ unresponsiveness to androgens. Deficient 5α-dihydrotestosterone binding in cultured skin fibroblasts. J Clin Invest 58:345–350

Kaufman M, Pinsky L, Baird PA, McGillivray BC (1979) Complete androgen insensitivity with a normal amount of 5α-dihydrotestosterone-binding activity in labium majus skin fibroblasts. Am J Med Genet 4:401–411

Kaufman M, Pinsky L, Gottlieb B, Schweitzer M, Brezezinski A, von Westarp C, Ginsberg J (1990) Androgen receptor defects in patients with minimal and partial androgen resistance classified according to a model of androgen-receptor complex energy states. Horm Res 33:87–94

Kazemi-Esfarjani P, Beitel LK, Trifiro M, Kaufman M, Rennie P, Sheppard P, Matusik RJ, Pinsky L. (1993) Substitution of valine-865 by methionine or leucine in the human androgen receptor causes complete or partial androgen insensitivity, respectively with distinct androgen receptor phenotypes. Mol Endocrinol 7:37–46

Keely EJ, Belsham DD, Wrogemann K, Faiman C (1993) Partial androgen insensitivity – are all tissues equal? Fertil Steril 60:366–368

Keenan BS, Meyer WJ, III, Hadjian AJ, Jones HW, Migeon CJ (1974) Syndrome of androgen insensitivity in man: absence of 5α-dihydrotestosterone binding protein in skin fibroblasts. J Clin Endocrinol Metab 38:1143–1146

Keenan BS, Meyer WJ, III, Hadjian AJ, Migeon CJ (1975) Androgen receptor in human skin fibroblasts. Characterization of a specific 17β-hydroxy-5α-androstan-3-one-protein complex in cell sonicates and nuclei. Steroids 25:535–552

Keenan BS, McNeel RL, Gonzales ET (1984) Abnormality of intracellular 5α-dihydrotestosterone binding in simple hypospadias: studies on equilibrium binding in sonicates of genital skin fibroblasts. Pediatr Res 18:216–220

Kemppainen JA, Lane MV, Sar M, Wilson EM (1992) Androgen receptor phosphorylation, turnover, nuclear transport, and transcriptional activation. Specificity for steroids and antihormones. J Biol Chem 267:968–974

Kennedy WR, Alter M, Sung JH (1968) Progressive proximal spinal and bulbar muscular atrophy of late onset. A sex-linked recessive trait. Neurology 18:671–680

Kimura N, Mizokami A, Oonuma T, Sasano H, Nagura H (1993) Immunocytochemical localization of androgen receptor with polyclonal antibody in paraffin-embedded human tissues. J Histochem Cytochem 41:671–678

Klocker H, Kaspar F, Eberle J, Überreiter S, Radmayr C, Bartsch G (1992) Point mutation in the DANN binding domain of the androgen receptor in two families with Reifenstein Syndrome. Am J Hum Genet 50:1318–1327

Knebelmann B, Boussin L, Guerrier D, Legeai L, Kahn A, Josso N, Picard J-Y (1991) Anti-Müllerian hormone Bruxelles: A nonsense mutation associated with the persistent Müllerian duct syndrome. Proc Natl Acad Sci USA 88:3767–3771

Koide R, Ikeuchi T, Onodera O, Tanaka H, Igarashi S, Endo K, Takahashi H, Kondo R, Ishikawa A, Hayashi T, et al. (1994) Unstable expansion of CAG repeat in hereditary dentatorubral-palliodoluysian atrophy (DRPLA). Nature Genet 6:9–13

Koivisto P, Visakorpi T, Kallioniemi O (1996) Androgen receptor gene amplification: a novel molecular mechanism for endocrine therapy resistance in human prostate cancer. Scand J Clinical Lab Invest 226:57–63

Koivisto P, Kononen J, Palmberg C, Tammela T, Hyytinen E, Isola J, Trapman J, Cleutjens K, Noordzij A, Visakorpi T, et al. (1997) Androgen receptor gene amplification: a possible molecular mechanism for androgen deprivation therapy failure in prostate cancer. Cancer Res 57:314–319

Krongrad A, Wilson CM, Wilson JD, Allman DR, McPhaul MJ (1991) Androgen increases androgen receptor protein while decreasing receptor mRNA in LNCaP cells. Mol Cell Endocrinol 76:79–88

Kuil CW, Mulder E (1994) Mechanism of antiandrogen action: conformational changes of the receptor. Mol Cell Endocrinol 102(1–2):R1–R5

Kuiper GG, Brinkmann AO (1995) Phosphotryptic peptide ananlysis of the human androgen receptor: detection of a hormone-induced phosphopeptide. Biochemistry 34(6):1851–1857

Kuiper GGJM, Faber PW, van Rooij HJC, et al. (1989) Structural organization of the human androgen receptor gene. J Mol Endocrinol 2:R1–R4

Kurz EM, Sengelaub DR, Arnold AP (1986) Androgens regulate the dendritic length of mammalian motoneurons in adulthood. Science 232:395–398

Kuttenn F, Mowszowicz I, Wright F, Baudot N, Jaffiol C, Robin M, Mauvais-Jarvis P (1979) Male pseudohermaphroditism: a comparative study of one patient with 5α-reductase activity and three patients with the complete form of testicular feminization. J Clin Endocrinol Metab 49:861–865

La Spada AR, Wilson EM, Lubahn DB, Harding AE, Fischbeck KH (1991) Androgen receptor gene mutations in X-linked spinal and bulbar muscular atrophy. Nature 352:77–79

La Spada AR, Roling DB, Harding AE, Warner CL, Spiegel R, Hausmanowa-Petrusewicz I, Yee WC, Fischbeck KH (1992) Meiotic instability and genotype-phenotype correlation of the trinucleotide repeat in X-linked spinal and bulbar muscular atrophy. Nature Genet 2:301–304

Langley E, Zhou Z, Wilson EM (1995) Evidence for an anti-parallel orientation of the ligand-activated human androgen receptor dimer. J Biol Chem 270(50):29983–29990

Lavedan C, Hofmann-Radvanyi H, Shelbourne P, Rabes J, Duros C, Savoy D, Dehaupas I, Luce S, Johnson K, Junien C (1993) Myotonic dystrophy: size- and sex-dependent dynamics of CTG meiotic instability, and somatic mosaicism. Am J Hum Genet 52:875–883

Lecca U, Parodo G, Fiore R, Martino E (1988) Embryonal carcinoma in two cases of androgen insensitivity syndrome: clinical, endocrinological and pathological features. Eur J Gynaecol Oncol 9:489–496

Lee MM, Donahoe PK (1993) Mullerian inhibiting substance: a gonadal hormone with multiple functions. Endocr Rev 14:152–164

Lee PA, Brown TR, LaTorre HA (1986) Diagnosis of the partial androgen insensitivity syndrome during infancy. JAMA 255:2207–2209

Lieberman BA, Bona BJ, Edwards DP, Nordeen SK (1993) The constitution of a progesterone response element. Mol Endocrinol 7:515–527

Lindzey J, Grossmann M, Kumar MV, Tindall DJ (1993) Regulation of the 5′-flanking region of the mouse androgen receptor gene by cAMP and androgen. Mol Endocrinol 7:1530–1540

Lobaccaro J-M, Lumbroso S, Belon C, Galtier-Dereure F, Bringer J, Lesimple T, Heron J-F, Pujol H, Sultan C (1993a) Male breast cancer and the androgen receptor gene. Nature Genet 5:109–110

Lobaccaro J-M, Lumbroso S, Ktari R, Dumas R, Sultan C (1993b) An exonic point mutation creates a *Mae*III site in the androgen receptor gene of a family with complete androgen insensitivity syndrome. Hum Mol Genet 2:1041–1043

Lobaccaro JM, Lumbroso S, Belon C, Chaussain JL, Toublanc JE, Leheup B, Sultan C (1993c) Androgen receptor (AR) gene mutations in 6 families with androgen insensitivity syndrome [Abstract]. Pediatr Res 33:S22-Abs #115

Lobaccaro JM, Lumbroso S, Belon C, Galtier-Dereure F, Bringer J, Lesimple T, Namer M, Cutuli BF, Pujol H, Sultan C (1993d) Androgen receptor gene mutation in male breast cancer. Hum Mol Genet 2:1799–1802

Lobaccaro JM, Lumbroso S, Berta P, Chaussain JL, Sultan S (1993e) Complete androgen insensitivity syndrome associated with a *de novo* mutation of the androgen receptor gene detected by single strand conformational polymorphism. J Steroid Biochem Mol Biol 44:211–216

Loda M, Fogt F, French FS, Posner M, Cukor B, Aretz HT, Alsaigh N (1994) Androgen receptor immunohistochemistry on paraffin-embedded tissue. Mod Pathol 7(3):388–391

Lubahn DB, Joseph DR, Sar M, Tan J-A, Higgs HN, Larson RE, French FS, Wilson EM (1988a) The human androgen receptor: complementary deoxyribonucleic acid cloning, sequence analysis and gene expression in prostate. Mol Endocrinol 2:1265–1275

Lubahn DB, Joseph DR, Sullivan PM, Willard HF, French FS, Wilson EM (1988b) Cloning of human androgen receptor complementary DNA and localization to the X chromosome. Science 240:327–330

Lubahn DB, Brown TR, Simental JA, Higgs HN, Migeon CJ, Wilson EM, French FS (1989) Sequence of the intron/exon junctions of the coding region of the human androgen receptor gene and identification of a point mutation in a family with complete androgen insensitivity. Proc Natl Acad Sci USA 86:9534–9538

Luboshitzky R, Dharan M, Goldman D, Hiss Y, Herer P, Lavie P (1997) Immunohistochemical localization of gonadotropin and gonadal steroid receptors in human pineal gland. J Clin Endocrin Metab 82(3):977–981

Luisi BF, Xu WX, Otwinowski Z, Freedman LP, Yamamoto KR, Sigler PB (1991) Crystallographic analysis of the interaction of the glucocorticoid receptor with DNA. Nature 352:497–505

Lukusa T, Fryns JP, Kleczkowska A, van den Berghe H (1991) Role of gonadal dysgenesis in gonadoblastoma induction in 46,XY individuals. The Leuven experience in 46,XY pure gonadal dygsenesis and testicular feminization syndromes. Genet Couns 2:9–16

Lumbroso S, Martin D, Lobaccaro JM, Chaussain JL, Belon C, Sultan C (1993) A new mutation within the deoxyribonucleic acid-binding domain of the androgen receptor gene in a family with complete androgen insensitivity syndrome. Fertil Steril 60:814–819

Lumbroso S, Lobaccaro JM, Belon C, Amram S, Bachelard B, Garandeau P, Sultan C (1994) Molecular prenatal exclusion of familial partial androgen insensitivity (Reifenstein syndrome). Eur J Endocrinol 130:327–332

Lustig RH, Hua P, Smith LS, Wang C-H, Chang C (1994) Androgen action on neurons *in vitro*: alterations in neuritic arborization and implications for developmental plasticity [Abstract]. Pediatr Res 35:71a-Abs #413

Lyon MF, Hawkes SG (1970) X-linked gene for testicular feminization in the mouse. Nature 227:1217 1219

MacDonald PC, Madden JD, Brenner PF, Wilson JD, Siiteri PK (1979) Origin of estrogen in normal men and in women with testicular feminization. J Clin Endocrinol Metab 49:905–916

Macke JP, Hu N, Hu S, Bailey M, King VL, Brown T, Hamer D, Nathans J (1993) Sequence variation in the androgen receptor gene is not a common determinant of male sexual orientation. Am J Hum Genet 53:844–852

MacLean HE, Chu S, Warne GL, Zajac JD (1993) Related individuals with different androgen receptor gene deletions. J Clin Invest 91:1123–1128

MacLean HE, Choi W, Warne GL, Zajac JD (1994) Abnormal androgen receptor binding affinity in five subjects with Kennedy's disease [Abstract]. Program of the 76th Annual Meeting of the Endocrine Society:633-Abs #1730

Mahtani MM, Lafrenière RG, Kruse TA, Willard HF (1991) An 18-locus linkage map of the pericentromeric region of the human X chromosome: genetic framework for mapping X-linked disorders. Genomics 10:849–857

Manuel M, Katayama KP, Jones HW, Jr (1976) The age of occurrence of gonadal tumors in intersex patients with a Y chromosome. Am J Obstet Gynecol 124:293–300

Marcelli M, Tilley WD, Wilson CM, Griffin JE, Wilson JD, McPhaul MJ (1990a) Definition of the human androgen receptor gene structure permits the identification of mutations that cause androgen resistance: premature termination of the receptor protein at amino acid residue 588 causes complete androgen resistance. Mol Endocrinol 4:1105–1116

Marcelli M, Tilley WD, Wilson CM, Wilson JD, Griffin JE, McPhaul MJ (1990b) A single nucleotide substitution introduces a premature termination codon into the androgen receptor gene of a patient with receptor-negative androgen resistance. J Clin Invest 85:1522–1528

Marcelli M, Tilley WD, Zoppi S, Griffin JE, Wilson JD, McPhaul MJ (1991a) Androgen resistance associated with a mutation of the androgen receptor at amino acid 772 (Arg→Cys) results from a combination of decreased messenger ribonucleic acid levels and impairment of receptor function. J Clin Endocrinol Metab 73:318–325

Marcelli M, Zoppi S, Grino PB, Griffin JE, Wilson JD, McPhaul MJ (1991b) A mutation in the DNA-binding domain of the androgen receptor gene causes complete testicular feminization in a patient with receptor-positive androgen resistance. J Clin Invest 87:1123–1126

Marschke KB, Tan J-A, Ho K, Wilson EM, French FS (1993) Androgen receptor DNA interactions. In: De Bellis A, Marschke KB (eds). New perspectives in endocrinology, Vol 49. New York:Serono Symposia Publications, Raven Press pp 41–50

Marschke KB, Ho K-C, Power SGA, Tan J-A, Kupfer SR, Wilson EM, French FS (1994) Response element diversity in androgen-receptor regulated genes. In: Dufau ML, Isidori A, Fabbri A (eds). Cell and molecular biology of the testis. New York, NY:Ares Serono Symposia, Raven Press pp 69–79

Matsuura T, Demura T, Aimoto Y, Mizuno T, Morikawa F, Tashiro K (1992) Androgen receptor abnormality in X-linked spinal and bulbar muscular atrophy. Neurology 42:1724–1726

Mauvais-Jarvis P, Bercovici JP, Crepy O, Gauthier F (1970) Studies on testosterone metabolism in subjects with testicular feminization syndrome. J Clin Invest 49:31–40

McEwan IJ, Gustafsson J (1997) Interaction of the human androgen receptor transactivation function with the general transcription factor TFIIF. Proc Natl Acad Sci USA 94(16):8485–8490

McKusick VA (1990) Testicular feminization syndrome. [Anonymous] Mendelian Inheritance in Man. Catalogs of autosomal dominant, autosomal recessive and sex-linked phenotypes (9th ed.). Baltimore:Johns Hopkins University Press pp 1724–1728

McPhaul MJ, Marcelli M, Tilley WD, Griffin JE, Isidro-Gutierrez RF, Wilson JD (1991a) Molecular basis of androgen resistance in a family with a qualitative abnormality of the androgen receptor and responsive to high-dose androgen therapy. J Clin Invest 87:1413–1421

McPhaul MJ, Marcelli M, Tilley WD, Griffin JE, Wilson JD (1991b) Androgen resistance caused by mutations in the androgen receptor gene. FASEB J 5:2910–2915

McPhaul MJ, Marcelli M, Zoppi S, Griffin JE, Wilson JD (1993) Genetic basis of endocrine disease 4. The spectrum of mutations in the androgen receptor gene that causes androgen resistance. J Clin Endocrinol Metab 76:17–23

McPhaul MJ, Marcelli M, Zoppi S, Wilson CM, Griffin JE, Wilson JD (1992) Mutations in the ligand-binding domain of the androgen receptor gene cluster in two regions of the gene. J Clin Invest 90:2097–2101

Mebarki F, Forest MG, Chatelain P, David M, Morel Y (1991) Analysis of exon 1 polymorphism and mutations of the androgen receptor gene makes possible the detection of heterozygote carriers and the prenatal diagnosis of the androgen insensitivity syndrome [Abstract]. Horm Res 35 (suppl 2):7-Abs #25

Mebarki F, Forest MG, Lauras B, Bertrand AM, Chatelain P, David M, Morel Y (1993) De novo mutation of the androgen receptor gene is not a rare event in androgen insensitivity syndrome as suggested by segregation of the trinucleotide repeat in the exon 1 and sequencing [Abstract]. 75th Annual Meeting of the Endocrine Society:201-Abst #602

Meyer WJ, III., Migeon BR, Migeon CJ (1975) Locus on human X chromosome for dihydrotestosterone receptor and androgen insensitivity. Proc Natl Acad Sci USA 72:1469–1472

Mhatre AN, Trifiro MA, Kaufman M, Kazemi-Esfarjani P, Figlewicz D, Rouleau G, Pinsky L (1993) Reduced transcriptional regulatory competence of the androgen receptor in X-linked spinal and bulbar muscular atrophy. Nature Genet 5:184–187

Migeon BR, Brown TR, Axelman J, Migeon CJ (1981) Studies of the locus for androgen receptor: Localization on the human X chromosome and evidence for homology with the *Tfm* locus in the mouse. Proc Natl Acad Sci USA 78:6339–6343

Mizokami A, Yeh S, Chang C (1994) Identification of 3', 5'-cyclic adenosine monophosphate response element and other *Cis*-acting elements in the human androgen receptor gene promoter. Mol Endocrinol 8:77–88

Moilanen A, Rouleau N, Ikonen T, Palvimo JJ, Janne OA (1997) The presence of a transcription activation function in the hromone-binding domain of androgen receptor is revealed by studies in yeast cells. FEBS Lett 412(2):355–358

Money J (1974) Long-term psychologic follow-up of intersexed patients. Clin Plast Surg 1(2):271–274

Morel Y, Mebarki F, Forest MG (1994) What are the indications for prenatal diagnosis in the androgen insensitivity syndrome? Facing clinical heterogeneity of phenotypes for the same genotype. Eur J Endocrinol 130:325–326

Morgan GE, Rowley G, Green PM, Chisholm M, Giannelli F, Brownlee GG (1997) Further evidence for the importance of an androgen response element in the factor IX promoter. Brit J Hemat 98(1):79–85

Morris JM (1953) The syndrome of testicular feminization in male pseudohermaphrodites. Am J Obstet Gynecol 65:1192–1211

Morris JM, Mahesh VB (1963) Further observations on the syndrome, "testicular feminization". Am J Obstet Gynecol 87:731–748

Morrow AF, Gyorki S, Warne GL, Burger HG, Bangah ML, Outch KH, Mirovics A, Baker HWG (1987) Variable androgen receptor levels in infertile men. J Clin Endocrinol Metab 64:1115–1121

Mowszowicz I, Riahi M, Wright F, Bouchard P, Kuttenn F, Mauvais-Jarvias P (1981) Androgen receptor in human skin cytosol. J Clin Endocrin Metab 52:338–344

Mowszowicz I, Melanitou E, Kirchhoffer M-O, Mauvais-Jarvis P (1983) Dihdrotestosterone stimulates 5α-reductase activity in pubic skin fibroblasts. J Clin Endocrinol Metab 56:320–325

Munoz-Torres M, Jodar E, Quesada M, Escobar-Jiminez F (1995) Bone mass in androgen insensitivity syndrome: response to hormonal replacement therapy. Calcified Tissue Int 57:94–96

Müller J (1984) Morphometry and histology of gonads from twelve children and adolescents with the androgen insensitivity (testicular feminization) syndrome. J Clin Endocrinol Metab 59:785–789

Müller J, Skakkebaek N (1984) Testicular carcinoma *in situ* in children with the androgen insensitivity (testicular feminisation) syndrome. Br Med J 288:1419–1420

Nagafuchi S, Yanagisawa H, Sato K, Shirayama T, Ohsaki E, Bundo M, Takeda T, Tadokoro K, Kondo I, Murayama N, et al. (1994) Dentatorubral and pallidoluysian atrophy expansion of an unstable CAG trinucleotide on chromosome 12p. Nature Genet 6:14–18

Nagel RA, Lippe BM, Griffin JE (1986) Androgen resistance in the neonate: use of hormones of the hypothalamic-pituitary-gonadal axis for diagnosis. J Pediatr 109:486–488

Nakagama H, Gunji T, Ohnishi S, Kaneko T, Ishikawa T, Makino R, Hayashi K, Shiga J, Takaku F, Imawari M (1991) Expression of androgen receptor mRNA in human hepatocellular carcinomas and hepatoma cell lines. Hepatology 14:99–102

Nakajima H, Kimura F, Nakagawa T, Furutama D, Shinoda K, Shimizu A, Oshawa N (1996) Transcriptional activation by the androgen receptor in X-linked spinal and bulbar muscular atrophy. J Neurol Sci 142(1–2):12–16

Nakamura M, Mita S, Matsuura T, Nagashima K, Tanaka H, Ando M, Uchino M (1997) The reduction of androgen receptor mRNA in motoneurons of X-linked spinal and bulbar muscular atrophy. J Neurol Sci 150(2):161–165

Nakao R, Yanase T, Sakai Y, Haji M, Nawata H (1993) A single amino acid substitution (Gly743→Val) in the steroid-binding domain of the human androgen receptor leads to Reifenstein syndrome. J Clin Endocrinol Metab 77:103–107

Nazareth LV, Weigel NL (1996) Activation of the human androgen receptor through a protein kinase A signaling pathway. J Biol Chem 271:19900–19907

Nemoto T, Ohara-Nemoto Y, Shimazaki S, Ota M (1994) Dimerization characteristics of the DNA- and steroid-binding domains of the androgen receptor. J Steroid Biochem Mol Biol 50:225–233

Neuschmid-Kaspar F, Gast A, Peterziel H, Schneikert J, Muigg A, Ransmayr G, Klocker H, Bartsch G, Cato ACB (1996) CAG-repeat expansion in androgen receptor in Kennedy's disease is not a loss of function mutation. Mol Cell Endocrinol 117(2):149–156

Newman K, Randolph J, Anderson K (1992) The surgical management of infants and children with ambiguous genitalia. Lessons learned from 25 years. Ann Surg 215(6):644–653

Newmark JR, Hardy DO, Tonb DC, Carter BS, Epstein JI, Isaacs WB, Brown TR, Barrack ER (1992) Androgen receptor gene mutations in human prostate cancer. Proc Natl Acad Sci USA 89:6319–6323

Nitsche EM, Moquin A, Adams PS, Guenette RS, Lakins JN, Sinnecker GHG, Kruse K, Tenniswood MP (1996) Differential display RT PCR of total RNA from human foreskin fibroblasts for investigation of androgen-dependent gene expression. Am J Med Genet 63:231–238

Nordeen SK, Suh BJ, Kühnel B, Hutchison CA, III (1990) Structural determinants of a glucocorticoid receptor recognition element. Mol Endocrinol 4:1866–1873

O'Connel MJ, Ramsey HE, Whang-Peng J, Wiernik PH (1973) Testicular feminization syndrome in three sibs: emphasis on gonadal neoplasia. Am J Med Sci 265:321–333

Ogata A, Matsuura T, Tashiro K, Moriwaka F, Demura T, Koyanagi T, Nagashima K (1994) Expression of androgen receptor in X-linked spinal and bulbar muscular atrophy and amyotrohic lateral sclerosis. J Neurol, Neurosur Psy 63:1274–1275

Ohno S, Lyon MF (1970) X-linked testicular feminization in the mouse as a non-inducible regulatory mutation of the Jacob-Monod type. Clin Genet 1:121–127

Oka M, Katabuchi H, Munemura M, Mizumoto J, Maeyama M (1984) An unusual case of male pseudohermaphroditism: complete testicular feminization associated with incomplete differentiation of the Müllerian duct. Fertil Steril 41:154–156

Orr HT, Chung M-y, Banfi S, Kwiatkowski TJ, Jr, Servadio A, Beaudet AL, McCall AE, Duvick LA, Ranum LPW, Zoghbi HY (1993) Expansion of an unstable trinucleotide CAG repeat in spinocerebellar ataxia type 1. Nature Genet 4:221–226

Palvimo JJ, Kallio PJ, Ikonen T, Mehto M, Jänne OA (1993) Dominant negative regulation of trans-activation by the rat androgen receptor: roles of the N-terminal domain and heterodimer formation. Mol Endocrinol 7:1399–1407

Palvimo JJ, Reinikainen P, Ikonen T, Kallio PJ, Moilanen A, Janne OA (1996) Mutual transcription interference between RelA and androgen receptor. J Biol Chem 271(39):24151–24156

Pelliniemi LJ, Dym M, Crigler JF, Retik AB, Fawcett DW (1980) Development of Leydig cells in human fetuses and in patients with androgen insensitivity. In: Steinberger A, Steinberger E (eds). Testicular development, structure and function. New York:Raven Press pp 49–54

Perutz MF, Johnson T, Suzuki M, Finch JT (1994) Glutamine repeats as polar zippers: their possible role in inherited neurodegenerative diseases. Proc Natl Acad Sci USA 91(12):5355–5358

Pettersson G, Bonnier G (1937) Inherited sex-mosaic in man. Hereditas 23:49–69

Pinsky L, Kaufman M, Summitt RL (1981) Congenital androgen insensitivity due to a qualitatively abnormal androgen receptor. Am J Med Genet 10:91–99

Pinsky L, Kaufman M, Gil-Esteban C, Sumbulian D (1983) Regulation of the androgen receptor in human genital skin fibroblasts, with a review of sex steroid receptor regulation by homologous and heterologous steroids. Can J Biochem Cell Biol 61:770–778

Pinsky L, Kaufman M, Killinger DW, Burko B, Shatz D, Volpe R (1984) Human minimal androgen insensitivity with normal dihydrotestosterone-binding capacity in cultured genital skin fibroblasts: evidence for an androgen-selective qualitative abnormality of the receptor. Am J Hum Genet 36:965–978

Pinsky L, Kaufman M, Chudley AE (1985) Reduced affinity of the androgen receptor for 5α-dihydrotestosterone but not methyltrienolone in a form of partial androgen resistance. J Clin Invest 75:1291–1296

Pinsky L, Trifiro M, Kaufman M, Beitel LK, Mhatre A, Kazemi-Esfarjani P, Sabbaghian N, Lumbroso R, Alvarado C, Vasiliou M, et al. (1992) Androgen resistance due to mutation of the androgen receptor. Clin Invest Med 15:456–472

Pion RJ, Dignam WJ, Lamb EJ, Moore JG, Frankland MV, Simmer HH (1965) Testicular feminization. Am J Obstet Gynecol 93:1067–1075

Poujol N, Lobaccaro JM, Chiche L, Lumbroso S, Sultan C (1997) Functional and structural analaysis of R607Q and R608K androgen receptor substitutions associated with male breast cancer. Mol Cell Endocrinol 130(1–2):43–51

Prader A (1954) Der Genitalbefund beim Pseudohermaphroditismus feminus des kongenitalen adrenogenitalen Syndroms. Helv Paediatr Acta 9:231–248

Pratt WB, Jolly DJ, Pratt DV, Hollenberg SM, Giguere V, Cadepond FM, Schweizer-Groyer G, Catelli M-G, Evans RM, Baulieu E-E (1988) A region in the steroid binding domain determines formation of the non-DNA-binding, 9 S glucocorticoid receptor complex. J Biol Chem 263:267–273

Prior L, Bordet S, Trifiro MA, Mhatre A, Kaufman M, Pinsky L, Wrogeman K, Belsham DD, Pereira F, Greenberg C, et al. (1992) Replacement of arginine 773 by cysteine or histidine in the human androgen receptor causes complete androgen insensitivity with different receptor phenotypes. Am J Hum Genet 51:143–155

Ptashne M (1988) How eukaryotic transcriptional activators work. Nature 335:683–689

Puck TT, Robinson A, Tjio JH (1960) Familial primary amenorrhea due to testicular feminization: a human gene affecting sex differentiation. Proc Soc Exp Biol Med 103:192–196

Puy L, MacLusky NJ, Becker L, Karsan N, Trachtenberg J, Brown TJ (1995) Immunocytochemical detection of androgen receptor in human temporal cortex. Characterization and application of polyclonal androgen receptor antibodies in frozen and paraffin-embedded tissues. J Steroid Biochem Mol Biol 55(2):197–209

Quarmby VE, Kemppainen JA, Sar M, Lubahn DB, French FS, Wilson EM (1990a) Expression of recombinant androgen receptor in cultured mammalian cells. Mol Endocrinol 4:1399–1407

Quarmby VE, Yarbrough WG, Lubahn DB, French FS, Wilson EM (1990b) Autologous down-regulation of androgen receptor messenger ribonucleic acid. Mol Endocrinol 4:22–28

Quigley CA, Evans BAJ, Simental JA, Marschke KB, Sar M, Lubahn DB, Davies P, Hughes IA, Wilson EM, French FS (1992a) Complete androgen insensitivity due to deletion of exon C of the androgen receptor gene highlights the functional importance of the second zinc finger of the androgen receptor *in vivo*. Mol Endocrinol 6:1103–1112

Quigley CA, Friedman KJ, Johnson A, Lafreniere RG, Silverman LM, Lubahn DB, Brown TR, Wilson EM, Willard HF, French FS (1992b) Complete deletion of the androgen receptor gene: definition of the null phenotype of the androgen insensitivity syndrome and determination of carrier status. J Clin Endocrinol Metab 74:927–933

Quigley CA, Simental JA, Marschke KB, Lubahn DB, Wilson EM, French FS (1992c) Naturally-occurring structural defects in the androgen receptor gene [Abstract]. J Cell Biochem Suppl 16C:38-Abs #L323

Quigley CA, De Bellis A, Marschke KB, El-Awady MK, Wilson EM, French FS (1995) Androgen receptor defects: historical, clinical, and molecular perspectives. Endocr Rev 16:271–321

Ramani P, Yeung CK, Habeebu SSM (1993) Testicular intratubular germ cell neoplasia in children and adolescents with intersex. Am J Surg Path 17:1124–1133

Ramaswamy G, Jagadha V, Tchertkoff V (1985) A testicular tumor resembling the sex cord with annular tubules in a case of the androgen insensitivity syndrome. Cancer 55:1607–1611

Reiner WG (1997) Sex assignment in the neonate with intersex or inadequate genitalia. Arch Pediat Adol Med 151:1044–1045

Renaud J, Rochel N, Ruff M, Vivat V, Chambon P, Gronemeyer H, Moras D (1995) Crystal structure of the RAR-g ligand-binding domain bound to all-*trans* retinoic acid. Nature 378:681–689

Rennie PS, Bruchovsky N, Leco KJ, Sheppard PC, McQueen SA, Cheng H, Snoek R, Hamel A, Bock ME, MacDonald BS, et al. (1993) Characterization of two *cis*-acting DNA elements involved in the androgen regulation of the probasin gene. Mol Endocrinol 7:23–36

Reyes FI, Winter JSD, Faiman C (1989) Endocrinology of the fetal testis. In: Burger H, de Kretser D, editors. The testis (2nd ed.). New York:Raven Press pp 119–142

Richards RI, Sutherland GR (1992) Dynamic mutations: a new class of mutations causing human disease. Cell 70:709–712

Ris-Stalpers C, Kuiper GGJM, Faber PW, Schweikert HU, van Rooij HCJ, Zegers ND, Hodgins MB, Degenhart HJ, Trapman J, Brinkmann AO (1990) Aberrant splicing of androgen receptor mRNA results in synthesis of a nonfunctional receptor protein in a patient with androgen insensitivity. Proc Natl Acad Sci USA 87:7866–7870

Ris-Stalpers C, Trifiro MA, Kuiper GGJM, Jenster G, Romalo G, Sai T, van Rooij HCJ, Kaufman M, Rosenfield RL, Liao S, et al. (1991) Substitution of aspartic acid-686 by histidine or asparagine in the human androgen receptor leads to a functionally inactive protein with altered hormone-binding characteristics. Mol Endocrinol 5:1562–1569

Ris-Stalpers C, de Blaeij TJP, Sleddens HFBM, Trapman J, Brinkmann AO (1992) Androgen receptor mutants and functional activity. [Abstract]. 74th Annual Meeting of the Endocrine Society:303-Abs #1008

Ris-Stalpers C, Verleun-Mooijman MCT, Trapman J, Brinkmann AO (1993) Threonine on amino acid position 868 in the human androgen receptor is essential for androgen binding specificity and functional activity. Biochem Biophys Res Commun 196:173–180

Ris-Stalpers C, Hoogenboezem T, Sleddens HFBM, Verleun-Mooijman MCT, Degenhart HJ, Drop SLS, Halley DJJ, Oosterwijk JC, Hodgins MB, Trapman J et al. (1994a) A practical approach to the detection of androgen receptor gene mutations and pedigree analysis in families with X-linked androgen insensitivity. Pediatr Res 36:227–234

Ris-Stalpers C, Verleun-Mooijman MCT, de Blaeij TJP, Degenhart HJ, Trapman J, Brinkmann AO (1994b) Differential splicing of human androgen receptor pre-mRNA in X-linked Reifenstein syndrome, because of a deletion involving a putative branch site. Am J Hum Genet 54:609–617

Ritzen EM (1992) Pubertal growth in genetic disorders of sex hormone action and secretion. Acta Paediatr Scand [Suppl] 383:22–25

Rodien P, Mebarki F, Mowszowicz I, Chaussain JL, Young J, Morel Y, Schaison G (1996) Different phenotypes in a family with androgen insensitivity caused by the same M780I point mutation in the androgen receptor gene. J Clin Endocrin Metab 81(8):2994–2998

Roehrborn CG, Lange JL, George FW, Wilson JD (1987) Changes in amount and intracellular distribution of androgen receptor in human foreskin as a function of age. J Clin Invest 79:44–47

Rosenfield RL, Lawrence AM, Liao S, Landau RL (1971) Androgens and androgen responsiveness in the feminizing testis syndrome. Comparison of complete and "incomplete" forms. J Clin Endocrinol 32:625–632

Ruizeveld de Winter JA, Janssen PJA, Sleddens HMEB, Verleun-Mooijman MCT, Trapman J, Brinkmann AO, Santerse AB, Schröder FH, van der Kwast TH (1994) Androgen receptor status in localized and locally progressive hormone refractory human prostate cancer. Am J Pathol 144:735–746

Rundlett SE, Wu X, Miesfeld RL (1990) Functional characterizations of the androgen receptor confirm that the molecular basis of androgen action is transcriptional regulation. Mol Endocrinol 4:708–714

Rutgers JL, Scully RE (1991) The androgen insensitivity syndrome (testicular feminization): a clinicopathologic study of 43 cases. Int J Gynecol Pathol 10:126–144

Sai T, Seino S, Chang C, Trifiro M, Pinsky L, Mhatre A, Kaufman M, Lambert B, Trapman J, Brinkmann AO, et al. (1990) An exonic point mutation of the androgen receptor gene in a family with complete androgen insensitivity. Am J Hum Genet 46:1095–1100

Sanborn BM, Caston LA, Chang C, Liao S, Speller R, Porter ID, Ku CY (1991) Regulation of androgen receptor mRNA in rat Sertoli and peritubular cells. Biol Reprod 45:634–641

Sar M, Lubahn DB, French FS, Wilson EM (1990) Immunohistochemical localization of the androgen receptor in rat and human tissues. Endocrinology 127:3180–3186

Sar M, Stumpf WE (1977) Androgen concentration in motor neurons of cranial nerves and spinal cord. Science 197:77–79

Sato N, Sadar MD, Bruchovsky N, Saatcioglu F, Rennie PS, Sato S, Lange PH, Gleave ME (1997) Androgenic induction of prostate-specific antigen gene is repressed by protein-protein interaction between the androgen receptor and AP-1/c-Jun in the human prostate cancer cell line LNCaP. J Biol Chem 272(28):17485–17494

Savage MO, Chaussain JL, Evain D, Roger M, Canlorbe P, Job JC (1978) Endocrine studies in male pseudohermaphroditism in childhood and adolescence. Clin Endocrinol 8:219–231

Schoenberg MP, Hakimi JM, Barrack ER (1994a) Allele size distribution of the androgen receptor (AR) CAG microsatellite in human prostate cancer [Abstract]. 8th Ann Mtg Soc Bas Urol Res

Schoenberg MP, Hakimi JM, Wang S, Bova GS, Epstein JI, Fischbeck KH, Isaacs WB, Walsh PC, Barrack ER (1994b) Microsatellite mutation (CAG24→18) in the androgen receptor gene in human prostate cancer. Biochem Biophys Res Commun 198:74–80

Schoenen J, Delwaide PJ, Legros JJ, Franchimont P (1979) Motoneuropathie héréditaire: la forme proximale de l'adulte liéé au sexe (ou maladie de Kennedy). J Neurol Sci 41:343–357

Scully RE (1981) Neoplasia associated with anomalous sexual development and abnormal sex chromosomes. [Anonymous] The intersex child. Pediatr Adol Endocr, Vol 8. Basel:Karger pp 203–217

Shan L, Rodriguez MC, Janne OA (1990) Regulation of androgen receptor protein and mRNA concentrations by androgens in rat ventral prostate and seminal vesicles and in human hepatoma cells. Mol Endocrinol 4:1636–1646

Shimada N, Sobue G, Doyu M, Yamamoto K, Yasuda T, Mukai E, Kachi T, Mitsuma T (1995) X-linked recessive bulbospinal neuronopathy: clinical phenotypes and CAG repeat size in androgen receptor gene. Muscle Nerve 18(12):1378–1384

Simental JA, Sar M, Lane MV, French FS, Wilson EM (1991) Transcriptional activation and nuclear targeting signals of the human androgen receptor. J Biol Chem 266:510–518

Sinclair AH, Berta P, Palmer MS, Hawkins JR, Griffiths BL, Smith MJ, Foster JW, Frischauf AM, Lovell-Badge R, Goodfellow PN (1990) A gene from the human sex-determining region encodes a protein with homology to a conserved DNA-binding motif. Nature 346:240–244

Sinnecker G, Kohler S (1989) Sex hormone-binding globulin response to the anabolic steroid stanozolol: evidence for its suitability as a biological androgen sensitivity test. J Clin Endocrinol Metab 68:1195–1200

Sinnecker GHG, Hiort O, Nitsche E, Holterhus PM, Kruse K (1997) Functional assessment and clinical classification of androgen sensitivity in patients with mutations of the androgen receptor gene. Europ J Pediatr 156(1):7–14

Smith DF, Toft DO (1993) Steroid receptors and their associated proteins. Mol Endocrinol 7:4–11

Smith DW, Marokus R, Graham JM, Jr (1985) Tentative evidence of Y-linked statural gene(s). Growth in the testicular feminization syndrome. Clin Pediatr 24:189–192

Song CS, Her S, Slomczynska M, Choi SJ, Jung MH, Roy AK, Chatterjee B (1993) A distal activation domain is critical in the regulation of the rat androgen receptor gene promoter. Biochem J 294:779–784

Soule SG, Conway G, Prelevic GM, Prentice M, Ginsburg J, Jacobs HS (1995) Osteopenia as a feature of the androgen insensitivity syndrome. Clin Endocrinol 43:671–675

Southren AL, Ross H, Sharma DC, Gordon G, Weingold AB, Dorfman RI (1965) Plasma concentration and biosynthesis of testosterone in the syndrome of feminizing testes. J Clin Endocrinol 25:518–525

Spencer JA, Watson JM, Lubahn DB, Joseph DR, French FS, Wilson EM, Graves JAM (1991) The androgen receptor gene is located on a highly conserved region of the X chromosomes of marsupial and monotreme as well as eutherian mammals. J Hered 82:134–139

Spencer JR, Torrado T, Sanchez RS, Vaughan ED, Jr., Imperato-McGinley J (1991) Effects of flutamide and finasteride on rat testicular descent. Endocrinology 129:741–748

Spencer JR, Vaughan ED, Jr, Imperato-McGinley J (1993) Studies of the hormonal control of postnatal testicular descent in the rat. J Urol 149:618–623

Strickland AL, French FS (1969) Absence of response to dihydrotestosterone in the syndrome of testicular feminization. J Clin Endocrinol Metab 29:1284–1286

Sultan C, Migeon BR, Rothwell SW, Maes M, Zerhouni N, Migeon CJ (1980) Androgen receptors and metabolism in cultured human fetal fibroblasts. Pediatr Res 14(1):67–69

Sultan C, Terraza A, Devillier C, Michel B, Descomps B, Jean R (1982) Androgen receptors in cultured human skin fibroblasts. British J Dermatol 107:40–46

Sultan C, Lumbroso S, Poujol N, Belon C, Boudon C, Lobaccaro JM (1993) Mutations of androgen receptor gene in androgen insensitivity syndromes. J Steroid Biochem Mol Biol 46:519–530

Sutherland RW, Wiener JS, Hicks JP, Marcelli M, Gonzales ETJ, Roth DR, Lamb DJ (1996) Androgen receptor gene mutations are rarely associated with isolated penile hypospadias. J Urol 156(2):828–831

Swanson ML, Coronel EH (1993) Complete androgen insensitivity with persistent Müllerian structures. A case report. J Reprod Med 38:565–568

Syms AJ, Norris JS, Panko WB, Smith RG (1985) Mechanism of androgen-receptor augmentation. Analysis of receptor synthesis and degradation by the density-shift technique. J Biol Chem 260:455–461

Takane KK, McPhaul MJ (1996) Functional analysis of the human androgen receptor promoter. Mol Cell Endocrinol 119(1):83–93

Tan J-A, Marschke KB, Ho K-C, Perry ST, Wilson EM, French FS (1992) Response elements of the androgen-regulated C3 gene. J Biol Chem 267:4456–4466

Tanaka F, Doyu M, Ito Y, Matsumoto M, Mitsuma T, Abe K, Aoki M, Itoyama Y, Fischbeck KH, Sobue G (1996) Founder effect in spinal and bulbar muscular atrophy (SBMA). Hum Mol Genet 5(9):1253–1257

The Huntington's Disease Collaborative Research Group (1993) A novel gene containing a trinucleotide repeat that is expanded and unstable on Huntington's disease chromosomes. Cell 72:971–983

Tilley WD, Marcelli M, Wilson JD, McPhaul MJ (1989) Characterization and expression of a cDNA encoding the human androgen receptor. Proc Natl Acad Sci USA 86:327–331

Tilley WD, Marcelli M, McPhaul MJ (1990) Expression of the human androgen receptor gene utilizes a common promoter in diverse human tissues and cell lines. J Biol Chem 265:13776–13781

Tincello DG, Hargreave TB, Wu FC, Padayachi T, Saunders PT (1992) Mutations in the androgen receptor gene of patients with androgen insensitivity [Abstract]. J Endocrinol 132 (suppl):Abs #87

Tommerup N, Schempp W, Meinecke P, Pedersen S, Bolund L, Brandt C, Goodpasture C, Guldberg P, Held KR, Reinwein H, et al. (1993) Assignment of an autosomal sex reversal locus (*SRA1*) and campomelic dysplasia (*CMPD1*) to 17q24.3–q25.1. Nature Genet 4:170–153

Trapman J, Klaassen P, Kuiper GGJM, van der Korput JAGM, Faber PW, van Rooij HCJ, Geurts van Kessel A, Voorhorst MM, Mulder E, Brinkmann AO (1988) Cloning, structure and expression of a cDNA encoding the human androgen receptor. Biochem Biophys Res Commun 153:241–248

Trifiro M, Gottlieb B, Pinsky L, Kaufman M, Prior L, Belsham DD, Wrogemann K, Brown CJ, Willard HF, Trapman J et al. (1991a) The 56/58 kDa androgen-binding protein in male genital skin fibroblasts with a deleted androgen receptor gene. Mol Cell Endocrinol 75:37–47

Trifiro M, Prior RL, Sabbaghian N, Pinsky L, Kaufman M, Nylen EG, Belsham DD, Greenberg CR, Wrogemann K (1991b) Amber mutation creates a diagnostic *Mael* site in the androgen receptor gene of a family with complete androgen insensitivity. Am J Med Genet 40:493–499

Trifiro MA, Lumbroso R, Beitel LK, Vasiliou DM, Bouchard J, Deal C, Van Vliet G, Pinsky L (1997) Altered mRNA expression due to insertion or substitution of thymine at position +3 of two splice-donor sites in the androgen receptor gene. Europ J Hum Gen 5(1):50–58

Truss M, Beato M (1993) Steroid hormone receptors: interaction with deoxyribonucleic acid and transcription factors. Endocr Rev 14:459–479

Ulloa-Aguirre A, Carranza-Lira S, Mendez JP, Angeles A, Chavez B, Perez-Palacios G (1990) Incomplete regression of Mullerian ducts in the androgen insensitivity syndrome. Fertil Steril 53:1024–1028

van der Schoot P (1992) Androgens in relation to prenatal development and postnatal inversion of the gubernacula in rats. J Reprod Fertil 95:145–158

van Gelderen HH (1986) Skeletal maturation in the XY female syndrome. Clin Genet 30:199–201

van Look PFA, Hunter WM, Corker CS, Baird DT (1977) Failure of positive feedback in normal men and subjects with testicular feminization. Clin Endocrinol 7:353–366

Varrela J, Alvesalo L, Vinkka H (1984) Body size and shape in 46,XY females with complete testicular feminization. Ann Hum Biol 11:291–301

Vasiliou M, Trifiro M, Pinsky L (1994) Mutations in the N-terminal domain of the human androgen receptor associated with androgen resistance syndrome [Abstract]. Program of the 76th Annual Meeting of the Endocrine Society:495-Abs #1179

Veldscholte J, Ris-Stalpers C, Kuiper GGJM, Jenster G, Berrevoets C, Claassen E, van Rooij HCJ, Trapman J, Brinkmann AO, Mulder E (1990) A mutation in the ligand binding domain of the androgen receptor of human LNCaP cells affects steroid binding characteristics and response to anti-androgens. Biochem Biophys Res Commun 173:534–540

Visakorpi T, Hyytinen E, Koivisto P, Tanner M, Keinanen R, Palmberg C, Palotie A, Tammela T, Isola J, Kallioniemi O (1995) In vivo amplification of the androgen receptor gene and progression of human prostate cancer. Nature Genet 9:401–406

Warne GL, Gyorki S, Risbridger GP, Khalid BAK, Funder JW (1984) Correlations between fibroblast androgen receptor levels and clinical features in abnormal male sexual differentiation and infertility. Aust N Z J Med 13:335–341

Warner CL, Griffin JE, Wilson JD, Jacobs LD, Murray KR, Fischbeck KH, Dickoff D, Griggs RC (1992) X-linked spinomuscular atrophy: a kindred with associated abnormal androgen receptor binding. Neurology 42:2181–2184

Weiss RE, Refetoff S (1992) Thyroid hormone resistance. Annu Rev Med 43:363–375

Wilkins L (1950) Heterosexual development. [Anonymous] The diagnosis and treatment of endocrine disorders in childhood and adolescence. Springfield:Thomas pp 256–279

Wilson CM, Griffin JE, Wilson JD, Marcelli M, Zoppi S, McPhaul MJ (1992) Immunoreactive androgen receptor expression in subjects with androgen resistance. J Clin Endocrinol Metab 75:1474–1478

Wilson CM, McPhaul MJ (1994) A and B forms of the androgen receptor are present in human genital skin fibroblasts. Proc Natl Acad Sci USA 91:1234–1238

Wilson CM, McPhaul MJ (1996) A and B forms of the androgen receptor are expressed in a variety of human tissues. Mol Cell Endocrinol 120(1):51–57

Wilson EM, French FS (1976) Binding properties of androgen receptors: Evidence for identical receptors in rat testis, epididymis and prostate. J Biol Chem 251:5620–5629

Wilson JD, Harrod MJ, Goldstein JL, Hemsell DL, MacDonald PC (1974) Familial incomplete male pseudohermaphroditism, type 1. Evidence for androgen resistance and variable clinical manifestations in a family with the Reifenstein syndrome. N Engl J Med 290:1097–1103

Wilson JD, George FW, Griffin JE (1981) The hormonal control of sexual development. Science 211:1278–1284

Wilson JD, Griffin JE, Russell DW (1993) Steroid 5α-reductase 2 deficiency. Endocr Rev 14:577–593

Wiren KM, Zhang X, Chang C, Keenan E, Orwoll ES (1997) Transcriptional up-regulation of the human androgen receptor by androgen in bone cells. Endocrinology 138(6):2291–2300

Wolf DA, Herzinger T, Hermeking H, Blaschke D, Horz W (1993) Transcriptional and post-transcriptional regulation of human androgen receptor expression by androgen. Mol Endocrinol 7:924–936

Wong C, Zhou Z, Sar M, Wilson EM (1993) Steroid requirement for androgen receptor dimerization and DNA binding. Modulation by intramolecular interactions between the NH2-terminal and steroid-binding domains. J Biol Chem 268:19004–19012

Wooster R, Mangion J, Eeles R, Smith S, Dowsett M, Averill D, Barrett-Lee P, Easton DF, Ponder BAJ, Stratton MR (1992) A germline mutation in the androgen receptor gene in two brothers with breast cancer and Reifenstein syndrome. Nature Genet 2:132–134

Word RA, George FW, Wilson JD, Carr BR (1989) Testosterone synthesis and adenylate cyclase activity in the early human fetal testis appear to be independent of human chorionic gonadotropin control. J Clin Endocrinol Metab 69:204–208

Wrange O, Eriksson P, Perlmann T (1989) The purified activated glucocorticoid receptor is a homodimer, J Biol Chem 264:5253–5259

Wurtz J, Bourguet W, Renaud J, Vivat V, Chambon P, Moras D, Gronemeyer H (1996) A canonical structure for the ligand-binding domain of nuclear receptors. Nature Struct Biol 3(1):87–94

Yapijakis C, Kapaki E, Boussiou M, Vassilopoulos D, Papageorgiou C (1996) Prenatal diagnosis of X-linked spinal and bulbar muscular atrophy in a Greek family. Prenatal Diag 16(3):262–265

Yeh S, Chang C (1996) Cloning and characterization of a specific coactivator, ARA70, for the androgen receptor in human prostate cells. Proc Natl Acad Sci USA 93:5517–5521

Yong EL, Roy A, Chua KL, Ratnam S, Yang M (1994) Complete androgen insensitivity due to a splice-site mutation in the androgen receptor gene and genetic screening with single-stranded conformation polymorphism. Fertil Steril 61:856–862

Youssoufian H, Kazazian HH, Jr, Phillips DG, Aronis S, Tsiftis G, Brown VA, Antonarakis SE (1986) Recurrent mutations in haemophilia A give evidence for CpG mutation hotspots. Nature 324:380–382

Yu W-hA, McGinnis MY (1986) Androgen receptor levels in cranial nerve nuclei and tongue muscles in rats. J Neurosci 6:1302–1307

Zachmann M, Prader A, Sobel EH, Crigler JF, Jr, Ritzen EM, Atares M, Ferrandez A (1986) Pubertal growth in patients with androgen insensitivity: indirect evidence for the importance of estrogens in pubertal growth of girls. J Pediatr 108:694–697

Zarate A, Canales ES, Soria J, Carballo O (1974) Studies on the luteinizing hormone- and follicle-stimulating hormone-releasing mechanism in the testicular feminization syndrome. Am J Obstet Gynecol 119:971–977

Zhang L, Fischbeck KH, Arnheim N (1995) CAG repeat length variation in sperm from a patient with Kennedy's disease. Hum Mol Genet 4(2):303–305

Zhang S, Murtha PE, Young CY (1997) Defining a functional androgen responsive element in the 5' far upstream flanking region of the prostate-specific antigen gene. Biochem Biophys Res Commun 231(3):784–788

Zhou Z, Sar M, Simental JA, Lane MV, Wilson EM (1994) A ligand-dependent bipartite nuclear targeting signal in the human androgen receptor. J Biol Chem 269:13115–13123

Zhou Z, Kemppainen JA, Wilson EM (1995) Identification of three proline-directed phosphorylation sites in the human androgen receptor. Mol Endocrinol 9:605–615

Zhou Z, Corden JL, Brown TR (1997) Identification and characterization of a novel androgen response element composed of a direct repeat. J Biol Chem 272:8227–8235

Zoppi S, Marcelli M, Deslypere J-P, Griffin JE, Wilson JD, McPhaul MJ (1992) Amino acid substitutions in the DNA-binding domain of the human androgen receptor are a frequent cause of receptor-binding positive androgen resistance. Mol Endocrinol 6:409–415

Zoppi S, Wilson CM, Harbison MD, Griffin JE, Wilson JD, McPhaul MJ, Marcelli M (1993) Complete testicular feminization caused by an amino-terminal truncation of the androgen receptor with downstream initiation. J Clin Invest 91:1105–1112

3 Behavioural correlates of testosterone

Kerrin Christiansen

Contents

3.1 Introduction

Behavioural endocrinology is the study of the interaction between hormones and behaviour. This interaction is bidirectional: hormones can affect behaviour, and behaviour can alter hormone levels. Thus, hormonal-behavioural correlations can be due to hormonal effects on behaviour, but certain behaviour (such as physical exercise, stress, sexual behaviour, alcohol consumption, and nutrition) is known to influence hormone levels as well (see below and Christiansen 1998).

Hormones do not cause behavioural changes per se; they can only alter the probability that particular behaviour will occur in the presence of a particular stimulus. Hormones can influence regions of the central nervous system (CNS) which contain hormone receptors by inducing changes in the rate of cellular function. The interaction of a hormone with its receptor begins a series of cellular events that lead to a genomic response wherein the hormone acts directly or indirectly to activate genes that regulate protein synthesis (e.g., Bixo et al. 1995; Chalepakis et al. 1990; Ford and Cramer 1982; Genazzani et al. 1992; Hutchinson 1991; McEwen 1992; McEwen et al. 1984; Sekeris 1990; Viru 1991).

Two decisive phases have been named in the discussion about the time of the effects of sex hormones on brain structures and consequently on behaviour. During fetal and neonatal life, relatively high concentrations of hormones, especially testosterone, are said to influence brain development by organizing the undifferentiated brain in a sex-specific manner. It has been shown, according to studies primarily of rodents, but also of primates and other mammals, that the hypothalamus, the hippocampus, the preoptic-septal region, and the limbic system (especially the amygdala) are important target areas for sex steroid action (Bettini et al. 1992; Brain and Haug 1992; Collaer and Hines 1995; Ellis 1982; Ford and Cramer 1982; Hutchinson 1991, 1993; Hutchinson and Steimer 1984; McEwen 1992; Michael and Bonsall 1990; Naftolin et al. 1990; Simon and Whalen 1987; Whalen 1982). These brain structures and hence the corresponding behavioural repertoires are then thought to be activated at the beginning of puberty when the production of sex hormones increases (Archer 1988; Beatty 1979; Becker et al. 1992; Schulkin et al. 1993). However, experimental studies of birds, rodents, and monkeys have demonstrated that an animal's previous experiences in aggressive encounters can sometimes be more important than the testosterone level in determining an individual's aggressiveness or dominance (Archer 1991; Gordon et al. 1979; Rejeski et al. 1988).

In humans, hormonal influences on behaviour are much less potent than in animals. Quantitative and qualitative behavioural differences in males and females are thought to result mainly from a combination of psychosocial factors which are the end product of differential experience and expectation produced by socialization. Moreover, hormonal influences on human behaviour are difficult to prove as pertinent research has to rely predominantly on correlational studies of endogenous hormone levels and behaviour which cannot ascertain hormonal *influences*. Only on very few occasions can scientists ethically manipulate hormone levels in humans to observe subsequent effects on brain and behaviour. But even from these studies on hormone substitution one cannot always draw firm conclusions regarding a particular hormonal-behavioural relation. If any, results from double-blind, placebo-controlled experimental designs can be interpreted as meaning that a particular hormone is the metabolic agent associated with behaviour. However, in behavioural endocrinology of humans, these are rare exceptions. Therefore, conclusions regarding hormonal effects on human behaviour have to be drawn with great care.

3.2 Sexuality

It is widely acknowledged that sexual behaviour in humans is multideter-
mined. Although no attempt will be made to deal with these issues here, it
should be pointed out that intrapsychic, social, somatic and cultural factors
can profoundly influence sexuality. The evidence presented here serves primarily to underline the contribution of sex hormones as a determinant of
sexual behaviour.

It has long been recognized that androgens play a critical role in human
male sexual behaviour. Prepubescent boys do not engage in sexual activity
outside the context of play. After puberty, when the testes begin to secrete
androgens, sex drive and the motivation to seek sexual contact become
powerful and are overtly expressed. Sexual performance and copulatory ability increase as well. The general pattern of age-dependent rise and decline of
androgen levels in men corresponds to average levels of male sexual activity
throughout the cycle of life. When blood levels of testosterone, especially
non SHBG-bound testosterone, diminish as men age, this mirrors their usually declining sexual interest and potency (Davidson et al. 1983). These observations suggest, but do not prove that male sexual behaviour is influenced by
androgens.

Less obvious and difficult to infer from everyday observation is the role of
testosterone in female sexual behaviour. Physiological testosterone levels in
women, which are one tenth of those in the normal male and to which males
are unresponsive, seemed to be negligible. Thus, the idea that androgens
could have enhancing effects on female sexual desire and arousal received little attention until synthezised testosterone was discovered to treat oophorectomized women.

3.2.1 Influence of testosterone on sexual behaviour in men

The physiological range of testosterone levels (3–12 ng/ml) is considerably
higher than that necessary to maintain normal sexual functions. Testosterone
levels found to be critical for sexual functions in males lie around 3 ng/ml
(Nieschlag 1979), and they show a clear intersubject variation. On the other
hand, levels at which a decline of androgen-related sexual behavior in individual subjects occurs appears to be reproducible (Gooren 1987).

Besides evidence from nonhuman primates and clinical case reports on effects of castration in human males (Nelson 1995), studies of hypogonadal
men on androgen replacement therapy provide convincing evidence of the
essential role of androgens in some aspects of male sexual behaviour (Table
3.1). In patients with induced or spontaneous hypogonadism either pathological withdrawal as well as reintroduction of exogenous androgens affected the
frequency of sexual phantasies, sexual arousal and desire, spontaneous erections during sleep and in the morning, ejaculation, sexual activities with and
without a partner, and orgasms through coitus or masturbation (Bancroft

Table 3.1. Significantly positive effects of androgens (testosterone, DHT) on various aspects of sexual behaviour in men

Behaviour	Endogenous testosterone/DHT level	Testosterone substitution
Sexual interest and phantasies	Nilsson et al. 1995	Anderson et al. 1992 Bancroft 1984 Carani et al. 1990 a Gooren 1987 O'Carroll and Bancroft 1984 Skakkebaek et al. 1981
Sexual arousal	–	Anderson et al. 1992 Bancroft 1984 Carani et al. 1990 a Gooren 1987 Su et al. 1993
Spontaneous erections (during sleep, in the morning)	Carani et al. 1992 Schiavi et al. 1988	Carani et al. 1990 a Luisi and Franchi 1980 Salmimies et al. 1982
Ejaculation	Schiavi et al. 1988	Gooren 1987 Salmimies et al. 1982 Skakkebaek et al. 1981
Sexual activities with partner	Schiavi et al. 1988	Carani et al. 1990 a
Orgasms in sexual activity (masturbation or coitus)	Knussmann et al. 1986 Mantzoros et al. 1995 Schiavi et al. 1988	Davidson et al. 1979

1984, 1986; Carani et al. 1990a, 1992; Davidson et al. 1979; Gooren 1987; Salmimies et al. 1982; Schiavi et al. 1988; Skakkebaek et al. 1981).

There is only limited evidence on the effects of testosterone adminstration to eugonadal men with or without sexual problems. In a controlled study of eugonadal men with diminished sexual desire O'Carroll and Bancroft (1984) produced a significant increase in sexual interest with injections of testosterone esters when compared to placebo injections. But in most of the men studied the increase in sexual interest was not translated into an improvement of their sexual relationship – perhaps because psychological problems with their partner had not been resolved with hormonal treatment only. When supraphysiological doses of testosterone used as potential hormonal male contraceptive agents were administered to healthy volunteers, this resulted in a significant increase in psychosexual stimulation or arousal during testosterone substitution. But there was no change in sexual activity or spontaneous erections (Anderson et al. 1992; Bagatell et al. 1994; Su et al. 1993).

As the healthy male produces much higher levels of androgens than necessary to maintain sexual function, lowering serum testosterone levels to the normal low range or increasing them to the high normal range in eugonadal men has no appreciable effect on sexual function (Buena et al. 1993). This led to the conclusion that androgens are only beneficial in those men whose endogenous levels are abnormally low. However, Bancroft (1984) pointed out that we cannot be certain on this point because with increasing levels of endogenous androgen supply it becomes more difficult to manipulate the circulating levels with exogenous hormones. The homeostatic mechanisms are powerful and the more testosterone is administered, the more the individual's own supply is suppressed or the metabolic clearance rate is increased. In a study by Benkert et al. (1979), who gave eugonadal men testosterone undecanoate daily to treat erectile dysfunction, no increase in circulating hormone levels was achieved. Their failure to produce any behavioural effect on erectile function therefore may not be due to ineffective androgens, but rather a result of their failure to alter hormone levels.

Indeed, in several studies a significant relation between physiological androgen levels and male sexual behaviour was observed. In a Swedish epidemiological investigation of 500 men aged 51 years low levels of non SHBG-bound testosterone were associated with low sexual interest (Nilsson et al. 1995). In young soldiers aged 18 to 22 years serum concentrations of 5α-dihydrotestosterone were a significant hormonal determinant of orgasmic frequency (Mantzoros et al. 1995). Knussmann et al. (1986) could ascertain in young healthy volunteers significantly positive correlations of salivary and total serum testosterone with the frequency of orgasms during 48 hours following the blood sampling. In their study, the majority of *intra*individual correlation coefficients (from 6 samples per subject) were also positive but some negative and insignificant ones were found as well. This finding points to the great *inter*individual variability of behavioural responses to hormones, and it could explain contradictory results from other pertinent studies on testosterone levels and frequency of orgasms (Buena et al. 1993; Kraemer et al. 1976; Persky et al. 1978; Raboch and Stárka 1972, 1973; Schwartz et al. 1980).

3.2.2 Influence of testosterone on sexual behaviour in women

A variety of models have been used to test the relationship between testosterone and sexuality in women. Because plasma testosterone levels peak around the time of ovulation (Ferin 1996), one investigational strategy has involved monitoring changes in several aspects of sexual behaviour at different points during the menstrual cycle. As plasma levels of estradiol also reach their highest point at the ovulatory phase, this research design makes it difficult to prove that testosterone alone induces the increase in sexual behaviour during the midcycle portion of the menstrual cycle observed in some studies (Adams et al. 1978; Dennerstein et al. 1994; Matteo and Rissman 1984). But several well-controlled correlational studies measuring circulating testosterone in women found evidence of an androgenic enhancement of sexual behaviour. Higher testosterone levels (midcycle peaks or average levels of plasma testosterone throughout the cycle) were associated with less sexual avoidance (Persky et al. 1982); more sexual gratification (Persky et al. 1978, 1982), sexual thoughts (Alexander and Sherwin 1993), and initiation of sexual activity (Morris et al. 1987); higher levels of sexual interest and desire (Alexander and Sherwin 1993; Alexander et al. 1990; Leiblum et al. 1983) and vasocongestive responses to erotic films (Schreiner-Engel et al. 1981); increased frequency of masturbation (Bancroft et al. 1983) and coitus (Morris et al. 1987); and a higher number of sexual partners (Cashdan 1995).

The positive relationship between testosterone and various measures of sexual interest and behaviour is intriguing; on the other hand, most studies failed to provide evidence of a peak of sexual behaviour at the time of the midcycle peak in testosterone. However, this is no argument against any testosterone-sexuality relationship in females, as an increase in testosterone does not have to produce an immediate behavioural response. For instance, the latency between androgen administration and increases in sexual desire in hypogonadal men ranges from days to several weeks.

The most powerful design for the study of the specificity of testosterone influence involves hormone replacement therapy in women who are oophorectomized. It is common clinical practise to treat these patients with estrogen replacement, but substitution of testosterone is also sensible as the women are deprived of ovarian androgen production as well. Some studies have shown – without contradictory evidence – that administration of testosterone, either alone or in addition to an estrogen replacement regimen, is more effective than estrogens alone or a placebo. In particular, an increase in sexual desire and phantasies was elicited, but also in sexual arousal and in coital or orgasmic frequency (Sherwin and Gelfand 1987; Sherwin et al. 1985).

Although there is converging evidence from correlational and experimental investigations that testosterone enhances male and female sexual behaviour, such as sexual desire or sexual phantasies, the underlying behavioural mechanism is not known. Testosterone might have direct effects on cognitive behaviour, e.g., influence the awareness of sexual cues (Alexander and Sherwin 1993), but it is also suggested that testosterone may act peripherally to enhance sexual pleasure and, thereby increase sexual desire (Davidson et al. 1982).

3.2.3 Influence of sexual behaviour on testosterone

The general concept that behaviour can feedback to hormone levels was first described with regard to sexual behaviour in an often cited publication (Anonymous 1970). A biologist working on an island attributed his increased beard growth immediately prior and during his visits to his girlfriend on the mainland to elevated androgen levels induced by sexual anticipation and sexual activity. Since then, numerous empirical studies dealt with effects of sexual behaviour (e.g., sexual stimulation, masturbation and coitus with or without orgasm) on testosterone levels. It could be demonstrated that almost any sexual behaviour can significantly alter sex hormone levels; however, cognitive factors and emotional involvement of the subjects produced mixed results. The majority of data on eugonadal men reports on effects of ejaculation. Orgasmic frequency in males, whether through masturbation or coitus,

correlated positively with free, non SHBG-bound testosterone and serum testosterone (Christiansen et al. 1984; Dabbs and Mohammed 1992; Knussmann
et al. 1986; Kraemer et al. 1976), while earlier investigations (Lee et al. 1974;
Monti et al. 1977; Stearns et al. 1973) could not ascertain behavioural-hor-
monal interactions. A significant rise in testosterone and DHT after masturbation was measured in blood samples of young males (Brown et al. 1978;
Monti et al. 1977; Purvis et al. 1976) while a case study by Fox et al. (1972)
found a testosterone increase in one male subject only after ejaculation during sexual intercourse but not after masturbation. This was explained by the
man's lack of emotional involvement with his autoerotic behaviour.

Endocrine effects of erotic stimulation were investigated by Christiansen
et al. (1984) who detected a significant increase of testosterone levels after
more or less accidental sexual stimulation through people, erotic pictures
and movies – not sexual activities – during 24 hours before the blood sampling. Even closer correlations were found in controlled laboratory experiments showing erotic or sexually neutral films (Carani et al. 1990b; Evans
and Distiller 1979; Hellhammer et al. 1985; Lincoln 1974; Pirke et al. 1974;
Stoléru et al. 1993).

Up to now, very little attention has been paid to behavioural-androgenic
effects in women. Only Dabbs and Mohammed (1992) measured salivary testosterone levels in women after sexual intercourse and detected a significant
increase of testosterone compared to a baseline value. Samples taken in the
evenings without preceding coitus did not show such an increase.

3.3 Stress

Activation of the hypothalamic-pituitary-adrenal axis and the subsequent release of cortisol is considered one of the major components of the physiological stress response in humans (Rose 1984). Stress responses of the pituitary-
gonadal axis are not as well known although their sensibility and specifity
are impressive.

Earliest studies investigated young military trainees in extremely stressful
situations during combat training (Kreuz et al. 1972; Rose et al. 1969) and
observed a significant decline of testosterone levels under psychologic and
somatic stress. Twenty years later, Opstad (1992) studied Norwegian military
cadets during five days of training involving strenuous exercise and almost
total deprivation of food and sleep. They confirmed the previous findings of
significantly decreased testosterone levels. Opstad attributed the hormonal response to extreme endurance training, sleep deficit and psychic stress (cf.
Christiansen et al. 1984; Cortés-Gallegos et al. 1983; Guezennec et al. 1994;
Singer and Zumoff 1992). In a similar study design, Bernton et al. (1995) investigated young male soldiers during eight weeks. They lived under extreme
psychosomatic stress with long exposure to rough environments, caloric deprivation, four hours of sleep per week and psychologic stressors including

constant risk of academic failure and the threat of simulated attack. The soldiers' testosterone levels decreased to clearly hypogonadal levels.

Even under less extreme conditions, psychosomatic stressors exert an influence on gonadal hormones: decreased concentrations of testosterone were found in males after surgery with anaesthesia (Carstensen et al. 1973; Matsumoto et al. 1970; Nakashima et al. 1975), driving heavygoods vehicles (Cullen et al. 1979), and routine flying missions in fighter type aircrafts (Leedy and Wilson 1985) as well as in female pilots flying a transport-plane (Dongyun and Yumin 1990).

Psychic stressors, e.g., financial difficulties, examinations, serious quarrels with other people, loss of close friends or relatives, dissatisfaction, boredom, or watching a movie with a stressful theme were generally followed by decreasing testosterone levels in males (Christiansen et al. 1985; Francis 1981; Hellhammer et al. 1985; Nilsson et al. 1995). Even anticipation of a stressful event, a final exam at the university, lead to a decrease of salivary testosterone in males and to an increase in females in the morning before the stressful situation (Christiansen and Hars 1995).

3.4 Physical exercise

A great deal of research on the effects of exercise upon the male and female reproductive system has taken place, and it has been demonstrated that some similarities, especially with regard to testosterone, exist between the sexes in the physiological outcomes of physical training when intrinsic gender differences in the endocrine system are acknowledged (Hackney et al. 1989; Shangold et al. 1984).

Despite the athletic appearance, male athletes have lower androgen levels than untrained men in a resting state. The results of comparative studies suggest significantly lower free and total testosterone concentations in chronically (several years) endurance-trained runners, weight lifters, rowers, cyclists, and swimmers (Arce et al. 1993; review in Arce and De Souza 1993; Hackney et al. 1988; Wheeler et al. 1984, 1991). In these studies, testosterone concentrations of trained subjects were only 60-85% of the age-matched untrained men.

With few exceptions (Elias et al. 1991, 1993; MacConnie et al. 1986) LH and FSH concentrations and pulsatile frequency-amplitude were unaffected by the training, even though testosterone was significantly reduced. Elias and Wilson (1993) emphasize that exercise effects on gonadotropic and gonadal hormones are independent of each other, and it is still uncertain what exact mechanisms cause the change in testosterone levels. Arce and De Souza (1993) attribute the decline in testosterone to alterations in hepatic and extrahepatic (muscles, skin) metabolism of testosterone (compare Cadoux-Hudson et al. 1985) which cannot be compensated by the athletes' gonads.

Acute effects of submaximal, prolonged (>60 min) exercise in marathon runners or cross-country skiers resemble hormonal changes found in endurance-trained men during resting state. After a 42-km marathon run or cross-country skiing over a distance of 75 km a highly significant decline in testosterone concentrations compared to pre-competition baselines was observed which could last as long as 5 days (Cook et al. 1986; Dessypris et al. 1976; Marinelli et al. 1994; Tanaka et al. 1986; Vasankari et al. 1993). After a long distance run (21 km or marathon), young sportswomen had significantly increased testosterone levels (Baker et al. 1982; De Crée et al. 1990; Hale et al. 1983). Even after a 30-min run significantly higher testosterone levels were observed (Shangold et al. 1981).

Regardless of the kind of sport, maximal or submaximal exercise (5–30 min) normally results in significant increases of testosterone levels in males independent of LH and FSH secretion. LH and FSH remain either relatively unchanged throughout the exercise (Sutton et al. 1973; Tegelman et al. 1988) or show an increase (Adlercreutz et al. 1976; Cumming et al. 1986; Kuoppasalmi et al. 1978).

As typical as the quick increase in testosterone concentrations is the rapid decline below baseline levels 15-60 minutes later (Adlercreutz et al. 1976; Elias et al. 1993). The reduction can last up to three days (Adlercreutz et al. 1976; Häkkinen and Pakarinen 1993; Kuoppasalmi et al. 1978; Tegelman et al. 1988) and its duration was found to depend on the intensity of the exercise (Häkkinen and Pakarinen 1993).

3.5 Aggression

For many decades, scientists have tried to capture and explain the phenomenon of human aggression and the observed sex differences (Eagly and Steffen 1986; Frodi et al. 1977; Gladue 1991a; Hyde 1984; Tieger 1980). Among biological factors, the endocrine system, especially testosterone, has been most intensively studied. A substantial body of data on subhuman primates has demonstrated a causative role of sex hormones in the development of sex-dimorphic aggressive behaviour (Archer 1988; Barfield 1984; Bernstein et al. 1983; Keverne 1979; Rose et al. 1971, 1975), and these results have focused attention on how endocrine activity and human aggression can interact with one another.

3.5.1 Prenatal hormones and aggression

Animal studies generally indicate that the presence of androgens in early life is important in establishing a biological readiness for future aggressive behaviour (Archer 1988). Much of the pertinent research on psychological effects of sex hormones in human studies consists of naturally occurring syndromes which result either from spontaneous endocrine excess or deficiency during fetal and early postnatal life (e.g., congenital adrenal hyperplasia) or prenatal sex hormone treatment of the pregnant mother. It is important to emphasize that true experimental design is precluded when studying potentially harmful hormone treatments in humans. Therefore, the researchers had to rely almost exclusively on these "experiments of nature" or existing clinical conditions to investigate how early exposure to hormones influences the potential for aggressive behavior in humans.

Overall, the results of pertinent studies show a slight effect of exposure to testosterone, progesterone with androgenic potential or diethylstilbestrol (a synthetic, nonsteroidal estrogen which exerts organizational effects similar to those of androgens converted to estrogens): an increase in physical aggressiveness and intense energy expenditure (e.g., vigorous play and athleticism) but not on verbal aggression during childhood years. The positive effects of early exposure to sex hormones are significant for both boys and girls, yet these influences seem subtle (Ehrhardt and Baker 1974; Ehrhardt and Meyer-Bahlburg 1981; Ehrhardt et al. 1989; Hines 1982; Jacklin et al. 1983; Reinisch 1981; Reinisch and Sanders 1984).

Exogenous estrogens, usually given in combination with progestagens and progestin-based progestagens can counteract endogenous androgenic effects and thus demasculinize certain aspects of aggressive behaviour in early childhood and adolescence (Ehrhardt et al. 1984; Meyer-Bahlburg and Ehrhardt 1982; Yalom et al. 1973; Zussman et al. 1975).

Archer (1991) explained the rather small or even insignificant effects of prenatal hormone excess with results obtained from animal studies. He suggests that organizing effects may only be detected clearly after puberty in the presence of adult sex hormone levels. In human studies the possibility that pubertal sex hormones are required to detect hormonal influences on behaviour remained untested except for the study by Yalom et al. (1973). He found stronger correlations of prenatal anti-androgen exposure to behaviour in adolescents after reaching puberty.

3.5.2 Adulthood testosterone levels

3.5.2.1 Aggressive behaviour

A number of studies provide evidence for an influence of circulating levels of androgens on aggression in males and females after puberty (Table 3.2). The hormonal effect is now referred to as *activational* as opposed to *organizational* effects in early life. In the beginning, several pertinent studies were carried out in prison where usually some of the inmates are highly aggressive. Here, the researchers expected to ascertain most likely a significant relationship between current testosterone levels and aggression and they actually did.

There is consistent data from five studies carried out on different types of violent male offenders who showed substantially higher testosterone levels than those found in selected samples of less violent prison inmates. Kreuz and Rose (1972) were the first to find that prisoners with a history of violent

Table 3.2. Endogenous testosterone and aggression in men and women

Nature of relationship between circulating testosterone and aggression	Studies based on behavioural measures	Studies based on self-report measures
Significant positive	Christiansen and Winkler 1992; Dabbs et al. 1987, 1988*, 1991; Ehlers et al. 1980*; Ehrenkranz et al. 1974; Elias et al. 1981; Gladue et al. 1989; Kedenburg 1977; Kreuz and Rose 1972; Lindman et al. 1987; Mattson et al. 1980; Olweus et al. 1980, 1988; Rada et al. 1976; Scaramella and Brown 1978	Christiansen and Knussmann 1987; Ehrenkranz et al. 1974; Gladue 1991b; Gray et al. 1991; Harris et al. 1996**; Houser 1979; Mattson et al. 1980; Olweus et al. 1980; Persky et al. 1971; Rada et al. 1983; Van Goozen et al. 1994*
Insignificant positive	Kedenburg 1977*; Lindman et al. 1992; Meyer-Bahlburg et al. 1974a, 1974b; Rada et al. 1983; Susman et al. 1987	Campbell et al. 1997; Dabbs et al. 1991; Doering et al. 1975; Meyer-Bahlburg et al. 1974b; Persky et al. 1977; Rada et al. 1976; Udry and Talbert 1988**
Insignificant negative	Susman et al. 1987*	Kreuz and Rose 1972; Monti et al. 1977; Persky et al. 1982*
Significant negative	–	Gladue 1991b*

Note: *These citations involved female subjects
**These citations involved female and male subjects

crime during adolescence showed higher testosterone levels than prisoners lacking such a history. Similar positive findings were reported by Ehrenkranz et al. (1974), Dabbs et al. (1987; 1991) as well as by Mattsson et al. (1980) who also found in their study of adolescent offenders that verbal aggression and impulsive behaviour in prison correlated significantly positively with testosterone levels.

The only study of female prison inmates (Dabbs et al. 1988) confirmed the results obtained from male offenders. Even the low testosterone concentrations in women, about 10 to 15% of circulating testosterone levels in men, seem to exert an influence on unprovoked violence in women.

With less severe offences the evidence for testosterone influence on aggressive behaviour is not as consistent. Neither Meyer-Bahlburg and his coworkers (1974a; 1974b), who investigated eight men with XYY-syndrome in comparison to normal XY-males nor Lindman et al. (1992) with their study on men who had been arrested for battering their wife under alcohol influence could find significantly higher testosterone levels in the study group in comparison to their controls, in the latter case non-violent pub patrons with a similar amount of alcohol consumption.

On the other hand, aggressive behaviour which was not a punishable offence also showed significant correlations with androgens in men and women. Under experimentally controlled alcohol intake, aggressively predisposed students were more dominant in a discussion and had higher free testosterone levels than nonaggressively predisposed students (Lindman et al. 1987). In male hockey players the pre-play testosterone levels correlated positvely with reactive aggression during the tournament (Scaramella and Brown 1978). Male patients in a clinic for nervous diseases showed more destructive aggression with higher levels of testosterone (Kedenburg 1977); in female patients he also detected a positive, however insignificant relation. The latter result was confirmed by Ehlers et al. (1980) who found that female outpatients of a neurobehavioural clinic who displayed more aggressive behaviour had significantly higher testosterone levels than women who were low in aggression.

In pubescent boys peer's and mothers' ratings of aggressive behaviour correlated positively with testosterone and several other androgens (Olweus et al. 1980, 1988; Susman et al. 1987).

In a group of traditionally living !Kung San hunter-gatherers (so-called bushmen) from the Kalahari desert in Namibia Christiansen and Winkler (1992) found that within a subgroup of physically aggressive San men violent behaviour correlated significantly positively with free testosterone and 5α-dihydrotestosterone (DHT) levels. As the physically aggressive men also exhibited higher mean values in body measurements of robustness of the face and trunk this finding may point to a possible pathway of *indirect* androgen action on human aggression, in addition to the widely accepted influence of testosterone on aggressive behaviour via its action on specific sites in the central nervous system.

3.5.2.2 Sexual aggression

Many aspects of sexual behaviour in the normal male are testosterone-dependent. With pathologically low serum testosterone levels a significant decrease in the frequency of sexual phantasies, sexual arousal and desire, spontaneous nocturnal or morning erections, ejaculations, sexual activity with and without a partner has been observed and also successfully treated with androgen replacement. Even among eugonadal men, some evidence for a positive relationship of endogenous testosterone with sexual behaviour has been found. This gave rise to the supposition that abnormally high androgen levels in men might elicit rape or other types of sexual aggression.

The first pertinent study was carried out on 52 sex offenders by Rada et al. (1976). A group of brutally violent rapists (according to clinical classification, police records and interviews) showed significantly higher levels of plasma testosterone than a combination of three other groups consisting of less overtly violent rapists, convicted child molesters and adult male volunteers. Taken together as one subgroup, rapists (both brutal and less violent) did not differ significantly from the child molesters in testosterone levels. Their mean serum testosterone values of 6.1 ng/ml (rapists) and 5.0 ng/ml (child molesters) both fell within the range of normal populations.

In a follow-up study Rada et al. (1983) failed to confirm differences in testosterone concentrations between a subsample of violent rapists and non-violent sex offenders or normal controls, while the group of child molesters had significantly lower testosterone levels than the rapists' group.

In a study of healthy normal young men Christiansen and Knussmann (1987a) found that interest in sexual aggression – assessed by measuring the viewing times of relevant slides – exhibited a low positive correlation with free testosterone and a significantly negative correlation with the hormone ratio DHT to testosterone. The correlation coefficients of free testosterone levels and interest in aggressive sexuality rose slightly as the aggressiveness illustrated in the slides increased.

3.5.2.3 Self-ratings of aggression

While the positive link between testosterone and past or present aggressive behaviour is fairly consistent, self-report measures of aggression, irritability, and hostility exhibit as many insignificant as significant relations with endogenous testosterone levels. Table 3.2 gives an overview of the distribution of significant and insignificant findings over the last twenty years, both for aggressive behaviour and self-ratings of aggression.

The inconsistent results for questionnaire data may be explained by several factors. First the selection of subjects and sample size: the age of the volunteers ranged between 10 to over 70 years; the sample sizes varied between 5 and 1709 subjects. Furthermore, great differences exist with regard to the number of serum or saliva samples collected for the hormone assays. In most studies only a single sample was used to determine the individual's sex hormone level; other investigators preferred to rely on up to 30 samples collected over a 2-month period in order to find *the* typical sex hormone level of a subject. Additional influence on the outcome of psychoendocrinological studies stems from the choice of the questionnaire. As the studies originate from countries all over the world, it was impossible always to use the same aggression inventory and thus quite diverse scales are supposed to represent the same personality trait, e.g., dominance, hostility or reactive aggression.

A positive relation of testosterone and questionnaire data of aggressiveness was found in both sexes from puberty to old age (Christiansen and Knussmann 1987a; Ehrenkranz et al. 1974; Gladue 1991b; Gray et al. 1991; Harris et al. 1996; Houser 1979; Mattsson et al. 1980; Olweus et al. 1980; Persky et al. 1971; Rada et al. 1983; Van Goozen et al. 1994a). Gladue (1991b) is the only one who found significantly negative correlations between testosterone and self-report measures of physical and verbal aggression in females – in contrast to the results of Van Goozen et al. (1994a) and Harris et al. (1996).

Several researchers could not detect any significant testosterone-aggressiveness relationship in men or women (Campbell et al. 1997; Dabbs et al. 1991; Doering et al. 1975; Kreuz and Rose 1972; Meyer-Bahlburg et al. 1974b;

Monti et al. 1977; Persky et al. 1977, 1982; Rada et al. 1976; Susman et al. 1987; Udry and Talbert 1988).

Thus it seems to be worth considering whether perhaps the relationship between testosterone and aggressiveness might be obscured by the interindividual variability in environmental, familial, and cognitive characteristics as well as personality traits that promote learning and emission of aggressive behaviour.

3.5.3 Testosterone administration

Correlational data reviewed in the previous chapters suggest that aggressive behaviour and presumably also aggressiveness in men and women are related to current, endogenous testosterone levels – but they do not prove a cause-effect relationship. In addition to studies on prenatal hormone treatment, research on the effects of testosterone intake in the adult female and male could possibly clarify the question whether aggression is actually testosterone-dependent.

Testosterone replacement therapy for hypogonadal males (Skakkebaek et al. 1981) or exogenous testosterone as potential hormonal male contraceptive (Anderson et al. 1992; Bagatell et al. 1994; Nieschlag 1992; Nieschlag and Behre in this volume) failed to show such an effect (Table 3.3). In a double-blind study by Björkqvist et al. (1994) male university students were given either testosterone, placebo, or no treatment for one week. After treatment the placebo group scored higher than both the control and testosterone group on self-evaluated anger, irritability, and impulsivity. Unfortunately, the authors of this well-controlled study did not assess aggressiveness with any of the standardized aggression questionnaires.

Table 3.3. Effects of exogenous testosterone on aggression in men and women

Nature of relationship between circulating testosterone and aggression	Studies based on behavioural measures	Studies based on self-report measures or interviews
Significant positive	van Goozen et al. 1995 a*	Brower et al. 1991; Hannan et al. 1991; Perry et al. 1990; Strauss et al. 1983, 1985* ; Van Goozen et al. 1991, 1994*, 1995 b
Insignificant positive	–	Anderson et al. 1992; Bagatell et al. 1994; Bahrke et al. 1990,1992; Nieschlag 1992; Skakkebek et al. 1981
Insignificant negative	–	Anderson et al. 1992; Björkqvist et al. 1994

Note: *These citations involved female subjects

With increasing anabolic-androgenic steroid abuse a new field of research opened up. Adverse behavioural effects such as increased irritability and aggressiveness have been reported in several field studies on men and women (Bahrke et al. 1992; Brower et al. 1991; Perry et al. 1990; Strauss et al. 1983, 1985). A double-blind study by Hannan et al. (1991) confirmed these findings. Increased aggressiveness (resentment, hostility and aggression subscales) occurred, even more so in high-dose anabolic steroid users. In a review of pertinent research Uzych (1992) concludes that the possibility of increased aggression after steroid abuse cannot be excluded. But to claim that aggression in anabolic-androgenic steroid users is only testosterone-dependent is too simple. Factors such as unstable personality may be the source of willingness to abuse steroids, as well as of aggressiveness. Bahrke et al. (1990) observed that irritability was slightly increased in many male steroid users but that only in a few, who were premorbid, might steroid use have been sufficient "to push them over the edge" and contribute to irrational or violent behaviour. Björkqvist et al. (1994) also conclude that steroid abuse may, for some, be a mediating factor enhancing aggressive tendencies by producing states of elated emotionality.

Further support of testosterone influence on aggression was published by van Goozen et al. (1994a, 1995a) and van Goozen et al. (1995b) who studied female-to-male and male-to-female transsexuals. After three months of cross-sex hormone treatment female-to-male transsexuals responded with more anger and aggression on a questionnaire describing hypothetical aversive situations. In male-to-female transsexuals anger and aggression proneness significantly decreased after androgen deprivation (van Goozen et al. 1995a; van Goozen et al. 1995b).

In order to understand the complexity of the relation between sex hormones and aggression one further aspect has to be considered: Testosterone and aggression seem to be mutually dependent. In addition to sex hormone influences on human aggression several studies have shown that assertive or aggressive behaviour followed by a rise in status leads to an increase in testosterone levels (Booth et al. 1989; Elias 1981; Gladue et al. 1989; Mazur and Lamb 1980; McCaul et al. 1992). Booth et al. (1989) also found that testosterone rose in tennis players 15 minutes before the next match – if the individual had won the previous match and probably anticipated winning again. Thus, the experience of winning and of a rise in status seemed to produce a rise in testosterone or to maintain an already elevated level, sustaining the winner's activation and readiness to enter subsequent competitions for higher status. Mazur's "Biosocial theory of status" (1985) incorporates these findings by hypothesizing a feedback loop between an individual's testosterone level and his or her assertiveness in attempting to achieve or maintain interpersonal status or dominance rank. This feedback loop may account for winning and losing "streaks" because each win reinforces a high testosterone level, which in turn reinforces further assertiveness or aggression.

This model of reciprocal effects between sex hormones and environment, although being more complex than simple hormone or experiential factors, does not fully explain the variation in aggressive behaviour between individuals. Much work still remains to be done before we will have reliable predictors of aggressive behaviour.

3.6 Mood

Depressive illness comprises a group of disorders that have different clinical symptoms and that respond to different treatment modalities. The symptoms of depression include reduced mood; low self-esteem; general fatigue; feelings of guilt; sleep and sex drive disturbances; absence of pleasure; agitated or retarded motor symptoms. Thus, depression covers a wide range of emotional and clinical states, and as a normal mood, depression is ubiquitous in human existence. It is an uncontradicted finding that depression is more prevalent in women than in men. This phenomenon could in part be explained by the hypothesis that sex hormones are involved in the etiology of some types of depression (Maggi and Perez 1985). The relationsship between cyclic hormonal changes and social behaviour has been examined in great detail (Bäckström 1992; Bäckström et al. 1983; Bains and Slade 1988; Bancroft 1993; Lewis 1990). Besides the contradictory, however significant roles of estradiol and progesterone, testosterone was found to be associated with higher levels of premenstrual dysphoria (Eriksson et al. 1992). Corresponding observations were made in female depressives who had higher testosterone levels than healthy controls (Hartmann et al. 1996; Vogel et al. 1978). In non-clinical studies on adolescent girls not suffering from depression either significantly positive (Paikoff et al. 1991) or insignificantly positive results (Susman et al. 1987) were found.

In males, the study of testosterone in depressive illness was prompted by the observation that in castrates and hypogonadal men treated with androgens the incidence of depressed mood and emotional instability appreciably decreased. Beach (1948) reported several cases of involutional melancholia in older men which were explained by a reduction of testicular hormones below the level necessary for mental stability. Improvement in mood and alertness was described in 65% of such cases when treated with testosterone propionate. However, more recent approaches to the psychoendocrinology of depression, using radioimmunoassays for determination of androgen levels in body fluids, were less successful in demonstrating anti-depressive properties of endogenous testosterone in male patients. Only Vogel et al. (1978) and Yesavage et al. (1985) found a significantly negative relation of testosterone with depression. The majority of clinical studies failed to ascertain significant effects of testosterone or testosterone production rate on depression (Levitt and Joffe 1988; Persky et al. 1968, 1971; Rubin et al. 1981, 1989; Sachar et al. 1973; Unden et al. 1988). Although some data suggested a dysfunction of the hypothalamo-pituitary-gonadal (HPG) axis, the majority of researchers concluded that there appears to be no major dysregulation of HPG axis activity in male endogenous depressives, even if dysregulation might be related to re-duced energy levels and sexual interest occurring in many depressive men (Angst 1983; Freedman and Carter 1982; Matussek 1980; Rubin et al. 1989; von Zerssen et al. 1984).

Parallel to clinical studies, research was extended to sex hormone levels and mood in healthy males with symptoms of severe depression. Besides nonconfirmatory evidence (Anderson et al. 1992; Persky et al. 1977; Susman et al. 1987), suprisingly, quite a number of significant, however controversial, relationships between mood and sex hormone status could be demonstrated, although hormonal and psychological variables lay within the normal range (Table 3.4). In a 2-month study of testosterone cycles and affective states among 20 healthy young males Doering et al. (1974) observed significant positive correlations between testosterone and self-ratings of depression. Houser (1979) tested five healthy men three times a week over a 10-week period and confirmed their findings: a significantly positve relationsship be-tween testosterone and self-ratings of depression and a negative correlation with elation. Christiansen and Knussmann (unpublished data) observed a corresponding trend in a large sample (n = 117) of unselected volunteers. Men with a high testosterone level had higher depression scores (p < 0,06) on a self-rating scale than low-testosterone males.

On the other hand, there also exists some evidence for an association of high testosterone levels with emotional well-being, as was found in early studies on testosterone substitution. A sample of 21 healthy young males was investigated and report was made of significantly negative correlations of salivary testosterone with depression and anxiety and a positive correlation with joyfulness (Hubert 1990; compare: Christiansen et al. 1984; Daitzman

Table 3.4. Testosterone and mood in the normal male or female

Study	Significant relationship of testosterone and self-ratings of mood
Doering et al. 1974	positive with depression
Houser 1979	positive with depression negative with elation
Daitzman and Zuckerman 1980	negative with depression
Christiansen et al. 1984	negative with anxiety
Diamond et al. 1989	negative with anxiety
Hubert 1990	negative with anxiety positive with joyfulness negative with depression
Paikoff et al. 1991*	negative with depression
Eriksson et al. 1992*	positive with premenstrual dysphoria
Christiansen and Knussmann (unpublished results)	positive with depression

Note: *These citations involved female subjects

and Zuckerman 1980; Diamond et al. 1989). Research in anabolic steroid abuse supported these findings. Pope and Katz (1988, 1989) contacted body building studios and offered members using steroids a cash payment to engage in confidential interviews about their steroid use. One third of 41 individuals reported to be manic or near manic. Most symptoms subsided when anabolic steroid use was discontinued (Pope and Katz 1989). Perry et al. (1990) investigated 20 competitive and noncompetitive weight lifters who consistently practised anbolic steroid abuse in cycles lasting 7 and 14 weeks. Accompanying the changes in physical parameters 70% of the men experienced becoming depressed more frequently while cycling. Clinical symptoms of depression were noted in 40% to 50% of the subjects, including low energy levels and excessive worrying.

According to Kashkin and Kleber (1989), anabolic steroids might have direct reward properties. Increasing numbers of reports on unexpected suicides in previously non-depressed young men who abruptly stopped using anabolic steroids have been noted. Thus, withdrawal symptoms manifested by anabolic steroid abusers and the symptoms of postpartum depression in women may result from a common underlying cause, dependence upon elevated steroid hormone levels (Nelson 1995).

3.7 Cognitive function

For many decades consistent sex differences in tests of cognitive abilities have been widely reported. During certain developmental stages – in particular during the first years of life and from puberty to early adulthood – girls surpass boys in several verbal skills. In contrast, males excel after about the tenth year and in adulthood at nonverbal skills, especially at spatial rotation and manipulation, at the related concept of field independence, and at mathematical reasoning (Halpern 1992; Hyde and Linn 1988; Hyde et al. 1990; Kimura 1996; Linn and Petersen 1985; Maccoby and Jacklin 1974; Masters and Sanders 1993; Wittig and Petersen 1979). In recent meta-analyses it was found that gender-differences in cognitive functioning have been decreasing since the seventies; however, they are still significant (Hyde and Linn 1988; Hyde et al. 1990; Linn and Petersen 1985; Stumpf and Klieme 1989).

Many sources of variance contribute to the gender differences which have been observed: culture, physical environment, socialization practices, X-chromosome linkage of spatial and partly of verbal skills as well as the degree of hemispheric lateralization of verbal and nonverbal processes (Blatter 1982; Hahn 1987; Vandenberg and Kuse 1979; Waber 1977). It is also generally accepted that sex hormones play a critical role in sex-typical cognitive abilities as well as in interindividual differences within the sexes.

3.7.1 Clinical studies and testosterone substitution

Evidence of a connection between sex hormones and spatial abilities came first from studies of phenotypic women with Turner's syndrome (X0 karyotype, no gonadal hormones) or testicular-feminization syndrome (XY karyotype; here the body tissues are unable to respond to normal levels of testosterone; see Chapter 2 by Quigley in this volume). These patients have female external genitalia, they are raised as girls, and develop a feminine gender identity. With regard to cognitive functioning such individuals' verbal skills surpass their spatial abilities, the typical pattern of cognitive abilities in women. Moreover, they are also on the average below healthy women in terms of their nonverbal performance but not of their verbal abilities (Collaer and Hines 1995; Dellantonio et al. 1984; Garron and van der Stoep 1969; Imperato-McGinley et al. 1991; Masica et al. 1969; Rovet and Netley 1979, 1982; Serra et al. 1978; Silbert et al. 1977). The strikingly poor nonverbal abilities are mainly explained by their low sex hormone levels.

Feminized prepubertal boys suffering from a kwashiorkor-induced endocrine dysfunction have been found to exhibit a more feminine cognitive style than a male control group. Dawson (1972) attributed this finding to a two-way relationship between sex hormones and socialization – because of their physical effeminateness, the boys tended to be raised into more feminine roles, including feminine cognitive behaviour.

Studies on men with idiopathic or aquired hypogonadotrophic hypogonadism appear to confirm the importance of testosterone for spatial abilities (Buchsbaum and Henkin 1980; Hier and Crowley 1982). A group of men with idiopathic hypogonadotrophic hypogonadism, and presumably a lifelong testosterone deficiency, performed significantly poorer than a group of men with late onset of pathologically reduced testosterone levels or normal controls on a number of spatial tests, but not on verbal tests. As short-term androgen therapy did not restore spatial function, these findings suggest that pre- and perinatal hormonal environments have lifelong effects on intellectual function in humans.

Although controversial, data from children with congenital adrenal hyperplasia (CAH) more or less support this hypothesis. Beginning in the third month of gestation, these patients are exposed to high levels of fetal androgens as a result of an enzymatic defect of the adrenal cortex. As a result, androgen production is greatly and continuously increased until treatment is initiated sometime in early post-natal life (Lauritzen 1987). While Lewis et al. (1968), Baker and Ehrhardt (1974), and McGuire et al. (1975) could not ascertain any effect of the prenatal exposure to higher-than-normal levels of androgens in CAH-children, Perlman (1973) and Resnick et al. (1986) found that CAH-girls were superior to their siblings or normal controls in spatial tasks. However, in boys with CAH the prenatal increase of androgens did not significantly influence their visual-spatial abilities (Resnick et al. 1986; Trautmann et al. 1995). This lack of androgen influence on cognition in boys is consistent with two studies measuring fetal or neonatal androgens. Neither

Table 3.5. Androgen treatment and cognitive functioning in men and women

Study	Sample	Cognitive task
Simonson et al. 1941	castrates, eunuchoids	critical flicker frequency [increase]
Simonson et al. 1944	older men	critical flicker frequency [increase]
Düker 1957	men in a state of mental exhaustion	arithmetical problem solving [increase]
Klaiber et al. 1971	college students	verbal and nonverbal repetitive tasks [increase]
Stenn et al. 1972	poorly androgenized male adolescents	verbal repetitive tasks [increase]
Sherwin 1988	oophorectomized women	short-term memory [increase] abstract reasoning [increase] perceptual speed [increase]
van Goozen et al. 1994 b	female-to-male transsexuals	visual-spatial skills [increase] verbal fluency [decrease]
Janowsky et al. 1994	older men	visual-spatial ability [increase]

[increase] increase, [decrease] decrease

Jacklin et al. (1988), who measured testosterone in umbilical cord blood obtained at birth, nor Finegan et al. (1992), who measured testosterone in second trimester amniotic fluid obtained via amniocentesis, detected a significant relationship to cognitive skills of boys at age six or age four.

Direct *manipulation of steroid hormones* supports the conclusion that androgens play a role in cognition. Studies on the effects of hormone treatment on cognitive functioning in adult males date back to 1941 and 1944 when Simonson and his co-workers published their experiments using methyl testosterone (Table 3.5). Administration of testosterone to castrated human males, eunuchoids, and older men improved their ability to perceive flicker (critical flicker frequency), a measure of attention and alertness, as long as the androgen treatment lasted. Düker (1957), using either testosterone, estradiol, or a combination of testosterone and estradiol produced a significant rise in concentration and speed solving simple arithmetical problems in a group of men with severe mental exhaustion. Similar results were reported by Stenn et al. (1972) who treated three poorly androgenized male adolescents. Intramuscular injections of testosterone enhanced their concentration and performance in a verbal fluency task.

A double-blind, cross-over, placebo-controlled study on substitution of sex steroids in oophorectomized women also demonstrated a causal link between testosterone and cognition (Sherwin 1988). Patients treated after surgery with either testosterone enanthate or a combination of testosterone enanthate, estradiol dienanthate, and estradiol benzoate showed stability both in sex hormone concentrations and in such aspects of cognitive performance which are typically the most prominent deficits in menopausal women, short- and long-term memory and perceptual speed.

Positive effects of testosterone treatment were also observed by Janowsky et al. (1994). When normal, aging men were given testosterone to enhance sexual functioning, as a side-effect they showed improved performance on visual-spatial tests. A similar result was obtained when female-to-male transsexuals received high doses of testosterone esters intramuscularly in preparation for sex therapy. Their spatial skills improved dramatically and their verbal fluency skills declined considerably within three months (van Goozen et al. 1994b). Due to ethical reasons, today manipulation of gonadal hormones is restricted to patients in clinical studies. Thus, it is over 25 years ago that Klaiber et al. (1971) tested effects of infused testosterone on mental performance in healthy male college students. After a 4-hour infusion of testosterone or a saline infusion in the control group, the post-infusion performances of the control group on a repetitive mental task had a significantly greater decline than in the testosterone-infused group. The result indicates that testosterone acted to prevent the typical decline from morning to afternoon in the performance of simple repetitive verbal or number tasks.

3.7.2 Endogenous testosterone levels

In the early seventies, determination of sex hormones using radioimmunoassays was established as a highly specific and reliable method. This enabled scientists to obtain quantitative data on circulating hormones in blood or saliva and to investigate the relationship of current sex steroids and cognitive abilities in normal, healthy subjects (Table 3.6). Komnenich et al. (1978) were the first to investigate healthy young women (n = 24) and men (n = 10). Four times within a month they measured the concentration of FSH, LH, testosterone, estradiol, and progesterone in plasma. On each of the days, simple repetitive tasks and a nonverbal test of field-independence were administered. Only the performance on the verbal tasks was positively related to estradiol in males. None of the other hormones exhibited a significant relation to cognitive performance in men or women.

In a relatively large sample of 43 men between 20 and 40 years Shute et al. (1983) detected a distribution of visual-spatial test scores as a function of androgen levels with the best-fitting third-order polynomial function describing the curve. Shute et al. reported that normal males selected for low plasma androgens were superior on certain spatial tests, while in their sample of 48 females the reverse was true, that is, highest-androgen females were superior to low-androgen women. However, due to the high cross-reactivity of the antibody used in their radioimmunoassays the authors speak of "general androgen" level instead of free, non SHBG-bound testosterone, which they originally intended to measure. Gouchie and Kimura (1991) found an effect similar to that of Shute and co-workers, using not the extremes of the group only, but a simple median split to divide all subjects on the basis of saliva testosterone levels in normal men and women. For one (paper folding test) of the two spatial tests there is a significant sex-by-hormone level interac-

Table 3.6. Serum and saliva androgens and cognitive abilities in normal males and females

Study	Cognitive task	Hormonal-cognitive relation
Komnenich et al. 1978	verbal visual-spatial visual-spatial visual-spatial	non significant (m) (f) curvilinear (m+f) non significant linear (m) positively linear (f)
Shute et al. 1983	visual spatial visual-spatial	curvilinear (m) positively linear (f)
Gordon and Lee 1986	visual-spatial	positively linear (m)
Christiansen and Knussmann 1987 b	visual-spatial field-independence verbal fluency	positively linear (m) positively linear (m) negatively linear (m)
McKeever and Deyo 1990	visual-spatial	positively linear (m)
Gouchie and Kimura 1991	visual-spatial (paper folding) visual-spatial (paper folding) visual-spatial + mathematical reasoning verbal perceptual speed	curvilinear (m+f) positively linear (f) curvilinear (m+f) non significant (m) (f) non significant (m) (f)
Tan and Akgün 1992	non-verbal	positively linear (m) non-significant (f)
Christiansen 1993	tactual-spatial field independence verbal fluency	positively linear (m) positively linear (m) negatively linear (m)

m = males; f = females; m+f = mixed group of males and females

tion, indicating that low levels of testosterone in males and high levels of testosterone in females are associated with superior performance. A composite score for tests on which males normally excel (paper folding, mathematical reasoning, mental rotations) shows a similar relation. On the other hand, multiple regression analyses of the male data alone revealed a trend in men for a positive linear relationship between saliva testosterone levels and visual spatial abilities (R = 0.29; p < 0.06). Gouchie and Kimura noted no evidence or consistent relationship between testosterone concentrations in men or women and their performance on perceptual speed tasks or vocabulary tests at which females usually excel over males.

The interpretation of such findings is complicated by the fact that testosterone may exert some of its effects through aromatization to estradiol in the brain. The suggestion has been made that it may in fact be the estrogen level which is related in a curvilinear fashion to spatial ability (Nyborg 1988). However, his hypothesis of a correlation between circulating serum estradiol and cognitive functioning could not be supported by the only two studies which, in addition to testosterone, measured serum estradiol (Christiansen 1993; Shute et al. 1983).

In contrast to Shute et al. (1983) and Gouchie and Kimura (1991) several studies have shown a significant linear testosterone-cognitive relationship.

Gordon and Lee (1986) investigated 32 men with four visual-spatial and four verbal tests and determined testosterone levels of their subjects. Testosterone concentrations correlated significantly positively with one spatial orientation task, but not with any of the other spatial or verbal tests. The study by Christiansen and Knussmann (1987b) attempted a broader investigation of the effects of androgens on spatial and nonspatial cognitive abilities in a larger sample of 117 men in their twenties. They collected blood and saliva samples to determine serum concentrations of testosterone, non SHBG-bound saliva testosterone, and 5α-dihydrotestosterone (DHT). The cognitive functioning was ascertained by 11 spatial and verbal ipsative test scores, reflecting intraindividual variance in the performance of these tasks, independent of the person's general level of achievement. The relationship between androgens and cognitive performance exhibited a clear pattern. All significant correlations between hormone values and verbal tests were negative. In contrast, significant correlations between androgens and spatial and field independence tests were all positive. Correspondingly, a more "masculine" cognitive pattern (superior skills on spatial tests as compared to verbal tests) was positively correlated to all three androgens. It should be noted that total serum testosterone clearly showed the greatest number of significant relations with verbal and spatial test scores. Three years later, McKeever and Deyo (1990) confirmed these findings of a positive correlation between androgen levels and spatial tasks with regard to DHT. Further evidence for a linear androgen-cognitive relationship come from Tan and Akgün (1992) who noted a positive correlation between nonverbal tasks and testosterone in a sample of Turkish university students. This was only the case for a subsample of right-handed men with right eye preference, but mixed-dominant males and young females showed no relationship.

As all the previous data were collected from individuals living in Western cultures, Christiansen (1993) tried to validate the findings in a non-Western group of healthy males. Althogether 256 !Kung San hunter-gatherers ("bushmen") and Kavango farmers from Namibia (Southern Africa) were investigated. They lived mainly on the subsistence level in their traditional lifestyle with a low degree of transition to Western culture. Testosterone, DHT, estradiol, and "free" salivary testosterone were determined. In order to make a sensible comparison of previous findings of hormone-related cognitive performance, spatial and verbal tests were the same or similar to those used in the study by Christiansen and Knussmann (1987b), however adapted for testing of illiterate subjects with no experience in paper and pencil tasks. The African data yielded the same hormonal cognitive pattern as was found in Western samples. Total and salivary testosterone showed the greatest number of significant relationships, a positive one with visual- or tactual-spatial tasks and a negative association with verbal tests.

The summary of sex hormone effects on cognitive abilities makes it reasonable to conclude that testosterone plays a role in cognitive functioning throughout life – from the prenatal period through adulthood till old age. But it has to be noted that explanations of inter- and intraindividual differences in cognitive abilities are complex and any causal model will have to recognize the reciprocal effects that environment and biology have on each other.

3.8 Key messages

- The interaction of testosterone with behaviour is bidirectional: testosterone can influence behaviour, and behaviour can alter testosterone levels.
- Testosterone affects brain development by organizing certain brain regions during fetal and neonatal life. At puberty, these brain structures and hence the behavioural repertoire are thought to be activated with increasing sex hormone concentrations. However, in humans behaviour is predominantly determined by intrapsychic, social and cultural factors; hormonal influences are less powerful than in animals.
- Male sexual behaviour is related to androgens. It becomes overt and powerful in puberty when the testes begin to secrete androgens.
- In hypogonadal males testosterone replacement provides convincing evidence of its role in some aspects of sexual behaviour. In eugonadal males testosterone substitution has only very limited influence.
- Endogenous testosterone levels in eugonadal men may be positively correlated to sexual interest and frequency of orgasms, but the findings are contradictory.
- In women, correlational and experimental studies show that testosterone is associated with an enhancement of sexual behaviour.
- Due to the endocrine effects of sexual stimulation and activities an increase in testosterone was observed in both sexes.
- Stress responses of the pituitary-gonadal axis show a remarkable sensibility. In males, testosterone levels decrease under psychosomatic and psychic stress, even under anticipation of stressful events, while testosterone concentrations in females rise.
- Compared to untrained men, well-trained athletes have lowered testosterone levels in a resting state.
- During and after prolonged submaximal exercise decreased testosterone levels in males and increased levels in females were measured. Acute effects of short exercise are a rise in testosterone concentrations, followed by a decline below baseline levels.
- Human aggression and endocrine activity are mutually dependant. Prenatal exposure to exogenous androgenic steroids results in slight increases of aggressive behaviour in boys and girls.
- Aggressive behaviour and self-report measures of aggression are predominantly positively correlated to endogenous testosterone levels although the aggressive act is generally removed in time from the hormone assessment.

- Testosterone replacement therapy or anabolic steroid abuse can result in increased aggressiveness, but not necessarily.
- Assertive or aggressive behaviour followed by a rise in status leads to a rise in testosterone levels.
- Androgens play a critical role in sex-typical cognitive functioning throughout life in normal men and women. Several studies have shown both linear and curvilinear effects of testosterone on visual-spatial abilities. Direct manipulation of testosterone supports the conclusion of its important role in cognition in females and males.
- The relation between mood and androgens is less clear. Early reports on castrates and hypogonadal men described an improvement in emotional stability following treatment with testosterone. However, neither clinical research on depressive men nor studies on mood and endogenous testosterone concentrations in normal males provided consistent results.

3.9 References

Adams DB, Gold AR, Burt AD (1978) Rise in female-initiated sexual activity at ovulation and its suppression by oral contraceptives. N Engl J Med 299:1145–1150

Adlercreutz H, Härkönen M, Kuoppasalmi K, Kosunen K, Näveri H, Rehunen S (1976) Physical activity and hormones. Adv Cardiol 18:144–157

Alexander GM, Sherwin BB (1993) Sex steroids, sexual behavior, and selection attention for erotic stimuli in women using oral contraceptives. Psychoneuroendocrinology 18:91–102

Alexander GM, Sherwin BB, Bancroft J, Davidson DW (1990) Testosterone and sexual behavior in oral contraceptive users and nonusers: A prospective study. Horm Behav 24:388–402

Anderson RA, Bancroft J, Wu FC (1992) The effect of exogenous testosterone on sexuality and mood of normal men. J Clin Endocrinol Metab 75:1505–1507

Angst J (1983) The origins of depression: Current concepts and approaches. Springer, Berlin

Anonymous (1970) Effects of sexual activity on beard growth in man. Nature 226:869–870

Arce JC, De Souza MJ (1993) Exercise and male factor infertility. Sports Med 15:146–169

Arce JC, De Souza MJ, Pescatello LS, Luciano AA (1993) Subclinical alterations in hormone and semen profile in athletes. Fertil Steril 59:398–404

Archer J (1988) The behavioral biology of aggression. Cambridge University Press, Cambridge

Archer J (1991) The influence of testosterone on human aggression. Brit J Psychol 82:1–28

Bäckström T (1992) Neuroendocrinology of premenstrual syndrome. Clin Obstet Gynecol 35:612–628

Bäckström T, Sanders D, Leask R, Davidson D, Warner P, Bancroft J (1983) Mood, sexuality, hormones, and the menstrual cycle. II. Hormone levels and their relationship to the premenstrual syndrome. Psychosom Med 45:503–507

Bagatell CJ, Heiman JR, Matsumoto AM, Rivier JE, Bremner WJ (1994) Metabolic and behavioral effects of high-dose, exogenous testosterone in healthy men. J Clin Endocrinol Metab 79:561–567

Bahrke MS, Wright JE, O'Connor JS, Strauss RH, Catlin DH (1990) Selected psychological characteristics of anabolic-androgenic steroid users. N Engl J Med 12:834–835

Bahrke MS, Wright JE, Strauss RH, Catlin DH (1992) Psychological moods and subjectively perceived bevavioral and somatic changes accompanying anabolic-androgenic steroid use. Am J Sports Med 20:717–724

Bains GK, Slade P (1988) Attributional patterns, moods, and the menstrual cycle. Psychosom Med 50:469–476

Baker ER, Mathur RS, Kirk RF (1982) Plasma gonadotropins, prolactin and steroid concentration in female runners immediately ofter a long distance run. Fertil Steril 38:38–41

Baker SW, Ehrhardt AA (1974) Prenatal androgen, intelligence, and cognitive sex differences. In: Friedman RC, Richart RM, Vande Wiele RL (eds) Sex differences in the brain. Wiley, New York, pp 53–76

Bancroft J (1984) Hormones and human sexual behavior. J Sex Mar Ther 10:3–21

Bancroft J (1986) Low sexual desire. In: Dennerstein L, Fraser I (eds) Hormones and behavior. Elsevier, Amsterdam, pp 396–405

Bancroft J (1993) The premenstrual syndrome – A reappraisal of the concept and the evidence. Psychol Med 24 (Suppl):1–47

Bancroft J, Sanders D, Davidson D, Warner P (1983) Mood, sexuality, and the menstrual cycle. III. Sexuality and the role of androgens. Psychosom Med 45:509–517

Barfield RJ (1984) Reproductive hormones and aggressive behavior. In: Flannelly KJ, Blanchard RJ, Blanchard DC (eds) Biological perspectives on aggression. Progress in clinical and biological research, Vol 169. Liss, New York, pp 105–134

Beach FA (1948) Hormones and behavior (2nd ed 1949) Hoeber, New York

Beatty WW (1979) Gonadal hormones and sex differences in nonreproductive behaviors in rodents: Organizational and activational influences. Horm Behav 12:112–163

Becker JB, Breedlove JM, Crews D (1992) Behavioral endocrinology. MIT Press, Cambridge

Benkert O, Witt W, Adam W, Leitz A (1979) Effects of testosterone undecanoate on sexual potency and the hypothalamic-pituitary-gonadal axis of impotent males. Arch Sex Behav 8:471–480

Bernstein I, Gordon TP, Rose RM (1983) The interaction of hormones, behavior and social context in nonhuman primates. In: Svare BB (ed) Hormones and aggressive behavior. Wiley, New York pp 535–561

Bernton E, Hoover D, Galloway R, Popp K (1995) Adaptation to chronic stress in military trainees. Ann NY Acad Sci 774:217–231

Bettini E, Pollio G, Santagati S, Maggi A (1992) Estrogen receptor in rat brain: Presence in the hippocampal formation. Neuroendocrinology 56:502–508

Bixo M, Bäckström T, Winblad B, Andersson A (1995) Estradiol and testosterone in specific regions of the human female brain in different endocrine states. J Steroid Biochem Mol Biol 55:297–303

Björkqvist K, Nygren T, Björklund A-C, Björkqvist S-E (1994) Testosterone intake and aggressiveness: Real effect or anticipation? Aggress Behav 20:17–26

Blatter P (1982) Sex differences in spatial ability: the X-linked gene theory. Percept Mot Skills 55:455–462

Booth A, Shelly G, Mazur A, Tharp G, Kittok R (1989) Testosterone, and winning and losing in human competition. Hormon Behav 23:556–571

Brain PF, Haug M (1992) Hormonal and neurochemical correlates of various forms of animal aggression. Psychoneuroendocrinology 17:537–551

Brower KJ, Blow FCX, Young YP, Hill EM (1991) Symptoms and correlates of anabolic-androgenic steroid dependence. Brit J Addict 86:759–768

Brown WA, Monti PM, Corriveau DP (1978) Serum testosterone and sexual activity and interest in men. Arch Sex Behav 7:97–103

Buchsbaum MS, Henkin RI (1980) Perceptual abnormalities in patients with chromatin negative gonadal dysgenesis and hypogonatropic hypogonadism. Int J Neurosci 11:201–209

Buena F, Swerdloff RS, Steiner BS, Lutchmansingh P, Peterson MA, Pandian MR, Galmarini M, Bhasin S (1993) Sexual function does not change when serum testosterone levels are pharmologically varied within the normal male range. Fertil Steril 59:1118–1123

Cadoux-Hudson TA, Few JD, Imms FJ (1985) The effects of exercise on the production and clearance of testosterone in well trained young men. Eur J Appl Physiol 54:321–325

Campbell A, Muncer S, Odber J (1997) Aggression and testosterone: Testing a bio-social model. Aggress Behav 23:229–238

Carani C, Zini D, Baldini A, Della Casa L, Ghizzani A, Marrama P (1990a). Effects on androgen treatment in impotent men with normal and low levels of free testosterone. Arch Sex Behav 19:223–234

Carani C, Bancroft J, Del Rio G, Granata ARM, Facchinetti F, Marrama P (1990b) The endocrine effects of visual erotic stimuli in normal men. Psychoneuroendocrinology 15:207–216

Carani C, Bancroft J, Granata A, Del Rio G, Marrama P (1992) Testosterone and erectile function, nocturnal penile tumescence and rigidity, and erectile response to visual erotic stimuli in hypogonadal and eugonadal men. Psychoneuroendocrinology 17:647–654

Carstensen, H, Amer I, Wide L, Amer B (1973) Plasma testosterone, LH and FSH during the first 24 hours after surgical operations. J Steroid Biochem 4:605–611

Casdan E (1995) Hormones, sex and status in women. Horm Behav 29:354–366

Chalepakis G, Schauer M, Slater EP, Beato M (1990) Gene regulation by steroid hormones. In: Alexis MN, Sekeris CE (eds) Activation of hormone and growth factor receptors. Kluver Academic Publishers, Dordrecht/Boston/London, pp 151–172

Christiansen K (1993) Sex-hormone related variations of cognitive performance in !Kung San hunter-gatherers of Namibia. Neuropsychobiology 27:97–107

Christiansen K (1998) Hypophysen-Gonaden-Achse Mann. In: Kirschbaum C, Hellhammer D (eds) Psychoneuroendokrinologie und Psychoimmunologie. Enzyklopädie der Psychologie, Biologische Psychologie. Hogrefe, Göttingen (in Druck)

Christiansen K, Hars O (1995) Effects of stress anticipation and stress coping strategies on salivary testosterone levels. J Psychophysiol 3:264

Christiansen K, Knussmann R (1987a) Androgen levels and components of aggression in men. Horm Behav 21:170–180

Christiansen K, Knussmann R (1987b) Sex hormones and cognitive functioning in men. Neuropsychobiology 18:27–36

Christiansen K, Knussmann R, Couwenbergs C (1984) Zusammenhänge zwischen Sexualhormonen des Mannes und Ernährung, Streß und Sexualverhalten. Homo 35:251–272

Christiansen K, Knußmann R, Couwenbergs C (1985) Sex hormones and stress in the human male. Horm and Behav 19:426–440

Christiansen K, Winkler EM (1992) Hormonal, anthropometrical, and behavioral correlates of physical aggression in !Kung San men of Namibia. Aggr Behav 18:271–280

Collaer ML, Hines M (1995) Human behavioral sex differences: A role for gonadal hormones during early development? Psychol Bull 118:55–107

Cook NJ, Read GF, Walker RF, Harris B, Riad-Fahmy D (1986) Changes in adrenal and testicular activity monitored by salivary sampling in males throughout marathon runs. Eur J Appl Physiol 55:634–638

Cortés-Gallegos V, Castaneda G, Alonso R, Sojo I, Carranco A, Cervantes C, Parra A (1983) Sleep deprivation reduces circulating androgens in healthy men. Arch Androl 10:33–37

Cullen J, Fuller R, Dolphin C (1979) Endocrine stress responses of drivers in a "real-life" heavygoods vehicle driving task. Psychoneuroendocrinology 4:107–115

Cumming DC, Brunsting LA, Strich G, Ries AL, Rebar RW (1986) Reproductive hormone increases in response to acute exercise in men. Med Sci Sports Exerc 18:369–373

Dabbs JM Jr, Mohammed S (1992) Male and female salivary testostrone concentrations before and after sexual activity. Physiol Behav 52:195–197

Dabbs JM Jr, Frady RL, Carr TS, Besch NF (1987) Saliva testosterone and criminal violence in young adult prison inmates. Psychosom Med 49:74–182

Dabbs JM Jr, Ruback RB, Frady R, Hopper CH, Sgoutas DS (1988) Saliva testosterin and criminal violence among women. Pers Indiv Differ 9:269–275

Dabbs JMJr, Jurkovic GJ, Frady RL (1991) Salivary testosterone and cortisol among late adolescent male offenders. J Abnorm Child Psychol 19:469–478

Daitzman R, Zuckerman M (1980) Desinhibitory sensation seeking, personality, and gonadal hormones. Pers Indiv Diff 1:103–110

Davidson JM, Camargo CA, Smith ER (1979) Effects of androgens on sexual behavior in hypogonadal men. J Clin Endocrinol Metab 48:955–958

Davidson JM, Kwan M, Greenleaf WJ (1982) Hormonal replacement and sexuality in men. Clin Endocrinol Metab 11:599–623

Davidson JM, Chen JJ, Crapo L, Gray GD, Greenleaf WJ, Catania JA (1983) Hormonal changes and sexual function in aging men. J Clin Endocrinol Metab 57:71–77

Dawson JML (1972) Effects of sex hormones on cognitive style in rats and men. Behav Genet 2:21–42

De Crée C, Lewin R, Ostyn M (1990) The monitoring of the menstrual status of female athletes by salivary steroid determination and ultrasonography. Eur J Appl Physiol 60:472–477

Dellantonio A, Lis A, Saviolo N, Rigon F, Tenconi R (1984) Spatial performance and hemispheric specialization in the Turner syndrome. Acta Med Auxol 16:193–203

Dennerstein L, Gotts G, Brown JB, Morse, CA, Farley TMM, Pinol A (1994) The relationship between the menstrual cycle and female sexual interest in women with PMS complaints and volunteers. Psychoneuroendocrinology 19:293–304

Dessypris A, Kuoppasalmi K, Adlercreutz H (1976) Plasma cortisol, testosterone, androstenedione and luteinizing hormone (LH) in a non-competitive Marathon run. J Steroid Biochem 7:33–37

Diamond P, Brisson GR, Caudas B, Péronnet F (1989) Trait anxiety, submaximal exercise and blood androgens. Eur J Applied Physiol Occup Physiol 58:699–704

Doering CH, Brodie HKH, Kraemer H, Becker H, Hamburg DA (1974) Plasma testosterone levels and psychologic measures in men over a 2-month period. In: Friedman RC, Richart RM, Vande Wiele RL, Stern LO (eds) Sex differences in behavior. Wiley, New York, pp 413–432

Doering CH, Brodie KH, Kraemer HC, Moos RH, Becker HB, Hamburg DA (1975) Negative affect and plasma testosterone: A longitudial human study. Psychosom Med 37:484–491

Dongyun S, Yumin W (1990) Flight influence on plasma levels of sex hormones of women pilots. Milit Med 155:262–264

Düker H (1957) Leistungsfähigkeit und Keimdrüsenhormone. Barth, München

Eagly AH, Steffen VJ (1986) Gender and aggressive behavior: A meta-analytic review of the social psychological literature. Psychol Bull 100:309–330

Ehlers CL, Rickler KC, Hovey JE (1980) A possible relationship between plasma testosterone and aggressive behavior in a female outpatient population. In: Girgis M, Kiloh LG (eds) Limbic epilepsy and dyscontrol syndrome. Elsevier/North Holland Biomedical Press, New York, pp 183–194

Ehrenkranz J, Bliss E, Sheard MH (1974) Plasma testosterone: Correlation with aggressive behavior and social dominance in man. J Psychosom Med 36:469–475

Ehrhardt AA, Baker SW (1974) Fetal androgens, human central nervous system differentiation and behavior sex differences. In: Friedman RC, Richart RM, Vande Wiele RL (eds) Sex differences in behavior. Wiley, New York, pp 33–51

Ehrhardt AA, Meyer-Bahlburg HFL (1981) Effects of prenatal sex hormones on gender related behavior. Science 211:1312–1318

Ehrhardt AA, Meyer-Bahlburg HFL, Feldman JF, Ince SE (1984) Sex-dimorphic behavior in childhood subsequent to prenatal exposure to exogenous progestagens and estrogens. Arch Sex Behav 13:457–478

Ehrhardt AA, Meyer-Bahlburg HFL, Rosen LR, Feldman JF, Veridiano NP, Elkin EJ, McEwen BS (1989) The development of gender-related behavior in females following prenatal exposure to diethylstilbestrol (DES). Horm Behav 23:526–541

Elias M (1981) Serum cortisol, testosterone, and testosterone-binding globulin responses to competitive fighting in human males. Aggress Behav 7:215–224

Elias AN, Wilson AF, Pandian MR, Chune G, James N, Stone SC (1991) CRH and gonadotropin secretion in physically active males after acute exercise. Eur J Appl Physiol 62:171–174

Elias AN, Wilson AF (1993) Exercise and gonadal function. Hum Reprod 8:1747–1761

Elias AN, Wilson AF, Pandian MR, Rojas FP, Kayalek R, Stone SC, James N (1993) Melatonin and gonadotropin after acute exercise in physically active males. Eur J Appl Physiol 66:357–361

Ellis L (1982) Developmental androgen fluctuations and the five dimensions of mammalian sex (with emphasis upon the behavioral dimension and the human species). Ethology Sociobiol 3:171–197

Eriksson E, Sundblad C, Lisjö P, Modigh K, Andersch B (1992) Serum levels of androgens are higher in women with premenstrual irritability and dysphoria than in controls. Psychoneuroendocrinology 17:195–204

Evans IM, Distiller LA (1979) Effects of luteinizing hormone-releasing hormone on sexual arousal in normal men. Arch Sex Behav 8:385–396

Ferin MJ (1996) The menstrual cycle: An integrative view. In: Adashi EY, Rock JA, Rosenwaks Z (eds) Reproductive endocrinology, surgery, and technology. Lippincott-Raven, Philadelphia, pp 103–121

Finegan JK, Niccols GA, Sitarenios G (1992) Relations between prenatal testosterone levels and cognitive abilities at age 4 years. Develop Psychol 28:1075–1089

Ford DH, Cramer EB (1982) The anatomical distribution of hormones in the central nervous system. In: Vernadakis A, Timiras PS (eds.) Hormones in development and aging. MTP Press, Lancaster, pp 151–180

Fox CA, Ismail AAA, Love DN, Kirkham KE, Loraine JA (1972) Studies on the relationship between plasma testosterone levels and human sexual activity. J Endocrinol 52:51–58

Francis KT (1981) The relationship between high and low trait psychological stress, serum testosterone, and serum cortisol. Experienta 37:1296–1297

Freedman R, Carter DB (1982) Neuroendocrine strategies in psychiatric research. In: Vernadakis A, Timiras PS (eds) Hormones in development and aging. MTP Press, Lancaster pp 619–636

Frodi A, Macaulay J, Thome PR (1977) Are women always less aggressive than men. A review of the experimental literature. Psychol Bull 84:634–660

Garron DC, Van der Stoep LP (1969) Personality and intelligence in Turner syndrome. A critical review. Arch Gen Psychiatry 21:339–346

Genazzani AR, Gastaldi M, Bidzinska B, Mercuri N, Genazzani AD, Nappi RE, Segre A, Petraglia F (1992) The brain as a target organ of gonadal steroids. Psychoneuroendocrinology 17:385–390

Gladue BA (1991 a) Qualitative and quantitative sex differences in self-reported aggressive behavioral characteristics. Psychol Reports 68:675–684

Gladue BA (1991 b) Aggressive behavioral characteristics, hormones, and sexual orientation in men and women. Aggress Behav 17:313–326

Gladue BA, Boechler M, McCaul KD (1989) Hormonal response to competition in human males. Aggress Behav 15:409–422

Gooren LJG (1987) Androgen levels and sex functions in testosterone-treated hypogonadal men. Arch Sex Behav 16:463–473

Gordon TP, Rose RM, Grady CL, Bernstein I (1979) Effects of increased testosterone secretion on the behavior of adult male rhesus monkeys living in a social group. Folia Primatol 32:149–160

Gordon HW, Lee PA (1986) A relationship between gonadotropins and visuospatial function. Neuropsychologia 24:563–576

Gouchie CT, Kimura D (1991) The relation between testosterone levels and cognitive ability patterns. Psychoneuroendocrinology 16:323–334

Gray A, Jackson DN, McKinlay JB (1991) The relation between dominance, anger, and hormones in normally aging men – Results from the Massachusetts male aging study. Psychosom Med 53:375–385

Guezennec CY, Satabin P, Legrand H, Bigard AX (1994) Physical performance and metabolic changes induced by combined prolonged exercise and different energy intakes in humans. Eur J Appl Physiol 68:525–530

Hackney AC (1989) Endurance training and testosterone. Sports Med 8:117–127

Hackney AC, Dolny DG, Ness RJ (1988) Comparison of resting reproductive hormonal profiles in select athletic groups. Biol Sport 4:200–204

Häkkinen K, Pakarinen A (1993) Acute hormonal responses to two different fatiguing heavy-resistance protocols in male athletes. J Appl Physiol 74:882–887

Hahn WK (1987) Cerebral lateralization of function: From infancy through childhood. Psychol Bull 101:376–392

Hale RW, Kosasa T, Krieger J (1983) A marathon: the immediate effect on female runners' luteinizing hormone, follicle-stimulating hormone, prolactin, testosterone, and cortisol levels. Am J Obstet Gynecol 146:550–554

Halpern DF (1992) Sex differences in cognitive abilities (2nd ed). Lawrence Erlbaum, Hillsdale (NJ)

Hannan CJ Jr, Friedl KE, Zold A, Kettler TM, Plymate SR (1991) Psychological and serum homovanillic acid changes in man administered androgenic steroids. Psychoendocrinology 16:335–343

Harris JA, Rushton JP, Hampson E, Jackson D (1996) Salivary testosterone and self-report aggressive and prosocial personality characteristics in men and women. Aggress Behav 22:321–331

Hartmann B, Baischer W, Koinig G, Albrecht A, Kirchengast S, Huber J, Langer G (1996) Disturbances in the hypothalamic-pituitary-gonadal axis of depressed fertile women compared to normal controls. In: Genazzani AR, Petraglia F, D'Ambrogio G, Genazzani AD, Artini PG (eds) Recent developments in gynecology and obstetrics. Parthenon, New York, pp 225–230

Hellhammer DH, Hubert W, Schürmeyer T (1985) Changes in saliva testosterone after psychological stimulation in men. Psychoneuroendocrinology 10:77–81

Hier DB, Crowley WF (1982) Spatial ability in androgen-deficient men. New Engl J Med 306:1202–1205

Hines M (1982) Prenatal gonadal hormones and sex differences in human behavior. Psychol Bull 92:56–80

Houser BB (1979) An investigation of the correlation between hormonal levels in males and mood, behavior and physical discomfort. Horm Behav 12:185–197

Hubert W (1990) Psychotropic effects of testosterone. In: Nieschlag E, Behre HM (eds) Testosterone: Action, deficiency, substitution. Springer, Berlin pp 51–71

Hutchinson JB (1991) Hormonal control of behaviour: Steriod action in the brain. Curr Opin Neurobiol 1:562–570

Hutchinson JB (1993) Sex hormone action on brain mechanisms of aggression. Aggress Behav 19:5

Hutchinson JB, Steimer T (1984) Androgen metabolism in the brain: Behavioural correlates. In: de Vries GJ, de Bruin J, Corner M (eds) Sex differences in the brain. Progress in brain research, Vol 61. Elsevier, Amsterdam, pp 23–51

Hyde JS (1984) How large are gender differences in aggression? A developmental meta-analysis. Develop Psychol 20:722–734

Hyde JS, Linn MC (1988) Gender differences in verbal ability: A meta-analysis. Psychol Bull 104:53–69

Hyde JS, Fennema E, Lamon SJ (1990) Gender differences in mathematics performance: A meta-analysis. Psychol Bull 107:139–155

Imperato-McGinley J, Pichardo M, Gautier T, Voyer D, Bryden MP (1991) Cognitive abilities in androgen-insensitive subjects: Comparison with control males and females from the same kindred. Clin Endocrinol 34:341–347

Jacklin CN, Maccoby EE, Doering CH (1983) Neonatal sex-steroid hormones and timidity in 6–18 month-old boys and girls. Dev Psychobiol 16:163–168

Jacklin CN, Wilcox KT, Maccoby EE (1988) Neonatal sex-steroid hormones and cognitive abilities at six years. Dev Psychobiol 21:567–574

Janowsky JS, Oviatt SK, Orwoll ES (1994) Testosterone influences spatial cognition in older men. Behav Neurosci 108:325–332

Kashkin KB, Kleber HD (1989) Hooked on hormones? An anabolic steroid addiction hypothesis. J Am Med Assn 262:3166–3170

Kedenburg HD (1977) Androgens and aggressive behavior in man. University Microfilms International, Ann Arbor (Michigan)

Keverne EB (1979) Sexual and aggressive behaviour in social groups of talapoin monkeys. In: Ciba Foundation Symposium 62, Sex hormones and behaviour. Excerpta Medica, Amsterdam, pp. 271–297

Kimura D (1996) Sex, sexual orientation and hormones influence human cognitive function. Curr Opin Neurobiol 6:259–263

Klaiber EL, Broverman DM, Vogel W, Abraham GE, Cone FL (1971) Effects of infused testosterone on mental performances and serum LH. J Clin Endocrinol Metab 32:341–349

Knussmann R, Christiansen K, Couwenbergs C (1986) Relations between sex hormone levels and sexual behavior in men. Arch Sex Behav 15:429–445

Komnenich P, Lane DM, Dickey RP, Stone SC (1978) Gonadal hormones and cognitive performance. Physiol Psychol 6:115–120

Kraemer HC, Becker HB, Brodie HKH, Doering CH, Moos RH, Hamburg DA (1976) Orgasmic frequency and plasma testosterone levels in normal human males. Arch Sex Behav 5:125–132

Kreuz LE, Rose RM (1972) Assessment of aggressive behavior and plasma testosterone in a young criminal population. J Psychosom Med 34:321–332

Kreuz LE, Rose RM, Jennings JR (1972) Suppression of plasma testosterone levels and psychological stress. Arch Gen Psychiatry 26:479–482

Kuoppasalmi K, Näveri H, Rehunen S, Härkönen M (1978) Effect of strenous anaerobic running exercise on plasma growth hormone, cortisol, LH, testosterone, androstenedione, estrone and estradiol. J Steroid Biochem 7:2029–2034

Lauritzen C (1987) Intersexualität. In: Wulf K-H, Schmidt-Matthiesen (eds) Gynäkologische Endokrinologie. Urban & Schwarzenberg, München, pp 35–94

Lee PA, Jaffe RB, Midgley AR Jr (1974) Lack of alteration of serum gonadotropins in men and women following sexual intercourse. Am J Obstet Gynecol 120:985–987

Leedy MG, Wilson MS (1985) Testosterone and cortisol levels in crewmen of U.S. Air Force fighter and cargo planes. Psychosom Med 47:333–338

Leiblum S, Bachmann G, Kemmann E, Colburn D, Schwartzman I (1983) Vaginal atrophy in the postmenopausal woman: The importance of sexual activity and hormones. J Am Med Assn 249:2195–2198

Levitt AJ, Joffe RT (1988) Total and free testosterione in depressed men. Acta Psychiatr Scand 77:346–348

Lewis JW (1990) Premenstrual syndrome as a criminal defense. Arch Sex Behav 19:425–441

Lewis VG, Money J, Epstein R (1968) Concordance of verbal and nonverbal ability in the adrenogenital syndrome. John Hopkins Med J 122:192–195

Lincoln GA (1974) Luteinizing hormone and testosterone in man. Nature 252:232–233

Lindman R, Järvinen P, Vidjeskog J (1987) Verbal interactions of aggressively and nonaggressively predisposed males in a drinking situation. Aggress Behav 13:187–196

Lindman R, von der Pahlen B, Öst B, Eriksson CJP (1992) Serum testosterone, cortisol, glucose, and ethanol in males arrested for spouse abuse. Aggress Behav 18:393–400

Linn MC, Petersen AC (1985) Emergence and characterization of sex differences in spatial ability: A meta-analysis. Child Develop 56:1479–1498

Luisi M, Franchi F (1980) Double-blind group comparative study of testosterone undecanoate and mesterolone in hypgonadal male patients. J Endocrinol Invest 3:305–308

Maccoby EE, Jacklin CN (1974) The psychology of sex differences. Stanford University Press, Stanford (CA)

MacConnie SE, Barkan A, Lampman RM, Schork MA, Beitins IZ (1986) Decreased hypothalamic gonadotropin-releasing hormone in male marathon runners. New Engl J Med 315:411–417

Maggi A, Perez J (1985) Role of female gonadal hormones in the CNS: Clinical and experimental aspects. Life Sci 37:893–906

Mantzoros CS, Georgiadis EI, Trichopoulos D (1995) Contribution of dihydrotestosterone to male sexual behaviour. Brit Med J 310:1289–1291

Marinelli M, Roi GS, Giacometti M, Bonini P, Banti G (1994) Cortisol, testosterone, and free testosterone in athletes performing a marathon in 4000 m altitude. Horm Res 41:225–229

Masica DN, Money J, Ehrhardt AA, Lewis VG (1969) IQ, fetal sex hormones and cognitive patterns: Studies in the testicular feminizing syndrome of androgen insensitivity. John Hopkins Med J 124:33–43

Masters MS, Sanders B (1993) Is the gender difference in mental rotation disappearing? Behav Genet 23:337–341

Matsumoto K, Takeyasu K, Mizutani S, Hamanaka Y, Uozomi T (1970) Plasma testosterone levels following surgical stress in male patients. Acta Endocrinol 65:11–17

Matteo S, Rissman EF (1984) Increased sexual activity during the midcycle portion of the human menstrual cycle. Horm Behav 18:249–255

Mattsson A, Schalling D, Olweus D, Löw H, Svensson J (1980) Plasma testosterone, aggressive behavior, and personality dimensions in young male delinquents. J Am Acad Child Psychiatry 19:476–490

Matussek N (1980) Zur Biochemie und Neuroendokrinologie der Depression. In: Heimann H, Giedke H (eds) Neue Perspektiven in der Depressionsforschung. Huber, Bern, pp 64–72

Mazur A (1985) A biosocial model of status in face-to-face primate groups. Soc Forces 64:377–402

Mazur A, Lamb TA (1980) Testosterone, status, and mood in human males. Horm Behav 14:236–246

McCaul K, Gladue B, Joppa M (1992) Winning, losing, mood, and testosterone. Horm Behav 26:486–505

McEwen BS (1992) Steroid hormones: Effect on brain development and function. Horm Res 37, Suppl 3:1–10

McEwen BS, Biegon A, Fischette CT, Luine VN, Parsons B, Rainbow TC (1984) Toward a neurochemical basis of steroid hormone action. In: Martini L, Ganong WF (eds) Frontiers in neuroendocrinology, Vol 8. Raven Press, New York, pp 153–176

McGuire LS, Ryan KO, Omenn GS (1975) Congenital adrenal hyperplasia: II. Cognitive and behavioral studies. Behav Genet 5:175–188

McKeever WF, Deyo RA (1990) Testosterone, dihydrotestosterone, and spatial task performances of males. Bull Psychonomic Soc 28:305–308

Meyer-Bahlburg HFL, Ehrhardt AA (1982) Prenatal sex hormones and human aggression: A review and new data on progestagen effects. Aggress Behav 8:39–62

Meyer-Bahlburg HFL, Boon DA, Sharma M, Edwards JA (1974a) Aggression und Androgene bei männlichen Individuen. Bericht über den 28. Kongreß der DGP, Bd. 4, Klinische Psychologie. Hogrefe, Göttingen pp 57–65

Meyer-Bahlburg HFL, Boon DA, Sharma M, Edwards JA (1974b) Aggressiveness and testosterone measures in man. J Psychosom Med 36:269–273

Michael RP, Bonsall RW (1990) Androgens, the brain and behavior in male primates. In: Balthazart J (ed) Hormones, brain and behavior in vertebrates, 2. Karger, Basel, pp 15–26

Monti P, Brown W, Corriveau D (1977) Testosterone and components of aggressive and sexual behavior in man. Am J Psychiatry 134:692–694

Morris MJ, Udry JR, Kahn-Dawood F, Dawood MY (1987) Marital sex frequency and midcycle female testosterone. Arch Sex Behav 16:27–37

Naftolin F, Perez J, Lerauth CS, Redmond DE, Garcia-Segura LM (1990) African green monkeys have sexually dimorphic and estrogen-sensitive hypothalamic neuronal membranes. Brain Res Bull 25:575–579

Nakashima A, Koshiyama K, Uozumi T, Monden J, Hamanaka Y, Kurachi K, Aono T, Mizutani S, Matsumoto K (1975) Effects of general anaesthesia and severity of surgical stress on serum LH and testosterone in males. Acta Endocrinol 78:258–269

Nelson RJ (1995) An introduction to behavioral endocrinology. Sinauer, Sunderland (Mass)

Nieschlag E (1979) The endocrine function of the human testis in regard to sexuality. In: Ciba Foundation Symposium: Sex, hormones and behavior. Experta Medica, Amsterdam, pp 182–208

Nieschlag E (1992) Testosteron, Anabolika und aggressives Verhalten bei Männern. Dtsch Ärzteblatt 89:1663–1666

Nilsson P, Moller L, Solstad K (1995) Adverse effects of psychosocial stress on gonadal function and insulin levels in middle-aged males. J Internal Med 237:479–486

Nyborg H (1988) Mathematics, sex hormones, and brain function. Behav Brain Sci 11:206–207

O'Carrol R, Bancroft J (1984) Testosterone therapy for low sexual interest and erectile dysfunction in men: A controlled study. Brit J Psychiatry 145:146–151

Olweus D, Mattsson A, Schalling D, Löw H (1980) Testosterone, aggression, physical and personality dimensions in normal adolescent males. Psychosom Med 42:253–269

Olweus D, Mattsson A, Schalling D, Löw H (1988). Circulating testosterone levels and aggression in adolescent males: A causal analysis. Psychosom Med 50:261–272

Opstad PK (1992) The hypothalamo-pituitary regulation of androgen secretion in young men after prolonged physical stress combined with energy and sleep deprivation. Acta Endocrinol 127:231–236

Paikoff RL, Brooks-Gunn J, Waren MP (1991) Effects of girls' hormonal status on depressive and aggressive symptoms over the course of one year. J Youth Adolescence 20:191–215

Perlman SM (1973) Cognitive abilities of children with hormone abnormalities: Screening by psychoeducational tests. J Lear Disabil 6:21–29

Perry PJ, Andersen KH, Yates WR (1990) Illicit anabolic steroid use in athletes. A case series analysis. Am J Sports Med 18:422–428

Persky H, Zuckerman M, Curtis GC (1968) Endocrine function in emotianally disturbed and normal men. J Nerv Ment Dis 146:488–497

Persky H, Smith KD, Basu GK (1971) Relation of psychological measures of aggression and hostility to testosterone production in man. Psychosom Med 33:265–277

Persky H, O'Brien C, Fine E, Howard W, Khan M, Beck R (1977) The effects of alcohol and smoking on testosterone function and aggression in chronic alcoholics. Am J Psychiatry 134:621–625

Persky H, Lief HI, Strauss D, Miller WR, O'Brien CP (1978) Plasma testosterone level and sexual behavior of couples. Arch Sex Behav 7:157–173

Persky H, Dreisbach L, Miller WR, O'Brien CP, Khan MA, Lief HI, Charney N, Strauss D (1982) The relation of plasma androgen levels to sexual behavior and attitudes of women. Psychosom Med 44:305–319

Pirke KM, Kockott G, Dittmar F (1974) Psychosexual stimulation and plasma testosterone in man. Arch Sex Behav 3:577–584

Pope HG, Katz DL (1988) Affective and psychotic symptoms associated with anabolic steroid use. Am J Psychiatry 145:487–490

Pope HG, Katz DL (1989) Homicide and near-homicide by anabolic steroid users. J Clin Psychiatry 51:28–31

Purvis K, Landgren B, Cekan Z, Diczfalusy E (1976) Endocrine effects of masturbation. J Endocrinol 70:439–444

Raboch J, Stárka L (1972) Coital activity of men and the levels of plasmatic testosterone. J Sex Res 8:219–224

Raboch J, Stárka L (1973) Reported coital activity of men and levels of plasma testosterone. Arch Sex Behav 2:309–316

Rada RT, Laws DR, Kellner R (1976) Plasma testosterone levels in the rapist. Psychosom Med 38:257–268

Rada RT, Laws DR, Kellner R, Stivastava L, Peake G (1983) Plasma androgens in violent and nonviolent sex offenders. Bull Am Acad Psychiatry Law 11:149–158

Reinisch JM (1981) Prenatal exposure to synthetic progestins increases potential for aggression in humans. Science 211:1171–1173

Reinisch JM, Sanders SA (1984) Prenatal gonadal steroidal influences on gender-related behavior. In: De Vries GH, de Bruin J, Corner M (eds) Sex differences in the brain. Progress in brain research, Vol 61. Elsevier, Amsterdam, pp 407–416

Rejeski WP, Brubaker PH, Herb RA, Kaplan JR, Korituik D (1988) Anabolic steroids and aggressive behavior in cynomolgus monkeys. J Behav Med 11:95–105

Resnick SM, Berenbaum SA, Gottesman II, Bouchard TJ Jr (1986) Early hormonal influences on cognitive functioning in congenital adrenal hyperplasia. Develop Psychol 22:191–198

Rose RM (1984) Overview of endocrinology of stress. In: Brown GM, Koslow SH, Reichlin S (eds) Neuroendocrinology and psychiatric disorder. Raven Press, New York, pp 95–122

Rose RM, Bourne PG, Poe RO, Mougey EH, Collins DR, Mason JW (1969) Androgen responses to stress II. Excretion of testosterone, epitestosterone, androsterone and eticholanolone during basic combat training and under threat of attack. Psychosom Med 31:418–436

Rose RM, Holaday JW, Bernstein IS (1971) Plasma testosterone, dominance rank and aggressive behaviour in male rhesus monkeys. Nature 231:366–368

Rose RM, Bernstein IS, Gordon TP (1975) Consequences of social conflict on plasma testosterone levels in rhesus monkeys. Psychosom Med 37:50–60

Rovet J, Netley C (1979) Phenotypic vs. genotypic sex and cognitive abilities. Behav Genet 9: 317–322

Rovet J, Netley C (1982) Processing deficits in Turner's syndrome. Develop Psychol 18:77–94

Rubin RT, Poland RE, Tower BB, Hart PA, Blodgett ALN, Forster B (1981) Hypothalamo-pituitary-gonadal function in primary endogenously depressed men: Preliminary findings. In: Fuxe K, Gustafsson JA, Wetterberg L (eds) Steroid hormone regulation of the brain. Pergamon, Oxford, pp 387–396

Rubin RT, Poland RE, Lesser IM (1989) Neuroendocrine aspects of primary endogenous depression. Pituitary-gonadal axis activity in male patients and matched control subjects. Psychoneuroendocrinology 14:217–229

Sachar EJ, Halpern F, Rosenfeld RS, Gallagher TF, Hellman L (1973) Plasma and urinary testosterone levels in depressed men. Arch Gen Psychiatry 28:15–18

Salmimies P, Kockott G, Pirke KM, Vogt HJ, Shill WB (1982) Effects of testosterone replacement on sexual behavior in hypogonadal men. Arch Sex Behav 11:345–353

Scaramella TJ, Brown WA (1978). Serum testosterone and aggressiveness in hockey players. Psychosom Med 40:262–265

Schiavi RC, Schreiner-Engel P, White D, Mandeli J (1988) Pituitary – gonadal function during sleep in men with hypoactive sexual desire and in normal controls. Psychosom Med 50:304–318

Schreiner-Engel P, Schiavi RC, Smith H, White D (1981) Sexual arousability and the menstrual cycle. Psychosom Med 43:199–214

Schulkin J (1993) Hormonally induced changes in mind and brain. Academic Press, San Diego

Schwartz MF, Kolodny RC, Masters WH (1980) Plasma testosterone levels of sexually functional and dysfunctional men. Arch Sex Behav 9:355–366

Sekeris CE (1990) The mitochondrial genome: A possible primary site of action of steroid hormones. In Vivo 4:317–320

Serra A, Pizzamiglio L, Boari A, Spera S (1978) A comparative study of cognitive traits in human sex chromosome aneuploids and sterile and fertile euploids. Behav Genet 8:143–154

Shangold MM (1984) Exercise and the adult female: Hormonal and endocrine effects. In: Terjung R (ed) Exercise and sports science reviews, Vol 12. Collamore Press, Lexington (MA), pp 53–79

Shangold MM, Gatz ML, Thysen B (1981) Acute effects of exercise on plasma concentrations of prolactin and testosterone in recreational women runners. Fertil Steril 35:699–702

Sherwin BB (1988) Estrogen and/or androgen replacement therapy and cognitive functioning in surgically menopausal women. Psychoneuroendocrinology 13:345–357

Sherwin BB, Gelfand MM (1987) The role of androgen in the maintenance of sexual functioning in oophorectomized women. Psychosom Med 49:397–409

Sherwin BB, Gelfand MM, Brender W (1985) Androgen enhances sexual motivation in females: A prospective, crossover study of sex steroid administration in the surgical menopause. Psychosom Med 47:339–351

Shute VJ, Pellegrino JW, Hubert L, Reynolds RW (1983) The relationship between androgen levels and human spatial abilities. Bull Psychonomic Soc 21:465–468

Silbert AR, Wolff PH, Lilienthal J (1977) Spatial and temporal processing in patients with Turner's syndrome. Behav Genet 7:11–21

Simon NG, Whalen RE (1987) Sexual differentiation of androgen-sensitive and estrogen-sensitive regulatory systems for aggressive behavior. Horm Behav 21:493–500

Simonson E, Kearns WM, Enzer N (1941) Effect of oral administration of methyltestosterone on fatigue in eunuchoids and castrates. Endocrinology 28:506–512

Simonson E, Kearns WM, Enzer N (1944) Effect of methyl testosterone treatment on muscular performance and the central nervous system of older men. J Clin Endocrinol 4:528–534

Singer F, Zumoff B (1992) Subnormal serum testosterone levels in male internal medicine residents. Steroids 57:86–89

Skakkebaek N, Bancroft J, Davidson D, Warner P (1981) Androgen replacement with oral testosterone undecanoate in hypogonadal men: A double blind controlled study. Clin Endocrinol 14:49–61

Stearns EL, Winter JSD, Faiman C (1973) Effects of coitus on gonadotropin, prolactin and sex steroid levels in man. J Clin Endocrinol Metab 37:687–691

Stenn PO, Klaiber EL, Vogel W, Broverman DM (1972) Testosterone effects on photic stimulation of the EEG and mental performances of humans. Percept Mot Skills 34:371–378

Stoléru SG, Ennaji A, Cournot A, Spira A (1993) LH pulsatile secretion and testosterone blood levels are influenced by sexual arousal in human males. Psychoneuroendocrinology 18:205–218

Strauss RH, Wright JE, Finerman GAM, Catlin DH (1983) Side effects of anabolic steroids in weight-trained men. Physician Sportsmed 11:86–98

Strauss RH, Liggett MT, Lanese RR (1985) Anabolic steroid use and perceived effects in ten weight-trained men. Physician Sportsmed 253:2871–2873

Stumpf H, Klieme E (1989) Sex-related differences in spatial ability: More evidence for convergence. Percep Mot Skills 69:915–921

Su TP, Pagliaro M, Pickar D, Wolkowitz W, Rubinow DR (1993) Neuropsychiatirc effects of anabolic steroids in male normal volunteers. J Am Med Assoc 269:2760–2764

Susman EJ, Inoff-Germain G, Nottelmann ED, Loriaux DL, Cutler GB Jr, Chrousos GP (1987) Hormones, emotional dispositions, and aggressive attributes in young adolescents. Child Develop 58:1114–1134

Sutton JR, Coleman MJ, Casey J, Lazarus L (1973) Androgen responses during physical exercise. Brit Med J 1:520–522

Tan Ü, Akgün A (1992) There is a direct relationship between nonverbal intelligence and serum testosterone level in young men. Int J Neurosci 64:213–216

Tanaka H, Cleroux J, de Champlain J, Ducharme J, Collu R (1986) Persistent effects of a marathon run on the pituitary-testicular axis. J Endocrinol Invest 9:97–101

Tegelman R, Carlström K, Pousette A (1988) Hormone levels in male ice hockey players during a 26-hour cup tournament. Int J Androl 11:361–368

Tieger T (1980) On the biological basis of sex differences in aggression. Child Develop 51:943–963

Toran-Allerand CD (1991) Organotypic culture of the developing cerebral cortex and hypothalamus: Relevance to sexual differentiation. Psychoneuroendocrinology 16:7–24

Trautmann PD, Meyer-Bahlburg HFL, Postelnek J, New MI (1995) Effects of early prenatal dexamethasone on the cognitive and behavioral development of young children: Results of a pilot study. Psychoneuroendocrinology 20:439–449

Udry JR, Talbert LM (1988) Sex hormone effects on personality at puberty. J Pers Soc Psychol 54:291–295

Unden F, Ljunggren JG, Beck-Friis J, Kjellman BF, Wetterberg L (1988) Hypothalamic-pituitary-gonadal axis in major depressive disorders. Acta Psychiatr Scand 78:138–146

Uzych L (1992) Anabolic-androgenic steroids and psychiatric-related effects: A review. Can J Psychiatry 37:23–28

Vandenberg SG, Kuse AR (1979) Spatial ability: A critical review of the sex-linked major gene hypothesis. In: Wittig MC, Petersen AC (eds) Determinants of sex-related differences in cognitive functioning. Academic Press, New York, pp 67–95

Van Goozen SHM, Frijda NH, Van de Poll NE (1994a) Anger and aggression in women: Influence of sports choice and testosterone administation. Aggress Behav 20:213–222

Van Goozen SHM, Cohen-Kettenis PT, Gooren LJG, Frijda NH, van de Poll NE (1994b) Activating effects of androgens on cognitive performance: Causal evidence in a group of female-to-male-transsexuals. Neuropsychologia 32:1153–1157

Van Goozen SHM, Frijda NH, Van de Poll NE (1995a) Anger and aggression during role-playing: Gender differences between hormonally treated male and female transsexuals and controls. Aggress Behav 21:257–273

Van Goozen SHM, Cohen-Kettenis PT, Gooren LJG, Frijda NH, Van de Poll NE (1995b) Gender differences in behavior: Activating effects of cross-sex hormones. Psychoneuroendocrinology 20:343–363

Vasankari TJ, Kujala UM, Heinonen OJ, Huhtaniemi JT (1993) Effects of endurance training on hormonal responses to prolonged physical exercise in males. Acta Endocrinol 129:109–113

Viru A (1991) Adaptive regulation of hormone interaction with receptor. Exp Clin Endocrinol 97:13–28

Vogel W, Klaiber EL, Broverman D (1978) Roles of the gonadal steroid hormones in psychiatric depression in men and women. Prog Neuro-Psych 2:487–503

Von Zerssen D, Berger M, Doerr P (1984) Neuroendocrine dysfunction in subtypes of depression. In: Shah NS, Donald AG (eds) Psychoneuroendocrine dysfunction in psychiatric and neurological illnesses: Influence of psychopharmacological agents. Plenum, New York, pp 357–382

Waber DP (1977) Sex differences in mental abilities, hemispheric lateralization, and rate of physical growth at adolescence. Develop Psychol 13:29–38

Whalen RE (1982) Current issues in the neurobiology of sexual differentiation. In: Vernadakis A, Timiras PS (eds) Hormones in development and aging. MTP Press, Lancaster, pp 273–304

Wheeler GD, Wall SR, Belcastro AN, Cumming DC (1984) Reduced serum testosterone and prolactin levels in male distance runners. J Am Med Assn 252:514–516

Wheeler GD, Singh M, Pierce WD, Epling WF, Cumming DC (1991) Endurance training decreases serum testosterone levels in men without change in luteinizing hormone pulsatile release. J Endocrinol Metab 72:422–425

Wittig MC, Petersen AC (1979) Determinants of sex-related differences in cognitive functioning. Academic Press, New York

Yalom ID, Green R, Fisk N (1973) Prenatal exposure to female hormones: Effect on psychosexual development in boys. Arch Gen Psychiatry 28:554–561

Yesavage JA, Davidson J, Widrow L, Berger PA (1985) Plasma testosterone levels, depression, sexuality, and age. Biol Psychiatry 20:222–225

Zussman JU, Zussman PP, Dalton K (1975) Postpubertal effects of prenatal administration of progesterone. Paper presented at the meeting of the Society for Research in Child Development, Denver (Col)

4 The role of testosterone in spermatogenesis

Gerhard F. Weinbauer and Eberhard Nieschlag

Contents

4.1 Introduction

The development of the male gamete is entirely dependent on the release of hypothalamic and pituitary factors (Fig. 4.1). Hypothalamic gonadotropin-releasing hormone (GnRH) binds to specific receptors on the pituitary gonadotrops and triggers the release of the gonadotropic hormones, luteinizing hormone (LH) and follicle-stimulating hormone (FSH). These hormones stimu-

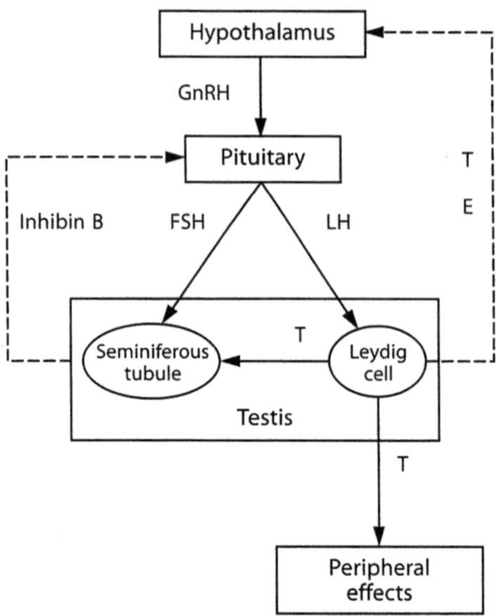

Fig. 4.1. Endocrine testicular feedback system of the male. Dashed lines represent inhibitory effects, full lines denote stimulatory actions. Testosterone (T) and estradiol (E) probably act via the hypothalamus whereas inhibin acts directly at the pituitary level. T and E can influence the secretion of both LH and FSH. Current evidence suggests that inhibin B has an important role in the control of FSH secretion in the male. GnRH = gonadotropin-releasing hormone

late testicular functions, i.e. synthesis of testosterone and production of spermatozoa. LH acts on Leydig cells in the interstitial compartment whereas FSH acts on Sertoli cells in the tubular compartment of the testis. Testosterone, synthesized and released by Leydig cells, acts on peritubular cells and on Sertoli cells within the testis. Hence, testosterone is a potent endocrine and paracrine stimulator of spermatogenesis (Schlatt et al. 1997; Weinbauer and Nieschlag 1996a). To ensure an adequate endocrine and local milieu compatible with male gametogenesis, testicular factors are secreted in the bloodstream and regulate GnRH and gonadotropin secretion.

Testosterone inhibits GnRH release which in turn interrupts LH and FSH secretion. Since testosterone is aromatized to estradiol or reduced to 5α-dihydrotestosterone (DHT) in the peripheral circulation and in the brain, testosterone acts either through DHT or estradiol. In fact, aromatase inhibitors or antiestrogens uncouple the hypothalamic-hyophyseal system from the testis and provoke an increase of gonadotropin serum concentrations. Recent work suggests that estradiol acts at the hypothalamic rather than on the pituitary level (Veldhuis et al. 1997). Although testosterone exerts a feedback on both gonadotropic hormones, an additional feedback loop exists between the testis and the brain for FSH. Inhibin, activin and follistatin are involved in this regulatory system. Inhibin functions to suppress FSH secretion and the homodimer activin stimulates FSH release (Mather et al. 1997), whereas follistatin antagonizes activin action. In the male, levels of inhibin B correlate well with FSH levels and it is suspected that inhibin B is produced by Sertoli cells under the control of FSH (Anderson et al. 1997).

4.2 Organization and kinetics of spermatogenesis

The germinal epithelium is composed of somatic Sertoli cells and the germ cells. On their way to becoming spermatozoa, germ cells experience three major events: the mitotic proliferation of spermatogonia, the production of haploid spermatids through meiosis and the morphological differentiation of round spermatids into elongated ones (testicular sperm). Sertoli cells divide during the prepubertal period until the establishment of the blood-testis barrier. Two major functions are assigned to the Sertoli cell: to enable and coordinate germ cell proliferation and development, and to determine adult testis size and sperm production. Evidence for the latter is compelling since it could be demonstrated under normal and pathological conditions that the number of Sertoli cells correlates precisely with the number of sperm produced (De Franca et al. 1995).

With regard to the role of the Sertoli cell in the control of germ cell evolution, substantial data is available to suggest that it could be the germ cells that – as a function of their developmental phase – control Sertoli cell functions. In fact, Sertoli cell secretions are influenced by the type of germ cells present in the seminiferous epithelium (Jegou 1993). Studies on heterologous germ cell transplantation showed that mouse Sertoli cells supported the development of rat germ cells, including the formation of spermatozoa (Russell and Brinster 1996). Since the time frames for germ cell production differ substantially (approx. 35 days in mouse vs 50 days in rats) it has been speculated that the germ cells might control Sertoli cell function. In that sense, the Sertoli cells would provide the required substances and factors upon request from germ cells. Whatever holds true, Sertoli cells and germ cells are intimately associated on a functional and morphological basis, and Sertoli cells possess specialized processes for interaction with germ cells.

4.2.1 Distribution of spermatogenic stages

The process of spermatogenesis not only covers cell division and differentiation but is also characterized by a particular topography and dynamics. In any given area of the seminiferous tubule, only certain associations of specific germ cell types are encountered. These associations are referred to as stages of spermatogenesis. The classification currently used relies primarily on the morphological appearance and development of the acrosome. The incidence of the different spermatogenic stages varies and it is believed that this variation reflects differences in the relative duration of each particular stage. Conventionally, a 12-stage system is used for mice and macaque monkeys, a 14-stage system for rats and a 6-stage system for men. These stages succeed each other along the length of the seminiferous tubule, and this phenomenon is known as the wave of spermatogenesis. In rodent testes, a defined section of the seminiferous tubule is occupied by a single spermatogenic stage and, in longitudinal sections, the succession of spermatogenic stages is readily apparent.

Among primates, cross-sections of seminiferous tubules may reveal the presence of more than one spermatogenic stage. For example, in men and chimpanzees, three and more different cellular associations can be identified in a single section (Johnson 1994; Schulze and Rehder 1984; Smithwick et al. 1996). It has been suggested that the presence of several stages results from a helical organisation of spermatogenesis (Schulze and Rehder 1984). This means that several spermatogenic waves run in parallel but are shifted spatially. The existence of complete spermatogenic waves in the human testis, however, has been questioned by others (Johnson 1994). The biological significance of this type of organization of the germinal epithelium is not known. It is interesting to note, however, that germ cell production per unit of testicular parenchyma is lower in primate testis than in rodent testis (Weinbauer and Nieschlag 1997).

4.2.2 Duration of the spermatogenic process

The time patterns of the spermatogenic process have been studied using incorporation of thymidine or replenishment of germ cells following irradiation (Clermont 1972). More recently, a non-radioactive approach based upon the incorporation of 5-bromodeoxyuridine (BrdU) into DNA during the S-phase of the cell cycle was established (Rosiepen et al. 1994). This approach has the advantage of avoiding radioactivity, ease of cellular localization of BrdU and applicability to nonhuman primate models (Rosiepen et al. 1997; Weinbauer et al. 1998a). The duration of the spermatogenic process, i.e. the succession of all stages, is species-specific. This process requires 12–14 days in rats, 7–9 days in mice, 17 days in the Chinese hamster, 9–11 days in macaques, 14 days in chimpanzees and 16 days in men. Although the duration of the cycle is remarkably constant under normal conditions, it is amenable to significant alterations under certain circumstances.

In mice, the first wave of spermatogenesis during puberty proceeds faster than the subsequent ones. Synchronization of spermatogenesis, i.e. the coordination of stages, can be achieved in the vitamin A-deficient and vitamin A-replaced rat model. In this model, spermatogenesis is halted at the level of pre-leptotene spermatocytes due to vitamin A deficiency, but is initiated in the majority of seminiferous tubules upon provision of vitamin A. As a consequence, at a given later time point, most of the germ cells are in a similar stage of spermatogenesis in contrast to the normal frequency distribution of the stages. After approximately ten spermatogenic cycles, however, the synchrony was lost and the frequency of the various spermatogenic stages had returned to normal indicating a change in the relative duration (Bartlett et al. 1990). Several reports described changes in the frequency of stages, implying an altered duration, following exposure of testes to heat or drugs. Administration of 2,5-hexanedione, a neurotoxin and Sertoli cell toxicant in the rat, changed stage frequencies and these alterations significantly prolonged the duration of the spermatogenic cycle by about one day (Rosiepen et al. 1995).

In contrast to these studies, deprivation of gonadotropic hormones and testosterone failed to affect the timing of the spermatogenic process (Clermont 1972 for review). However, our recent studies on the gonadotropin control of the kinetics of germ cell proliferation in the macaque model (cynomolgus monkey) yielded an unexpected finding (Weinbauer et al. 1998a). Administration of a GnRH antagonist throughout a period of 25 days slightly reduced the progression of BrdU-labelled germ cells during meiosis and significantly retarded spermatid progression during spermiogenesis. This is the first indication that lack of gonadotropic hormones might influence dynamic aspects of spermatogenesis, at least in the primate. Further investigations, however, are required to substantiate this view.

4.3 Initiation of spermatogenesis by testosterone

Initiation refers to the first completion of the spermatogenic process during puberty. In rodents germ cell proliferation commences around birth or shortly thereafter, whereas in nonhuman primates and men a sustained period of reproductive quiescence precedes the onset of puberty. During this period the secretion of reproductive hormones ceases. In general, though, serum testosterone levels show a peak around birth. Blockade of the neonatal increase of androgens by means of a GnRH antagonist leads to compromised ejaculatory behaviour and fertility in the adult rat (Kolho and Huhtaniemi 1989). On the other hand, prevention of the neonatal rise of testosterone in a primate species, the common marmoset, had no discernible effect on the reproductive system once the animal has attained adulthood (Lunn et al. 1997). At present, the biological significance of the perinatal testosterone peak remains unclear.

Earlier investigations on the effects of hCG, LH or testosterone on the initiation of spermatogenesis in the rat model suggested that germ cell proliferation could be initiated but not sustained beyond the meiotic stages (Weinbauer and Nieschlag 1993). Notwithstanding this, it is clear that gonadotropic hormones are important for the initation of spermatogenesis since hypophysectomy in prepubertal rats markedly increases the number of degenerating germ cell (Russell et al. 1987). Administration of GnRH antagonist to 20-day old rats blocked the formation of spermatids and this effect could be overcome by concomitant administration of testosterone (Ganguly et al. 1994). It has also been reported that LH/androgens are involved in the quantitative achievement of sperm numbers during puberty: immunization of 18-day old rats against the LH receptor provoked a 50% reduction of testicular sperm counts at the age of 88 days (Graf et al. 1997). In the hypogonadotropic (*hpg*) mouse, which lacks gonadotropins due to a major deletion in the GnRH gene, administration of testosterone to weaning animals throughout eight weeks induced the formation of spermatozoa (Singh et al. 1995). These spermatozoa even proved fertile in an in-vitro fertilization as-

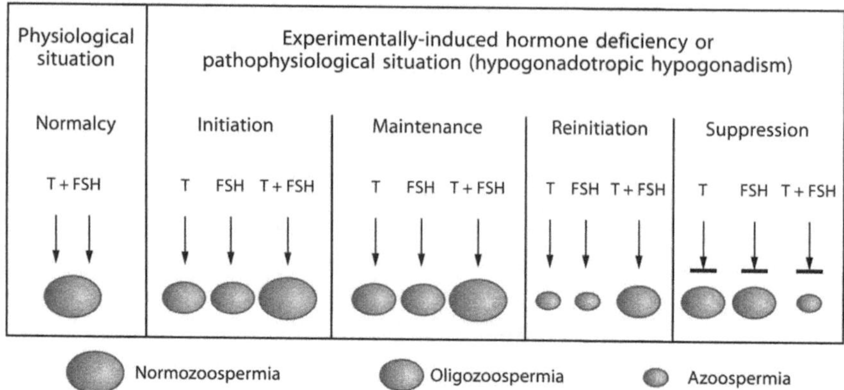

Fig. 4.2. Representation of the role and effects of testosterone (T) and follicle-stimulating hormone (FSH) alone and in combination on the spermatogenic process in the primate. For quantitatively normal sperm production (normozoospermia) both hormones are necessary. Conversely, the complete inhibition of spermatogenesis (azoospermia) usually requires that both T and FSH are suppressed. Although it has been shown that spermatogenesis can be reinitiated with T alone, in clinical practice, both hormones are needed to re-establish fertility

say. Estradiol has also been claimed to stimulate stem cell multiplication during the initial spermatogenic cycle in the rat (Kula 1988).

In primates it has been established beyond doubt that testosterone alone is capable of inducing the production of sperm (Fig. 4.2) since administration of high amounts of exogenous testosterone to immature cynomolgus monkeys led to the induction of complete spermatogenesis (Marshall et al. 1984). However, the number of spermatozoa in the ejaculates was rather low compared to those of adult animals. The evidence for men is equally compelling and derives from studies in boys bearing Leydig cell tumours and or mutations of the LH receptor. In the first instance, it was found that spermatogenesis was complete in those tubules adjacent to the Leydig cell tumors but was incomplete in more remote tubules (Weinbauer and Nieschlag 1996a). Secondly, activating mutations of the LH receptor were associated with precocious puberty and premature maturation of the male gonad (Gromoll et al. 1998; Shenker et al. 1993; Simoni et al. 1997). Conversely, an inactivating mutation of the LH receptor abolished ligand-induced cAMP production and was associated with Leydig cell hypoplasia and male pseudohermaphroditism (Kremer et al. 1995).

These data clearly demonstrate that testosterone, if present, exerts a stimulatory effect on germ cell development. This, however, does not prove that testosterone is indispensable for the commencment of spermatogenesis. Patients with isolated LH deficiency ("fertile eunuchs") have atrophied Leydig cells but complete spermatogenesis and normal testis (Behre et al. 1997). Recently a patient has been described with normal LH and marginally elevated FSH levels but markedly reduced testosterone concentrations (de Roux et al. 1997). This patient was phenotypically androgen-deficient and had testes of subnornal size. However, spermatogenesis was complete (approx. 40×10^6 sperm/ml ejaculate). These observations demonstrate that initiation of spermatogenesis can take place despite substantially reduced testosterone levels (Fig. 4.2).

4.4 Maintenance of spermatogenesis by testosterone

Maintenance refers to the requirements of the testis during ongoing spermatogenesis. Testosterone is an essential component for the maintenance of the spermatogenic process since selective elimination of Leydig cells or immunization against LH were accompanied by a reduction of germ cell numbers and an increase in the number of morphologically abnormal germ cells (Weinbauer and Nieschlag 1990, 1993). The adverse effects of androgen deficiency on the testicular system can be reversed by testosterone supplementation (Kerr et al. 1993). Immunization against the LH receptor in 90-day old rats reduced testicular sperm numbers by 50% until day 150 of age (Graf et al. 1997) and in the mouse model, LH receptor immunization abolished male fertility but it is not clear whether this effect was due to lack of spermatozoa or lack of appropriate mating behaviour (Remy et al. 1993, 1996). In the rabbit, immunization against LH induced a loss and eventually disappearance of spermatids (Jeyakumar et al. 1995). Detailed studies on the effects of testosterone withdrawal suggested that the initial effect of androgen deprivation in the rat testis is a specific loss of step 7 and 8 round spermatids (O'Donnell et al. 1996). Provision of testosterone at the time of induction of experimental hypogonadotropism revealed that testosterone prevented testicular involution and maintained germ cell numbers. Awoniyi et al. (1992) reported that testosterone alone can maintain quantitatively normal numbers of testicular spermatozoa in the GnRH-immunized rat. However, it has also been suggested that FSH is still present in this experimental paradigm (McLachlan et al. 1994).

Immunization against testosterone in the rhesus monkey reduced peripheral testosterone levels but failed to lower intratesticular concentrations of testosterone (Wickings and Nieschlag 1978). As one would expect, testis size and spermatogenesis remained unaffected. More recently, adult bonnet monkeys were immunized against LH and this intervention caused blockade of sperm production (Suresh et al. 1995). Conversely, supplementation of testosterone to gonadotropin-deficient monkeys prevented or delayed the involution of the germinal epithelium and of sperm production (Weinbauer and Nieschlag 1993). The magnitude of effects is related to the administered dose of testosterone. Our knowledge about the role of testosterone for maintenance of human spermatogenesis is incomplete but it appears that hCG/testosterone can maintain the spermatogenic process to a qualitatively normal extent (Fig. 4.2). Anecdotal evidence derives from earlier case reports on patients with hypogonadotropic hypogonadism in whom spermatogenesis could be maintained by hCG once it had been restarted using gonadotropin therapy (Johnsen 1978). Application of hCG has been used for suppression of testicular function, taking advantage of the elevation of testosterone concentrations. However, complete suppression of sperm production could not be achieved suggesting that testicular testosterone maintained spermatogenesis.

4.5 Reinitiation of spermatogenesis by testosterone

Reinitiation refers to the restart of spermatogenesis once it has been interrupted. In the rat model, testosterone re-induced the formation of sperm (Weinbauer and Nieschlag 1990) and the sperm produced were fertile (Awoniyi et al. 1992). It is of interest to note that the outcome of efforts to reinitiate spermatogenesis by testosterone was related to the experimental approach for induction of gonadotropin deficiency. Following hypophysectomy-induced testicular regression, testosterone stimulated germ cell production but quantitatively normal numbers were not restored. In contrast, in estradiol or GnRH-suppressed animals, the quantitative restoration of the spermatogenic process was achieved (Awoniyi et al. 1992). It must be stated, however, that others were unable to confirm this effect in the GnRH-immunized rat (McLachlan et al. 1994). It is conceivable that the differing effects of testosterone on spermatogenesis in the hypophysectomized vs gonadotropin-suppressed animal model relate to the levels of FSH since it has been suggested that FSH secretion is not totally abolished or re-induced by testosterone following immunization against GnRH (McLachlan et al. 1994). The latter circumstance has been observed in rats exposed to GnRH antagonist and testosterone (Sharma et al. 1990): testosterone restored FSH bioactivity and hence quantitatively normal spermatogenesis.

Testosterone has also been shown to reinitiate spermatogenesis in gonadotropin-deficient monkeys but not to a quantitatively normal extent (Weinbauer and Nieschlag 1996b). A number of studies reported the reinitiation of complete spermatogenesis either with testosterone or hCG alone (Matsumoto 1994; Weinbauer and Nieschlag 1993). It must be realized, however, that in the majority of patients sperm counts remained small (Fig. 4.2). It has been suggested that the success of gonadotropin therapy might depend on whether therapy is initiated prior to or after puberty has occurred. Another confounding variable relates to the fact that the spermatogenic status, i.e. the precise spermatogenic lesion, and the previous history of spermatogenic activity are frequently unknown. The question also remains whether these patients are really FSH-deficient or whether they have been exposed to some stimulatory effects prior to therapy. The success of gonadotropin therapy is associated with testis size at start of therapy (Kliesch et al. 1994) and it seems that patients responding to hCG monotherapy have larger testes prior to therapy than those who do not respond (Kung et al. 1994). Although testosterone can reinitiate spermatogenesis in men, this effect, if achieved, is usually not sufficient to restore fertility. Moreover, stimulation of Leydig cells by hCG along with hMG proved better than the combination of testosterone and FSH (Schaison et al.

1993) and, clinically, the combination of LH and FSH activity or GnRH proved most successful (Burgues et al. 1997; Kliesch et al. 1995).

The Djungarian hamster (*Phodopus sungorus*) represents an interesting exception among the animal models investigated so far. Testicular function in this species is regulated by seasonal cues and shortened exposure to light triggers complete testicular involution and disruption of spermatogenesis at the level of early spermatocytes. Treatment of out-of-season and aspermatogenic animals with LH restored Leydig cell function and testicular testosterone production but had no discernible effects on the seminiferous epithelium (Milette et al. 1988; Niklowitz et al. 1989). This observation is remarkable since activation and restoration of Leydig cell activity exerted a pronounced stimulatory effect on germ cell numbers in primates and other rodent species (Schaison et al. 1993; Schlatt et al. 1995; Singh and Handelsman 1996). Apparently testosterone is not required to re-establish the spermatogenic process in this species and the production of sperm is entirely under the control of FSH (Lerchl et al. 1993; Niklowitz et al. 1997).

4.6 Testicular distribution and transport of testosterone

Testosterone is synthesized in the interstitial compartment and acts on the tubular compartment. Because of its lipophilic properties testosterone can be expected to diffuse freely throughout the testis. Notwithstanding this, specific transport mechanisms cannot be excluded entirely (Fig. 4.3). Androgen-

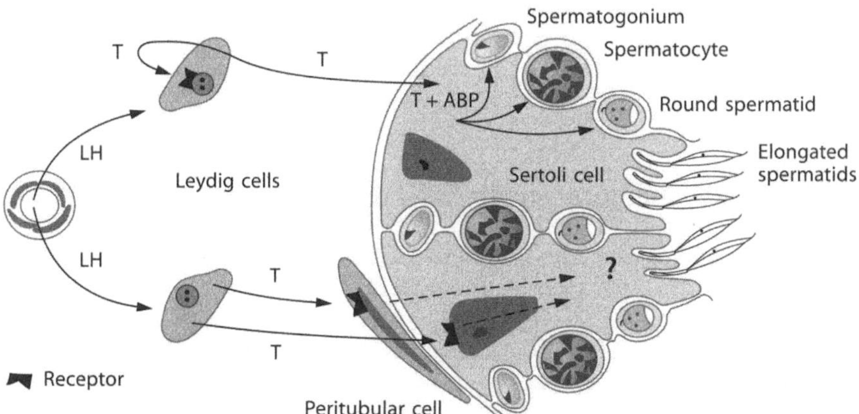

Fig. 4.3. Testicular production, distribution and action of testosterone (T). T is synthesized in the Leydig cells and acts through specific nuclear receptors on Leydig cells, peritubular cells and Sertoli cells. Germ cells lack androgen receptors. Within the seminiferous tubule T can be bound to androgen binding protein (ABP) and distributed within the germinal epithelium. The T/ABP complex was found to be internalized by all types of germ cells. Whether peritubular and Sertoli cells, upon testosterone stimulation, produce specific androgen-dependent pro-spermatogenic factors, could not be clarified to date

binding protein (ABP) is produced within the seminiferous tubules by Sertoli cells and is secreted into the interstitial fluid and the seminiferous tubule fluid (Danzo and Eller 1985). Hence, ABP may influence the testicular and intratubular distribution of androgens. Testicular ABP production is influenced by testosterone and FSH, and gonadotropin deficiency lowers testicular ABP contents (Danzo et al. 1990; Huang et al. 1991, 1992). In the rodent testis, testosterone levels and androgen receptor expression were maximal prior to the onset of spermatid elongation (stages VII–VIII) and ABP levels were highest during the final stages of spermatid maturation (stages VIII–XII). Sex hormone-binding acitivity has also been shown for human testis tissue and human ABP cDNA has been isolated (Reventos et al. 1988).

The precise localization of androgens within the tubular compartment is not known. Immunocytochemical attempts to localize testosterone must be viewed with caution since fixation and dehydration will strongly influence the presence and distribution of steroids in general. Androgens might be transported into germ cells because it has been shown that germ cells bind ABP and testosterone-ABP complexes (Felden et al. 1992). In fact, in isolated seminiferous tubules, when exposed to testosterone-ABP complexes, the complexes are taken up by the Sertoli cells and passed on to the adluminal germ cells (Gerard et al. 1994) and endocytosis of ABP by germ cells has been demonstrated in rats, monkeys and men (Gerard 1995). Whether this mechanism has physiological relevance under in-vivo conditions still awaits proof. However, an association between spermiogenesis and the availability of ABP has been postulated (Pogach et al. 1993). To date, the strongest hint for a direct involvement of testicular ABP in the control of spermatogenesis comes from transgenic mice overexpressing rat ABP (Joseph et al. 1997; Larriba et al. 1995). These animals exhibit reduced fertility and histological analysis revealed focal spermatogenic dysfunction. It could be speculated that elevated levels of ABP might capture androgens, thereby preventing them from acting on germ cells or that the rat ABP competes with endogenous mouse ABP.

4.7 Testicular androgen receptor and its regulation

The androgen receptor is a ligand-dependent transcription factor whose molecular properties are well characterized (see Chapter 2 of this volume by Quigley). Within the testis, androgen receptors are found on Leydig cells, peritubular cells and Sertoli cells (Fig. 4.3). Germ cells seemingly lack androgen receptors (Bremner et al. 1994; van Roijen et al. 1995), although the presence of receptors on elongating spermatids of steps X–XII has also been suggested (Vornberger et al. 1994). The current view, however, is that the nuclear receptor-mediated androgen action is conveyed by the somatic cells of the testis. Sertoli cell expression of the androgen receptor is stage-dependent with maximal levels in stage VII of spermatogenesis and nadir values thereafter (Bremner et al. 1994). Both FSH and testosterone influence the expression of the androgen receptor. FSH stimulated both mRNA and protein expression (Blok et al. 1992b; Verhoeven and Cailleau 1988) whereas deprivation of testosterone partly reduced or eliminated testicular androgen receptor expression (Blok et al. 1992a; Bremner et al. 1994). The developmental expression of the androgen receptor in Leydig cells and Sertoli cells appears under differential control in the rat (Shan et al. 1995, 1997). GnRH antagonist-induced gonadotropin and testosterone deficiency caused a reduction of androgen receptor expression in prepubertal but not in adult Leydig cells. Concomitant administration of testosterone significantly increased androgen receptor expression only in prepubertal Leydig cells. Conversely, androgen receptor expression in Sertoli cells was highly dependent on the androgenic milieu in adult animals but less so in immature animals.

4.8 Testosterone and differentiation of somatic testicular cells

Apart from the trophic actions of testosterone on the germinal epithelium during initiation of spermatogenesis, testosterone also plays an important role in the morphological and presumably functional differentiation of the somatic testicular cell types prior to the onset of spermatogenic activity. Formation of the male gonad is dependent on testosterone and is discussed elsewhere in this book (see Chapter 2 of this volume by Quigley). It is well established that the differentiation of Leydig cells from their fibroblast precursors is under the control of LH/androgens (Teerds et al. 1994). This view is supported by the demonstration that exogenous testosterone increases androgen receptor expression on Leydig cells (Shan et al. 1997) and that the functional maturation of Leydig cells is controlled by LH/hCG (Schlatt et al. 1995). The morphological and functional maturation of peritubular cells in the primate is induced by androgens, as shown from investigation of the expression of α-smooth muscle actin in these cells. Peritubular cells are myoid cells endowed with the ability of contraction. Presumably these contractions serve to transport released germ cells and possibly to extrude fluid towards the epididymis. Peritubular cells in the immature monkey do not express detectable amounts of this type of actin. However, upon treatment with either testosterone (Schlatt et al. 1993) or hCG (Schlatt et al. 1995) actin expression was encountered. Interestingly, FSH potentiated the effects of androgens. These observations suggest an FSH-dependent local interaction between Sertoli cells and peritubular cells during androgen action. Unlike in immature animals, deprivation of testicular androgens in the adult animal did not alter actin expression in peritubular cells.

With regard to the maturation of Sertoli cells, testosterone but not FSH appears to be the key factor. Testosterone, but not FSH induced an adult-type pattern of distribution of filamentous and monomeric actin within the Sertoli cells, i.e. arrangement of this form of actin in the peripheral cytoplasm (Schlatt et al. 1993). Appropriate morphological and functional organization by the Sertoli cell is necessary to maintain contact with the developing spermatids during the process of spermiogenesis. It has been shown in the rat model that FSH established the competency of the Sertoli cells to bind round spermatids (Cameron et al. 1993). In these studies, testosterone only stimulated spermatid binding once Sertoli cells had been primed with FSH. These cooperative mechanisms may operate during reinitiation of spermatogenesis. Gonadotropin-deprived Sertoli cells lose the competency to bind spermatids, have abnormally shaped ectoplasmic specializations and altered distribution patterns of f-actin and vinculin (Muffly et al. 1994). Testosterone restored spermatid binding only in the presence of FSH. More recently, N-cadherin has been suggested as a candidate mediating the endocrine-controlled interaction between Sertoli cells and round spermatids (Perryman et al. 1996).

4.9 Follicle-stimulating hormone and spermatogenesis

4.9.1 The follicle-stimulating hormone receptor

The FSH receptor (and LH receptor) belong to the family of G-coupled receptors and are characterized by a large extracellular domain (Simoni et al. 1997). Unlike the LH receptor, that is also expressed in extragonadal tissues, expression of the FSH receptor is confined to the testis (Dankbar et al. 1995). Within the testis, the FSH receptor is only expressed in Sertoli cells (Böckers et al. 1994; Vannier et al. 1996). Hence, the Sertoli cell is the target of FSH action. Expression of the FSH receptor is regulated by FSH but not by testosterone in the immature rat (Maguire et al. 1997). Alternative splicing generates several transcripts of the FSH receptor (Simoni et al. 1997) but the physiological significance of this event is unknown. The onset of FSH receptor expression coincides with the onset or puberty (Laborde et al. 1996) and with the appearance of spermatocytes (Rannikko et al. 1996). During the process of spermatogenesis, FSH receptor expression and ligand binding is related to the stage of spermatogenesis. Expression levels and FSH binding activity were highest during stages XII–II and lowest in stages VI–VII (Kangasniemi et al. 1990; Heckert and Griswold 1991; Ranniko et al. 1996).

4.9.2 Initiation, maintenance and reinitiation of spermatogenesis by FSH

Initiation of spermatogenesis is influenced by FSH in rodent models. Immunization against FSH decreased germ cell numbers (Shetty et al. 1996) and immunization against the FSH receptor reduced the production of sperm (evaluated at the age of 88 days) by about 50% (Graf et al. 1997) in the rat. Administration of FSH promoted germ cell proliferation and development in intact (Matikainen et al. 1994) and hypophysectomized (Russell et al. 1993; Vihko et al. 1991) immature rats. In mice, FSH is not obligatory for the achievement of fertility as revealed from animals deficient in the FSH β-subunit (Kumar et al. 1997). It must be pointed out, however, that the testes were smaller in FSH-deficient animals, indicating a role for FSH in Sertoli cell development. In contrast to the lack of effect of FSH on fertility, immunization against the FSH receptor was found to affect fertility in male mice (Remy et al. 1996). In the *hpg* mouse, FSH stimulated spermatogonial and spermatocyte numbers but had no additive effects beyond the stimulatory action of testosterone (Singh and Handelsman 1996). Although the importance of FSH for spermatogenesis in the adult rat had been questioned earlier, it seems clear by now that FSH has a role to play: immunization against the FSH receptor lowered sperm production (Graf et al. 1997) and administration of FSH to adult LH/FSH-deprived animals was beneficial for spermatogenesis (Weinbauer and Nieschlag 1993 for review). FSH has also been shown to partially restore spermatogenesis after suppression following immunization against GnRH (McLachlan et al. 1995).

The essential role of FSH for spermatogenesis in the primate is evident from a number of studies (Fig. 4.2). FSH stimulates spermatogonial and Sertoli cells numbers in immature macaque monkeys (Arslan et al. 1993; Schlatt et al. 1995). In adult monkeys, immunization against FSH (Moudgal et al. 1992; Weinbauer and Nieschlag 1993) or the FSH receptor (Moudgal et al. 1997; Suresh et al. 1995) had dramatic effects on the spermatogenic function of the testis: animals became either azoospermic or severly oligozoospermic, and in the latter case were even infertile. These data strongly suggest that FSH is a crucial and indispensable factor for normal spermatogenesis in the primate. This view is further supported by clinical observations from male contraceptive trials since the achievement of complete suppression of sperm production necessitates adequate suppression of FSH levels. Another piece of evidence for the promoting action of FSH on spermatogenesis was gathered from a hypohysectomized patient bearing an activating mutation of the FSH receptor (Gromoll et al. 1996). This patient, although having subnormal testosterone levels, had normal testis size, sperm numbers in the normal range and was fertile upon supplementation with testosterone.

The role of FSH during initiation of human spermatogenesis has recently been questioned since it was found that an inactivating mutation of the FSH receptor was compatible with sperm production and fertility (Tapanainen et al. 1997). This was surprising since the same mutation was associated with gonadal dysgenesis in women (Aittomäki et al. 1995). It must be stated, however, that in the study by Tapanainen et al. (1997), serum levels of inhibin B, an indicator of endogenous FSH activity (Anderson et al. 1997) was not totally abolished in four out of the five men studied. Interestingly, only in the fifth patient, who was infertile, was inhibin B undetectable. Hence, the question arises whether this FSH receptor mutation completely abolished all FSH activity in all men studied.

4.10 Mechanism of testosterone and FSH action during spermatogenesis

The precise mode and mechanism of action of both testosterone and FSH on the spermatogenic process have not been disclosed until now. Thus far, it would appear that both hormones exert a general trophic effect on the cellular constitutents of the germinal epithelium. The possibility must be considered that these hormones not only stimulate germ cell proliferation but also act as survival factors for germ cells. For example, lack of gonadotropic stimulation or selective androgen withdrawal are associated with pronounced loss of germ cells via cell death (Brinkworth et al. 1995; Russell and Clermont 1977; Troiano et al. 1994) and hormone supplementation remedies this situation (Russell et al. 1993). It has been extremely difficult to identify testosterone- or FSH-specific factors that would allow the study of specific endocrine-related testicular events. An orphan homeobox gene, *Pem*, is expressed in mouse Sertoli cells and the expression correlated with the onset of meiosis. It was shown that testosterone and LH control the expression of this transcription factor (Lindsey and Wilkinson 1996). Whether this factor is entirely dependent on androgens remains to be seen since the effects of FSH were not investigated.

cAMP responsive element modulator (CREM), is a transcription factor involved in the cAMP-related signalling cascade (Monaco et al. 1996). By means of alternative splicing and alternate promotor usage, CREM activators and repressors are generated. CREM repressors are expressed in premeiotic and meiotic cells whereas activators are predominantly expressed in round spermatids (Delmas et al. 1993; Foulkes et al. 1992). It was observed that inactivation of the CREM gene arrested spermatogenesis at the level of round spermatids (Blendy et al. 1996; Nantel et al. 1996). Thus, CREM seemingly plays a crucial role during spermatid maturation and differentiation. The transition from repressor to an activator was found to be due to activator transcript stabilisation mediated by FSH (Foulkes et al. 1993). Moreover, hypophysectomy abolished testicular CREM expression and this effect could be reversed by FSH but not by LH or prolactin (Foulkes et al. 1993). If CREM is indeed a specifically FSH-regulated transcription factor, another factor must be postulated to explain how FSH, acting on Sertoli cells, can regulate specific gene expression in germ cells. Furthermore, the suggested dependency of CREM on FSH must be reconciled with the findings that the FSH-β-deficient mouse, although having smaller testis, displays complete spermatogenesis and normal fertility (Kumar et al. 1997). If CREM expression is entirely FSH-controlled, a spermatogenic defect similar to that seen in CREM-deficient mice would be expected in FSH-deficient mice. Notwithstanding this, CREM might be an important regulator of spermatid maturation in the human testis since we found that CREM expression was diminished or undetectable in patients with round spermatid maturation arrest (Weinbauer et al. 1998 b).

The fact that the receptors are only present on somatic testicular cells implies the existence of putative yet unidentified mediators of hormone action on Sertoli cells. Alternatively, testosterone and FSH might simply enable the

Sertoli cells to establish proper contact with germ cells and might stimulate the Sertoli cells to provide the required nutrients for germ cell development. The observations that both hormones can essentially exert similar effects on spermatogenesis would be compatible with this assumption. In the case of testosterone, the possibility of a direct effect on germ cells mediated via ABP (Section 4.6) or membrane steroid binding sites (see Chapter 1 of this volume by Rommerts) must also be considered. It has been shown in numerous studies that the combination of LH/testosterone and FSH is most effective in stimulating testicular function. Moreover, it is proven beyond doubt that the effects of either hormone are additive when administered in combination, highlighting the fact that both hormones act through different mechanisms.

It is assumed that testosterone acts within the testis following conversion into DHT by 5α-reductase activity. At comparable doses, DHT is as effective as testosterone in stimulating spermatogenesis (Singh et al. 1995). Testicular testosterone concentrations substantially surpass those measured in the circulation. This might be necessary to provide peripheral testosterone pulses upon stimulation with LH or might serve to generate sufficient amounts of DHT. The latter view is questionable, however, since the affinity of DHT for the androgen receptor exceeds that of testosterone by about tenfold (Grino et al. 1990). Recent studies revealed another facet of steroid hormone action on spermatogenesis: mice lacking the estrogen receptor-α exhibited severe impairment of spermatogenesis and were infertile (Eddy et al. 1996; Lubahn et al. 1993). The diameter of the seminiferous tubules was increased and the architecture of the germinal epithelium was disrupted. The underlying cause for this spermatogenic defect is that fluid resorption in the efferent ducts is lacking in these animals, causing back pressure and extension of tubules (Hess et al. 1997 a).

These observations demonstrate a pivotal role of estrogen for spermatogenesis. Two forms of the estrogen receptor, α and β, have been described (see Kuiper and Gustaffson 1997 for details), and both receptor forms are present in the efferent ducts and in the epididymis (Hess et al. 1997 b). It has also been suggested that P450 aromatase activity is present in germ cells, probably in round spermatids, and that germ cells are a source of testicular estrogens (Janulis et al. 1996). Stimulatory effects of estrogen on germ cell numbers in the immature rat have been suggested previously (Kula 1988) and recent studies revealed stimulation of rat gonocyte proliferation in-vitro (Li et al. 1997). Recent work indicates that aromatase inhibitors impair the development of spermatids in monkeys (Shetty et al. 1997). Altogether, these observations suggest that the mode of testosterone action on spermatogenesis/sperm transport is mediated not only by DHT but also by estradiol. The deleterious effects of estrogen receptor targetting on testicular function identifies estradiol/estradiol receptor as an essential factor in the control of male reproductive physiology.

4.11 Testosterone or follicle-stimulating hormone: which is the key player?

A synopsis of the reported effects and efficacy of LH/testosterone and FSH during spermatogenesis reveals that both hormones exert a beneficial effect on germ cell development and that these effects are comparable and additive. At least in the primate it would appear that both endocrine factors share a common target, the spermatogonial population (van Alphen et al. 1988; Marshall et al. 1995; Weinbauer et al. 1991). The importance of both factors is also suggested by the observations that immunization, against either LH or FSH, affected spermatogenesis in the monkey (Suresh et al. 1995). It has been argued that testosterone but not FSH influences spermatid maturation. However, evidence has been provided that FSH can influence chromatin condensation in spermatids (Moudgal et al. 1997). In relative terms it would appear that testosterone might eventually be more important for spermatogenesis in rats and mice, whereas FSH seem to be a prime regulator of gametogenesis in primates. Overall, though, it would appear that neither LH/testosterone nor FSH have primacy regarding the control of spermatogenesis. It is also important to recognize that neither hormone, in most instances, is sufficient to support spermatogenesis quantitatively. This achievement normally requires the combined action of both factors. An exception is the Djungarian hamster, whose spermatogenic process is under FSH but not testosterone control. For quantitatively normal spermatogenesis both hormones are necessary and of equal importance. The production of sperm as such, however, can be achieved with either hormone alone.

4.12 Relevance of the experimental design and the animal model for the study of testosterone action on spermatogenesis

The experimental paradigm that has been adopted by most investigators and which is still in use to investigate the roles of testosterone/LH in the spermatogenic process comprises the administration of the hormone of interest to immature animals, to adult animals with experimentally induced pituitary deficiency and inactivation of the hormones by selective immunization. More recently, immunization against hormones has been extended by immunization against their receptors. Selective elimination of the Leydig cells as the source of androgens using ethane dimethane sulphonate (EDS) has also been used, although the possibility of EDS effects on Sertoli and germ cells cannot be entirely ruled out. As far as animal species are concerned, the rat has been used most extensively and mice are being used increasingly since the advent of gene-targetting. Nonhuman primates have been used to a lesser extent for obvious reasons.

Solid evidence is available to prove that the choice of the experimental model for the study of the endocrine control of spermatogenesis can influence the outcome. For example, restoration of the spermatogenic process by testosterone is more difficult in hypophysectomized rats than in rats immunized against GnRH (Zirkin 1993). These observations suggest that selective elimination of gonadotropic hormones could have consequences other than abolishing all anterior pituitary hormones. It has been proposed from the GnRH-immunized rats that testosterone can quantitatively maintain and restore spermatogenesis (Zirkin 1993). On the other hand, it has been argued that testosterone stimulates FSH in the GnRH-immunized animals (McLachlan et al. 1994), thus proposing that the observed effects result from a combined effect of FSH and testosterone. The GnRH and gonadotropin-deficient rat appears to represent a rather particular animal model: blockade of the pituitary GnRH receptors by a GnRH antagonist revealed that testosterone, under these circumstances, paradoxically stimulates pituitary FSH levels and also FSH release (Arslan et al. 1989; Sharma et al. 1990). Hence, such experimental models are not suited for addressing the selective role of testosterone in spermatogenesis. A "clean" and unambigous rat paradigm for studying the role of testosterone is not available.

Although the availability of nonhuman primates as animals models is somewhat restricted, these models appear quite superior to the rat for analysis of the hormonal regulation of spermatogenesis. In primates given GnRH antagonists, unlike in the rat, testosterone does not stimulate FSH release (Khurshid et al. 1991; Pavlou et al. 1991). Hence the GnRH antagonist-treated and testosterone-supplemented nonhuman primate provides a suitable model for unravelling the role of testosterone in spermatogenesis. Another substantial difference between rats and nonhuman primates is suggested from analysing the initial response of the germinal epithelium to withdrawal of gonadotropic hormones.

Detailed histological studies in the rat identified a specific loss of spermatocytes and spermatids as the predominant lesion following gonadotropic hormone deprivation (McLachlan et al. 1994; O'Donnell et al. 1996; Sinha-Hikim and Swerdloff 1993; Sinha-Hikim et al. 1997) although Russell and Clermont (1977) also observed some abnormal earlier germ cells after hypophysectomy. In the nonhuman primate, loss of spermatogonia (Weinbauer et al. 1991) and meiotic arrest (Aravindan et al. 1993) have been suggested. Prolonged withdrawal of gonadotropic hormones eliminates spermatids, spermatocytes and even differentiated spermatogonia in the monkey (Marshall et al. 1986; Weinbauer et al. 1991). In contrast to this, some spermatocytes and round spermatids are retained despite continued elimination of LH and FSH in the rat (Chandolia et al. 1991a, b; McLachlan et al. 1995). We observed recently that GnRH antagonist treatment provoked a dramatic loss of the spermatogonial proliferation (assessed from S-phase labelling studies, Weinbauer et al. 1998a) and in the production/survival of B-type spermatogonia (Zhengwei et al. 1998, Fig. 4.4). These observations suggest a substantial difference between the rat and the nonhuman primate in the testicular response to endocrine disturbances.

Immunization against testosterone, LH or LH receptor has been used in rats and monkeys (see above). Although this approach yielded interesting re-

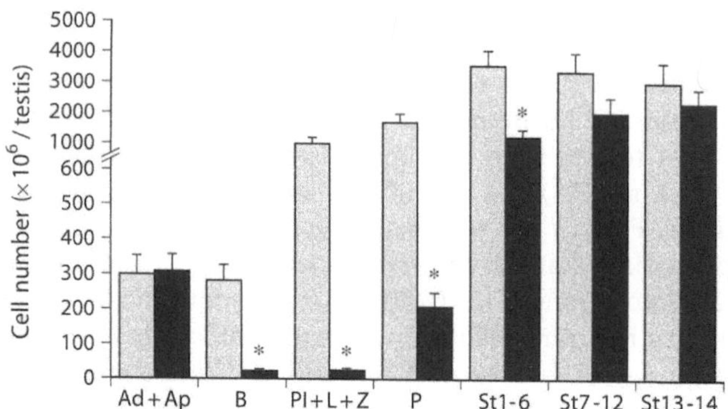

Fig. 4.4. Effects of withdrawal of LH and FSH by administration of the GnRH antagonist cetrorelix throughout 25 days in the adult cynomolgus monkey. Note that under cetrorelix treatment the production/survival of B-type spermatogonia is specifically compomised (modified from Zhengwei et al. 1998). Grey bars denote vehicle-treated animals; black bars represent gonadotropin-deficient animals. Values are mean±sem of 4–5 animals/group; * significantly different from vehicle-treated animals

sults, it has the disadvantage that such manipulation will severely interfere with the balance of the feedback system. Any drop of serum testosterone levels markedly elevates not only LH but also FSH secretion. Since FSH is a pronounced stimulator of spermatogenesis, it becomes difficult to assess the role of testosterone in germ cell formation precisely from these experimental approaches. Overall, it appears that, given our current knowledge, the GnRH antagonist-treated nonhuman primate is the most appropriate model to investigate the role of testosterone in spermatogenesis.

4.13 Key messages

- Testosterone is able to initiate, maintain and reinitiate the formation of spermatozoa.
- For establishment of quantitatively normal numbers of germ cells FSH is needed in addition.
- Testosterone either acts indirectly via the somatic testicular cells or, possibly, directly via binding to ABP and internalization by germ cells.
- The testicular effects of testosterone are mediated by DHT and estradiol. In fact, estradiol became a "male hormone" during the past few years.
- Neither testosterone nor FSH have a primacy during regulation of gametogenesis but rather cooperate and act via different mechanisms.
- At present, the nonhuman primate appears to be the most appropriate model for analysis of the role of testosterone in human spermatogenesis.

4.14 References

Aittomäki K, Lucena JLD, Pakarinen P, Sistonen P, Tapanainen J, Gromoll J, Kaskikari R, Sankila EM, Lehväslaiho H, Engel AR, Nieschlag E, Huhtaniemi I, de la Chapelle A (1995) Mutation in the follicle-stimulating hormone receptor gene causes hereditary hypergonadotropic ovarian failure. Cell 82:959–968

Anderson RA, Wallace EM, Groome NP, Bellis AJ, Wu FCW (1997) Physiological relationship between inhibin B, follicle stimulating hormone secretion and spermatogenesis in normal men and response to gonadotrophin suppression by exogenous testosterone. Hum Reprod 12:746–751

Aravindan GR, Gopalakrishnan K, Ravindranath N, Moudgal NR (1993) Effect of altering endogenous gonadotrophin concentrations on the kinetics of testicular germ cell turnover in the bonnet monkey (*Macaca radiata*). J Endocrinol 137:485–495

Arslan MA, Weinbauer GF, Schlatt S, Shahab M, Nieschlag E (1993) FSH and testosterone, alone or in combination, initiate testicular growth and increase the number of spermatogonia and Sertoli cells in a juvenile non-human primate (*Macaca mulatta*). J Endocrinol 136:235–243

Arslan M, Weinbauer GF, Khan SA, Nieschlag E (1989) Testosterone and dihydrotestosterone, but not estradiol, selectively maintain pituitary and serum follicle-stimulating hormone in gonadotropin-releasing hormone antagonist treated male rats. Neuroendocrinology 49:395–401

Awoniyi CA, Zirkin BR, Chandrashekar V, Schlaff WD (1992) Exogenously administered testosterone maintains spermatogenesis quantitatively in adult rats actively immunized against gonadotropin-releasing hormone. Endocrinology 130:3283–3288

Bartlett JMS, Weinbauer GF, Nieschlag E (1990) Stability of spermatogenic synchronization achieved by depletion and restoration of vitamin A in rats. Biol Reprod 42:603–612

Behre HM, Nieschlag E, Meschede D, Partsch CJ (1997) Diseases of the hypothalamus and pituitary gland. In: Nieschlag E, Behre HM (eds) Andrology – Male reproductive health and dysfunction. Springer Verlag, Berlin, pp 115–132

Blendy JA, Kaestner KH, Weinbauer GF, Nieschlag E, Schütz G (1996) Severe impairment of spermatogenesis in mice lacking the CREM gene. Nature 380:162–165

Blok LJ, Bartlett JMS, Bolt-de-Vries J, Themmen APN, Brinkmann AO, Weinbauer GF, Nieschlag E, Grootegoed JA (1992a). Effect of testosterone deprivation on expression of the androgen receptor in rat prostate, epididymis and testis. Int J Androl 15:182–198

Blok LJ, Hoogerbrugge JW, Themmen APN, Baarends WN, Post M, Grootegoed JA (1992b) Transient down-regulation of androgen receptor messenger ribonucleic acid (mRNA) expression in Sertoli cells by follicle-stimulating hormone is followed by up-regulation of androgen receptor mRNA and protein. Endocrinology 131:1243–1349

Böckers TM, Nieschlag E, Kreutz MR, Bergmann M (1994) Localization of follicle-stimulating hormone (FSH) immunoreactivity and hormone receptor mRNA in testicular tissue from infertile men. Cell Tiss Res 278:595–600

Bremner WJ, Millar MR, Sharpe RM, Saunders PTK (1994) Immunohistochemical localization of androgen receptors in the rat testis: evidence for stage-dependent expression and regulation by androgens. Endocrinology 135:1227–1234

Brinkworth MH, Weinbauer GF, Schlatt S, Nieschlag E (1995) Identification of male germ cell undergoing apoptosis in adult rats. J Reprod Fertil 105:25–33

Burgues S, Calderon MD, and the Spanish Collaborative Group on Male Hypogonadotropic Hypogonadism (1997) Subcutaneous self-administration of highly purified follicle stimulating hormone and human chorionic gonadotrophin for the treatment of male hypogonadotrophic hypogonadism. Hum Reprod 12:980–986

Cameron DF, Muffly KE, Nazian SJ (1993) Testosterone stimulates spermatid binding to competent Sertoli cells in vitro. Endocr J 1:61–65

Chandolia RK, Weinbauer GF, Simoni M, Behre HM, Nieschlag E (1991a) Comparative effects of chronic administration of the non-steroidal antiandrogens flutamide and Casodex on the reproductive system of the adult male rat. Acta Endocrinol 125:547–555

Chandolia RK, Weinbauer GF, Fingscheidt U, Bartlett JMS, Nieschlag E (1991b) Effects of flutamide on testicular involution induced by an antagonist of gonadotrophin-releasing hormone and on stimulation of spermatogenesis by follicle-stimulating hormone. J Reprod Fertil 93:313–323

Clermont Y (1972) Kinetics of spermatogenesis in mammals: Seminiferous epithelium cycle and spermatogonial renewal. Physiol Rev 52:198–236

Dankbar B, Brinkworth MH, Schlatt S, Weinbauer GF, Nieschlag E, Gromoll J (1995) Ubiquitous expression of the androgen receptor and testis-specific expression of the FSH receptor in the cynomolgus monkey (*Macaca fascicularis*) revealed by a ribonuclease protection assay. J Steroid Biochem Mol Biol 55:35–41

Danzo BJ, Eller BC (1985) The ontogeny of biologically active androgen biding protein in rat plasma, testis and epididymis. Endocrinology 117:1380–1388

Danzo BJ, Pavlou SN, Anthony HL (1990) Hormonal regulation of androgen binding protein in the rat. Endocrinology 127:2829–2838

De Franca LR, Hess RA, Cooke PS, Russell D (1995) Neonatal hypthyroidism causes delayed Sertoli cell maturation in rats treated with propylthiouracil: evidence that the Sertoli cell controls testis growth. Anat Rec 242:57–69

de Roux N, Young J, Misrahi M, Genet R, Chanson P, Schaison G, Milgrom E (1997) A family with hypogonadotropic hypogonadism and mutations in the gonadotropin-releasing hormone receptor. New Engl J Med 337:1597–1602

Delmas V, van der Hoorn F, Mellström B, Jegou B, Sassone-Corsi P (1993) Induction of CREM activator proteins in spermatids: down-stream targets and implications for haploid germ cell differentiation. Mol Endocrinol 7:1502–1514

Eddy EM, Washburn TF, Bunch DO, Goulding EH, Gladen BC, Lubahn DB, Korach KS (1996) Targeted disruption of the estrogen receptor gene in male mice causes alteration of spermatogenesis and fertility. Endocrinology 137:4796–4805

Felden F, Gueant JL, Ennya A, Gerard A, Fremont S, Nicolas JP, Gerard H (1992) Photoaffinity labelled rat androgen-binding protein and human sex hormone steroid-binding protein bind specifically to rat germ cells. Mol Endocrinol 9:36–46

Foulkes NS, Mellström B, Benusiglio E, Sassone-Corsi P (1992) Developmental switch of CREM function during spermatogenesis: from antagonist to transcriptional activator. Nature 355:80–84

Foulkes NS, Schlotter F, Pevet P, Sassone-Corsi P (1993) Pituitary hormone FSH directs the CREM functional switch during spermatogenesis. Nature 362:264–267

Ganguly A, Misro MM, Das RP (1994) Roles of FSH and testosterone in the initiation of spermatogenesis in preppubertal rats medically hypophysectomized by a GnRH antagonist. Arch Androl 32:111–210

Gerard A (1995) Endocytosis of androgen-binding protein (ABP) by spermatogenic cells. J Steroid Biochem Mol Biol 53:533–542

Gerard H, Gerard A, Ennya A, Felden F, Gueant JL (1994) Spermatogenic cells do internalize Sertoli androgen-binding protein: a transmission electron microscopy autoradiographic study in the rat. Endocrinology 134:1515–1527

Graf KM, Dias JA, Griswold MA (1997) Decreased spermatogenesis as the result of an induced autoimmune reaction directed against the gonadotropin receptors in male rats. J Androl 18:174–185

Grino PB, Griffin JE, Wilson JD (1990) Testosterone at high concentrations interacts with the human androgen receptor similarly to dihydrotestosterone. Endocrinology 126:1165–1172

Gromoll J, Simoni M, Nieschlag E (1996) An activating mutation of the FSH receptor autonomously sustains spermatogenesis in a hypophysectomized man. J Clin Endocrinol Metab 81:1367–1370

Gromoll J, Partsch CJ, Simoni M, Nordhoff V, Sippell WG, Nieschlag E, Saxena B (1998) A mutation in the first transmembrane domain of the lutropin receptor causes male precocious puberty. J Clin Endocrinol Metab 83:476–480

Heckert L, Griswold MD (1991) Expression of follicle-stimulating hormone receptor mRNA in rat testes and Sertoli cells. Mol Endocrinol 5:670–677

Hess RA, Bunick D, Lee KH, Bahr J, Taylor JA, Korach KS, Lubahn DB (1997a) A role for oestrogens in the male reproductive system. Nature 390:509–512

Hess RA, Gist DH, Bunick D, Lubahn DB, Farrell A, Bahr J, Cooke PS, Greene GL (1997b) Estrogen receptor (α&ß) expression in the excurrent ducts of the adult male rat reproductive tract. J Androl 18:602–611

Huang HFS, Pogach LM, Nathan E, Giglio W, Seebode JJ (1991) Synergistic effects of follicle-stimulating hormone and testosterone on the maintenance of spermiogenesis in hypophysectomized rats: relationship with the androgen-binding protein status. Endocrinology 128:3152–3161

Huang HFS, Pogach L, Giglio W, Nathan E, Seebode J (1992) GnRH antagonist induced arrest of spermiogenesis in rats is associated with altered androgen binding protein distribution in the testis and epididymis. J Androl 13:153–159

Janulis L, Bahr JM, Hess RA, Bunick D (1996) P450 aromatase messenger ribonucleic acid expression in male rat germ cells: detection by reverse transcription-polymerase chain reaction amplification. J Androl 17:651–658

Jegou B (1993) The Sertoli-germ cell communication network in mammals. Int Rev Cytol 147:25–96

Jeyakumar M, Suresh R, Krishnamurthy HN, Mougal NR (1995) Changes in testicular function following specific deprivation of LH in the adult rabbit. J Endocrinol 147:111–120

Johnsen SG (1978) Maintenance of spermatogenesis induced by hMG treatment by means of continuous hCG treatment in hypogonadotophic men. Acta Endocrinol 89:763–769

Johnson L (1994) A new approach to study the architectural arrangement of spermatogenic stages revealed little evidence of a partial wave along the length of the human seminiferous tubules. J Androl 15:435–441

Joseph DR, O'Brien DA, Sullivan PM, Bcchis M, Tsuruta JK, Petrusz P (1997) Overexpression of androgen binding protein/sex hormone-binding globulin in male transgenic mice: tissue distribution and phenotypic disorders. Biol Reprod 56:21–32

Kangasniemi M, Kaipia A, Toppari J, Perheentupa A, Huhtaniemi I, Parvinen M (1990) Cellular regulation of follicle-stimulating hormone (FSH) binding in rat seminiferous tubules. J Androl 11:336–343

Kerr JB, Millar M, Maddocks S, Sharpe RM (1993) Stage-dependent changes in spermatogenesis and Sertoli cells in relation to the onset of spermatogenic failure following withdrawal of testosterone. Anat Rec 235:547–559

Khurshid S, Weinbauer GF, Nieschlag E (1991) Effects of testosterone and gonadotropin-releasing hormone (GnRH) antagonist on basal and GnRH-stimulated gonadotrophin secretion in orchidectomized monkeys. J Endocrinol 129:363–370

Kliesch S, Behre HM, Nieschlag E (1994) High efficacy of gonadotropin or pulsatile gonadotropin-releasing hormone treatment in hypogonadotropic hypogonadal men. Eur J Endocrinol 131:347–354

Kliesch S, Behre, Nieschlag E (1995) Recombinant human follicle-stimulating hormone and human chorionic gonadotropin for induction of spermatogenesis in a hypogonadotropic male. Fertil Steril 63:1326–1328

Kohlo KL, Huhtaniemi I (1989) Neonatal treatment of male rats with a gonadotropin-releasing hormone antagonist impairs ejaculation and fertility. Physiol Behav 46:373–377

Kremer H, Kraaij R, Toledo SPA, Psot M, Fridman JB, Hayashida CY, van Reen M, Milgrom E, Ropers HH, Marimann E, Themmen APN, Brunner HG (1995) Male pseudohermaphroditism due to a homozygous missense mutation of the luteinizing hormone receptor gene. Nature Gen 9:160–164

Kuiper CG, Gustafsson JA (1997) The novel estrogen receptor-beta subtype: potential role in the cell- and promoter-specific actionss of estrogens and anti-estrogens. FEBS Lett 410:87–90

Kula K (1988) Induction of precocious maturation of spermatogenesis in infant rats by human menopausal gonadotropin and inhibition by simultaneous administration of gonadotropins and testosterone. Endocrinology 122:34–39

Kumar TF, Wang Y, Lu N, Matzuk MM (1997) Follicle stimulating hormone is required for ovarian follicle maturation but not male fertility. Nature Gen 15:201–204

Kung AWC, Zhong YY, Lam KSL, Wang C (1994) Induction of spermatogenesis with gonadotrophins in Chinese men with hypogonadotrophic hypogonadism. Int J Androl 17:241–247

Laborde P, Barkey RJ, Belair L, Remy JJ, Dijane J, Salesse R (1996) Ontogenesis of LH and FSH receptors in postnatal rabbit testes: age-dependent differential expression of long and short RNA transcripts. J Reprod Fertil 108:25–30

Larriba S, Esteban C, Toran N, Gerard A, Audi L, Gerard H, Reventos J (1995) Androgen binding protein is tissue-specifically expressed and biologically active in transgenic mice. J Steroid Biochem Mol Biol 53:573–578

Lerchl A, Sotiriadu S, Behre HM, Pierce J, Weinbauer GF, Kliesch S, Nieschlag E (1993) Restoration of spermatogenesis by follicle-stimulating hormone despite low intratesticular testosterone in photoinhibited hypogonadotropic Djungarian hamsters (*Phodopus sungorus*). Biol Reprod 49:1108–1116

Li H, Papadopoulos V, Vidic B, Dym M, Culty M (1997) Regulation of rat testis gonocyte proliferation by platelet-derived growth factor and estradiol: identification of signaling mechanisms involved. Endocrinology 138:1289–1298

Lindsey S, Wilkinson MF (1996) *Pem*: a testosterone- and LH-regulated homeobox gene expressed in mouse Sertoli cells and epididymis. Dev Biol 179:471–484

Lubahn DB, Moyer JS, Golding TS, Couse JF, Korach KS, Smithies O (1993) Alteration of reproductive function but not prenatal sexual development after insertional disruption of the mouse estrogen receptor gene. PNAS 90:11162–11166

Lunn SF, Cowen GM, Fraser HM (1997) Blockade of the neonatal increase in testosterone by a GnRH antagonist: the free androgen index, reproductive capacity and postmortem findings in the male marmoset monkey. J Endocrinol 154:125–131

Maguire SM, Tribley WA, Griswold MD (1997) Follicle-stimulating hormone (FSH) regulates the expression of FSH receptor messenger ribonucleic acid in cultured Sertoli cells and in hypophysectomized rat testis. Biol Reprod 56:1106–1111

Marshall GR, Wickings EJ, Nieschlag E (1984) Testosterone can initiate spermatogenesis in an immature nonhuman primate, Macaca fascicularis. Endocrinology 114:2228–2233

Marshall GR, Jockenhövel F, Lüdecke D, Nieschlag E (1986) Maintenance of complete but quantitatively reduced spermatogenesis in hypophysectomized monkeys by testosterone alone. Acta Endocrinol 113:424–431

Marshall GR, Zorub DS, Plant TM (1995) Follicle-stimulating hormone amplifies the population of differentiated spermatogonia in the hypophysectomized testosterone-replaced adult rhesus monkey (*Macaca mulatta*). Endocrinology 136:3504–3511

Mather JP, Moore A, Li RH (1997) Activins, inhibins, and follistatins: further thoughts on a growing family of regulators. Proc Soc Exp Biol Med 215:209–222

Matikainen T, Toppari J, Vihko KK, Huhtaniemi I (1994) Effects of recombinant human FSH in immuature hypophysetomized male rats: evidence for Leydig cell-mediated action on spermatogenesis. J Endocrinol 141:449–457

Matsumoto AM (1994) Hormonal therapy of male hypogonadism. Endocrinol Metab Clin North Am 23:857–875

McLachlan RI, Wreford NG, Meachem SJ, de Kretser DM, Robertson DM (1994) Effects of testosterone on spermatogenic cell populations in the adult rat. Biol Reprod 51:945–955

McLachlan RI, Wreford NG, deKretser DM, Robertson DM (1995) The effects of recombinant follicle-stimulating hormone on the restoration of spermatogenesis in the gonadotropin-releasing hormone-immunized adult rat. Endocrinology 136:4035–4043

Milette JJ, Schwartz NB, Turek FW (1998) The importance of follice stimulating hormone in the initiation of testicular growth in photostimulated Djungarian hamsters. Endocrinology 122:1060–1066

Monaco L, Nantel F, Foulkes NS, Sassone-Corsi P (1996) CREM: A transcriptional master switch governing the cAMP response in the testis. In: Hansson V, Levy FO, Tasken K (eds) Signal transduction in testicular cells. Springer Verlag, Berlin, pp 69–94

Moudgal NR, Ravindranath N, Murthy GS, Dighe RR, Aravindan GR, Martin F (1992) Long-term contraceptive efficacy of vaccine of ovine follicle-stimulating hormone in male bonnet monkeys (*Macaca radiata*). J Reprod Fertil 96:91–102.

Moudgal NR, Sairam MR, Krishnamurthy HN, Sridhar S, Krishnamurthy H, Khan H (1997) Immunization of male bonnet monkeys (*M. radiata*) with a recombinant FSH receptor preparation affects testicular function and fertility. Endocrinology 138:3065–3068

Muffly KE, Nazian SJ, Cameron DF (1994) Effects of follicle-stimulating hormone on the junction-related Sertoli cell cytoskeleton and daily sperm production in testosterone-treated hypophysectomized rats. Biol Reprod 51:158–166

Nantel F, Monaco L, Foulkes NS, Masquilier D, LeMeur M, Henriksen K, Dierich A, Parvinen M, Sassone-Corsi P (1996) Spermiogenesis deficiency and germ cell apoptosis in CREM-mutant mice. Nature 380:159–162

Niklowitz P, Khan S, Bergmann M, Hoffman K, Nieschlag E (1989) Differential effects of follicle-stimulating hormone and luteinizing hormone on Leydig cell function and restoration of spermatogenesis in hypophysectomized and photoinhibited Djungarian hamsters (*Phodopus sungorus*). Biol Reprod 41:871–880

Niklowitz P, Lerchl A, Nieschlag E (1997) *In vitro* fertilizing capacity of sperm from FSH-treated photoinhibited Djungarian hamsters (*Phodopus sungorus*). J Endocrinol 154:475–481

O'Donnell L, McLachlan RI, Wreford NG, de Kretser DM, Robertson DM (1996) Testosterone withdrawal promotes stage-specific detachment of round spermatids from the rat seminiferous epithelium. Biol Reprod 55:895–901

Pavlou SN, Brewer K, Farley MG, Lindner J, Bastias MC, Rogers BJ, Swift LL, Rivier JE, Vale WV, Conn PM, Herbert CM (1991) Combined administration of a gonadotropin-releasing hormone antagonist and testosterone in men induces reversible azoospermia without loss of libido. J Clin Endocrinol Metab 73:1360–1369

Perryman KJ, Stanton PG, Loveland KL, McLachlan RI, Robertson DM (1996) Hormonal binding dependency of neural cadherin in the binding of round spermatids to Sertoli cells *in vitro*. Endocrinology 137:3877–3883

Pogach L, Giglio W, Nathan W, Huang HFS (1993) Maintenance of spermiogenesis by exogenous testosterone in rats treated with a GnRH antagonist: relationship with androgen-binding protein status. J Reprod Fertil 98:415–422

Rannikko A, Penttila TL, Zhang FP, Toppari J, Parvinen M, Huhtaniemi I (1996) Stage-specific expression of the FSH receptor gene in the prepubertal and adult rat seminiferous epitelium. J Endocrinol 151:29–35

Remy JJ, Bozon V, Couture L, Goxe B, Salesse R, Garnier J (1993) Suppression of fertility in male mice by immunization against LH receptor. J Reprod Immunol 25:63–79

Remy JJ, Couture L, Rabesona H, Haertle T, Salesse R (1996) Immunization against exon 1 decapeptides from lutropin/choriogonadotropin receptor or the follitropin receptor as potential male contraceptive. J Reprod Immunol 32:37–54

Reventos J, Hammond GL, Crozat A, Brooks DE, Gunsalus GL, Bardin CW, Musto NS (1988) Hormonal regulation of rat androgen-binding protein (ABP) messenger ribonucleic acid and homology of human testosterone-estradiol-binding globulin and ABP complementary deoxyribonucleic acids. Mol Endocrinol 2:125–132

Rosiepen G, Arslan M, Nieschlag E, Weinbauer GF (1997) Estimation of the duration of the cycle of the seminiferous epithelium in the non-human primate *Macaca mulatta* using the 5-bromodeoxyuridine technique. Cell Tiss Res 288:365–369

Rosiepen G, Chapin RE, Weinbauer GF (1995) The duration of the cycle of the seminiferous epithelium is altered by administration of 2,5-hexanedione in the adult Sprague-Dawley rat. J Androl 16:127–135

Rosiepen G, Weinbauer GF, Schlatt S, Behre HM, Nieschlag E (1994) Duration of the cycle of the seminiferous epithelium, estimated by the 5-bromodeoxyuridine technique, in laboratory and feral rats. J Reprod Fertil 100:299–306

Russell LD, Brinster RL (1996) Ultrastructural observations of spermatogenesis following transplantation of rat testis cells into mouse seminiferous tubules. J Androl 17:615–627

Russell LD, Clermont Y (1977) Degeneration of germ cells in normal, hypophysectomized and hormone treated hypophysectomized rats. Anat Rec 187:347–366

Russell LD, Alger LE, Nequin LG (1987) Hormonal control of pubertal spermatogenesis. Endocrinology 120:1615–1632

Russell LD, Corbin TJ, Borg KE, de Franca LR, Grasso P, Bartke A (1993) Recombinant human follicle-stimulating hormone is capable of exerting a biological effect in the adult hypophysectomized rat by reducing the numbers of degenerating germ cells. Endocrinology 133:2062–2070

Schaison G, Yaoung J, Pholsena M, Nahoul K, Couzinet B (1993) Failure of combined follicle-stimulating hormone-testosterone administration to initiate and/or maintain spermatogenesis in men with hypogonadotropic hypogonadism. J Clin Endocrinol Metab 77:1545–1549

Schlatt S, Weinbauer GF, Arslan M, Nieschlag E (1993) Appearance of a smooth muscle actin in peritubular cells of monkey testes is induced by androgens, modulated by follicle-stimulating hormone, and maintained after hormonal withdrawal. J Androl 14:340–350

Schlatt S, Arslan M, Weinbauer GF, Behre HM, Nieschlag E (1995) Endocrine control of testicular somatic and premeiotic germ cell development in the immature testis of the primate *Macaca mulatta*. Eur J Endocrinol 133:235–247

Schlatt S, Meinhardt A, Nieschlag E (1997) Paracrine regulation of cellular interactions in the testis: factors in search of a function. Eur J Endocrinol 137:107–117

Schulze W, Rehder U (1984) Organization and morphogenesis of the human seminiferous epithelium. Cell Tiss Res 237:395–407

Shan LX, Zhu LJ, Bardin CW, Hardy MP (1995) Quantitative analysis of androgen receptor messenger ribonucleic acid in developing Leydig cells and Sertoli cells by in situ hybridization. Endocrinology 136:3856–3862

Shan LX, Bardin CW, Hardy MP (1997) Immunohistochemical analysis of androgen effects on androgen receptor expression in developing Leydig and Sertoli cells. Endocrinology 138:1259–1266

Sharma OP, Khan SA, Weinbauer GF, Arslan M, Nieschlag E (1990) Bioactivity and immunoreactivity of androgen stimulated pituitary FSH in GnRH antagonist-treated male rats. Acta Endocrinol 122:168–174

Shenker A, Laue L, Kosugi S, Merendino L, Minegishi T, Cutler GN (1993) A constitutively activating mutation of the luteinizing hormone receptor in familial precocious puberty. Nature 365:652–654

Shetty G, Krishnamurthy H, Krishnamurthy HN, Bhatnagar AS, Moudgal RN (1997) Effect of estrogen deprivation on the reproductive physiology of male and female primates. J Steroid Biochem Mol Biol 61:157–166

Shetty J, Marathe GK, Dighe RR (1996) Specific immunoneutralization of FSH leads to apoptotic cell death of the pachytene spermatocytes and spermatogonial cells in the rat. Endocrinology 137:2179–2182

Simoni M, Gromoll J, Nieschlag E (1997) The follicle-stimulating hormone receptor: biochemistry, molecular biology, physiology, and pathophysiology. Endocr Rev 18:739–773

Singh J, Handelsman DJ (1996) The effects of recombinant FSH on testosterone-induced spermatogenesis in gonadotrophin-deficient (*hpg*) mice. J Androl 17:382–393

Singh J, O'Neill C, Handelsman DJ (1995) Induction of spermatogenesis by androgens in gonadotropin-deficient (*hpg*) mice. Endocrinology 136:5311–5321

Sinha Hikim AP, Swerdloff RS (1993) Temporal and stage-specific changes in spermatogenesis of rat after gonadotropin deprivation by a potent gonadotropin-releasing hormone antagonist treatment. Endocrinology 133:2161–2170

Sinha Hikim AP, Rajavashisth TB, Sinha Hikim I, Lue Y, Bonavera JJ, Leung A, Wang C, Swerdloff RS (1997) Significance of apoptosis in the temporal and stage-specific loss of germ cells in the adult rat after gonadotropin deprivation. Biol Reprod 57:1193–1201

Smithwick EB, Young LG, Gould KG (1996) Duration of spermatogenesis and relative frequency of each stage in the seminiferous epithelial cycle of the chimpanzee. Tiss Cell 28:357–366

Suresh R, Medhamurthy R, Moudal NR (1995) Comparative studies on the effects of specific immunoneutralization of endogenous FSH or LH on testicular germ cell transformations in the adult bonnet monkey (*Macaca radiata*). Am J Reprod Immunol 34:35–43

Tapanainen JS, Aittomäki K, Min J, Vaskivuo T, Huhtaniemi IT (1997) Men homozygous for an inactivating mutation of the follicle-stimulating hormone (FSH) receptor gene present variable suppression of spermatogenesis and fertility. Nature Gen 15:205–206

Teerds KJ, Veldhuizen-Tsoerkan MB, Rommerts FFG, de Rooij DG, Dorrington DJ (1994) Proliferation and differentiation of testicular interstitial cells: aspects of Leydig cell development in the (pre)pubertal and adult testis. In: Verhoeven G, Habenicht UF (eds) Molecular and cellular endocrinology of the testis. Springer, Berlin Heidelberg New York Tokyo, pp 37–66

Troiano L, Fustini MF, Lovato E, Frasoldati A, Malorni W, Capri M, Grassilli E, Marrama P, Franceschi C (1994) Apoptosis and spermatogenesis: evidence from an in vivo model of testosterone withdrawal in the adult rat. Biochem Biophys Res Comm 202:1315–1321

van Alphen MAA, van de Kant HJG, de Rooij DG (1988) Follicle-stimulating hormone stimulates spermatogenesis in the adult monkey. Endocrinology 123:1449–1455

van Roijen JH, van Assen S, van der Kwast TH, de Rooij DG, Boersma JA, Vreeburg JTM, Weber RFA (1995) Androgen receptor immunoexpression in the testis of infertile men. J Androl 16:510–516

Vannier B, Loosfelt, Meduri G, Pichon C, Milgrom E (1996) Anti-human FSH receptor monoclonal antibodies: immunochemical and immunocytochemical characterization of the receptor. Biochemistry 35:1358–1366

Veldhuis JD, Iranmanesh A, Samojlik E, Urban RJ (1997) Differential sex steroid negative feedback regulation of pulsatile follicle-stimulating hormone secretion in healthy older men: deconvolution analysis and steady-state sex-steroid hormone infusions in frequently sampled healthy older individuals. J Clin Endocrinol Metab 82:1248–1254

Verhoeven G, Cailleau J (1988) Follicle-stimulating hormone and androgens increase the concentration of the androgen receptor in Sertoli cells. Endocrinology 122:1541–1550

Vihko KK, LaPolt PS, Nishimori K, Hsueh AJW (1991) Stimulatory effects of recombinant follicle-stimulating hormone on Leydig cell function and spermatogenesis in immature hypophysectomized rats. Endocrinology 129:1926–1932

Vornberger W, Prins G, Musto NA, Suarez-Quian CA (1994) Androgen receptor distribution in rat testis: new implications for androgen regulation of spermatogenesis. Endocrinology 123:2307–2316

Weinbauer GF, Nieschlag E (1990). The role of testosterone in spermatogenesis. In: Nieschlag E, Behre HM (eds) Testosterone-action, deficiency, substitution. Springer-Verlag, Berlin, pp 23–50

Weinbauer GF, Nieschlag E (1993) Hormonal control of spermatogenesis. In: de Kretser DM (ed) Molecular biology of the male reproductive system. Academic Press, New York, pp 99–142

Weinbauer GF, Nieschlag E (1996a) Hormonal regulation of reproductive organs. In: Greger R, Windhorst U (eds) Comprehensive human physiology – from cellular mechanisms to integration. Springer-Verlag, Berlin, pp 2231–2252

Weinbauer GF, Nieschlag (1996b) The Leydig cell as a target for male contraception. In: Payne AH, Hardy MP, Russell LD (eds) The Leydig cell. Cache River Press, Clearwater, pp 629–662

Weinbauer GF, Nieschlag E (1997) Germ cell proliferation and germ cell loss in the primate testis. In: Waites GMH, Frick J, Baker GWH (eds) Current advances in andrology. Monduzzi Editore, Bologna, pp 211–220

Weinbauer GF, Behre HM, Fingscheidt U, Nieschlag E (1991) Human follicle stimulating hormone exerts a stimulatory effect on spermatogenesis, testicular size, and serum inhibin levels in the gonadotropin-releasing hormone antagonist-treated non-human primate (Macaca fascicularis). Endocrinology 129:1831–1839

Weinbauer GF, Schubert J, Yeung CH, Rosiepen G, Nieschlag E (1998a) Gonadotrophin-releasing hormone (GnRH) antagonist arrests premeiotic germ cell proliferation but does not inhibit meiosis in the male monkey: A quantitative analysis using 5-bromodeoxyuridine and dual parameter flow cytometry. J Endocrinol 156:23–34

Weinbauer GF, Behr R, Bergmann M, Nieschlag E (1998b) Testicular cAMP response element modulator (CREM) protein is expressed in round spermatids but is absent or reduced in men with round spermatid maturation arrest. Mol Human Reprod (in press)

Wickings EJ, Nieschlag E (1978) The effect of active immunization with testosterone on pituitary-gonadal feedback in the male rhesus monkey (*Macaca mulatta*). Biol Reprod 18:602–607

Zhengwei Y, Wreford NG, Schlatt S, Weinbauer GF, Nieschlag E, McLachlan RI (1998) GnRH antagonist-induced gonadotropin withdrawal acutely and specifically reduces type B spermatogonial number in the adult macaque. J Reprod Fertil (in press)

Zirkin BR (1993) Regulation of spermatogenesis in the adult mammal: gonadotropins and androgens. In: Desjardins C, Ewing LL (eds) Cell and molecular biology of the testis. Oxford University Press, Oxford, pp 166–188

5 Androgens and hair

Valerie Anne Randall

<table>
<tr><td colspan="3">Contents</td></tr>
</table>

5.1 Introduction

Human hair growth plays a major role in social and sexual communication. Often subconsciously people all over the world classify a person's state of health, sex, sexual maturity and age by assessing their scalp and body hair.

The importance of hair can be seen in the many social customs associated with it in different cultures such as never cutting the scalp hair by Sikhs or shaving the head of Buddhist monks. Body hair is also involved; for example the widespread custom of shaving men's beards and the habit in Northern Europe and the USA, but frequently not in Southern Europe, of shaving women's axillary hair. This is why abnormalities of hair growth, both greater or less than "normal", even including the common male pattern baldness, cause widespread psychological distress.

Androgens are the most obvious regulators of human hair growth. Although hair with a major protective role such as the eyelashes, eyebrows and scalp hair is produced by children in the absence of androgens, the formation of long pigmented hair on the axillae, pubis, face etc. needs androgens in both sexes. In contrast, androgens may also inhibit hair growth on the scalp, causing baldness. How one type of hormone can cause these contradictory effects in the same tissue in different body sites simultaneously in one person is an endocrinological paradox. The hair follicle has another exciting characteristic. It is the only tissue in the adult body which can regenerate, often producing a new hair with different features. This is how androgens can stimulate such major changes.

Recently, there has been a great deal of interest in the hair follicle promoted by the discovery that the antihypertensive drug, minoxidil, could sometimes stimulate hair growth. However, relatively little is known about the precise functioning of this complex cell biological system at the biochemical level. Nevertheless, our knowledge about the mechanism of androgens in the follicle has enabled the treatment of hirsutism in women with antiandrogens, such as cyproterone acetate, and currently the 5α-reductase type II inhibitor, finasteride, developed to regulate prostate disorders, is under clinical trial for use in male pattern baldness. Greater understanding of hair follicle biology may also enable the development of further treatments in the future.

This chapter will cover our current knowledge of the structure and function of hair follicles, their responses to androgens, the mechanism of action of androgens in the follicle and current modes of control of androgen-potentiated disorders.

5.2 Structure and function of the hair follicle

5.2.1 Functions of hair

Hairs cover almost all the body surface of human beings except for the soles of the feet, palms of the hands and the lips. They are fully keratinised tubes of dead epithelial cells where they project outside the skin. They taper to a point, but otherwise are extremely variable in length, thickness, colour and cross-sectional shape. These differences occur between individuals e.g. blonde, red or dark haired people and between specific body areas within one individual such as the long thick scalp and adult male beard hairs and the short, fine ones on the back of the hand. They also occur on the same parts of the same person at different stages of their life e.g. the fine, short almost colourless hairs on a boy's face are replaced by darker, thicker and longer beard hairs in adulthood.

The main functions of mammalian hair are insulation and camouflage. These are no longer necessary for the "naked ape," although vestiges of this remain in the seasonal patterns of our hair growth (Randall and Ebling 1991) and the erection of our body hairs when shivering with cold. Mammals often have specialised hairs as neuroreceptors e.g. whiskers and this remains slightly in human body hair with its good nerve supply, but the main functions of human hair are protection and communication. Eyelashes and eyebrow hairs prevent substances entering the eyes and scalp hair may protect the scalp and back of the neck from sun damage during our upright posture. During puberty the development of axillary and pubic hair signals the beginning of sexual maturity in both sexes (Marshall and Tanner 1969, 1970; Winter and Faiman 1972, 1973) while the male beard readily distinguishes the sexes like the mane of the lion.

5.2.2 Structure of the hair follicle

Each hair is produced by a hair follicle. Hair follicles are cylindrical epidermal down-growths into the dermis and subcutaneous fat (Fig. 5.1). Each enlarges at its base into a hair bulb where it surrounds the tear-shaped, mesenchyme-derived dermal papilla. The dermal papilla, which contains specialised fibroblast-like cells embedded in an extracellular matrix and separated from the epithelial components by a basement membrane, regulates many aspects of hair growth.

The hair is produced by epithelial cell division in the bulb; the keratinocytes move upwards, undergoing differentiation into the various layers of the follicle. The central portion forms the hair itself whose colour is produced by pigment donated by the follicular melanocytes. By the time it reaches the surface the cells are fully keratinised and dead. The hair is surrounded by two multi-layered epithelial sheaths: the inner root sheath, which helps it move through the skin and which disintegrates when level with the sebaceous gland, and the outer root sheath, which becomes continuous with the epidermis completing the skin's protective barrier (Fig. 5.1).

5.2.3 The hair follicle growth cycle

Cell division continues until the hair reaches the appropriate length for its body site. The length of this period of hair growth, or *anagen*, can range from two years or more on the scalp (Kligman 1959) to only about two months on the finger (Saitoh and Sakamoto 1970).

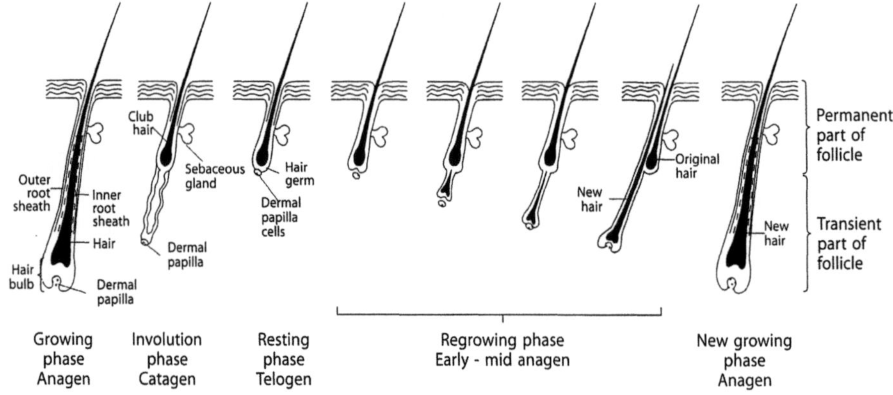

Fig. 5.1. Diagram of the hair follicle growth cycle.
Hair follicles pass through regular cycles of growth and rest during which the lower part of the follicle is regenerated. This enables the follicle to produce a different type of hair in response to hormonal stimuli to co-ordinate changes in the body's development e.g. sexual maturity or seasonal climate changes

At the end of anagen, cell division stops and the lower follicle regresses, entering a transient stage known as *catagen*. The hair itself becomes fully keratinised with a swollen or "club" end and moves up in the skin, resting below the level of the sebaceous gland. The dermal papilla also regresses, losing the extracellular matrix and the cells become inactive. The dermal papilla cells rest below the *club hair* associated with epithelial cells (Fig. 5.1) and the follicle then enters a variable period of rest termed *telogen*. At the end of telogen the dermal papilla cells reactivate, epithelial cells recommence cell division and a lower follicle is regenerated growing back down into the dermis and producing a new hair (Fig. 5.1). The new hair grows up into the permanent part of the hair follicle alongside the old hair which is shed (Fig. 5.1).

The origin of the epithelial cells which give rise to the new lower follicle is currently the subject of some debate. Comparatively recently epithelial stem cells were identified in the bulge region of the outer root sheath below the sebaceous gland (Cotsarelis et al. 1990), but the traditional view of stem cells in the epithelial germ, known as germinative epithelial cells, is supported by elegant cell co-culture experiments of the various follicular cell types (Jahoda and Reynolds 1996). It seems likely that both stem cell types are involved in the hair follicle, with the bulge cells as less specialised, higher order stem cells in line with the haemopoietic system.

Although the hair follicle growth cycle has been well documented (Kligman 1959), the control mechanisms are complex and still not understood (Stenn et al. 1996). It is clear that the early stages of anagen at least partially recapitulate the embryogenesis of the hair follicle to an unique extent in the adult. The processes of the hair growth cycle allow the follicle to replace the hair with a different type to correlate with changes in the environment or maturity of the individual. These changes are co-ordinated by the pineal-hypophysis-pituitary system (Ebling et al. 1991). Co-ordination to the environment is particularly important for some mammals, such as arctic hares, which need a longer, warmer and white coat in the snowy winter but a shorter, brown coat in the summer to increase their chances of survival. Human beings in the temperate regions also exhibit seasonal changes in both scalp (Orentreich 1969, Randall and Ebling 1991) and body hair (Randall and Ebling 1991). The main change in human hair growth is the production of adult patterns of body hair growth after puberty, like the male lion's mane, in response to androgens; seasonal fluctuations in human body hair growth may also co-ordinate to those of androgens (discussed in Randall and Ebling 1991).

5.3 The effects of androgens on human hair growth

5.3.1 Hair types in children

In utero the human body is covered with quite long, colourless *lanugo* hairs. These are shed before birth and at birth, or shortly after, babies normally exhibit pigmented, quite thick protective hairs on the eyebrows and eyelashes and variable amounts on the scalp; by the age of three or four the scalp hair is usually quite well developed, though it will not yet have reached its maximum length. These readily visible pigmented hairs are known as *terminal* hairs and are formed by large deep *terminal follicles* (Fig. 5.2). This emphasises that terminal hair growth on the scalp, eyelashes and eyebrows is not androgen-dependent. The rest of the body is often considered hairless but, except for the glabrous skin of the lips, palms and sole of the feet, is normally covered with fine, short, almost colourless *vellus hairs* produced by small, short *vellus follicles* (Fig. 5.2). The molecular mechanisms involved in the distribution and formation of the different types of follicles during embryogenesis are not clear, but the transcription factor LEF-1, various homeobox genes and retinoic acid have all been implicated (Stenn et al. 1996).

5.3.2 Androgen-stimulated increases in terminal hair growth

5.3.2.1 Increase in terminal hair after puberty

One of the first signs of puberty is the gradual appearance of a few larger and more pigmented *intermediate* hairs, firstly in the pubic region and later in the axillae. These are replaced by longer and darker hairs (Fig. 5.2) and the area spreads. The terminal hair appearance parallels the rise in circulating androgen levels and occurs later in boys than girls (Marshall and Tanner 1969, 1970; Winter and Faiman 1972, 1973). In boys, similar changes occur gradually on the face, starting above the mouth and on the central chin,

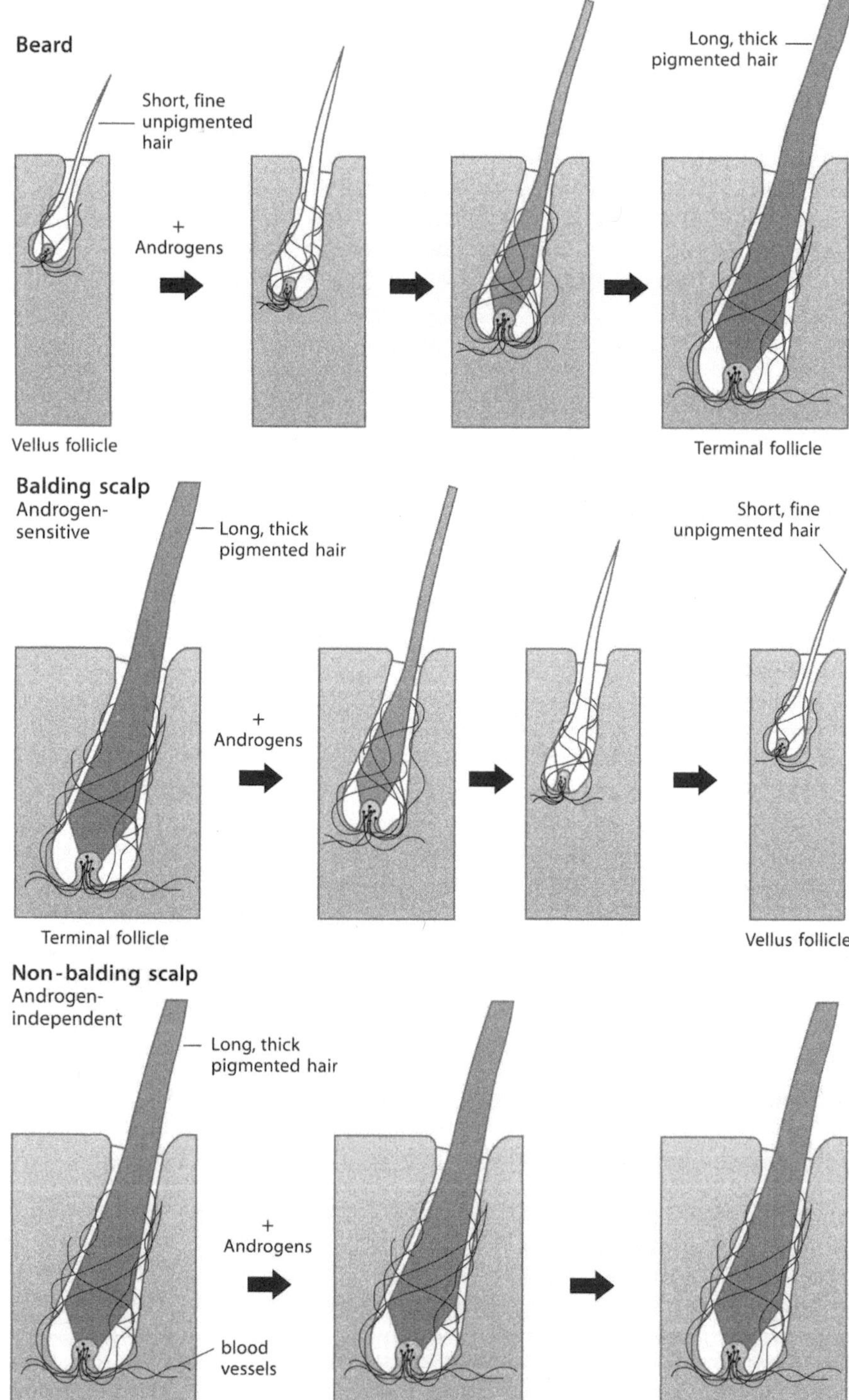

Beard
Short, fine unpigmented hair
+ Androgens
Long, thick pigmented hair
Vellus follicle
Terminal follicle

Balding scalp
Androgen-sensitive
Long, thick pigmented hair
+ Androgens
Short, fine unpigmented hair
Terminal follicle
Vellus follicle

Non-balding scalp
Androgen-independent
Long, thick pigmented hair
+ Androgens
blood vessels

eventually generally spreading over the lower part of the face and parts of the neck, readily distinguishing the adult male. The adult man's pubic hair distribution also differs from the woman, extending in a diamond shape up to the navel in contrast to the woman's inverted triangle. Terminal hair on the chest and sometimes the back is also normally restricted to men, though both sexes may also develop intermediate terminal hairs on their arms and legs, with terminal hairs normally restricted to the lower limbs in women (Fig. 5.3).

In all areas the responses are gradual, often taking many years. Beard weight increases dramatically during puberty, but continues to rise until the mid-thirties (Hamilton 1958), while terminal hair growth on the chest and in the external ear canal may first be seen many years after puberty (Hamilton 1946).

The amount of body hair is very variable and differs both between families within one race and between races, with Caucasians generally exhibiting more than Orientals (Hamilton 1958). This implicates a genetically-determined response to circulating triggers. The responses of the follicles themselves also vary with female hormone levels, being sufficient to stimulate terminal hair growth in the pubis and axillae, but male hormones required for other areas, such as the beard and chest. Beard hair growth also remains high, right into a man's seventies, while axillary growth is maximal in the mid-twenties and falls quite rapidly then in both sexes. These contrasts are presumably due to differential gene expression within follicles from the various body sites. The intrinsic response of individual follicles is retained when follicles are transplanted to other skin sites (Ebling and Johnson 1959); this is the basis of corrective hair follicle transplant surgery (Orentreich and Durr 1982).

5.3.2.2 Evidence for the role of androgens

Although various circulating factors affect human hair growth, androgens are the clearest regulators in contrast to most mammals (Ebling et al. 1991). Other circulating factors (reviewed in Randall 1994a) which have effects are an adequate diet due to the follicles' high metabolic demands (Bradfield 1971), the hormones of pregnancy, which cause a prolonged anagen resulting in a synchronised shedding of a proportion of scalp hairs post-partum (Lynfield 1960), and hypothyroidism, which inhibits hair growth (Jackson et al. 1972). Growth hormone is also necessary in combination with androgens for

Fig. 5.2. Summary of various paradoxical effects of androgens on hair follicles. Androgens have no effects on some follicles (lower diagram), stimulate the gradual transformation of vellus follicles to terminal ones producing large pigmented hairs in many regions (upper diagram), while causing the reverse effect on the scalp in genetically-disposed individuals (middle diagram). The hair follicle undergoes several hair cycles (see Fig. 5.1) between producing vellus and terminal hairs (modified from Randall 1996)

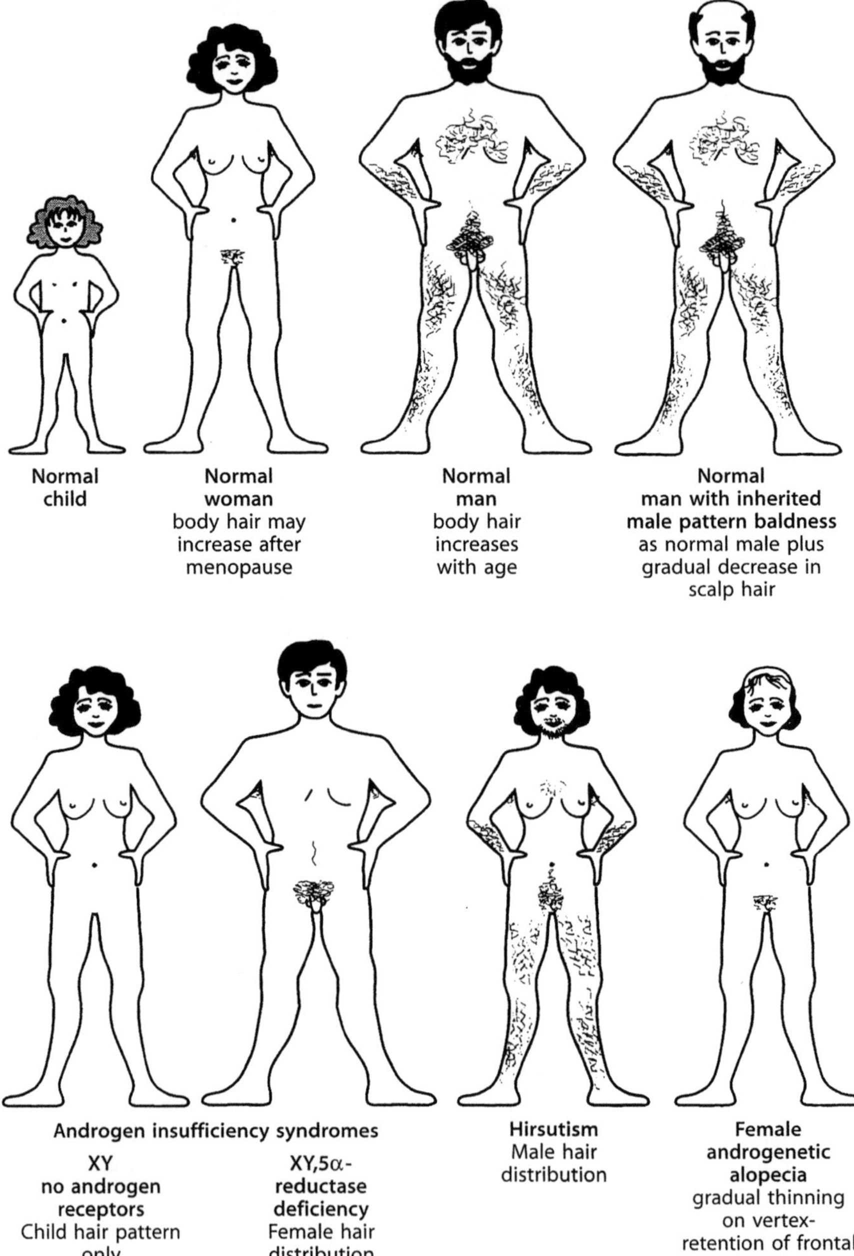

Fig. 5.3. Terminal hair distribution in people under differing endocrine conditions

normal body hair development in boys (Zachmann and Prader 1970). There is a range of evidence supporting the importance of androgens which fits in well with the concept of much terminal hair growth being a secondary sexual characteristic. As well as the correlation between the pubertal rise of androgens and the appearance of terminal hairs, testosterone will stimulate beard growth in eunuchs and elderly men (Chieffi 1949) and increased beard growth noted by an isolated endocrinologist is ascribed to his rising androgens on anticipating his girlfriend's arrival! An extensive study in the USA also showed that castration before puberty prevented beard and axillary hair growth and after puberty reduced them (Hamilton 1951a, 1958). Nevertheless, the strongest evidence for the essential nature of androgens is the lack of any body hair, even the female pubic and axillary pattern, in adult XY complete androgen insensitivity patients with absent or dysfunctional androgen receptors (Quigley, this volume, Chapter 2).

5.3.3 Inhibition of scalp hair growth by androgens (androgenetic alopecia)

5.3.3.1 In men

There is a generalised loss of hair follicles from the scalp in both sexes by the seventh or eighth decade known as *senile balding*. This differs from the progressive baldness seen in *androgenetic alopecia,* also known as *male pattern baldness* or *alopecia, common baldness* or *androgen-dependent alopecia.* This is characterised by a gradual inhibition of terminal follicles to smaller vellus follicles (Fig. 5.2) with the length of anagen decreasing and that of telogen increasing. The connective tissue sheath left in the dermis when the follicle becomes miniaturised frequently becomes subject to chronic inflammation; this may prevent terminal hair regrowth in long-term baldness. Balding occurs in a precise pattern described by Hamilton (1951b), starting with regression of the frontal hairline in two wings and balding in the centre of the vertex. These areas gradually expand and coalesce, exposing large areas of scalp; generally the back and sides of the scalp retain terminal hair even in extreme cases (Fig. 5.3). Hamilton's scale was later modified by Norwood (1975) to include a wider range of patterns.

Male pattern baldness is androgen-dependent, since unless they are given testosterone, it does not occur in castrates (Hamilton 1942) nor XY individuals with androgen insensitivity due to non-functional androgen receptors (Quigley this volume, Chapter 2). There is also a marked inherited tendency to develop it (Hamilton 1942). The incidence of androgenetic alopecia in Caucasians is high with estimates varying widely. Other races exhibit it to a lesser extent (Hamilton 1951b; Setty 1970) and it is also seen in some primates, being well studied in the stump-tailed macaque. Male pattern baldness primarily causes concern amongst those who develop marked loss before their forties and early balding has been linked to myocardial infarction (Lesko et al. 1993). Whether this indicates a dual end-organ sensitivity or reflects the psychological stress early balding induces in the youth-orientated American culture is unknown. No relationship between the incidence of balding and prostatic carcinoma has been seen in men between fifty and seventy (Demark-Wahnefried et al. 1997).

5.3.3.2 In women

Androgenetic alopecia has also been described in women, but the pattern of expression is normally different. Women generally do not show the frontal recession, but retain the frontal hairline and exhibit thinning on the vertex which may lead to balding (Ludwig 1977) (Fig. 5.3). Post-menopausal women may exhibit the masculine pattern (Venning and Dawber 1988). The progression of balding in women is normally slow and a full endocrinological investigation is recommended if a rapid onset is seen (Dawber and Van Neste 1995).

5.3.4 Hirsutism

Hirsutism is the development of male pattern body hair growth in women. This causes marked psychological distress because the person erroneously feels that they are changing sex. The extent of body hair growth which causes a problem varies and depends on the amount of normal body hair amongst her race or sub-group. Normally hirsutism would include terminal hair on the face, chest or back. Ferriman and Gallwey (1961) introduced a scale for grading hirsutism which is widely used, especially to monitor hirsutism progression with, or without, treatment.

Hirsutism is often associated with an endocrine abnormality of the adrenal or ovary causing raised androgens and is frequently associated with polycystic ovarian (PCO) syndrome. Some women have no obvious underlying disorder and are termed "idiopathic". The proportion of these is larger in older papers as modern methods increase the range of abnormalities that can be detected e.g. low sex hormone building globulin. The assumption that idiopathic hirsutism is due to a greater sensitivity to normal androgens is given credence by the report of hirsutism on only one side of a woman by Jenkins and Ash (1973).

5.4 The mechanism of androgen action in the hair follicle

5.4.1 Hair growth in androgen insufficiency syndromes

As described in Chapters 1 and 2 of this volume, androgens from the blood stream enter the cell and bind to specific, intracellular androgen receptors, usually in the form of testosterone or its more potent metabolite, 5α-dihydrotestosterone. The hormone-receptor complex then activates the appropriate gene transcription for that cell type.

The importance of androgen receptors within hair follicles for the development of androgen-dependent hair growth is shown by androgen-insensitivity patients without functional androgen receptors (see this volume, Chapter 2). These individuals produce no body hair at puberty, even with high circulating androgen levels, nor do they go bald (Fig. 5.3).

Men with 5α-reductase deficiency also contribute to our understanding because they exhibit axillary and female pattern pubic hair, but very little beard growth; they are not reported to have male pattern baldness either (Griffin and Wilson 1989) (Fig. 5.3). This suggests that the formation of terminal pubic and axillary hair can be mediated by testosterone itself, while that of the secondary sexual hair of men requires the presence of 5α-dihydrotestosterone. Since androgens are stimulating the same transformation, presumably via the same receptor, this is currently difficult to understand although it is further evidence of the intrinsic differences within hair follicles. It suggests that some more specific aspect of androgen action is involved in male pattern hair follicles which require 5α-dihydrotestosterone, such as interaction with a specific transcription factor. Interestingly, androgen-dependent sebum production by the sebaceous glands attached to hair follicles is also normal in 5α-reductase deficiency (Imperato-McGinley et al. 1993). The identification of two forms of 5α-reductase, type 1 and type 2, has made the situation more complex, but all individuals with 5α-reductase deficiency so far have been shown to be deficient in 5α-reductase type 2 (reviewed by Randall 1994b) which appears to be the important form for much androgen-dependent hair growth.

5.4.2 The current model for androgen action in the hair follicle

5.4.2.1 The role of the dermal papilla

The mesenchyme-derived dermal papilla plays a major role in determining the type of hair produced by a follicle as shown by an elegant series of experiments involving the rat whisker by Oliver, Jahoda, Reynolds and colleagues (reviewed by Jahoda and Reynolds 1996). Whisker dermal papillae transplanted into ear or glabrous skin stimulated the production of whisker follicles and hair growth could also be stimulated by cultured dermal papilla cells reimplanted *in vivo*.

In many embryonic steroid-regulated tissues, including the prostate and the breast, steroids act via the mesenchyme (Cunha et al. 1987). Since during their growth cycles hair follicles recapitulate the stages of embryogenesis and ^{3}H-testosterone had been localised to rat follicle dermal papillae, the hypothesis was proposed that androgens would act on the other components of the hair follicle via the dermal papilla (Randall et al. 1991). In this hypothesis androgens would alter the ability of the dermal papilla cells to syn-

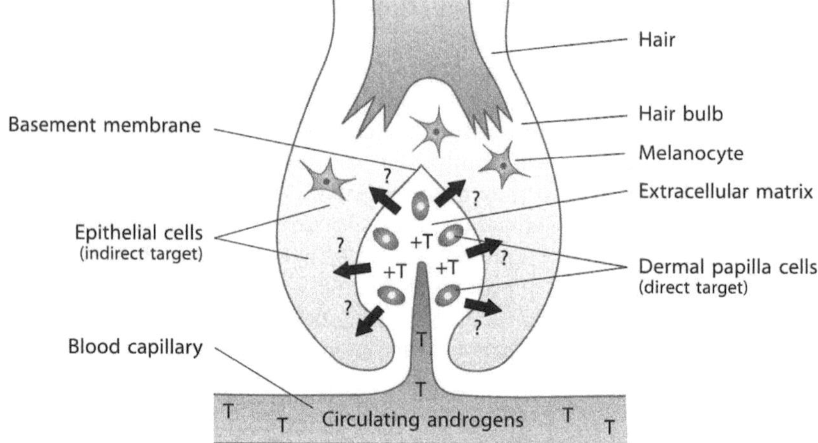

Fig. 5.4. The current model of androgen action in the hair follicle.
Androgens from the blood enter the hair follicle via the dermal papilla's blood supply. They are bound by androgen receptors in the dermal papilla cells which then alter their production of regulatory paracrine factors; these then alter the activity of follicular keratinocytes and melanocytes. T = Testosterone; ? = unknown paracrine factors (modified from Randall 1994)

thesise or release controlling factors which would affect follicular keratinocytes, melanocytes and connective tissue sheath cells and also probably the dermal endothelial cells to alter the follicles' blood supply in proportion to its change in size (Fig. 5.4). These factors could be growth factors and/or extracellular matrix proteins.

This hypothesis has now received a great deal of experimental support. Androgen receptors have been localised by immunohistochemistry in the dermal papilla and not the keratinocyte cells (Choudhry et al. 1992). Cultured dermal papilla cells derived from androgen-sensitive follicles such as beard (Randall et al. 1992) and balding scalp (Hibberts and Randall 1998) contain higher levels of androgen receptor than androgen-insensitive non-balding scalp. Importantly, metabolism of testosterone by cultured dermal papilla cells also reflects hair growth in 5α-reductase deficiency patients with beard, but not pubic or non-balding scalp, cells forming 5α-dihydrotestosterone (Hamada et al. 1996; Itami et al. 1990; Thornton et al. 1993). All these results have led to wide acceptance of the hypothesis.

5.4.2.2 Paracrine factors implicated in mesenchyme-epithelial interactions in the hair follicle

The production of growth factors by cultured dermal papilla cells derived from human and rat hair follicles has been investigated by several groups on the basis of the primary role of the dermal papilla, its potential probable role in androgen action and the retention of hair growth-promoting ability by cultured rat cells (discussed above). Bioassays have shown that human dermal papilla cells secrete soluble proteinaceous factors which stimulate the growth of other dermal papilla cells (Randall et al. 1991; Thornton et al. 1998), outer root sheath cells (Itami et al. 1995), transformed epidermal keratinocytes (Hibberts and Randall 1996) and endothelial cells (Hibberts et al. 1996b) with testosterone *in vitro* stimulating greater mitogenic capacity of beard cells to affect beard but not scalp dermal papilla cells (Thornton et al. 1998), outer root sheath cells (Itami et al. 1995) and keratinocytes (Hibberts and Randall 1996). Interestingly, testosterone decreased the mitogenic capacity of androgenetic alopecia dermal papilla cells from both men (Hibberts and Randall 1996) and stump-tailed macaques (Obana and Uno 1996).

To date only IGF-1 has been identified as androgen-regulated *in vitro* (Itami et al. 1995), but stem cell factor, the ligand for the melanocyte receptor *c-kit*, is secreted in greater quantities by beard dermal papilla cells than non-balding scalp cells (Hibberts et al. 1996a), unlike vascular endothelial growth factor (Hibberts et al. 1996b). Other factors which have been implicated in the follicular dermal papilla include keratinocyte growth factor and hepatocyte growth factor, though many more have been located in the epidermis (reviewed by Stenn et al. 1996). Further study of this area should increase our understanding of the complex hair follicle and lead to better treatments for hair follicle disorders.

5.5 The treatment of androgen-potentiated hair disorders

5.5.1 Androgenetic alopecia

Currently, the most effective treatment for male pattern baldness is the transplant of follicles from non-balding sites into the balding region. This has significant disadvantages; not only is it very invasive and heavily reliant on the skill of the operator for a good cosmetic result, but the alopecia continues to progress behind the transplanted area so that further transplants are often required.

Antiandrogen therapy is not a practical option for men due to the side-effects, but cyproterone acetate, in combination with estrogen to ensure contraception and prevent the potential feminisation of a male fetus, has been used in women. It increased the percentage of hair follicles in anagen and may cause some regrowth, but is probably most effective in preventing further progression (Dawber and Van Neste 1995; Peereboom-Wynia et al. 1989). Since cyproterone acetate is unavailable in the USA, spirolactone and high-dose cimetidine have been used as alternative antiandrogens.

Minoxidil, a vasodilator used for hypertension, stimulated excessive hair growth as a side-effect. This provoked major interest in hair follicle biology because it demonstrated that vellus follicles could be stimulated to form terminal hairs. Topical application of minoxidil has been used in both male and female androgenetic alopecia. It stimulates regrowth in up to thirty per cent with only about ten per cent obtaining complete regrowth. Most success occurs with younger men and with the early stages of balding, i.e. Hamilton stage V or less (Dawber and Van Neste 1995).

Recently, finasteride, a 5α-reductase type II inhibitor, has been developed to treat androgen-potentiated prostate disorders. Oral finasteride is currently undergoing carefully controlled clinical trials around the world with promising early results (Kaufman 1996). All of these treatments need to be used continually because they are opposing a natural process which, if treatment is discontinued, retains all the components to continue to progress.

5.5.2 Hirsutism

Cosmetic treatments such as bleaching, depilatory measures such as shaving or waxing or electrolysis are common. Indeed, electrolysis with the aim of permanent removal by killing the dermal papilla and germinative epithelium/stem cells is the only potentially permanent treatment, but it is expensive, time consuming and may cause scarring; removal by laser treatment has recently been introduced (Lucky 1997). The most common endocrine treatment, outside the USA, is the antiandrogen, cyproterone acetate, given with estrogen if the woman is premenopausal; spirolactone or flutamide can be used as an alternative (Fruzzetti 1997). Patients have to be well-motivated because hair growth on the face generally takes at least nine months before a noticeable effect occurs, although any acne will be cleared in a couple of months and effects on thigh hair growth will be seen in four to six months (Sawers et al. 1982). Facial responses are seen first on the sides of the face and last on the upper lip, in reverse order to the appearance of facial hair in men (personal observations).

Recently finasteride has also been used for hirsutism with some success (Fruzzetti 1997). This seems logical as 5α-reductase type II is necessary for male pattern body hair growth (see Section 5.4.1). Contraception is still required with finasteride as it also has the potential to affect the development of a male fetus.

Overall, there have been major changes in the treatment of androgen-potentiated disorders over the last ten years. The ideal treatment of a uniformly effective, topical treatment which is inactivated on contact with the blood or is specific for hair follicles is not yet available. Further research on the biology of androgen action in the hair follicle may facilitate its development.

> ## 5.6 Key messages
>
> - Androgens are the main regulators of human hair growth.
> - Androgens have paradoxically different effects on hair follicles depending on their body site. They can stimulate the formation of large hairs e.g. beard, axilla, have no effect e.g. eyelashes or inhibit follicles on the scalp.
> - All effects are gradual.
> - Androgen-potentiated disorders of hair growth are common including hirsutism in women and androgenetic alopecia in both sexes.
> - Androgen receptors are necessary for all androgen-dependent hair growth and 5α-reductase type II for most, but not female patterns of axillary and pubic hair.
> - The current model for androgen action in the hair follicle proposes that androgens act via the cells of the dermal papilla, altering their production of regulatory paracrine factors such as growth factors which then influence the activity of other follicular components, e.g. keratinocytes, melanocytes and endothelial cells.
> - Antiandrogens, generally cyproterone acetate, and a 5α-reductase type II inhibitor, finasteride, are being used to control androgenetic alopecia and hirsutism.
> - Endocrine treatments may need several months to show their effects and will need to be used continually.
> - Further understanding of the mechanism of androgens in the hair follicle is necessary to enable the development of better treatments, preferably working topically and specific to the hair follicle.

5.7 Acknowledgements

The personal research mentioned in this review could not have been carried out alone. I would like to thank all my colleagues and collaborators over the years, particularly the late Professor John Ebling, and Kazu Hamada, Nigel Hibberts, Tracey Jenner, S. Kate, Andrew Messenger, Michael Nutbrown and Julie Thornton. I should also like to acknowledge the expert preparation of this manuscript by Mrs Christine Dove and Mr Chris Bowers and Mrs Jenny Braithwaite for the figures.

5.8 References

Bradfield RB (1971) Protein deprivation: comparative response of hair roots, serum protein and urinary nitrogen. Amer J Clin Nutrit 24:405–410

Chieffi M (1949) Effect of testosterone administration on the beard growth of elderly males. J Geront 4:200–204

Choudhry R, Hodgins MB, Van Der Kwast TH, Brinkman AO, Boersma, WJA (1992) Localisation of androgen receptors in human skin by immunohistochemistry: implications for the hormonal regulation of hair growth, sebaceous glands and sweat glands. J Endocr 133:467–475

Cotsarelis C, Sun T, Lavker R (1990) Label-retaining cells reside in the bulge area of the pilosebaceous unit; implications for follicular stem cells, hair cycle and skin carcinogenesis. Cell 61:1329–1337

Cunha GR, Donjacour AA, Cook PS, Mee S, Bigby RH, Higgins S, Sugimura Y (1987) The endocrinology and developmental biology of the prostate. Endocr Rev 8:338–362

Dawber R, Van Neste D (1995) Hair and scalp disorders. Martin Dunitz, London

Demark-Wahnefried W, Lesko SM, Conaway MR, Robertson CN, Clark RV, Lobaugh B, Mathias BJ, Strigo TS, Paulsa DF (1997) Serum androgens: associations with prostate cancer risk and hair patterning. J Androl 18:495–500

Ebling FJG, Johnson E (1959) Hair growth and its relation to vascular supply in rotated skin grafts and transposed flaps in the albino rat. J Embry Exp Morph 9:285–293

Ebling FJG, Hale PA, Randall VA (1991) Hormones and hair growth. In: Goldsmith LA (ed) Biochemistry and physiology of the skin, Second ed. Clarenden Press, Oxford pp 660–696

Ferriman D, Gallwey JD (1961) Clinical assessment of body hair growth in women. J Clin Endocr Metab 21:1440–1447

Fruzetti F (1997) Treatment of hirsutism: antiandrogen and 5α-reductase inhibitor therapy. In: Azziz R, Nestler JE, Dewailly D (eds) Androgen excess disorders in women. Lippincott-Raven, Philadelphia pp 787–797

Griffin JE, Wilson JD (1989) The resistance syndromes: 5α-reductase deficiency, testicular feminisation and related disorders. In: Baudet AL, Sly WS, Valle D (eds) The metabolic basis of inherited disease. Scriver CR, McGraw-Hill, New York pp 1919–1944

Hamada K, Thornton MJ, Liang I, Messenger AG, Randall VA (1996) Pubic and axillary dermal papilla cells do not produce 5α-dihydrotestosterone in culture. J Invest Derm 106:1017–1022

Hamilton JB (1942) Male hormone stimulation is a prerequisite and an incitant in common baldness. Amer J Anat 71:451–480

Hamilton JB (1946) A secondary sexual character that develops in men but not in women upon aging of an organ present in both sexes. Anat Rec 94:466–467

Hamilton JB (1951a) Quantitative measurement of a secondary sex character and axillary hair. Ann N Y Acad Sci 53:585–599

Hamilton JB (1951b) Patterned loss of hair in man; types and incidence. Ann N Y Acad Sci 53:708–728

Hamilton JB (1958) Age, sex and genetic factors in the regulation of hair growth in man: a comparison of Caucasian & Japanese populations. In: Montagna W, Ellis RA (eds) The biology of hair growth. Academic Press, New York pp 399–433

Hibberts NA, Randall VA (1996) Testosterone inhibits the capacity of cultured balding scalp dermal papilla cells to produce keratinocyte mitogenic factors. In: Van Neste D, Randall VA (eds) Hair research for the next millenium. Elsevier Science, Amsterdam pp 303–306

Hibberts NA, Randall VA (1998) Dermal papilla cells from human balding scalp hair follicles contain higher levels of androgen receptors than those from non-balding scalp. J Endocr (in press)

Hibberts NA, Messenger AG, Randall VA (1996a) Dermal papilla cells derived from beard hair follicles secrete more stem cell factor (SCF) in culture than scalp cells or dermal fibroblasts. Biochem Biophys Res Commun 222:401–405

Hibberts NA, Sato K, Messenger AG, Randall VA (1996b) Dermal papilla cells from human hair follicles secrete factors (eg VEGF) mitogenic for endothelial cells. J Invest Derm 106:341

Imperato-McGinley J, Gautier T, Cai L, Yee B, Epstein J, Pochi P (1993) The androgen control of sebum production. Studies of subjects with dihydrotestosterone deficiency and complete androgen insensibility. J Clin Endocr Metab 76:524–528

Itami, S, Kurata S, Takayasu S (1990) 5α-reductase activity in cultured human dermal papilla cells from beard compared with reticular dermal fibroblasts. J Invest Derm 94:150–152

Itami S, Kurata S, Takayasu S (1995) Androgen induction of follicular epithelial cell growth is mediated via insulin-like growth factor I from dermal papilla cells. Biochem Biophys Res Commun 212:988–994

Jackson D, Church RE, Ebling FJG (1972) Hair diameter in female baldness. Br J Derm 87:361–367

Jahoda CAB, Reynolds AJ (1996) Dermal-epidermal interactions; adult follicle-derived cell populations and hair growth. In: Whiting DA (ed) Dermatologic clinics 14. Update on hair disorders. WB Saunders, Philadelphia pp 573–583

Jenkins JS, Ash S (1973) The metabolism of testosterone by human skin in disorders of hair growth. J Endocr 59:345–351

Kaufman KD (1996) Clinical studies on the effects of oral finasteride, a type II 5α-reductase inhibitor, on scalp hair in men with male pattern baldness. In: Van Neste D, Randall VA (eds) Hair research for the next millenium. Amsterdam: Elsevier pp 363–365

Kligman AM (1959) The human hair cycle. J Invest Derm 33:307–316

Lesko SM, Rosenberg L, Shapiro S (1993) A case-control study of baldness in relation to myocardial infarction in men. J Amer Med Ass 269:998–1003

Lucky AW (1997) Physical treatments of unwanted hair. In: Azziz R, Nestler JE, Dewailly D (eds) Androgen excess disorders in women. Lippincott-Raven, Philadelphia pp 779–786

Ludwig E (1977) Classification of the types of androgenetic alopecia (common baldness) occurring in the female sex. Br J Derm 97:247–254

Lynfield YL (1960) Effect of pregnancy on the human hair cycle. J Invest Derm 35:323–327

Marshall WA, Tanner JM (1969) Variations in pattern of pubertal change in girls. Arch Dis Child 44:291–303

Marshall WA, Tanner JM (1970) Variations in the pattern of pubertal changes in boys. Arch Dis Child 45:13–23

Norwood OT (1975) Male pattern baldness, classification and incidence. South Med J 68:1359–1365

Obana N, Uno H (1996) Dermal papilla cells in macaque alopecia trigger a testosterone-dependent inhibition of follicular cell proliferation. In: Van Neste D, Randall VA (eds) Hair research for the next millenium. Elsevier Science, Amsterdam pp 307–310

Orentreich N (1969) Scalp hair replacement in man. In: Montagna W, Dobson RL (eds) Adv Biol Skin Vol 9. Hair growth. Pergamon Press, Oxford pp 99–108

Orentreich N, Durr N P (1982) Biology of scalp hair growth. Clin Plas Surg 9:197–205

Peereboom-Wynia JDR, Van Der Willigen AH, Stolz E, Van Joost TH (1989) The effect of cyproterone acetate on hair roots and hair shaft diameter in androgenetic alopecia in females In: Van Neste D, Lachapelle JM, Antoine JL (eds) Trends in human hair growth and alopecia research. Kluwer, Academic Publishers, Dordrecht pp 207–213

Randall VA (1994a) Androgens and human hair growth. Clin Endocr 40:439–457

Randall VA (1994b) The role of 5α-reductase in health and disease. In: Sheppard M, Stewart P (eds) Hormones, enzymes and receptors. Baillière's Clinical endocrinology and metabolism 8:405–431

Randall VA, Ebling FJG (1991) Seasonal changes in human hair growth. Br J Derm 124:146–151

Randall VA, Thornton MJ, Hamada K, Redfern CPF, Nutbrown M, Ebling FJG, Messenger AG (1991) Androgens and the hair follicle: cultured human dermal papilla cells as a model system. Ann N Y Acad Sci 642:355–375

Randall VA, Thornton MJ, Messenger AG (1992) Cultured dermal papilla cells from androgen-dependent human follicles (e.g. beard) contain more androgen receptors than those from non-balding areas. J Endocr 133:141–147

Saitoh M, Sakamoto M (1970) Human hair cycle. J Invest Derm 54:65–81

Sawers RA, Randall VA, Iqbal MJ (1982) Studies on the clinical and endocrine aspects of antiandrogens. In: Jeffcoate SL (ed) Androgens and antiandrogen therapy. Current topics in endocrinology. John Wiley, Chichester pp 145–168

Setty LR (1970) Hair patterns of the scalp of white and negro males. Amer J Phys Anthrop 33:49–55

Stenn KS, Combates NJ, Eilertson KJ, Gordon JS, Poardinas JR, Parimoo S, Prouty S (1996) Hair follicle growth controls. In: Whiting DA (ed) Dermatologic clinics 14. Update on hair disorders. WB Saunders, Philadelphia pp 543–558

Thornton MJ, Laing I, Hamada K, Messenger AG, Randall VA (1993) Differences in testosterone metabolism by beard and scalp hair follicle dermal papilla cells. Clin Endocr 39:633–639

Thornton MJ, Hamada K, Messenger AG, Randall VA (1998) Beard, but not scalp, dermal papilla cells secrete autocrine growth factors in response to testosterone *in vitro*. J Invest Derm (in press)

Venning VA, Dawber R (1988) Patterned androgenic alopecia. J Amer Acad Derm 18:1073–1077

Winter JSD, Faiman C (1972) Pituitary-gonadal relations in male children and adolescents. Paed Res 6:125–135

Winter JSD, Faiman C (1973) Pituitary-gonadal relations in female children and adolescents. Paed Res 7:948–953

Zachmann M, Prader A (1970) Anabolic and androgenetic effect of testosterone in sexually immature boys and its dependency on GH. J Clin Endocr Metab 30:85–95

Zachmann M, Aynsley-Green A, Prader A (1976) Interrelations of the effects of growth hormone and testosterone in hypopituitarism. In: Pecile A, Muller EE (eds) Growth hormone and related peptides. Excerpta Medica, Amsterdam and Oxford pp 286–296

 6 Androgens and bone metabolism

Joel S. Finkelstein

Contents

6.1 Introduction

Osteoporosis is one of the leading causes of morbidity and mortality in the elderly. Osteoporosis affects 20 million Americans and leads to approximately 1.5 million fractures each year (Finkelstein 1996). The annual cost of health care and lost productivity attributed to osteoporosis exceeds $13 billion in the United States. Even though osteoporosis is less common in men than in women, men lose about 30% of their trabecular bone and 20% of their cortical bone during their lifetime. Thirty percent of all hip fractures occur in men (Cooper et al. 1992) and the incidence of hip fractures in men over the age of 65 is 4 to 5 per 1000 (Jacobsen et al. 1990). By the age of 90 one of every six men will have fractured their hip. Hypogonadism has been identified as a probable risk factor for hip fractures in men (Jackson et al. 1992). Case series of men with osteoporotic fractures suggest that hypogonadism is present in between 7 and 30% of such individuals (Jackson et al. 1992; Jackson and Kleerekoper 1990; Kelepouris et al. 1995; Seeman et al. 1983; Stanley et al. 1991). This review will examine the roles of androgen in bone metabolism.

6.2 Mechanism of action of androgens on bone

6.2.1 Effects of androgens on osteoblasts in vitro

The mechanism(s) whereby androgens affect bone density is still unclear. Some data suggest that androgens may affect osteoblast function directly. Several observations are consistent with this notion. First, androgen receptors have been found on normal human osteoblasts (Colvard et al. 1989), in human osteosarcoma cell lines (Orwoll et al. 1991) and in bone marrow-derived stromal cells (Bellido et al. 1995). Second, both aromatizable and non-aromatizable androgens stimulate proliferation of human osteoblasts *in vitro* (Gray et al. 1992; Kasperk et al. 1989; Kasperk et al. 1997; Vaishnav et al. 1988), a process that appears to require adequate stores of vitamin D (Somjen et al. 1989). Third, dihydrotestosterone (DHT), a non-aromatizable androgen, and DHEA stimulate differentiation of human osteoblasts *in vitro* (Kasperk et al. 1989, 1997) although this effect has not been seen consistently in all studies. The ability of 1,25-dihydroxyvitamin D to stimulate alkaline phosphatase activity *in vitro* is enhanced by DHT (Gray et al. 1992). DHT also stimulates collagen production *in vitro* (Gray et al. 1992). The effects of androgens on osteoblast proliferation and differentiation might be due to increased local production of TGF-β or increased sensitivity to the mitogenic effects of fibroblast growth factor and IGF-II (Kasperk et al. 1990, 1997).

6.2.2 Effects of androgens on osteoclasts in vitro

Although it appears likely that androgens stimulate osteoblast activity, it also appears that androgens inhibit osteoclast activity. Because androgen receptors are not expressed on osteoclasts, effects of androgens on osteoclastic activity are likely indirect and may involve local production or action of cytokines in bone. Both testosterone and DHT inhibit the production of interleukin-6 (IL-6) by bone marrow-derived stromal cells by inhibiting expression of the IL-6 gene, an effect which requires the androgen receptor (Bellido et al. 1995). Targeted disruption of the IL-6 gene prevents the increase in osteoclastogenesis and the bone loss that normally occurs after orchiectomy (Bellido et al. 1995). Both testosterone and DHT inhibit PTH-stimulated accumulation of cAMP by osteoblasts *in vitro* (Fukayama and Tashjian 1989) and partly inhibit parathyroid hormone (PTH) and interleukin-1 (IL-1) induced bone resorption in mouse calvarial cultures (Pilbeam and Raisz 1990). Because IL-1 and IL-6 promote osteoclast activation and differentiation and stimulate bone resorption, decreased local production or activity of IL-1, IL-6, or other osteoclast-stimulating cytokines likely plays in important role in the mechanism whereby androgens inhibit bone resorption.

6.2.3 Effects of androgens on bone in laboratory animals

Although there has been a proliferation of studies examining the impact of androgens on bones in men, studies in humans have important limitations. First, populations of hypogonadal men are often heterogeneous with respect to the cause of hypogonadism, time of onset, severity, and therapy. Second, detailed studies of bone histomorphometry, which are needed to address many important questions, are difficult to perform in humans. Third, some experimental manipulations, such as estrogen administration, cannot be performed in men, thereby limiting the ability to investigate some key pathophysiological issues. Because of these and other limitations, animal models have been important in examining the effects of androgens on bone.

Almost all studies of the effects of androgens on bone in laboratory animals have been performed in rats. The effect of androgens on the rat skeleton appears to depend on the age of the animal and the type of bone being examined. In young, growing male rats, castration reduces calcium content of the long bones and increases the number of tibial osteoclasts (Saville 1969; Schoutens et al. 1984). Orchiectomy does not alter the cross-sectional area, medullary area or cortical area of long bones in young rats but it does reduce the periosteal cortical bone formation rate (Gunness and Orwoll 1995; Turner et al. 1989). In trabecular bone, both osteoclast number and the extent of osteoclast-covered bone surfaces increase after orchiectomy in young rats resulting in trabecular bone loss (Gunness and Orwoll 1995; Turner et al. 1989; Wakley et al. 1991). All of these skeletal changes in young castrated rats can be prevented by both aromatizable and non-aromatizable androgens. These findings suggest that, in young growing rats, androgens stimulate periosteal growth of cortical bone and inhibit resorption of trabecular bone.

In aged rats, orchiectomy reduces calcium content of the femur and tibia (Vanderschueren et al. 1992; Verhas et al. 1986; Wink and Felts 1980). This decrease is due to a reduction in cortical thickness with no change in cortical bone density (Vanderschueren et al. 1992; Verhas et al. 1986). Orchiectomy also reduces trabecular bone mass in aged rats (Vanderschueren et al. 1992; Wink and Felts 1980). This decrease is associated with an increase in the percentage of trabecular surfaces covered by osteoblasts and osteoclasts. As in young rats, these skeletal changes in old castrated rats can be prevented by both aromatizable and non-aromatizable androgens (Vanderschueren et al. 1992).

6.2.4 Potential roles of testosterone metabolites

It is still unclear whether the effects of testosterone on bone are due to testosterone itself or one of its metabolites such as estradiol or DHT. Because testosterone can be converted to DHT by ground spongiosa from human bone *in vitro*, some investigators have hypothesized that DHT is the active intracellular androgen in human bone (Schweikert et al. 1980). However, the conversion rate of testosterone to DHT, which is catalyzed by the enzyme 5α-reductase, was very low. Furthermore, ground spongiosa probably contains cartilage cells which could be responsible for conversion of testosterone to DHT. More recently, it has been demonstrated that osteoblast-like cells cultured from postmenopausal women can convert androstenedione into testosterone and DHT (Bruch et al. 1992). Still, administration of finasteride, an inhibitor of 5α-reductase activity, had no effect on bone density in young male rats (Rosen et al. 1995). Although this latter finding, in particular, suggests that it is likely that bone cells can convert testosterone to DHT, a role for DHT in bone metabolism remains to be established.

Several observations suggest that aromatization of testosterone into estrogens is crucial for many of the effects of testosterone on bone. First, estrogen receptors have been demonstrated in human osteoblast-like cells (Eriksen et al. 1988). Second, aromatase activity has been demonstrated in human osteosarcoma cells (Tanaka et al. 1993). Third, estrogen administration can maintain bone mass in castrated male rats (Cruess and Hong 1978) and in male-to-female transsexuals (Van Kesteren et al. 1996). Fourth, and most important, a man with complete estrogen resistance due to a genetic defect in the estrogen receptor and a man with estrogen deficiency due to a mutation in aromatase P450 both had osteopenia in the lumbar spine and proximal radius despite normal or high testosterone levels and complete virilization (Carani et al. 1997; Morishima et al. 1995; Smith et al. 1994). These findings provide the most compelling evidence to date that estrogens are required to establish normal peak bone mass and/or maintain bone density in men. Further studies are needed to assess the relative roles of androgens and estrogens in bone metabolism in men.

6.2.5 Effects of androgens on calcium regulatory hormones and IGF-1 in humans

Androgens may also affect bone metabolism by effects on calcium regulatory hormones. It has been reported that serum calcitonin levels are lower in hypogonadal men than in normal men (Foresta et al. 1983); that testosterone administration increases calcitonin levels in hypogonadal men (Foresta et al. 1985); and that testosterone administration enhances the hypocalcemic effect of calcitonin in orchiectomized rats (Ogata et al. 1970). One group of investigators reported that serum $1,25\text{-}(OH)_2$ vitamin D levels were low, whereas other investigators reported that serum $1,25\text{-}(OH)_2$ vitamin D levels were normal in hypogonadal men (Francis et al. 1986; Jackson et al. 1987). The reported effects of testosterone therapy on serum $1,25\text{-}(OH)_2$ vitamin D levels have also been variable and may be related to the duration or mode of therapy. Francis et al. (1986) reported that testosterone therapy increased serum $1,25\text{-}(OH)_2$ vitamin D levels but their patients were only followed for 6 weeks. Katznelson et al. (1996) also found that serum $1,25\text{-}(OH)_2$ vitamin D levels increased during the first several months of testosterone therapy but then returned to normal. We did not detect any change in $1,25\text{-}(OH)_2$ vitamin D levels in GnRH-deficient men before and after long-term physiological androgen replacement (Finkelstein et al. 1989) and others found no changes in $1,25\text{-}(OH)_2$ vitamin D or PTH levels of older men with borderline hypogonadism when testosterone levels were raised into the mid-portion of the normal range (Tenover 1992). Recently, two reports, one of which clearly utilized supraphysiological doses of testosterone, indicated that serum calcium levels decrease and serum PTH levels increase when testosterone is administered to hypogonadal men (Katznelson et al. 1996; Wang et al. 1996). Because consistent alterations have not been observed, it seems unlikely that effects on calcium regulatory hormones play an important role in the effects of androgens on bone mass.

It is, however, quite possible that important effects of testosterone are mediated by changes in growth hormone (GH) or insulin like growth factor-1 (IGF-1). GH administration increases bone density in adult men with acquired GH deficiency (Baum et al. 1996). Administration of IGF-1 increases bone formation in humans (Grinspoon et al. et al. 1995). Aromatizable androgens increase growth hormone and IGF-1 secretion in boys with delayed puberty (Link et al. 1986; Ulloa-Aguirre et al. 1990). Similar effects have been reported in some, but not all studies, with use of non-aromatizable androgens (Link et al. 1986; Ulloa-Aguirre et al. 1990). Thus, it is important to assess effects of testosterone on GH and IGF-1 production when evaluating the effects of testosterone administration on bone metabolism.

6.2.6 Effects of androgens on bone turnover in humans

Bone remodeling is a continuous process characterized by two opposing activities: formation of new bone by osteoblasts and resorption of old bone by osteoclasts. Bone remodeling can be assessed either by quantitative histomorphometry of bone biopsies after tetracycline labeling or by measuring biochemical markers of bone turnover in blood and urine. Although quantitative histomorphometry can distinguish the cortical and trabecular contributions to bone remodeling, it is invasive, expensive, and poorly reproducible. Therefore, measurements of biochemical markers of bone formation and resorption in blood and urine are used routinely to assess bone turnover. Osteoblast activity can be assessed by measuring serum levels of type I procollagen extension peptides (amino and carboxy-terminal propeptides that are cleaved during the formation of type I collagen) or other non-collagenous proteins made by osteoblasts including osteocalcin, and alkaline phosphatase, particularly the skeletal fraction of the latter. Bone resorption can be assessed by measuring urinary excretion of degradation products of type I collagen including hydroxyproline, pyridinium cross links (pyridinoline and deoxypyridinoline), and collagen type 1 crosslinked N-telopeptide (NTX).

Histomorphometric analyses of iliac crest bone biopsies from hypogonadal men have produced variable results. Finkelstein et al. (1989) reported that bone turnover is decreased in most men with isolated GnRH deficiency receiving long-term gonadal steroid replacement therapy. Jackson et al. (1987) reported that indices of bone formation were modestly increased in six hypogonadal men with osteoporosis compared to eight eugonadal osteoporotic controls. Some histomorphometric data suggest that testosterone therapy increases bone formation in hypogonadal men (Baran et al. 1978; Francis et al. 1986) while other data suggest that testosterone decreases bone formation in such patients (Jackson et al. 1987).

Most data derived from measurements of markers of bone turnover suggest that bone turnover is increased in untreated hypogonadal men. Serum osteocalcin and bone-specific alkaline phosphatase levels and urinary hydroxyproline excretion are increased in men who have undergone castration and levels of all markers fall when bone resorption is inhibited with calcitonin administration (Stepan et al. 1989). Similarly, serum osteocalcin and alkaline phosphatase levels increase in men rendered hypogonadal by administration of a long-acting GnRH analog (Goldray et al. 1993). Urinary hydroxyproline excretion and serum alkaline phosphatase levels were also increased in the heterogenous group of hypogonadal men studied by Jackson et al. (1987).

Several studies have examined the effects of testosterone administration on bone turnover. Serum osteocalcin levels increased but urinary excretion of hydroxyproline was unchanged in normal young men treated with pharmacological doses of a parenteral testosterone ester for six months (Young et al. 1993). A small increase in serum alkaline phosphatase levels was also detected in female-to-male transsexuals treated with high doses of testosterone esters (Van Kesteren et al. 1996). In contrast, urinary excretion of deoxypyridinoline and NTX and serum bone-specific alkaline phosphatase levels all decreased in middle-aged eugonadal men with vertebral crush fractures treated with more physiologic doses of a parenteral testosterone ester (Anderson et al. 1997). Because treatment with testosterone esters increases both testosterone and estradiol levels, it is difficult to determine if the changes in these studies are due to androgens, estrogens, or both. In elderly men with borderline low serum testosterone levels, testosterone administration reduced urinary hydroxyproline excretion whereas serum osteocalcin levels did not change (Tenover 1992). Two recent studies have examined the effects of testosterone on bone turnover indices in hypogonadal men. In these studies, testosterone replacement therapy reduced indices of bone resorption including urinary excretion of pyridinium cross links and NTX (Katznelson et al. 1996; Wang et al. 1996). The effects of testosterone administration on markers of bone formation in hypogonadal men are less clear. Wang et al. (1996) reported that serum osteocalcin and type I procollagen levels increased with no change in bone-specific alkaline phosphatase levels after 6 months of sublingual testosterone replacement in doses that produced supraphysiological serum testosterone concentrations. We also failed to detect a change in bone-specific alkaline phosphatase levels after six months of testosterone replacement therapy but found that bone-specific alkaline phosphatase levels were significantly decreased after 12 and 18 months of testosterone administration (Katznelson et al. 1996). Overall, these findings suggest that physiological doses of testosterone primarily inhibit bone turnover but that pharmacological doses may additionally stimulate bone formation. In addition, the variability in the results of these studies underscores the importance that patient populations, dosage of testosterone administration, mode of testosterone administration, and use of differing biochemical markers may make when evaluating the effects of androgens on bone turnover.

6.2.7 Effects of androgens on bone in normal men

In adults, bone density at any point in time is determined both by the peak bone mass achieved during development and the subsequent amount of bone loss. Androgens, by affecting both of these processes, are an important determinant of bone mass in men. Both cortical and trabecular bone density increase dramatically during puberty in boys and girls (Bonjour et al. 1991; Gilsanz et al. 1988; Krabbe and Christiansen 1984; Mazess and Cameron 1974; McCormick et al. 1991; Theintz et al. 1992). The pubertal rise in testosterone is followed closely by an increase in serum alkaline phosphatase and, subsequently, bone density increases (Krabbe et al. 1984; Riis et al 1985). These data strongly suggest that testosterone, or one of its metabolites, is responsible for the pubertal rise in bone mineral density. Peak trabecular bone density is usually achieved by the age of 18 years in males (Bonjour et al. 1991; McCormick et al. 1991) though peak cortical bone density may not be reached until a few years later (Mazess and Cameron 1974; Theintz et al. 1992). Furthermore, peak cortical bone mineral density is higher in normal men than in women (Bonjour et al. 1991; Mazess and Cameron 1974), suggesting that androgens have an independent effect on peak cortical bone mass. Bone density remains relatively stable in young adults males before it declines slowly in later life. Although some cross-sectional studies have suggested that a decline in gonadal function may be responsible for the decrease in bone density as men age (Foresta et al. 1984; Rudman et al. 1994), most studies have been unable to demonstrate a correlation between serum androgens and bone mass in aging men (McElduff et al. 1988; Meier et al. 1987). Longitudinal studies of bone density in older adult men are needed to assess the relationship between changes in gonadal function and bone mass more precisely.

6.3 Bone density in men with disorders of androgen metabolism

The observation that androgen deficiency could produce osteoporosis in men was first made by Albright, who noted that eunuchs often developed osteoporosis. With the advent of improved methods for measuring bone density, many groups have now reported that bone density is decreased in men with hypogonadism of many different etiologies. In some of these reports, however, coexisting disorders make it difficult to determine the effects of hypogonadism, per se, on the skeleton. For example, a series of reports involving a small number of patients suggested that cortical bone density is decreased in men with Klinefelter syndrome (Foresta et al. 1983; Smith and Walker 1977). It is difficult, however, to determine whether the osteopenia of these men is due to their hypogonadism or is an independent effect related to their genetic defect. Furthermore, the small number of patients studied, variability in serum testosterone levels, and use of outdated techniques for determining bone density limits the utility of these reports. Osteopenia has also been reported in hypogonadal men with hemochromatosis (Diamond et al. 1991). In these men, it is difficult to determine whether other problems, such as concomitant liver disease, might also be contributing to their low bone mineral density.

Most studies of bone density in hypogonadal men have been performed in heterogeneous groups of patients including men with both primary and secondary hypogonadism. Detailed studies of the effects of androgens on bone mass in men have been performed, however, in several more homogeneous groups of men: men with primary hypogonadism due to castration; men with secondary hypogonadism due to isolated GnRH deficiency, hyperprolactinemia, or GnRH analog therapy; men receiving treatment with inhibitors of 5α-reductase; patients with complete androgen insensitivity, male-to-female and female-to-male transsexuals; and men with histories of constitutionally delayed puberty. Additionally, some studies have examined the effects of androgens on bone mass in elderly men, the group that is probably of the greatest clinical importance. These studies provide prospective data for the effects of androgens on bone mass in men and will be discussed in detail.

6.3.1 Bone density in men after castration

In contrast to women, in whom there is a large amount of data relating primary hypogonadism to bone mass, little such data exist in men. Stepan et al.

(1989) measured bone mineral density of the lumbar spine using dual photon absorptiometry in 12 men who had undergone bilateral orchiectomy because of sexual delinquency up to 11 years previously (mean 5.6 years). In nine of these men, measurements were repeated one to three years later. Bone density decreased progressively with increasing years since castration. Although it appeared that the rate of bone loss was greater in the first several years after orchiectomy, the number of observations was too small to demonstrate such a relationship conclusively. As noted previously, biochemical indices of bone resorption and bone formation were increased compared to normal men and there was a significant association between urinary hydroxyproline excretion and the rate of bone loss.

Recently, a retrospective study examined the effect of castration on bone mineral density and fractures in men with non-stage A prostate cancer (Daniell 1997). Osteoporotic fractures were identified in 13.6% of men treated with orchiectomy and 1.1% of patients who did not undergo orchiectomy. Six of 16 men who survived for at least five years after orchiectomy sustained osteoporotic fractures that were not attributable to their maligancy. Bone mineral density of the femoral neck, measured an average of 65 months after castration, was 20% lower than in controls without prostate cancer and 13% lower than in non-castrate controls with prostate cancer. These findings demonstrate that severe gonadal steroid deficiency is associated with clinically significant bone loss in men but do not indicate whether androgen deficiency, estrogen deficiency, or both are the primary cause of bone loss in hypogonadal men.

6.3.2 Bone density in men with hyperprolactinemic hypogonadism

Greenspan et al. (1986) measured cortical bone density in the forearm by single photon absorptiometry (SPA) and trabecular bone density in the spine by quantitative computed comography (QCT) in 18 men between the ages of 30 and 79 with secondary hypogonadism caused by prolactin-secreting tumors. Five men had secondary hypothyroidism and/or secondary adrenal insufficiency and were receiving physiological hormone replacement. Both cortical and trabecular bone mineral density were significantly decreased in the hyperprolactinemic men compared to age-matched controls. Cortical bone density was associated with the duration of hyperprolactinemia. There was no significant correlation between cortical and trabecular bone density, suggesting that cortical and trabecular bone respond differently to androgen deficiency. These findings suggest that hypogonadism in hyperprolactinemic men leads to osteopenia but do not rule out the possibility that other hormonal abnormalities in these men may have an adverse effect on skeletal integrity.

To assess the effects of restoration of gonadal steroid secretion on bone density in men with hyperprolactinemic hypogonadism, Greenspan et al. (1989) performed serial measurements of bone density for 6 to 48 months in

20 such men who were treated with bromocriptine, transsphenoidal surgery, and/or cranial radiation. In those men whose serum testosterone levels normalized, cortical bone density increased significantly and there was a significant association between the change in serum testosterone levels and the change in bone density. In the men who remained hypogonadal, cortical bone density did not change. Trabecular bone density of the lumbar spine did not change significantly in either group.

6.3.3 Bone density in men with idiopathic hypogonadotropic hypogonadism

Men with idiopathic hypogonadotropic hypogonadism (IHH) are hypogonadal due to an isolated absence of hypothalamic gonadotropin-releasing hormone (GnRH) secretion. Otherwise, pituitary function is intact in these men. Thus, they provide a useful model to examine the effects of severe gonadal steroid deficiency on bone mass in men in the absence of other hormonal abnormalities. Because IHH is almost always a congenital abnormality, these patients also provide a valuable model to assess the effects of gonadal steroid deficiency on pubertal bone development (i.e. the attainment of peak bone mass). We measured cortical bone density of the radial shaft by SPA and trabecular bone density of the lumbar spine by QCT in 23 young men with IHH (Finkelstein et al. 1987). Because bone density increases dramatically during puberty, patients with open epiphyses were compared to adolescent controls matched for bone age whereas patients with fused epiphyses were compared to age-matched adult men. Both cortical and trabecular bone mineral density were markedly decreased in IHH men and the osteoporosis was equally severe in the men with open and fused epiphyses. The average bone density value was below the 1st percentile for age-matched men and eight men had trabecular bone density values below the fracture threshold despite their young age. Because severe osteopenia was already present in men who were skeletally immature, these data suggest that the osteopenia of IHH men is due to inadequate pubertal bone accretion rather than post-maturity adult bone loss. Histomorphometric analyses of iliac crest bone biopsy specimens from 10 IHH men receiving long-term androgen replacement therapy suggested that bone turnover is low in most such patients (Finkelstein et al. 1989).

Guo et al. (1997) recently assessed bone mineral density in ten adult IHH men who had been treated with hCG or testosterone esters for 2 to 22 years. As in the patients in our study, both cortical and trabecular bone density were markedly decreased. In addition, they reported that the severity of osteopenia was related to the age of initial therapy so that the patients whose therapy was delayed the longest had the lowest bone mineral density. Surprisingly, biochemical markers of bone formation and bone resorption were increased suggesting that bone turnover is above normal in IHH men despite therapy that produced normal serum testosterone levels, a finding that is difficult to reconcile with the histomorphometric data presented above.

To assess the effects of androgen replacement on bone mass in IHH men, we made longitudinal measurements of cortical and trabecular bone density in 21 of these 23 men while serum testosterone levels were maintained in the normal range with either pulsatile GnRH, human chorionic gonadotropin, or intramuscular testosterone therapy for an average of two years (Finkelstein et al. 1989). In the men who were initially skeletally mature, there was a small but significant increase in cortical bone density whereas trabecular bone density did not change. Both cortical and trabecular bone density increased significantly in the men who were skeletally immature at the beginning of the study. Furthermore, cortical bone density increased more in the skeletally immature men, suggesting that much of the increase in bone density in men with open epiphyses is due to a completion of the process of pubertal bone accretion. Despite these increases, bone density remained well below the levels of normal men. This finding suggests either that factors other than gonadal steroid deficiency are involved in the pathogenesis of the osteopenia of IHH men or that there may be a critical period in development when gonadal steroid secretion must be normal in order to achieve a normal peak bone density.

Similar results were obtained by Guo et al. (1997) who reported that total body bone mineral, a measure that is predominantly cortical bone, increased during four months of high-dose hCG therapy in skeletally mature IHH men, whereas spinal bone mineral density did not change. Serum bone specific alkaline phosphatase and urinary excretion of NTX both decreased from their previously elevated baseline levels, suggesting that bone turnover decreased. Because high dose hCG therapy increased serum levels of both testosterone and estradiol, it is difficult to determine whether these changes in bone turnover and bone density were due to increases in androgens or estrogens.

6.3.4 Bone density in men receiving long-acting GnRH analogs

The effects of secondary hypogonadism on bone mass have also been studied by examining the effects of daily administration of a long-acting GnRH analog to men with benign prostatic hyperplasia (Goldray et al. 1993). GnRH analog therapy produced severe testosterone deficiency in all men. In 10 of 17 men, bone density of the lumbar spine decreased significantly over a period of 6 to 12 months. Serum alkaline phosphatase and osteocalcin levels increased significantly, suggesting that bone turnover was increased. Serum levels of calcium, vitamin D metabolites, and parathyroid hormone did not change significantly. In general, the effects of GnRH analog administration on bone metabolism in men were similar to those previously described in women. Thus, it is likely that long-term GnRH analog use in men will lead to clinically significant bone loss. Strategies to prevent GnRH analog-induced bone loss, which have been employed successfully in women (Finkelstein et al. 1994), need to be tested in men so that safe long-term use of these agents will be feasible.

6.3.5 Bone density in men receiving 5α-reductase inhibitors

As noted above, conversion of testosterone to DHT has been demonstrated in spongiosa from normal and osteoporotic human bone (Schweikert et al. 1980). Because the conversion of testosterone to DHT is catalyzed by the enzyme 5α-reductase, it is possible that inhibition of 5α-reductase activity might affect skeletal integrity. Bone age was delayed substantially in a boy with 5α-reductase deficiency suggesting that DHT is important for normal skeletal maturation (Fisher et al. 1978). However, when adult men with benign prostatic hypertrophy were treated with finasteride for 12 months, no changes in vertebral bone mineral density, serum osteocalcin, or urinary calcium excretion were observed despite an 80% decrease in serum DHT levels (Matzkin et al. 1992). Similar results were obtained in a retrospective study in which patients with benign prostatic hypertrophy who had received finasteride for an average of ten months were compared with matched controls (Tollin et al. 1996). Although these results might suggest that DHT does not play an important physiological role in maintaining skeletal integrity, it is possible that different effects would be seen with more complete suppression of DHT production. Furthermore, there are two steroid 5α-reductase enzymes, only one of which is inhibited by finasteride. The skeletal consequences of inhibiting both 5α-reductase enzymes are unknown.

6.3.6 Bone density in patients with androgen insensitivity

In normal adults, peak cortical bone density is approximately 15% higher in men than in women. Patients with complete androgen insensitivity syndrome are genetic males who are phenotypic females due to the absence of androgen receptor activity. Because they are genetic males with no response to androgens, they provide a valuable model to assess whether the sexual diphormism in peak bone density is genetically or hormonally determined. We measured cortical and trabecular bone density in eight adults with complete androgen insensitivity to determine whether their bone density would be similar to that expected for their genetic sex (Finkelstein et al. 1992a). Cortical bone mineral density of the radial shaft was lower than that of normal men but similar to that of normal women. This finding suggests that androgen action contributes to the normal sexual diphormism in cortical bone mineral density and that the Y chromosome, per se, is not sufficient to guarantee the higher cortical bone density observed in normal men. Cortical thickness and bone formation rates were also lower in androgen-resistant rats and normal female rats compared to male controls (Vanderschueren et al. 1993). Interestingly, trabecular bone density of the lumbar spine was lower than expected for either men or women of the same age in patients with complete androgen insensitivity. In contrast, trabecular bone density is maintained in androgen-resistant rats, possibly by increased production of estrogens (Vanderschueren et al. 1993). Soule et al. (1995) reported that bone density of the femoral neck and lumbar spine were decreased in five adult patients with androgen insensitivity. However, two of the patients in their study did not receive hormone replacement therapy for many years after castration so that their osteopenia could easily be attributed to estrogen deficiency and two patients were castrated prepubertally making it difficult to establish the diagnosis of androgen insensitivity with certainty. Recently, it was reported that estrogen therapy increased bone density of the spine and femoral neck in a patient with complete androgen insensitivity (Vered et al. 1997).

6.3.7 Bone density in transsexuals

The effects of gonadal steroids on bone have also been examined in both male-to-female and female-to-male transsexuals (Lips et al. 1989, 1996; Van Kesteren et al. 1996). In genetic males treated with high-dose estrogens plus antiandrogens, spinal bone mineral density increased even though serum testosterone levels were markedly suppressed. This finding suggests that estrogens, in supraphysiological doses, can maintain bone density in androgen-deficient men. Biochemical markers of bone turnover and some histomorphometric parameters of bone formation were suppressed suggesting that, as in women, estrogens suppress bone turnover in men. In female-to-male transsexuals treated with either parenteral testosterone esters or oral testosterone undecanoate, bone mineral density did not change. Serum alkaline phosphatase levels increased but serum osteocalcin levels did not change significantly. Thus, no consistent effect was observed on biochemical markers of bone formation. Histomorphometric analyses of iliac crest bone biopsies showed that the extent of eroded bone surfaces and the bone formation rate were both decreased, consistent with an antiresorptive effect of testosterone administration while mean cortical thickness was increased suggesting an additional anabolic effect of testosterone on bone (Lips et al. 1996). Although studies in transsexuals may provide useful information regarding the effects of sex steroids on bone metabolism, the pharmacological doses of gonadal steroids administered and the heterogeneity of the treatment programs make interpretation of the data somewhat difficult.

6.3.8 Bone density in men
with histories of constitutionally-delayed puberty

As noted above, the observation that bone density failed to normalize during prolonged gonadal steroid replacement in IHH men suggested that there may be a critical period in development during which puberty must occur in order to achieve a normal peak bone mineral density. To test this hypothesis, we measured bone mineral density of the radial shaft using SPA and bone mineral density of the lumbar spine and proximal femur using dual-energy xray absorptiometry (DXA) in adult men with histories of constitutionally delayed puberty (mean age=26 years) and compared their values with a well-matched group of normal men (Finkelstein et al. 1992b, 1996). None of the patients had ever received any form of hormonal therapy and all subsequently had undergone complete spontaneous pubertal development. Bone mineral density was significantly lower in men with histories of delayed puberty than in normal controls at all sites. On average, bone mineral density was 1.4, 0.9 and 0.7 standard deviations below normal at the radial shaft, lumbar spine, and femoral neck, respectively. In fact, bone mineral density of the radial shaft was at least 2 SD below the normal mean in 8 of 23 men. No changes were observed in bone mineral density of the radial shaft or lumbar spine were observed after two years of follow-up. Bertelloni et al. reported that cortical BMD of the radial shaft was markedly decreased in adolescent boys (age 13.1 to 15.8 years) with delayed puberty, even when compared to bone-age matched controls. Bone density increased significantly in the boys who were treated with testosterone for 6 months but not in untreated boys (Bertelloni et al. 1995). These findings demonstrate that radial, femoral, and spinal bone mineral density are decreased in men with histories of constitutionally delayed puberty and suggest that the timing of puberty may be an important determinant of peak bone mass. Because the peak bone density achieved during development is an important determinant of bone density in later life, men with histories of delayed puberty may be at increased risk for osteoporotic fractures. Finally, a history of delayed puberty may be an important clue to the etiology of low bone density in men with "idiopathic" osteoporosis.

6.4 Therapy of androgen-deficiency bone loss

6.4.1 Effects of androgen replacement on bone density

The effects of androgen replacement therapy on bone mass has been examined in several studies. As noted above, bone density increases during long-term treatment of men with idiopathic hypogonadotropic hypogonadism, particularly in those men who are still skeletally immature at the time that therapy is initiated (Finkelstein et al. 1989). Most studies of androgen replacement in hypogonadal men have been conducted on heterogeneous populations of hypogonadal men, including men with both primary and secondary hypogonadism and men with both congenital and acquired hypogonadism. We investigated bone density and the effects of testosterone replacement therapy in 36 men with acquired primary (n=7) or secondary (n=29) hypogonadism, none of whom had received prior testosterone therapy, and compared their results with 44 age-matched normal men (Katznelson et al. 1996). Spinal bone density was measured using both DXA, a technique that assesses both trabecular and cortical bone in the spine, and QCT, a technique that assesses exclusively trabecular bone. Spinal bone density was significantly lower than that of age-matched normal men using both techniques. Spinal bone density increased significantly (5% by DXA and 14% by QCT) during 12 to 18 months of testosterone replacement therapy. Serum bone-specific alkaline phosphatase and urinary excretion of deoxypyridinoline, markers of bone formation and resorption, both decreased significantly suggesting that the beneficial effects of testosterone replacement are due to its ability to inhibit bone turnover.

Wang et al. (1996) evaluated the effects of six months of sublingual testosterone replacement therapy in 67 hypogonadal men (26 men with primary hypogonadism, 27 men with secondary hypogonadism, and 14 undetermined). Some men had congenital hypogonadism whereas hypogonadism was acquired in others. Although there was a small decrease in urinary NTX and calcium excretion, suggesting a suppression of bone resorption, and an increase in serum osteocalcin and type I procollagen, suggesting an increase in bone formation, bone mineral density did not change significantly, possibly due to the short period of observation.

Recently, Behre et al. (1997) measured trabecular bone density of the spine using QCT in 72 hypogonadal men (37 men with primary and 35 men with secondary hypogonadism) receiving long-term gonadal steroid replacement therapy (using either parenteral testosterone, hCG, pulsatile GnRH, transdermal testosterone, or oral testosterone undecanoate). Once again, men with both congenital and acquired hypogonadism were included. For the group as a whole, bone density was low at baseline. Baseline bone density was lower in men with secondary hypogonadism than in men with primary hypogonadism and in men who had not received gonadal steroid therapy prior to their initial bone density measurement. Bone density increased significantly in the group as a whole, particularly in men who had not received gonadal steroid therapy prior to their initial measurement (mean increase of 15% in previously-treated men and 39% in previously-untreated men). Most of the increases in bone density occurred during the first year of gonadal steroid therapy and bone density normalized in many patients, particularly in men who had received androgen therapy for at least one year before entering the study. These findings suggest that in this heterogeneous group of patients, gonadal steroid therapy increases, and may normalize bone density. Although the response to therapy may, at first glance, appear to be superior to that reported by Katznelson et al. (1996), the final bone densities of the men who were clinically comparable (i.e. those without histories of prior treatment) were virtually identical in the two studies and were in the lower portion of the normal range (i.e. 125 to 130 mg/cc). Controlled studies, in which patients with primary, secondary, congenital, and acquired hypogonadism are evaluated separately, are needed to determine whether bone density truly normalizes during long-term therapy of hypogonadal men.

6.4.2 Other therapies of androgen-deficiency bone loss

To date, gonadal steroid replacement is the only form of therapy that has been evaluated for treatment of bone loss in hypogonadal men. Many hypogonadal men, however, (e.g. men with benign prostatic hypertrophy or prostate cancer receiving GnRH analog therapy or other forms of androgen ablation) have contraindications to androgen therapy. Because bone resorption is increased in such men, anti-resorptive therapy would seem to be logical. In addition, we have recently demonstrated that daily parathyroid hormone administration, an agent that stimulates bone formation, prevents GnRH analog-induced bone loss in women with endometriosis, in whom estrogen therapy may be contraindicated (Finkelstein et al. 1994; Finkelstein et al. in press). Thus, it seems likely that parathyroid hormone might also prevent GnRH analog-induced bone loss in men. The use of anabolic agents like parathyroid hormone, or antiresorptive agents, such as bisphosphonates or calcitonin, to prevent bone loss in hypogonadal men with contraindications to testosterone therapy needs further evaluation.

6.5 Areas for future investigation

A number of important issues related to the effects of androgens on bone in men need further investigation. Our understanding of the cellular and molecular mechanisms whereby androgens act on the skeleton is limited. In particular, little is known about the mechanisms whereby androgens stimulate osteoblast proliferation and differentiation or about the mechanisms that may be responsible for differential effects of androgens on cortical and trabecular bone. The role of testosterone metabolites, such as DHT and estrogens, remains an important area for future studies. The observation that trabecular bone mineral density is decreased in patients with estrogen resistance or aromatase deficiency suggests an important role for estrogens but is difficult to reconcile with studies demonstrating that non-aromatizable androgens can maintain bone mass in castrate male animals. From a clinical standpoint, the degree of testosterone deficiency that leads to bone loss remains to be established. Studies are also needed to examine the effects of anti-resorptive agents such as bisphosphonates, or anabolic agents such as parathyroid hormone, on bone density in hypogonadal men, particularly in individuals who may have contraindications to androgen therapy. Finally, the effects of testosterone administration on bone density and other organ systems in aging men with borderline low testosterone levels need to be investigated thoroughly as this group of individuals is far and above the largest potential target for androgen therapy.

6.6 Key messages

- Osteoporosis is common in men.
- Hypogonadism is one of the major risk factors for osteoporosis in men.
- Androgens inhibit osteoclastic bone resorption and may stimulate osteoblastic bone formation, particularly subperiostial cortical bone.
- Androgen metabolites, particularly estrogens, may be key mediators of the effects of androgen on bone.

6.7 References

Anderson FH, Francis RM, Peaston RT, Wastell HJ (1997) Androgen supplementation in eugonadal men with osteoporosis: effects of six months' treatment on markers of bone formation and resorption. J Bone Miner Res 12:472–478

Baran DT, Bergfeld MA, Teitelbaum SL, Avioli LV (1978) Effect of testosterone therapy on bone formation in an osteoporotic hypogonadal male. Calcif Tissue Res 26:103–106

Baum HBA, Biller BMK, Finkelstein JS, Cannistraro KB, Oppenheim DS, Schoenfeld DA, Michel TH, Wittink H, Klibanski A (1996) Effects of physiologic growth hormone therapy on bone density and body composition in patients with adult-onset growth hormone deficiency. Ann Intern Med 125:883–890

Behre HM, Kliesch S, Leifke E, Link TM, Nieschlag E (1997) Long term effect of testosterone therapy on bone mineral density in hypogonadal men. J Clin Endocrinol Metab 82:2386–2390

Bellido T, Jilka RL, Boyce BF, Girasole G, Broxmeyer H, Dalrymple SA, Murray R, Manolagas SC (1995) Regulation of interleukin-6, osteoclastogenesis, and bone mass by androgens. The role of the androgen receptor. J Clin Invest 95:2886–2895

Bertelloni S, Baroncelli GI, Battini R, Perri G, Saggese G (1995) Short-term effect of testosterone treatment on reduced bone density in boys with constitutional delay of puberty. J Bone Miner Res 10:1488–1495

Bonjour J-P, Theintz G, Buchs B, Slosman D, Rizzoli R (1991) Critical years and stages of puberty for spinal and femoral bone mass accumulation during adolescence. J Clin Endocrinol Metab 73:555–563

Bruch H-R, Wolf L, Budde R, Romalo G, Schweikert H-U (1992) Androstenedione metabolism in cultured human osteoblast-like cells. J Clin Endocrinol Metab 75:101–105

Carani C, Simoni M, Faustini-Fustini M, Serpente S, Boyd J, Korach KS, Simpson ER (1997) Aromatase deficiency in the male: effect of testosterone and estradiol treatment. N Engl J Med 337:91–95

Colvard DS, Eriksen EF, Keeting PE, Wilson EM, Lubahan DB, French FS, Riggs BL, Spelsberg TC (1989) Identification of androgen receptors in normal human osteoblast-like cells. Proc Natl Acad Sci 86:854–857

Cooper C, Campion G, Melton LD (1992) Hip fractures in the elderly: a world-wide projection. Osteoporos Int 2:285–289

Cruess RL, Hong KC (1978) The long term effect of estrogen administration on the metabolism of male rat bone. Proc Soc Exp Biol Med 159:368–373

Daniell HW (1997) Osteoporosis after orchiectomy for prostate cancer. J Urol 157:439–444

Diamond T, Stiel D, Posen S (1991) Effects of testosterone and venesection on spinal and peripheral bone mineral in six hypogonadal men with hemochromatosis. J Bone Miner Res 6:39–43

Eriksen EF, Colvard DS, Berg NJ, Graham ML, Mann KG, Spelsberg TC, Riggs BL (1988) Evidence of estrogen receptors in normal human osteoblast-like cells. Science 241:84–86

Finkelstein JS (1996) Osteoporosis. In: Bennett JC, Plum F (eds) Cecil Textbook of Medicine, 20th Edition. W.B. Saunders Co., Philadelphia, pp 1379–1384

Finkelstein JS, Klibanski A, Neer RM, Greenspan SL, Rosenthal DI, Crowley WF (1987) Osteoporosis in men with idiopathic hypogonadotropic hypogonadism. Ann Intern Med 106:354–461

Finkelstein JS, Klibanski A, Neer RM, Doppelt SH, Rosenthal DI, Segre GV, Crowley WF (1989) Increases in bone density during treatment of men with idiopathic hypogonadotropic hypogonadism. J Clin Endocrinol Metab 69:776–783

Finkelstein J, Klibanski A, Neer, RM. (1992a) Cortical and trabecular bone density in patients with the complete androgen insensitivity syndrome. International Congress of Endocrinology, Nice (abstract)

Finkelstein JS, Neer RM, Biller BMK, Crawford JD, Klibanski A (1992b) Osteopenia in adult men with a history of delayed puberty. N Engl J Med 326:600–604

Finkelstein JS, Klibanski A, Schaefer EH, Hornstein MD, Schiff I, Neer RM (1994) Parathyroid hormone for the prevention of bone loss induced by estrogen deficiency. N Engl J Med 331:1618–1623

Finkelstein JS, Klibanski A, Neer RM (1996) A longitudinal evaluation of bone mineral density in adult men with histories of delayed puberty. J Clin Endocrinol Metab 81:1152–1155

Fisher LK, Kogut MD, Moore RJ, Goebelsmann U, Weitzmann JJ, Isaacs H, Griffin JE, Wilson JD (1978) Clinical, endocrinological, and enzymatic characterisation of two boys with 5α-reductase deficiency: evidence that a single enzyme is responsible for the 5α-reduction of cortisol and testosterone. J Clin Endocrinol Metab 47:653–665

Foresta C, Ruzza A, Mioni R, Meneghello A, Baccichetti C (1983) Testosterone and bone loss in Klinefelter syndrome. Horm Metab Res 15:56–57

Foresta C, Ruzza A, Mioni R, Guarneri G, Gribaldo R, Meneghello A, Mastrogiacomo I (1984) Osteoporosis and decline of gonadal function in the elderly male. Horm Res 19:18–22

Foresta C, Zanatta GP, Busnardo B, Scanelli G, Scandellari C (1985) Testosterone and calcitonin plasma levels in hypogonadal osteoporotic young men. J Endocrinol Invest 8:377–379

Francis RM, Peacock M, Aaron JE, Selby PL, Taylor GA, Thompson J, Marshall DH, Horsman A (1986) Osteoporosis in hypogonadal men: role of decreased plasma 1,25-dihydroxyvitamin D, calcium malabsorption, and low bone formation. Bone 7:261–268

Fukayama S, Tashjian AH (1989) Direct modulation by androgens of the response of human bone cells (SaOS-2) to human parathyroid hormone (PTH) and PTH-related protein. Endocrinology 125:1789–1794

Gilsanz V, Gibbens DT, Roe TF, Carlson M, Senac MO, Boechat MI, Huang HK, Schulz EE, Libanati CR, Cann CC (1988) Vertebral bone density in children: effect of puberty. Radiology 166:847–850

Goldray D, Weisman Y, Jaccard N, Merdler C, Chen J, Matzkin H (1993) Decreased bone density in elderly men treated with the gonadotropin-releasing hormone agonist decapeptyl (D-Trp⁶-GnRH). J Clin Endocrinol Metab 76:288–290

Gray C, Colston KW, Mackay AG, Taylor ML, Arnett TR (1992) Interaction of androgen and 1,25-dihydroxyvitamin D_3: effects on normal rat bone cells. J Bone Miner Res 7:41–46

Greenspan SL, Neer RM, Ridgway EC, Klibanski A (1986) Osteoporosis in men with hyperprolactinemic hypogonadism. Ann Intern Med 104:777–782

Greenspan SL, Oppenheim DS, Klibanski A (1989) Importance of gonadal steroids to bone mass in men with hyperprolactinemic hypogonadism. Ann Intern Med 110:526–531

Grinspoon SK, Baum HBA, Peterson S, Klibanski A (1995) Effects of rhIGF-1 administration on bone turnover during short-term fasting. J Clin Invest 96:900–906

Gunness M, Orwoll E (1995) Early induction of alterations in cancellous and cortical bone histology after orchiectomy in mature rats. J Bone Miner Res 10:1735–1744

Guo C-Y, Jones TH, Eastell R (1997) Treatment of isolated hypogonadotropic hypogonadism effect on bone mineral density and bone turnover. J Clin Endocrinol Metab 82:658–665

Jackson JA, Kleerekoper M (1990) Osteoporosis in men: diagnosis, pathophysiology and prevention. Medicine 69:137–152

Jackson JA, Kleerekoper M, Parfitt AM, Rao DS, Villanueva AR, Frame B (1987) Bone histomorphometry in hypogonadal and eugonadal men with spinal osteoporosis. J Clin Endocrinol Metab 65:53–58

Jackson JA, Riggs MW, Spiekerman AM (1992) Testosterone deficiency as a risk factor for hip fractures in men: a case-control study. Am J Med Sci 304:4–8

Jacobsen SJ, Goldberg J, Miles TP, Brody JA, Stiers W, Rimm AA (1990) Hip fracture incidence among the old and very old: a population-based study of 745–435 cases. Am J Public Health 80:871–873

Kasperk CH, Wergedal JE, Farley JR, Linkhart TA, Turner RT, Baylink DJ (1989) Androgens directly stimulate proliferation of bone cells in vitro. Endocrinology 124:1576–1578

Kasperk C, Fitzsimmons R, Strong D, Mohan S, Jennings J, Wergedal J, Baylink D (1990) Studies of the mechanism by which androgens enhance mitogenesis and differentiation in bone cells. J Clin Endocrinol Metab 71:1322–1329

Kasperk CH, Wakley GK, Hierl T, Ziegler R (1997) Gonadal and adrenal androgens are potent regulators of human bone cell metabolism in vitro. J Bone Miner Res 12:464–471

Katznelson L, Finkelstein JS, Schoenfeld DA, Rosenthal DI, Anderson EJ, Klibanski A (1996) Increase in bone density and lean body mass during testosterone administration in men with acquired hypogonadism. J Clin Endocrinol Metab 81:4358–3465

Kelepouris N, Harper KD, Gannon F, Kaplan FS, Haddad JG (1995) Severe osteoporosis in men. Ann Intern Med 123:452–460

Krabbe S, Christiansen C (1984) Longitudinal study of calcium metabolism in male puberty. I. Bone mineral content, and serum levels of alkaline phosphatase, phosphate and calcium. Acta Paediatr Scand 73:745–749

Krabbe S, Hummer L, Christiansen C (1984) Longitudinal study of calcium metabolism in male puberty. II. Relationship between mineralization and serum testosterone. Acta Paediatr Scand 73:750–755

Link K, Blizzard RM, Evans WS, Kaiser DL, Parker MW, Rogol AD (1986) The effect of androgens on the pulsatile release and the twenty-four hour mean concentration of growth hormone in peripubertal males. J Clin Endocrinol Metab 62:159–164

Lips P, Asscheman H, Uitewaal P, Netelenbos JC, Gooren L (1989) The effect of cross-gender hormonal treatment on bone metabolism in male-to-female transsexuals. J Bone Miner Res 4:657–662

Lips P, Van Kesteren PJM, Asscheman H, Gooren LJG (1996) The effect of androgen treatment on bone metabolism in female-to-male transsexuals. J Bone Miner Res 11:1769–1773

Matzkin H, Chen J, Weisman Y, Goldray D, Pappas F, Jaccard N, Braf Z (1992) Prolonged treatment with finasteride (a 5α-reductase inhibitor) does not affect bone density and metabolism. Clin Endocrinol (Oxf) 37:432–436

Mazess RB, Cameron JR (1974) Bone mineral content in normal U.S. whites. In: Mazess RB (ed). Proceedings, International Conference on Bone Mineral Measurement. National Institute of Arthritis, Metabolism, and Digestive Diseases, (DHEW publication no. NIH 75-683) Washington, DC pp 228–238

McCormick DP, Ponder SW, Fawcett HD, Palmer JL (1991) Spinal bone mineral density in 335 normal and obese children and adolescents: evidence for ethnic and sex differences. J Bone Miner Res 6:507–513

McElduff A, Wilkinson M, Ward P, Posen S (1988) Forearm mineral content in normal men: relationship to weight, height and plasma testosterone concentrations. Bone 9:281–283

Meier DE, Orwoll ES, Keenan EJ, Fagerstrom RM (1987) Marked decline in trabecular bone mineral content in healthy men with age: lack of association with sex steroid levels. J Am Geriatr Soc 35:189–197

Morishima A, Grumbach MM, Simpson ER, Fisher C, Qin K (1995) Aromatase deficiency in male and female siblings caused by a novel mutation and the physiological role of estrogens. J Clin Endocrinol Metab 80:3689–3698

Ogata E, Shimazawa E, Suzuki H, Yoshitoshi Y, Asano H, Ando H (1970) Androgens and the enhancement of hypocalcemic response to thyrocalcinonin in rats. Endocrinology 87:421–426

Orwoll ES, Stribrska L, Ramsey EE, Keenan EJ (1991) Androgen receptors in osteoblast-like cell lines. Calcif Tissue Int 49:183–187

Pilbeam CC, Raisz LG (1990) Effects of androgens on parathyroid hormone and interleukin-1 stimulated prostaglandin production in cultured neonatal mouse calvariae. J Bone Miner Res 5:1183–1188

Riis BJ, Krabbe S, Christiansen C, Catherwood BD, Deftos LJ (1985) Bone turnover in male puberty: a longitudinal study. Calcif Tissue Int 37:213–217

Rosen HN, Tollin S, Balena R, Middlebrooks VL, Moses AC, Yamamoto M, Zeind AJ, Greenspan SL (1995) Bone density in normal male rats treated with finasteride. Endocrinology 136:1381–1387

Rudman D, Drinka PJ, Wilson CR, Mattson DE, Scherman F, Cuisinier MC, Shultz S (1994) Relations of endogenous anabolic hormones and physical activity to bone mineral density and lean body mass in elderly men. Clin Endocrinol (Oxf) 40:653–661

Saville PD (1969) Changes in skeletal mass and fragility with castration in the rat: a model of osteoporosis. J Am Geriatr Soc 17:155–166

Schoutens A, Verhas M, L'Hermite-Baleriaux M, L'Hermite M, Verschaeren A, Dourov N, Mone M, Heilporn A, Tricot A (1984) Growth and bone haemodynamic responses to castration in male rats. Reversibility by testosterone. Acta Endocrinol 107:428–432

Schweikert HU, Rulf W, Niederle N, Schafer HE, Keck E, Kruck F (1980) Testosterone metabolism in human bone. Acta Endocrinol 95:258–264

Seeman E, Melton LJ, O'Fallon WM, Riggs BL (1983) Risk factors for spinal osteoporosis in men. Am J Med 75:977–983

Smith DAS, Walker MS (1977) Changes in plasma steroids and bone density in Klinefelter's syndrome. Calcif Tissue Res 22:225–228

Smith EP, Boyd J, Frank GR, Takahashi H, Cohen RM, Specker B, Williams TC, Lubahn DB, Korach KS (1994) Estrogen resistance caused by a mutation in the estrogen-receptor gene in a man. N Engl J Med 331:1056–1061

Somjen D, Kaye AM, Harell A, Weisman Y (1989) Modulation by vitamin D status of the responsiveness of rat bone to gonadal steroids. Endocrinology 125:1870–1876

Soule SG, Conway G, Prelevic GM, Prentice M, Ginsburg J, Jacobs HS (1995) Osteopenia as a feature of the androgen insensitivity syndrome. Clin Endocrinol (Oxf) 43:671–675

Stanley HL, Schmitt BP, Poses RM, Deiss WP (1991) Does hypogonadism contribute to the occurrence of minimal trauma hip fracture in elderly men? J Am Geriatr Soc 39:766–771

Stepan JJ, Lachman M, Zverina J, Pacovsky V, Baylink DJ (1989) Castrated men exhibit bone loss: effect of calcitonin treatment on biochemical indices of bone remodeling. J Clin Endocrinol Metab 69:523–527

Tanaka S, Haji M, Nishi Y, Yanase T, Takayanagi R, Nawata H (1993) Aromatase activity in human osteoblast-like osteosarcoma cell. Calcif Tissue Int 52:107–109

Tenover JS (1992) Effects of testosterone supplementation in the aging male. J Clin Endocrinol Metab 75:1092–1098

Theintz G, Buchs B, Rizzoli R, Slosman D, Clavien H, Sizonenko PC, Bonjour J-P (1992) Longitudinal monitoring of bone mass accumulation in healthy adolescents: evidence for a marked reduction after 16 years of age at the levels of lumbar spine and femoral neck in female subjects. J Clin Endocrinol Metab 75:1060–1065

Tollin SR, Rosen HN, Zurowski K, Saltzman B, Zeind AJ, Berg S, Greenspan SL (1996) Finasteride therapy does not alter bone turnover in men with benign prostatic hyperplasia – a clinical research center study. J Clin Endocrinol Metab 81:1031–1034

Turner RT, Hannon KS, Demers LM, Buchanan J, Bell NH (1989) Differential effects of gonadal function on bone histomorphometry in male and female rats. J Bone Miner Res 4:557–563

Ulloa-Aguirre A, Blizzard, RM, Garcia-Rubi E, Rogol AD, Link K, Christie CM, Johnson ML, Veldhuis JD (1990) Testosterone and oxandrolone, a nonaromatizable androgen, specifically amplify the mass and rate of growth hormone (GH) secreted per burst without altering GH secretory burst duration or frequency or the GH half-life. J Clin Endocrinol Metab 71:846–854

Vaishnav R, Beresford JN, Gallagher JA, Russell RGG (1988) Effects of the anabolic steroid stanozolol on cells derived from human bone. Clin Sci 74:455–460

Van Kesteren P, Lips P, Deville W, Popp-Snijders C, Asscheman H, Megens J, Gooren L (1996) The effect of one-year cross-sex hormonal treatment on bone metabolism and serum insulin-like growth factor-1 in transsexuals. J Clin Endocrinol Metab 81:2227–2232

Vanderschueren D, Van Herck E, Suiker AMH, Visser WJ, Schot LPC, Bouillon R (1992) Bone and mineral metabolism in aged male rats: short and long term effects of androgen deficiency. Endocrinology 130:2906–2916

Vanderschueren D, Van Herck E, Suiker AMH, Visser WJ, Schot LPC, Chung K, Lucas RS, Einhorn TA, Bouillon R (1993) Bone and mineral metabolism in the androgen-resistant (testicular feminized) male rat. J Bone Miner Res 8:801–809

Vered I, Kaiserman I, Sela B-A, Sack J (1997) Cross genotype sex hormone treatment in two cases of hypogonadal osteoporosis. J Clin Endocrinol Metab 82:576–578

Verhas M, Schoutens A, L'Hermite-Baleriaux M, Dourov N, Verschaeren A, Mone M, Heilporn A (1986) The effects of orchidectomy on bone metabolism in aging rats. Calcif Tissue Int 39:74–77

Wakley GK, Schutte HD, Hannon KS, Turner RT (1991) Androgen treatment prevents loss of cancellous bone in the orchidectomized rat. J Bone Miner Res 6:325–330

Wang C, Eyre DR, Clark R, Kleinberg D, Newman C, Iranmanesh A, Veldhuis J, Dudley RE, Berman N, Davidson T, Barstow TJ, Sinow R, Alexander G, Swerdloff RS (1996) Sublingual testosterone replacement improves muscle mass and strength, decreases bone resorption, and increases bone formation markers in hypogonadal men – a clinical research center study. J Clin Endocrinol Metab 81:3654–3662

Wink CS, Felts WJL (1980) Effects of castration on the bone structure of male rats: a model of osteoporosis. Calcif Tissue Int 32:77–82

Young NR, Baker HWG, Liu G, Seeman E (1993) Body composition and muscle strength in healthy men receiving testosterone enanthate for contraception. J Clin Endocrinol Metab 77:1028–1032

7 Androgens and muscles

Shalender Bhasin, Rachelle Bross, Thomas W. Storer
and Richard Casaburi

Contents

7.1 Introduction

Four thousand years ago, Sushruta described the use of testicular extracts for the treatment of impotence in the Indian scriptures, Ayurveda. In modern times, Charles Eduard Brown-Sequard resurrected interest in the use of testicular extracts for the treatment of aging-related illnesses and symptoms. Brown-Sequard claims to have rejuvenated himself by injecting the extracts of guinea pig testes. It is debatable whether the testicular extracts did him any good; however, for his insight that the secretions of one organ might affect other organs of the body, he is known as the father of modern endocrinology. Ever since the chemical synthesis of testosterone in the 1930's, there has been great interest in exploring the anabolic applications of androgens for augmenting muscle size and strength. The widespread abuse of androgenic steroids by athletes and recreational body builders is based on the premise that testosterone produces muscle hypertrophy and improves performance (Cowart 1987; Strauss and Yesalis 1991); this premise remained unsubstantiated for over 50 years, fueling one of the longest running, and highly polarizing debates in endocrinology (American College of Sports Medicine 1984; Elashoff et al. 1991; Haupt and Rovere 1984; Mooradian et al. 1987; Strauss and Yesalis 1991; Wilson 1988). It is only in the last two to three years that well-controlled studies have unequivocally demonstrated that testosterone induces muscle hypertrophy under specific experimental conditions. We still do not know whether replacement or supraphysiologic doses of androgens improve muscle performance, or produce clinically meaningful changes in body composition in sarcopenic states such as HIV-infection, cancer cachexia, and aging. Furthermore, the anabolic effects of androgens on muscles vary with the age of the individual, the prevalent testosterone levels, the exercise stimulus, and the modulating influence of nutritional factors, growth hormone and other muscle-specific growth factors.

7.2 Testosterone effects on body composition in men

Testosterone effects on body composition will be discussed in three contexts:
- Effects of testosterone replacement on fat-free mass in hypogonadal men
- Effects of modulating serum testosterone concentrations in the broad normal range
- Effects of supraphysiologic doses of testosterone in eugonadal men

7.2.1. Effects of testosterone replacement on body composition in hypogonadal men

Testosterone replacement increases nitrogen retention in castrated males of several animal species (Kochakian 1950), eunuchoidal men, boys before pu-

Table 7.1. Effects of testosterone replacement on body composition in hypogonadal men

Study	Age (years)	Testosterone regimen	Change in fat-free mass	Change in fat mass	Change in muscle strength
Bhasin et al. 1997b	19–47	Testosterone enanthate 100 mg weekly for 10 weeks	5.0±0.7 kg (9.9±1.4%) increase by underwater weight and D_2O	No change in fat mass by underwater weight and D_2O	+22±3%
Katznelson et al. 1996	22–69	Testosterone enanthate or cypionate 100 mg weekly for 18 months	7±2% increase by bioelectrical impedance	14±4% decrease in percent body fat, 13±4% decrease in SC fat	Not measured
Brodsky et al. 1996	33–57	Testosterone cypionate 3 mg/kg/2 weeks for 6 months	+15% by DXA scan	11% decrease in fat mass	Not measured
Wang et al. 1996	19–60	Sublingual testosterone 5 mg three times a day for 6 months	+0.9 kg (+2%) by DXA scan	No change in fat mass	No change in arm press, 8.7 kg increase in leg press

berty, and in women (Kenyon et al. 1940). Several recent studies (Bhasin et al. 1997b; Brodsky et al. 1996; Katznelson et al. 1996; Wang et al. 1996) have re-examined the effects of testosterone on body composition and muscle mass in more detail (Table 7.1). We administered 100 mg testosterone enanthate weekly for ten weeks to seven hypogonadal men (Bhasin et al. 1997b). All subjects underwent a 10–12 week washout and were maintained on a constant eumetabolic diet. Body weight increased by 4.5±0.6 kg (P=0.005) after ten weeks of testosterone replacement. Fat-free mass, estimated from underwater weight, increased by 5.0±0.8 kg (P=0.004) (Fig. 7.1);

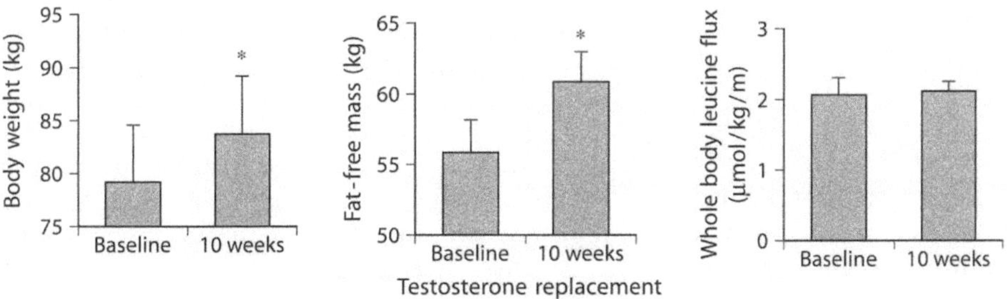

Fig. 7.1. Effects of testosterone replacement on body weight, fat-free mass (by underwater weight) and whole body leucine flux in hypogonadal men. Mean (±SEM) body weight, fat-free mass and whole body leucine flux before (*Baseline*) and after ten weeks of weekly i.m. 100 mg testosterone enanthate in seven hypogonadal men. *p < 0.005 vs baseline. (Adapted with permission from Bhasin et al. 1997b)

body fat did not change. Similar increases in fat-free mass were observed using the deuterium water dilution method. Arm and leg muscle cross sectional areas, assessed by magnetic resonance imaging, increased significantly. Substantial increases in muscle strength were also noted after treatment. These results demonstrated that testosterone replacement augments fat–free mass, muscle size and strength in hypogonadal men.

Using dual-energy X-ray absorptiometry, Brodsky et al. (1996) reported a 15% increase in fat-free mass and a 11% decrease in fat mass in hypogonadal men. The muscle mass increased by 20% and accounted for 65% of the increase in fat-free mass. The muscle accretion during testosterone treatment was associated with a 56% increase in fractional muscle protein synthesis.

Percent body fat is significantly greater in hypogonadal than eugonadal men (Katznelson et al. 1996). Testosterone replacement of androgen-deficient men is associated with a significant decrease in body fat (Katznelson et al. 1996).

A sublingual, cyclodextrin-complexed, testosterone formulation produced a modest increase in fat-free mass (+0.9 kg) and muscle strength (+8.7 kg) in hypogonadal men (Wang et al. 1996); however, the testosterone dose used in this study was smaller than the doses used in previous studies. Thus, the published studies are in agreement that in hypogonadal men replacement doses of testosterone increase fat-free mass.

7.2.2 Effect of modulating serum testosterone levels in the broad normal range

The range of serum testosterone levels in healthy men is wide and extends from 275 to 1,100 ng/dL. We do not know what serum levels of testosterone are optimum for maintaining muscle mass. Although there are some data that sexual function can be maintained by serum testosterone levels at the low end of the normal male range, we do not know whether these low normal levels are sufficient to maintain muscle mass. To examine the effects of modulating serum testosterone in the broad normal range, we used a long-acting GnRH agonist, leuprolide (Lupron Depot, TAP Pharmaceutical Company, Chicago, IL) to suppress endogenous testosterone secretion in healthy, young men (Byerley et al. 1993). Two different levels of serum testosterone concentrations were created by testosterone replacement at doses of four mg/day or eight mg/day by means of a testosterone microsphere formulation. This combined treatment resulted in serum testosterone levels that were in the low-normal range (287±40 ng/dl) in the four mg/d group and the high-normal range (803±46 ng/dl) in eight mg/d group; serum LH levels were markedly suppressed in both the groups. Protein dynamics, measured by a primed, constant infusion of $1-(1-{}^{13}C)$ leucine, nitrogen balance and urinary 3-methyl histidine excretion, did not significantly change as testosterone levels were pharmacologically manipulated within the broad normal range. Similarly, insulin sensitivity, glucose appearance, and lactate production from glucose measured by D $(1-{}^{13}C)$ glucose infusion did not reveal significant treatment (before and after treatment), or testosterone dose (4 vs. eight mg/d) effects. Total cholesterol, HDL cholesterol, HDL subfractions 2 and 3, LDL, VLDL, triglycerides, and apolipoprotein B levels did not differ significantly between the groups either before or after testosterone treatment. Within the constraints imposed by a small sample size and a 9-week treatment duration, these pilot data indicate that as long as the testosterone levels are maintained by replacement therapy in the broad normal range, a eumetabolic state with regard to protein, carbohydrate and lipid metabolism is sustained.

7.2.3 Effect of supraphysiologic doses of testosterone on body composition

Intense controversy persisted until recently with respect to the effects of supraphysiologic doses of androgenic steroids on body composition and muscle strength (American College of Sports Medicine 1988; Bardin 1996; Casaburi et al. 1996; Elashoff et al. 1991; Haupt and Rovere 1984; Strauss and Yesalis 1991; Wilson 1988). Many of the previous studies were neither blinded, nor placebo-controlled. The doses of androgens used in most studies were relatively low and it is surprising that any effects were seen at all. In some studies, the energy and protein intake was not controlled and in other, the exercise stimulus was not standardized so that the effects of androgen could not be evaluated independently of the effects of strength training (Bardin 1996; Casaburi et al. 1996). Another confounding factor in some studies was the inclusion of competitive athletes whose desire to win might preclude compliance with standardized regimens of diet, exercise, and drug administration (Wilson 1988). We conducted a placebo-controlled, double-blind, randomized, clinical trial to separately assess the effects of supraphysiologic doses of testosterone and resistance exercise on fat–free mass, muscle size and strength (Bhasin et al. 1996). Healthy men, 19–40 years of age, and within 15% of their ideal weight, were randomly assigned to one of four groups: placebo but no exercise; testosterone but no exercise; placebo plus exercise; and testosterone plus exercise. The men received 600 mg testosterone enanthate or placebo weekly for ten weeks. To assure compliance, testosterone injections were administered by the nursing staff in the Clinical Study Center. Serum total and free testosterone levels, measured seven days after each injection, increased five-fold; these were nadir levels and serum testosterone levels at other times must have been higher. Serum LH levels were markedly suppressed in the two testosterone-treated but not placebo-treated men, providing additional evidence of compliance. Men in the exercise groups underwent weight lifting exercises three times weekly; the training stimulus was standardized based on the subjects' initial 1-repetition maximum and supervised. Fat-free mass by underwater weighing, muscle size by magnetic resonance imaging, and muscle strength of the arms and legs in bench press and squat exercises were measured before and after ten weeks of treatment (Fig. 7.2).

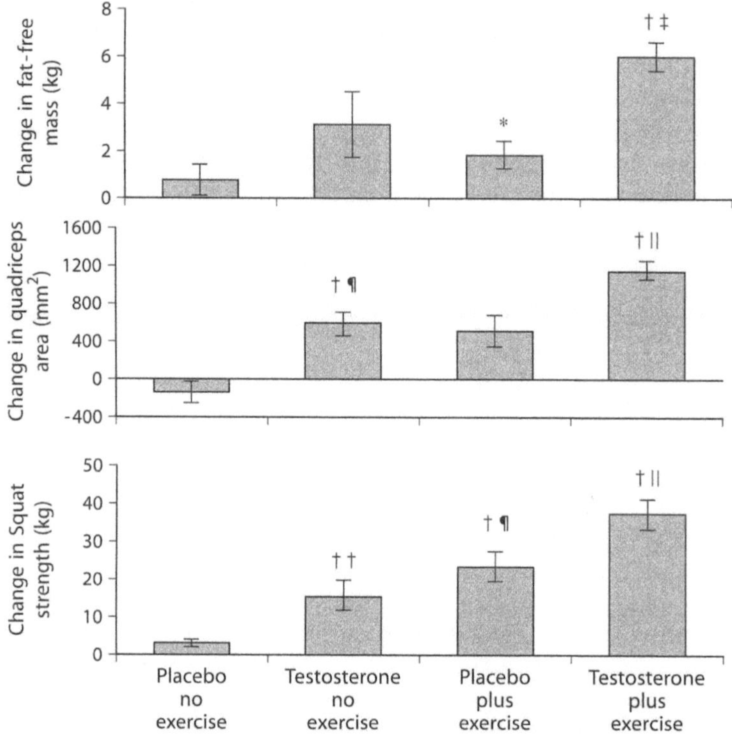

Fig. 7.2. Effect of a supraphysiologic dose of testosterone on fat-free mass, muscle size, and strength in healthy, eugonadal men. Changes from baseline in mean±SE fat-free mass, quadriceps area, and muscle strength in the squat exercise over the ten weeks of treatment. *, change significantly different from zero P=0.017; †, change significantly different from zero P<0.001; ‡, change significantly greater than that in either no exercise group, P<0.05; ¶, change significantly greater than that in placebo, no exercise group P<0.05; ‖, change significantly greater than in all other groups P<0.05; ††, change significantly different from zero P=0.004. (Adapted with permission from Bhasin et al. 1996)

Compared to placebo-treated subjects, the men given testosterone alone had greater gains in muscle size in the arm (mean (± SE) change in triceps area 13.2±3.3 vs. −2.1±2.9%, p<0.05) and leg (change in quadriceps area 6.5±1.3 vs. −1.0±1.1%, p<0.05), and strength in the bench press (increase 10±4 vs. −1±2%, p<0.05) and squat exercise capacity (increase 19±6 vs. 3±1%, p<0.05). The two interventions together produced greater gains in fat–free mass (+9.5±1.0%), and muscle size (+14.7±3.1% in triceps area and +14.1±1.3% in quadriceps area) than either placebo or exercise alone, and greater gains in muscle strength (+24±3% in bench press strength, and +39±4% in squat exercise capacity) than either non-exercising group. We did not observe any significant changes in red blood cell counts or liver enzymes in any treatment group. Serum PSA levels did not change during treatment and no abnormalities were detected in the prostate on digital rectal examination during the ten-week treatment period. Two men in the testosterone-group and one man receiving placebo injections developed acne. These results demonstrate that supraphysiologic doses of testosterone, especially when combined with strength training, increase fat-free mass, muscle size and strength in normal men.

In an open-label study that was not placebo-controlled, Griggs et al. (1989a) administered testosterone enanthate at a dose of three mg/kg/week to healthy men, 19–40 years of age. Muscle mass, estimated from creatinine excretion, increased by a mean of 20% and ^{40}K mass increased 12% after 12 weeks of testosterone treatment. In a separate study (Griggs et al. 1989b; Welle et al. 1992), these investigators administered a similar dose of testosterone enanthate for 12 months to men with muscular dystrophy. Lean body mass increased 4.9 kg (approximately 10%) at three months; these gains were maintained for 12 months, although no further gains in lean body mass were observed with continued testosterone treatment beyond three months.

Young et al. (1993) examined fat-free mass by DXA scan in 13 non-athletic men treated with 200 mg testosterone enanthate weekly for six months during the course of a male contraceptive study. This was an open-label study that included untreated men as controls. Testosterone treatment increased serum testosterone levels by 90% and was associated with 9.6% increase in fat-free mass and 16.2% decrease in fat mass. Changes in muscle strength varied across different muscle groups; most consistent changes were reported in hip abduction which increased 19.2%.

Collectively, these data (Bhasin et al. 1996; Forbes et al. 1992; Griggs et al. 1989a; Welle et al. 1992; Young et al. 1993) demonstrate that when dietary intake and exercise stimulus are controlled, supraphysiologic doses of testosterone produce further increases in fat-free mass and strength in eugonadal men.

7.3 Effects of androgenic steroids on strength, athletic performance and endurance

Excellent reviews of the effects of anabolic steroids on strength (Casaburi et al. 1996; Elashoff et al. 1991; Wilson 1988) have emphasized the flaws in the design of published studies. Approximately one half of the published studies report an increase in muscle strength after androgen administration while others find no change. All the studies in which a beneficial effect on strength was shown used athletes who were previously trained, consumed high protein diet, and used an orally active anabolic steroid, methandrostenolone (dianabol), (Ariel 1972; Bowers and Reardon 1972; Casaburi et al. 1996; Freed et al. 1975; Hervey et al. 1976, 1981; Johnson et al. 1972; O'Shea 1971; Stamford and Maffatt 1974; Ward 1973; Wilson 1988). The doses of androgenic steroids used in the studies prior to our recent study were low and it is surprising that any effects were seen at all at daily doses of ten mg methandrostenolone or less. Our study (Bhasin et al. 1996) demonstrated that supraphysiologic doses of testosterone increase effort-dependent muscle strength and that these effects are augmented by resistance training.

Apart from weight lifting and occasional examination of endurance exercise, little is known about the effects of testosterone on athletic performance. An exception is a study by O' Shea and Winkler (1970) which found no changes in swim time. The literature with respect to sprint times, throwing distances and jumping abilities is sparse and no study has examined the effects on performance in sports, muscle fatigue, power, task–based performance, and effort–independent performance. One study of prepubertal boys (Golding et al. 1974) reported improved performance in vertical jump with increasing testosterone levels, but the training level was not controlled. The published literature is consistent in finding no improvement in either aerobic performance or maximal oxygen uptake following the use of anabolic steroids (Casaburi et al. 1996; Fahey et al. 1973; Fowler et al. 1965; Hervey et al. 1976; Johnson et al. 1975; Loughton and Ruhling 1977; O'Shea 1971).

To evaluate if the testosterone-induced hypertrophy is associated with improved "quality" of muscle function, we computed the strength gains for a unit increase in muscle cross-sectional area in men who had received testosterone alone or strength training alone. Resistance exercise and testosterone both produced similar increases in quadriceps muscle cross-sectional area; however, the strength gains in the squat exercise for each unit change in muscle cross-sectional area tended to be greater in the men undergoing exercise alone than in those receiving testosterone. Testosterone treatment was associated with significantly greater weight gain than exercise alone. Therefore, it is conceivable that in sports that require the athletes to lift their body weight during the course of the athletic event, such as high jump, exercise alone would be more beneficial because it would produce similar gains in strength without as much weight gain.

7.4 Construction of a testosterone dose response curve from the published data

Forbes (1985) collected data from published reports in which body composition was measured before and after administration of anabolic steroids. In some studies, body composition was not directly measured but lean body mass was estimated from the nitrogen balance data. The author concluded that increment in lean body mass was directly related to the cumulative dose of testosterone (Forbes 1985). Forbes' hypothesis predicts that continued treatment with the same dose of androgen for a longer duration would produce greater muscle accretion than treatment for a shorter duration. However, the studies conducted in the author's own laboratory (Griggs et al. 1989b; Welle et al. 1992) do not support the proposal that the gains in muscle mass are proportional to the cumulative androgen dose. For instance, patients with muscular dystrophy (Welle et al. 1992) treated with a supraphysiologic dose of testosterone for three months gained 4.9 kg of lean body mass; continued treatment for 12 months did not produce further gains although the cumulative testosterone dose was four times larger. Our studies in hypogonadal and eugonadal men also demonstrate that most of the weight gain during testosterone treatment occurs within the first six to eight weeks at which time body weight tends to reach a new plateau.

We propose that the testosterone dose response curve is log-linear; each log unit increment in serum testosterone levels is associated with an 8–9% increase in fat-free mass. Thus the magnitude of muscle accretion in response to androgen treatment depends on the increment in serum testosterone levels rather the cumulative dose. Studies are currently underway to test this hypothesis prospectively. How can we reconcile with this hypothesis the data from our pilot study (Byerley et al. 1993) in which modulation of serum testosterone levels in the broad normal range did not produce significant changes in measures of protein dynamics? Two explanations are possible. First, it is conceivable that we failed to detect small but physiologically significant effects because of the small sample size of our pilot study and the relative insensitivity of the methods used (whole body leucine flux, and nitrogen balance). Alternatively, it is possible that two dose response curves may exist, as has been proposed (Wilson 1988), one in the hypogondal range with V_{max} response at the lower end of the normal range and a second dose response curve in the supraphysiologic range.

7.5 Mechanisms of androgen-induced muscle hypertrophy

7.5.1. Testosterone effects on protein synthesis

The biochemical processes that mediate androgen-induced accretion of muscle mass are not well understood. Based on observations that androgenic steroids increase nitrogen retention, we hypothesized that testosterone produces a generalized increase in whole body protein synthesis. Protein synthetic activity is decreased in gonadectomized adult male rats; conversely, testosterone increases incorporation of amino acids into protein and RNA polymerase activity. Testosterone effects on protein degradation are less well understood. Androgen administration alone or in combination with exercise decreases degradation rate of myofibrillar proteins in heart and skeletal muscles of female guinea pigs, but there are no human studies on androgen effects on protein degradation.

We measured whole body protein synthesis in seven hypogonadal men by using a primed, steady-state, infusion of L-$[^{13}C_1]$ leucine before and after ten weeks of testosterone replacement (Bhasin et al. 1997b). We observed no significant changes in whole body leucine flux, leucine oxidation or non-oxidative leucine disappearance rates. This appears paradoxical because accretion of new muscle mass must involve an increase in protein synthesis. Several explanations are possible. First, increase in protein synthesis may occur early in the course of testosterone treatment and once a new steady state is achieved, the leucine flux may return to baseline. Because we measured leucine turnover only at the end of the ten week treatment period, it is possible that we may have missed this early increase. Indeed, most of the weight gain occurred in the first six weeks of testosterone treatment. Second, it is possible that changes in whole body protein synthesis may be small and therefore, were not detected because of the relative insensitivity of the methods used and the small sample size of this study. Third, testosterone may selectively increase protein synthesis within the muscle compartment. Because fractional muscle protein synthesis contributes only 20 − 25% to whole body protein synthesis, small but significant changes in the muscle compartment may not be reflected in measurements of whole body protein synthesis. Brodsky et al. (1996) demonstrated that testosterone replacement of young hypogonadal men increases fractional muscle protein synthesis rate by over 50%. Similarly, Urban et al. (1995) reported significant increases in fractional muscle protein synthesis rates in older men given testosterone replacement for six weeks. Thus, testosterone increases muscle mass by stimulating muscle protein synthesis. Muscle protein breakdown was not measured in any of these studies. All the protein turnover studies have been performed in the post-absorptive state in which protein breakdown exceeds protein synthesis; it is possible that protein turnover studies performed during the fed state may more accurately reflect the anabolic effects of testosterone.

7.5.2 Testosterone effects on growth hormone and insulin-like growth factor-1 axis

The biochemical mechanisms that mediate testosterone-induced increase in muscle protein synthesis are not known. Testosterone increases circulating IGF-1 levels by stimulating GH secretion. However, testosterone produces muscle hypertrophy even in hypogonadal men who have undergone hypophysectomy and have GH deficiency (Bhasin et al. 1997b). It is likely that testosterone directly upregulates IGF-1 expression within the muscle. Anabolic interventions such as exercise and testosterone increase IGF-1 mRNA within the skeletal muscle (Urban et al. 1995). Testosterone downregulates IGFBP-4 mRNA. Because IGFBP-4 attenuates the physiologic effects of IGF-1, reciprocal changes in IGF-1 and its binding protein provide one potential mechanism for amplifying the anabolic signal.

7.5.3 Does testosterone-induced increase in muscle size represent muscle hypertrophy or hyperplasia?

It is generally believed that the increase in muscle volume during testosterone treatment represents muscle hypertrophy; however, this issue has not been firmly established. A combined regimen of exercise and methandienone given to young men for six weeks increased muscle dimensions and strength (Alen et al. 1984). These changes in muscle volume were associated with an increase in muscle fiber diameter consistent with the hypothesis that hypertrophy mediates testosterone-induced increase in muscle size. Testosterone administration to animals is associated predominantly with skeletal muscle hypertrophy. However, Griggs et al. (1989a) found no significant change in muscle fiber diameter in healthy men treated with testosterone enanthate at a dose of three mg/kg/week.

7.5.4 Are testosterone effects mediated through an androgen-receptor mediated pathway?

Androgen receptor is expressed in the skeletal muscle and its binding characteristics are similar to those in other tissues (Dahlberg et al. 1981; Krieg 1976; Michel and Bauheu 1980; Saartok et al. 1984). However, we do not know whether anabolic effects of supraphysiologic doses of testosterone are mediated through an androgen receptor mediated pathway. Androgen receptor in the skeletal muscle of the adult animal is saturated at physiologic testosterone concentrations (Bartsch et al. 1980). Therefore, it is possible that the anabolic effects of supraphysiologic doses of testosterone are mediated through an androgen receptor-independent mechanism, such as through an anti-glucocorticoid effect (Wilson 1988).

Several androgenic steroids inhibit binding of dexamethasone to glucocorticoid receptor in the skeletal muscle (Mayer and Rosen 1977). Glucocorticoid antagonists can retard the atrophy of levator ani muscle after castration (Konagaya and Max 1986). Testosterone can reverse, in part, the atrophy caused by glucocortioids in some experimental paradigms (Reid et al. 1996). Androgen administration to normal men increases urinary free cortisol (Hervey et al. 1981). On the contrary, androgenic steroids have very low binding affinity to the glucocorticoid receptor. Furthermore, adrenalectomy in female rabbits does not attenuate the myotrophic effects of androgens (Salmons 1983), suggesting that the anti-glucocorticoid action of androgens may be limited to its supraphysiologic doses. If there are two separate mechanisms mediating testosterone's anabolic effects, there might also exist two separate dose-response curves, one in the hypogonadal range with V_{max} corresponding to the lower end of the male range and a second dose-response curve in the supraphysiologic range representing the anti-glucocorticoid effect. This speculation remains to be tested.

7.5.5 Is 5α-reduction of testosterone obligatory for mediating its effects?

We do not know whether 5α-reduction of testosterone to its metabolite, dihydrotestosterone, is required for mediating its anabolic effects. Patients with benign prostatic hypertrophy (BPH) who have been treated with 5α-reductase inhibitor, finasteride, do not experience muscle wasting. Similarly, patients with congenital deficiency of 5α-reductase enzyme have normal muscle mass. Sattler et al. (1996) have proposed that a defect in DHT generation contributes to sarcopenia in a subset of HIV-infected men with wasting. However, we do not know whether decreased DHT levels can contribute to loss of lean tissue, independent of the decrease in testosterone levels. The bulk of evidence suggests that conversion of testosterone to DHT is not obligatory for maintaining muscle mass.

7.5.6 Testosterone effects on neuromuscular transmission

Testosterone may affect muscular function by its effects on neuromuscular transmission (Blanco et al. 1997; Leslie et al. 1991). The expression of choline acetyltransferase mRNA in spinal cord motoneurons in the male rat is testosterone-dependent (Blanco et al. 1997). The spinal cord motor nuclei that innervate the foot muscle, flexor digitorum brevis, is sexually dimorphic and its size is regulated by testosterone levels (Leslie et al. 1991). The neuronal nitric oxide synthase enzyme may be an additional site for androgen regulation of muscle function.

7.6 Testosterone effects on regional fat metabolism

It is not clear why some studies of testosterone administration report a decrease in fat mass, while others show only an increase in fat-free mass and no change in fat mass. Marin et al. (1992, 1995) have proposed that testosterone is one of the determinants of regional fat distribution. Visceral obesity in men is associated with lower testosterone levels (Seidell et al. 1990). In middle aged men, serum testosterone levels correlate inversely with mid-segment obesity and directly with plasma HDL levels and insulin sensitivity (Barrett-Connors and Khaw 1988; Seidell et al. 1990). Physiologic testosterone replacement decreases waist to hip ratios, blood glucose and insulin levels (Marin et al. 1992). Testosterone inhibits lipid uptake and enhances lipid mobilization from adipocytes in a region-specific manner, thus determining the proportion of fat in central and peripheral adipose tissues in man (Marin et al. 1995).

7.7 Modulation of testosterone effects by other growth factors

Testosterone is only one of the many factors regulating body composition, other determinants include genetic factors, nutrition, growth hormone and

IGF-1 levels, growth and differentiation factor-8, and cytokines. Resistance exercise augments testosterone effects. The growth hormone axis may modulate the anabolic response to androgens (Fryburg et al. 1997). The interactions of nutrition, particularly protein intake, growth and differentiation factor-8, and cytokines with testosterone have not been studied.

7.8 Testosterone effects on body composition and protein dynamics in other physiologic states

7.8.1. Effects of testosterone replacement in pre-pubertal boys

Boys and girls before their entry into puberty have similar body composition; however, during the course of pubertal development, the boys accrue substantially greater amounts of muscle mass. These gender differences in muscle accretion have been attributed in part to the difference in sex steroid milieu in boys and girls. The boys experience over a hundred-fold increase in serum testosterone levels during the pubertal transition to adulthood. Testosterone increases nitrogen retention in immature castrated animals, and in pre-pubertal boys (Table 7.2; Arslanian and Suprasongsin 1997; Gregory et

Table 7.2. Effects of testosterone administration in prepubertal boys

Study	Subjects	Testosterone regimen	Body composition changes	Metabolic changes
Arslanian and Suprasongsin 1997	14.9–16.5 years, delayed puberty	Testosterone enanthate 50 mg every 2 weeks for 4 months	7.6±1.5 kg (20.7±4.2 %) increase in fat-free mass and 7.8% decrease in body fat	Decreased whole body proteolysis and oxidation, increased fat oxidation
Gregory et al. 1992	Boys with constitutional delay of puberty	Testosterone undecanoate 40 mg daily vs. placebo	Greater increases in height velocity and fat-free mass in testosterone-treated boys	No differences in energy expenditure and muscle strength between two groups
Soliman et al. 1995	Boys with constitutional delay of puberty, 14.3+0.7 years	100 mg testosterone enanthate monthly or placebo	Increased height velocity, weight gain, and IGF-1 levels in testosterone-treated boys	
Mauras et al. 1994	Healthy, pre-pubertal, short boys	Testosterone enanthate 3 mg/kg every 2 weeks for 4–6 weeks	Body composition not reported	Increases in whole body protein synthesis and calcium retention

al. 1992; Mauras et al. 1994; Soliman et al. 1995). Gregory et al. (1992) administered 40 mg oral testosterone undecanoate daily for three months to 18 boys with constitutional delay of puberty. Testosterone treatment was associated with increased height velocity and greater gains in fat-free mass than placebo-treatment. In another study (Arslanian and Suprasongsin 1997), 50 mg testosterone enanthate given every two weeks to seven healthy boys with constitutional delay of puberty was associated with an eight kg increase in fat-free mass and a 3.3 kg decrease in body fat. Whole body proteolysis and leucine oxidation were decreased after testosterone treatment. Mauras et al. (1994) also reported increase in non-oxidative leucine disappearance rates in pre-pubertal boys after 4–6 weeks of testosterone treatment. Oxandrolone, an orally administered androgenic steroid, increases linear growth velocity in boys with constitutional delay of puberty (Stanhope and Brook 1985; Stanhope et al. 1988). Collectively, these data suggest that increasing serum testosterone levels in prepubertal boys are associated with marked increases in protein synthesis and lean body mass accretion.

7.8.2 Effects of testosterone supplementation on body composition and muscle function in older men

Changes in gonadal function during the aging process are discussed in detail in another chapter. As a group, serum testosterone levels are statistically lower in older men than those seen in younger men even though most older men have normal or low-normal testosterone levels. These observations have led to speculation that older men might be relatively insensitive to the end organ effects of testosterone. There is evidence that androgen receptor number and affinity are decreased in many organs of the aging rat (Greenstein 1979; Haji et al. 1981). However, in the human, older men have increased rather than decreased sensitivity to androgen feedback effects on pituitary LH and FSH secretion (Winters et al. 1984). This issue of androgen insensitivity of muscle and bone to testosterone effects in older men has not been studied.

The anabolic effects of testosterone replacement on body composition in older men have been examined in several short-term studies (Table 7.3; Janowsky et al. 1994; Morley et al. 1993; Sih et al. 1997; Tenover 1992; Urban et al. 1995). In one such study, Tenover (1992) administered testosterone enanthate, 100 mg per week, or placebo, for 12 weeks to older men with testosterone levels of less than 400 ng/dl. In this double blind, placebo-controlled, fat-free mass increased after testosterone replacement, but the increments were relatively modest.

Morley et al. (1993) studied the effects of 200 mg testosterone cypionate given every two weeks to older men with bioavailable testosterone levels of less than 60 ng/dl. These investigators noted a modest increase in hand grip in testosterone-treated men, although body composition did not change.

Several clinical trials are currently in progress to examine the long-term effects of testosterone replacement in older men. In a recently completed

Table 7.3. Effects of testosterone supplementation in older men

Study	Subjects	Treatment regimen	Changes in body composition	Changes in muscle function	Comments
Tenover 1992	60–75 years, serum testosterone <400 ng/dL	Testosterone enanthate 100 mg weekly for 3 months	1.8 kg increase in fat-free mass; no change in fat mass	No change in grip strength	Mild increases in PSA and hematocrit
Morley et al. 1993	69–89 years, bioavailable testosterone less than 75 ng/dL	Testosterone enanthate 200 mg every 2 weeks for 3 months	No change in fat mass or body weight	Increase in grip strength	
Sih et al. 1997	healthy men, 51–79 years, serum bio-available testosterone <60 ng/dL	Testosterone cypionate 200 mg every 2 weeks for 12 months	0.9 (+3%) cm increase in mid-arm circumference, no change in fat mass	4–5 kg increase in grip strength	No change in PSA, increase in hematocrit
Urban et al. 1995	Healthy elderly, 67+2 years, testosterone <480 ng/dL	Testosterone enanthate weekly for 4 weeks to increase testosterone to 500–1000 ng/dL	Body composition not reported	Increase in hamstring and quadriceps work per repetition; no change in endurance	Approximately 2 fold increase in fractional muscle protein synthesis rate

study, Sih et al. (1997) treated older men over the age of 50 with bioavailable testosterone of less than 60 ng/dl with placebo or testosterone cypionate, 200 mg/2 weeks, for a period of one year. There was a modest improvement in grip strength after testosterone treatment. The fat free mass did not change significantly. Three men in the testosterone group dropped out because of marked increases in hematocrit. There was no change in prostate-specific antigen.

Urban et al. (1995) treated five older men with replacement doses of testosterone for six weeks. Testosterone replacement increased fractional muscle protein synthesis and muscle IGF-1 mRNA expression in the treated older men. Thus short-term studies in older men demonstrate that physiologic testosterone replacement produces only modest increases in lean body mass and grip strength. It remains uncertain whether physiologic replacement can induce meaningful changes in muscle function in older men.

7.9 Key messages

- There is consensus that replacement doses of testosterone in hypogonadal men and supraphysiologic doses given to eugonadal men increase fat-free mass, muscle size and strength.
- We do not know whether testosterone supplementation can produce clinically significant improvement in muscle function in older men or in patients with wasting disorders such as HIV-infection or cancer cachexia.
- Testosterone selectively stimulates fractional muscle protein synthesis, thus producing muscle hypertrophy.
- The molecular mechanisms by which testosterone exerts its anabolic effects are not well understood.

7.10 References

Alen M, Hakkinen K, Komi PV (1984) Changes in neuromuscular performance and muscle fiber characteristics of elite power athletes self-administering androgenic and anabolic steroids. Acta Physiol Scand 122:535–544

American College of Sports Medicine (1984) Position Stand on the use of anabolic-androgenic steroids in sports. Sports Med Bull 19:13–15

Ariel G (1972) The effect of anabolic steroid (methandrostenolone) upon selected physiologic parameters. Ath Training 7:190–196

Arslanian S, Suprasongsin C (1997) Testosterone treatment in adolescents with delayed puberty: changes in body composition, protein, fat and glucose metabolism. J Clin Endocrinol Metab 82:2313–3220

Bardin CW (1996) The anabolic action of testosterone. N Engl J Med 335:52–53

Barrett-Connors E, Khaw K-T (1988) Endogenous sex-hormones and cardiovascular disease in men. A prospective population-based study. Circulation 78:539–545

Bartsch W, Krieg M, Voigt KD (1980) Quantitation of endogenous testosterone, 5-alpha-dihydrotestosterone and 5-alpha-androstane-3-alpha, 17-beta-diol in subcellular fractions of the prostate, bulbocavernosus/levator ani muscle, skeletal muscle, and heart muscle of the rat. J Steroid Biochem 13:259–267

Bhasin S, Storer TW, Berman N, Callegari C, Clevenger BA, Phillips J, Bunnell T, Tricker R, Shirazi A, Casaburi R (1996) The effects of supraphysiologic doses of testosterone on muscle size and strength in men. N Engl J Med 335:1–7

Bhasin S, Bremner WJ (1997a) Emerging issues in androgen replacement therapy. J Clin Endocrinol Metab 82:3–8

Bhasin S, Storer TW, Berman N, Yarasheski K, Phillips J, Clevenger B, Lee WP, Casaburi R (1997b) A replacement dose of testosterone increases fat-free mass and muscle size in hypogonadal men. J Clin Endocrinol Metab 82:407–413

Bhasin S, Tenover JS (1997c) Sarcopenia: Issues in testosterone replacement of older men. J Clin Endocrinol Metab 82:1659–1660

Blanco CE, Popper P, Micevych P (1997) Anabolic-androgenic steroid induced alterations in choline acetyltransferase messenger RNA levels of spinal cord motoneurons in the male rat. Neuroscience 78:973–882

Bowers RW, Reardon JP (1972) Effect of dianabol on strength development and aerobic capacity. Med Sci Sports Exer 4L:54

Brodsky IG, Balagopal P, Nair KS (1996) Effects of testosterone replacement on muscle mass and muscle protein synthesis in hypogonadal men – a Clinical Research Center Study. J Clin Endocrinol Metab 81:3469–3475

Byerley LO, Lee WNP, Swerdloff RS, Buena F, Nair KS, Buchanan T, Goldberg R, Bhasin S (1993) Effect of modulating serum testosterone levels in normal male range on protein, glucose, and lipid metabolism in men: implications for testosterone replacement therapy. Endocr J 1:253–262

Casaburi R, Storer T, Bhasin S (1996) Androgen effects on body composition and muscle performance. In: Bhasin S, Gabelnick H, Spieler JM, Swerdloff RS, Wang C (eds) Pharmacology, Biology, and Clinical Applications of Androgens: Current Status and Future Prospects. Wiley-Liss, New York, NY, pp 283–288

Cowart V (1987) Steroids in sports: after four decades, time to return these genies to the bottle. JAMA 257:421–423

Dahlberg E, Snochowski M, Gustaffsson JA (1981) Regulation of the androgen and glucocorticoid receptors in the rat and mouse muscle cytosol. Endocrinology 108:1431–1436

Elashoff JD, Jacknow AD, Shain SG, Braunstein GD (1991) Effects of anabolic/androgenic steroids on muscle strength. Ann Intern Med 115:387–393

Fahey TD, Brown CH (1973) The effects of an anabolic steroid on the strength, body composition, and endurance of college males when accompanied by a weight training program. Med Sci Sports Exer 5:272–277

Forbes GB (1985) The effect of anabolic steroids on lean body mass: the dose response curve. Metabolism 34:571–573

Forbes GB, Porta CR, Herr B, Griggs RC (1992) Sequence of changes in body composition induced by testosterone and reversal of changes after the drug is stopped. JAMA 267:397–399

Freed DLJ, Banks AJ, Longson D, Burley DM (1975) Anabolic steroids in athletics: crossover double-blind trial on weight lifters. Br Med J 2:471–474

Fowler WM, Gardner GW, Egstron GH (1965) Effect of an anabolic steroid on physical performance of young men. J Appl Physiol 20:2038–2044

Fryburg DA, Weltman A, Jahn LA, Weltman J, Samojlik E, Hintz RL, Veldhuis JD (1997) Short-term modulation of the androgen milieu alters pulsatile, but not exercise- or growth hormone (GH)-releasing hormone-stimulated GH secretion in healthy men: impact of gonadal steroid and GH secretory changes on metabolic outcomes. J Clin Endocrinol Metab 82:3710–3719

Golding LA, Freydinger JE, Fishel SS (1974) The effect of an androgenic-anabolic steroid and a protein supplement on size, strength, weight, and body composition in athletes. In: Craig TT (ed). The medical aspects of sports: 15. Chicago, American Medical Association, pp 25–28

Greenstein BD (1979) Androgen receptors in the rat brain, anterior pituitary glans and ventral prostate gland: effects of orchiectomy and aging. J Endocrinol 81:75–81

Gregory JW, Greene SA, Thompson J, Scrimgeour CM, Rennie MJ (1992) Effects of oral testosterone undecanoate on growth, body composition, strength, and energy expenditure of adolescent boys. Clin Endocrinol (Oxford) 37:207–213

Griggs RC, Kingston W, Josefowicz RF, Herr BE, Forbes G, Halliday D (1989a) Effect of testosterone on muscle mass and muscle protein synthesis. J Appl Physiol 66:498–503

Griggs RC, Pandya S, Florence JM, Brooke MH, Kingston W, Miller JP, Chutkow J, Herr BE, Moxley RT (1989b) Randomized controlled trial of testosterone in myotonic dystrophy. Neurology 39:219–222

Haupt HA, Rovere GD (1984) Anabolic steroids: a review of the literature. Am J Sports Med 12:469–475

Hervey GR, Hutchison I, Knibbs AV, Burkinshaw L, Jones PR, Norgan NG, Levell MJ (1976) Anabolic effects of mathandienone in men undergoing athletic training. Lancet 2:699–702

Hervey GR, Knibbs AV, Burkinshaw L, Burkinsahw L, Morgan DB, Jones PRM, Chette DR, Vartsky D (1981) Effects of methandienone on the performance and body composition of men undergoing athletic training. Clin Sci Lond. 60:457–461

Janowsky JS, Oviatt SK, Orwoll ES (1994) Testosterone influences spatial cognition in older men. Behav Neurosci 108:625–632

Johnson LC, Roundy ED, Allsen PE, Fisher AG, Slivester LF (1975) Effect of anabolic steroid treatment on endurance. Med Sci Sports 7:287–289

Johnson LC, Fisher G, Silvester LJ, Hofheins CC (1972) Anabolic steroid: effects on strength, body weight, oxygen uptake and spermatogenesis upon mature males. Med Sci Sports 4:43–45

Katznelson L, Finkelstein JS, Schoenfeld DA, Rosenthal DI, Anderson EJ, Klibanski A (1996) Increase in bone density and lean body mass during testosterone administration in men with acquired hypogonadism. J Clin Endocrinol Metab 81:4358–4365

Kenyon AT, Knowlton K, Sandiford I, Kock FC, Lotwin G (1940) A comparative study of the metabolic effects of testosterone propionate in normal men and women and in eunuchoidism. Endocrinology 26:26–45

Kochakian CD (1950) Comparison of protein anabolic properties of various androgens in the castrated rat. Am J Physiol 60:553–558

Konagaya M, Max SR (1986) A possible role for endogenous glucocorticoid in orchiectomy-induced atrophy of the rat levator ani muscle: studies with RU38486, a potent glucocorticoid antagonist. J Steroid Biochem 25:305–311

Krieg M (1976) Characterization of the androgen receptor in the skeletal muscle of the rat. Steroids 28:261–268

Leslie M, Forger NG, Breedlove SM (1991) Sexual dimorphism and androgen effects on spinal motoneurons innervating the rat flexor digitorum brevis. Brain Res 561:269–273.

Loughton SJ, Ruhling RO (1977) Human strength and endurance responses to anabolic steroid and training. J Sports Med Phys Fitness 17:285–296

Marin P, Krotkiewski M, Bjorntorp P (1992) Androgen treatment of middle-aged, obese men: effects on metabolism, muscle, and adipose tissues. Eur J Med 1:329–336.

Marin P, Oden B, Bjorntorp P (1995) Assimilation and mobilization of triglycerides in subcutaneous abdominal and femoral adipose tissue in vivo in men: effects of androgens. J Clin Endocrinol Metab 80:239–243

Mauras N, Haymond MW, Darmaun D, Vieira NE, Abrams SA, Yergey AL (1994) Calcium and protein kinetics in pre-pubertal boys. J Clin Invest 93:1014–1019

Mayer M, Rosen F (1975) Interaction of anabolic steroids with glucocorticoid receptor sites in rat muscle cytosol. Am J Physiol 229:1381–1386

Michel G, Bauheu EE (1980) Androgen receptor in skeletal muscle: characterization and physiologic variations. Endocrinology 107:2088.

Mooradian AD, Morley JE, Korenman SG (1987) Biological actions of androgens. Endocr Rev 8:1–28

Morley JE, Perry HM III, Kaiser FE, Kraenzle D, Jensen J, Houston K, Mattammal M, Perry M (1993) Effects of testosterone replacement therapy in old hypogonadal males: a preliminary study. J Am Geriatr Soc. 41:149–152

O'Shea JP (1971) The effects of an anabolic steroid on dynamic strength levels of weight lifters. Nutr Rep Int 4:363–367

O'Shea JP, Winkler W (1970) Biochemical and physical effects of an anabolic steroid in competitive swimmers and weightlifters. Nutr Rep Int 2:351–355

Reid IR, Wattie DJ, Evans MC, Stapleton JP (1996) Testosterone therapy in glucocorticoid-treated men Arch Intern Med 156:1173–1177

Saartok T, Dahlberg E, Gustaffsson JA (1984) Relative binding affinity of anabolic-androgenic steroids, comparison of the binding to the androgen receptors in skeletal muscle and in prostate as well as sex hormone binding globulin. Endocrinology 114:2100–2107

Salmons S (1983) Myotrophic effects of anabolic steroids. Vet Res Commun 7:19–26

Sattler FR, Antonipillai I, Allen J, Horton R (1996) Wasting and sex hormones: evidence for the role of dihydrotestosterone in AIDS patients with weight loss. Abstract #376 presented at the XI International Conference on AIDS. Vancouver, Canada

Seidell J, Bjorntorp P, Sjostrom L, Kvist H, Sannerstedt R (1990) Visceral fat accumulation in men is positively associated with insulin, glucose and C-peptide levels, but negatively with testosterone levels. Metabolism 39:897–901

Sih R, Morley JE, Kaiser FE, Perry HM, Patrick P, Ross C (1997) Testosterone replacement in older hypogonadal men: a 12 month randomized controlled trial. J Clin Endocrinol Metab 82:1661–1667

Soliman AT, Khadir MM, Asfour M (1995) Testosterone treatment in adolescent boys with constitutional delay of growth and development. Metabolism 44:1013–1016

Stamford BA, Moffatt R (1974) Anabolic steroid: effectiveness of an ergogenic aid to experienced weight lifters. J Sports Med Phys Fitness 14:191–197

Stanhope R, Buchanan CR, Fenn GC, Preece MA (1988) Double blind placebo-controlled trial of low-dose oxandrolone in the treatment of boys with constitutional delay of growth and puberty. Arch Dis Child 63:501–505

Stanhope R, Brook CG (1985) Oxandrolone in low dose for constitutional delay of growth and puberty in boys. Arch Dis Child 60:379–381

Strauss RH, Yesalis CE (1991) Anabolic steroids in the athlete. Ann Rev Med 42:449–460

Tenover JS (1992) Effects of testosterone supplementation in the aging male. J Clin Endocrinol Metab 75:1092–1098

Urban RJ, Bodenburg YH, Gilkison C, Foxworth J, Coggan AR, Wolfe RR, Ferrando A (1995) Testosterone administration to elderly men increases skeletal muscle strength and protein synthesis. Am J Physiol 269:E820–E826

Wang C, Eyre DR, Clark R, Kleinberg D, Newman C, Iranmanesh A, Veldhuis J, Dudley RE, Berman N, Davidson T, Barstow TJ, Sinow R, Alexander G, Swerdloff RS (1996) Sublingual testosterone replacement improves muscle mass and strength, decreases bone resorption, and increases bone resorption markers in hypogonadal men – a Clinical Research Center Study. J Clin Endocrinol Metab 81:3654–3662

Ward P (1973) The effects of an anabolic steroid on strength and lean body mass. Med Sci Sports 5:277–282

Welle S, Josefowicz R, Forbes G, Griggs RC (1992) Effect of testosterone on metabolic rate and body composition in normal men and men with muscular dystrophy. J Clin Endocrinol Metab 74:332–335

Wilson JD (1988) Androgen abuse by athletes. Endocr Rev 9:181–200

Winters SF, Sherins RJ, Troen P (1984) The gonadotropin-suppressive activity of androgen is increased in elderly men. Metabolism 33:1052–1059

Young NR, Baker HWG, Liu G, Seeman E (1993) Body composition and muscle strength in healthy men receiving testosterone enanthate for contraception. J Clin Endocrinol Metab 77:1028–1032

8 Androgens, cardiovascular risk factors and atherosclerosis

Arnold von Eckardstein

Contents

8.1 Introduction: Sex-dependent differences in the risk for atherosclerotic vessel diseases

Atherosclerotic coronary, cerebral, and peripheral vessel diseases are the most frequent cause of death and disability in the industrialized countries for both men and women. Before menopause both morbidity and mortality is much lower in women than in men because of atherosclerotic vessel diseases. After menopause the incidence of atherosclerotic vessel diseases increases steeply (Gorodeski and Utian 1994). In Germany, for example, before age of 65 four times as many as women die because of myocardial infarction (MI). After age 75, coronary heart disease (CHD) mortality is much higher in women than in men so that finally the overall cardiovascular mortality does not differ among sexes (Statistisches Bundesamt 1997). The approximately 10-year gap in the clinical manifestation of atherosclerotic vessel diseases between men and women is usually explained by the anti-atherogenic effects of estrogens. The anti-atherogenicity of estrogens is also indicated by the high incidence of cardiovascular disease in ovariectomized women who do not substitute estrogens, and by the lower cardiovascular event rates of postmenopausal women who take estrogens (Barrett-Connor and Bush 1991; Gorodeski and Utian 1994). In addition to these clinical studies, many animal studies also provided evidence for the anti-atherogenicity of estrogens (Bourassa et al. 1996; Bruck et al. 1997). Moreover, compared to men of the same age, premenopausal women have a more favourable cardiovascular risk factor profile with higher levels of high density lipoprotein cholesterol (HDL-C) and lower levels of low density lipoprotein cholesterol (LDL-C), lipoprotein(a) (Lp(a)), homocysteine, and fibrinogen. After menopause, women present a more adverse risk factor profile. Hormone replacement therapy increases levels of HDL-C and decreases levels of LDL-C and Lp(a) (Barrett-Connor and Bush 1991; Lobo 1991; Sacks and Walsh 1994; Sacks et al. 1995). Beyond their beneficial effects on many cardiovascular risk factors, estrogens were shown to have various other anti-atherogenic effects such as endothelium-dependent and independent vasodilation, inhibition of smooth muscle (SMC) migration and proliferation, and inhibition of lipid accumulation in macrophages (MΦ) (Hutchison et al. 1997a; Sacks et al. 1995; St. Clair 1997).

Much less is known on the role of androgens in atherosclerosis. Prevalences of 20 to 30% hypoandrogenism in men over 65 years and hyperandrogenism in pre-menopausal women (Barnes and Rosenfeld 1989; Vermeulen and Kaufmann 1995), however, show the importance of this issue. More and more indications for the administration of exogenous androgens and suppression of endogenous testosterone are being introduced into medical practice or are under active discussion (see Chapters 10 and 18).

8.2 Epidemiological and clinical evidence for the role of androgens in atherosclerosis

8.2.1 Studies in men

Two prospective cohort studies of 1009 men aged 40 to 79 years who were followed up for 12 years, and of 2512 men aged 45 to 59 years who were followed up for 5 years, did not reveal any significant association between serum levels of testosterone measured at baseline and the occurrence of coronary events (Barrett-Connor and Khaw 1988; Yarnell et al. 1995). Confirming these studies, three nested prospective case control studies of 344 men with MI and 344 age-matched controls with follow-up times of 6–8, 9.5, and 19–20 years did not find significant associations between serum levels of testosterone at baseline and future coronary events (Cauley et al. 1987; Contoreggi et al. 1990; Philips et al. 1988).

In two prospective 12 year follow-up studies of 242 and 510 men, respectively, aged 50 to 79 years, Barrett-Connor et al. (1986, 1987) investigated the association between the weak androgen dehydroepiandrosterone sulfate (DHEAS) and the occurrence of CHD events. Baseline DHEAS levels below the median were associated with a threefold increased risk of cardiovascular death independent of other cardiovascular risk factors (Barrett-Connor and Khaw 1987; Barrett-Connor et al. 1986). A nested prospective case control also found significantly lower DHEAS levels in 238 Japanese men with myocardial infarction (MI) as compared to 476 matched controls without MI during an 18 year follow-up. However, the inverse correlation between DHEAS and coronary risk was not independent of other risk factors (LaCroix et al. 1992). Two other nested case-control studies did not demonstrate significant associations between DHEAS levels and cardiovascular disease in men (Contoreggi et al. 1990; Newcomer et al. 1994).

Alexandersen and colleagues (1996) reviewed the outcomes of 30 cross-sectional case-control studies in men looking at the relationships between the serum levels of various androgens (testosterone, free testosterone, DHEA, DHEAS, and/or androstendione) and different coronary end points. 18 studies found significant inverse associations, 11 no significant association and only one a significant positive association between adrenal androgens and angiographically assessed CHD (Alexandersen et al. 1996). It is noteworthy that all studies which measured levels of free testosterone found an inverse association between this bioavailable androgen and CHD. In one angiographic study of 55 men without prior MI, serum levels of free testosterone were negatively correlated with the degree of CHD (Philips et al. 1994). Alexandersen and colleagues therefore concluded that low levels of androgens increase the risk of CHD in men (1996).

Few interventional studies have been performed on the effect of exogenous androgens on atherosclerosis, only one of which was randomized (Wu et al. 1993). In all cases no coronary end points but symptoms of myocardial ischemia were recorded. Treatment of male CHD patients with testosterone reduced the severity and frequency of angina pectoris events (Hamm 1942; Lesser 1943; Levine and Likoff 1943; Sigler and Tulgan 1943; Wu et al. 1993) as well as the occurrence of exercise-induced ST-segment depression (Jaffé 1977; Rosano et al. 1993). On the other hand, however, there are several case reports of illegal use of anabolic androgens in athletes who then suffered from premature MI (Mammen 1993; McNutt et al. 1988).

8.2.2 Studies in women

By contrast to the neutral or even beneficial role of testosterone and DHEAS for cardiovascular risk for men, the few studies in women revealed pro-atherogenic associations of androgens with CHD. However, only few prospective data are available on the importance of testosterone and other androgens as cardiovascular risk factors in women. Barrett-Connor and Goodman-Gruen (1995) reported a 19 year follow-up of 651 postmenopausal women. Serum levels of testosterone, bioavailable testosterone, and androstendione did not differ between those women with and those without a heart disease history at baseline. Cardiovascular mortality during follow-up was not associated with any androgen serum level (Barrett-Connor and Goodman-Gruen 1995). An earlier 12 year follow-up study of 289 women 60 to 79 years of age revealed a positive association between serum levels of DHEAS and cardiovascular mortality (Barrett-Connor and Khaw 1987).

In an angiographic study of 60 postmenopausal women, serum levels of free testosterone were correlated with the maximum percentage reduction of the luminal diameter of coronary arteries. This correlation was independent of age, body mass index (BMI), systolic blood pressure, smoking, as well as levels of cholesterol, insulin, and estradiol (Philips et al. 1997).

Indirect evidence for the atherogenicity of androgens in women comes from clinical studies in which women with CHD were affected more frequently than control women by clinical symptoms of androgen excess such as hirsutism and polycystic ovaries (Wild et al. 1990; Birdsall et al. 1997). In a combined angiography and pelvic ultrasound study of 143 women aged 60 years or less, the presence of polycystic ovaries was associated with an increased number of stenosed coronary arteries (Birdsall et al. 1997). The presence of polycystic ovaries was also associated with elevated levels of insulin and triglycerides and lower levels of HDL-C. Moreover, a multivariate analysis showed that the presence of CHD and a family history of MI were independent predictors of polycystic ovaries (Birdsall et al. 1997). In another case-control study of 16 premenopausal women older than 40 years with polycystic ovary syndrome (PCOS) and increased testosterone levels and 16 matched controls, B-mode ultrasound of the carotid arteries revealed a significantly increased intima-media thickness in women with PCOS (Guzick et al. 1996). Since carotid intima-media thickness is a preclinical form of atherosclerosis and a strong prognostic marker for future MI, this study also indicates a pro-atherogenic role of androgens in women. However, as discussed in more detail below, it is important to note that hyperandrogenemia in women is confounded by a broad scope of atherogenic metabolic disturbances including obesity, insulin resistance, and dyslipidemia. Since it has not been conclusively shown whether obesity and hyperinsulinemia precipitate hyperandrogenism in women or vice versa, the presently available clinical data does not allow any conclusion on the etiological contribution of hyperandrogenism to atherosclerosis in women.

8.3 Animal studies on the role of androgens in atherosclerosis

The effects of testosterone on atherosclerosis have been investigated in only few animal studies which yielded contradictory results. Larsen et al. (1993) investigated the effects of intramuscular injections of testosterone-enanthate or placebo on atherosclerosis in castrated male rabbits. To avoid testosterone-related differences in serum cholesterol levels between treated and untreated animals, cholesterol levels were titrated by a cholesterol-rich diet. After 17 weeks of treatment, 19 rabbits in the verum group and 17 rabbits of the placebo group did not differ in the cholesterol content of the aorta (Larsen et al. 1993). Likewise, in another study of cholesterol-fed rabbits treatment with the artificial androgen stanozolol did not lead to significant changes in the extent of atherosclerotic lesions (Fogelberg et al. 1990). Bruck et al. (1997) found sex-specific effects of testosterone on the development of atherosclerotic lesions in rabbits. 32 castrated male rabbits and 32 ovariectomized female rabbits were divided into four groups of eight animals which were fed a cholesterol-rich diet and which received either a placebo, or estradiol-valerate alone, testosterone-enanthate alone, or both testosterone-enanthate and estradiol-valerate. After 12 weeks of treatment, morphometry of the proximal aorta detected significantly reduced intimal thickening in rabbits of both sexes which received the combination therapy, in female rabbits which were treated with estradiol alone, and in male rabbits which were treated with testosterone. The sex-specific anti-atherogenic effects of testosterone and estradiol were independent of changes in lipoprotein levels (Bruck et al. 1997). In three studies, 5 to 12 weeks long, treatment of male rabbits with DHEA reduced the extent of atherosclerotic lesions. Interestingly, the beneficial effects of DHEA were independent of changes in lipoprotein levels (Arad et al. 1989; Eich et al. 1993; Gordon et al. 1988).

By contrast to the beneficial or neutral effects of testosterone and other androgens on atherosclerosis in male rabbits, treatment of male chicks with testosterone resulted in a dose-dependent increase of aortic atherosclerosis (Toda et al. 1984).

Treatment of four female ovariectomized cynomolgus monkeys with testosterone for 24 months increased the size of atherosclerotic plaques in the coronary artery by a factor 2 as compared to both four untreated animals without ovariectomy and four untreated animals with ovariectomy. The gain in atherosclerosis occurred independently of changes in lipid levels and, surprisingly, despite improved endothelial reactivity (Adams et al. 1995). Interestingly, in the study by Bruck et al. (1997), treatment of female rabbits with testosterone alone also tended to increase the extent of atherosclerotic lesions. Thus, by contrast to the neutral or beneficial effects of androgens observed in male animals, experimentally induced hyperandrogenism may favour atherosclerosis in female animals.

8.4 Effect of androgens on cardiovascular risk factors

Arteriosclerosis is a multifactorial disease. Its pathogenesis is favoured by various risk factors. In addition to the non-modifiable risk factors age, male sex, and positive family history, classic modifiable risk factors include smoking, diabetes mellitus, arterial hypertension, and high blood levels of LDL-C as well as low levels of HDL-C (International Task Force for the Prevention of Cardiovascular Disease 1992; The Expert Panel 1993). More recent studies have also identified increased plasma levels of triglycerides, Lp(a), insulin, homocysteine, fibrinogen, clotting factor VII, tissue plasminogen activator antigen or plasminogen activator inhibitor type 1 (PAI-1) as independent cardiovascular risk factors (Assmann et al. 1996; Boushey et al. 1995; Després et al. 1996; Ernst and Resch 1993; NIH Consensus Development Panel 1993, Stein and Rosensen 1997; Thompson et al. 1995). Androgens have profound effects on several of these risk factors.

8.4.1 Regulation of lipid metabolism by androgens

8.4.1.1 Evidence from cross-sectional population and clinical studies

The effects of testosterone on lipid metabolism are contradictory (Table 8.1). Prepubertal boys and girls do not differ significantly in their serum lipid and lipoprotein levels. By contrast to girls, in whom levels of HDL-C and LDL-C change little with puberty, sexually maturing boys experience a decrease in HDL-C and increases in LDL-C and triglycerides (Bagatell and Bremner 1995). In conflict with these longitudinal studies, several cross-sectional population studies *in men* found a positive correlation between plasma levels of testosterone and HDL-C and inconsistent relationships of testosterone with triglycerides, total cholesterol, and LDL-C (Barrett-Connor 1995; Freedman et al. 1991; Hämäläinen et al. 1986; Kiel et al. 1989; Phillips et al. 1994; Yarnell et al. 1993). Two very recent studies took into account the confounding effects of obesity and insulin levels on the relationships between androgens and plasma levels of lipids and lipoproteins (Simon et al. 1997; Tchernof et al. 1997; see also 4.2 in this chapter). Upon univariate analysis, Tchernof et al. (1997) found significant negative correlations between plasma levels of testosterone (and DHEAS and androstendione) with triglycerides, total cholesterol, and LDL-C. Upon multivariate analysis, however, which adjusted for the associations of both androgens and lipids with visceral obesity as well as with serum levels of insulin and free fatty acids (FFA), the correlations between the androgens and the lipid parameters lost their statistical significance (Tchernof et al. 1997). Likewise, in a case-control study of 50 men who were matched by age and ethnic background but differed by testosterone levels, hypoandrogenemia was associated with significantly higher BMI, higher waist-hip-ratio (WHR), higher systolic blood pressure, higher fasting and 2 hour glucose and insulin levels, and higher levels of total cholesterol, LDL-C, triglycerides, and apoB as well as with lower levels of HDL-C and apoA-I. After adjustment for BMI and WHR, only the negative associations of testosterone with levels of insulin and triglycerides remained significant (Simon et al. 1997). In conclusion, the association of testosterone levels with a more favourable lipoprotein profile observed in cross-sectional studies of men does not appear to reflect direct regulatory effects of testosterone on lipoprotein metabolism but rather the inverse correlation of testosterone levels with body fat and insulin resistance, which in turn have great impact especially on levels of triglycerides and HDL-C (Table 8.1).

A typical condition associated with androgen excess *in premenopausal women* is the polycystic ovary syndrome (PCOS) (Barnes and Rosenfeld 1989). Various investigators reported that the PCOS is associated with low levels of HDL-C and elevated serum concentrations of triglycerides, total cholesterol, and LDL-C (Conway et al. 1992; Dahlgren et al. 1992a; Mattson et al. 1984; Talbot et al. 1995; von Eckardstein et al. 1996; Wild et al. 1985; Wortsman and Soler 1982). In a retrospective study, Dahlgren and coworkers (1992b) observed that this adverse lipid profile is also maintained after menopause. However, dyslipidemia in women with PCOS is confounded with a high prevalence of obesity and hyperinsulinemia, i.e. insulin resistance. Thus, it is a matter of debate whether or not low HDL-C and hypertriglyceridemia in women with PCOS are independent of overweight and insulin resistance (Chang et al. 1983; Conway et al. 1993; Dahlgren et al. 1992a; Dunaif et al. 1988; Franks 1989; Graf et al. 1990; Holte et al. 1994; Mattson et al. 1984; von Eckardstein et al. 1996; Wild et al. 1990; Wortman and Soler 1982). Moreover, hyperandrogenemia itself contributes to obesity, insulin resistance, and hyperinsulinemia (Björntorp 1993, 1994; Haffner 1996; Tchernoff et al. 1996). Therefore the question is as yet unresolved whether the frequently observed dyslipidemias in women with PCOS is attributed directly to hyperandrogenemia or to confounding metabolic disturbances (Table 8.1).

8.4.1.2 Evidence from interventional studies

8.4.1.2.1 Effects on HDL-cholesterol

Substitution of testosterone in hypogonadal men was associated with increases, decreases or no change of HDL-C levels (Asscheman et al. 1994; Bagatell et al. 1994; Hromadova et al. 1989; Kirkland et al. 1987; Mendoza et al. 1981; Merrigiola et al. 1995; Ozata et al. 1996; Sorva et al. 1988; Zgliczynski et al. 1996). Application of androgen-like anabolic steroids or supraphysiological amounts of testosterone were consistently found to decrease HDL-C, especially in HDL-subclasses HDL2 and LpA-I (Friedel et al. 1990; Glazer et al. 1991; Marcovina et al. 1996; Thompson et al. 1989; Zmuda et al. 1996) (Table 8.1). Clinical experiences from studies on the suppression of endogenous testosterone production through antagonists of the gonadotropin releasing hormone (GnRH) also indicate that testosterone lowers HDL-C since these drugs cause increases of HDL-C dependent on dosage and time (Baga-

Table 8.1. Outcomes of clinical and epidemiological studies on relationships between testosterone and various metabolic parameters

		Men	Women
Associations of endogenous testosterone levels with metabolic variables	*Lipoproteins* *Glucose metabolism* *Adipose tissue* *Hemostasis*	HDL[a], (LDL[b], Triglycerides[b]) Glucose tolerance[a], Insulin[b] Body fat[b], Leptin[b] Fibrinogen[b], PAI-1[b]	HDL[b], LDL[a], Triglycerides[a] Glucose tolerance[b], Insulin[b] Body fat[a], Leptin[a] Fibrinogen[a], PAI-1[a]
Effects of testosterone substitution in hypo-gonadism on metabolic variables	*Lipoproteins* *Glucose metabolism* *Adipose tissue* *Hemostasis*	Lp(a)[b] Insulin[b] Leptin[b] Fibrinogen[b]	
Effects of supra-physiological androgen dosages on metabolic variables	*Lipoproteins* *Glucose metabolism* *Adipose tissue* *Hemostasis*	HDL[b], Lp(a)[b] Fibrinogen[b], PAI-1[b]	HDL[b, c] Fibrinogen[b, c], PAI-1[b, c]
Effects of supressed endogenous testosterone on metabolic variables	*Lipoprotein* *Glucose metabolism* *Adipose tissue* *Hemostasis*	HDL[a], Lp(a)[a] Insulin[a] Leptin[a]	

[a] Designates a positive correlation of testosterone with or an increasing effect of testosterone on the respective parameter.

[b] Designates a negative correlation of testosterone with or a decreasing effect of testosterone on the respective parameter

[c] Designates effects observed after administration of the androgen-like steroid tibolone

Parameters in parentheses represent weak or inconsistent correlations.

T = testosterone; TG = triglycerides, HDL-C = HDL-cholesterol, LDL-C = LDL-cholesterol, PAI-1 = plasminogen activator inhibitor type 1

tell et al. 1992; Behre et al. 1994; Goldberg et al. 1985; von Eckardstein et al. 1997) (Table 8.1). Suppression of testosterone by four week treatment with the GnRH antagonist cetrorelix caused mean increases of HDL-C and apoA-I by 20% which were predominantly due to the increase in levels of HDL subclass LpA-I (von Eckardstein et al. 1997). Since GnRH antagonists suppress gonadotropins it is theoretically possible that the lack of luteinizing and/or follicle-stimulating hormones is responsible for their HDL raising effects. Since testosterone is aromatized to estradiol, one may also argue that lack of estradiol rather than lack of testosterone caused the increase in HDL. However, application of GnRH-antagonists did not increase levels of HDL-C if testosterone-enanthate was injected in parallel to maintain normal testosterone serum concentrations (Bagatell et al. 1992; Goldberg et al. 1985). Thus, the absence of testosterone rather than gonadotropins or estradiol is the most likely reason for the increase in HDL-C in men treated with GnRH-antagonists.

In women, exogenous androgens also appear to decrease HDL-C levels. Hormone replacement therapy of postmenopausal women with the synthetic

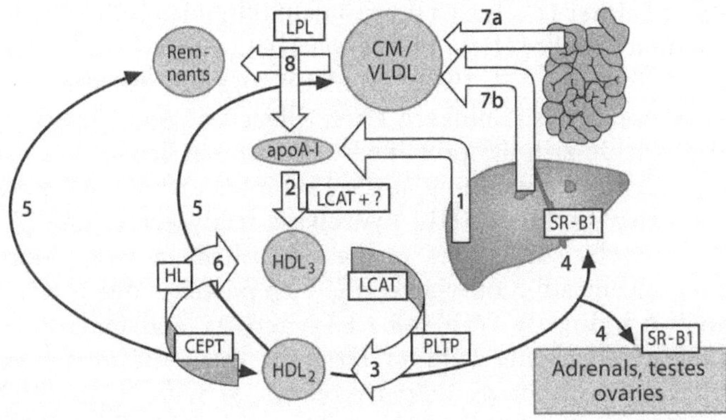

Fig. 8.1. Model of HDL metabolism.
The major protein constituent of HDL, apoA-I, is secreted by the liver as a lipid-free protein or lipid-poor particle (1). This precursor is maturated to HDL by the acquisition of lipids from cells or plasma lipoproteins (*not shown*) and by the esterification of cholesterol through lecithin:cholesterol acyltransferase (LCAT) (2). The small HDL₃ particles thus generated grow to larger HDL₂ through ongoing LCAT-mediated cholesterol esterification. Moreover, phospholipid transfer protein (PLTP) fuses HDL₃ into HDL₂ (3). Cholesteryl esters, especially of HDL₂, are targets either for SR-B1-mediated selective uptake into the liver and into steroido-genic organs (4) or for the exchange with triglycerides in apoB-containing particles by cholesteryl ester transfer protein (CETP) (5). Triglycerides and phospholipids in HDL₂ are substrate for hepatic lipase (HL) (6). Removal of cholesteryl esters by SR-B1 and CETP as well as lipolysis though HL decrease the size of HDL2. Moreover, the interconversion of HDL₃ and HDL₂ by PLTP, CETP, and HL releases lipid-free apoA-I or lipid-poor HDL-precursors which re-enter the aforementioned cycle. HDL precursors are also generated by the li-polysis of chylomicrons, which carry nutritional lipids (7a), and of VLDL, which contain endogenously synthe-sized lipids (7b) (8). Additionally, the concentration of these triglyceride-rich lipoproteins is also an important determinant of HDL-C levels since this is an important determinant of CETP activity (5). Dark arrows indicate lipid transfers, open arrows metabolic pathways of entire particles

steroid tibolone, which exerts estrogenic, gestagenic, and androgenic effects, caused 20% decreases of HDL-C and apoA-I. This is in contrast to experiences with postmenopausal hormone replacement therapies with estrogens alone or with estrogens and progestins in combination which both cause significant increases of HDL-C (Bjarnasson et al. 1997) (Table 8.1).

The mode of action by which testosterone regulates HDL-C levels is not resolved. Figure 8.1 summarizes important steps in HDL metabolism. Changes in the concentration and/or activity of the involved proteins, enzymes, and receptors will probably result in changes of HDL-C levels. However, little is known on the regulation of their genes by testosterone. The production rate of HDL is determined by the synthesis of apoA-I which is its main protein constituent (Brinton et al. 1994). In mice testosterone was found to increase the synthesis of apoA-I (Tang et al. 1991). The catabolism of HDL is regulated by various factors including plasma activities of the cholesterol esterifying enzyme lecithin:cholesterol acyltransferase (LCAT) and of cholesteryl ester transfer protein (CETP) which exchanges cholesteryl esters (CE) and triglycerides between HDL and apoB-containing lipoproteins (Jonas 1991; Tall 1993, 1995). Suppression of testosterone by cetrorelix did not cause consistent changes in the activities of either LCAT, CETP, or phospholipid transfer protein which mediates the fusion of smaller HDL$_3$ into larger HDL$_2$ (Albers et al. 1996; Tall 1995; von Eckardstein et al. 1997, 1998). Lipolytic enzymes are also important determinants of HDL-C levels (Goldberg 1996; Olivecrona and Olivecrona 1995). Lipolysis of triglyceride-rich lipoproteins by lipoprotein lipase (LPL) generates apoA-I and phospholipid containing surface remnants which are considered HDL-precursors. Hepatic lipase (HL) hydrolyses triglycerides and phospholipids of HDL and thereby contributes to the catabolism of HDL. Moreover, differences in the concentration of triglyceride-rich particles due to variation in lipolytic activities indirectly regulates CETP activity and thereby HDL-C concentration (Tall 1993). Some authors reported that administration of testosterone to men increases the postheparin plasma activity of HL which would be in agreement with an HDL-lowering effect of exogenous testosterone (Glueck et al. 1976; Sorva et al. 1988; Zmuda et al. 1993). However, we observed only a moderate decrease in postheparin plasma HL activity when testosterone was suppressed by the GnRH antagonist cetrorelix (von Eckardstein et al. 1998). Likewise, castration of male rats did not cause significant changes in postheparin plasma activity of HL or in hepatic HL mRNA levels. Subsequent substitution of testosterone raised HL-activity without changing hepatic HL-mRNA expression (Peinado-Onsurbe 1993). Very recently, the HDL receptor SR-B1 was shown to be an important regulator of HDL-C levels, at least in mice, by mediating the selective uptake of cholesteryl esters from HDL into liver and steroidogenic organs (Rigotti et al. 1997). The regulation of this receptor by testosterone has not been studied. However, since SR-B1 is involved in the supply of cholesterol to steroidogenic organs, it is an interesting target for regulation by testosterone.

8.4.1.2.2 Effects on lipoprotein(a)

Levels of triglycerides and LDL-C were not affected by either administration of exogenous androgens or suppression of endogenous testosterone. However, testosterone, like estrogens and progestins (Sacks and Walsh 1994), exerts strong regulatory effects on serum levels of Lp(a). This particle resembles LDL by the presence of one molecule apoB-100 and by its high content of CE. In Lp(a), apoB is covalently linked with a glycoprotein termed apo(a) by a disulfide bridge. Lp(a) levels vary considerably in the population between 0 and up to 300 mg/dl with a frequency distribution which is skewed to lower concentrations. Twin and sib pair studies have explained 90% of the interindividual variability in Lp(a) levels through variation in the apo(a) gene (Boerwinkle et al. 1992). Although Lp(a) levels are generally assumed to remain stable throughout life, estrogens, progestins, growth hormone, and thyroxin were shown to lower Lp(a) levels (Angelin 1997). Results of many case-control studies and most prospective population studies demonstrated that Lp(a) levels higher than 30 mg/dl are an independent risk factor for coronary, cerebrovascular, and peripheral atherosclerotic vessel diseases (Assmann et al. 1996; Maher and Brown 1995; Stein and Rosenson 1997).

Administration of testosterone to orchidectomized patients with prostate cancer (Berglund et al. 1996, Henrikson et al. 1992) as well as administration of supraphysiological doses of testosterone-enanthate to healthy men were associated with significant 25% to 59% decreases of Lp(a) which were correlated to baseline levels of Lp(a), i.e. those men with the highest initial Lp(a) levels experienced the greatest changes in Lp(a) levels (Marcovina et al. 1996; Ozata et al. 1996; Zmuda et al. 1996). In agreement with these observations suppression of endogenous testosterone by treatment with cetrorelix for four weeks caused a pronounced increase of Lp(a) levels by 40% to 60% which was also correlated with baseline values of Lp(a) (von Eckardstein et al. 1997). Likewise, in patients with prostate cancer, orchidectomy or suppression of endogenous testosterone by the GnRH analogue buserelin resulted in a highly significant increase of Lp(a) levels (Arrer et al. 1996; Berglund et al. 1996; Henrikson et al. 1992). Animal studies have also provided evidence for the involvement of testosterone in the regulation of Lp(a) levels. Frazer and colleagues observed that Lp(a) levels decrease in male but not in female apo(a)-transgenic mice after sexual maturation. Castration of male animals restored initial Lp(a) levels which decreased again upon application of dihydrotestosterone (DHT) (Frazer et al. 1995).

Since testosterone is aromatized to estradiol and since estradiol is well known to decrease Lp(a), one may argue that lack of estradiol rather than lack of testosterone caused the increase in Lp(a). However, testosterone also suppressed Lp(a) levels when it was administered in combination with aromatization inhibitor testolactone (Zmuda et al. 1996).

8.4.2 Interdependences of androgens, obesity, and insulin sensitivity

Numerous observations point to mutual relationships between androgens, body fat distribution, and insulin sensitivity, of which the latter two are also involved in the regulation of HDL and triglyceride metabolism (Björntorp 1993, 1994, 1996; Haffner 1996; Tchernof et al. 1996). It is, however, not clear whether androgens regulate adipose tissue and insulin sensitivity or whether vice versa adipocytes and insulin regulate testosterone levels. Probably both forms of causal relationships exist (Figs. 8.2 and 8.3).

In men, overweight and insulin resistance are frequently associated with hypotestosteronemia (Seidell et al. 1990). Hypogonadal men are frequently obese and have increased levels of the adipocyte-derived hormone leptin which signals to the hypothalamus the size of energy storage in adipose tissue and thereby regulates food intake (Lönnquist and Schalling 1997; Rosenbaum et al. 1997; Vogel 1996). Frequently also serum levels of insulin, a marker for insulin resistance, are increased (Björntorp 1996; Haffner and Valdez 1995; Marin et al. 1995; Vermeulen 1996). Substitution of testosterone in hypogonadal men reduced levels of leptin and insulin (Behre et al. 1997; Sih et al. 1996). Treatment of obese men with testosterone was associated with a decrease of visceral fat tissue mass and, in parallel, with a rise in insulin sensi-

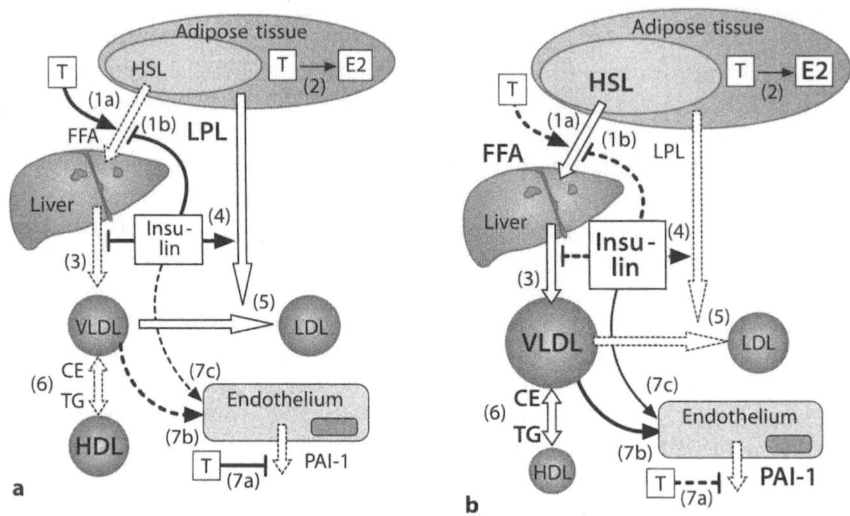

Fig. 8.2. Model of metabolic effects of testosterone in men.
Testosterone (T) activates hormone sensitive lipase (HSL) in adipocytes and thereby decreases body fat mass (1a). This implies little aromatization of testosterone into estradiol (2), i.e helps to maintain normal testosterone levels in non-obese men (**a**). Hydrolysis of body fat by HSL produces free fatty acids (FFA) which stimulate hepatic VLDL production (3a). However, this hypertriglyceridemic effect is balanced by improved insulin sensitivity in lean individuals with the result of reduced FFA release from adipocytes (1b), inhibited VLDL production (2b), and stimulated secretion of lipoprotein lipase (LPL) by the adipose tisse (3). Normal VLDL production and regular lipolysis of VLDL (and chylomicrons) by LPL (4) lead to normotriglyceridemia and, via low CETP-mediated exchange of cholesteryl esters (CE) and triglycerides (TG) between VLDL and HDL (6, see Fig. 6.1), to normal HDL-C levels. Taken together, normal testosterone levels, low insulin levels, and normotriglyceridemia help to suppress PAI-1 production in the endothelium (7). In hypogonadal men (**b**), low testosterone levels impair lipolysis in adipocytes and favour obesity (1a). Enhanced aromatization of testosterone into estradiol in obese men (2) further decreases testosterone levels. Obesity causes insulin resistance with the result of increased FFA release from adipocytes (1b), disinhibited VLDL production (3), and decreased LPL secretion (4). Both increased VLDL secretion (3) and decreased lipolysis of triglycerides-rich lipoproteins (5) cause hypertriglyceridemia which stimulates the CETP-mediated removal of cholesteryl esters from HDL and thereby causes low HDL-C (6). Finally, hypotestosteronemia (7a), hypertriglyceridemia (7b), and hyperinsulinemia (7c) stimulate the production of PAI-1 in endothelial cells.

tivity and an improvement of dyslipidemia (Marin et al. 1992, 1993; Rebuffé-Scrive et al. 1991). In the opposite experiment, suppression of testosterone by the GnRH-antagonist cetrorelix increased serum levels of leptin and insulin (von Eckardstein et al. 1998). These data indicate that, at least in men, testosterone reduces the mass of fat tissue especially in the abdomen and (thereby?) insulin resistance. On the other hand, however, weight loss in obese men is associated with an increase of serum levels of testosterone and a decrease of serum levels of estradiol so that obesity may cause hypotestosteronemia (Leenen et al. 1994; Stanick et al. 1981) (Fig. 8.2).

Interestingly, women present the opposite relationships between androgens, insulin, and obesity. Testosterone is positively correlated with BMI and leptin levels (Björntorp 1993, 1994; Mantzoros et al. 1997; Rouru et al. 1997).

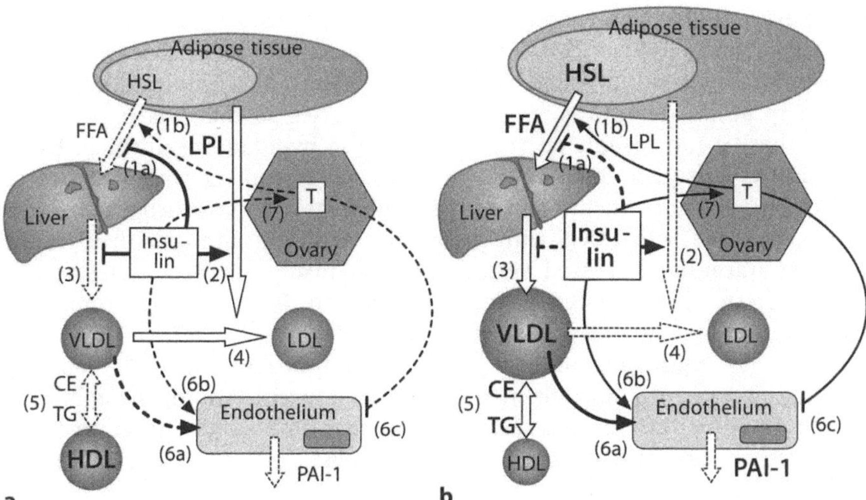

Fig. 8.3. Model of metabolic effects of testosterone in women.
In non-obese women with normal insulin sensitivity (**a**), adipocytes release limited amounts of free fatty acids (FFA) (1) and regular amounts of lipoprotein lipase (LPL) (2). VLDL are secreted at regular amounts by the liver (3) and properly hydrolysed by LPL (4). Normotriglyceridemia is associated with a low exchange of triglycerides (triglycerides) and cholesteryl esters (CE) between HDL and VLDL (5) so that HDL-C levels stay normal. Normotriglyceridemia (6a) and low insulin levels (6b) inhibit the release of PAI-1 from endothelial cells. Low insulin levels also limit the ovarian production of testosterone (T) (7). Low testosterone levels support insulin in suppressing FFA release from adipocytes (1b) and thereby hepatic VLDL production as well as in inhibiting PAI-1 release from endothelial cells (6c). Obese women, by contrast (**b**), have insulin resistance and hyperinsulinemia. In the adipose tissue, insulin resistance increases the production of FFA (1) and inhibits LPL secretion (2). Enhanced hepatic VLDL secretion (3) and low LPL activity (4) cause hypertriglyceridemia and, indirectly via enhanced CE/triglycerides exchange between VLDL and HDL (5), low HDL-C. Hypertriglyceridemia (6a) and hyperinsulinemia (6b) both stimulate PAI-1 secretion from endothelial cells. Hyperinsulinemia also stimulates the production of testosterone in the ovary (7). Hyperandrogenemia then aggravates the detrimental effects of insulin resistance on FFA release from adipocytes (1b) and thereby on hepatic VLDL production (as well as of hyperinsulinemia on PAI-1 secretion from endothelial cells (6c)

Hyperinsulinemia and insulin resistance in women are associated with low serum levels of sexual hormone binding globulin (SHBG), an indirect measure of female hyperandrogenism. Women with PCOS are frequently characterized by abdominal obesity and high serum levels of leptin and insulin, both basal and after glucose intake (Chang et al. 1983; Conway et al. 1993; Dunaif et al. 1988; Graf et al. 1990; Haffner 1996; Laughlin et al. 1997). In a prospective study, 20% of women with SHBG-levels below the 5th percentile developed diabetes mellitus type 2 within the 12 year follow-up period (Lindstedt et al. 1991). These studies do not allow any conclusion as to whether obesity causes hyperandrogenemia or vice versa. Administration of testosterone to transsexual women increased insulin resistance in skeletal muscle (Björntorp 1993, 1994). Likewise, treatment of female cynomolgus monkeys with testosterone increased BMI and the mass of both visceral fat and muscle and caused insulin resistance (Adams et al. 1995). Administration of low doses of testosterone to female rats resulted in reduced glucose transport and glycogen synthesis in muscle which both indicate insulin resistance (Björntorp 1993, 1994). These data suggest that in women hyperandrogenism preceeds obesity and insulin resistance. However, insulin was also shown to stimulate androgen synthesis in ovaries so that, via hyperinsulinemia, obesity and insulin resistance may cause hyperandrogenism in women with PCOS (Pasquali et al. 1995) (Fig. 8.3). In agreement with a causal role of insulin resistance in the pathogenesis of female hyperandrogenism, treatment of women with PCOS with metformin or the insulin sensitizer troglitazone significantly decreased serum levels of insulin as well as testosterone and free testosterone, independent of changes in BMI or gonadotropin levels (Ehrmann et al. 1997, Velazquez et al. 1997).

How may androgens influence obesity and insulin sensitivity and how may obesity and insulin influence testosterone levels? Androgens enhance lipolysis in adipocytes by increasing the expression of β-adrenergic receptors, adenylate cyclase, proteinkinase A and hormone sensitive lipase (HSL) (Figs. 8.2 and 8.3). Interestingly, testosterone up-regulates the expression of androgen receptors in adipocytes which may further enhance lipolysis (Björntorp 1993, 1994). Increased lipolysis reduces fat storage in adipocytes, which may explain the weight reduction of obese and hypogonadal men treated with androgens (Tchernof et al. 1996). Lipolysis, however, also increases the concentration of FFA in the circulation. FFA stimulate hepatic VLDL synthesis and glucose production and decrease hepatic insulin utilisation. Thereby androgens may induce hypertriglyceridemia, insulin resistance, and impaired glucose tolerance which are frequently observed in hyperandrogenemic women (Fig. 8.3). On the other hand, adipocytes express testosterone aromatase which converts testosterone into estradiol so that the increased amount of adipose tissue may cause hypotestosteronemia and increased levels of estradiol in obese men (Tchernof et al. 1996) (Fig. 8.2). Finally, insulin was shown to stimulate androgen synthesis in the ovaries, so that obesity may cause hyperandrogenism in women with PCOS via hyperinsulinemia (Pasquali et al. 1995) (Fig. 8.3).

Whether androgens determine body fat distribution and insulin sensitivity or vice versa, obesity and insulin resistance have strong effects on lipid metabolism. Insulin inhibits lipolysis in adipose tissue, suppresses hepatic VLDL-synthesis and activates lipoprotein lipase. In insulin-resistant individuals, increased release of FFA, increased hepatic VLDL production and disturbed lipolysis cause hypertriglyceridemia which sometimes becomes obvious only in the postprandial phase. The increased concentration of triglyceride-rich lipoproteins activates CETP which exchanges triglycerides of VLDL and LDL against CE of HDL so that finally HDL-C levels decrease (Després and Marette 1994; Frayn 1993; Reaven 1988) (Figs. 8.2 and 8.3). Hypertriglyceridemia, however, also interferes with the coagulation and fibrinolytic systems and increases, for example, plasma levels of fibrinogen, factor VII or PAI-1 (Juhan-Vague et al. 1993; Hamsten et al. 1994).

8.4.3 Effects of androgens on the hemostatic system

Since thrombosis plays an important role in acute cardiovascular events like MI and stroke it is not surprising that various hemostatic parameters have been associated with risk for cardiovascular disease (Smith 1996). The risk of MI increases with plasma levels of the thrombogenic factors fibrinogen and factor VII, as well as with plasma levels of the fibrinolysis inhibitor PAI-1 or tissue plasminogen activator antigen which represents the inactivated form (Ernst and Resch 1993; Thompson et al. 1995). Platelet aggregability is another important factor which determines the thrombophilicity of the hemostatic system and thereby cardiovascular risk.

Plasma levels of fibrinogen are inversely correlated with endogenous testosterone in men (Glueck et al. 1993; Yang et al. 1993). In agreement with a fibrinogen lowering effect of testosterone, administration of supraphysiological dosages testosterone to 32 healthy men participating in a trial on male contraception, led to a sustained decrease of fibrinogen by 15 to 20% over 52 weeks of treatment (Andersen et al. 1995a). In this study the doubling of endogenous testosterone levels initially also led to significant decreases of PAI-1, protein S, and protein C as well as to increases of antithrombin III and β-thromboglobulin. Suppression of testosterone in patients with prostate cancer or benign prostate hypertrophy, however, by treatment with the nonsteroidal anti-androgen casodex or the GnRH agonist leuprolide exerted no significant effects on plasma fibrinogen levels (Eri et al. 1995a, 1995b).

In hypogonadal men fibrinolytic plasma activity is reduced (Bennet et al. 1987), probably due to increased PAI-1 activity. In both normolipidemic and hyperlipidemic men levels of endogenous testosterone correlated inversely with plasma levels PAI-1 (Glueck et al. 1993; Yang et al. 1993). Administration of supraphysiological amounts of testosterone to men in a contraception study as well as adminsitration of the anabolic androgen stanozolol decreased PAI-1 (Andersen et al. 1995b; Verheijen et al. 1984). In vitro, testosterone inhibited the secretion of PAI-1 from bovine aortic endothelial cells (Sobel et al. 1995) (Fig. 8.2).

By contrast to these apparantly anti-thrombotic effects of testosterone in men, androgens were also shown to exert prothrombotic activities. High dosages of androgens increase platelet aggregability by decreasing cyclooxygenase activity and thereby decreasing the anti-aggregatory effects of prostaglandins (Pilo et al. 1981; Winkler 1996).

Like the relationships between testosterone, obesity, and insulin, the relationships between testosterone and fibrinogen or testosterone and PAI-1 observed in women are opposite of those observed in men. In women, testosterone correlates positively with PAI-1. Moreover, women with PCOS have higher levels of PAI-1 and fibrinogen than normoandrogenemic women (Dahlgren et al. 1994; Sampson et al. 1996). Probably, however, since insulin is a potent stimulator of PAI-1 expression (Hamsten et al. 1994), the positive association of testosterone with PAI-1 does not reflect a direct effect of androgens but rather the presence of hyperinsulinemia in hyperandrogenic women (Fig. 8.3). In agreement with this model, both body weight reduction and insulin lowering therapy of women with PCOS with metformin or troglitazone led to significant decreases of PAI-1 levels (Andersen et al. 1995; Ehrmann et al. 1997; Velazquesz et al. 1997). Moreover, treatment of endometriosis with the weak androgen danozolol as well as postmenopausal hormone replacement therapy with tibolone, which also has androgenic properties, led to significant decreases of PAI-1 levels (Bjarnasson et al. 1997; Cortees-Prieto 1987; Winkler 1996). Tibolone treatment of postmenopausal women also caused a significant decrease in fibrinogen levels (Bjarnasson et al. 1997).

In conclusion, by lowering fibrinogen and PAI-1, androgens appear to exert beneficial effects on the pro-coagulatory and fibrinolytic activities of plasma. However, these effects may be opposed by pro-aggregatory effects on thrombocytes.

8.5 Effects of androgens on cells of the arterial wall and vascular function

The pathogenesis of atherosclerosis involves various cell types which either form the intact arterial wall, namely endothelial cells and smooth muscle cells (SMC), or migrate into the arterial wall in the course of atherogenesis, e.g. MΦs, lymphocytes, or mast cells (Fuster 1994; Libby 1995; Ross 1990, 1995). The intact endothelium protects the arterial wall from atherosclerosis by many mechanisms. It is a barrier towards atherogenic lipoproteins, it regulates vascular tone by the production of nitric oxide, and it exerts a broad scope of anti-thrombotic features. Endothelial cells also express receptors to which circulating blood cells bind before they can migrate into the arterial wall and exert their pro-atherogenic activities (Busse and Fleming 1996; Griedling and Alexander 1996; Jang et al. 1994; Lüscher 1993; Vogel 1997; Zimmerman et al. 1996).

In the intact arterial wall, vascular SMCs regulate the arterial tone and produce the extracellular matrix. Proliferation and migration of SMCs are important steps in the formation of neointima and stenoses. Apoptosis of vascular SMCs contribute to plaque instability and rupture (Owens 1996).

Among the non-resident cells of the arterial wall, MΦs appear to play a key role in atherosclerosis. By contrast to most other cells, in which the intracellular accumulation of cholesterol is self-limited by down-regulation of cholesterol biosynthesis and of LDL-receptor mediated lipoprotein uptake, MΦs can internalize large amounts of exogenous lipids by various unregulated scavenger receptors and phagocytosis. Lipid accumulation transforms MΦs into foam cells, which form the initial atherosclerotic lesions, i.e. fatty streaks, but also are responsible for late atherosclerotic events such as plaque rupture. Moreover, MΦ foam cells express a broad variety of cytokines and growth factors which attract other cells into the arterial wall and which stimulate proliferation and migration of SMCs (Schmitz et al. 1997; Tabas 1996, 1997). Stimulated by the gender difference in risk for premature atherosclerosis, research on sex hormone effects on vascular function and vascular cell biology has started only recently but has focused on estrogens (Hutchison et al. 1997a; St. Clair 1997).

8.5.1 Effects of androgens on vascular reactivity

An early hallmark of atherosclerosis is decreased vascular responsiveness to various hormonal stimuli due to endothelial dysfunction and due to disturbances in vascular SMC physiology. As a result, decreased vasodilation and enhanced vasoconstriction can lead to vasospasm and angina pectoris. Moreover, endothelial dysfunction also contributes to coronary events by promoting plaque rupture and thrombosis (Busse and Fleming 1996; Griedling and Alexander 1996; Libby 1995; Lüscher 1993; Vogel 1997).

Many clinical studies as well as *in vivo* and *in vitro* experiments have documented the protective effects of estrogens on vascular function (Sacks et al. 1995; Hutchison et al. 1997a). By contrast, little and contradictory data is available on the effects of testosterone on vascular reactivity. Herman and colleagues (1997) conducted a study in patients with prostate cancer, who were deprived of endogenous androgens either surgically or pharmacologically, and in healthy men or men with non-prostate cancers as controls. Brachial artery diameter was measured by external vascular ultrasound at rest, during reactive hyperemia (i.e. endothelium-dependent dilation), and after administration of nitroglycerin (i.e. endothelium-independent dilation). Androgen-deprived patients had a greater endothelium-dependent vasodilation than normoandrogenemic controls. Upon multivariate analysis this association was independent of differences in cholesterol levels or of the presence of malignancy. The endothelium-dependent vasodilation by nitroglycerin was not different between the groups (Herman et al. 1997). Male-to-female transsexuals who received estrogens also had greater endothelium-dependent vasodilation than male control subjects. The two studies in transsexuals, however, could not answer the question whether adminstration of estrogens or deprivation of androgens was responsible for the improved vascular function (McCrohon et al. 1997; New et al. 1997).

By contrast to this clinical data on men, which suggest a detrimental effect of androgens on vascular responsiveness, two *in vivo* studies in female monkeys and dogs of both sexes as well as most *in vitro* studies with animal vessels have indicated the opposite. After treatment for two years with testosterone, ovariectomized female cynomolgus monkeys underwent coronary angiography. Intracoronary injections of acetylcholine caused significant endothelium-dependent vasodilation in testosterone-treated but not in untreated animals. By contrast endothelium-independent vasodilation after administration of nitroglycerin occurred regularly in both treatment groups (Adams et al. 1995). Chou and colleagues (1996) infused increasing concentrations of testosterone into both male and female dogs to investigate by intracoronary ultrasound the effects of testosterone on the reactivity of coronary arteries both to endothelium-dependent and independent stimuli. Testosterone induced vasodilation by endothelium-dependent and independent mechanisms. Most *in vitro* studies with isolated rings of coronary arteries and/or aortas from rats, rabbits, and pigs also found that in both sexes testosterone improves both endothelium-dependent and/or independent vascular responsiveness (Costarella et al. 1996; Farhat et al. 1995; Yue et al. 1995). One study, however, observed that testosterone worsens the endothelium-dependent relaxation of aortic rings from rabbits which were either made hypercholesterolemic or exposed to tobacco smoke (Hutchison et al. 1997b).

The molecular mechanisms by which testosterone (and estradiol) regulate vascular tone are unknown. Typically steroid hormones regulate cell function by binding to specific intracellular receptors which then regulate gene expression (Evans 1988). Androgen receptor expression has been found in rat aortic SMCs (Higashiura et al. 1997). Moreover, dihydrotestosterone increased the density of thromboxan receptors on aortic SMCs from rats and guinea pigs (Masuda et al. 1991). The involvement of prostaglandins in the regulation of vascular tone is also suggested by the observation that testosterone increases the response of coronary arteries to prostaglandin F2α (Farhat et al. 1995). However, in some *in vivo* and *in vitro* animal studies, pretreatment with the prostaglandin synthesis inhibitor indomethacin had no effect on testosterone-induced vasodilation. This contradicts the mediation of testosterone effects on the arterial wall by eicosanoids such as thromboxan (Chou et al. 1996; Yue et al. 1995).

In addition, steroid hormones can regulate cellular function by non-genomic mechanisms which involve, for example, the modulation of cell membrane channels (McEwen 1991). Several observations speak against genomic pathways and for non-genomic pathways to be involved in the vascular effects of androgens. First, the anti-androgen flutamide did not affect the relaxation of rabbit coronary arteries by testosterone (Yue et al. 1995). Second, since endothelial cells can convert testosterone into estradiol by aromatase, it is possible that testosterone exerts its vasoactive effects via estradiol. However, neither the aromatase inhibitor aminogluthemide nor the estrogen-receptor antagonist ICI 182,780 prevented testosterone-induced vasodilation (Chou et al. 1996; Yue et al. 1995). Third, barium chloride attenuated the testosterone-induced vasorelaxation of rabbit aortas and coronary arteries, indicating that testosterone modulates the conductance of potassium channels of SMCs (Yue et al. 1995).

Most studies found both endothelium-dependent and independent effects of testosterone on vascular reactivity. Moreover, several studies indicate the involvement of endothelial nitric oxide (Adams et al. 1995; Chou et al. 1996; Costarella et al. 1996; Farhat et al. 1995; Herman et al. 1997). In one *in vivo* study of dog coronary arteries and one *in vitro* study of rat aorta, the nitric oxide synthase inhibitor L-NAME prevented testosterone-induced vasodilation (Chou et al. 1996; Costarella et al. 1996). In another *in vitro* study of rabbit aortas and coronary arteries, however, L-NAME had no effect on testosterone-induced vasodilation (Yue et al. 1995). In agreement with the latter, *in vitro* expression of nitric oxide synthase in human aortic endothelial cell was found stimulated by estradiol but not by testosterone (Hishikawa et al. 1995).

In conclusion, the design of the few clincial studies on men does not allow any conclusion about the role of testosterone in the regulation of vascular tone. Most animal studies suggest that testosterone like estradiol improves vascular reactivity, probably through both endothelium-dependent and independent mechanisms. Data on the molecular basis of this action is limited and controversial.

8.5.2 Effects of androgens on smooth muscle cell proliferation and macrophage functions

Recently estradiol was found to inhibit proliferation and migration of SMCs, which are important steps in atherosclerosis. In control experiments, testosterone did not exert these effects on SMCs (Akishita et al. 1997; Kolodgie et al. 1996). Moreover, the protection of female rabbits by estradiol but not the protection of male rabbits by testosterone from atherosclerosis was associated with decreased incorporation of 5′-bromo-2′-deoxyuridine into DNA of neointimal cells, an *in vivo* marker of arterial SMC proliferation (Bruck et al. 1997).

Unregulated uptake of oxidatively modified lipoproteins via scavenger receptors leads to intracellular accumulation of cholesteryl esters in MΦs and thereby to foam cell formation (Ross 1990, 1995; Tabas 1996, 1997). Estradiol inhibits oxidation of LDL both in the presence and absence of cells including MΦs (St. Clair 1997). By contrast, testosterone was found to increase the oxidation of LDL by placenta MΦs in vitro (Zhu et al. 1997).

After internalization, oxidized LDL are directed via endosomes to lysosomes for degradation. Cholesteryl esters are hydrolysed by lysosomal acid lipase. Thereby liberated cholesterol molecules leave the lysosome membrane to be re-esterified by the microsomal enzyme acylCoA-cholesterolacyltransferase. Thus generated cholesteryl esters are stored in the cytosol and responsible for the foamy appearance of lipid-laden MΦs (Tabas et al. 1996, 1997). The transport of cholesterol from lysosomes to the site of re-esterification is inhibited *in vitro* by various steroids with an oxo-group at the C17 or C20 position such as progesterone, pregnenolone, and androstendione. 17-hydroxy-steroids including testosterone were less effective (Aikawa et al. 1994). Cytosolic cholesteryl esters can be hydrolysed by neutral cholesterol esterase (NCEH), which is activated by cylclic AMP. In adipose tissue of female rats NCEH is more active than in adipose tissue of male rats. Moreover, exogenous estradiol increases NCEH activity in male rats and in female rats which have been ovariectomized. *In vitro* estradiol but not testosterone increased the activity of NCEH in the murine MΦ cell line J774, probably by increasing the activity of a cyclic AMP dependent protein kinase A (St. Clair 1997; Tomita et al. 1996).

Lipid-loaded MΦs produce various cytokines including chemotactic protein 1, interleukin 1, and tumor necrosis factor 1α, as well as growth factors such as platelet derived growth factor 1. These bioactive molecules induce various processes which contribute to atherosclerosis, e.g. recruitment of

MΦs into the vascular wall and SMC proliferation and migration (Libby 1994; Ross 1990, 1995; Schmitz et al. 1997). Effects of testosterone on the production of cytokines and growth factors have not been studied in foam cell MΦ models but in MΦs which have been activated by lipopolysacharides. Whether these results are also valid for MΦs in the arterial wall is not known: estradiol but not testosterone inibited the migration of monocytes in response to chemotactic protein 1. Interleukin 1 production in rat testicular MΦs and hamster peritoneal MΦs was also independent of testosterone (Hayes et al. 1996; St. Clair 1997; Yoshida et al. 1996).

In conclusion unlike estradiol, testosterone does not exert the anti-proliferative effects on SMCs and does not appear to inhibit the accumulation of cholesteryl esters in MΦs.

8.6 Key messages

- Associations of testosterone and DHEA(S) with CHD unravelled in epidemiological and clinical studies suggest a neutral or even anti-atherogenic effect of endogenous androgens in men and a pro-atherogenic effect in women. Data from interventional studies are lacking.
- Likewise, exogenous testosterone and DHEAS exert neutral or beneficial effects on the development of atherosclerosis in male animals and detrimental effects in female animals.
- In men serum levels of endogenous testosterone are correlated positively with HDL-C and negatively with LDL-C, triglycerides, fibrinogen, and PAI-1. In women these relationships are the opposite. These associations, however, are confounded with obesity and insulin resistance which both are associated with hypotestosteronemia in men and with hyperandrogenemia in women and with low HDL-C as well as with high LDL-C, triglycerides, fibrinogen, and PAI-1 in both sexes.
- Administration of exogenous androgens to men decreases serum levels of HDL-C, Lp(a), PAI-1, fibrinogen, insulin, and leptin as well as visceral body fat mass. Suppression of endogenous testosterone in men results in opposite changes. Vice versa, weight loss and/or pharmacological sensitizing to insulin raises testosterone levels in men and decreases androgen levels in women. Thus mutual relationships exist between androgens, body fat distribution, and insulin sensitivity all of which are involved in the regulation of HDL and triglyceride metabolism and the hemostatic system.
- The majority of data from animal studies indicate that testosterone improves vascular reactivity to humoral stimuli. It is not clear whether effects of testosterone are dependent or independent of the endothelium and whether they are exerted by genomic or non-genomic pathways. Informative clinical studies in humans are missing.

8.7 References

Adams MR, Williams JK, Kaplan JR (1995) Effects of andogens on coronary artery atherosclerosis and atherosclerosis-related impairment of vascular responsiveness. Arterioscler Thromb Vasc Biol 15:562–570

Aikawa K, Furuchi T, Fujimoto Y, Arai H, Inoue K (1994) Structure-specific inhibiton of lysosomal cholesterol transport in macrophages by various steroids. Biochim Biophys Acta 1213:127–134

Akishita M, Ouchi Y, Miyoshi H, Kozaki K, Inoue S, Ishikawa M, Eto M, Toba K, Orimo H (1997) Estrogen inhibits cuff-induced intimal thickening of rat femoral artery: effects on migration and proliferation of vascular smooth muscle cells. Atherosclerosis 130:1–10

Albers JJ, Tu A-Y, Wolfbauer G, Cheung MC, and Marcovina SM (1996) Molecular biology of phospholipid transfer protein. Curr Opin Lipidol 7:88–93

Alexandersen P, Haarbo J, Christiansen C (1996) The relationship of natural androgens to coronary heart disease in males: a review. Atherosclerosis 125:1–13

Andersen P, Seljeflot I, Abdelnoor A, Arnesen M, Dale PO, Lovik A, Birkeland K (1995a) Increased insulin sensitivity and fibrinolytic capacity in obese women with polycystic ovary syndrome. Metabolism 44:611–616

Andersen RA, Ludlam CA, Wu CW (1995b) Haemostatic effects of supraphysiological levels of testosterone in normal men. Thromb Haemost 74:693–697

Angelin B (1997) Therapy for lowering lipoprotein(a) levels. Curr Opin Lipidol 8:337–341

Arad Y, Badimon JJ, Badimon L, Hembree WC, Ginsberg HN (1989) Dehydorepiandrosterone feeding prevents aortic fatty streak formation and cholesterol accumulation in cholesterol-fed rabbits. Arteriosclerosis 9:159–165

Arrer E, Jungwirth A, Mack D, Frick J, Patsch W (1996) Treatment of prostate cancer with gonadotropin releasing hormone analogue: effect on lipoprotein(a). J Clin Endocrinol Metab 81:2508–2511

Asscheman H, Gooren LJG, Megens JAJ, Nauta J, Klosterboer HJ, Eikelboom F (1994) Serum testosterone level is the major determinant of the male-female differences in serum levels of high-densitiy lipoprotein (HDL) cholesterol and HDL 2 cholesterol. Metabolism 43:935–939

Assmann G, Schulte H, von Eckardstein A (1996) Hypertriglyceridemia and elevated levels of lipoprotein(a) are risk factors for major coronary events in middle-aged men. Am J Cardiol 77:1179–1184

Bagatell CJ, Bremner WJ (1995) Androgen and progestagen effects on plasma lipids. Prog Cardiovasc Dis 38:255–271

Bagatell CJ, Knopp RH, Vale WW, Rivier JE, Bremner WJ (1992) Physiologic testosterone levels in normal men suppress high-density lipoprotein cholesterol levels. Ann Intern Med 116:967–973

Bagatell CJ, Heiman JR, Matsumoto AM, Rivier JE, Bremner WJ (1994) Metabolic and behavioural effects of high dose exogenous testosterone in healthy men. J Clin Endocrin Metab 79:561–567

Barnes R, Rosenfeld R (1989) The polycystic ovary syndrome: Pathogenesis and treatment. Ann Intern Med 110:386–397

Barret-Connor EL (1995) Testosterone and risk factors for cardiovascular disease in men. Diabetes Metab 21:156–161

Barrett-Connor EL, Bush TL (1991) Estrogen and coronary heart disease in women. JAMA 265:1861–1867

Barrett-Connor EL, Goodman-Gruen D (1995) Prospective study of endogenous sex hormones and fatal cardiovascular disease in postmenopausal women. Br Med J 311:1193–1196

Barrett-Connor EL, Khaw KT, Yen SSC (1986) A prospective study of dehydroepiandrosteronesulfate, mortality, and cardiovascular disease. N Engl J Med 315:1519–1524

Barrett-Connor EL, Khaw KT (1987) Absence of an inverse relation of dehydroepiandrosteronesulfate with cardiovascular mortality in postmenopausal women. N Engl J Med 317:711

Barrett-Connor EL, Khaw KT (1988) Endogenous sex hormones and cardiovascular disease in men: a prospective population based study. Circulation 78:539–545

Behre HM, Böckers A, Schlingheider A, Nieschlag E (1994) Sustained suppression of serum LH, FSH, and testosterone and increase of high-density lipoprotein cholesterol by daily injections of the GnRH antagonist cetrorelix over 8 days in normal men. Clin Endocrinol (Oxf) 40:241–248

Behre HM, Simoni M, Nieschlag E (1997) Strong association between serum levels of leptin and testosterone in men. Clin Endocrinol 47:237–240

Bennet A, Sié P, Caron P, Boneu B, Bazex J, Pontonnier F, Barret A, Louvet JP (1987) Plasma fibrinolytic activity in a group of hypogonadic men. Scand J Clin Lb Invest 47:23–27

Berglund L, Carlström K, Stege R, Gottlieb C, Eriksson M, Angelin B, Henriksson P (1996) Hormonal regulation of serum lipoprotein(a) levels: effects of parenteral administration of estrogen or testosterone in males. J Clin Endocrinol Metab 81:2633–2637

Birdsall M, Farquhar C, White H (1997) Association between polycystic ovaries and extent of coronary artery disease in women having cardiac catheterization. Ann Intern Med 126:32–35

Bjarnasson NH, Bjarnason K, Haarbo J, Coelingh Bennink HJT, Christiansen C (1997) Tibolone: Influence on markers of cardiovascular disease. J Clin Endocrinol Metab 82:1752–1756

Björntorp P (1993) Hyperandrogenicity in women – a prediabetic condition? J Intern Med 234:579–83

Björntorp P (1994) Fatty acids, hyperinsulinemia, and insulin resistance: which comes first? Curr Opin Lipidol 5:166–174

Björntorp P (1996) The regulation of adipose tissue distribution in humans. Int J Obesity 20:291–302

Boerwinkle E, Leffert CC, Lin J, Lackner C, Chiesa G, Hobbs HH (1992) Apolipoprotein(a) gene accounts for greater than 90% of the variation in plasma lipoprotein(a) concentrations. J Clin Invest 90:52–60

Bourassa PAK, Milos MM, Gaynor BJ, Breslow JL, Aielllo, RJ (1996) Estrogen reduces atherosclerotic lesion development in apolipoprotein E deficient mice. Proc Natl Acad Sci USA 93:10022–10027

Boushey CJ, Beresford SA, Omenn GS, Motulsky AG (1995) A quantitative assessment of plasma homocysteine as a risk factor for vascular disease. Probable benefits of increasing folic acid intakes. JAMA. 274:1049–57

Brinton EA, Eisenberg S, Breslow JL (1994) Human HDL-cholesterol levels are determined by apoA-I fractional catabolic rate, which correlates inversely with estimates of HDL size. Arterioscler Thromb 14:707–720

Bruck B, Brehme B, Gugel N, Hanke S, Finking G, Lutz C, Benda N, Schmahl FW, Haasis R, Hanke H (1997) Gender-specific differences in the effects of testosterone and estrogen on the development of atherosclerosis in rabbits. Arterioscler Thromb Vasc Biol 17:2192–2199

Busse R, Fleming I (1996) Endothelial dysfunction in atherosclerosis. J Vasc Res 33:181–94

Caron P, Bennet A, Camare R, Louvet JP, Boneu S, Sie P (1989) Plasminogen activator inhibitor in plasma is related to testosterone in men. Metabolism 38:1010–1015

Cauley JA, Gutai JP, Kuller LD, Dai WS (1987) Usefulness of sex steroid hormone levels in predicting coronary artery disease in men. Am J Cardiol 60:771–777

Chang RJ, Nakamura RM, Judd HL, Kaplan SA (1983) Insulin resistance in nonobese patients with polycystic ovarian disease. J Clin Endocrinol Metabol 57:356–359

Chou TM, Sudhir K, Hutchison SJ, Ko E, Amidon TM, Collins P, Chatterjee K (1996). Testosterone induces dilatation of canine coronary conductance and resistance arteries in vivo. Circulation 94:2614–2619

Contoreggi CS, Blackman MR, Andres R, Muller DC, Lakatta EG, Fleg JL, Harman SM (1990) Plasma levels of estradiol, testosterone, and DHEAS do not predict risk of coronary artery disease in men. J Androl 11:460–470

Conway GS, Agrawal R, Betteridge DJ, Jacobs HS (1992) Risk factors for coronary artery disease in lean and obese women with the polycystic ovary syndrome. Clin Endocrinol 37:119–25

Conway GS, Clark PM, Wong D (1993) Hyperinsulinaemia in the polycystic ovary syndrome confirmed with a specific immunoradiometric assay for insulin. Clin Endocrinol 38:219–22

Cortees-Prieto J (1987) Coagulation and fibrinolysis in postmenopausal women treated with Org OD 14. Maturitas 1:67–83

Costarella CE, Stallone JN, Rutecki GW, Whittier FC (1996) Testosterone causes direct relaxation of rat thoracic aorta. J Pharmacol Experiment Therap 277:34–39

Dahlgren E, Johannson S, Lindstedt G, Knutsson F, Oden A, Janson P O, Mattson L A, Crona N and Lundberg PA (1992a) Woman with polycystic ovary syndrome wedge resected in 1956 to 1965:A long-term follow-up focussing on natural history and circulating hormones Fertil Steril 57:505–513

Dahlgren E, Janson PO, Johansson S, Lapidus L, Oden A (1992b) Polycystic ovary syndrome and risk for myocardial infarction: evaluated from a risk factor model based on a prospective population study of women. Acta Obstet Gynecol Scand 71:599–604

Dahlgren E, Lapidus A, Janson P, Lindstedt G, Johannson S, Tengborn L (1994) Hemostatic and metabolic variables in women with polycystic ovary syndrome. Fertil Steril 61:455–460

Després JP, Marette A (1994) Relation of components of insulin resistance syndrome to coronary disease risk. Curr Opin Lipidol 5:274–289

Després JP, Lamarche B, Mauriege P, Cantin B, Dagenais GR, Moorjani S, Lupien PJ (1996) Hyperinsulinemia as independent risk factor for ischemic heart disease. N Eng J Med 334:952–957

Dunaif A, Mandeli J, Fluhr H, Dobrjansky A (1988) The impact of obesity on chronic hyperinsulinemia on gonadotropin release and gonadal steroid secretion in the polycystic ovary syndrome. J Clin Endocrinol Metabol 66:131–139

Ehrmann DA, Schneider DJ, Sobel BE, Cavaghan MK, Imperial J, Rosenfield RL, Polonsky KS (1997) Troglitazone improves defects in insulin action, insulin secretion, ovarian steroidogenesis, and fibrinolysis in women with polycystic ovary syndrome. J Clin Endocrinol Metab 82:2108–2116

Eich DM, Nestler JE, Johnson DE, Dworkin GH, Ko D, Wechsler AS, Hess ML (1993) Inhibition of accelerated coronary atherosclerosis with dehydroepinadrosterone in the heterotopic rabbit model of cardiac transplantation. Circulation 87:261–265

Eri LM, Urdal P (1995a) Effects of the nonsteroidal androgen casodex on lipoproteins, fibrinogen and plasminogen activator inhibitor in patients with benign prostatic hyperplasia. Eur J Urol 27:274–279

Eri LM, Urdal P, Bechensteen AG (1995b) Effects of the luteinizing hormone-releasing hormone agonist leuprolide on lipoproteins, fibrinogen and plasminogen activator inhibitor in patients with benign prostatic hyperplasia. J Urol 154:100–104

Ernst E, Resch KL (1993) Fibrinogen as a cardiovascular risk factor: A meta-analysis and review of the literature. Ann Intern Med 118:956–965

Evans RM (1988) The steroid and thyroid hormone receptor superfamily. Science 240:889–895

Farhat MY, Wolfe R, Vargas R, Foegh ML, Ramwell (1995) Effect of testosterone treatment on vasoconstrictor response of left anterior descending coronary artery in male and female pigs. J Cardiovasc Pharmacol 25:495–500

Fogelberg M, Björkhem I, Diczfalusy U, Henriksson P (1990) Stanozolol and experimental atherosclerosis: atherosclerosis development and blood lipids during anabolic steroid therapy of New Zealand white rabbits. Scand J Clin Lab Invest 50:693–700

Franks S (1989) Polycystic ovary syndrome: a changing perspective. Clin Endocrinol 31:87–120

Frayn KN (1993) Insulin resistance and lipid metabolism. Curr Opin Lipidol 4:197–204

Frazer KA, Narla G, Zhang JL, Rubin EM (1995). The apolipoprotein(a) gene is regulated by sex hormones and acute phase inducers in YAC transgenic mice. Nature Genetics 9:424–431

Freedman DS, O'Brien TR, Flanders WD (1991) Relation of serum testosterone levels to high density lipoprotein cholesterol and other characteristics in men. Arterioscler Thromb 11:307–315

Friedel DKE, Hannan CJ, Jones RE, Kettler TM, Plymate SR (1990) High density lipoprotein cholesterol is not decreased if an aromatizable androgen is administered. Metabolism 39:69–77

Fuster V (1994) Mechanisms leading to myocardial infarction: insights from studies in vascular biology. Circulation 90:2126–2146

Glazer G (1991) Atherogenic effects of anabolic steroids on serum lipid levels. Steroids 151:1925–1933

Glueck CJ, Gartside P, Fallat RW, Mendoza S (1976) Effect of sex hormones on protamine inactivated and resistant postheparin plasma lipase. Metabolism 25:625–630

Glueck CJ, Glueck HI, Stroop D, Speirs J, Hamer T, Tracy T (1993) Endogenous testosterone, fibrinolysis, and coronary heart disease risk in hyperlipidemic men. J Lab Clin Med 122:412–420

Goldberg IJ (1996) Lipoprotein lipase and lipolysis: Central roles in lipoprotein metabolism and atherogenesi. J Lipid Res 37:693–707

Goldberg RB, Rabin D, Alexander AN, Coelle GC, Getz GS (1985) Suppression of plasma testosterone leads to increase in serum total and high density lipoprotein cholesterol and apoproteins A-I and B. J Clin Endocrinol Metab 60:203–207

Gordon GB, Bush DE, Weisman HF (1988) Reduction of atherosclerosis by administration of dehydroepiandrosterone: A study in the hypercholesterolemic New Zealand white rabbit with aortic intimal injury. J Clin Invest 82:58–64

Gorodeski GI, Utian WH (1994) Epidemiology and risk factors of cardiovascular disease in postmenopausal women. In: Lobo RA (ed) Treatment of the postmenopausal woman: basic and clinical aspects. Raven Press New York NY pp 199–221

Graf MJ, Richards CJ, Brown V, Meissner L, Dunaif A (1990) The independent effects of hyperandrogenaemia, hyperinsulinaemia, and obesity on lipid and lipoprotein profiles in women. Clin Endocrinol 33:119–131

Griendling KK, Alexander RW (1996) Endothelial control of the cardiovascular system: recent advances. FASEB J 10:283–292

Guzick DS, Talbott EO, Sutton-Tyrell K, Herzog HC, Kuller LH, Wolfson SK (1996). Carotid atherosclerosis in women with polycystic ovary syndrome: Initial results from a case control study. Am J Obstet Gynecol 174:1224–1232

Haffner SM (1996) Sex hormone-binding protein, hyperinsulinemia, insulin resistance and noninsulin-dependent diabetes. Horm Res 45:233–237

Haffner SM, Valdez RA (1995) Endogenous sex hormones: impact on lipds, lipoproteins, and insulin. Am J Med 98 (Suppl):S40–S47

Hämäläinen E, Adlercreutz H, Ehnholm C (1986) Relationship of serum lipoproteins and apoproteins to the binding capacity of sex hormone binding globulin in healthy Finnish men. Metabolism 35:535–541

Hamm L (1942) Testosterone propionate in the treatment of angina pectoris. J Clin Endocrinol 2:325–328

Hamsten A, Karpe F, Bavenholm P, Silveira A (1994) Interactions amongst insulin, lipoproteins, and hemostatic function relevant to coronary heart disease. J Int Med 236 (Suppl 736):75–88

Hayes R, Chalmers SA, Nikolic-Paterson DJ, Atkins RC, Hedger MP (1996) Secretion of bioactive interleukin 1 by rat testicular macrophages in vitro. J Androl 17:41–49

Henriksson P, Angelin B, Berglund L (1992) Hormonal regulation of serum Lp(a) levels. J Clin Invest 89:1166–1171

Herman SM, Robinson JTC, McCredie RJ, Adams MR, Boyer MJ, Celermajer DS (1997) Androgen deprivation is associated with enhanced endothelium-dependent dilatation in adult men. Arterioscler Thromb Vasc Biol 17:2004–2009

Higashiura K, Mathur RS, Halushka PV (1997) Gender-related differences in androgen regulation of thromboxan A2 receptors in rat aortic smooth muscle cells. J Cardiovasc Pharmacol 29:311–315

Hishikawa K, Nakaki T, Marumo T, Suzuki H, Kato R, Saruta T (1995) Up regulation of nitric oxide synthase by estradiol in human aortic endothelial cells. FEBS Lett 360:291–295

Holte J, Bergh T, Berne C, Lithell H (1994) Serum lipoprotein lipid profile in women with the polycystic ovary syndrome: relation to anthropometric, endocrine and metabolic variables. Clin Endocrinol 41:463–471

Hromadova M, Hacik T, Malatinsky E, Sklovsky A, Cervenakov I (1989) Some measures of lipid metabolism in young sterile men before and after testosterone treatment. Endoc Exp 23:205–211

Hutchison SJ, Sudhir K, Chou TM, Chatterjee K (1997a) Sex hormones and vascular reactivity. Herz 22:141–150

Hutchison SJ, Sudhir K, Chou TM, Sievers RE, Zu BQ, Sun YP, Deedwania PC, Glantz SA, Parmely WW, Chatterjee K (1997b) Testosterone worsens endothelial dysfunction associated with hypercholesterolemia and environmental tobacco smoke exposure in male rabbit aorta. J Am Coll Cardiol 29:800–807

International Task Force for the Prevention of Coronary Heart Disease (1992) Prevention of coronary heart disease: scientific background and new clinical guidelines. Nutr Metab Cardiovasc Dis 2:113–156

Jaffe MD (1977) Effect of testosterone on on postexercise ST segment depression. Br Heart J 39:1217–1222

Jang Y, Lincoff AM, Plow EF, Topol EJ (1994) Cell adhesion molecules in coronary artery disease. J Am Coll Cardiol 24:1591–1601

Jonas A (1991) Lecithin:cholesterol acyltransferase in the metabolism of high density lipoproteins. Biochim Biophys Acta 1084:205–225

Juhan-Vague I, Alessi MC, Nalbone G (1993) Fibrinolysis and atherothrombosis. Curr Opin Lipidol 4:477–483

Kiel DP, Baron CA, Plymate SR (1989) Sex hormones and lipoproteins in men. Am J Med 87:35–39

Kirkland RT, Keenan BS, Probstfield JL, Patsch W, Tsai-Lien L, Clayton GW, Insull W Jr (1987) Decrease in plasma high density lipoprotein cholesterol levels at puberty in boys with delayed adolesence: Correlation with plasma testosterone levels. JAMA 257:502–507

Kolodgie FD, Jacob A, Wilson PS, Carlson GC, Farb A, Verma A, Virmani R (1996) Estradiol attenuates directed migration of vascular smooth muscle cells in vitro. Am J Pathol 148:969–976

LaCroix AZ, Yano K, Reed DM (1992) Dehydroepiandrosteronsulfate, incidence of myocardial infarction, and extent of atherosclerosis in men. Circulation 86:1529–1534

Larsen BA, Nordestgaard B, Stender S, Kjeldsen K (1993) Effect of testosterone on atherogenesis in cholesterol-fed rabbits with similar plasma cholesterol levels. Atherosclerosis 99:79–86

Laughlin GA, Morales AJ, Yen SSC (1997) Serum leptin levels in women with polycystic ovary syndrome: The role of insulin resistence hyperinsulinemia. J Clin Endocrinol Metab 82:1692–1696

Leenen R, van der Kooy K, Seidell JC, Deurenberg P, Koppeschaar HP (1994) Visceral fat accumulation in relation to sex hormones in obese men and women undergoing weight loss therapy. J Clin Endocrinol Metab 78:1515–1520

Lesser MA (1943) The treatment of angina pectoris with testosterone propionate: Further observation. N Eng J Med 228:185–188

Levine S, Likoff W (1943) The therapeutic value of testosterone propionate in angina pectoris. N Eng J Med 228:770–773

Libby P (1995) Molecular bases of the acute coronary syndromes. Circulation 91:2844–2850

Lindstedt G, Lundberg P, Lapidus L, Lundgren H, Bengtsson C, Björntorp P (1991) Low sex hormone binding globulin concentration as an independent risk factor for development of NIDDM: 12 yr follow up of population study of women in Gothenburg. Diabetes 40:123–128

Lobo RA (1991) Clinical Review 27: effects of hormonal replacement on lipids and lipoproteins in postmenopausal women. J Clin Endocrinol Metab 73:925–930

Lönnquist F, Schalling M (1997) Role of leptin and its receptor in human obesity. Curr Opin Endocrinol Diabetes 4:164–171

Lüscher TF (1993) The endothelium as a target and mediator of cardiovascular disease. Eur J Clin Invest 23:670–685

Maher VMG, Brown BG (1995) Lipoprotein(a) and coronary heart disease. Curr Opin Lipidol 6:229–235

Mammen EF (1993) Androgens and antiandrogens. Gynecol Endocrinol 7 (Suppl) 79–86

Mantzoros CS, Dunaif A, Flier JS (1997) Leptin concentrations in the polycystic ovary syndrome. J Clin Endocrinol Metab 82:1687–1691

Marcovina SM, Lippi G, Bagatell CJ, Bremner WJ (1996) Testosterone-induced suppression of lipoprotein(a) in normal men: relation to basal lipoprotein(a) level. Atherosclerosis 122:89–95

Marin P, Holmäng S, Jönsson L, Sjöstrom L, Kvist H, Holm G, Lindstedt G, Björntorp P (1992) The effects of testosterone treatment on body composition and metabolism in middle aged men. Int J Obesity 16:991–997

Marin P, Holmäng S, Gustafsson C, Jönsson L, Kvist H, Elander A, Eldh J, Sjöstrom L, Holm G, Björntorp P (1993) Androgen treatment of abdominally obese men. Obesity Res 1:245–251

Marin P, Odén B, Björntorp P (1995) Assimilation and mobilization of triglycerides in subcutaneous abdominal and femoral adipose tissue in vivo in men: effects of androgens. J Clin Endocrinol Metab 80:239–243

Masuda A, Mathur A, Halushka PV (1995) Testosterone increases thromboxane A2 receptors in cultured rat smooth muscle cells. Circ Res 69:638–643

Mattson LA, Cullberg G, Hamberger L, Samsioe G, Silverstolpe G (1984) Lipid metabolism in women with polycystic ovary syndrome: possible implications for an increased risk for coronary heart disease. Fertil Steril 42:579–84

McCrohon JA, Walters WAW, Robinson JC, McCredie RJ, Turner L, Adams MR, Handelsman DJ, Celermajer DS (1997) Arterial reactivity is enhanced in genetic males taking high dose estrogens. J Am Coll Cardiol 29:1432–1436

McEwen (1991) Non-genomic and genomic effects of steroids on neural activity. Trends Pharmacol Sci 12:141–147

McNutt RA, Ferenchick GS, Kirlin PA, Hamlin NJ (1988) Acute myocardial infarction in a 22 year old world class weightlifter abusing anabolic steroids. Am J Cardiol 62:164–167

Mendoza SG, Osuna A, Zerpa A, Gartside PS, Glueck CJ (1981) Hypertriglyceridemia and hypoalphalipoproteinemia in azoospermic and oligospermic young men: relationships of endogenous testosterone to triglyceride and high density lipoprotein metabolism. Metabolism 30:481–486

Meriggiola MC, Bremner WJ, Paulsen CA (1995) Testosterone enanthate at a dose of 200 mg/week decreases HDL-cholesterol levels in healthy men. Int J Androl 18:237–242

New G, Timmins KL, Duffy SJ, Tran BT, O'Brien RC, Harper RW, IT Meredith IA (1997) Long-term estrogen therapy improves vascular function in male to female transsexuals. J Am Coll Cardiol 29:1437–1444

Newcomer LM, Manson JE, Barbierir JL, Hennekens CH, Stampfer MJ (1994) Dehydroepinandrosterone sulfate and the risk of myocardial infarction in US male physicians: a prospective study. Am J Epidemiol 140:870–877

NIH Consensus Development Panel on Triglyceride, High Density Lipoprotein, and Coronary Heart Disease (1993). JAMA 269:505–510

Olivecrona G, Olivecrona T (1995) Triglyceride lipases and atherosclerosis. Curr Opin Lipidol 6:291–305

Owens GK (1995) Regulation and differentiation of vascular smooth muscle cells. Physiol Rev 75:487–517

Ozata M, Yildrimkaya M, Bulur M, Yilmaz K, Bolu E, Corakci A, Gundogan MA (1996) Effects of gonadotropin and testosterone treatment on lipoprotein(a), high density lipoprotein particles, and other lipoprotein levels in male hypogonadism. J Clin Endocrinol Metab 81:3372–3378

Pasquali R, Casimiri F, De Iasio R, Mesini P, Boschi S, Chierici R, Flamia R, Biscotti M, Vicennati V (1995) Insulin regulates testosterone and sex-hormone binding globulin concentrations in adult normal weight and obese men. J Clin Endocrinol Metab 80:654–658

Peinado-Onsurbe J, Staels B, Vanderschueren D, Bouillon R, Auwerx J (1993) Effects of sex steroids on hepatic and lipoprotein lipase activity and mRNA in the rat. Horm Res 40:184–188

Philips GB, Yano K, Stemmermann GN (1988) Serum sex hormones and myocardial infarction in the Honolulu Heart Program: pitfalls in prospective studies on sex hormones. J Clin Epidemiol 41:1151–1156

Phillips GB, Pinkernell BH, Jing TY (1994) The association of hypotestosteronemia with coronary artery disease in men. Arterioscler Thromb 14:701–706

Phillips GB, Pinkernell BH, Jing TY (1997) Relationship between serum sex hormones and coronary artery disease in postmenopausal women. Arterioscler Thromb Vasc Biol 17:695–701

Pilo R, Aharony D, Raz A (1981) Testosterone potentiation of ionophere and ADP induced platelet aggregation: relationship to arachidonic acid metabolism. Thromb Haemost 46:538–542

Reaven GM (1988) Role of insulin resistance in human disease. Diabetes 37:1495–1607

Rebuffé-Scrive M, Marin P, Bjorntorp P (1991) Effect of testosterone on abdominal adipose tissue in men. Int J Obes 15:791–5

Rigotti A, Trigatti B, Babitt J, Penman M, Xu S, Krieger M (1997) Scavenger receptor B1 – a cell surface receptor for high density lipoproteins. Curr Opin Lipidol 8:181–188

Rosano GM, Sarrel PM, Poole WP, Collins P (1993) Beneficial effect of oestrogen on exercise induced myocardial ischemia in women with coronary artery disease. Lancet 342:133–136

Rosenbaum M, Leibel RL, Hirsch JH (1997) Obesity. N Engl J Med 337:396–406

Ross R (1990). The pathogenesis of atherosclerosis: a perspective for the 1990s. Nature 362:801–809

Ross R (1995) Cell biology of atherosclerosis. Ann Rev Physiol 57:791–804

Rouru J, Anttila L, Koskinen P, Penttila TA, Irjala K, Huupponen R, Koulu M (1997) Serum leptin concentrations in women with polycystic ovary syndrome. J Clin Endocrinol Metab 82:1697–1700

Sacks FM, Walsh BW (1994) Sex hormones and lipoprotein metabolism. Curr Opin Lipidol 5:236–240

Sacks FM, Gerhard M, Walsh BW (1995) Sex hormones, lipoproteins, and vascular reactivity. Curr Opin Lipidol 6:161–166

Sampson M, Kong C, Patel A, Unwin R, Jacobs HS (1996) Ambulatory blood pressure profiles and plasminogen activator inhibitor (PAI-1) activity in lean women with and without the polycystic ovary syndrome. Clin Endocrinol (Oxf) 45:623–9

Schmitz G, Orsó E, Rothe G, Klucken J (1997) Scavenging, signalling and adhesion coupling in macrophages: implications for atherosclerosis. Curr Opin Lipidol 8:287–300

Seidell JC, Björntorp P, Sjöström L, Kvist H, Sannerstedt R (1990) Visceral fat accumulation in men is positively associated with insulin, glucose, and C-peptide levels but negatively with testosterone levels. Metabolism 39:897–901

Sigler LH, Tuglan J (1943) Treatment of angina pectoris with testosterone propionate. NY St J Med 43:1424–1428

Sih R, Morley JE, Kaiser FE, Perry HM, Patrick P, and Ross C (1997) Testosterone replacement in older hypogonadal men: a 12 month randomized controlled trial. J Clin Endocinol Metab 82:1661–1667

Simon D, Charles MA, Nahoul K, Orssaud G, Kremski J, Hully V, Joubert E, Papoz L, Eschwege E (1997) Association between plasma total testosterone and cardiovascular risk factors in healthy adult men: the Telecom Study. J Clin Endocrinol Metab 82:682–685

Smith EB (1996) Haemostatic factors and atherogenesis. Atherosclerosis 124:137–143

Sobel MI, Winkel CA, Macy LB, Liao P, Bjornsson TD (1995) The regulation of plasminogen activators and plasminogen activator inhibitor type 1 in endothelial cells by sex hormones. Am J Obstetr Gynecol 173:801–808

Sorva R, Kuusi T, Taskinene MR, Perheentupa J, Nikkila EA (1988) Testosterone substitution increases the activity of lipoprotein lipase and hepatic lipase in hypogonadal men. Atherosclerosis 69:191–197

Stanick S, Dornfeld LP, Maxwell MH, Viosca A, Korenman SG (1981) The effect of weight loss on reproductive hormones in obese men. J Clin Endocrinol Metab 53:828–832

Statistisches Bundesamt (ed) (1997) Statistical year book 1997. Wiesbaden

St. Clair RW (1997) Effects of estrogens on macrophage foam cells: a potential target for the protective effects of estrogens on atherosclerosis. Curr Opin Lipidol 8:281–286

Stein JH, Rosenson RS (1997) Lipoprotein Lp(a) excess and coronary heart disease. Arch Intern Med 157:1170–1176

Tabas I (1996) The stimulation of the cholesterol esterifcation pathway by atherogenic lipoproteins in macrophages. Curr Opin Lipidol 7:260–268

Tabas I (1997) Phospholipid metabolism in cholesterol loaded macrophages. Curr Opin Lipidol 8:266–274

Talbott E, Guzick D, Clerici A, Berga S, Detre K, Weimer K and Kuller L (1995) Coronary heart disease risk factors in women with polycystic ovary syndrome. Arterioscl Thromb Vasc Biol 15:821–6

Tall AR (1993) Plasma cholesteryl ester transfer protein. J Lipid Res 34:1255–1274

Tall A (1995) Plasma lipid transfer enzymes. Ann Rev Biochem 64:235–25

Tang J, Srivastava RAK, Krul ES, Baumann D, Pfleger B A, Kitchens RT, Schonfeld G (1991) In vivo regulation of apolipoprotein A-I gene expression by estradiol and testosterone occurs by different mechanisms in inbred strains of mice. J Lipid Res 32:1571–1585

Tchernof A, Labrie F, Bélanger A, Després JP (1996) Obesity and metabolic complications: contribution of dehyroepiandrosterone and other steroid hormones. J Endocrinol 150 (Suppl):155–164

Tchernof A, Labrie F, Bélanger A, Prud'homme D, Bouchard C, Tremblay A, Nadeau A, Després JP (1997) Relationships between endogenous steroid hormone, sex hormone binding globulin and lipoprotein levels in men: contribution of visceral obesity, insulin levels and other metabolic variables. Atherosclerosis 133:235–244

The Expert Panel (1993) Summary of the second report of the National Cholesterol Education Programme on detection, evaluation, and treatment of high blood cholesterol in adults. (Adult Treatment Panel II). JAMA 269:3015–3023

Thompson PD, Cullinane EM, Sady SP, Chenevery C, Saritelli AL, Sady MA (1989) Contrasting effects of testosterone and stanozolol on serum lipoprotein levels JAMA 261:1165–1168

Thompson SG, Kienast J, Pyke SDM, Haverkate F, van de Loo JCW for the European Concerted Action on Thrombosis and Disabilities Angina Pectoris Study Group (1995). Hemostatic factors and the risk of myocardial infarction or sudden death in patients with angina pectoris. N Engl J Med 332:635–641

Toda T, Toda Y, Cho BH, Kummerow FA (1984) Ultrastructural changes in the comb and aorta of chicks fed excess testosterone. Atherosclerosis 51:47–53

Tomita T, Sawamura F, Uetsuka R, Chiba T, Miura S, Ikeda M, Tomita I (1996) Inhibition of cholesteryl ester accumulation by 17β-estradiol in macrophges through activation of neutral cholesterol esterase. Biochim Biophys Acta 1300:210–218

Velazquez EM, Mendoza SG, Wang P, Glueck CJ (1997) Metformin therapy is associated with a decrease in plasma plasminogen activator inhibitor 1, lipoprotein(a), and immunoreactive insulin levels in patients with the polycystic ovary syndrome. Metabolism 46:454–457

Verheijen JH, Rijken DC, Chang GT, Preston FE, Kluft C (1984) Modulation of rapid plasminogen activator inhibitor in plasma by stanozolol. Thromb Haemost 51:396–397

Vermeulen A (1996) Decreased androgen levels and obesity in men. Ann Medi 28:13–15

Vermeulen A, Kaufman JM (1995) Ageing of the hypothalamo-pituitary-testicular axis in men. Hor Res 43:25–28

Vogel G (1996) Leptin: a trigger for puberty? Science 274:1466–1467

Vogel RA (1997) Coronary risk factors, endothelial function, and atherosclerosis: a review. Clin Cardiol 20:426–432

von Eckardstein S, von Eckardstein A, Bender HG, Schulte H, Assmann G (1996) Elevated low density lipoprotein-cholesterol in women with polycystic ovary syndrome. J Gynecol Endocrin 10:311–318

von Eckardstein A, Kliesch S, Nieschlag E, Chirazi A, Assmann G, Behre HM (1997) Suppression of endogenous testosterone in young men increases serum levels of HDL-subclass LpA-I and lipoprotein(a) J Clin Endocrinol Metab 82:3367–3372

von Eckardstein A, Büchter D, Kliesch S, Chirazi A, Nieschlag E, Assmann G, Behre HM (1998) Effects of testosterone suppression in young men by cetrorelix on plasma lipids, lipolytic enzymes, lipid transfer proteins, insulin, and leptin (submitted)

Wild RA, Painter PC, Coulson PB, Carruth BK, Ranney GB (1985) Lipoprotein lipid concentrations and cardiovascular risk in women with the polycystic ovary syndrome. J Clin Endocrinol Metabol 61:946–51

Wild RA, Grubb B, Hartz MD, van Nort JJ, Bachman W, Bartholomew M (1990) Clinical signs of androgen excess as risk factors for coronary artery disease. Fertil Steril 54:255–260

Winkler UH (1996) Effects of androgens on haemostasis. Maturitas 24:147–155

Wortsman J, Soler NG (1982) Abnormalities of fuel metabolism in the polycystic ovary syndrome. Obstet Gynecol 60:342–345

Wu SZ, Weng XZ (1993) Therapeutic effects of an androgenic preparation on myocardial ischemia and cardiac function in 62 elderly male coronary heart disease patients. Chin Med J 106:415–418

Yang XC, Jing TY, Gesnick LM, Phillips GB (1993) Relation of hemostatic factors to other risk factors for coronary artery disease and to sex hormones in men. Arterioscler Thromb 13:467–471

Yarnell JWG, Beswick AD, Swetnam PM, Riad-Fahmy D (1993) Endogenous sex hormones and ischemic heart disease in men. The Caerphilly Prospective study. Arterioscler Thromb 13:517–523

Yoshida Y, Nakamura Y, Sugino N, Shimamura K, Ono M, Kato H (1996) Changes in interleukin 1 production of peritoneal macrophages during estrous cycle in golden hamsters. Endocrine J 43:151–156

Yue P, Chatterjee K, Beale C, Poole-Wilson PA, Collins P (1995) Testosterone relaxes rabbit coronary arteries and aorta. Circulation 91:1154–1160

Zgliczynski S, Ossowski M, Slowinska-Szrednicka J, Brzenzinska A, Zgliczynski W, Soszynski P, Chotowska E, Szrednicki, M Sadowski, Z (1996) Effect of testosterone replacement therapy on lipids and lipoproteins in hypogonadal and elderly men. Atherosclerosis 121:35–43

Zhu XD, Bonet B, Knopp RH (1997) 17β estradiol, progesterone, and testosterone inversely modulate low density lipoprotein oxidation and cytotoxicity in cultured placental trophoblast and macrophages. Am J Obstet Gynecol 177:196–209

Zimmerman GA, McIntyre TM, Prescott SM (1996) Adhesion and signaling in vascular cell-cell interactions. J Clin Invest 98:1699–1702

Zmuda JM, Fahrenbach MC, Younkin BT (1993) The effect of testosterone aromatization on high density lipoprotein cholesterol levels and post-heparin lipolytic activity. Metabolism 39:69–77

Zmuda JM, Thompson PD, Dickenson R, Bausserman LL (1996) Testosterone decreases lipoprotein(a) in men. Am J Cardiol 77:1244–1247

9 Androgens and the prostate

Julian Frick, Andreas Jungwirth and Erwin Rovan

Contents

9.1 Introduction

The prostate is androgen-dependent, requiring testosterone for its growth, development, differentiation and function. The involution of the prostate gland is initiated by testosterone depriviation (orchidectomy, medical castration). Subsequent administration of exogenous androgen and/or cessation of medical androgen blockade, however, results in re-growth of the prostate gland. However, it only attains its original size, and the response to androgen does not cause prostatic overgrowth.

Within the prostate testosterone, as the principle plasma androgen, is converted to 5α-dihydrotestosterone (DHT) by the enzyme 5α-reductase which is localised on the nuclear membrane. DHT preferentially binds to the androgen receptor (AR) in all the target cells. The formation of the DHT-AR complex is essential to activate the growth-regulatory pathways and those related to the synthesis of secretory proteins. The DHT-AR complex is directly responsible for modulating gene expression in the prostate. Under the influence of androgens a series of protein-like substances is expressed from the epithelial cells of the prostate and represent an important part of the seminal fluid. Although benign prostatic hypertrophy (BPH) and/or prostate cancer have never developed in men castrated early in life, DHT would still appear to be involved in the pathogenesis of prostatic disease. BPH and prostate cancer are both characterised by uncontrolled growth, with the effect of androgen possibly promotional rather than causal.

This chapter deals with androgens and the prostate and includes all recently studied aspects. In a certain percentage of men of increasing age a slight to marked decrease in testosterone and its bioavailability will be found. These hormonal changes may be involved in alterations of sexual function, prostate function and growth, muscle mass, bone formation and mineralization. There is an increasing demand for treatment of this hypogonadal stage in an older male population, as quality of life is already a real issue. Androgen therapy in this age group is fraught with hazard and has to be carried out very carefully in regard to side-effects, overtreatment, possible induction of prostate growth and cancer development (see also Chapter 16 by Kaufman and Vermeulen in this volume).

Two areas of concern summarize the risks of androgen therapy in older men: cardiovascular disease and accleration of benign and/or malignant prostate disease. Other less pronounced areas of concern include water retention (including exacerbation of hypertension), hepatotoxicity, development of gynecomastia and polycythemia. Androgen replacement in an older man requires careful consideration of these issues and caution when applying therapy.

9.2 Anatomy and histopathology

The prostate is a glandulomuscular organ which surrounds the urethra at the bladder outlet and is enclosed in a fibrous capsule. The adult prostate is about the size and shape of a horse chestnut with the apex positioned forwards. The bulk of the gland is located laterally to the urethra; these are joined above by a thin bridge of glandular substance – the anterior lobe – and medially just below the vesical outlet is the glandular group of the median lobe. The posterior lobe constitutes the anatomic apex of the prostate, as palpated rectally. At the base of the prostatic urethra are the verumontanum and the ejaculatory ducts.

The prostate, apparently so named due to its location anterior to the bladder and seminal vesicles, is present in essentially all mammalian species, but has widely varying morphology.

The prostate is a distinct tubuloalveolar gland. It is frequently referred to as "lobed", although some confusion surrounds the use of this term. It may refer to its separate and distinct anatomical structure, to histologically discrete areas within a given structure, to zones that respond differentially to hormones, or to zones with a differential tendency to metastasis.

The human prostate is frequently described as a compact gland about the size of a chestnut, weighing approximately 20 g. It surrounds the urethra at the base of the bladder with the two ejaculatory ducts penetrating the gland through its posterior surface close to its upper border. The organ contains between 30 and 50 tubuloalveolar glands which empty into the prostatic urethra via 15 to 30 ducts. Calcified concretions which may exceed 1 mm in diameter exist in the secretion within the lumen of acini. The gland was considered by early anatomists to be largely homogeneous, but more recent studies have shown this is not correct. There is, nevertheless, still some dispute as to the precise subdivision of the gland. A detailed anatomical study recognizes three glandular regions, the peripheral zone, central zone, and preprostatic region surrounded by a thick anterior fibromuscular stroma (McNeal 1981). Benign prostatic hyperplasia is specifically restricted to the preprostatic region (Fig. 9.1).

The dog prostate has been a popular experimental system for studies of prostatic function for several reasons. It is the only accessory gland in this species, and when stimulated with pilocarpine hydrocholoride, large quantities of pure prostatic secretion can be obtained. Moreover, the canine pros-

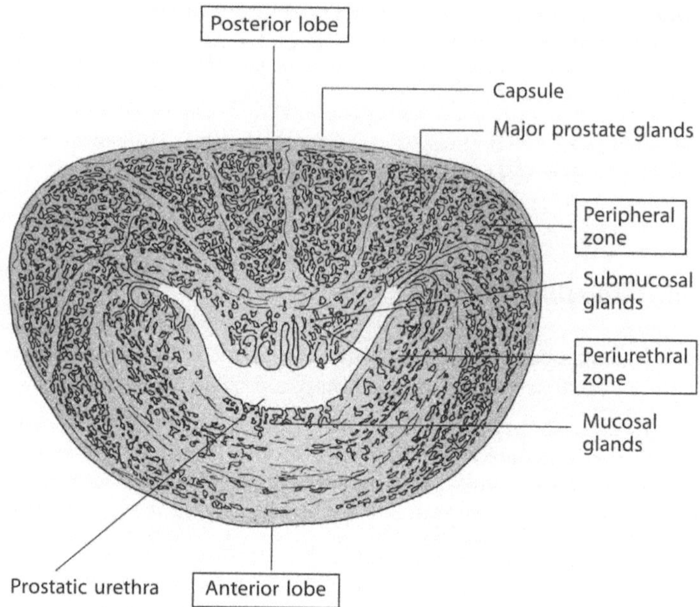

Fig. 9.1. Schematic presentation of the zonal arrangement of the prostate

tate has a tendency to hypertrophy, and thus provides a suitable model system in which to study this aspect of human pathology.

The rat prostate is also a popular experimental system, particularly for studies of androgenic control of male accessory glands. The rat prostate is a complex structure with several distinct anatomical lobes. Some confusion surrounded the classification of these lobes in early descriptions. The currently accepted classification encompasses the ventral prostate, which is a bilobed structure situated ventral to the urethra, and the dorsolateral aspect of the urethra. The coagulating gland is an additional lobe of the prostate and was previously known as the anterior prostate. It lies adjacent to the seminal vesicles, with which it shares a common peritoneal sheath. The ventral, lateral, and dorsal lobes of the prostate are each drained into the urethra by multiple ducts, whereas each coagulating gland is drained by single duct.

Arterial supply to the seminal vesicles and prostate of man has been the subject of a number of studies. There is considerable variation between specimens in the arrangement of the blood vessels and some confusion has arisen due to differences in the nomenclature. It seems that the main arterial supply is derived from the anterior division of the internal iliac (hypogastric) artery. From this vessel arise the umbilical, vesiculodeferential and prostatovesical arteries. The seminal vesicles and vasa deferentes are supplied by branches from the vesiculodeferential artery, whereas the prostatovesical artery divides to form the inferior vesical and prostatic arteries. The prostatic artery further subdivides into a urethral and capsular group of blood vessels. The urethral group supplies the periurethral regions of the prostate and the urethra, whereas the capsular group supplies the ventral and dorsal regions of the prostatic capsule and approximately two-thirds of the parenchyma. A particular feature of the prostatic arteries is a tortuosity of "corkscrew" pattern on the surface of the organ and also in the stroma. Venous blood leaves the prostate via the plexus of Santorini situated at the base of the gland, particularly on the anterior and lateral surface (Netter 1976).

9.2.1 Cell types

The prostatic epithelium in the human is composed of three types of cells: the secretory epithelial cells, the basal cells, and/or the stem cells. In any cell-renewing population cells flow from quiescent reserve stem cells to a more rapidly dividing transient proliferating population and finally to the formation of fully mature, nondividing terminally differentiated secretory cells that will then die off. In the prostate the most common tall columnar epithelial cells are terminally differentiated and are easily distinguished by their morphology and abundant secretory granules and enzymes that stain abundantly with acid phosphatase and other enzymes such as leucine amino peptidase and prostate-specific antigen. These tall columnar secretory cells appear like rows of a picket fence resting next to each other, with their base attached to a basement membrane and with their nuclei located at their base. Above the nucleus is a clear zone of abundant Golgi apparatus, and the upper cellular periphery is rich in secretory granules and enzymes. The apical plasma membrane facing the lumen possesses microvilli, and secretions move out into the open collecting spaces of the acinus. These epithelial cells encircle the periphery of the acinus, and the acini drain into ducts that connect to the urethra.

Much smaller and less abundant in number are the basal cells, which are present in less than 10% of the number of secretory epithelial cells. These small cells are not columnar and rounder in shape, with little cytoplasm and large irregularly shaped nuclei. They are less differentiated and almost devoid of secretory products such as acid phosphatase. The basal cells are rich in tonofilaments and stain brightly with fluorescent antibodies to keratin. It was mistakenly believed that these cells were myoepithelial, but this is not true because they are not rich in actin or myosin. It is believed that these undifferentiated basal cells give rise to secretory epithelial cells and, as such, function as a type of stem cells. The importance of understanding the biology of these basal cells lies in the growing evidence that many neoplasias, both benign and malignant, are really stem-cell diseases (Isaacs 1987).

9.2.2 Stroma and tissue matrix

The noncellular and connective tissue of the prostate compose what is termed the extracellular matrix. The extracellular matrix has long been recognized as one of the important inductive components during normal development of many different types of cells. A matrix system is defined as a biological scaffolding or residual skeleton structure that organizes cells as well as subcellular particles. The cellular matrix system is one of the most active areas of modern cell biology (i.e., extracellular matrix, cytoskeleton, and nuclear matrix). The extracellular matrix is far more than a supporting scaffolding because it has been shown to play a central role in the development and the control of cellular function. It now appears that the extracellular matrix is just one of the three major matrix systems that interact and form the overall tissue matrix system of the prostate. According to this concept, the epithelial cell rests upon the basement membrane, which is connected by an extracellular matrix to the stromal cells. It is believed that phase shifts and communication through these structural matrix elements may play a central role in controlling prostatic development and function (Isaacs et al. 1981).

9.3 Regulation of prostatic growth

The most important androgen in the human male is testosterone, and as soon as testosterone reaches the prostate cell by diffusion, it is metabolized into other steroids by a series of enzymatic processes. More than 75% of testosterone is converted into the most important intraprostatic androgen, namely 5α-DHT. This transformation of testosterone into DHT takes place under the influence of 5α-reductase (Bruchovsky et al. 1968). DHT then binds to the activated androgen receptors (AR) in the cell. This hormone receptor complex (HRC) is finally transformed and transferred into the cell nucleus, where the RNA polymerase is activated and followed by subsequent mRNA synthesis. Thereafter, mRNA is transported into the cytoplasmatic compartment where it is converted into secretory proteins. At a neurologic command, these proteins are secreted into the lumen; this takes place during ejaculation. Naturally, there are also other factors that possibly modify the testosterone effect on the prostate cell (Bruchovsky et al. 1975). Interaction between estrogen and androgen might induce growth of the prostate cell, so that a certain synergistic effect may occur. This has also been seen in several experiments, for example, when castrated dogs were treated with estrogen and dihydrotestosterone, the prostate gland grew to enormous size. Particularly estradiol doubles the prostate's androgen-induced increase in volume. This synergistic effect, however, is not found in the rat. Just as there are great differences in the prostate of various species, so there are also great differences in the pathology of the prostate.

In the normal adult prostate, the epithelial cells are continuously undergoing change. In this process of self-renewal, the rate of prostatic cell death is balanced by an equal rate of prostatic cell qualification so that neither involution nor overgrowth of the gland normally occurs with time. If an adult male is castrated, the plasma testosterone levels rapidly decrease to below 1.0 ng/ml. As a

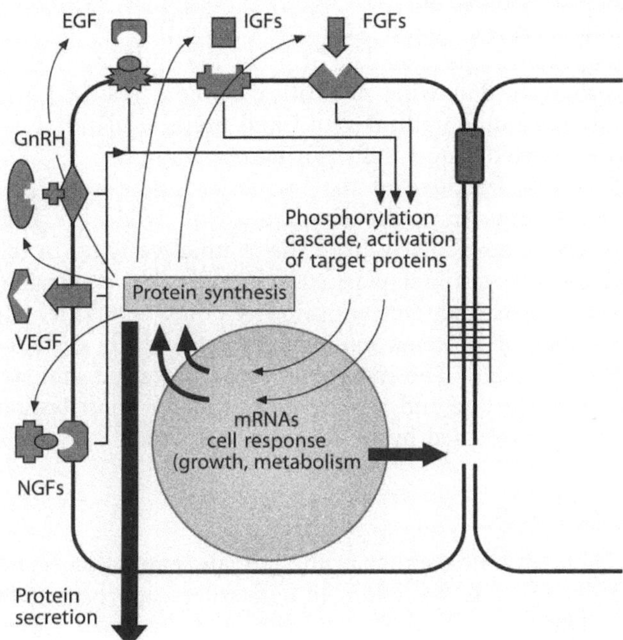

Fig. 9.2. Autocrine/paracrine growth factors known to be active in the prostate

result, the prostate rapidly decreases in size and loses significant amounts of its volume due to a major loss of glandular epithelial cells but not of the stromal or basal epithelial cells. Only the glandular epithelial cells are androgen-dependent and undergo apoptosis following castration and/or androgen deprivation (Mooradian et al. 1987). Prostate epithelial growth is additionally modulated by a series of growth factors i.e. epidermis growth factors (EGF), transforming growth factor-β (TGF-β, fibroblast growth factors (FGF), nerve growth factors (NGF) and insulin-like growth factors (IGF I/II) (Cohen et al. 1991; Martikainen et al. 1990; Sherwood et al. 1992). The mechanism of action of these possible growth factors may be either paracrine (between populations), autrocrine (within populations) or intracrine (within individual cells) (Fig. 9.2).

9.4 Normal function

9.4.1 Prostate secretion

Ejaculate volume in the human male is normally approximately 3.0–3.5 ml but ranges from 2 to 6 ml. The ejaculate is composed of two important components. The first is the spermatozoa and the second the seminal plasma. Both fractions are easily separated by centrifugation. The seminal plasma consists of the secretions from the accessory glands of the male genital tract, namely epididymis, vas deferens, ampullae, seminal vesicles, prostate gland, bulbourethral gland (Cowper's gland) and the urethral glands (Littré's glands). In the ejaculate volume the individual glands contribute approximately the following quantities: seminal vesicle 1.5–2 ml, prostate gland 0.5 ml, and bulbourethral gland and urethral glands 0.1–0.2 ml. These fractions are secreted sequentially during ejaculation.

The first fractions of the human ejaculate are rich in spermatozoa and in components from the prostate gland, while the later fractions are rich in products particularly from the seminal vesicles. In relation to the other body fluids, the seminal plasma is a somewhat unusual secretion, primarily because of its high concentration of potassium, zinc, citric acid, fructose, and a number of proteins and enzymes to which certain functions are ascribed (Coffey 1985). The individual ions, important amines and proteins expressed in the prostate and representing a major contribution to the seminal plasma will be discussed in the following section.

9.4.1.1 Zinc

The high zinc content in the human seminal plasma stems almost exclusively from prostate secretion and measures approximately 50 mg per 100 g dry weight.

As early as 1921 Bertrand and Vladesco reported that the human prostate gland has the highest zinc content of all tested organs. Various animal experiments have also shown that the zinc content in the prostate gland of many animals is very high, just as in man, and that a very specific zinc absorption occurs in the prostate. Zinc was localized in the human prostate by radioautographic tests performed in the epithelial cells. It is also known that zinc levels seen in benign prostate hyperplasia are elevated, while prostate malignancies are accompanied by a reduction in zinc levels (Byar 1974; MacKenzie et al. 1962).

Many physiologic roles have been postulated for zinc. A specific zinc-binding protein was described, particularly in malignant prostate disease, but it has also been repeatedly emphasized that zinc is a very considerable antibacterial factor in the prostate, as already suggested by the studies by Fair and Wehner (1976). Numerous studies have shown zinc to be one of the important factors effecting a local antibacterial mechanism. One study conducted in 36 healthy men with no bacterial prostatitis showed a zinc content of approximately 350 µg/ml in the prostate secretion, with a very wide range of 150–1000 µg/ml. In contrast, studies have been reported in 61 patients, 15 of whom had documented chronic bacterial prostatitis. In these patients the zinc content had dropped to an average of 50 µg/ml, and the range (0–139 µg/ml) was also depressed compared with healthy study subjects. In vitro studies have also shown that zinc actually evokes an antibacterial effect on gram-positive as well as gram-negative bacteria (Madsen et al. 1978).

9.4.1.2 Citric acid

One of the main sources of citrate secretion is the prostate gland, and citrate is one of the most important and most essential anions in human seminal plasma with an average content of 376 mg per 100 ml. This observation was already made by Huggins (1947), who examined nine human specimens of prostate secretion and found a citric acid content of 480–2688 mg per 100 ml. In contrast, he found a very minor citric acid content in the secretion of the seminal vesicle, namely maximum 20 mg citric acid per 100 ml. Citric acid, and very often also the acid phosphatases (see below), have been considered chemical indicators of prostate function. But it has also been seen that the citrate level in the prostate gland is not directly correlated with plasma testosterone content. Under normal or elevated plasma testosterone the citric acid content in the prostate gland would also have to be normal or elevated, and when testosterone is very low, the prostate cells would also have to secrete very little citric acid.

9.4.1.3 Cholesterol

Scott (1945) reported the cholesterol content in human seminal plasma to be approximately 103 mg per 100 ml, with about 83 mg phospholipids per 100 ml. The range for cholesterol content in seminal plasma is, however, relatively large, from 11 to 103 mg per 100 ml. Furthermore, a certain relationship is promoted between cholesterol and the phospholipids in the seminal plasma. White et al. (1976) found that this protects the spermatozoa from temperature shock. The prostate gland is certainly one of the most important sources of the cholesterol found in seminal plasma. It is also believed that cholesterol synthesis in the prostate is elevated in cases of benign prostate hyperplasia.

9.4.1.4 Spermine

The prostate gland is the main site of secretion of spermine, and the content of this protein in human seminal plasma ranges between 50 and 350 mg per 100 ml. Great interest has always surrounded spermine and other related polyamines in man, for example, spermidine and putrescine. These peptides have been repeatedly associated with growth. The role of spermine has not yet been precisely established, but it is felt that spermines play an important role by protecting the genital tract from infectious agents (Fair and Parrish 1981; Stamey et al. 1968).

9.4.1.5 Seminin

Seminin ia a proteolytic enzyme with a molecular weight of 30 kDa. Seminin usually appears in the first fraction of a split ejaculate, and it is therefore believed to come from the prostate. This seminal proteinase has no certain fibrinolytic properties, as was once presumed. Today it is believed that seminin particularly influences liquefaction and coagulation of the seminal plasma (Amelar 1962; Tauber and Zaneveld 1976).

9.4.1.6 Prostaglandin

The most important source of prostaglandins is not the prostate gland but the seminal vesicle. The term prostaglandin is certainly a misnomer, since its discoverer (von Euler 1934) believed that these compounds come from the prostate gland and not from the seminal vesicle. This view was later corrected by Eliasson (1959) in his very precise studies. Thus, this compound should be called seminoglandin instead of prostaglandin, but the original name has survived and has not changed despite clear scientific evidence. The prostaglandin content of the seminal plasma is about 100–300 µg/ml. Later, it was primarily Bergström et al. (1986) who discovered that a large number of prostaglandins exist. These authors purified the individual prostaglandins and identified their chemical structure.

There are some 15 different prostaglandins in the seminal fluid alone. The prostaglandins are divided into the main groups A, B, E, and F, according to the structure of the cyclopentane ring. Each of these groups is subdivided according to the position and number of double bonds in the side chains. As we know, these compounds are very potent pharmacologic substances that entail a number of biologic processes. Prostaglandins certainly play an important role in spermatozoa motility and spermatozoa transport, as well as in testicular and penile contraction. Moreover, it is believed that the prostaglandins from the seminal plasma, which are released into the vagina, influence the cervical mucus, increase vaginal secretion, and influence spermatozoa transport in the female genital tract. To date, these effects have not been established with 100% certainty.

9.4.1.7 Acid phosphatase and acid prostate phosphatase

Acid prostate phosphatase is a glycoprotein with a molecular weight of 102 kDa (Babson and Read 1959). Both of these enzymes, namely acid phosphatase and prostate acid phosphatase, are expressed in the prostate gland. The seminal plasma contains approximately 800–1500 U/ml acid phosphatase. Both enzymes have long been used to monitor prostate carcinomas, above all the course of very advanced prostate malignancies. Neither of these enzymes play a major role with regard to inflammatory changes in the prostate region. Various studies to date have failed to show that the phosphatases in the prostate gland inhibit or promote inflammation. For many years both factors were decisive for monitoring the course of advanced prostate carcinomas, until prostate-specific antigen assumed this role.

9.4.1.8 Prostate-specific antigen

In 1979 Wang and coworkers reported for the first time on this protein that is expressed by the epithelial cells of the prostate gland. This parameter is a very specific one, and although it is not always successful in early detection of prostate cancer, today it is the standard parameter for monitoring the course of a prostate malignancy, either after radical prostatectomy or during endocrine therapy of an advanced, already metastasizing malignancy (Wang et al. 1981). Prostate-specific antigen (PSA) is also a glycoprotein with a molecular weight of 33kDa. The entire amino acid sequence of this protein is known. Bahnson (1991) reported elevated PSA levels in a patient with a granulomatous prostatitis that developed following instillation of Bacille Calmette-Guérin for a superficial bladder tumor. Elevated serum PSA has been reported in isolated cases of patients with prostatitis or a prostate infarction (Speights and Brawn 1996; Tchetgen and Oesterling 1997). It has been shown that PSA is helpful in the early diagnosis of prostate cancer and in this respect is superior to other available tests like examination and transrectal ultrasonography. PSA is also helpful in staging of locally confined disease. It can be used to identify or exclude local extension of disease, if combined with T category and grade of differentiation determined on biopsy. The same parameters also give an indication of the presence of lymph node metastases, which may prevent unnecessary and invasive staging procedures in certain groups of patients with favourable prognostic factors and a low PSA-value. Progression of untreated prostate cancer in various stages can be monitored by PSA. The true value of the marker in this respect is still underexplored. It may be possible that PSA will be shown to differentiate effectively between aggressive and non-progressive disease. In this respect, it could become an essential tool to identify those patients that may not require treatment at all. PSA is also a useful marker for therapy response. An elevation of PSA after radical prostatectomy indicates local or metastatic progression, which will occur within 1–2 years. PSA is an androgen-dependent enzyme and decreases under endocrine treatment. It is unexplained why, in spite of its endocrine-dependent character, PSA rises with endocrine-independent progression of prostate cancer. (Arcangeli et al. 1997; Bangma et al. 1995).

9.4.1.9 Immunoglobulins

It is known that immunoglobulins can be demonstrated in human seminal plasma. IgG values of 7 mg/100 ml and 22 mg/100 ml and IgA values of 0 mg/100 ml and 6 mg/100 ml in seminal plasma have been reported. To date, however, no IgM has been discovered in seminal plasma. The source of these immunoglobulins is not precisely known. Whether or not part of these immunoglobulins is produced in the prostate gland cannot be reported with certainty. Immunoglobulin levels in the seminal plasma are lower than in plasma, but the possibility exists that these diffuse through the blood-seminal-plasma barrier, so that this path must also be taken into consideration.

What role the immunoglobulins play, and whether they help prevent infections in the region of the accessory glands of the male genital apparatus is not known for sure (Coffey 1985).

9.4.1.10 Prostate proteins

In addition to PSA, other proteins are also expressed in the prostate gland. One of these proteins is prostatein, which is composed of several chains, with various molecular weights ranging from 6 to 14 kDa. It is known that the gene for these proteins is under strict androgen control, but the precise role of prostatein has not yet been discovered.

9.4.2 Transport of compounds into the prostate secretion

These transport mechanisms into the prostate secretion are important because they may contribute to possible new therapies for acute and chronic recurrent prostatitis, and also because new possibilities for chemotherapy are needed, above all for prostate carcinoma. Certain compounds reach concentrations in the prostate secretion similar to those in blood, or are sometimes even enriched in the prostate secretion, i.e., the levels in the prostate secretion are higher than in blood. In general, the transport of compounds into the prostate secretion can be described as follows: the substances pass the cell membrane by nonionic diffusion or through the membrane by fat solubility. An important role is also played by the pH value of the seminal plasma or the pH value of prostate secretion. This pH value varies from 6 to 8 in humans, with a mean of 6.6. However, when the prostate gland becomes inflamed, the pH value of the prostate secretion rises to about 7 or even higher. This must be taken into consideration when using certain medications, namely particular circumstances in the prostate gland create particular pH values. It must also be remembered that the prostate gland contains a very considerable transport force, and that prostatic transport systems proceed under particular hormonal influence. Thus the transport capacity for various substances might be stimulated by androgens and inhibited by estrogens (Frick 1994).

9.5 Epidemiology and natural history of benign prostatic hyperplasia (BPH)

BPH appears to be a prostatic condition that occurs with aging (Berry et al. 1984). Autopsy studies show 50% of men in their fifth decade have pathologic proof of BPH and that prevalence increases to 90% by the ninth decade. Symptoms of prostatic outlet obstruction are coincident with the development of BPH. A survey of men 60 years and older who never had previous prostate surgery reported a 35% prevalence of one or more symptoms of prostatism (hesitancy, straining, weak stream, intermittency, or use of a catheter). Approximately 25% of men aged 55 note a decreased urinary flow rate and this increases to 50% at age 75. As a result of such a high incidence of symptomatic BPH, it has been estimated that a 40- to 50-year old male has a 20% to 30% chance of requiring prostatectomy for symptomatic BPH in his lifetime (Boyle et al. 1991). Though BPH appears to increase with age, the symptoms caused by BPH do not necessarily increase with age in any particular individual. Watchful waiting and placebo can result in improvement of BPH symptoms in approximately 40% of patients. Moreover, dangerous complications of BPH, such as acute urinary retention, renal failure from obstructive uropathy and/or urosepsis occur in only a small percentage of patients. In most instances relief of symptoms and quality of life issues are the main reason for seeking therapy (Garraway et al. 1991).

In a study carried out by Lytton and coworkers (1968), the incidences of prostatectomy was 0.2 per 1000 patient years at age 40 through 49, 1.2 at 50 through 59, 5.7 at 60 through 69, 10 for 70 through 79, and 10.9 after age 80. These data suggest that a 40-year-old man who survives to age 80 has about a 10% likelihood of undergoing prostatectomy. Such estimates are made by assuming a 40-year-old man instantly ages and is exposed to the incidence of surgery in each age group. Using 1987 rates, the probability of a prostatectomy for a 40-year-old man surviving to age 80 would be about 37%. It also is important to note, given geographic variations in rates, that the risk might vary considerably amoung communities, countries and continents.

Fang-Liu (1993) described a relatively low incidence of histological BPH among Chinese men at autopsy compared to non-Chinese. This was a study performed in China in the twenties of this century. In a recent analysis of autopsies of Bejing Medical University between 1989–1992, the prevalence of histological BPH at autopsy was actually more compatible with other series: about 20% for men in their 50's, 50% for men in their 60's and 70's and 80% for men over 80 (Barry 1990; Guess 1992).

There are numerous international mortality rates described for BPH. Countries with Asian populations seem to have low death rates from BPH. However, the many inconsistencies in these data suggest that recording and reporting practices may have a greater influence than BPH occurrence in deriving the reported rates. For example, two medically advanced Western countries, the US and Norway, reported BPH death rates in the late 1970s of 23 per 100000 person-years and 119 per 100000 person-years, respectively, for men over age 75. Because the greater majority of US men have BPH by this older age, it is hard to explain the radically different death rates in these groups on the basis of a higher incidence of BPH. Moreover, during this same era, rates of prostatectomy were roughly comparable in Norway and New England, making it unlikely that undertreatment of BPH led to higher death rates in the former region (Guess 1995).

9.6 Endocrinology and pathophysiology of BPH

9.6.1 Plasma hormone levels in patients with BPH

Normal functioning Leydig cells require a normal functioning hormonal axis with pulsatile expression of GnRH in the hypothalamus and, consequently, of LH in the pituitary. In the normal human male, the major circulating serum androgen is testosterone, which is almost exclusively (more than 95%) of testicular origin. Under normal physiological conditions the Leydig cells of the testis are the major source of testicular androgens. The Leydig cells are stimulated by the gonadotropins (primarly LH) to synthesize testosterone from acetate and cholesterol. The spermatic vein concentration of testosterone is 40 to 50 µg/100 ml. Other androgens also leave the testes by the spermatic vein, and these include androstanediol, androstenedione (3 µg/100 ml), dehydroepiandrosterone (7 µg/100 ml), and dihydrotestosterone (at 0.4 µ/100 ml); therefore, the concentrations of these androgens are much lower than testosterone, all being less than 15% of that of testosterone (Table 9.1).

The total testosterone that enters the plasma is referred to as the testosterone blood production rate and is 6 to 7 mg/day in the human. Although other steroids, such as androstenedione from the adrenals, can be converted by peripheral metabolism to testosterone, they probably account for less than 5% of the overall production of plasma testosterone.

The average testosterone concentration in the adult human male plasma is approximately 611 ng/100 ml±186 and is not greatly related to age between 25 and 70 years, although it does decline gradually to approximately 500 mg/100 ml after 70 years of age. It is recognized that plasma concentrations of testosterone can vary widely in an individual during the course of the day

Table 9.1. Plasma levels of sex steroids in healthy human males

Steroid	Plasma concentration ng/100ml	Daily blood production rate (mg/day)
Testosterone	611±186	6.6±0.5
Dihydrotestosterone (DHT)	56±20	0.3±0.06
3α-Androstanediol	14±4	0.2±0.03
Androstenediol	161±52	
Androsterone	54±32	0.28
Androstenedione	150±54	1.4
Dehydroepiandrosterone (DHEA)	501±98	29
Dehydroepiandrosterone sulfate (DHEAS)	135.925±48.000	
Progesterone	30	0.75
17β-Estradiol	2.5±0.8	0.045

and may reflect both episodic and diurnal variations in the production rate of testosterone.

Only 2% of the total serum testosterone is not protein bound; this is termed the free testosterone in the plasma, and it is at a concentration of approximately 15 ng/100 ml or less than 1 nM. Only this free testosterone is available for metabolism by the liver and intestines, primarily to 17-keto-steroids, which are then secreted into the urine as final water-soluble conjugates of sulfuric acid and glucuronic acid. Total 17-ketosteroids in the urine of adult males ranges from 4 to 25 mg per 24 hrs and is not an accurate index of testosterone production, since other steroids from the adrenals as well as non-androgenic steroids can be metabolized to 17-ketosteroids. Only small (25–160 µg/day) amounts of testosterone enter the urine without metabolism, and this urinary testosterone represents less than 2% of the daily testosterone production.

Under normal conditions aging in human males is accompanied by a gradual decline in levels of all fractions of serum testosterone. These include total testosterone, free testosterone, and testosterone not bound to sex-hormone binding globulin (SHBG). There is great inter-individual variability in plasma testosterone levels among healthy older men (Tenover 1997; Vermeulen 1991). Fig. 9.3 shows the course of plasma testosterone and 17-β-estradiol over the lifespan of men. In the fifth decade plasma testosterone levels begin to decline, whereas 17-β-estradiol levels remain fairly constant throughout the recording period (Frick 1974). Not all men are destined to become hypogonadal with increasing age, and the criteria for defining an older adult male as testosterone-deficient is not really established yet. Serum gonadotropin levels are not helpful in this regard, since older men, even those with subnor-

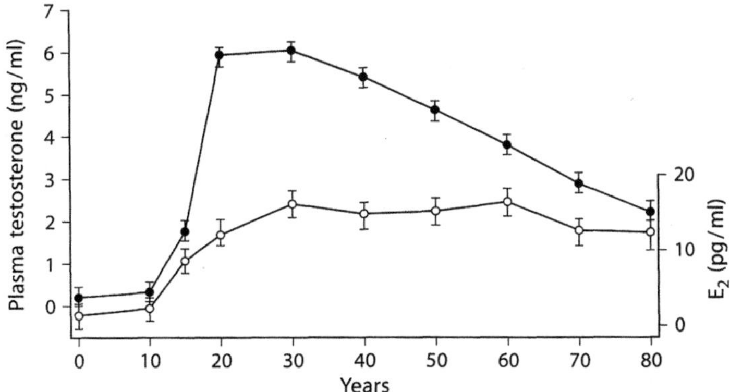

Fig. 9.3. This figure demonstrates mean testosterone and 17β-estradiol plasma levels during the lifetime of the normal male; 10 individuals in every decade were investigated. As can be seen from the testosterone curve, the levels rise at the age of 14 and remain in the same range for about 30 years. After age 45 on average a slight decrease in the testosterone plasma levels begins. The 17β-estradiol curve also shows an increase of the plasma levels after age 15 and remains more or less at the same level during the whole lifetime of a male (Frick unpublished)

mal plasma testosterone levels, usually have gonadotropin values within normal range. There is agreement that the free testosterone content decreases with age. This special testosterone fraction is generally considered to be a parameter of the biologically available testosterone. That the free testosterone component shows a greater decrease than total testosterone might be caused by the age-dependent increase of the binding capacity of sex-hormone binding globulin (Deslypere and Vermeulen 1981).

Plasma DHT levels in elderly human males are either unchanged or slightly decreased. In elderly men with manifest BPH plasma DHT values are somewhat higher compared with a normal control group of similar age (Hammond et al. 1978; Partin et al. 1991). Androstanediol glucuronide plasma and urine levels decrease significantly with age. This steroid is supposed to be an important parameter to measure peripheral androgen action (Walsh and Wilson 1976).

9.6.2 Androgen metabolism

Androgens are synthetized and secreted by Leydig cells of the testis and by cells of the zona reticularis of the adrenals.

Androgen biosynthesis commonly starts from cholesterol, but synthesis from acetyl-coenzyme A is also possible. For transport to target organs cholesterol, as a small and lipophilic molecule, has to be bound to a carrier lipoprotein complex (LDL: low density lipoprotein) to avoid uncontrolled passive diffusion into each cell of all organs. Specific integral membrane receptor proteins (LDL receptors) bind the LDL-cholesterol complex, which is internalized together with the receptor by endocytosis. After fusion with lysosomal vesicles, cholesterol is separated from the complex and the receptor is recycled back to the plasma membrane. A smaller proportion of cholesterol is stored as cholesterol ester in lipid droplets, while the major part is transported as free cholesterol to mitochondria (Träger 1977).

Cholesterol is a molecule of 27 C atoms, consisting of three hexane rings, one pentane ring and an aliphatic side chain with three methyl groups. Two steric isoforms are known: a transconfiguration (α-form) and a cis-configuration (β-form).

The first steps of androgen biosynthesis occur at the inner membrane of mitochondria. A special enzyme complex (cytochrome P 450 scc) transforms cholesterol into pregnenolone by oxidative cleaving of the side chain and by hydroxylation of C 20 and C 22. Pregnenolone is transported to the endoplasmic reticulum by specific cytoskeleton proteins. Integral membrane enzymes (3-β-hydroxy-steroid-dehydrogenase) convert pregnenolone into progesterone. In Leydig cells a further hydroxylase (cytochrome 450c 17) and a cytochrome reductase transform progesterone first to 17-α-hydroxy-progesterone and then to androstendione, which is converted to testosterone by a membrane bound keto-reductase (17-β-oxy-hydroxy-steroid-dehydrogenase) (Andersson and Moghrabi 1997; Labrie et al. 1997).

pregnenolone   **testosterone**

$\downarrow$ 3β-HSD $\uparrow$ 17β-OHSD

progesterone $\rightarrow$ **androstendione**

P 450 c 17

\+ cytochrome reductase

In the prostate and particularly also in the testes and the epididymis, testosterone is converted to the more active dihydrotestosterone by membrane-bound isoforms of 5α-reductase. In contrast to peptide hormones steroids are not stored within synthetizing cells. They are continuously transported through the plasma membrane by specific transport proteins and thus reach extracellular space and blood capillaries. In peripheral blood plasma most androgens are bound to steroid-hormone binding globulins (SHBG). The plasma concentration is a balance between secretion and metabolism which can be estimated by the metabolic evaluation of steroid kinetics (Luke and Coffey 1994).

The rate of secretion can be calculated according to the formula:

production rate (nmol/d) = MCR (metabolic clearance rate: 1/d) × C
(plasma concentration: nmol/l)

The major portion of circulating testosterone (60 to 70%) is metabolized to $\Delta 4$-androstenedion. This steroid initiates the degradation of inactive androgens. There are two possible degradation pathways: one proceeds predominantly in the liver via androstendion to ethiocholanolone and androsterone with further glucurination and sulfation, the other in Leydig cells and Sertoli cells via 17-β hydroxylized substrates to 3β-5α androstendiols. Therefore DHT can be seen as the first step of testosterone degradation, although it reveals a higher biological activity than testosterone itself. In contrast, DHT synthetized in hepatocytes is immediately converted to androstendion and is biologically inactive. Only a small amount of testosterone ($<1\%$) is directly excreted as glucuronide (Ishimaru et al. 1977).

A considerable proportion of androgens is provided by the adrenal cortex. Dihydroepiandrosterone (DHEA), dihydroepiandrosteronsulfate (DHEA-S) and $\Delta 4$-androstendion are precursors showing weak biological activity (Ebeling and Koivisto 1994). They are partially metabolized to testosterone in liver cells. The existence of a second source for testosterone should be taken into consideration if complete testosterone down-regulation for prostate cancer treatment is intended.

9.6.3 5α-reductase and additional enzymes

After free testosterone in plasma has entered the prostatic cells through diffusion, it is rapidly metabolized to other steroids by a series of prostatic enzymes. Over 90% of the testosterone is irreversibly converted to the main prostatic androgen, DHT, through the action of the enzyme 5α-reductase located on the endoplasmic reticulum and on the nuclear membrane. The enzyme 5α-reductase reduces the unsaturated bond in testosterone molecule between the 4 and 5 position to form 5α-reduced DHT. The enzyme is primarily localized to the stroma (Bruchovsky and Dunstan-Adams 1985). 5α-Reductase can also convert androstenedione or progesterone to the 5α-reduced form. After DHT is formed from testosterone, it is subjected to a series of reversible reactions to form 3α-diol (5α-androstane 3α, 17β-diol) and 3β-diol (5α-androstane 4β- 17β-diol). The enzymes that perform this transformation of DHT are 3α- or 3β-hydroxysteroid oxidoreductases (3α-HSOR or 3β-HSOR). These enzymes utilize NADP as a cofactor, but, in contrast to 5α-reductase, they can also utilize NAD. The equilibrium for the metabolism of DHT favors the formation of DHT. On the other hand, 3β-diol is not very effective as an androgen because it is rapidly and irreversibly converted to the triol form by hydroxylation in the 6α or 7α-position. The triols are very water soluble and inactive as androgens and cannot reform DHT.

9.6.4 Adrenal androgens

There is evidence that overstimulation of the adrenal cortex causes the production of adrenal steroids that stimulate the growth of the prostate gland. For example, in humans, abnormal virilisation has been observed in immature males with hyperfunction of the adrenal cortex resulting from neoplasia or hyperplasia of the adrenal gland. In rodents, overstimulation of the adrenals can also induce limited prostate growth, even in the absence of testicular androgens (Mobbs et al. 1973). For example, administration of exogenous ACTH to castrated animals significantly increases the growth of sex accessory organs. Such ACTH stimulation of prostatic growth has been observed in both castrated and castrated-hypophysectomized animals, but does not occur in animals that are adrenalectomized.

The effect of normal levels of adrenal androgens on the prostate in intact human and adult male rats may not be significant because adrenalectomy has very little effect on prostate size and/or morphology of epididymis and seminal vesicles. Furthermore, following castration in animals, the prostate diminshes to a very small size (90% reduction in cell mass) without concomitant adrenalectomy. Finally, the small involuted ventral prostate in the castrated rat cannot be significantly reduced further by performing additional adrenalectomy or hypophysectomy. In castrated rats, the DHT levels in the prostatic tissue are approximately 30% of that in normal intact animals. Adrenalectomy lowers DHT to nondetectable levels without further diminution in prostate growth. This indicates that a threshold level of DHT is required in the prostate to stimulate growth and that the castrate level is below this threshold. It has also been concluded similarly that the human prostate does not restore itself following castration, indicating that adrenal androgens are insufficient to compensate for the loss of testicular function. Quantitative morphometry of the human prostate also confirms that the adrenal gland has little effect on the prostate (Oesterling et al. 1986).

The adrenal steroids (DHA, DHA-S and androstenedione) are synthesized from acetate and cholesterol and are secreted by the normal human adrenal glands. Essentially all of the DHA in the male plasma is of adrenal cortex origin, and the production rate is 10 to 30 mg/day. Less than 1% of the total testosterone plasma is derived from dehydroepiandrosterone (Endoh et al. 1996).

The prostate and seminal vesicles of the rat, as well as the human prostate, can slowly hydrolyze DHEA-S to free steroids through a prostate sulfatase enzymatic activity, but the degree of 5α-conversion is low, thereby explaining why DHEA-S is not a very potent androgen.

A second adrenal androgen is androstenedione, and its plasma concentration in adult males is approximately 150 ± 54 ng/100 ml. The blood production rate of androstenedione in human males is about 2 to 6 mg/day. Androstenedione cannot be converted directly to DHT and is therefore considered as a weak androgen. An important role for androstenedione in the male may be its peripheral conversion to estrogen through aromatization.

The adrenal gland also produces C21 steroids (e.g., progesterone). The plasma production rate (0.75 mg/day) is low, producing a low plasma progesterone concentration of 0,3 ng/ml (Coffey 1985). Although progesterone is weakly androgenic, it does not exert a significant effect on the prostate at the low concentrations present in normal male plasma. In summary, under normal conditions the adrenals do not support significant growth of prostatic tissue.

9.6.5 Androgen receptors

The androgen receptor (AR) is a member of the steroid receptor superfamily which plays an important role in male sexual differentiation and prostate cell proliferation. Mutations or abnormal expression of AR in prostate cancer can play a key role in the process that changes prostate cancer from an androgen-dependent to an androgen-independent stage (Newmark et al. 1992). Generally AR concentrations in BPH have been found to be higher than in normal human prostate, mostly due to increased values in the nuclear fraction, presumably reflecting the high 5α-reductase activity in this tissue, and possibly indicating a high degree of androgen stimulation. This may be a factor that sensitizes this tissue and facilitates accelerated growth. Chang and coworkers (1988a, 1988b, 1995) have recently suggested that AR may need coactivators such as AR-associated protein (ARA_{70}) for optimal androgen activity. In addition to this effect ARA_{70} might also act as enhancer to increase androgenic activity (Yeh and Chang 1996).

9.6.6 Estrogen receptors and the role of estrogens

In the dog, estrogen is involved in the induction of the androgen receptor. Furthermore, the canine prostate contains abundant quantities of high-affinity estrogen receptor. Patients with larger volumes of BPH have higher levels of estradiol. This means that in the human male androgen may also act synergistically with estrogen. In the human prostate multiple sites bound for estrogen binding were found in the cytosol as well as in nuclear preparations (Ekman et al. 1983). There are two classes of binding sites, one class of binding sites corresponding to the classic high-affinity estrogen receptor and the second class to a lower-affinity binding protein. Although high-affinity estrogen receptors are present in human BPH, the levels are significantly lower than in normal peripheral tissue. If estrogen is indeed involved in the induction of BPH in man, higher levels of receptors would be expected (Trachtenberg et al. 1980).

There is enough evidence from multiple studies that BPH is under endocrine control (Huggins and Stevens 1940; Isaacs 1983). As reports from the literature show, a reversible regression of BPH could be achieved with potent LHRH-agonists, which confirms the assumption that androgens at least have an important supportive role in development of BPH (Peters and Walsh 1987). It could also be shown that larger volumes of BPH have higher levels of free testosterone and estrogen. These findings demonstrate that plasma androgens, acting synergistically with increased estrogen, are responsible for persistent stimulation of BPH. Increased nuclear androgen receptor may support the view that estrogen induction of the androgen receptor in man is responsible for the synergism associated with increased BPH volume with age. Coffey and Walsh (1990) were unable to demonstrate increased estrogen receptors in human BPH.

Based on these observations, it is reasonable to assume that persistent androgenic stimulation, possibly coupled with estrogen synergism, is involved in the growth of BPH with age. If so, therapeutic attempts at lowering plasma testosterone levels, reducing estrogen levels (without increasing androgen levels), or interfering with androgenic stimulation of the prostate through other mechanisms, such as inhibitors of androgen metabolism within the prostate, may prevent progression of BPH with age.

9.7 Approaches to hormonal treatment of BPH

9.7.1 GnRH analogues and progestins

Gonadotropin-releasing hormone agonists (GnRH), antiandrogens, estrogen preparations, and bilateral orchiectomy all affect the prostate by diminishing serum testosterone or blocking its effect on the prostate (Berry et al. 1986). These agents have a definite place in the management of prostate cancer and prostatic outlet obstruction from cancer. Currently, however, they have no role in the routine management of BPH (Geller et al. 1967; Lepor and Stone 1995; Peters and Walsh 1987).

9.7.2 5α-Reductase inhibitors

As mentioned earlier, the prostate is hormonally sensitive and androgen deprivation as well as blockage of 5α-reductase can decrease the size of the prostate. This led to the development of finasteride, a 5α-reductase inhibitor. Within prostate tissues 5α-reductase is responsible for conversion of testosterone to dihydrotestosterone.

Blockage of this enzyme results in DHT deprivation within the prostate (Geller 1990; Stoner 1992). As a result, the prostate decreases in size. Currently, finasteride is the only drug that acts through hormonal pathways approved for treatment of BPH in many countries around the world. A recent report of the 3-year efficacy and safety of finasteride provides the best long-term data. Finasteride appears to be a remarkably safe drug with no dangerous side-effects. The most common side-effect, impotence, occurs in less than 5% of patients. The major disadvantage of finasteride may be efficacy, which is difficult to interpret. In the three-year study by the Finasteride Study Group (Stoner 1994), 543 patients initially were placed on finasteride for prostatism; the dosage was 5 mg/day. After 36 months 297 patients were evaluable. A statistically significant improvement from baseline was shown in maximum urinary flow rate and total symptom score. After three years, those followed and still taking 5 mg of finasteride daily in the study, had a 2,4 ml/second mean increase in maximum flow rate compared with baseline before initiation of treatment. At least 40% of patients demonstrated a 3 ml/second increase in the maximum flow rate. Likewise, after three years, the symptom scores improved by 3,6 points, and 40% of patients reported a 50% or greater improvement in symptom scores. After three years prostate volume had decreased approximately 27%. Despite objective statistically significant data showing finasteride indeed improves prostatism, the same studies suggest that it may take up to 12 months of medication to see any benefit and clearly a significant percentage of patients do not respond to the medication. Prospective, randomized, double-blind and placebo-controlled multicenter studies with finasteride were terminated after a two-year treatment period using 5 mg finasteride/day or placebo. The studies (SCARP, PROSPECT and PROWESS) included a total of 4222 patients and showed a 57% decreased risk of acquiring retention and a 34% decreased chance of requiring surgery because of clinical and symptomatic BPH. Finasteride may be most beneficial in preventing progression of prostatism. The prostate-specific antigen (PSA) level diminishes by approximately 50% during the first year of finasteride therapy and this reduction persisted throughout the three-year study. Finasteride, therefore, is remarkably safe but has very little effect on acute obstructive symptoms and patients must realize that it may take three to six months on the medication before benefits may be realized, if at all. It is likely that other medications with similar mechanisms of action will be developed. Whether they will be more effective compared than finasteride has yet to be seen (Marberger 1997; Nickel et al. 1996).

9.8 Effect of androgens in the initiation and promotion of prostatic cancer

The mechanism of action of androgens and other factors in the development of prostatic cancer is not yet exactly unterstood (Henderson et al. 1991). However, androgens play a distinct role. This is evidenced by the fact that prostate cancer almost never develops in eunuchs, in hypogonadotropic patients and in patients with 5α-reductase deficiency. In autopsy studies, patients with hepatic cirrhosis, a disease that decreases serum testosterone levels, have lower rates of latent prostatic cancer compared with controls. Furthermore, the regression of prostate cancer after androgen deprivation is well recognized and indicates that the majority of prostate cancer cells are partially androgen-dependent, at least initially.

Dietary fat may modulate prostate cancer risk via its effect on hormonal status (Graham et al. 1983). Dietary fat produces a promotional effect on sex steroids and the development of prostate cancer in genetically susceptible rats by reducing the induction interval for prostate cancer in response to testosterone (Pollard and Luckert 1986). Reduction of fat intake reduces the incidence of prostate cancer in free-living rats but not in germ-free rats. An age-associated decrease in testosterone concentration has been observed in free-living rats fed a reduced-calorie diet, suggesting that diet influences the reduction of testosterone associated with aging. It was observed that growth induction of the peripheral zone of the prostate by sex steroids in rats increases the incidence of prostate cancer in response to a carcinogen. These findings suggest that proliferation of the gland induced by androgens increases the incidence of prostate cancer in genetically susceptible experimental animals. Dietary fat and carcinogens enhance the induction process. Available data in humans are consistent with this observation from animal studies (Meikle and Smith 1990).

9.9 Influence of androgens in prostatic cancer growth

9.9.1 Epidemiology of prostate cancer

Age-adjusted incidence rates for carcinoma of the prostate have risen steadily in Europe and the United States over the last decades (Blair and Fraumeni 1978). Although hormonal, environmental, and genetic factors may contribute to the development of either incidental or clinically significant prostate cancer, specific epidemiologic causes have not been identified. Often carcinoma of the prostate is diagnosed at an advanced stage: nearly 30% to 40% of patients have metastatic disease at the time of diagnosis. Almost 80% of these men can be expected to die from carcinoma of the prostate. At least in these patients, early detection of prostate cancer may lead to therapy that could reduce mortality rates.

Epidemiologic studies of prostate cancer are able to provide significant information about distribution of the disease, age, race, nationality, and other factors of influence (Morton 1994). Evaluation of these data may identify variables important for the development of the disease, as well as potential preventive measures. Equally important, recognition of factors that imply a higher risk for the development of prostate cancer may allow identification of certain groups that should be targeted for early detection through screening studies.

After 50 years of age, both mortality and incidence rates from prostate cancer increase almost exponentially. Ninety-five per cent of cases of prostate cancer are dignosed in men between 45 and 89 years of age, with a median age at diagnosis of 72 years. Prostate cancer is the most frequent carcinomatous disease in men over 65 (Winkelstein and Ernster 1979).

Often, it is difficult to separate racial factors from other influences in epidemiologic studies of cancer. Culture, diet, and environment vary among ethnic or racial groups even within the same country. Nevertheless, there are striking differences in the incidence of prostatic cancer that are notable on a worldwide basis and among ethnic groups within the same country. Curiously, most studies suggest that the prevalence of clinically occult prostatic cancer found at autopsy is similar throughout all countries and racial groups. However, the rate of clinically apparent disease and prostatic cancer mortality rates differ widely (Fig. 9.4) (Meikle and Smith 1990).

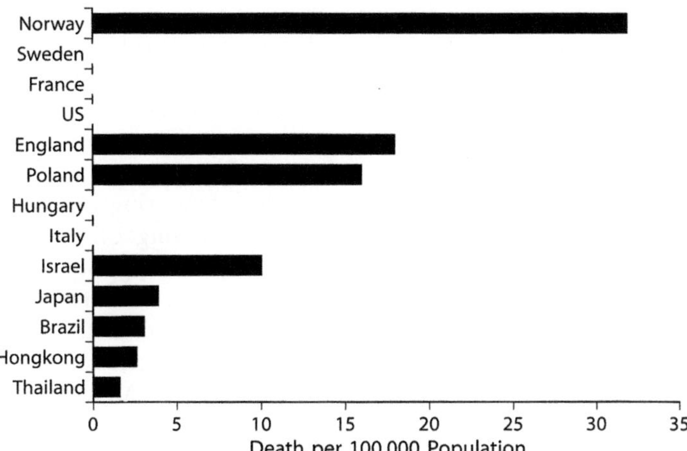

Fig. 9.4. Death rates of prostate cancer patients in different countries around the world. There is a significant difference in death rates between European and North American countries and countries in the far east

Migration studies have been used as evidence that environmental factors are important in the pathogenesis of prostate cancer. These studies indicate that if individuals migrate from a host population with a low rate of prostate cancer to an area with a high rate, the migrants finally reach the same rate of prostate cancer present in the new country. Studies showing the reverse effect are lacking because migrations from host countries with high rates to countries with low rates have not been reported.

There are reports on the incidence of prostate cancer in families. In first-order-relatives (father and brother) of men with proven prostate cancer there is a three-fold higher risk of dying from this tumor. Brothers of younger men with prostate cancer have a higher risk for the development of the disease than brothers of older men (Meikle et al. 1982, 1985).

As already mentioned in Section 9.8, dietary fat intake is particularly implicated in the pathogenesis of prostate cancer. According to international studies, fat consumption is directly correlated with mortality and incidence rates of prostate cancer. Particularly high correlation coefficients between prostate cancer mortality rates and per capita consumption of meat, milk and fats have been reported. Prostate cancer incidence rates among ethnic groups in different regions of the US correlate positively with dietary fat intake (Kolonel et al. 1981).

9.9.2 Age and androgen levels in relation to prostate cancer

There is a general agreement that plasma testosterone values decline with age. In regard to androgen plasma concentrations in patients with prostate cancer compared with age-matched controls, there are inconsistent reports claiming the testosterone plasma levels in prostate cancer patients can be elevated, unchanged or reduced (Meikle et al. 1989). Some of these inconsistences may be explained by circadian variations of testosterone values in individuals, by incorrect age-matching of control groups and/or by some effects of the tumor itself on the hormonal axis. Men with advanced stages of prostate cancer frequently present low plasma testosterone concentrations (Drafta et al. 1982). There is a correlation between testosterone values and prognosis, as men with low testosterone have a less pronounced response to hormonal treatment.

There are also reports from studies comparing hormone plasma levels of different races. Prostate cancer patients among US blacks and controls had higher plasma testosterone than a comparable group in Nigeria. In healthy college-age black, white and Japanese men in California higher total and free plasma testosterone concentrations were found in the young blacks than in whites or Japanese. These differences of androgen content might be found in the young population but diminish with age (Ahluwalia et al. 1981; Meikle and Smith 1990).

Genetic factors might also have a distinct influence on the plasma levels of sex steroids in men. These factors may influence the production rate of testosterone and DHT.

9.9.3 Androgen receptor and androgen sensitivity of cancer-related tissue

The increasing interest in androgen receptor (AR) research is based on the knowledge that androgen is required for the development and growth of prostate cancer. The process that changes prostate cancer from an androgen-dependent to an androgen-independent stage is mainly based on mutations or abnormal expression of AR in prostate cancer (Barrack and Tindall 1987). Recent studies have shown that mutant ARs can be generated that bind steroid but are nonfunctional or that do not bind but are in fact active. The hypothetic assumption might be very realistic that human prostate cancer contains AR gene mutations that change AR function and/or inactivate it which may affect the cancer cell phenotype. AR mutation in newly diagnosed and untreated organ-defined prostate cancer may be an indicator for progression to an androgen-independent stage (Trachtenberg and Walsh 1982). In a recent study presented by Chang et al. (1995) it has been shown that the AR transcriptional activity is enhanced in the presence of an AR co-activator (ARA_{70}). This fact may expand androgen activity in the prostate that AR alone cannot reach. This protein may also have the ability to interact with AR change during the progression of prostate cancer from an androgen-dependent to an androgen-independent stage. Interactions of co-activators with AR as well as mutations in the AR and/or AR gene might be involved in the development, growth and/or progression of prostate cancer (Yeh and Chang 1996).

9.9.4 Role of estrogens in prostatic cancer

Indirect evidence suggests that estrogens do not play a role in prostate cancer. The systemic administration of estrogens as treatment for metastatic disease exerts its effect primarily through inhibition of LH release from the pituitary gland. Alcoholic liver disease is associated with hyperestrogenism but lower incidence of prostate cancer. After the discovery of androgen deprivation as a therapeutic approach to prostate cancer by Huggins and Hodges in 1941, estrogen preparations were used over decades for therapeutic purposes in patients with advanced stages of prostate cancer (Veterans Administration Cooperative Urological Research Group 1970).

9.10 Increasing androgen insensitivity of prostate cancer and progression

The growth of any tissue, whether normal or cancerous, depends upon the quantitative relationship between the rate of cell proliferation and apoptosis. In normal adult tissues a balanced environment is responsible for non-continuous growth. It has been shown in different tissues that this balance is mainly regulated by a series of tissue-specific growth factors. Cancer cells have one fundamental feature insofar as they lose their normal responsiveness to these tissue-specific growth factors, with the result that continuous growth and apoptosis occur. Certain types of human cancers respond normally to growth factors (Aaronson 1991). This response might be the same to that of normal tissue where the tumor originates. In the prostate one of these growth factors regulating cell growth and cell death is androgen. Androgens stimulate cell proliferation and inhibit apoptosis. Prostate cancers may behave like normal prostatic tissue insofar as they retain normal ability for androgenic regulation of cell proliferation and cell death. These are then the tumors which respond to androgen deprivation initially. Around 70%–75% of all men with prostate cancer treated by androgen deprivation respond to this kind of therapy to varying degrees. The therapeutic response may differ from individual to individual, finally all subjects may come to a stage resistant to further anti-androgen treatment. In most prostate cancer patients increasing androgen insensitivity may lead to a tumor progression with evidence of metastatic disease (Baley et al. 1995; Ruizeveld de Winter et al. 1994; Sato et el. 1993; Tilley et al. 1995).

9.11 Hormone substitution in elderly men and prostate cancer

Adverse effects of testosterone therapy in older men have been assessed on the basis of data obtained over relatively short periods of study and in a limited number of men. The two areas of most concern in regard to adverse effects in the older age group are potentiation of cardiovascular disease and acceleration of benign and/or malignant prostate disease (Gooren and Poldermann 1990). Other areas of concern include water retention (including exacerbation of hypertension), hepatotoxicity, development of gynecomastia and polycythemia.

A portion of the increased cardiovascular disease risk in men compared to women may be due to androgens. Androgen effects on serum lipoproteins may play a role, but other factors, such as direct androgen effects on atherogenesis or modulation of vasoconstriction, may also be important. Studies of androgen therapy in older men have utilized parenteral therapy and have demonstrated that the major serum cholesterol profiles are not adversely affected (WHO 1992).

Prostatic cancer and benign protatatic hyperplasia are diseases of the aging male (Lunglmayer 1997). Androgens have been shown to have a role in the promotion of, and androgen deprivation has been used as treatment for, both of these diseases. Almost all short-term studies of androgen therapy in older men evaluating serum prostate specific antigen (PSA), prostate size, and/or urine flow parameters have shown no change in these parameters with therapy (Behre et al. 1994). Although short-term androgen therapy may seem to have no major immediate impact on the prostate in older men, experience may be too limited to infer long-term safety. However, reports from the literature show that short-term testosterone supplementation can also have some effect on lipoproteins, hematological parameters and PSA, so that older men should be screened carefully for testosterone supplementation and followed periodically throughout testosterone treatment (Nieschlag 1996). The mode of administration of the so-called right testosterone preparation is also important. This means a long-acting testosterone substance given in a form that guarantees constant release rates within the normal range, avoiding unnecessary and/or harmful testosterone peaks (Behre and Nieschlag 1992; Meikle et al. 1992; Partsch et al. 1995 and Chapter 10 by Nieschlag and Behre in this volume). Testosterone supplementation should be avoided in patients with prostate cancer independently of the stage of the disease as this steroid has an impact on cell growth in prostate cancer, even if the complete mechanism of action still requires some clarification (Tenover 1996).

9.12 Key messages

- The most important intraprostatic androgen is DHT arising from transformation of testosterone under the influence of 5α-reductase. DHT binds to the activated AR, the ARC is then transferred to the cell nucleus where the complex is prepared for mRNA synthesis and specific protein expression.
- Following castration only the androgen-dependent epithelial cells undergo apoptosis. Growth factors (EGF, TGF-β, FGF, IGF I/II) modulate prostate epithelial growth.
- The incidence of BPH may show regional differences. Symptoms caused by BPH do not necessarily increase with age. At present dangerous complications of BPH are fairly rare.
- Overproduction of adrenal androgens (DHEA-S, DHEA, androstenedione) may stimulate prostate growth, whereas under normal conditions the adrenals do not contribute significantly to prostatic growth. Following castration intraprostatic DHT decreases by 30% and additional adrenalectomy lowers DHT to nondectable values.
- Larger volumes of BPH are associated with higher levels of estradiol. In the human prostate there are multiple sites for estrogen receptors in the cytosol as well as in nuclear preparations. Plasma androgens, in synergism with increased estrogen, are responsible for persistent stimulation of BPH.
- 5α-reductase inhibitors induce androgen deprivation within prostate tissue. Multicenter clinical trials using such an inhibitor showed a 40% increase in maximum flow rate, 40%–50% improvement of symptom score, a 57% decreased risk of urinary retention and 34% decreased incidence of surgery.

- Androgens have a distinct role in the development of prostate cancer. Prostate cancer regresses after androgen deprivation. Dietary fat via sex steroids affects the development of prostate cancer
- Risk of prostate cancer in families (father-brother, brother-brother) is 3–4 times higher compared with an age-matched population.
- There are inconsistent reports from the literature indicating that testosterone levels in prostate cancer patients can be elevated, unchanged or reduced.
- The change from an androgen-dependent to an androgen-independent tumor is mainly based on mutations or abnormal expression of AR in prostate cancer. The transcriptional activity of AR is enhanced in the presence of AR co-activator (ARA_{70}).
- During androgen deprivation therapy the response declines and finally results in a stage resistant to anti-androgen treatment. Increasing androgen insensitivity may lead to tumor progression.
- Testosterone supplementation in older men provokes concern in regard to acceleration of benign and/or malignant prostate disease: Older men should be carefully screened and followed concerning testosterone supplementation therapy.

9.13 References

Aaronson SA (1991) Growth factors and cancer. Science 254:1146–1149

Ahluwalia B, Jackson MA, Jones GW (1981) Blood hormone profiles in prostate cancer patients in high-risk and low-risk populations. Cancer 48:2267–2273

Amelar RD (1962) Coagulation, liquefaction and viscosity of human semen. J Urol 87:187–190

Andersson S, Moghrabi N (1997) Physiology and molecular genetics of 17β-hydroxysteroid dehydrogenases. Steroids 62:143–147

Arcangeli CG, Ornstein DK, Keetch DW, Andriole GL (1997) Prostate specific antigen as a screening test for prostate cancer. The United States experience. Urol Clin North Am 24 (2):299–206

Babson AL, Read PA (1959) A new assay for prostatic acid phosphatase in serum. Am J Clin Pathol 32:88–91

Bahnson RR (1991) Elevation of prostate specific antigen from bacillus Calmette-Guérin-induced granulomatous prostatitis. J Urol 146:1368–1369

Baley PA, Yoshida K, Qian W, Sehgal I, Thompson TC (1995) Progression to androgen insensitivity in a novel in vitro mouse model for prostate cancer. J Steroid Biochem Mol Biol 52(5):403–413

Bangma CH, Blijenberg BG, Schoder FH (1995) Prostate-specific antigen: its clinical use and application in screening for prostate cancer. Scand J Clin Lab Invest Suppl (Norway) 221:35–44

Barrack ER, Tindall DJ (1987) A critical evaluation of the use of androgen receptor assays to predict the androgen responsiveness of prostatic cancer. Prog Clin Biol Res 239:155–187

Barry MJ (1990) Epidemiology and natural history of benign prostatic hyperplasia. Urol Clin North Am 17:495–499

Behre HM, Nieschlag E (1992) Testosterone buciclate (20-Aet-1) in hypogonadal men: Pharmacokinetics and pharamcodynamics of the new long-acting testosterone ester. J Clin Endocrinol Metab 75:1204–1210

Behre HM, Bohmeyer J, Nieschlag E (1994) Prostate volume in testosterone treated and untreated hypogonadal men in comparison to age-matched normal controls. Clin Endocrinol 40:341–349

Bergström S, Carlson LA, Weeks LR (1986) The prostaglandins: a family of biologically active lipids. Pharmacol Rev 20:1–48

Berry SJ, Coffey DS, Walsh PC, Ewing LL (1984) The development of human benign prostatic hyperplasia with age. J Urol 132:474–478

Berry SJ, Coffey DS, Strandberg JD, Ewing LL (1986) Effect of age, castration and testosterone replacement on the development and restoration of canine benign prostatic hyperplasia. Prostate 9:295–302

Bertrand G, Vladesco R (1921) Prostatic zinc concentration. C R Acad Sci [III] 173:176–179

Blair A, Fraumeni JF (1978) Geographic patterns of prostate cancer in the United States. JNCI 61:1379–1384

Boyle P, McGinn R, Maisonneuve P, La Vecchia C (1991) Epidemiology of benign prostatic hyperplasia: present knowledge and studies needed. Eur Urol 20 (1):3–10

Bruchovsky N, Dunstan-Adams E (1985) Regulation of α-reductase activity in stroma and epithelium of human prostate. In: Bruchovsky N, Chapdelaine A, Neumann F (eds) Regulation of androgen action. Bruckner, Berlin, pp 31–34

Bruchovsky N, Wilson JF (1968) The conversion of testosterone to 5α androstan-17β-ol-3-one by rat prostate in vivo and in vitro. J Biol Chem 243:2012–2121

Bruchovsky N, Lesser B, van Doorn E, Craven S (1975) Hormonal effects on cell proliferation in rat prostate. Vitam Horm 33:61–66

Byar DP(1974) Zinc in male sex accessory organs: distribution and hormonal response. In: Brandes D (ed) Male sex accessory organs, structure and function in mammals. Academic Press, New York, pp 161–171

Chang CS, Kokontis J, Liao ST (1988a) Molecular cloning of human and rat complementary DANN encoding androgen receptors. Science 240:324–326

Chang CS, Kokontis J, Liao ST (1988b) Structural analysis of complementary DANN and amino acid sequences of human and rat androgen receptors. Proc Natl Acad Sci USA 85(19):7211–7215

Chang CS, Saltzmann A, Yeh S, Young W, Keller E, Lee HJ, Wang C, Mizokami A (1995) Androgen receptor: an overview. Crit Rev Eukaryot Gene Expr 5 (2):97–125

Coffey DS (1985) Biochemistry and physiology of prostate and seminal vesicles. In: Walsh PC (ed) Campell's urology, 5th ed, Saunders, Philadephia, pp 1081–1121

Coffey DS, Walsh PC (1990) Clinical and experimental studies of benign prostatic hyperplasia. Urol Clin North Am 17:461–480

Cohen P, Peehl DM, Lamson G, Rosenfeld RG (1991) Insulin-like growth factors (IGFs), IGF receptors, and IGF-binding proteins in primary cultures of prostate epithelial cells. J Clin Endocrinol Metab 401–407

Deslypere JP, Vermeulen A (1981) Aging and tissue androgens. J Clin Endocrinol Metab 53:430–434

Drafta D, Proca E, Zamfir V (1982) Decreased steroids in benign prostatic hypertrophy and carcinoma of the prostate. J Steroid Biochem 17:689–693

Ebeling P, Koivisto VA (1994) Physiological importance of dehydroepiandrosterone. Lancet 343:1479–1481

Ekman P, Barrack ER, Greene GL, Jensen EV, Walsh PC (1983) Estrogen receptors in human prostate: Evidence for multiple binding sites. J Clin Endocrinol Metab 57:166–170

Eliasson R (1959) Studies on prostaglandins. Occurrence; formation and biological actions. Acta Physiol Scand 158 (Suppl) 46:1–9

Endoh A, Kristiansen SB, Casson RP, Buster JE, Hornsby PJ (1996) The zona reticularis is the site of biosynthesis of dehydroepiandrosterone and dehydroepiandrosterone sulfate in the adult human adrenal cortex resulting from its low expression of 3 β-hydroxysteroid dehydrogenase. J Clin Endocrinol Metab 81 (10):3558–3565

Fair WR, Parrish RT (1981) Antibacterial substance in prostatic fluid. In: Murphy GP, Sandberg AA, Karr JP (eds) Prostatic cell: Structure and function, part A. Liss, New York, pp 247–264

Fair WR, Wehner N (1976) The prostatic antibacterial factor: identity and significance. Prog Clin Biol Res 6:383–403

Fang Liu Gu (1993) The incidence of benign prostatic hyperplasia and prostatic cancer in China. Chin J Surg 31:323–326

Frick J (1974) Improved plasma testosterone assay by competitive protein binding and clinical availability. Invest Urol 12:27–29

Frick J (1994) Physiology and pathophysiology of prostate infection. In: Weidner W, Madsen PO, Schiefer HG (eds.): Prostatitis: etiopathology, diagnosis and therapy. Heidelberg, Springer Verlag, pp 11–22

Garraway WM, Collins GN, Lee RJ (1991) High prevalence of benign prostatic hypertrophy in the community. Lancet 338:469–471

Geller J (1990) Effect of finasteride, a 5α-reductase inhibitor on prostate tissue androgens and prostate-specific antigen. J Clin Endocrinol Metab 71 (6):1552–1555

Geller J, Fruchtmann B, Meyer C, Newman H (1967) Effect of progestational agents on gonadal and adrenal cortical function in patients with benign prostatic hypertrophy and carcinoma of the prostate. J Clin Endocrinol Metab 27 (4):556–560

George FW (1997) Androgen metabolism in the prostate of the finasteride treated adult rat: a possible explanation for the differential action of testosterone and 5α-dihydrotestosterone during development of the male urogenital tract. Endocrinology 13:871–877

Gooren LJG, Poldermann KH (1990) Safety aspects of androgen therapy. In: Nieschlag E, Behre HM (eds) Testosterone – action, deficiency, substitution. Springer, Berlin Heidelberg pp 182–197

Graham S, Harighey B, Marshall J, Priore R, Byres T, Rzepka T, Mettlin C, Pontes JE (1983) Diet in the epidemiology of carcinoma of the prostate gland. JNCI 70:687–692

Guess HA (1992) Benign prostatic hyperplasia antecedents and natural history. Epidemiol Rev 14:131–140

Guess HA (1995) Epidemiology and natural history of benign prostatic hyperplasia. Urol Clin of North Am 22:247–261

Hammond GL, Kontturi M, Vihko P, Vihko R (1978) Serum steroids in normal males and patients with prostatic diseases. Clin Endocrinol 9:113–121

Henderson BE, Ross RK, Pike MC (1991) Toward the primary prevention of cancer. Science 254:1131–1138

Huggins C, Hodges CV (1941) Studies on prostatic cancer I: The effect of castration, estrogen and androgen injection on serum phosphatases in metastatic carcinoma of the prostate. Cancer Res I:293–297

Huggins C (1947) The prostatic secretion. Harvey Lect 42:148–151

Huggins C, Stevens RA (1940) The effect of castration on benign hypertrophy of the prostate in man. J Urol 43:705–707

Isaacs JT (1983) Changes in dihydrotestosterone metabolism and the development of benign prostatic hyperplasia in the aging beagle. J Steroid Biochem 18:749–757

Isaacs JT (1987) Control of cell proliferation and cell death in the normal and neoplastic prostate: A stem cell model In: Rodgers CH, Coffey DS, Cunha G (eds): Benign prostatic hyperplasia, vol 2. Washington, DC, US Department of Health and Human Services, NIH Publication 87-2881, pp 85–94

Isaacs JT, Barrak ER, Isaacs WB, Coffey DS (1981) The relationship of cellular structure and function. The matrix systems. Prog Clin Biol Res 75A:1–24

Ishimaru T, Pages L, Horton R (1977) Altered metabolism of androgens in elderly men with benign prostatic hyperplasia. J Clin Endocrinol Metab 45:695–701

Kolonel LN, Hankin JH, Lee J (1981) Nutrient intakes in relation to cancer incidence in Hawaii. Br J Cancer 44:332–229

Labrie F, Luu-The V, Lin SX, Labrie C, Simard J, Breton R, Belanger A (1997) The key role of 17β-hydroxysteroid dehydrogenases in sex steroid biology. Steroids 62:148–158

Lepor H, Stone E (1995) Long-term results of medical therapies for benign prostatic hyperplasia. Curr Op Urol 5:18–21

Luke MC, Coffey DS (1994) The male sex accessory tissues: structure, androgen action and physiology In: Knobil E, Neil JD (eds) The physiology of reproduction, Raven Press, New York pp 1435–1480

Lunglmayer G (1997) Androgen deficiency in aging men. In: Geoffrey MH, Frick J, Baker GWH (eds) Current advances in andrology, Proceedings of the VIth International Congress of Andrology 1997, pp 289–292

Lytton B, Emery JM, Harvard BM (1968) The incidence of benign prostatic obstruction. J Urol 99(5):639–645

MacKenzie A, Hall T, Whitemore WF Jr (1962) Zinc content of expressed human prostatic fluid. Nature 193:72–73

Madsen PO, Baumueller A, Hoyne U (1978) Experimental models for determination of antimicrobials in prostatic tissue, interstitial fluid and secretion. Scand J Infect Dis (Suppl) 14:145–150

Marberger M (1997) PROWESS study, SIU-Congress Montreal, Canada

Martikainen P, Kyprianou N, Isaacs JT (1990) Effect of transforming growth factor-β 1 on proliferation and death of rat prostatic cells. Endocrinology 2963–2968.

McNeal JE (1981) The zonal anatomy of the prostate. Prostate 2:35–49

Meikle AW, Smith JA (1990) Epidemiology of prostate cancer. Urol Clin North Amer 17:709–718

Meikle AW, Stanish WM (1982) Familial prostatic cancer risk and low testosterone. J Clin Endocrinol Metab 54:1104–1108

Meikle AW, Smith JA, West DW (1985) Familial factors affecting prostatic cancer risk and plasma sex steroid levels. Prostate 6:121–128

Meikle AW, Smith JA, Stringham JD (1989) Estradiol and testosterone metabolism and production in men with prostate cancer. J Steroid Biochem 33:19–24

Meikle AW, Mazer NA, Moellner JF, Stringham JD, Tolman KG, Sander SW (1992) Enhanced transdermal delivery of testosterone across non scrotal skin produces physiological concentrations of testosterone and its metabolites in hypogonadal men. J Clin Endocrinol Metab 74:623–628

Mobbs BJ, Johnson LE, Connolly JG (1973) Influence of the adrenal gland on prostatic activity in adult rats. J Endocrinol 59:335–343

Mooradian AD, Morley JE, Korenman SG (1987) Biological actions of androgens. Endocr Rev 8:1–27

Morton RA Jr (1994) Racial differences in adenocarcinoma of the prostata in North American men. Urology (United States) 1994; 44/5:637–645

Netter FH (1976) The Ciba collection of medical illustration: genital organs. Thieme Verlag, Stuttgart pp 2–282

Newmark JR, Hardy DO, Tonb DC, Carter BS, Epstein JL, Isaacs WB, Brown TR, Barrack ER (1992) Androgen receptor gene mutations in human prostate cancer. Proc Natl Acad Sci USA 89:6319–6323

Nickel JC, Fradet Y, Boake RC, Pommerville PJ, Perreault JP, Afridi SK, Elhilali MM (1996) Efficacy and safety of finasteride therapy for benign prostatic hyperplasia. Results of a two-year randomized controlled trial (the PROSPECT-Study). Can Med Assoc J 155(9):1251–1259

Nieschlag E (1996) Testosterone replacement therapy: something old, something new ... Clin Endocrinol (Oxf) 45 1996; 45/3:261–262

Oesterling JE, Epstein JI, Walsh PC (1986) The inability of adrenal androgens to stimulate the adult prostate – An autopsy evaluation of men with hypogonadotropic hypogonadism and panhypopituitarism. J Urol, 136:103–104

Partin AW, Oesterling JE, Epstein JI (1991) Influence of age and endocrine factors on the volume of benign prostatic hyperplasia. J Urol 145:405–409

Partsch CJ, Weinbauer GF, Fang R, Nieschlag E (1995) Injectable testosterone undecanoate has more favorable pharmacokinetics and pharmacodynamics than testosterone enanthate. Eur J Endocrinol 132:514–519

Peters CA, Walsh PC (1987) The effect of nafarelin acetate, a luteinizing hormone-releasing hormone agonist, on benign prostatic hyperplasia. N Engl J Med 317:599–604

Pollard M, Luckert PH (1986) Promotional effects of testosterone and high fat diet on the development of autochthonous prostate cancer in rats. Cancer Lett 32:223–227

Ruizeveld de Winter JA, Janssen PJ, Sleddens HM, Verleun-Mooijman MC, Trapman J, Brinkman AO, Santerse AB, Schröder FH, van der Kwast TH (1994) Androgen receptor status in localized and locally progressive hormone refractory human prostate cancer. Am J Pathol 144 (4):735–746

Sato N, Suzuki H, Shimazaki J (1993) Mechanism of loss of androgen dependency in androgen-dependent tumor. Hum Cell 6(3):170–175

Scott WW (1945) Lipids of prostatic fluid, seminal plasma and enlarged prostate gland of man. J Urol 53:712–718

Sherwood ER, Fong CJ, Lee C, Kozlowski JM (1992) Basic fibroblast growth factor: a potential mediator of stromal growth in the human prostate. Endocrinology 2955–2963

Speights VO Jr, Brawn PN (1996) Serum prostate specific antigen levels in non-specific granulomatous prostatitis. Br J Urol 77(3):408–410

Stamey TA, Fair WR, Timothy MM, Wehner N (1968) Antibacterial nature of prostatic fluid. Nature 218:44–415

Stoner E (1992) The clinical effects of a 5α-reductase inhibitor, finasteride, on benign prostatic hyperplasia. The Finasteride Study Group. J Urol 147 (5):1298–1302

Stoner E (1994) Three year safety and efficacy data on the use of Finasteride in the treatment of benign prostatic hyperplasia. Urology 43:284–289

Tauber PF, Zaneveld LJD (1976) Coagulation and liquefaction of human semen. In: Hafez ESE (ed) Human semen and fertility regulation in men. Mosby, St Louis, pp 153–166

Tchetgen MB, Oesterling JE (1997) The effect of prostatitis, urinary retention, ejaculation, and ambulation on the serum prostate-specific antigen concentration. Urol Clin North Am 24(2):283–291

Tenover JS (1996) Androgen therapy in aging men. In: Bahasin S, Gabelnick HL, Spieler JM, Swerdloff RS, Wang C (eds). Pharmacology, biology, and clinical applications of androgens: current status and future prospects. Wiley Liss, New York, Chichester pp 309–319

Tenover JS (1997) Androgen deficiency in aging men. In: Geoffrey MH, Frick J, Baker GWH (eds) Current advances in andrology, Proceedings of the VI [th] International Congress of Andrology 1997, pp 285–288

Tilley WD, Bentel MJ, Aspinall JO, Hall RE, Horsfall DJ (1995) Evidence for a novel mechanism of androgen resistence in the human prostate cancer cell line, PC-3. Steroids 60 (1):180–186

Trachtenberg J, Hicks LL, Walsh PC (1980) Androgen and estrogen receptor content in spontaneous and experimentally induced canine prostatic hyperplasia. J Clin Invest 65:1051–1059

Trachtenberg J, Walsh PC (1982) Correlation of prostatic nuclear androgen receptor content with duration of response and survival following hormonal therapy in advanced prostatic cancer. J Urol 127:466–470

Träger L (1977) Steroidhormone: Biosynthese, Stoffwechsel, Wirkung. Springer Verlag Berlin, pp 164–197

Vermeulen A (1991) Clinical review 24: Androgens in the aging male. J Clin Endocrinol Metab 73:221–224

Veterans Administration Cooperative Urological Research Group (1970) Estrogen treatment for cancer of the prostate: Early results with three doses of diethylstilbestrol and placebo. Cancer 26:257–262

von Euler US (1934) Zur Kenntnis der pharmakologischen Wirkungen von Nativsekreten und Extracten männlicher accessorischer Geschlechtsdrüsen. Arch Pathol Pharmakol 175:78–84

Walsh PC, Wilson JD (1976) The induction of prostatic hypertrophy in the dog with androstanediol. J Clin Invest 57:1093–1097

Wang MC, Valenzuela LA, Murphy GP, Chu TM (1979) Prostate antigen: a new potential marker for prostatic cancer. Prostate 2:89–96

Wang MC, Papsidero LM, Kuriyama M, Valenzuela LA, Murphy GP, Chu TM (1981) Prostate antigen: a new potential marker for prostatic cancer. Prostate 2:89–96

White IG, Darin-Bennett A, Poulos A (1976) Lipids of human semen. In: Hafez ESE (ed) Human semen and fertility regulation in men. Mosby, St Louis, pp 144–152

Winkelstein W Jr, Ernster VL (1979) Epidemiology and etiology. In Murphy GP (ed) Prostatic cancer. Littleton, Massachusetts, PSG Publishing, pp 1–17

World Health Organization (1992) Nieschlag E, Wang C, Handelsman DJ, Swerdloff RS, Wu FCW, Einer-Jensen N, Khanna J, Waites GMH (eds) Guidelines for the use of androgens. WHO, Geneva.

Yeh S, Chang CS (1996) Cloning and characterization of a specific coactivator, ARA70 for the androgen receptor in human prostate cells. Proc Natl Acad Sci USA 93:5517–5521

10 Pharmacology and clinical uses of testosterone

Eberhard Nieschlag and Hermann M. Behre

Contents

10.1 Historical development of testosterone therapy

The first experimental proof that the testes produce a substance responsible for virility was provided by Berthold (1849). He transplanted testes from roosters into the abdomen of capons and recognized that the animals with the transplanted testes behaved like normal roosters: "They crowed quite considerably, often fought among themselves and with other young roosters and showed a normal inclination to hens". Berthold concluded that the virilizing effects were exerted by testicular secretions reaching the target organs via the bloodstream. Berthold's investigation is generally considered the origin of experimental endocrinology (Simmer and Simmer 1961). Following his observation various attempts were made to use testicular preparations for therapeutic purposes. The best known experiments are those by Brown-Séquard (1889), who tried testis extracts on himself (which can at best have had placebo effects). The first testicular extracts with demonstrable biological activity were prepared by Loewe and Voss (1930) using the seminal vesicle as a test organ. Finally, the groundstone for modern androgen therapy was laid when steroidal androgens were first isolated from urine by Butenandt (1931), testosterone was obtained in crystalline form from bull testes by David et al. (1935) and testosterone was chemically synthesized by Butenandt and Hanisch (1935) and Ruzicka and Wettstein (1935).

Immediately after its chemical isolation and synthesis testosterone was introduced into clinical medicine (unthinkable if it had happened today) and used for the treatment of hypogonadism. Since testosterone was ineffective orally it was either compressed into pellets and applied subcutaneously (see Chapter 12 by Handelsman, this volume) or was used as 17α-methyltestosterone. In the 1950s longer acting injectable testosterone esters (Junkmann et al. 1957) became the preferred therapeutic modality. In the 1950s and 1960s chemists and pharmacologists concentrated on the chemical modification of androgens in order to emphasize their erythropoietic or anabolic effects (Kopera 1985). These preparations never played an important role in the treatment of hypogonadism. In the late 1970s the orally effective testosterone undecanoate was added to the spectrum of testosterone preparations clinically used (Coert et al. 1975; Nieschlag et al. 1975). Finally, transdermal testosterone applied either through scrotal skin (Bals-Pratsch et al. 1986) or non-scrotal skin (Mazer et al. 1992) was introduced into clinical practice in 1994 and 1995 respectively, first in the USA and later also in other countries.

10.2 General considerations

Although testosterone has been in clinical use for over 60 years, it has never received much interest from clinical research. This is partly due to the fact that hypogondal men requiring testosterone treatment constitute only a small minority of all patients and hypogonadism is not a life-threatening disease. Since development of new preparations is mainly a task of the pharmaceutical industry and hypogonadal patients did not promise to contribute a substantial economic profit, development of testosterone preparations was slow. Only recently has the question of testosterone treatment of senescent men and, to a certain extent also the search for a hormonal male contraceptive spurred interest in the pharmacology and application of testosterone

Today oral, injectable and transdermal testosterone preparations are available for clinical use. There are practically no studies available comparing the various preparations with the goal of identifying the optimal preparation for substitution purposes. While the older injectable preparations, which are still the predominant form for substitution, produce supraphysiological serum testosterone levels, the newer preparations achieve levels in the physiological range. We are only beginning to understand which serum levels are required to achieve the various biological effects of testosterone and to avoid untoward side-effects. In particular, very little is known about long-term effects of testosterone therapy caused by different preparations. Under these circumstances it appears that the consensus reached by a Workshop Conference on Androgen Therapy organised jointly by WHO, NIH and FDA in 1990 still provides the best therapeutic guidelines: "The consensus view was that the major goal of therapy is to replace testosterone levels at as close to physiologic concentrations as is possible" (WHO 1992). Until other evidence is provided, all testosterone preparations will best be judged by this principle.

Another important question is which androgen preparation should be used for clinical purposes. Numerous androgenic steroids have been synthesized and used clinically in the past. The synthetic androgens were produced with the aim to enhance selectively certain aspects of testosterone activity e.g. the anabolic effect on muscles or the hematopoietic effect. Some of these molecules proved to have toxic side-effects in particular upon long-term use (as required for substitution of hypogonadism) or the desired effects were never proven in controlled clinical trials (as advocated by evidence-based medicine). In addition, some of these steroids could not be converted to 5α-DHT or estrogen as is testosterone and therefore cannot develop the full spectrum of acitivities of testosterone. The important biological significance of these conversions is described in Chapter 1 (by Rommerts) and 2 (by Quigley) of this volume. For these reasons, the synthetic preparations have almost disappeared from the market and testosterone as produced naturally is the prevailing androgen used in clinical medicine. In its various preparations testosterone has been on the market for over 6 decades and as one of the oldest "drugs" in clinical use has demonstrated its high safety. However, new insights into the molecular mechanisms of androgen action may lead to the development of steroids suited for specific purposes. As an example, 7α-methyl-19-nortestosterone may be quoted, as it is experiencing a renaissance due to its high androgenicity combined with low prostatotropic effects shown in animal experiments (Kumar et al. 1997; Suvisaari et al. 1997). Whether such steroids may become useful and safe for clinical use remains to be seen.

This chapter provides an overview of the various conventional and new testosterone preparations and discusses the clinical use of testosterone. Some aspects will be expanded in the following chapters.

10.3 Pharmacology of testosterone preparations

As all other androgens, testosterone derives from the basic structure of androstane. This molecule consists of three cyclohexane and one cyclopentane ring (perhydrocyclopentanephenanthrene ring) and a methyl group each in position 10 and 13. Androstane itself is biologically inactive and obtains activity through oxygroups in position 3 and 17. Testosterone, the quantitatively most important androgen synthesized in the organism, is characterized by an oxo group in position 3, a hydroxy group in position 17 and a double bond in position 4 (Fig. 10.1).

To make testosterone therapeutically effective three approaches have been used:
1) different routes of administration,
2) esterification in position 17, and
3) chemical modification of the molecule.

In addition, these approaches have been combined. Since of practical clinical relevance, the route of administration is used here for categorizing the various testosterone preparations (overview in Table 10.1).

10.3.1 Oral administration

10.3.1.1 Free testosterone

Free unesterified testosterone as physiologically secreted by the testes would appear to be the first choice when considering substitution therapy. When ingested orally in the free form testosterone is absorbed well from the gut but is effectively metabolized and inactivated in the liver before it reaches

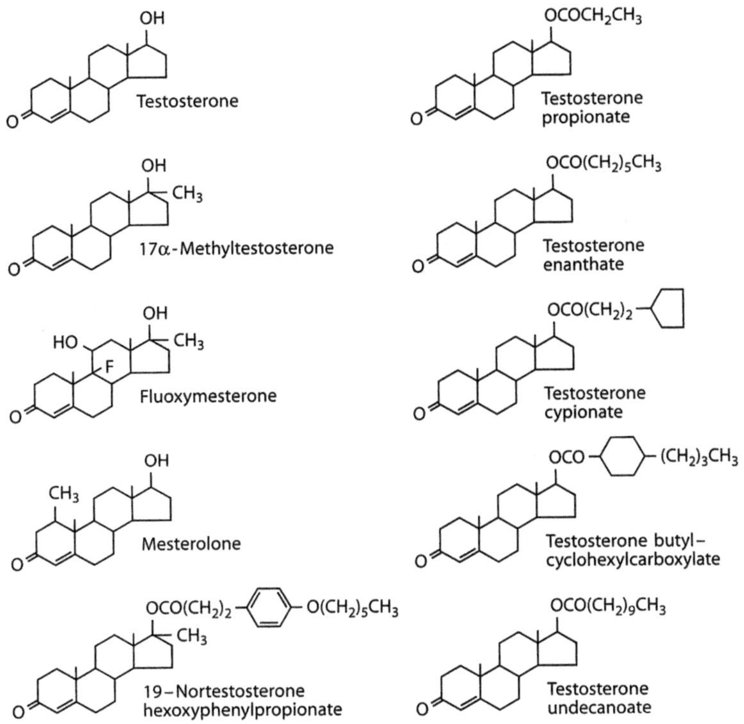

OH
Testosterone

OH
CH_3
17α-Methyltestosterone

OH
HO
CH_3
F
Fluoxymesterone

OH
CH_3
Mesterolone

$OCO(CH_2)_2$—O$(CH_2)_5CH_3$
CH_3
19–Nortestosterone
hexoxyphenylpropionate

$OCOCH_2CH_3$
Testosterone
propionate

$OCO(CH_2)_5CH_3$
Testosterone
enanthate

$OCO(CH_2)_2$
Testosterone
cypionate

OCO—$(CH_2)_3CH_3$
Testosterone butyl–
cyclohexylcarboxylate

$OCO(CH_2)_9CH_3$
Testosterone
undecanoate

Fig. 10.1. Molecular structure of testosterone and clinically used testosterone esters and derivatives

the target organs ("first-pass-effect"). Only when a dose of 200 mg is ingested, which exceeds 30 fold the amount of testosterone produced daily by a normal man, the metabolizing capacity of the liver is overruled. With such doses an increase in peripheral testosterone blood levels becomes measurable and clinical effects can be observed (Daggett et al. 1978; Johnsen et al. 1974; Nieschlag et al. 1975, 1977). The testosterone-metabolizing capacity of the liver, however, is age- and sex-dependant. An oral dose of 60 mg free testosterone does not affect peripheral testosterone levels in normal adult men, but produces a significant rise in prepubertal boys and women (Nieschlag et al. 1977). This demonstrates that testosterone induces liver enzymes responsible for its own metabolism (Johnsen et al. 1976). When the liver is severely damaged its metabolizing capacity decreases. Thus, in patients with liver cirrhosis a dose of 60 mg testosterone (ineffective in normal men) produces high serum levels (Nieschlag et al. 1977).

Since hypogonadal men usually have normal liver function 400–600 mg testosterone must be administered daily if the patient is to be substituted by oral testosterone (Johnsen 1978; Johnsen et al. 1974). Besides the fact that such high steroid concentrations are uneconomical, the possibility of toxic side-effects of such huge testosterone doses cannot be excluded, especially when given over long periods of time as required for substitution therapy.

Table 10.1. Mode of application and dosage of various testosterone preparations

Preparation	Route of application	Full substitution dose
In clinical use		
Testosterone enanthate	Intramuscular injection	200–250 mg every 2-3 weeks
Testosterone cypionate	Intramuscular injection	200 mg every 2 weeks
Testosterone undecanoate	Oral	2–4 capsules à 40 mg per day
Transdermal testosterone patch	Scrotal skin	1 membrane per day
Transdermal testosterone patch	Non-scrotal skin	1 or 2 systems per day
Testosterone implants	Implantation under the abdominal skin	3–6 implants à 200 mg every 6 months
Under development		
Testosterone cyclodextrin	Sublingual	2,5–5,0 mg twice daily
Testosterone undecanoate	Intramuscular injection	1000 mg every 8–10 weeks
Testosterone buciclate	Intramuscular injection	1000 mg every 12–16 weeks
Testosterone microspheres	Intramuscular injection	315 mg for 11 weeks
Obsolete		
17α-Methyltestosterone	Oral	(25–50 mg per day)
Fluoxymesterone	Sublingual	(10–25 mg per day)
	Oral	(10–20 mg per day)

However, in a small group of patients treated for as long as seven years with oral testosterone such side-effects have not been observed (Johnson 1978). Nevertheless, oral administration of free testosterone has not become a generally accepted method for therapeutic purposes.

As a relict of experiments performed last century (see 10.1), preparations containing animal testis extracts or dried organ powder are still being manufactured and are available on the market. Although synthesized in the testis, the testosterone content of these preparations is fairly low since the testis, in contrast to other endocrine glands (such as the thyroid), does not store its hormonal products. Moreover, the testosterone in these orally consumed products cannot become effective for the reasons described above. Such preparations may at best exert placebo effects and do not belong to a rational therapeutic repertoire.

10.3.1.2 17α-Methyltestosterone

Several attempts have been made to modify the testosterone molecule by chemical means in order to render it orally effective, i.e. to delay metabolism in the liver. In this regard, the longest known testosterone derivative is 17α-methyltestosterone (Ruzicka et al. 1935) which is a fully effective oral androgen preparation. 17α-methyltestosterone is quickly absorbed and maximal blood levels are observed 90 to 120 minutes after ingestion. The half-life in blood amounts to approximately 150 minutes (Alkalay et al. 1973).

Ever since this steroid was introduced for clinical use, repeated reports about toxic side-effects such as an increase in serum liver enzymes (Carbone et al. 1959), cholestasis of the liver (de Lorimer et al. 1965; Werner et al. 1950), and peliosis of the liver (Westaby et al. 1977) have appeared. It is of interest that humans are more susceptible to the hepatotoxic effects of methyltestosterone than rats (Heywood et al. 1977a) or dogs (Heywood et al. 1977b). Later on, an association between long-term methyltestosterone treatment and liver tumors was found (Farrell et al. 1975; Goodman and Laden 1977; Boyd and Mark 1977; Paradinas et al. 1977; Coombes et al. 1978; Falk et al. 1979; Bird et al. 1979; McCaughan et al. 1985). While these side-effects appear to be clearly related to methyltestosterone application, the isolated observation of a seminoma in a 36-year old man on high-dose methyltestosterone seems rather incidental (Vogelzang et al. 1986).

The side-effects are due to the alkyl group in the 17α-position and have also been reported for other steroids with this configuration (Krüskemper and Noell 1967). Because of the side-effects methyltestosterone should no longer be used therapeutically, in particular since effective alternatives are available (Nieschlag 1981). The German Endocrine Society declared methyltestosterone obsolete in 1981 and the German Federal Health Authority ruled that methyltestosterone should be withdrawn from the market (Methyltestosterone 1988). In other countries, however, methyltestosterone is still in use, a practice which should be terminated.

10.3.1.3 Fluoxymesterone

The androgenic activity of fluoxymesterone was enhanced over that of testosterone by the introduction of fluorine and the addition of a hydroxy group into the steroid skeleton of testosterone. This substance also contains a 17α-methyl group and accordingly there is a risk of hepatotoxicity with long-term use. Therefore, this androgen has disappeared from the market.

10.3.1.4 Mesterolone

Mesterolone can be considered a derivative of the 5α-reduced testosterone metabolite 5α-dihydrotestosterone (DHT) which is protected from fast metabolism in the liver by a methyl group in position 1 (Gerhards et al. 1966) and thus becomes orally active. It is free of liver toxicity. Unlike testosterone mesterolone cannot be metabolized to estrogens (Breuer and Gütgemann 1966) and acts, like DHT, on a molecular level. It has only limited effects in suppressing pituitary gonadotrophin secretion (Aakvaag and Stomme 1974; Gordon et al. 1975). It can only be considered a weak or partially active androgen. Altogether, mesterolone is not suited for the substitution of hypogonadism.

10.3.1.5 Testosterone undecanoate

When testosterone is esterified in the 17β-position with a long aliphatic side chain such as undecanoic acid and given orally, its route of absorption from the gastrointestinal tract is shifted from the vena portae to the lymph and reaches the circulation via the ductus thoracicus (Coert et al. 1975; Horst et al. 1976). Absorption is improved if the ester is taken in arachis oil (Nieschlag et al. 1975) and with a meal (Frey et al. 1979). After oral ingestion of a 40 mg capsule, of which 63% i.e. 25 mg is testosterone, maximum serum levels are reached 2 to 6 hours later (Nieschlag et al. 1975; Schürmeyer et al. 1983). Thus, with 2 to 4 capsules (80 to 160 mg) per day substitution of hypogonadism can be achieved. (For further pharmacokinetic considerations see Chapter 11 by Behre and Nieschlag in this volume).

Along with injectable testosterone esters, oral testosterone undecanoate belongs to the standard repertoire for the treatment of hypogonadism, although the widely fluctuating serum levels and the relavtively short-lived serum testosterone peaks make this type of therapy less than ideal.

10.3.2 Sublingual application

17α-methyltestosterone was found to be more effective when applied sublingually than when ingested orally (Escamilla 1949). This type of substitution should, however, not be practised because of the liver toxicity of methyltestosterone summarized above.

The solubility of the hydrophobic testosterone molecule can be enhanced by incorporation into hydroxypropyl-β-cyclodextrins (Pitha et al. 1986) which are macro-ring structures consisting of cyclic oligosaccharides. When testosterone incorporated into such cyclodextrins is administered sublingually steep increases in serum testosterone occur lasting for one or two hours (Stuenkel et al. 1991). Hypogonadal men treated with three daily doses for 60 days showed improvement of their condition (Salehian et al. 1995; Wang et al. 1996b). This is an interesting new approach to testosterone substitution, but unless more constant serum levels can be achieved this therapy would require repeated daily applications and would have the same disadvantages as conventional oral testosterone undecanoate therapy.

10.3.3 Rectal application

In order to avoid the first-pass effect of the liver, testosterone can be applied rectally in suppositories (Hamburger 1958). Administration of a suppository containing 40 mg testosterone results in an immediate and steep rise of serum testosterone lasting for about four hours. Effective serum levels can be achieved by repeated applications (Nieschlag et al. 1976). This therapy, however, never gained much popularity probably because the patients find it unacceptable to use suppositories three times daily on a long-term routine basis.

10.3.4 Nasal application

The first-pass effect of the liver can also be avoided by applying testosterone to the nasal mucosa (Danner and Frick 1980). However, unreliable absorption patterns and short-lived serum peaks prevent this form of application from becoming a desirable option for long-term substitution therapy and it has never passed the experimental state.

10.3.5 Intramuscular application

10.3.5.1 Testosterone esters

The most widely-used testosterone substitution therapy is the intramuscular injection of testosterone esters. While free unesterified testosterone has a half-life of only ten minutes and would have to be injected very frequently, testosterone esters have a prolonged half-life. To a certain degree the length of the ester moiety side chain determines the duration of action. Pharmacokinetic details of the esters will be discussed in Chapter 11 by Behre and Nieschlag. In short, for substitution purposes *testosterone propionate* must be injected every two to three days while *testosterone enanthate* when given in doses of 200 to 250 mg allows spacing of the injections at about two-week intervals. Two other clinically available testosterone esters, *testosterone cypionate* and *testosterone cyclohexanecarboxylate* have very similar kinetic properties to enanthate so that they can be used in the same doses and intervals (Gooren 1987; Nieschlag et al. 1976; Schulte-Beerbühl and Nieschlag 1980; Snyder and Lawrence 1980; Schürmeyer and Nieschlag 1984; Sokol et al. 1982).

The disadvantage of all these esters is that they produce initially supraphysiological testosterone levels which may exceed normal levels severalfold and then slowly decline, so that before the next injection pathologically low levels may be reached. The patients recognize these ups and downs of testosterone levels in parallel variations of the general well-being, sexual activity and emotional stability. Despite these disadvantages testosterone enanthate and cypionate are still the standard therapy for male hypogonadism.

Because of these shortcomings of the available esters the World Health Organization (WHO) initiated a steroid synthesis programme (Crabbé et al. 1980) out of which a series of new testosterone esters was developed. When tested in small laboratory rodents a specific ester was identified that showed greatly prolonged activity, namely *testosterone-trans-4-n-butylcyclohexyl-carboxylate*, free name *testosterone buciclate*. This ester was then tested in castrated monkeys and found to raise testosterone serum levels of the animals into the normal range for about four months when a single dose of 40 mg was injected. The same amount of testosterone given as testosterone enanthate produced supraphysiological serum testosterone levels for eight days which returned into the subnormal range by three weeks (Rajalakshmi and Ramakrishnan 1989; Weinbauer et al. 1986). In a further preclinical study in monkeys testosterone buciclate in combination with a GnRH antagonist proved very successful in achieving azoospermia (Weinbauer et al. 1989). In a phase-I clinical study with hypogonadal men single injections of 600 mg testosterone buciclate produced serum testosterone levels in the normal range for 12 weeks (Behre and Nieschlag 1992). In a first clinical trial for hormonal male contraception single injections of 1000 mg showed a similarly long duration of action and were well tolerated by normal volunteers (Behre et al. 1995). Supraphysiologic levels were not observed at any point. Therefore, testosterone buciclate has the longest duration of action of any injectable ester tested so far and appears to have a great potential for clinical use. Further studies are awaited with great interest.

In China *testosterone undecanoate* dissolved in teaseed oil was first used for intramuscular injections and satisfactory results for the substitution of hypogonadism were reported (Wang et al. 1991). When tested in monkeys a half-life considerably longer than that for testosterone enanthate was found (Partsch et al. 1995). Ensuing phase-I and phase-II trials confirmed these findings so that testosterone undecanoate injections of 1000 mg (in castor oil) may allow injection intervals of up to 8 weeks (see Chapter 11 by Behre and Nieschlag, this volume). Thus, next to testosterone buciclate, the undecanoate ester appears to be a promising injectable preparation currently undergoing clinical testing.

10.3.5.2 Testosterone microspheres

Drugs can be incorporated into biodegradable microspheres. Such drug-loaded microspheres when injected intramuscularly provide controlled release of the substance for several weeks or even months. As an example, microencapsulated GnRH agonists have become a valuable modality in the treatment of prostatic carcinoma. Testosterone has been incorporated into poly (DL-lactide-co-glycolide) microspheres. When first tested in castrated monkeys single injections resulted in an elevation of serum levels above the lower limit of normal for several months (Asch et al. 1986). When similar microsphere injections containing 315 mg of testosterone were given to eight hypogonadal men serum testosterone levels slowly increased to peak levels at about eight weeks and fell thereafter to reach pathological levels again by 11 weeks (Burris et al. 1988). In a later study the size-range and the testosterone loading of the microspheres were adjusted so that in hypogonadal men single intramuscular injections resulted in relatively constant serum levels within the normal range for about 70 days (Bhasin et al. 1992). These two clinical studies demonstrated that the microspheres can be adapted to the required needs and the results were encouraging. However, no further development has occurred since, probably due to problems with stability and reproducibility of the microspheres (Bhasin and Swerdloff 1996).

10.3.6 Subdermal implants

Shortly after the chemical synthesis of testosterone it was used clinically in the form of implants. For this purpose testosterone was compressed into short rods which were then implanted subcutaneously and lasted for several weeks or months. With the advent of other modalities they went out of general use. However, recent investigations found favourable pharmacokinetic profiles with these implants (Handelsman et al. 1990; Jockenhövel et al. 1996) so that there is renewed interest (Nieschlag 1996). In this volume, this kind of substitution is discussed in Chapter 12 by Handelsman.

The use of testis-shaped testosterone-loaded implants were described by Japanese investigators (Kaetsu et al. 1988). The implants were prepared from 10 g vinyl monomer and 6.4 g testosterone by radiation-induced polymerisation. These artificial testes were placed in the scrotum of orchidectomized patients and provided serum testosterone levels sufficient for substitution for over a year. The implants serve the double function of a cosmetic and an endocrine prothesis. An analysis of the kinetic profiles more detailed than provided in the publications would be desirable to further explore this modality. However, further publications could not be traced and it proved impossible to contact the authors.

10.3.7 Transdermal testosterone

The skin easily absorbs steroids and other drugs and transdermal drug delivery has become a widely used therapeutic modality. Transdermal substitution of estradiol is now one of the general methods for treating ovarian insufficiency. For this purpose, however, only micrograms of estradiol are required, whereas milligrams of testosterone are necessary for treatment of male hypogonadism. Such large amounts of androgens can only be administered through the skin of the trunk if large areas are covered with androgens. Moreover, the dosage is difficult to control and accidental person-to-person transfer may occur as shown by androgenization of the female partners of men using androgen-containing creams (Delanoe et al. 1984). This kind of androgen treatment using DHT-containing creams is presented in Chapter 15 of this book by Schaison and Conzinet.

Different areas of the skin, however, show different rates of steroid absorption – the scrotum has the highest rate, about 40-fold higher than the forearm (Feldmann and Maibach 1967). This difference in absorption rates has been exploited for the development of a transdermal therapeutic system (TTS) to deliver testosterone. 40 and 60 cm^2 large polymeric membranes loaded with 10 or 15 mg testosterone, when attached to the scrotal skin, deliver sufficient amounts of the steroid to provide hypogonadal men with serum levels in the physiological range (Bals-Pratsch et al. 1986; Findlay et al. 1987; Korenmann et al. 1987). The membranes need to be renewed every day. When applied in the morning and worn until the next morning the resulting serum testosterone levels resemble the normal diurnal variations of serum testosterone in normal men without supraphysiological peaks (Bals-Pratsch et al. 1988). More pharmacokinetic details and clinical experience with transscrotal testosterone therapy are provided in Chapter 13 by Atkinson in this book.

While testosterone is readily absorbed by genital skin, transdermal systems for use on non-genital skin require enhancers to facilitate sufficient testosterone passage through the skin. Such systems have recently become available for clinical use and are described in detail in Chapter 14 by Meikle and in Chapter 13 by Atkinson in this volume. If one or two such systems are worn for 24 hours physiologic serum testosterone levels can be mimicked, as with transscrotal patches (Brocks et al. 1996; Meikle et al. 1996). However, skin reactions occur at a higher rate due to the alcoholic enhancer used and the occlusive nature of the systems (Jordan 1997).

Nevertheless, both transdermal modalities through either scrotal or non-genital skin provide the most physiologic serum testosterone levels in comparison with all other available preparations.

10.4 Use of testosterone in male hypogonadism

The prime indication for testosterone is substitution therapy of male hypogonadism. An overview of the syndromes is provided in Table 10.2. Hypogonadism may be caused by hypothalamic, pituitary, testicular or target organ lesions (for a detailed description the reader is referred to the textbook by Nieschlag and Behre 1997). The clinical symptoms of all syndromes and disease entities are predominantly due to a lack of testosterone or its action. The most frequent disorders requiring testosterone substitution are Klinefelter syndrome, Kallman syndrome, idiopathic hypogonadotropic hypogonadism (IHH), anorchia and pituitary insufficiency. Some disorders such as varicocele, orchitis, maldescended testes and Sertoli-cell-only syndrome may not, or only eventually require testosterone substitution. Although discrete endocrine alterations may be noted by laboratory tests in these patients, the endocrine capacity of the Leydig cells remains high enough to maintain serum testosterone in the lower physiological range. In order to achieve fertility in patients with hypothalamic (IHH) or pituitary insufficiency, treatment with gonadotropins (hCG/hMG) or pulsatile GnRH may be required temporarily (e.g. Kliesch et al. 1994). Once a pregnancy has been induced these patients will go back on testosterone substitution. Cases with hypogonadism of testicular origin in whom infertility cannot be treated require testosterone substitution continuously. In all these patients testosterone substitution is a lifelong therapy.

Table 10.2. Overview of disorders with male hypogonadism (from Nieschlag and Behre 1997)

Hypothalamic-pituitary origin (hypogonadotropic syndromes = secondary hypogonadism)
- Idiopathic hypogonadotrophic hypogonadism (IHH) including Kallman syndrome
- Prader-Labhart-Willi syndrome
- Laurence-Moon-Biedl syndrome
- Constitutional delay of puberty
- Pituitary insufficiency/adenomas
- Pasqualini syndrome
- Hyperprolactinemia
- Hemochromatosis

Testicular origin (hypergonadotropic syndromes = primary hypogonadism)
- Congenital anorchia
- Acquired anorchia
- Klinefelter syndrome
- XYY syndrome
- XX male
- Noonan syndrome
- Gonadal dysgenesis
- Leydig cell tumours
- Maldescended testes
- Varicocele
- Sertoli-cell-only syndrome
- General disease e.g. renal failure, liver cirrhosis, diabetes, myotonia dystrophica
- Male pseudohermaphroditismus due to enzyme defects in testosterone biosynthesis
  or LH-receptor defects

Target organ resistance to androgens
- Testicular feminization
- Reifenstein syndrome
- Perineoscrotal hypospadia with pseudovagina
- Infertility with androgen resistance
- Receptor positive androgen resistance
- Undervirilised fertile male syndrome

There is general agreement that patients with "classical" disorders of primary or secondary hypogonadism should receive testosterone substitution therapy. However, there is a relatively large group of patients in whom hypogonadism develops as a corollary of other acute or chronic diseases. Although these patients lack testosterone and show symptoms of hypogonadism, testosterone is usually not administered to them. Just why substitution is withheld is not quite clear. Probably in many physicians' minds testosterone is predominantly associated with sexual functions. However, the better the general effects of testosterone on well-being, mood, bones, muscles and red blood are understood, the more testosterone substitution will be considered. Similarly, male senescence may be associated with symptoms of hypogonadism and, again, there is no general consensus whether testosterone substitution should be provided or not. Because of the controversies and unresolved problems surrounding these areas, two chapters of this volume are devoted to these topics, Chapter 16 by Kaufman and Vermeulen, "Androgens in male senescence" and Chapter 17 by Liu and Handelsman, "Androgen therapy in non-gonadal disease".

For the time being the principle may be followed that any type of hypogonadism documented by decreased serum testosterone concentrations deserves testosterone substitution, unless there is a clear contraindication, of which there are only few.

10.4.1. Symptoms of hypogonadism

The time of the onset of testosterone deficiency is of greater importance for the clinical symptoms than the localisation of the cause. Lack of testosterone or testosterone action during weeks 8 to 14 of *fetal life*, the period of sexual differentiation, leads to the development of intersexual genitalia (see Chapter 2 by Quigley in this volume). Lack of testosterone at the end of fetal life results in maldescended testes and small penis size. In later life the onset of testosterone deficiency before or after the completion of puberty determines the clinical appearance (Table 10.3).

If testosterone is lacking from the time of normal onset of *puberty* onwards, eunuchoidal proportions will develop, i.e. arm span exceeds the standing height and lower length of body (from soles to symphysis) exceeds upper length (from symphysis to top of the cranium) and the bone mass will not

Table 10.3. Symptoms of hypogonadism in relation to the time of onset of testosterone deficiency (from Nieschlag and Behre 1997)

Organ/function	Testosterone deficiency	
	Before completion of puberty	After completion of puberty
Bones	Eunuchoidal proportions, osteoporosis	Decrease of bone mass, osteoporosis
Larynx	Lack of voice mutation	No change
Secondary hair	Horizontal pubic hair line, straight frontal hairline, sparse beard	First, no change in pattern, decrease in density
Skin	Lack of sebum and acne, fine wrinkles	Atrophy, paleness, fine wrinkles
Bone marrow	Anemia	Anemia
Muscles	Underdeveloped	Atrophic
Penis	Infantile	No change in size
Prostate	Underdeveloped	Atrophic
Spermatogenesis	Not initiated	Regression
Ejaculate	Anejaculation or small volume	Decreasing volume
Libido	Not developed	Loss
Potency	Not developed	Erectile dysfunction

develop to its normal level. The distribution of fat will remain prepubertal and feminine, i.e. emphasis of hips, buttocks and lower belly. Voice mutation will not occur. The frontal hairline will remain straight without lateral recession, beard growth is absent or scanty, the pubic hairline remains straight. Hemoglobin and erythrocytes will be in the lower normal to subnormal range. Early development of fine perioral and periorbital wrinkles are characteristic. Muscles remain underdeveloped. The skin is dry due to lack of sebum production and remains free of acne. The penis remains small, the prostate is underdeveloped. Spermatogenesis will not be initiated and the testes remain small. If an ejaculate can be produced it will have a very small volume. Libido and potency will not develop.

A lack of testosterone occuring in *adulthood* cannot change body proportions, but will result in decreased bone mass and osteoporosis. Early-on lower backache and, in an advanced stage, vertebral fractures may occur. Once mutation has occurred the voice will not change again. Lateral hair recession and baldness when present will persist, the secondary sexual hair will become scanty and in advanced cases, a female hair pattern may again develop. Mild anemia may develop. Muscle mass and power decreases. The skin will become atrophied and wrinkled. The prostate will decrease in volume while the penis will not change its size. Spermatogenesis will decrease and as a consequence, also the size of the testes, which will become softer. Libido and sexual arousability will decrease or disappear while potency will be less affected.

10.4.2 Clinical practice of substitution therapy

The symptoms of androgen deficiency can be prevented or reversed with testosterone treatment. Of all testosterone preparations and routes of application described above the intramuscular injection or oral ingestion of testosterone esters used to be the most widely accepted and practiced modalities for the treatment of all forms of hypogonadism. More recently, transdermal testosterone has become a valuable alternative. Testosterone substitution is started when the diagnosis is established and serum testosterone levels are found below the normal range, taking into account the various influences on serum testosterone levels including diurnal variations. In order to establish the diagnosis by documenting low serum testosterone levels, usually determination of testosterone in a serum sample taken between 08.00 and 10.00 in the morning is sufficient (Vermeulen and Verdonck 1992). Pooled sera will not improve diagnostic accuracy.

For full substitution pharmacokinetic and clinical studies show that 200 to 250 mg testosterone enanthate or testosterone cypionate must be injected every two weeks (Cunningham et al. 1990; Davidson et al. 1979; Gooren 1987; Nieschlag et al. 1976; Schulte-Beerbühl and Nieschlag 1980; Snyder and Lawrence 1980; Sokol et al. 1982). If oral substitution is preferred, 40 mg testosterone undecanoate must be given 2 to 4 times daily. These doses have been shown to be effective in the majority of hypogonadal men in either open (Franchi et al. 1978; Franchimont et al. 1978; Gooren 1987; Maisey et al. 1981; Morales et al. 1997) or double-blind controlled studies (Luisi and Franchi 1980; Skakkebaek et al. 1981) when libido and potency as well as physical and mental activity were taken as parameters. Although relatively high testosterone doses are consumed with this regimen, liver function is not negatively affected, as could be shown in 35 men taking 80 to 200 mg testosterone undecanoate over ten years (Gooren 1994).

Transdermal testosterone preparations may be used as a first choice and are specifically suited for patients suffering from fluctuating symptoms caused by testosterone enanthate injections. Another advantage is the self-applicability of the system. Abuse is easily avoided as testosterone is quickly eliminated after removal of the patch. In cases of suspected prostate diseases this offers an advantage over injectable preparations. Transscrotal preparations (Testoderm) consist of a film containing 10 or 15 mg natural testosterone. These are applied daily and lead to physiologic serum testosterone levels (Ahmed et al. 1988; see Chapter 13 by Atkinson; Bals-Pratsch et al. 1988; Carey et al. 1988; Cunningham et al. 1989; Findley et al. 1989). In our own experience of over ten years adequate long-term substitution can be achieved without serious side-effects under regular use. Serum testosterone levels are maintained in the lower normal range which is sufficient to induce e.g. normal bone density (Behre et al. 1997 b). Transdermal delivery systems on non-scrotal skin (Androderm, Andropatch, Testotop) also result in physiological serum levels, depending on the number of systems applied (Brocks et al. 1996). When the systems are applied to different body areas e.g. back, abdomen, thigh or upper arm rather similar pharmacokinetic patterns of serum testosterone are achieved (Meikle et al. 1996).

If a patient has pronounced androgen deficiency, has never received testosterone and has passed the age of puberty he is immediately treated with a full maintenance dose of testosterone. In cases of secondary hypogonadism when fertility is requested, testosterone therapy can be interrupted and GnRH or hCG/hMG-therapy can be implemented until sperm counts increase and a pregnancy has been induced. Testosterone therapy does not prevent the chance of initiating or reinitiating spermatogenesis with releasing or gonadotropic hormones. In some cases when the stimulatory therapy had been terminated testosterone alone may maintain spermatogenesis for some time (Baranetsky and Carlson 1980; see also Chapter 4 by Weinbauer and Nieschlag in this book).

Patients with residual testosterone production may not require a full maintenance dose, e.g. Klinefelter patients in an early phase of testosterone deficiency. In these cases injection intervals of testosterone esters may be extended beyond the two-week period; these cases may also be suited for low-dose testosterone undecanoate therapy (i.e. one or two times 40 mg daily) or intermittent transdermal treatment. This dose would not entirely suppress the residual endogenous testosterone production and would supplement the lacking hormone. Low-dose testosterone undecanoate or transdermal treatment may also be used in patients in whom therapy is started at the age of the onset of puberty. Full substitution at this stage would accelerate pubertal development above the normal rate and could cause premature epiphyseal closure leading to short stature. 40 mg testosterone undecanoate per day or testosterone enanthate 250 mg every 6 to 8 weeks over a 1 to 3 year period and followed by increases of the dose and shortening of the injection intervals respectively is the proper regimen in these cases. Definitive therapeutic regimens using transdermal testosterone for this purpose have not yet been established. In patients with secondary hypogonadism treatment with GnRH or gonadotropins can be started at the expected age of puberty, but this therapy involving portable pumps or injections every other day is much more demanding and cumbersome for the patient than testosterone therapy.

10.4.3 Surveillance of testosterone substitution therapy

The physiological effects of testosterone (Mooradian et al. 1987) can be used for monitoring the efficacy of testosterone substitution therapy (Table 10.3). Since therapy aims at replacing the testosterone endogeneously lacking and since physiological serum concentrations are well known, serum testosterone levels provide also a good parameter for therapy surveillance. Guidelines for monitoring testosterone therapy have been compiled by WHO (1992).

10.4.3.1 Behaviour and mood

The general well-being of a patient is a good parameter to monitor the effectiveness of replacement therapy. Under sufficient testosterone replacement the patient feels physically and mentally active, vigorous, alert and in good spirits; low testosterone levels will be accompanied by lethargy, inactivity and depressed mood (Burris et al. 1992; Wang et al. 1996a).

10.4.3.2 Sexuality

The presence and frequency of sexual thoughts and fantasies correlate with appropriate testosterone substitution, while loss of libido and sexual desire are a sign of subnormal testosterone values. Spontaneous erections such as those during sleep will not occur if testosterone replacement is inadequate; however, erections due to visual erotic stimuli may be present even with low testosterone levels. The frequency of ejaculations and sexual intercourse correlate with serum testosterone levels in the normal to subnormal range. Therefore, detailed psychological exploration or a diary on sexual activity are useful adjuncts in assessing testosterone substitution. For objective evaluation of psychosexual effects weekly questionnaires on sexual thoughts and fantasies, sexual interest and desire, satisfaction with sexuality, frequency of erections and number of morning erections and ejaculations may be used. These clinical experiences are substantiated by studies on androgen replacement in hypogonadal men (Bals-Pratsch et al. 1988; Behre et al. 1992; Burris et al. 1992; Carani et al. 1992; Clopper et al. 1993; Cunningham et al. 1990; Morales et al. 1997) and by recent findings in normal men treated with GnRH analogues (Bagatell et al. 1994; Behre et al. 1994; Buena et al. 1993).

Priapism has been reported to occur in individual cases at the beginning of testosterone substitution (Endres et al. 1987; Ruch and Jenny 1989; Zelissen and Stricker 1988). This is an extremely rare effect; in our experience of 30 years of substitution therapy only one case is recollected. Decreasing the testosterone dose is the rational consequence, but intervention by aspirating blood from the corpora cavernosa or administration of sympathomimetic drugs may be acutely necessary.

10.4.3.3 Phenotype

Muscles and physical strength grow under testosterone treatment and the patient develops a more vigorous appearance. Due to its anabolic effects body weight increases by about 5%. Therefore, accurate recording of body weight belongs to the routine control of the patient. The increase in lean body mass at the expense of body fat can be measured by sophisticated techniques but has not yet been applied routinely to the monitoring of androgen replacement (Young et al. 1993). Moreover, the distribution of subcutaneous fat that shows feminine characteristics in hypogonadism (hips, lower abdomen, nates) may change with increasing muscle mass. In particular, testosterone appears to reduce abdominal fat (Rebuffé-Scrive et al. 1991).

The appearance and maintenance of a male sexual hair pattern is a good parameter for monitoring testosterone replacement (see Chapter 5 by Randall in this volume). In particular, beard growth and frequency of shaving can easily be recorded. Hair growth in the upper pubic triangle is an important indicator of sufficient androgen substitution. While women, boys and untreated hypogonadal patients have a straight frontal hairline, androgenization is accompanied by temporal recession of the hairline and – if a predisposition exists – by the development of baldness. The pattern of male sexual sexual hair is of greater importance than the intensity of hair growth since no correlation could be found between the intensity of body hair growth and serum testosterone levels in the normal range (Knussmann et al. 1992). A well-substituted patient may have to shave daily. However, if there is no genetic disposition for dense beard growth, additional testosterone will not increase facial hair.

Sebum production correlates with circulating testosterone levels and hypogonadal men may suffer from dry skin. In an early phase of treatment patients may even complain about the necessity of shampooing more frequently; they have to be informed that this is a part of normal maleness. The occurrence of acne may be a sign of supraphysiological testosterone levels and the dose should be reduced accordingly.

Gynecomastia may be caused by increased estradiol levels during testosterone therapy, especially under testosterone enanthate. After adaption of androgen therapy and consecutive decrease of estradiol serum levels gynecomastia usually disappears. If gynecomastia preexists due to an increased estradiol/testosterone ratio in hypogonadal men, it may reduce during adequate testosterone therapy.

Patients who have not undergone pubertal development will experience voice mutation soon after initiation of testosterone therapy. During normal pubertal development the voice begins to break when serum testosterone levels reach about 10 nmol/l and SHBG drops (Pedersen et al. 1986). Mutation of the voice is very assuring for the patient and helps him to adjust to his environment by closing the gap between his chronological and biological age. It is specifically important for the patient to be recognized as an adult male on the phone. Once the voice has mutated it is no longer a useful parameter for monitoring the replacement therapy since the size of the larynx, the vocal chords and thus the voice achieved will be maintained without requiring further androgens.

In prepubertal patients penis growth will be induced by testosterone treatment and normal erectile function will develop. Since penile androgen receptors diminish during puberty, growth will cease even under continued testosterone treatment (Shabsigh 1997).

Patients who did not undergo puberty before the onset of hypogonadism may also develop eunuchoidal body proportions because of retarded closure of the epiphyseal lines of the extremities. Testosterone treatment will briefly stimulate growth, but will then lead to closure of the epiphysis and will arrest growth. In these patients, an X-ray of the left hand and distal end of the lower arm should be made before treatment. The epiphyseal closure may be followed by further X-rays during the course of treatment. In addition, body height and armspan – as measured from the tip of the right to the tip of the left middle finger – should be measured until no furthter growth occurs. Continued growth, in particular of the armspan, indicates inadequate androgen substitution.

10.4.3.4 Blood pressure

Inadequate dosage of androgens, as it can be observed during misuse of testosterone and anabolic steroids, may increase the blood pressure by increasing blood electrolytes and water retention, leading to edema. During effective testosterone substitution therapy in hypogonadal men such side-effects are not observed (e.g. Whitworth et al. 1992). However, all androgens cause some degree of sodium retention and a small expansion of extracellular fluid volume that may contribute to weight increase in healthy individuals. Regular blood pressure measurement should be performed during testosterone therapy, especially during inception of treatment when the testosterone dosages have to be adapted, and in men with additional problems of the heart and kidneys (Gooren 1994).

10.4.3.5 Serum testosterone

When serum testosterone levels are used to judge the quality of testosterone substitution it is necessary to be aware of the pharmacokinetic profiles of the different testosterone preparations (see above). Moreover, in longitudinal surveillance of testosterone therapy it is important to use assay systems that strictly undergo internal and external quality control. Generally, testosterone serum levels should be measured just before the injection or implantation of the next dose of long-acting preparations. The time point of the last injection or administration of oral or transdermal testosterone must be recorded to interpret the serum levels measured. Levels below the lower normal limit at the end of a three week injection interval after testosterone enanthate injection should lead to an increase of injection frequency to two week intervals. If the levels are in the high physiological range at the end of the injection interval, the dosing intervals may be extended. Low serum testosterone levels two to four hours after ingestion of oral testosterone undecanoate should prompt counselling of the patient so that the capsule is taken together with a meal and testosterone is better absorbed. However, monitoring of treatment with oral testosterone undecanoate is difficult to base on serum testosterone levels and other parameters are of more importance if this mode of therapy is chosen. Inadequate testosterone serum levels after scrotal testosterone application should give rise to careful investigation whether the delivery system is placed correctly or if, e.g. because of increased genital hair growth, the contact between skin and patch is not optimal. Transdermal systems applied to non-genital skin may show poor adhesiveness, in particular when the patient sweats. Repeated measurements should be extended until the best application scheme is found for the individual patient. If other parameters indicate inadequate substitution, monitoring of testosterone serum levels is also of importance. During regular check-ups testosterone serum levels should be measured every 6 to 12 months.

10.4.3.6 Free testosterone, SHBG

In blood, testosterone is bound to sex hormone binding globulin (SHBG). Only 2% of testosterone is not bound and is available for biological action of testosterone (free testosterone). Since total testosterone correlates well with free testosterone, separate determination of free testosterone is not necessary for routine monitoring. Since accurate testosterone measurements are still difficult, additional problems should be circumvented by avoiding the use of available assays for very low free testosterone concentrations. Exceptions occur in the case of hyperthyreoidism or with ingestion of antiepileptic drugs that may cause a significant increase of SHBG, leading to a lowering of free testosterone levels. In these cases, measurement of SHBG may be indicated.

10.4.3.7 Saliva testosterone

Testosterone can be determined in saliva. The concentrations correlate to free testosterone concentrations in serum. This monitoring procedure can be easily applied without the help of medical staff (Navarro et al. 1994; Schürmeyer et al. 1983; Tschöp et al. 1998). However, since the available assays are not very robust, measurement of saliva testosterone has not become a widespread methodology.

10.4.3.8 Dihydrotestosterone

Determination of dihydrotestosterone does not play a role in routine monitoring of testosterone replacement therapy, but may be of importance in experimental use of testosterone preparations and monitoring biological effects of androgens. In the rare case that a patient does not respond properly to substitution, DHT measurement may be helpful for explanation.

10.4.3.9 Serum estradiol

In sensitive patients very high serum testosterone levels, as they may occur under testosterone enanthate, may be converted to estrogens and cause gynecomastia. This is an indication to reduce the dose or switch to another testosterone preparation. In this case monitoring serum estradiol levels is useful.

10.4.3.10 Gonadotropins

The determination of LH and FSH plays a key role in establishing the diagnosis of hypogonadotropic (i.e. secondary) or hypergonadotropic (i.e. primary) hypogonadism. However, during surveillance of testosterone therapy they are of secondary importance. By negative feed-back regulation between hypothalamus, pituitary and testes there is negative correlation between serum testosterone and LH, as well as to some extent to FSH levels in normal men.

In cases with primary hypogonadism (e.g. intact hypothalamic and pituitary function) FSH and in particular LH increase with decreasing testosterone levels and may normalize under testosterone, especially with injectable testosterone esters and thus indicate sufficient therapy. However, in the most frequent form of primary hypogonadism, in patients with Klinefelter's syndrome, LH and FSH often do not show significant suppression during testosterone substitution. Moreover, oral or transdermal testosterone may have only little effects on gonadotropins. Therefore LH is not the best indicator of sufficient testosterone replacement therapy.

10.4.3.11 Erythropoiesis

Since erythropoiesis is androgen-dependant, hypogonadal patients usually have a mild anemia (with values in the female normal range) which normalizes under testosterone treatment. Therefore, hemoglobin, red blood cell count and hematocrit are good parameters for surveillance of replacement therapy. If too much testosterone is administered, supraphysiological levels of hemoglobin, erythrocytes and hematocrit as a sign of polycythemia can develop, indicating that the testosterone dose should be scaled down (Hajjar et al. 1997; Matsumoto et al. 1985; Sih et al. 1997). In some cases phlebotomy may be required acutely. If adequate stimulation is lacking despite adequate testosterone therapy, lack of iron should be ruled out and treated if necessary. At the beginning of therapy we check red blood cell values every three months, later on annually.

Testosterone has been claimed to potentiate sleep apnea; however, only case reports about incidence of sleep apnea during testosterone treatment have been published (Matsumoto et al. 1985). The two men who demonstrated worsening of obstructive apnea on testosterone replacement therapy had pathologically elevated erythrocyte counts and hematocrit ($>59\%$), sufficient to require therapeutic phlebotomy. The increased hematocrit, increased mass of pharyngeal muscle bulk, as well as neuroendocrine effects of testosterone during testosterone therapy in these patients were discussed as possible reasons.

10.4.3.12 Liver function

The testosterone preparations proposed for testosterone replacement do not have negative side-effects on liver function (Gooren 1994). Nevertheless, many physicians believe that testosterone may disturb liver function. This impression derives from 17α-methyltestosterone and other 17α-alkylated anabolic steroids which indeed are liver toxic and which should no longer be used in the clinic (see above). However, there may be ethnic differences since the weekly application of 200 mg testosterone enanthate, i.e. double the dose used for substitution, for several months led to a slight increase of liver transaminases while this effect was not seen in non-Chinese men (Wu et al. 1996). Under more physiologic testosterone doses this phenomenon is not observed (Wang et al. 1991).

Monitoring liver function is of special interest in hypogonadal patients with concommitant diseases that affect liver function, or in patients whose hypogonadism is induced by general diseases. In such cases additional medication is necessary that may influence liver function and thus influence testosterone metabolism, e.g. by increasing SHBG production. We usually determine liver enzymes once per year routinely.

10.4.3.13 Lipid metabolism

Whether cardiovascular risk factors are affected by testosterone therapy remains a matter of debate (see Chapter 8 by von Eckardstein in this volume). In adult male hypogonadism beneficial effects such as an increase in HDL-cholesterol as well as adverse effects such as a decrease in HDL-cholesterol or an increase in LDL-cholesterol have been demonstrated (Ozata et al. 1996;

Sorva et al. 1988). In a one-year study, in which especially older, hypogonadal men were recruited, improvement in LDL cholesterol without effects on HDL cholesterol was reported (Zgliczynski et al. 1996).

Under testosterone replacement therapy, changes of lipid metabolism appear to occur within the physiological ranges and to date monitoring of these parameters is not part of routine surveillance of hypogonadal patients on testosterone treatment. With increasing knowledge, however, it may be necessary to try and select untreated and treated hypogonadal patients by additional screening procedures to identify those who may be at risk of atherosclerotic complications.

Besides lipid profiles other metabolic parameters such as overweight and especially an accumulation of abdominal fat predispose men for cardiovascular diseases and diabetes. This condition is more frequent in men with low testosterone and SHBG levels (Tchernof et al. 1996). The newly discovered hormonal product of adipocytes leptin is a candidate link between these different metabolic systems. Substitution therapy of male hypogonadism normalizes initially elevated leptin concentrations and improves obesity and therefore is a useful marker of therapeutic effectiveness (Behre et al. 1997a; Jockenhövel et al. 1997; Sih et al. 1997).

10.4.3.14 Prostate and seminal vesicles

The prostate and seminal vesicles are androgen-sensitive organs and are small in hypogonadal patients. They increase under testosterone therapy. Testosterone induces their normal functions, as indicated by the appearance of seminal fluid. Well-substituted patients should have ejaculate volumes in the normal range (≥ 2 ml).

There is much concern about the effects of testosterone with regard to the development of benign prostatic hyperplasia (BPH) and carcinoma of the prostate and this issue is specifically dealt with in Chapter 9 by Frick et al. in this book. A widely accepted theory on the pathogenesis of BPH suggests that prostatic enlargement is mediated through the action of 5α-DHT and that these alterations are related to intraprostatic events rather than to increases in serum concentrations of testosterone or 5α-DHT (Meikle et al. 1997; Morgentaler et al. 1996; Nomura et al. 1988). Furthermore, estrogens may be involved in hormonal regulation in prostatic tissue (Thomas et al. 1994). Testosterone therapy increases prostate volume in hypogonadal men, but only to the prostate size seen in age-matched controls (Behre et al. 1994). This is also the case in patients on scrotal testosterone treatment leading to somewhat elevated serum DHT-levels. PSA levels increase slightly during therapy but remain within the normal range in a younger population (Behre et al. 1994; Meikle et al. 1997). Though limited in accuracy, sensitivity and specificity, rectal palpation of the prostate for size, surface and consistency belongs to the regular check-up of patients under testosterone treatment.

Because of the incidence of benign prostatic hyperplasia and prostate carcinoma increasing with age and the risk of stimulating the growth of a pre-existing carcinoma by testosterone, patients should be examined carefully before onset of testosterone therapy and those over 45 years of age under testosterone treatment should be monitored at least yearly. In addition to rectal palpation, transrectal ultrasonography should be applied as the entire organ can be imaged non-invasively. Serum PSA and uroflow measurements should be performed in these patients at the same time intervals. It is reassuring that a recent study on long-term testosterone treatment of men around 70 years old found a higher rate of BPH in the control than in the treatment group over a two-year observation period (Hajjar et al. 1997).

10.4.3.15 Bone mass

Hypogonadism is associated with decreased bone density by increased bone resorption and decreased mineralization, resulting in premature osteoporosis and increased risk of fractures (see Chapter 6 by Finkelstein in this volume). Testosterone replacement in hypogonadal patients results in an increase in bone density (Behre et al. 1997b; Devogelaer et al. 1992). Since estrogens play an important role in bone metabolism and structure it is important that the testosterone preparation used for substitution can be converted to estrogens. This is the case with all testosterone preparations described above, but not with DHT or some testosterone derivatives such as mesterolone which are therefore not suited for medium- and long-term substitution purposes (see also Chapter 15 by Schaison and Conzinet in this volume).

Only advanced changes in bone density can be recognized by usual X-ray. For monitoring early signs of inadequate bone density different methods are available, e.g. dual photon absorptiometry (DPA), dual energy X-ray absorptiometry (DXA) or volumetric methods such as quantitative computer tomography of the lumbar spine (QCT) or the peripheral quantitative computer tomography of radial or tibial bone (pQCT). These methods work with high accuracy and reproducibility. However, the clinical relevance in the surveillance of testosterone therapy is not fully established yet. We prefer monitoring of lumber bone density by quantitative computer tomography of the lumber spine (QCT) in hypogonadal men.

Since osteoporosis and the possible risk of fractures influence the quality of life significantly, we now subject all patients prior to and during testosterone substitution therapy to bone density measurements regularly every two years and take the result into consideration for adaptation of testosterone regimen. If not all patients can be monitored we would at least strongly recommend measuring bone density in patients with long-standing untreated hypogonadism and in older patients when hypogonadism is first diagnosed. Further monitoring should then be implemented if values for bone density are subnormal at the inception of testosterone therapy.

10.5 Constitutional delay of puberty

Testosterone enanthate can be used to initiate puberty in boys with constitutional delay of pubertal development (Albanese et al. 1995; de Lange et al. 1979; Kaplan et al. 1973; Rosenfeld et al. 1982). Different regimens have been tested. In order not to cause early closure of the epiphyses we prefer a short-term treatment with three injections of 250 mg testosterone enanthate given in monthly intervals. After the third injection a pause of three months follows; if pubertal development does then not proceed spontaneously another three-month treatment course is added. Since the differential diagnosis between constitutional delay of puberty and idiopathic hypogonadotropic hypogonadism may be difficult, this type of treatment allows reassessment of the diagnosis at the end of the treatment-free interval. The treatment does not influence final height, but usually leads to relatively rapid virilization of the boys, the effect hoped for by such treatment.

More recently, low dose oral testosterone undecanoate has been tested for the treatment of constitutional delay of puberty (Albanese et al. 1994; Brown et al. 1995; Butler et al. 1992). For example, treatment of 11–14 year old prepubertal boys with 20 mg testosterone undecanoate per day for six months resulted in an increase in growth velocity without advancing bone age and pubertal development (Brown et al. 1995). Such "mild" treatment appears to be suited for an early phase when virilisation is not yet requested.

10.6 Overtall stature

The effect of testosterone on epiphysial closure may be used to treat boys who may be dissatisfied with their prospective final overtall body height. Treatment has to start before the age of 14 (Zachmann et al. 1976). Doses of 500 mg testosterone enanthate have to be administered every two weeks for at least a year to produce effects (Brämswig et al. 1981). This treatment should be reserved for special cases since tall stature is not a disease but rather a cosmetic and psychological problem. However, social and psychological conflicts caused by this condition should not be underestimated.

An additional reservation comes from the possible effects of such high-dose testosterone treatment at this early age on fertility, the prostate, the cardiovascular system, on bones and other organs. Long-term follow-up of men treated on average ten years earlier with high-dose testosterone for tall stature revealed no negative effects on sperm parameters and reproductive hormones in comparison to controls (de Waal et al. 1995; Lemcke et al. 1996). Prostate morphology as evaluated by ultrasonsography did not show any abnormalities and serum lipids were not different from the control group. Slightly lower sperm motility was rather attributable to a higher incidence of varicocele and maldescended testes in the treated men than to the treatment as such. Thus it appears that as far as evaluated, the high-dose treatment has no long-term negative side-effects.

10.7 Unproven use of testosterone in male infertility

Since testosterone has been used so effectively in the treatment of the endocrine insufficiency of the testes, its use has also been attempted in the treatment of idiopathic male infertility. Testosterone rebound was one of the earliest modalities in this regard. The published success rate in terms of pregnancies varied considerably from centre to centre, but remained low overall (Charny and Gordon 1978; Getzoff 1955; Lamensdorf et al. 1975; Rowley and Heller 1972). All studies were uncontrolled trials without placebo and double-blinding and therefore inconclusive. Testosterone rebound therapy cannot be recommended for treatment of infertility and is no longer practiced.

More recently, testosterone undecanoate has been tested in the treatment of idiopathic male infertility. However, a significant increase in pregnancy rates could not be demonstrated (Comhaire et al. 1995; Kloer et al. 1980; Push et al. 1989). When testosterone undecanoate was given in combination with tamoxifen and/or hMG, an improvement of semen parameters was observed (Adamopoulos et al. 1995, 1997). However, in these studies no pregnancy rates were reported. The therapeutic goal for every infertility treatment should be an increase in pregnancy rates, therefore, studies in which only improved semen parameters are reported, without examining the pregnancy rates, must be considered as inconclusive in terms of infertility treatment.

Similarly, after many years of clinical use no significant effect of mesterolone on pregnancy rates could be demonstrated in an extensive WHO-sponsored multicentre trial (WHO 1989). Thus, to date testosterone and other androgens have no place in evidence-based treatment of idiopathic male infertility (Nieschlag and Leifke 1997).

10.8 Contraindications to testosterone treatment

Effects and side-effects of testosterone therapy have been described in detail above. Here the major reasons for not initiating or for interrupting a testosterone therapy should be summarised briefly.

The major contraindications to a testosterone therapy is a *prostate carcinoma*. A patient with an existing prostate carcinoma should not receive testosterone. A carcinoma has to be excluded before starting therapy and the patient on testosterone should regularly be checked for a prostate carcinoma (digital exploration, PSA, transrectal sonography and biopsy, if necessary).

Breast cancer cells often are hormone-sensitive, especially estrogen-sensitive, and therefore, for reasons of safety, breast cancer is considered a contraindication to testosterone treatment. However, breast cancer is a relatively rare cancer in men and no cases of testosterone substitution and occurence of breast cancer have been published, as an extended literature search revealed. Thus, this warning cannot be substantiated.

In some countries *sexual offenders* may be treated by castration or antian-drogenic therapy. It would be a serious mistake to administer testosterone to such patients. Relapses and renewed crimes could be the consequence.

Testosterone suppresses spermatogenesis, a phenomenon exploited for hormonal male contraception (see Chapter 18 of this volume). In hypogonad-al patients with reduced spermatogenetic function testosterone administra-tion will also decrease sperm production. Such patients who wish to father children e.g. by techniques of artificial fertilisation, should not receive full testosterone substitution therapy, at least not for the time their sperm are ne-cessary for fertilisation of eggs. This may become of increasing importance as it has been shown that not only residual sperm in patients with secondary hypogonadism but also with Klinefelter syndrome may be able to fertilise eggs via intracytoplasmatic sperm injection (ICSI) and induce pregnancies (e.g. Bourne et al. 1997).

10.9 Overall effect of testosterone

Testosterone has many biological functions and, as demonstrated in this chapter, testosterone is a safe medication. There are only very few reasons why testosterone should be withheld from a hypogonadal patient (see 10.8). Nevertheless, to date many hypogonadal men do still not receive the benefit of testosterone therapy, because they are not properly diagnosed and the therapeutic consequences are not drawn. Some physicians even believe that the shorter life expectancy of men compared to women could be attributed to effects of testosterone, possibly mediated through changes in lipid metabo-lism. Hence it may be asked whether testosterone may have a life-shortening effect on patients with hypogonadism under testosterone treatment. Appro-priate controlled studies to answer this question directly are not available and are unlikely to be performed since it would appear unethical to withhold testosterone lifelong from a hypogonadal control group. However, there are two retrospective historical studies available addressing the problem.

A retrospective analysis of the life expectancy of inmates of an institution for the mentally handicapped in the USA came to the conclusion that early castration would lead to a longer life expectancy (Hamilton and Mestler 1969). However, this could be explained by the preference of castration as treatment for the physically more active inmates, whereas lack of mobility is the major predictor of shortened life expectancy among institutionalised men. In contrast, the retrospective comparison of the life expectancy of sing-ers born between 1581 and 1858 and castrated prepubertally in order to pre-serve their high voices to intact singers born at the same time did not reveal a significant difference between the lifespan of intact and castrated singers (Nieschlag et al. 1993).

Since neither the inmates nor the historical singers can be considered representative for the present population, these contradictory studies can only provide hints but no conclusive answer. There is, however, no proof that testosterone is a life-shortening agent. To prevent a hypogonadal patient from receiving the necessary substitution would force him to continue a miserable life of low quality. If testosterone in physiological doses should cause "side" effects these would indeed be the normal biological effects. The risks inherent to testosterone, be it of endogenous or exogenous origin, would then appear to be the tribute men have to pay for being men.

10.10 Key messages

- The principle indications for testosterone therapy are the various forms of male hypogonadism. For substitution testosterone preparations should be used that can be converted to 5α-dihydrotestosterone as well as to estrogens in order to develop the full spectrum of testosterone action.
- Injectable, oral and transdermal testosterone preparations are available for clinical use. The best preparation is the one that replaces testosterone serum levels at as close to physiologic concentrations as is possible.
- In six decades of clinical use testosterone has proven to be a very safe medication. No toxic effects are known. The only important contraindication is the presence of a prostate carcinoma.
- Testosterone therapy should be monitored by patients' well-being, alertness and sexual activity, by occasional serum testosterone levels, hemoglobin and hematocrit, by bone density measurements and prostate parameters (rectal examination, PSA and transrectal sonography).
- Testosterone can be used to initiate puberty in boys with constitutional delay of pubertal development. Careful dosing does not lead to premature closure of the epiphyses and reduced height.
- High-dose testosterone treatment in early puberty may prevent expected overtall stature in boys. Negative long-term effects of this treatment have not become evident to date.
- No evidence has been provided that testosterone treatment leads to higher pregnancy rates in male idiopathic infertility and should therefore not be used for this indication.
- The risks inherent to testosterone, be it of endogenous or exogenous origin, appear to be the tribute men have to pay for being men.

10.11 References

Aakvaag A, Stromme SB (1974) The effect of mesterolone administration to normal men on the pituitary-testicular function. Acta Endocrinol 77:380–386

Adamopoulos DA, Nicopoulou S, Kapolla N, Vassipoulos P, Karamertzanis M, Kontogeorgos L (1995) Endocrine effects of testosterone undecanoate as a supplementary treatment to menopausal gonadotropins or tamoxifen citrate in idiopathic oligozoospermia. Fertil Steril 64:818–24

Adamopoulos DA, Nicopoulou St, Kapolla N, Karamertzanis M, Andreou E (1997) The combination of testosterone undecanoate with tamoxifen citrate enhances the effects of each agent given independently on seminal parameters in men with idiopathic oligozoospermia. Fertil Steril 67:756–62

Ahmed SR, Boucher AE, Manni A, Santen RJ, Bartholomew M, Demers LM (1988) Transdermal testosterone therapy in the treatment of male hypogonadism. J Clin Endocrinol Metab 66:546–551

Albanese A, Stanhope R (1995) Predictive factors in the determination of final height in boys with constitutional delay of growth and puberty. J Pediatr 126:545–550

Albanese A, Kewley GD, Long A, Pearl KN, Robins DG, Stanhope R (1994) Oral treatment for constitutional delay of growth and puberty in boys: a randomised trial of an anabolic steroid or testosterone undecanoate. Arch Dis Child 71:315–317

Alkalay D, Khemani L, Wagner WE, Bartlett MF (1973) Sublingual and oral administration of methyltestosterone. A comparison of drug bioavailability. J Clin Pharmac 13:142–151

Asch RH, Heitman TO, Gilley RM, Tice TR (1986) Preliminary results on the effects of testosterone microcapsules. In: Zatuchni GI, Goldsmith A, Spieler JM, Sciarra JJ (eds) Male contraception: advances and future prospects, Harper & Row, Philadelphia, pp 347–360

Bagatell CJ, Heimann JR, Rivier JE, Bremner WJ (1994) Effects of endogeneous testosterone and estradiol on sexual behaviour in normal young men. J Clin Endocrinol Metab 78:711–716

Bals-Pratsch M, Knuth UA, Yoon YD, Nieschlag E (1986) Transdermal testosterone substitution therapy for male hypogonadism. Lancet ii:943–946

Bals-Pratsch M, Langer K, Place VA, Nieschlag E (1988) Substitution therapy of hypogonadal men with transdermal testosterone over one year. Acta Endocrinol 118:7–13

Baranetsky NG, Carlson HE (1980) Persistence of spermatogenesis in hypogonadotropic hypogonadism treated with testosterone. Fertil Steril 34:477–482

Behre HM, Nieschlag E (1992) Testosterone buciclate (20-Aet-1) in hypogonadal men: Pharmacokinetics and pharmacodynamics of the new long-acting testosterone ester. J Clin Endocrinol Metab 75:1204–1210

Behre HM, Nashan D, Hubert W, Nieschlag E (1992) Depot gonadotropin-releasing hormone-agonist blunts the androgen-induced suppression of spermatogenesis in a clinical trial of male contraception. J Clin Endocrinol Metab 74:84–90

Behre HM, Böckers A, Schlingheider A, Nieschlag E (1994) Sustained suppression of serum LH, FSH testosterone and increase of high-density lipoprotein cholesterol by daily injections of the GnRH antagonist cetrorelix over 8 days in normal men. Clin Endocrinol 40:241–248

Behre HM, Simoni M, Nieschlag E (1997a) Strong association between leptin and testosterone. Clin Endocrinol 47:237–240

Behre HM, Kliesch S, Leifke E, Link TM, Nieschlag E (1997b) Long-term effect of testosterone therapy on bone mineral density in hypogonadal men. J Clin Endocrinol Metab 82:2386–2390

Berthold AA (1849) Transplantation der Hoden. Archiv für Anatomie, Physiologie und wissenschaftliche Medicin, Berlin, 42–46

Bhasin S, Swerdloff RS (1996) Sustained delivery of testosterone by a long acting, biodegradable testosterone microsphere formulation in hypogonadal men. In: Pharmacology, biology, and clinical application of androgens. Current status and future prospects. Bhasin S (ed) Wiley-Liss, New York, pp 481–486

Bhasin S, Swerdloff RS, Steiner B, Peterson MA, Meridores T, Galmirini M, Pandian MR, Goldberg R, Berman N (1992) A biodegradable testosterone microcapsule formulation provides uniform eugonadal levels of testosterone for 10–11 weeks in hypgonadal men. J Clin Endocrin Metab 74:75–83

Bird D, Vowles K, Anthony PP (1979) Spontaneous rupture of a liver cell adenoma after long term methyltestosterone: report of a case successfully treated by emergency right hepatic lobetomy. Br J Surg 66:212–213

Bourne H, Stern K, Clarke G, Pertile M, Speirs A, Baker GHW (1997) Delivery of normal twins following the intracytoplasmatic injection of spermatozoa from a patient with 47,XXY Klinefelter's syndrome. Hum Reprod 12:2447–2450

Boyd PR, Mark GJ (1977) Multiple hepatic adenomas and a heptatocellular carcinoma in a man on oral methyl testosterone for eleven years. Cancer 40:1765–1770

Brämswig JH, Schellong G, Borger HJ, Breu H (1981) Testosteron-Therapie hochwüchsiger Jungen. Dtsch Med Wschr 106:1656–1661

Breuer H, Gütgemann D (1966) Wirkung von 1α-Methyl-5α-androstan-17β-ol 3-on (Mesterolon) auf die Steroidausscheidung beim Menschen. Arzneimittelforschung 16:759–762

Brocks DR, Meikle AW, Boike SC, Mazer NA, Zariffa N, Audet PR, Jorkasky DK (1996) Pharmacokinetics of testosterone in hypogonadal men after transdermal delivery: influence of dose. J Clin Pharmacol 36:732–739

Brown D, Butler CGE, Kelnar CJH, Wu FCW (1995) A double blind, placebo controlled study of the effects of low dose testosterone undecanoate on the growth of small for age, prepubertal boys. Archi Dis Child 73:131–135

Brown-Séquard CE (1889) Des effets produits chez l'homme par des injections souscutanées d'un liquide retiré des testicules frais de cobaye et de chien. Cr Séanc Soc Biol 1:420–430

Buena F, Swerdloff RS, Steiner BS, Lutchmansingh P, Peterson MA, Pandian MR, et al. (1993) Sexual function does not change when serum testosterone levels are pharmacologically varied within the normal male range. Fertil Steril 59:1118–1123

Burris AS, Ewing LL, Sherins RJ (1988) Initial trial of slow-release testosterone microspheres in hypogonadal men. Fertil Steril 50:493–497

Burris AS, Banks SM, Carter CS, Davidson JM, Sherins RJ (1992) A long-term prospective study of the physiologic and behavioural effects of hormone replacement in untreated hypogondadal men. J Androl 13:297–304

Butenandt A (1931) Über die chemische Untersuchung des Sexualhormons. Z angew Chem 44:905–908

Butenandt A, Hanisch G (1935) Über Testosteron. Umwandlung des Dehydroandrosterons in Androstendiol und Testosteron; ein Weg zur Darstellung des Testosterons aus Cholesterin. Hoppe-Seyler's Z Physiol Chem 237:89–98

Butler GE, Sellar RE, Hendry RF, Qalker M, Kelnar CJH, Wu FCW (1992) Oral testosterone undecanoate in the management of delayed puberty in boys: pharmacokinetics and effects on sexual maturation and growth. J Clin Endocrin Metabol 75:37–44

Carani C, Bancroft J, Granata A, Del Rio G, Marrama P (1992) Testosterone and erectile function: nocturnal penile tumescence and rigidity, and erectile response to visual erotic stimuli in hypogonadal men. Psychoneuroendocrinology 17:647–654

Carbone JV, Grodsky GM, Hjelte V (1959) Effect of hepatic dysfunction on circulating levels of sulphobromopthalein and its metabolites. J Clin Invest 38:1989–1996

Carey PO, Howards SS, Vance ML (1988) Transdermal testosterone treatment of hypogonadal men. J Urol 140:76–79

Charny CW, Gordon JA (1978) Testosterone rebound therapy: a neglected modality. Fertil Steril 29:64–68

Clopper RR, Voorhess ML, MacGillivray MH, Lee PA, Mills B (1993) Psychosexual behavior in hypopituitary men: a controlled comparison of gonadotropin and testosterone replacement. Psychoneuroendocrinology 18:149–161

Coert A, Geelen J, de Visser J, van der Vies J (1975) The pharmacology and metabolism of testosterone undecanoate (TU), a new orally active androgen. Acta Endocrinol 79:789–800

Comhaire F, Schoonjans F, Abelmassih R, Gordts S, Campo R, Dhont M, Milingos S, Gerris J (1995) Does treatment with testosterone undecanoate improve the in-vitro fertilizing capacity of spermatozoa in patients with idiopathic testicular failure? (Results of a double blind study). Hum Reprod 10:2600–2602

Coombes GB, Reiser J, Paradinas FJ, Burn I (1978) An androgen-associated hepatic adenoma in a trans-sexual. Br J Surg 65:869–870

Crabbé P, Diczfalusy E, Djerassi C (1980) Injectable contraceptive synthesis: an example of international cooperation. Science 209:992–994

Cunningham GR, Cordero E, Thornby JI (1989) Testosterone replacement with transdermal therapeutic systems. JAMA 261:2525–2530

Cunningham GR, Hirshkowitz M, Korenman SG, Karacan I (1990) Testosterone replacement therapy and sleep-related erections in hypogonadal men. J Clin Endocrinol Metab 70:792–797

Daggett PR, Wheeler MJ, Nabarro JDN (1978) Oral testosterone: a reappraisal. Horm Res 9:121–129

Danner C, Frick G (1980) Androgen substitution with testosterone-containing nasal drops. Int J Androl 3:429–431

David K, Dingemanse E, Freud J, Laquer E (1935) Über krystallinisches männliches Hormon aus Hoden (Testosteron), wirksamer als aus Harn oder aus Cholesterin bereitetes Androsteron. Hoppe-Seyler's Z physiol Chem 233:281–282

Davidson JM, Camargo CA, Smith ER (1979) Effects of androgen on sexual behavior in hypogonadal men. J Clin Endocrinol Metab 48:955

de Lorimer AA, Gordan GS, Löwe RC, Carbone JV (1965) Methyltestosterone: related steroids and liver function. Arch Intern Med 116:289–294

deLange WE, Snoep MC, Doorenbos H (1979) The effect of short-term testosterone treatment in boys with delayed puberty. Acta endocrinol 91:177–183

Delanoe D, Fougeyrollas B, Meyer L, Thonneau P (1984) Androgenisation of female partners of men on medroxy-progesterone acetate/percutaneous testosterone contraception. Lancet i:276

Devogelaer JP, de Cooman S, Nagant de Deuxchaisnes C (1992) Low bone mass in hypogonadal males. Effect of testosterone substitution therapy, a densitometric study. Maturitas 15:17α23

De Waal WJ, Vreeburg JTM, Bekkering F, de Jongt FH, de Muinck Keizer-Schrama SMPF, Drop SLS, Weber RFA (1995) High dose testosterone therapy for reduction of final height in constitutionally tall boys: does it influence testicular function in adulthood? Clin Endocrinol 43:87–95

Endres W, Shin YS, Rieth M, Block T, Schmiedt E, Knorr D (1987) Priapism in Fabry's disease during testosterone treatment. Klin Wschr 65:925

Escamilla RF (1949) Treatment of preadolescent eunuchoidism with (methyl)testosterone linguets. Amer Pract 3:425

Falk H, Thomas LB, Popper H, Ishak KG (1979) Hepatic angiosarcoma associated with androgenic-anabolic steroids. Lancet ii:1120–1123

Farrell GC, Joshua DE, Uren RF, Baird PJ, Perkins KW, Kronenberg H (1975) Androgen-induced hepatoma. Lancet i:430–432

Feldmann RJ, Maibach HI (1967) Regional variation in percutaneous penetration of ^{14}C cortisol in man. J Invest Dermatol 48:181–83

Findlay JC, Place VA, Snyder PJ (1987) Transdermal delivery of testosterone. J Clin Endocrinol Metab 64:266–268

Findlay JC, Place VA, Snyder PJ (1989) Treatment of primary hypogonadism in men by the transdermal administration of testosterone. J Clin Endocrinol Metab 68:369–373

Franchi F, Luisi M, Kicovic PM (1978) Long-term study of oral testosterone undecanoate in hypogonadal males. Int J Androl 1:270–278

Franchimont P, Kicovic PM, Mattei A, Roulier R (1978) Effects of oral testosterone undecanoate in hypogonadal male patients. Clin Endocrinol 9:313–320

Frey H, Aakvaag A, Saanum D, Falch J (1979) Bioavailability of oral testosterone in males. Eur J Clin Pharmacol 16:345–349

Gerhards E, Gibian H, Kolb KH (1966) Zum Stoffwechsel von 1α-Methyl-5α-androstan-17β-ol-3-on (Mesterolon) beim Menschen. Arzneimittelforschung 16:458–463

Getzoff PL (1955) Clinical evaluation of testicular biopsy and the rebound phenomenon. Fertil Steril 6:465:474

Goodman MA, Laden AMJ (1977) Hepatocellular carcinoma in association with androgen therapy. Med J Aust i:220–221

Gooren LJG (1987) Androgen levels and sex functions in testosterone-treated hypogonadal men. Arch Sex Behav 16:463–473

Gooren LJG (1994) A ten year safety study of the oral androgen testosterone undecanoate. J Androl 15:212–215

Gordon RD, Thomas MJ, Poyntin JM, Stocks AE (1975) Effect of mesterolone on plasma LH, FSH and testosterone. Andrologia 7:287–296

Hajjar RR, Kaiser FE, Morley JE (1997) Outcomes of long-term testosterone replacement in older hypogonadal males: a retrospective analysis. J Clin Endocrinol Metab 82:3793–3796

Hallagan JB, Hallagan LF, Snyder MB (1989) Anabolic-androgenic steroid use by athletes. New Engl J Med 321:1042–1045

Hamburger C (1958) Testosterone treatment and 17-ketosteroid excretion. Acta Endocrinol 28:529–536

Handelsman DJ, Conway AJ, Boylan LM (1990) Pharmacokinetics and pharmacodynamics of testosterone in man. J Clin Endocrinol Metab 70:216–222

Heywood R, Hunter B, Green OP, Kennedy SJ (1977a) The toxicity of methyl testosterone in the rat. Toxicology Letters 1:27–31

Heywood R, Chesterman H, Ball SA, Wadsworth PF (1977b) Toxicity of methyl testosterone in the beagle dog. Toxicology 7:357–365

Horst HJ, Höltge WJ, Dennis M, Coert A, Geelen J, Voigt KD (1976) Lymphatic absorption and metabolism of orally administered testosterone undecanoate in man. Klin Wschr 54:875–879

Jockenhövel F, Vogel E, Kreutzer E, Reinhard W, Lederbogen S, Reinwein D (1996) Pharmacokinetics and pharmacodynamics of subcutaneous testosterone implants in hypogonadal man. Clin Endocrinol 45:61–71

Jockenhövel F, Blum WF, Vogel E, Englaro P, Müller-Wieland D, Reinwein D, Rascher W, Krone W (1997) Testosterone substitution normalizes elevated serum leptin levels in hypgonadal men. J Clin Endocrinol Metab 82:2510–2513

Johnsen SG (1978) Long-term androgen therapy with oral testosterone. In: Patanelli DJ (ed) Hormonal control of male fertility. DHEW publication No (NIH) 78–1097. Bethesda, 123–143

Johnsen SG, Bennet EP, Jensen VG (1974) Therapeutic effectiveness of oral testosterone. Lancet 2:1473–1475

Johnsen SG, Kampmann JP, Bennet EP, Jörgensen F (1976) Enzyme induction by oral testosterone. Clin Pharmacol Ther 20:233–237

Jordan WP (1997) Allergy and topical irritation associated with transdermal testosterone administration: a comparison of scrotal and nonscrotal transdermal systems. Am J Cont Dermat 8:108–113

Junkmann K (1957) Long-acting steroids in reproduction. Rec Progr Horm Res 13:389–419

Kaetsu I, Yoshida M, Asano M, Yamanaka H, Imai K, Nakai K, Mashimo T, Yuasa H (1988) Controlled release for hormone therapies by LHRH analogue containing polymer needles and testosterone containing artificial testis. Drug Develop Ind Pharm 14:2519–2533

Kaplan JG, Monshant T, Bernstein R, Parks JS, Bingiovanni AM (1973) Constitutional delay of growth and development: Effect of treatment with androgens. J Pediatr 82:38–44

Kliesch S, Behre HM, Nieschlag E (1994) High efficacy of gonadotropin or pulsatile GnRH treatment in hypogonadotropic hypogonadal men. Europ J Endocr 131:347–354

Kloer H, Hoogen H, Nieschlag E (1980) Trial of high-dose testosterone undecanoate in treatment of male infertility. Int J Androl 3:121

Knussmann R, Christiansen K, Kannmacher J (1992) Relations between sex hormone levels and characters of hair and skin in healthy young men. Am J Phys Anthropol 88:59–67

Knuth UA, Behre HM, Belkien L, Bents H, Nieschlag E (1985) Clinical trial of 19 nortestosterone-hexoxyphenylpropionate (Anadur) for male fertility regulation. Fertil Steril 44:814–821

Knuth UA, Maniera H, Nieschlag E (1989) Anabolic steroids and semen parameters in bodybuilders. Fertil Steril 52:1041–1047

Kopera H (1985) The history of anabolic steroids and a review of clinical experience with anabolic steroids. In: Eikelboom FA, van der Vies J (eds) Anabolics in the '80s. Acta endocrinol (Suppl) 271:11–18

Korenman SG, Viosca S, Garza D, Guralnik M, Place V, Campbell P, Stanik Davis S (1987) Androgen therapy of hypogonadal men with transscrotal testosterone system. Am J Med 83:471–478

Krüskemper HI, Noell G (1967) Steroidstruktur und Lebertoxizität. Acta endocrinol 54:73–80

Kumar N, Suvisaari J, Tsong Y-Y, Aguillaume C, Bardin WC, Lähteenmaki P, Sundaram K (1997) Pharmacokinetics of 7α-methyl-19-nortestosterone in men and cynomolgus monkeys. J Androl 18:352–358

Lamensdorf H, Compere D, Begley G (1975) Testosterone rebound therapy in the treatment of male infertility. Fertil Steril 26:469–472

Lemcke B, Zentgraf J, Behre HM, Kliesch S, Brämswig JH, Nieschlag E (1996) Long-term effects on testicular function of high-dose testosterone treatment for excessively tall stature. J Clin Endocrinol Metab 81:296–301

Loewe S, Voss HE (1930) Der Stand der Erfassung des männlichen Sexualhormons (Androkinins). Klin Wschr 9:481–487

Luisi M, Franchi F (1980) Double-blind group comparative study of testosterone undecanoate and mesterolone in hypogonadal male patients. J Endocrinol Invest 3:305–308

Maisey NM, Bingham J, Marks V, English J, Chakraborty J (1981) Clinical efficacy of testosterone undecanoate in male hypogonadism. Clin Endocrinol 14:625–629

Matsumoto AM, Sandblom RE, Schoene RB, Lee KA, Giblin EC, Pierson DJ, Bremner WJ (1985) Testosterone replacement in hypogonadal men: effects on obstructive sleep apnea, respiratory drives, and sleep. Clin Endocrinol 22:713–721

Mazer NA, Heiber WE, Moellmer JF, Meikle AW, Stringham JD, Sanders SW, Tolman KG, Odell WD (1992) Enhanced transdermal delivery of testosterone: a new physiological approach for androgen replacement in hypogonadal men. J Contrld Release 19:347–362

McCaughan GW, Bilous MJ, Gallagher ND (1985) Long-term survival with tumor regression in androgen-induced liver tumors. Cancer 56:2622–2626

Meikle AW, Arver S, Dobs AS, Sanders SW, Rajaram L, Mazer N (1996) Pharmacokinetics and metabolism of a permeation-enhanced testosterone transdermal system in hypogonadal men: influence of application site – a clinical research center study. J Clin Endocrinol Metab 81:1832–1840

Meikle AW, Arver S, Dobs AS, Adolfsson J, Sanders SW, Middleton RG, Stephenson RA, Hoover DR, Rajaram L, Mazer NA (1997) Prostate size in hypogonadal men treated with a nonscrotal permeation-enhanced testosterone transdermal system. Urology 49:191–196

Methyltestosteron-Monographie (Anonymous) (1988) Bundesanzeiger 40:5140–5141

Mooradian AD, Morley JE, Korenman SG (1987) Biological actions of androgens. Endocr Rev 8:1–28

Morales A, Johnston B, Heaton JPW, Lundie M (1997) Testosterone supplementation for hypogonadal impotence: assessment of biochemical measures and therapeutic outcomes. J Urol 157:849–854

Morgentaler A, Carl MD, Bruning MD, DeWolf WC (1996) Occult prostate cancer in men with low serum testosterone levels. JAMA 276:1904–1906

Navarro MA, Villabona CM, Blanco A, Gomez JM, Bonnin RM, Soler J (1994) Salivary exretory pattern of testosterone in substitutive therapy with testosterone enanthate. Fertil Steril 61:125–128

Neff MS, Goldberg J, Slifkin RF, Eiser AR, Calamia V, Kaplan M, Baez A, Gupta S, Matoo N (1981) A comparison of androgens for anemia in patients on hemodialysis. N Engl J med 304:871–875

Nieschlag E (1981) Ist die Anwendung von Methyltestosteron obsolet? Dtsch med Wschr 106:11123

Nieschlag E (1996) Testosterone replacement therapy: something old, something new... (Commentary). Clin Endocrinol 45:261–262

Nieschlag E, Behre HM (eds) (1997) Andrology: Male reproductive health and dysfunction. Springer, Heidelberg, New York

Nieschlag E, Leifke E (1997) Empirical therapies for male idiopathic infertility. In: Nieschlag E, Behre HM (eds) Andrology: Male reproductive health and dysfunction, Springer Publisher, Heidelberg, pp 311–319

Nieschlag E, Mauss J, Coert A, Kicovic P (1975) Plasma androgen levels in men after oral administration of testosterone or testosterone undecanoate. Acta endocrinol 79:366

Nieschlag E, Cüppers HJ, Wiegelmann W, Wickings EJ (1976) Bioavailability and LH suppressing effect of different testosterone preparations in normal and hypogonadal men. Hormone Research 7:138

Nieschlag E, Cüppers EJ, Wickings EJ (1977) Influence of sex, testicular development and liver function on the bioavailability of oral testosterone. Europ J Clin Invest 7:145

Nieschlag E, Nieschlag S, Behre HM (1993) Life expectancy and testosterone. Nature 366:215

Nomura A, Heilbrunn LK, Stemmermann GN, Judd HL (1988) Prediagnostic serum hormones and the risk of prostate cancer. Cancer Res. 48:3515–3517

Ozata M, Yildirimkaya M, Bulur M, Yilmaz K, Bolu E, Corakci A et al (1996) Effects of gonadotropin and testosterone treatments on lipoprotein (a), high density lipoprotein particles and other lipoprotein levels in male hypogonadism. J Clin Endocrinol Metab 81:3372–3378

Paradinas FJ, Bull TB, Westaby D, Murray-Lyon IM (1977) Hyperplasia and prolapse of hepatocytes into hepatic veins during longterm methyltestosterone therapy: possible relationships of these changes to the development of peliosis hepatis and liver tumours. Histopathology 1:225–246

Partsch CJ, Weinbauer GF, Fang R, Nieschlag E (1995) Injectable testosterone undecanoate has more favourable pharmacokinetics and pharmacodynamics than testosterone enanthate. Eur J Endocrinol 132:514–519

Pedersen MF, Moller S, Krabbe S, Bennett P (1986) Fundamental voice frequency measured by electroglottography during continuous speech. A new exact secondary sex characteristic in boys in puberty. Int J Ped Otohinol 11:21–27

Pitha J, Harman SM, Michel ME (1986) Hydrophilic cyclodextrin derivatives enable effective oral administration of steroidal hormones. J Pharm Sci 75:165–167

Pusch HH (1989) Oral treatment of oligozoospermia with testosterone-undecanoate: results of a double-blind-placebo-controlled trial. Andrologia 21:76–82

Rajalakshmi M, Ramakrinshnan PR (1989) Pharmacokinetics and pharmacodynamics of a new long-acting androgen ester: maintenance of physiological androgen levels for 4 months after a single injection. Contraception 40:399–412

Rebuffé-Scrive M, Marin P, Björntorp P (1991) Short communication: effect of testosterone on abdominal adipose tissue in men. Int J Obes 15:791–795

Rosenfeld RG, Northcraft GB, Hintz RL (1982) A prospective, randomized study of testosterone treatment of constitutional delay of growth and development in male adolescents. Pediatrics 69:681–687

Rowley MJ, Heller CG (1972) The testosterone rebound phenomenon in the treatment of male infertility. Fertil Steril 23:498–504

Ruch W, Jenny P (1989) Priapism following testosterone administration for delayed male puberty. Am J Med 86:256

Ruzicka L, Wettstein A (1935) Synthetische Darstellung des Testishormons, Testosteron (Androsten 3-on-17-ol). Helv chim Acta 18:1264–1275

Ruzicka L, Goldberg MW, Rosenberg HR (1935) Herstellung des 17-Methyl-testosterons und anderer Androsten- und Androstanderivate. Zusammenhänge zwischen chemischer Konstitution und männlicher Hormonwirkung. Helv Chim Acta 18:1487–1498

Salehian B, Wang C, Alexander G, Davidson T, McDonald V, Berman N, Dudley RE, Ziel F, Swerdloff RS (1995) Pharmacokinetics, bioefficacy, and safety of sublingual testosterone cyclodextrin in hypogonadal men: comparison to testosterone enanthate – a clinical research center study. J Clin Endocrinol Metab 80:3567–3575

Schulte-Beerbühl M, Nieschlag E (1980) Comparison of testosterone dihydrotestosterone, luteinizing hormone, and follicle-stimulating hormone in serum after injection of testosterone enanthate or testosterone cypionate. Fertil Steril 33:201–203

Schürmeyer T, Wickings EJ, Freischem CW, Nieschlag E (1983) Saliva and serum testosterone following oral testosterone undecanoate administration in normal and hypogonadal men. Acta Endocrinol 102:456–462

Schürmeyer T, Nieschlag E (1984) Comparative pharmacokinetics of testosterone enanthate and testosterone cyclohexanecarboxylate as assessed by serum and salivary testosterone levels in normal men. Int J Androl 7:181–187

Schürmeyer T, Jung K, Nieschlag E (1984) The effect of an 1100 km run on testicular, adrenal and thyroid hormones. Int J Androl 7:276–282

Shabsigh R (1997) The effects of testosterone on the cavernous tissue and erectile function. World J Urol 15:21–26

Sih R, Morley JE, Kaiser FE, Perry HM, Patrick P, Ross C (1997) Testosterone replacement in older hypogonadal men: a 12-month randomized controlled trial. J Clin Endocrinol Metab 82:1661–1667

Simmer H, Simmer I (1961) Arnold Adolph Berthold (1803–1861). Zur Erinnerung an den hundertsten Todestag des Begründers der experimentellen Endokrinologie. Dtsch med Wschr 86:2186–2192

Skakkebaek NE, Bancroft J, Davidson DW, Warner P (1981) Androgen replacement with oral testosterone undecanoate in hypogonadal men: a double blind controlled study. Clin Endocrinol 14:49–61

Snyder PJ, Lawrence DA (1980) Treatment of male hypogonadism with testosterone enanthate. J Clin Endocrinol Metab 51:1335

Sokol RZ, Palacios A, Campfield LA, Saul C, Swerdloff RS (1982) Comparison of the kinetics of injectable testosterone in eugonadal and hypogonadal men. Fertil Steril 37:425–430

Sorva R, Kuusi T, Taskinen MR, Perheentupa J, Nikkila EA (1988). Testosterone substitution increases the activity of lipoprotein lipase and hepatic lipase in hypogonadal males. Atherosclerosis 69:191–197

Stuenkel CA, Dudley RE, Yen SC (1991) Sublingual administration of β-cyclodextrin inclusion complex stimulates episodic androgen release in hypogonadal men. J Clin Endocrinol Metab 72:1054–1059

Tchernof A, Labrie F, Belanger A, Despres JP (1996) Obesity and metabolic complications: contribution of dehydroepiandrosterone and other steroid hormones. J Endocrinol 150:S155–S164

Thomas JA, Keenan EJ (1994) Effects of estrogens on the prostate. J Androl 15:97–99

Tschöp M, Behre HM, Nieschlag E, Dressendörfer RA, Strasburger CJ (1998) A time-resolved fluorescence immunoassay for the measurement of testosterone in saliva: monitoring of testosterone replacement herapy with testosterone buciclate. Clin Endocrinol (in press)

Vermeulen A, Verdonck G (1992) Representatives of a single point plasma testosterone level for the long term hormonal milieu in men. J Clin Endocrinol Metab 74:939–942

Vogelzang NJ, Arnold JL, Chodak GJ, Schoenberg H (1986) Androgen and germ cell testicular cancers. JAMA 255:906

Wang C, Alexander G, Berman N, Salehian B, Davidson T, McDonald V, Steiner B, Hull H, Callegari C, Swerdloff R (1996a) Testosterone replacement therapy improves mood in hypogonadal men – a clinical research center study. J Clin Endocrinol Metab 81:3578–3583

Wang C, Eyre R, Clark D, Kleinberg C, Newman I, Iranmanesh A, Veldhuis R, Dudley RE, Berman N, Davidson T, Barstow TS, Sinow R, Alexander G, Swerdloff R (1996b) Sublingual testosterone replacement improves muscle mass and strength, decreases bone resorption and increases bone formation markers in hypogonadal men – a clinical research center study. J Clin Endocrinol Metab 81:3654–3662

Wang L, Shi DC, Lu SY, Fang RY (1991) The therapeutic effect of domestically produced testosterone undecanoate in Klinefelter syndrome. New Drugs Mark 8:28–32

Weinbauer GF, Marshall GR, Nieschlag E (1986) New injectable testosterone ester maintains serum testosterone of castrated monkeys in the normal range for four months. Acta endocrinol 113:128–132

Weinbauer GF, Kurshid S, Fingscheidt U, Nieschlag E (1989) Sustained inhibition of sperm production and inhibin secretion, induced by a gonadotropin-releasing hormone (GnRH) antagonist and delayed testosterone substitution in non-human primates (Macaca fascicularis). J Endocrinol 123:303–310

Weinbauer GF, Jackwerth B, Yoon YD, Behre HM, Yeung CH, Nieschlag E (1990) Pharmacokinetics and pharmacodynamics of testosterone enanthate and dihydrotestosterone enanthate in non-human primates. Acta endocrinol 22:432–442

Werner SC, Hamger FM, Kritzler RA (1950) Jaundice during methyltestosterone therapy. Am J Med 8:325–331

Westaby D, Ogle SJ, Paradinas FJ, Randell JB, Murray-Lyon IM (1977) Liver damage from long-term methyltestosterone. Lancet ii:261–263

Whitworth JA, Scoggins BA, Andrews J, Williamson PM, Brown MA (1992) Haemodynamic and metabolic effects of short term administration of synthetic sex steroids in humans. Clin Exp Hypertens 14:905–922

World Health Organization, Nieschlag E, Wang C, Handelsman DJ, Swerdloff RS, Wu F, Einer-Jensen N, Waites G (1992) Guidelines for the use of androgens. WHO, Geneva

World Health Organization Task Force on the Diagnosis and Treatment of Infertility (1989) Mesterolone and idiopathic male infertility: a double-blind study. Int J Androl 12:254–264

Wu FCW, Farley TMM, Peregoudov A, Waites GMH, World Health Organisation Task Force on Methods for the Regulation of Male Fertility (1996) Effects of testosterone enanthate in normal men: experience from a multicenter contraceptive efficacy study. Fertil Steril 65:626–636

Young NR, Baker HWG, Liu G, Seeman E (1993) Body composition and muscle strength in healthy men receiving testosterone enanthate for contraception. J Clin Endocrinol Metab 77:1028–1032

Zachmann MA, Murset G, Gnehm HE, Prader A (1976) Testosterone treatment of excessively tall boys. J Pediatr 88:116–123

Zelissen PMJ, Stricker BHC (1988) Severe priapism as a complication of testosterone substitution therapy. Am J Med 85:273

Zgliczynksi S, Ossowski M, Slowinska-Srzednicka J, Brzezinska A, Zgliczynski W, Soszynksi P (1996) Effects of testosterone replacement therapy on lipids and lipoproteins in hypogonadal and elderly men. Atherosclerosis 121:35–43

11 Comparative pharmacokinetics of testosterone esters

Hermann M. Behre and Eberhard Nieschlag

Contents

11.1 Introduction

In replacing an endogenous hormone a safe general principle appears to be to mimic, as closely as possible, the normal concentrations of that hormone or its active metabolites. Following this principle testosterone treatment of male hypogonadism should avoid unphysiologically high testosterone serum concentrations to prevent possible side-effects or low concentrations to prevent androgen deficiency.

This chapter compares the pharmacokinetics of different testosterone esters widely used for substitution therapy and reevaluates them using computer analysis and simulation. For pharmacokinetics of special testosterone preparations the reader is referred to separate chapters: Chapter 12 for Testosterone Implants and Chapters 13 and 14 for Transdermal Testosterone.

11.2 Principles of pharmacokinetic evaluation

11.2.1 Polyexponential functions

The analysis of pharmacokinetics of different testosterone esters for clinical substitution therapy is based on the testosterone serum concentrations measured in the general circulation. Applying non-linear least squares regression analysis it is possible to obtain the best fitting n-term polyexponential function describing the pharmacokinetic profile of the drug according to the data actually measured. The use of polyexponential functions does not require a particular compartmental model to be used, which is fortunate since often there is no firm basis for choosing one such model over another. Different commercially accessible computer programs are available for calculating the parameters of the best-fitting polyexponential functions. The parameters of the polyexponential functions can be used to calculate pharmacokinetic parameters such as area-under-the-concentration-versus-time-curve (AUC), time to reach maximum serum levels (t_{max}), maximum serum concentration (c_{max}) and terminal elimination half-life ($t_{1/2}$) (Wagner 1975). The pharmacokinetic single-dose analysis presented in this chapter was performed applying the computer program RSTRIP (RSTRIP Ver. 5.0, MicroMath Scientific Software, Salt Lake City, Utah, USA).

11.2.2 Mean residence time (MRT)

In recent years more and more noncompartmental methods have been used for pharmacokinetic analysis. Twenty years ago statistical moment theory was introduced to pharmacokinetic analysis (Yamaoka et al. 1978). The times for the individual molecules to be eliminated can be described in terms of a statistical distribution function, i.e. the individual molecules can be eliminated just by chance within the first minutes or might still reside in the body weeks later. The mean residence time is a characteristic of this collective behaviour and is the mean of the residence times of individual molecules (Cutler 1987). The mean residence time can be regarded as the statistical moment analogy to half-life ($t_{1/2}$) (Gibaldi and Perrier 1982).

Assuming linear pharmacokinetics, the mean residence time (MRT) is characteristic for a special drug, independent of the administered dose. However, mean residence time is a function of how a drug is administered. The

MRT of a drug after non-instantaneous administration, e.g. intramuscular injection, will always be greater than the MRT after intravenous administration and can be regarded as approximately the sum of the MRT of the drug at the intramuscular depot and the MRT in the general circulation (Collier 1989; Mayer and Brazzell 1988; Yamaoka et al. 1978).

11.2.3 Pharmacokinetic simulation

Assuming linear pharmacokinetics, the parameters of the n-term polyexponential function describing single-dose pharmacokinetics of a drug can be used for prediction of drug concentrations applying multiple dosing (Wagner 1975). Maximal, minimal and average serum concentrations and the time to reach maximal serum concentrations at steady state can be predicted by computer simulation (Fig. 11.1). Area from zero to infinity under the single dose concentration-versus-time-curve is equal to the area within a dosage interval at the steady state (Gibaldi and Perrier 1982).

Computer simulation can also be applied to calculate an appropriate loading dose to shorten the period of time required to attain steady state plasma concentrations and to calculate expected serum concentrations for different loading doses, maintenance doses and injection intervals (Gibaldi and Perrier 1982; Gladtke and von Hattingberg 1977; Wagner 1975). Projections of multiple dose serum concentrations have the advantage that the design of a clinical study or therapy can be simulated and it is not necessary to perform multiple experiments for varying doses and injection intervals in humans. Even when a multiple dose study is performed, peak concentrations are often missed because of the blood sampling scheme used (Wagner 1975). Usually, drug serum concentrations are measured just before the next administration, but the serum concentration at this time point is just the minimal concentra-

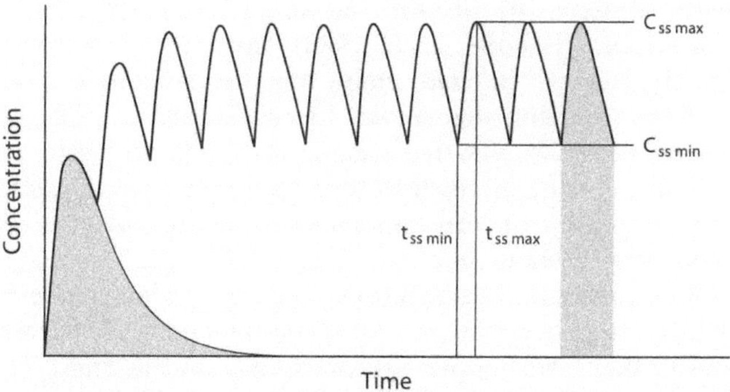

Fig. 11.1. Application of computer simulation based on pharmacokinetic parameters of single dose kinetics for prediction of multiple dose serum concentrations. $C_{ss\,max}$ maximal serum concentration at steady state; $C_{ss\,min}$ minimal serum concentration at steady state; $t_{ss\,max}$ time to reach maximal serum concentration at steady state; $t_{ss\,min}$ time to reach minimal serum concentration at steady state

tion at steady state and does not reflect the true drug concentration time course during the administration interval. To obtain the proper pharmacokinetic information, very frequent blood sampling would be necessary which is often neither feasible nor acceptable for patients or study volunteers. In this chapter some examples of dosage regimen calculations and prediction of multiple dose kinetics will be compared to the values of actually performed studies.

11.3 Pharmacokinetics of testosterone esters

Esterification of the testosterone molecule at position 17, e.g., with propionic or enanthic acid, prolongs the activity of testosterone in proportion to the length of the side chain when administered intramuscularly (Junkmann 1952, 1957). Studies applying gas chromatography-mass spectrometry that allow discrimination between endogenous testosterone and exogenously administered deuterium-labelled testosterone propionate-19,19,19-d_3 and its metabolite testosterone-19,19,19-d_3 were able to show that after intramuscular administration, the testosterone ester is slowly absorbed into the general circulation and then rapidly converted to the active unesterified metabolite (Fujioka et al. 1986). The observation that the time at the injection site is the major factor determining the residence time of the drug in the body agrees with pharmacokinetic studies in rats showing that the androgen ester 19-nortestosterone decanonate, when injected into the musculus gastrocnemius of the rat *in vivo*, is absorbed unchanged from the injection depot in the muscle into the general circulation according to first-order kinetics with a long half-life of 130 h (van der Vies 1965). Comparisons of the absorption kinetics of different testosterone esters clearly show that the half-lives of the absorption of the esters increase when the esterified fatty acids have a longer chain (van der Vies 1985). The ester is then rapidly hydrolysed in plasma, as could be shown by *in vitro* rat studies (van der Vies 1970) and *in vivo* human studies (Fujioka et al. 1986). The rate of hydrolysis again depends on the structure of the acid chain, but this process is much faster than release from the injection depot (van der Vies 1985). Similarly, the duration of action of the orally effective ester testosterone undecanoate seems to be dependent on the time of absorption of the uncleaved lipophilic testosterone undecanoate via the ductus thoracicus from the gut (Maisey et al. 1981; Schürmeyer et al. 1983).

The unesterified testosterone is the active substance for substitution therapy of male hypogonadism. As the metabolism of the testosterone ester to the unesterified testosterone occurs rapidly, and it could be shown that e.g. after intravenous injection of testosterone enanthate or testosterone these compounds have parallel pharmacokinetics (Sokol and Swerdloff 1986), the unesterified testosterone molecule determined by specific assay is regarded as the parameter suitable for evaluating the pharmacokinetics of different testosterone esters for substitution therapy of male hypogonadism (Cantrill et al. 1984; Conway et al. 1988; Sokol et al. 1982; Sokol and Swerdloff 1986; Snyder and Lawrence 1980).

It is known from clinical studies for male contraception that testosterone esters suppress the endogenous LH and testosterone secretion (Nieschlag and Behre 1997). If pharmacokinetics of testosterone esters are studied in normal male volunteers, the testosterone concentration measurable in the serum is the sum concentration resulting from the endogenous testosterone and the serum concentration of the exogenous testosterone hydrolysed from the ester (Anderson et al. 1997). Because the endogenous testosterone is suppressed to hypogonadal values during the first days after administration of the testosterone ester, the changes in testosterone serum concentrations after administration represent the combined pharmacokinetics of the endogenous and exogenous testosterone. Hypogonadal patients are characterised by impaired or absent endogenous testosterone secretion; exogenous testosterone administration can further suppress endogenous testosterone secretion only to a limited degree, if at all. Accordingly, in hypogonadal patients the serum concentration versus time profile is mainly a reflection of the pharmacokinetics of the exogenously administered testosterone ester alone. In this chapter the evaluation of pharmacokinetic parameters for different testosterone esters is based on the increases of testosterone serum concentrations over basal levels in hypogonadal patients.

11.3.1 Testosterone propionate

11.3.1.1 Single dose pharmacokinetics

Single dose pharmacokinetics of testosterone propionate were studied in seven patients with secondary hypogonadism due to chromophobe adenomas of the pituitary, aged 19–58 years (Nieschlag et al. 1976). Five patients were investigated prior to, two patients six months after surgical removal of the adenoma. No patients had received testosterone treatment previously. 50 mg of testosterone propionate were injected at 18.00 h on the control day. Blood samples were obtained at 8.00 h on the following test days. Basal testosterone levels, measured at 8.00 h on the control day, were subtracted from the testosterone concentrations measured on the test days to evaluate the pharmacokinetics of the exogenously administered testosterone. Measured testosterone concentrations±SEM and the best-fitted pharmacokinetic profile of testosterone propionate kinetics are shown in Fig. 11.2. Maximal testosterone levels in the supraphysiological range were seen shortly after injection (40.2 nmol/l, $t_{max} = 14$ h). Testosterone levels below the normal range were observed following day 2 (57 h) after injection. The calculated values for AUC were 1843 nmol × h/l, for MRT 1.5 d and 0.8 d for terminal half-life (Table 11.1).

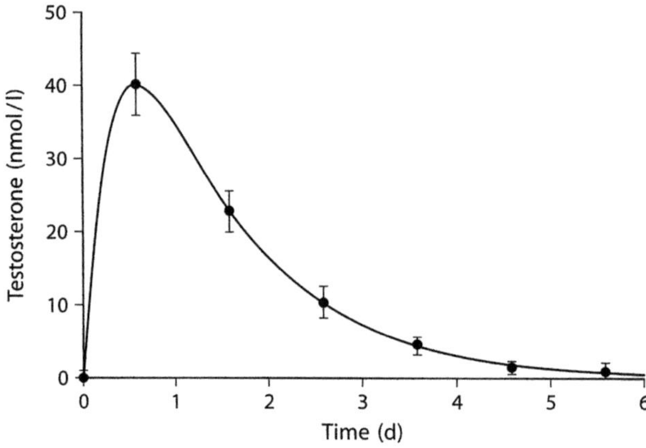

Fig. 11.2. Single dose pharmacokinetics of testosterone propionate in seven hypogonadal patients. *Closed circles,* mean ± SEM of testosterone serum concentrations actually measured; *curve,* best-fitted pharmacokinetic profile

Table 11.1 Comparative pharmacokinetics of different testosterone esters after intramuscular injection to hypogonadal patients. (MRT = mean residence time)

Testosterone ester	Terminal half-life (d)	MRT (d)
Testosterone propionate	0.8	1.5
Testosterone enanthate	4.5	8.5
Testosterone undecanoate	20.9	34.9
Testosterone buciclate	29.5	65.0

11.3.1.2 Multiple dose pharmacokinetics

Based on the single dose pharmacokinetic parameters, a multiple dose pharmacokinetic simulation was performed. Expected testosterone serum concentrations after multiple dosing of 2 times 50 mg testosterone propionate twice per week (e.g. injections Mondays and Thursdays, 8.00 h) are shown in Fig. 11.3. Shortly after injection high supraphysiological testosterone serum concentrations up to 45 nmol/l are observed. At the end of the injection interval (three and four days, respectively) testosterone serum concentrations below the lower range of normal testosterone values are projected (7 nmol/l and 3 nmol/l, respectively).

Judged by the data from pharmacokinetic analysis and simulation, administration of testosterone propionate is not suitable for substitution therapy of male hypogonadism because of resulting wide fluctuations of testosterone serum concentrations and maximal injection intervals of three days for the 50 mg dose.

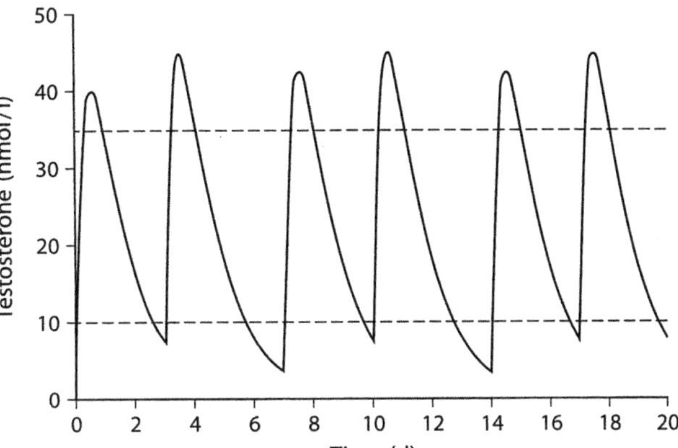

Fig. 11.3. Multiple dose pharmacokinetics of testosterone propionate after injection of 50 mg testosterone propionate twice per week (e.g. Mondays and Thursdays). *Solid curve,* pharmacokinetic simulation; *broken lines,* range of normal testosterone values

11.3.2 Testosterone enanthate

11.3.2.1 Single dose pharmacokinetics

Single dose pharmacokinetics of testosterone enanthate were studied in seven patients with primary hypogonadism, three castrates and four patients with Klinefelter syndrome, aged 20–58 years (Nieschlag et al. 1976). The usual testosterone substitution therapy in these patients was discontinued at least six weeks before the investigation. 250 mg of testosterone enanthate were injected at 18.00 h on the control day. Blood samples were obtained at 8.00 h on the following test days. Increases of testosterone serum concentrations (mean±SEM) over basal values (measured on control day) and the best fitted pharmacokinetic profile of testosterone enanthate kinetics are shown in Fig. 11.4. Maximal testosterone levels in the supraphysiological range were seen shortly after injection (39.4 nmol/l, $t_{max} = 10$ h). Testosterone levels below the normal range were observed following day 12 after injection. The calculated values were 9911 $nmol \times h/l$ for AUC, 8.5 d for MRT and 4.5 d for terminal half-life (Table 11.1).

11.3.2.2 Multiple dose pharmacokinetics

Based on the pharmacokinetic parameters of single dose pharmacokinetics multiple dose pharmacokinetic simulations for equal doses of 250 mg testosterone enanthate and injection intervals of one to four weeks were performed. With weekly injection intervals supraphysiological maximal testosterone serum concentrations up to 78 nmol/l are observed at steady state shortly after injection and supraphysiological minimal testosterone serum concentrations up to 40 nmol/l just before the next injection (Fig. 11.5). Injecting 250 mg of testosterone enanthate every two weeks results in maximal supraphysiological testosterone serum concentrations up to 51 nmol/l shortly after injection and testosterone serum levels at the lower range for normal

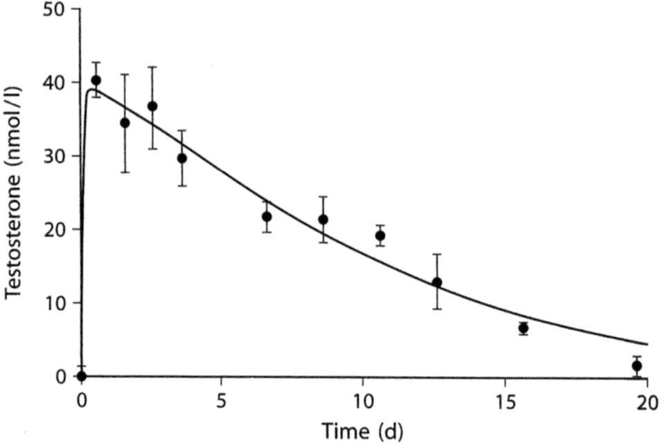

Fig. 11.4. Single dose pharmacokinetics of testosterone enanthate in seven hypogonadal patients. *Closed circles,* mean ± SEM of testosterone serum concentrations actually measured; *curve,* best-fitted pharmacokinetic profile

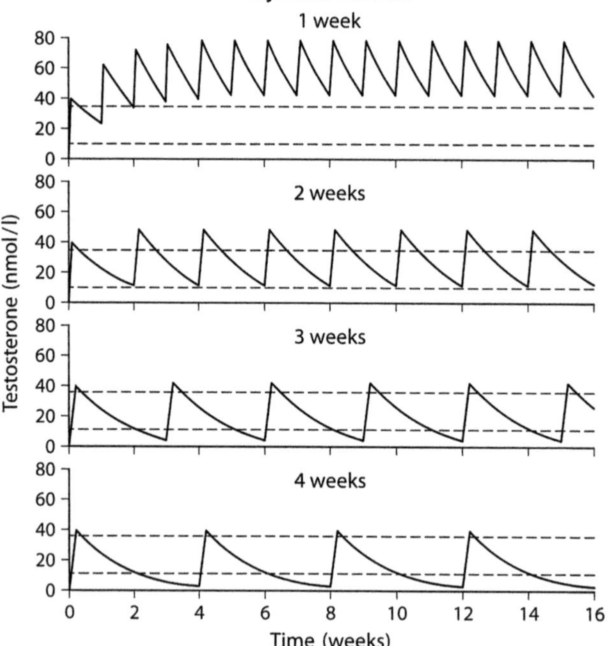

Fig. 11.5. Multiple dose pharmacokinetics of testosterone enanthate after injection of 250 mg testosterone enanthate every week (*upper panel*), every second week (*upper middle panel*), every three weeks (*lower middle panel*) and every four weeks (*lower panel*). *Solid curves,* pharmacokinetic simulations; *broken lines,* range of normal testosterone values

testosterone serum concentration shortly before the next injection. If the injection interval is extended to three weeks, testosterone serum concentrations below the normal range are observed 14 days after injection. With injection intervals of four weeks, testosterone serum concentrations are in the subnormal range at week three and four and effective testosterone substitution is not guaranteed (Fig. 11.5).

The calculated testosterone serum concentrations at steady state obtained by computer simulation correspond well to the results of published studies describing multiple dose testosterone enanthate pharmacokinetics. In a clinical trial for male contraception 20 healthy men were injected with 200 mg/wk of testosterone enanthate for 12 weeks (Cunningham et al. 1978). Minimal serum concentrations of testosterone at steady state, i.e. the testosterone serum concentration just before the next injection, were measured at 31.2 nmol/l to 39.5 nmol/l after weekly injection of 200 mg testosterone enanthate. Very similar data were obtained in recent contraceptive studies when normal men received 200 mg/wk testosterone enanthate injections for 18 months (Anderson and Wu 1996; Wu et al. 1996). The data of these studies fit well with the computer-calculated minimal testosterone serum concentrations of 40 nmol/l and maximal testosterone levels of 78 nmol/l after multiple injections of testosterone enanthate at a dosage of 250 mg/wk.

Snyder and Lawrence (1980) administered 100 mg/wk (n=12), 200 mg/2 wks (n=10), 300 mg/3 wks (n=9) and 400 mg/4 wks (n=6) testosterone enanthate to hypogonadal patients during a study period of three months. Blood was drawn during the last injection period, when steady state had been reached, every day (100 mg/wk) up to every fourth day (400 mg/4 wks). Similar to the computer simulation described above for 250 mg testosterone enanthate and injections intervals of one to four weeks, initial supraphysiological testosterone serum levels were seen shortly after injection. In the 100 mg/wk treatment group, where daily blood sampling was performed, mean peak serum concentrations were seen 24 h after injection. Comparable to the results of the computer simulation, after injection of 200 mg/2 wks testosterone enanthate, following initial supraphysiological testosterone serum levels, values fell to progressively lower values before the next injection, eventually reaching the lower normal limit (Snyder and Lawrence 1980). Similar results were described after injection of 300 mg/3 wks or 400 mg/4 wks testosterone enanthate. The authors conclude that the testosterone enanthate doses of 200 mg have to be injected every two weeks or of 300 mg every 3 weeks to guarantee an effective substitution therapy.

Demisch and Nickelsen (1983) deduce from their studies with testosterone enanthate for testosterone replacement therapy that if a dose of 250 mg testosterone enanthate once every three weeks is used, the concentration of both total and "free" testosterone are sufficiently high in the first and second week. In the third week, however, total testosterone moves to the lower limit of the male range and erectile disturbances were reported by the patients.

11.3.3 Testosterone cypionate and testosterone cyclohexanecarboxylate

Testosterone cypionate (cyclopentylpropionate) pharmacokinetics were compared with those of testosterone enanthate in a cross-over study involving six healthy men aged 20–29 years. Three subjects received 194 mg of testosterone enanthate, followed seven weeks later by 200 mg of testosterone cypio-

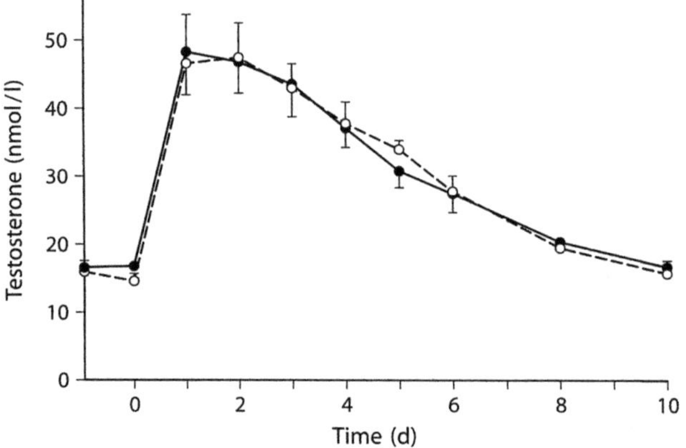

Fig. 11.6. Comparative pharmacokinetics of 194 mg of testosterone enanthate and 200 mg of testosterone cypionate after intramuscular injection to 6 normal volunteers. *Closed circles,* mean ± SEM of testosterone enanthate kinetics; *open circles,* mean ± SEM of testosterone cypionate kinetics

nate and vice versa (amount of unesterified testosterone 140 mg in both preparations). The serum testosterone profiles were identical after injection of both preparations in equivalent doses, both in terms of maximal concentrations and in terms of duration of elevation above basal levels (Fig. 11.6, Schulte-Beerbühl and Nieschlag 1980).

In a subsequent clinical study the pharmacokinetics of testosterone cyclohexanecarboxylate were compared to the pharmacokinetics of testosterone enanthate in a single blind cross-over study in seven healthy young men, aged 22–31 years. Four volunteers chosen at random first received 200 mg testosterone cyclohexanecarboxylate, followed five weeks later by a 200 mg testosterone enanthate injection. The other three volunteers received testosterone enanthate first and testosterone cyclohexanecarboxylate five weeks later (Schürmeyer and Nieschlag 1984).

After injection of either testosterone enanthate or testosterone cyclohexanecarboxylate, testosterone concentrations in serum increased sharply and reached maximum levels, 4–5 times above basal, 8–24 h after injection. During following days a parallel decay of testosterone levels occurred after injection of both ester preparations, with testosterone serum concentrations slightly, but significantly lower after testosterone cyclohexanecarboxylate injection compared to testosterone enanthate injection two, three and seven days after administration. Basal serum levels were reached seven days after testosterone cyclohexanecarboxylate administration and nine days after injection of testosterone enanthate.

Because testosterone cypionate, testosterone cyclohexanecarboxylate and testosterone enanthate had comparable suppressing effects on LH and consequently on endogenous testosterone secretion, it can be concluded from these studies in normal volunteers that all three esters with similar molecular structure possess comparable pharmacokinetics of exogenous testosterone serum concentrations. Testosterone cypiónate or testosterone cyclohexanecarboxylate do not provide a more advantageous pharmacokinetic profile than testosterone enanthate.

This observation is in agreement with a clinical study of replacement therapy with single dose administration of 200 mg of testosterone cypionate in 11 hypogonadal patients (Nankin 1987). Pharmacokinetics of testosterone cypionate were very similar to the pharmacokinetics of testosterone enanthate described above, with peak testosterone serum concentrations shortly occurring after injection and decay to baseline values by day 13 to 14 after injection.

11.3.4 Testosterone ester combinations

Testosterone ester mixtures have been widely used for substitution therapy of male hypogonadism (e.g. *Testoviron Depot 50*: 20 mg testosterone propionate and 55 mg testosterone enanthate; *Testoviron Depot 100*: 25 mg testosterone propionate and 110 mg testosterone enanthate; *Sustanon* 250: 30 mg testosterone propionate, 60 mg testosterone phenylpropionate, 60 mg testosterone isocaproate and 100 mg testosterone decanoate). These combinations are used following the postulate that the so-called short-acting testosterone ester (e.g testosterone propionate) is the effective testosterone for substitution during the first days of treatment and the so-called long-acting testosterone (e.g. testosterone enanthate) warrants effective substitution for the end of injection interval. However, this assumption is not supported by the pharmacokinetic parameters of the single testosterone esters. Both testosterone propionate and testosterone enanthate cause highest testosterone serum concentrations shortly after injection (Fig. 11.2 and Fig. 11.4). Accordingly, addition of testosterone propionate to testosterone enanthate only increases the initial undesired testosterone peak and worsens the pharmacokinetic profile that ideally should follow zero-order kinetics (Fig. 11.7). The computer simulation agrees well with the limited published single dose testosterone values that have been measured in hypogonadal patients treated with the combination of testosterone propionate and testosterone enanthate. Maximal increases

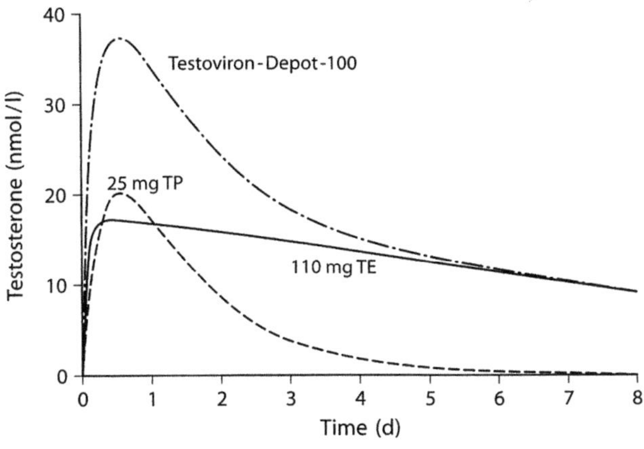

Fig. 11.7. Pharmacokinetic profile of Testoviron Depot 100 (110 mg testosterone enanthate and 25 mg testosterone propionate) in comparison to the pharmacokinetics of the individual testosterone esters of the mixture. *Curves*, pharmacokinetic simulations

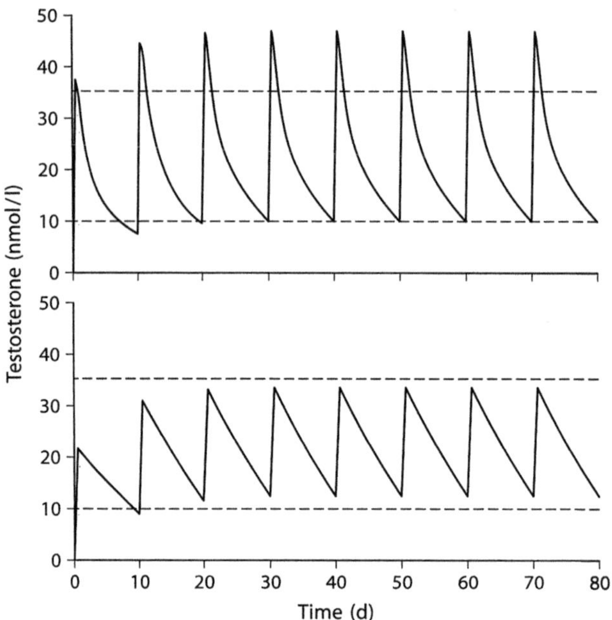

Fig. 11.8. Multiple dose pharmacokinetics of the testosterone ester mixture Testoviron Depot 100 (110 mg testosterone enanthate and 25 mg testosterone propionate = 100 mg unesterified testosterone, upper panel) every 10 d in comparison to 139 mg testosterone enanthate (= 100 mg unersterified testosterone, lower panel) every 10 d. *Solid curves,* pharmacokinetic simulations; *broken lines,* range of normal testosterone values

of approximately 40 nmol/l testosterone over basal values are described one day after intramuscular administration of a testosterone ester combination of 115.7 mg testosterone enanthate and 20 mg testosterone propionate to three hypogonadal patients (Fukutani et al. 1974).

A comparison of computer-simulated testosterone serum concentrations after multiple dose injections of Testoviron Depot 100 (110 mg testosterone enanthate and 25 mg testosterone propionate = 100 mg unesterified testosterone) every 10 d and 139 mg testosterone enanthate (= 100 mg unersterified testosterone) every 10 d is shown in Fig. 11.8. As can be expected by the single dose kinetics of the individual esters, injection of the testosterone ester mixture (upper panel) produces a much wider fluctuation of testosterone serum concentrations relative to injection of testosterone enanthate alone (lower panel). This simulation shows that the injections of testosterone enanthate alone produce a more favourable pharmacokinetic profile in comparison to injections of testosterone propionate and testosterone enanthate ester mixtures in comparable doses. For treatment of male hypogonadism there is no advantage in combining short- and long-acting testosterone esters.

11.3.5 Testosterone undecanoate

11.3.5.1 Oral administration

Testosterone undecanoate pharmacokinetics after single dose administration were tested in eight hypogonadal patients and twelve normal men (Schürmeyer et al. 1983). Directly before and at hourly intervals after oral application of three times 40 mg of testosterone undecanoate in arachis oil (Andriol) taken together with a standardized breakfast, matched saliva samples, as a parameter for free testosterone at the tissue level, and blood samples were collected at hourly intervals for up to 8 h. After administration of testosterone undecanoate serum and saliva testosterone always showed a parallel rise and fall, as demonstrated by a constant saliva/serum testosterone ratio. On average maximum levels could be observed 5 h after testosterone undecanoate administration. However, the serum testosterone profile showed a high interindividual variability of the time when maximum concentrations were reached, as well as of the maximum levels themselves that ranged from 17 to 96 nmol/l. Due to the high interindividual variability of testosterone serum levels achieved, no meaningful n-term polyexponential function describing the pharmacokinetics of testosterone undecanoate in normal men or hypogonadal patients could be calculated by computer analysis. When the individual serum concentration versus time curves were centralized about the time of maximal serum concentrations, serum concentrations significantly different from basal values were seen only two hours before and one hour after the time of maximal serum concentrations in hypogonadal patients (Fig. 11.9, Schürmeyer et al. 1983). Based on this observation it can be deduced that even with administration of testosterone undecanoate 3 times daily, only short-lived testosterone peaks resulting in high fluctuations can be obtained.

This judgment is in agreement with the data of a two month multiple dose study with testosterone undecanoate for replacement therapy in hypogonadal men (Skakkebaek et al. 1981). Applying a double blind cross-over design,

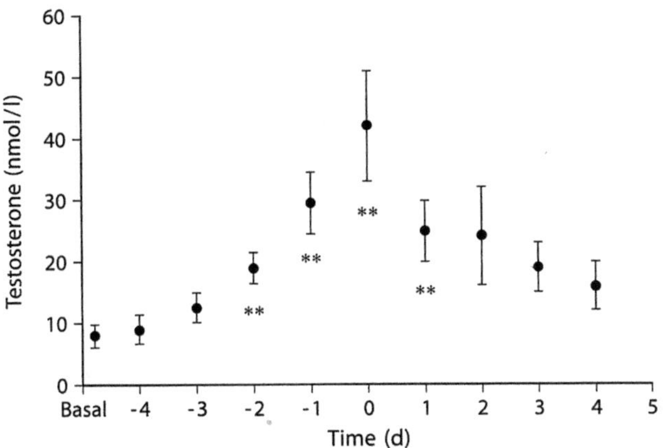

Fig. 11.9. Single dose pharmacokinetics of testosterone undecanoate after oral administration of 120 mg of the ester to 8 hypogonadal patients. Because of high interindividual variability of testosterone serum concentrations after administration of testosterone undecanoate, individual curves were all centralized about the time of maximal serum concentrations (time 0). *Asterisks* indicate significant higher testosterone serum concentrations compared to pretreatment values (basal) (mean ± SEM)

serum testosterone levels were studied in 12 hypogonadal patients to whom 80 mg of testosterone undecanoate had been administered twice per day 12 h apart. Whereas 4 h after administration of testosterone undecanoate a significant increase of testosterone serum levels was observed compared to the placebo group, no significant difference in testosterone serum levels between treatment and placebo control group was seen 12 h after administration. Even 4 h after administration, in four of twelve patients testosterone levels were still below the lower level of the normal range after both one month and two months of treatment. A significant marked variability between subjects as well as within the same subjects has also been observed in other clinical studies (Cantrill et al. 1984; Conway et al. 1988). From these and our own studies on the pharmacokinetics of testosterone undecanoate it can be concluded that, because of its unpredictable pharmacokinetics due to high inter- and intraindividual variability, the marked fluctuations of testosterone serum levels and the unreliable rise of testosterone levels to the normal range, testosterone undecanoate is not a satisfactory drug for substitution therapy of male hypogonadism.

11.3.5.2 Intramuscular administration

11.3.5.2.1 Preclinical studies

While testosterone undecanoate has been available for oral substitution for more than two decades, it was recently demonstrated in China that intramuscular administration of testosterone undecanoate in teaseed oil (125 mg/ml) has a prolonged duration of action (Wang 1991). We therefore tested the pharmacokinetics of testosterone undecanoate in comparison to testosterone enanthate in two groups of orchiectomized cynomolgus monkeys (Partsch et al. 1995). After injection of 10 mg/kg body weight of the respective esters serum levels of testosterone remained above the lower limit of normal for 108 days, compared to 31 days after testosterone enanthate injection. Pharmacokinetic analysis revealed a terminal half-life of 25.7 ± 4.0 days and mean residence time of 40.7 ± 4.1 days for testosterone undecanoate, compared to 10.3 ± 1.1 days and 11.6 ± 1.1 days for testosterone enanthate. The maximal testosterone concentration of 72.6 ± 11.7 nmol/l after testosterone undecanoate injection was significantly lower than 177.0 ± 21.3 nmol/l after testosterone enanthate injection.

11.3.5.2.2 Clinical studies

Because of the favourably long duration of action of intramuscular testosterone undecanoate in monkeys we subsequently performed a clinical phase I-study on the single dose pharmacokinetics of testosterone undecanoate in man. Hypogonadal patients were given intramuscular injections of 250 mg (n=7) or 1000 mg testosterone undecanoate (n=7). Follow-up examinations were performed 1, 2, 3, 5 and 7 days after injection and then weekly up to

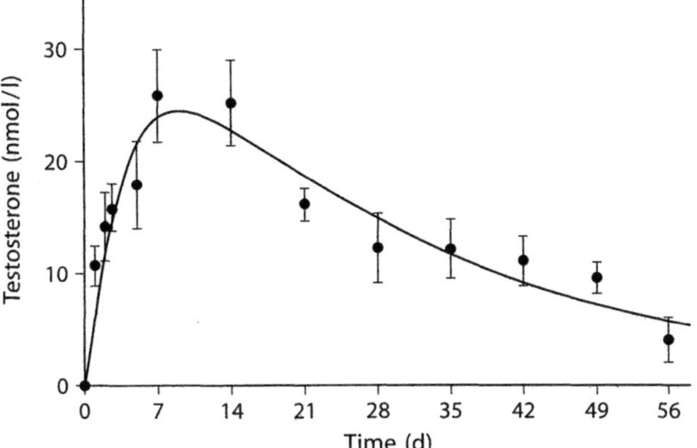

Fig. 11.10. Single dose pharmacokinetics of testosterone undecanoate after intramuscular injection of 1000 mg of the ester to seven hypogonadal patients. *Closed circles,* mean±SEM of testosterone serum concentrations actually measured; *curve,* best-fitted pharmacokinetic profile

study week 8. Whereas no prolonged increase of testosterone was observed in the 250 mg-group, serum levels of testosterone in the higher dose group increased from 4.8±0.9 nmol/l (mean±SEM) to maximum levels of 30.5±4.3 nmol/l at day 7 (t_{max}). Testosterone levels remained within the normal range up to week 7 (13.5±1.2 nmol/l). Non-linear least squares regression analysis revealed a terminal elimination half-life for intramuscular testosterone undecanoate of 20.9±6.0 days and a mean residence time of 34.9±8.2 days (Fig. 11.10) (Table 11.1).

Similar to the preclinical study in monkeys, the clinical study in hypogonadal men demonstrated favourable pharmacokinetics of intramuscular testosterone undecanoate. Because of the relatively low concentration of 125 mg testosterone undecanoate per milliliter teaseed oil, however, administration of the 1000 mg dose requires an injection volume of 8 ml which renders intramuscular administration impracticable. In collaboration with Jenapharm (Jena, Germany) we reformulated the intramuscular testosterone undecanoate preparation and performed a clinical study with testosterone undecanoate dissolved in castor oil at a higher concentration of 250 mg/ml. 14 hypogonadal patients received an intramuscular injection of 1000 mg of the reformulated testosterone undecanoate preparation. Maximal serum levels were lower than in the study with the Chinese preparation and did not exceed 25 nmol/l. This observation is in agreement with a elegant study on a possible influence of injection volume on the pharmacokinetics of nandrolone esters (Minto et al. 1997). As in the first study the duration of action of intramuscular testosterone undecanoate was six to eight weeks. Follow-up studies with multiple injections of 1000 mg testosterone undecanoate every six to eight weeks are currently being performed which are based on pharmacokinetic computer simulation.

11.3.6 Testosterone buciclate

11.3.6.1 Preclinical studies

A first study on the pharmacokinetics of the new WHO/NIH androgen ester testosterone buciclate was performed in two groups of four long-term orchiectomized cynomolgus monkeys, *Macaca fascicularis*, weighing 2.8–4.6 kg (Weinbauer et al. 1986). One group received a single intramuscular injection of 40 mg testosterone buciclate in aqueous suspension or 32.8 mg testosterone enanthate dissolved in sesame oil. Both preparations contained equal amounts of testosterone, namely 23.6 mg. Testosterone enanthate injections resulted in supraphysiological serum levels of testosterone for eight days, followed by a rapid decline with levels lower than the physiological limit after three weeks. In contrast, testosterone buciclate produced a moderate increase of serum testosterone levels into the physiological range with a peak level (c_{max}) of 29.9 ± 5.9 nmol/l on day 14. Serum levels of testosterone remained in the physiological range for a period of four months. The levels never exceeded the physiological range of testosterone.

These favourable results on the pharmacokinetics of testosterone buciclate were confirmed in castrated rhesus monkeys. After a single intramuscular injection of 40 mg testosterone buciclate in 1 ml to nine monkeys, serum levels of testosterone remained in the normal physiological range for 80–136 days (Rajalakshmi and Ramakrishnan 1989). In addition, it could be demonstrated in castrated rhesus monkeys that significantly higher testosterone levels can be achieved when the intramuscular injection of 80 mg testosterone buciclate is given as four injections of 0.5 ml at four different sites compared to a single injection of 2 ml (Rajalakshmi and Ramakrishnan 1989).

11.3.6.2 Clinical studies

To assess the pharmacokinetics of testosterone buciclate in men the first clinical study was performed in eight men with primary hypogonadism under the auspices of the WHO Male Task Force on Methods for the Regulation of Male Fertility (Behre and Nieschlag 1992). The men were randomly assigned to two study groups and were given either 200 (group I) or 600 mg (group II) testosterone buciclate intramuscularly. Whereas in group I serum androgen levels did not rise to normal values, in group II androgens increased significantly and were maintained in the normal range up to 12 weeks with maximal serum levels (c_{max}) of 13.1 ± 0.9 nmol/l (mean $\pm$ SEM) in study week 6 (t_{max}). No initial burst release of testosterone was observed in either study group. Pharmacokinetic analysis revealed a terminal elimination half-life of 29.5 ± 3.9 days and a mean residence time of 65.0 ± 9.9 days (Fig. 11.11) (Table 11.1).

Because of the promising results of the first clinical study with testosterone buciclate, a follow-up study was initiated in six patients with primary hypogonadism. After complete wash-out from previous therapy all men re-

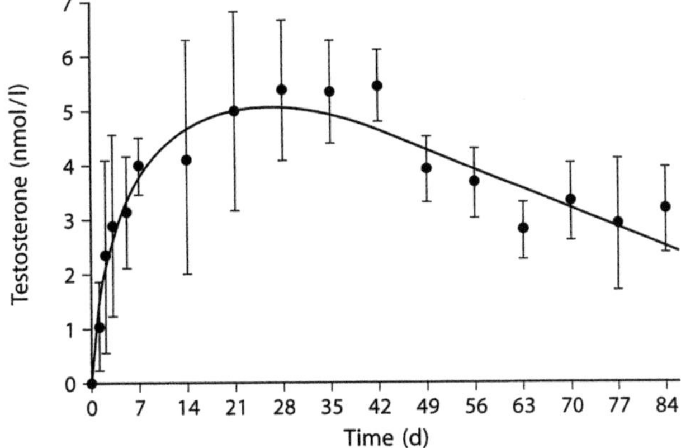

Fig. 11.11. Single dose pharmacokinetics of testosterone buciclate after intramuscular injection of 600 mg of the ester to four hypogonadal patients. *Closed circles,* mean ± SEM of testosterone serum concentrations actually measured; *curve,* best-fitted pharmacokinetic profile

ceived a single intramuscular injection of 1000 mg testosterone buciclate. As in the previous study with lower doses, no initial burst release of testosterone was observed. Maximal testosterone serum levels were observed nine weeks (t_{max}) after injection with a mean value of 13.1 ± 1.8 nmol/l (c_{max}). Following peak concentrations, testosterone serum levels gradually declined and remained within the normal range up to week 16. This study demonstrated that an increase of the injected dose of testosterone buciclate from 600 to 1000 mg prolongs the duration of action significantly, but does not lead to significantly higher maximal serum levels of testosterone.

The long duration of action of testosterone buciclate was recently also demonstrated in the first contraceptive study with this new testosterone ester. After a single injection of 1200 mg testosterone buciclate at a concentration of 400 mg/ml to eight normal men, serum levels of testosterone remained within the normal range, whereas gonadotropins and spermatogenesis was significantly suppressed for at least 18 weeks (Behre et al. 1995). These studies demonstrate that the long-acting testosterone buciclate is well suited for substitution therapy of male hypogonadism as well as for male contraception.

11.4. Key messages

- Results of clinical studies demonstrate that computerized pharmacokinetic analysis and simulation can be applied with advantage to predict multiple-dose serum concentrations from single dose kinetics and are valuable tools for planning clinical therapy.
- The available testosterone esters for intramuscular injection (testosterone propionate, testosterone enanthate, testosterone cypionate, testosterone cyclohexanecarboxylate) are not satisfactory for the treatment of male hypogonadism. Doses and injection intervals most frequently used

in the clinic lead to initial supraphysiological testosterone levels and subnormal values before the next injection. To obtain testosterone serum concentrations continuously in the normal range, unacceptably frequent small doses would have to be injected.

- Oral administration of testosterone undecanoate results in high interindividual and intraindividual variability of serum testosterone values. The testosterone elevations, only short-lived, result in wide fluctuations of serum concentrations.

- Intramuscular injection of 1000 mg testosterone undecanoate to hypogonadal men maintains serum levels of testosterone within the normal range for six to eight weeks weeks. If multiple dose studies currently performed confirm these initial results, intramuscular testosterone undecanoate could become a valuable preparation for substitution therapy of male hypogonadism and for male contraception.

- The longest duration of action can be achieved with testosterone buciclate. A single intramuscular injection of 1000 mg maintains serum testosterone in the physiological range for 16 weeks. Unfortunately, testosterone buciclate is not yet available for clinical therapy.

11.5. References

Anderson RA, Wu FC (1996) Comparison between testosterone enanthate-induced azoospermia and oligozoospermia in a male contraceptive study. II. Pharmacokinetics and pharmacodynamics of once weekly administration of testosterone enanthate. J Clin Endocrinol Metab 81:896–901

Anderson RA, Wallace AM, Kicman AT, Wu FC (1997) Comparison between testosterone enanthate-induced azoospermia and oligozoospermia in a male contraceptive study. IV. Suppression of endogenous testicular and adrenal androgens. Hum Reprod 128:1657–1662

Behre HM, Nieschlag E (1992): Testosterone buciclate (20 Aet-1) in hypogonadal men: pharmacokinetics and pharmacodynamics of the new long-acting androgen ester. J Clin Endocrinol Metab 75:1204–1210

Behre HM, Baus S, Kliesch S, Keck C, Simoni M, Nieschlag E (1995) Potential of testosterone buciclate for male contraception: endocrine differences between responders and nonresponders. J Clin Endocrinol Metab 80:2394–2403

Cantrill AJ, Dewis P, Large DM, Newman M, Anderson DC (1984) Which testosterone replacement therapy? Clin Endocrinol 21:97–107

Collier PS (1989) Mean residence time: some further considerations. Biopharm Drug Dispos 10:443–451

Conway AJ, Boylan LM, Howe C, Ross G, Handelsman DJ (1988) Randomized clinical trial of testosterone replacement therapy in hypogonadal men. Int J Androl 11:247–264

Cunningham GR, Silverman VE, Kohler PO (1978) Clinical evaluation of testosterone enanthate for induction and maintenance of reversible azoospermia in man. In: Patanelli DJ (ed) Hormonal control of male fertility. Department of Health, Education and Welfare. National Institutes of Health, Bethesda, Md, Publ No (NIH) 78–1097, pp 71–87

Cutler DJ (1987) Definition of mean residence times in pharmacokinetics. Biopharm Drug Dispos 8:87–97

Demisch K, Nickelsen T (1983) Distribution of testosterone in plasma proteins during replacement therapy with testosterone enanthate in patients suffering from hypogonadism. Andrologia 15:536–541

Fujioka M, Shinohara Y, Baba S, Irie M, Inoue K (1986) Pharmacokinetic properties of testosterone propionate in normal men. J Clin Endocrinol Metab 63:1361–1364

Fukutani K, Isurugi K, Takayasu H, Wakabayashi K, Tamaoki B-I (1974) Effects of depot testosterone therapy on serum levels of luteinizing hormone and follicle-stimulating hormone in patients with Klinefelter's syndrome and hypogonadotropic eunuchoidism. J Clin Endocrinol Metab 39:856–864

Gibaldi M, Perrier D (1982) Pharmacokinetics. Marcel Dekker, In., New York

Gladtke E, von Hattingberg HM (1977) Pharmakokinetik. 2nd ed Springer, Berlin, Heidelberg, New York

Junkmann K (1952) Über protrahiert wirksame Androgene. Arch Path Pharmacol 215:85–92

Junkmann K (1957) Long-acting steroids in reproduction. Recent Prog Horm Res 13:389–419

Maisey NM, Bingham J, Marks V, English J, Chakraborty J (1981) Clinical efficacy of testosterone undecanoate in male hypogonadism. Clin Endocrinol 14:625–629

Mayer PR, Brazzell RK (1988) Application of statistical moment theory to pharmacokinetics. J Clin Pharmacol 28:481–483

Minto CF, Howe C, Wishart S, Conway AJ, Handelsman DJ (1997). Pharmacokinetics and pharmacodynamics of nandrolone esters in oil vehicle: effects of ester, injection site and injection volume. J Pharmacol Exp Ther 281:93–102.

Nankin HR (1987) Hormone kinetics after intramuscular testosterone cypionate. Fertil Steril 47:1004–1009

Nieschlag E, Behre HM (1997) Male contribution to contraception – experimental approaches. In: Nieschlag E, Behre HM (eds) Andrology – Male reproductive health and dysfunction. Berlin, Heidelberg, New York: Springer-Verlag, pp 386–393

Nieschlag E, Cüppers HJ, Wiegelmann W, Wickings EJ (1976) Bioavailability and LH-suppressing effect of different testosterone preparations in normal and hypogonadal men. Horm Res 7:138–145

Partsch CJ, Weinbauer GF, Fang R, Nieschlag E (1995) Injectable testosterone undecanoate has more favourable phamacokinetics and pharmacodynamics than testosterone enanthate. Eur J Endocrinol 132:514–519

Rajalakshmi M, Ramakrishnan PR (1989): Pharmacokinetics and pharmacodynamics of a new long-acting androgen ester: maintenance of physiological androgen levels for 4 months after a single injection. Contraception 40:399–412

Schulte-Beerbühl M, Nieschlag E (1980) Comparison of testosterone, dihydrotestosterone, luteinizing hormone, and follicle-stimulating hormone in serum after injection of testosterone enanthate or testosterone cypionate. Fertil Steril 33:201–203

Schürmeyer T, Nieschlag E (1984) Comparative pharmacokinetics of testosterone enanthate and testosterone cyclohexanecarboxylate as assessed by serum and salivary testosterone levels in normal men. Int J Androl 7:181–187

Schürmeyer T, Wickings EJ, Freischem CW, Nieschlag E (1983) Saliva and serum testosterone following oral testosterone undecanoate administration in normal and hypogonadal men. Acta endocrinol 102:456–462

Skakkebaek NE, Bancroft J, Davidson DW, Warner P (1981) Androgen replacement with oral testosterone undecanoate in hypogonadal men: a double blind controlled study. Clin Endocrinol 14:49–61

Snyder PJ, Lawrence DA (1980) Treatment of male hypogonadism with testosterone enanthate. J Clin Endocrinol Metab 51:1335–1339

Sokol RZ, Swerdloff RS (1986) Practical considerations in the use of androgen therapy. In: Santen JR, Swerdloff RS (eds) Male reproductive dysfunction. Marcel Dekker, New York, pp 211–225

Sokol RZ, Palacios A, Campfield LA, Saul C, Swerdloff RS (1982) Comparison of the kinetics of injectable testosterone in eugonadal and hypogonadal men. Fertil Steril 37:425–430

van der Vies J (1965) On the mechanism of action of nandrolone phenylpropionate and nandrolone decanoate in rats. Acta endocrinol 49:271–282 van der Vies J (1970) Model studies in vitro with long-acting hormonal preparations. Acta endocrinol 64:656–669

van der Vies J (1985) Implications of basic pharmacology in the therapy with esters of nandrolone. In: Eikelboom FA, van der Vies J (eds) Anabolics in the '80s. Acta endocrinol, suppl 271, 110:38–44

Wagner JG (1975) Fundamentals of clinical pharmacokinetics. Drug Intelligence Publications, Inc, Hamilton, Illinois

Wang LZ (1991) The therapeutic effect of domestically produced testosterone undecanoate in Klinefelter syndrome. New Drugs Market 8:28–32

Weinbauer GF, Marshall GR, Nieschlag E (1986) New injectable testosterone ester maintains serum testosterone of castrated monkeys in the normal range for four months. Acta endocrinol 113:128–132

Wu FCW, Farley TMM, Peregoudov A, Waites GMH, WHO (1996) Effects of testosterone enanthate in normal men: experience from a multicenter contraceptive efficacy study. Fertil Steril 65:626–636

Yamaoka K, Nakagawa T, Uno T (1978) Statistical moments in pharmacokinetics. J Pharmacokin Biopharm 6:547–558

12 Clinical pharmacology of testosterone pellet implants

David J. Handelsman

Contents

12.1 Introduction

Androgen therapy may be divided into physiological and pharmacological applications. The pharmacological applications usually involve non-physiological doses of synthetic androgens as second-line, empirical treatment where more specific medical therapy is not yet available. The physiological applications consist of androgen replacement therapy, the treatment of androgen deficiency in hypogonadal men. Androgen replacement therapy aims to replicate physiological actions of endogenous testosterone by steadily maintaining physiological blood levels of testosterone. Since the underlying disorders are virtually always irreversible, this requires life-long administration of testosterone, making it desirable that the testosterone formulations be long-acting. Reliable therapeutic compliance over the lifetime of the patient depends heavily on a convenient formulation which ensures the continuity of treatment. The pharmacological properties of testosterone, notably its rapid hepatic metabolism and very low oral bioavailability, dictate the need for development of depot, sustained-release testosterone formulations (Parkes 1938; Wilson 1980). The perfect depot would be safe, effective, inexpensive, convenient, and long-acting with a reproducible, zero-order release profile. Not surprisingly, even six decades after entry of testosterone into clinical use (Foss 1939; Hamilton 1937), this ideal has not been achieved. Nevertheless one of the oldest testosterone formulations, the subdermal testosterone implant, provides a very close approximation to this ideal in providing stable blood testosterone levels lasting 4–6 months after a single implantation. Curiously this cheap, safe and effective treatment modality was neglected for decades despite its many advantages for androgen replacement therapy but is now undergoing a revival of interest, particularly since its desirable pharmacological properties have been outlined (Cantrill et al. 1984; Conway et al. 1988; Handelsman et al. 1990, 1997; Jockenhövel et al. 1996; Nieschlag 1996; Zacharin and Warne 1997).

12.2 History

Remarkably by modern standards, testosterone entered clinical usage (Hamilton 1937) within two years of its chemical identification (David et al. 1935) and synthesis (Butenandt and Hanisch 1935; Ruzicka and Wettstein 1935) in 1935. During its purification it became apparent that testosterone had negligible oral bioavailability and a very short duration of action parenterally (Foss 1939; Parkes 1938), later shown to be due to rapid hepatic metabolism (Frey et al. 1979; Hellman et al. 1956; Nieschlag et al. 1975, 1977). These pharmacological features led to an early recognition of the need for long-acting depot testosterone (Foss 1939; Parkes 1938). Within the few years until the hiatus created by World War II, numerous formulations of testosterone and its derivatives were reported. Subdermal pellet implantation was among the earliest effective modalities employed for clinical application of testosterone (Deansley and Parkes 1937; Howard and Vest 1939; Vest and Howard 1939). The experimental observation that subdermal implants showed the most potent, lasting effects of any steroid formulation (Deansley and Parkes 1938; Hamilton and Dorfman 1939) was quickly applied by clinical investigators (Foss 1939; Howard and Vest 1939; Vest and Howard 1939) and testosterone implants became an established form of androgen replacement therapy by 1940. Similar effects were observed for other bioactive steroids including androgens, estrogens, progestins and mineralocorticoids with depots lasting up to 1 year (Emmens 1941; Forbes 1941; Loeser 1940; Thorn and Firor 1940). Despite their long availability for clinical usage, few clinical trials (Reiter 1963; Swyer 1953) and only a single pharmacological study (Bishop and Folley 1951) of testosterone implants was reported until recently when pharmacological (Handelsman et al. 1990; Jockenhövel et al. 1996) and clinical (Cantrill et al. 1984; Conway et al. 1988; Handelsman et al. 1997; Zacharin and Warne 1997) studies redefined their clinical pharmacology and applications. Thus subdermal testosterone implantation, already established by 1940 as a highly effective and near-optimal depot form of androgen replacement, was curiously neglected until the recent revival of interest.

12.3 Formulation and physical features

The original testosterone implants were manufactured by high-pressure tableting of crystalline steroid with a cholesterol excipient (Loeser 1940; Bishop and Folley 1951). These proved brittle, hard to standardize or sterilize and exhibited surface unevenness and fragmentation during *in-vivo* absorption to produce an uneven late release rate (Emmens 1941). These limitations were overcome in the 1950's by switching to high-temperature moulding whereby molten testosterone (without excipients) was cast into cylindrical moulds to produce more robust implants. These have more uniform composition, resulting in a more steady and prolonged release and reduced tissue reaction. Sterilisation is achieved by a combination of high-temperature exposure during fabrication together with surface sterilisation or, more recently, terminal gamma-irradiation. The testosterone pellet implants (Fig. 12.1) currently are available in two sizes with a common diameter (4.5 mm), 100 mg (length 6 mm, surface area 117 mm^2) and 200 mg (length 12 mm, surface area 202 mm^2).

12.4 Implantation procedure

Pellets are implanted under sterile conditions for routine minor or office-type surgery using a trochar and cannula. The usual site is under the skin of the lower abdominal wall level with, and lateral to, the umbilicus but other possible sites include the buttock, deltoid, gluteal, and upper thigh, although there is less experience with these (Zacharin and Warne 1997). Disposable sterile plastic instruments as well as re-usable stainless steel ones that require sterilisation are available commercially. For the conventional lower abdominal site, a small incision (~1 cm) is made under local anaesthetic at

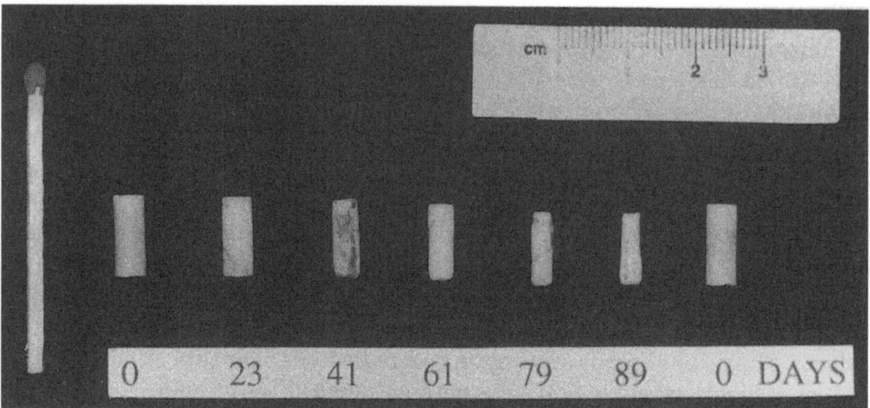

Fig. 12.1. Testosterone pellet implants made by melting crystalline steroid without excipient and moulding into a cylindrical shape (Organon Pty Ltd). Seven 200 mg testosterone implants are illustrated. The two un-implanted 200 mg pellets (far left and right) with length of 12 mm and diameter of 4.5 mm are shown with a centimeter scale and a matchstick for size comparison. Testosterone implants are absorbed by an uniform surface erosion mechanism that preserves their original cylindrical shape, as illustrated by the middle five implants which were extruded after 23 to 89 days carriage in the body. Time in days of residence in the body after implantation is indicated on the legend below the implants

least 5 cm from the mid-line at the level of the umbilicus to allow introduction of the trocar. Pellets are distributed in individual tracks fanning out from the puncture site and discharged from the trocar by an obturator at 5–10 cm distal to the puncture site. The puncture is closed without suture by using adhesive strips and covered with a waterproof dressing left in situ for a week. Antibiotics are not required routinely. Implantation may be more difficult in men with little subcutaneous fat and occasionally in areas with unusually heavy subdermal fibrosis from previous implantations.

An alternative implantation procedure uses an incision on the lateral aspect of the buttocks corresponding to the skin covered by underwear (Zacharin and Warne 1997). This site is cosmetically preferable both for being less visible externally as well as the incisions being reusable, i.e. so that pellets can be implanted alternate sides, repeatedly re-using previous insertion incisions and tracks. The implantation procedure is similar at this site except that two pellets are placed in each track and a single silk suture is used to close the wound; the patient is instructed to remove the suture himself about one week later.

12.5 Absorption

12.5.1 Mechanism of absorption

Absorption of testosterone from subdermal pellets occurs via uniform erosion of the pellet's surface from which the steroid leeches out according to the solubility of testosterone in the extracellular fluid. This is supported by the observation that pellets recovered up to 3 months after implantation retain their cylindrical shape (Fig. 12.1). A mathematical model incorporating a uniform rate of surface erosion (Forbes 1941) accurately fits direct measurements of release rate (Bishop and Folley 1951).

Deviations from the simple surface-erosion model may occur late in the time-course of absorption if the surface area from which absorption occurs enlarges unpredictably. This may happen where surface irregularities eventually lead to pitting or fragmentation as the pellet size decreases. Although pellet geometry (especially surface area) is the rate-limiting factor in absorption from subdermal testosterone pellets (Emmens 1941), other relevant factors include

1) steroid chemistry notably hydrophobicity,
2) pellet hardness, smoothness and size,
3) the site of implantation, its local blood flow and trauma and
4) the tissue reaction and encasing of the pellet (Bishop and Folley 1951; Emmens 1941; Forbes 1941; Foss 1939).

Circulating sex steroid levels, however, are not important (Emmens 1941). Few of these factors have been tested systematically in humans and their relative importance is unclear.

12.5.2 Absorption kinetics and bioavailability

Empirical estimates of the effective testosterone release rate can be made directly by measuring residue in extruded pellets according to time in situ as well as indirectly from the % absorbed-time plots, and these independent estimates are in remarkable agreement. For the 200 mg testosterone implants, a direct estimate of absorption rate is 1.3 (95% confidence interval 1.22–1.37) mg/day based on the weighed remnants of 59 extruded pellets which exhibited linear rate of release with time for over 100 days after implantation (Fig 12.2). This is corroborated independently by Wagner-Nelson plots which are also virtually linear (Fig 12.3) consistent with a good approximation to zero-order release of testosterone. The % absorbed-time plots provide an estimate of 2.5 months for the effective half-time of absorption and an independent estimate of 1.3 mg/day testosterone release rate for the 200 mg pellet. The number of pellets did not influence the testosterone absorption rate. These estimates are ~40% lower (despite 66% greater initial surface area) than the only previous estimate of 1.1 mg/day for a 100 mg pellet arrived at from weight of pellets removed at intervals after implantation in the antecubital or subscapular region (Bishop and Folley 1951). This discrepancy suggests either differences in pellet composition or possibly site-specific absorption characteristics. These testosterone release rates are also consistent with indirect estimates based on blood testosterone concentrations produced by implantation of single 100 mg and 200 mg pellets in women (Dewis et al. 1986;

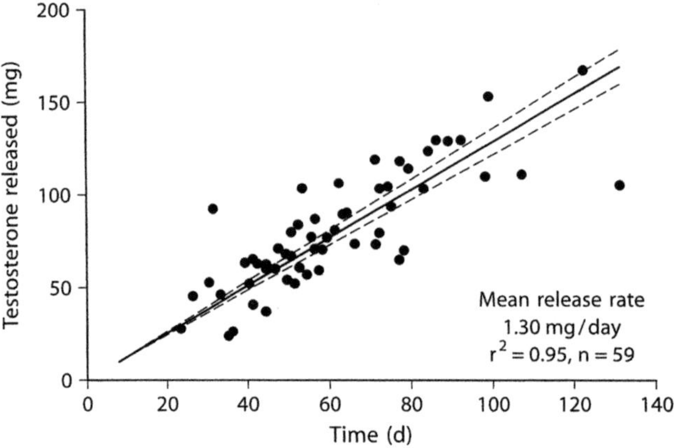

Fig 12.2. Direct estimates of testosterone release rate from plotting the weights of pellets extruded against time in days from implantation to extrusion. Extruded pellets were cleaned, dried and weighed to determine the mass of testosterone released by comparison with unimplanted 200 mg pellets (mean 202.5 mg). The amount of testosterone released was a linear function of time ($r^2 = 0.95$, n = 59) for up to 120 days. The testosterone release rate is estimated from the slope of the linear regression (solid line) and 95% prediction band (dotted lines). This was 1.3 mg/day per 200 mg pellet with 95% confidence interval of 1.22–13.7 mg/day

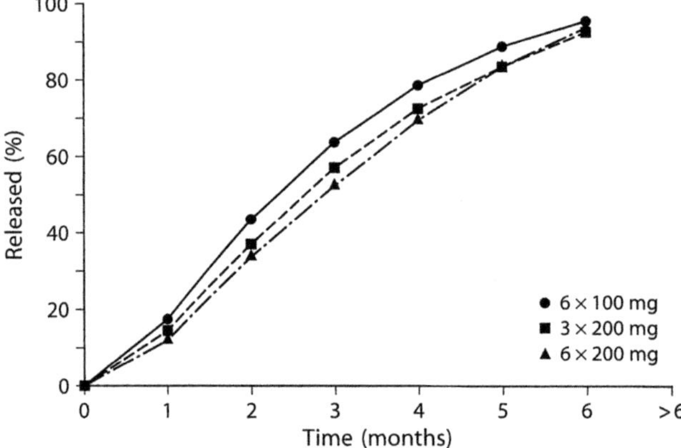

Fig 12.3. Percent absorbed-time plots of testosterone release over 6 months in 43 hypogonadal men on one of the 3 pellet regimens 6 × 200 mg (filled circles, n = 32), 6 × 100 mg (open circles, n = 28) and 3 × 200 mg (filled triangles, n = 51). The near linear Wagner-Nelson plots indicate a virtual zero-order release rate and the similarity of the 3 curves indicate that the release rates for the 100 mg and 200 mg pellets are similar

Thom et al. 1981) after correcting for gender differences in testosterone clearance rates (Gandy 1977; Southren et al. 1968).

Empirical confirmation of the pharmacological significance of the surface-area limited model of testosterone absorption from subdermal implants is provided by comparing the time-course of blood testosterone and gonadotrophin levels after implantation of either three 200 mg or six 100 mg pellets. While both regimens provide the same total testosterone dose (600 mg), the 16% higher initial surface area of the six 100 mg implants produced higher free testosterone levels and greater gonadotropin suppression in the first (but not the second) 3 months after implantation. This pharmacokinetic and pharmacodynamic evidence supports the surface area-limited model of testosterone absorption at steady-state.

Like other depot steroid formulations (Burris et al. 1988; Diaz-Sanchez et al. 1989), testosterone implants demonstrate a minor and transient accelerated initial (or "burst") release. This lasts for 1–2 days and involves only ~1.5% of total testosterone release reaching mean testosterone concentrations of <50 nmol/L (Jockenhövel et al. 1996). Although these testosterone concentrations exceed the eugonadal reference range (<35 nmol/L), they are less than the maximal blood testosterone concentrations produced by intramuscular testosterone ester injections which routinely peak at 40–80 nmol/L (Behre and Nieschlag 1998).

The bioavailability of testosterone (defined by appearance in bloodstream) from subdermal pellets is virtually complete. Full biodegradability is expected for a parenterally delivered steroid absorbed into the systemic circulation and avoiding first-pass hepatic inactivation. Following testosterone pellet implantation, blood SHBG concentrations, the major determinant of testosterone metabolic clearance rate (Petra et al. 1985), remains unaltered, making it reasonable to assume a constant testosterone clearance rate of 540 l/m^2/day (Southren et al. 1968; Gandy 1977) throughout the lifespan of an implant. Using the time-course of plasma testosterone, it can be calculated that virtually all the testosterone from a 200 mg implant would be absorbed within 6 months which corresponds to the calculation based on the testosterone release rate estimated independently.

12.6 Pharmacokinetics

12.6.1 Pharmacological studies

The clinical pharmacology of testosterone implants has been described in several studies (Conway et al. 1988; Handelsman et al. 1990; Jockenhövel et al. 1996). The most comprehensive involves a random sequence, cross-over clinical study of 43 androgen-deficient men with primary (hypergonadotropic, n=22) or secondary (hypogonadotropic, n=21) hypogonadism (Handelsman et al. 1990). Androgen-deficient men were treated sequentially with 3 regimens – six 100 mg, three 200 mg or six 200 mg testosterone implants – at intervals of ≥6 months when blood testosterone concentrations returned to baseline. Blood samples for testosterone and gonadotropin assay were obtained prior to and at 4-week intervals after implantation of 111 pellet implantations (6×100 mg–28 implants; 6×200 mg–32 implants; 3×200 mg–51 implants). Systematic study of various implant sites would be valuable to define the clinical role of alternative implantation sites and any site-specific variations in release rate and duration of action of pellets which are suspected (Bishop and Folley 1951; Handelsman et al. 1990).

12.6.2 Total and free testosterone levels

Implantation of testosterone pellets gives highly reproducible and dose-dependent time-course for circulating total and free testosterone (Fig. 12.4). Total testosterone levels on the 1200 mg dose are higher ($p < 0.0001$) than those of either 600 mg combinations which in turn produced similar ($p = 0.95$) circulating testosterone concentrations and time course. Plasma testosterone levels peaked at the first month and gradually declined to return to baseline. Testosterone concentrations reached baseline by 6 months after either of the 600 mg dose regimens ($3 \times 200mg$, $6 \times 100mg$) but remained significantly elevated after 6 months following the 1200 mg dose. Plasma free testosterone exhibited virtually the same time-course as total testosterone except that free (but not total) testosterone levels were significantly higher in the first 3 months after the 6×100 mg regimen (which has a higher initial surface area) compared with the 3×200 mg regimen. This was consistent pellet surface area being a major determinant of testosterone release rates.

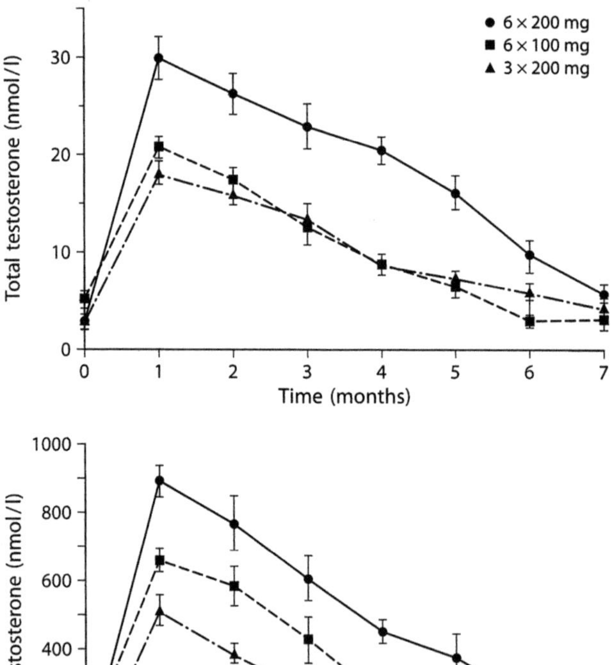

Fig 12.4. Total (upper panel) and free (lower panel) testosterone levels over 6 months in 43 hypogonadal men on one of the 3 pellet regimens 6×200 mg (filled circles, $n = 32$), 6×100 mg (filled squares, $n = 28$) and 3×200 mg (filled triangles, $n = 51$). Data is plotted as mean and standard error of mean

Weekly blood sampling for one month after implantation of 6×100 mg testosterone pellets in 15 hypogonadal men demonstrated a gradual rise of total and free testosterone and suppression of gonadotropins (Conway et al. 1988). Disregarding the minor "burst" release over the first day (see section 12.5.2), peak blood testosterone concentrations are attained between two and four weeks after implantation.

12.7 Pharmacodynamics

12.7.1 Clinical effects

Maintenance of libido, potency and well-being is very consistent on all three dose regimens lasting for 4–5 months with either of the 600 mg dose regimens and 6 months with the 1200 mg dose. After completion of cross-over most men (30/43) preferred to continue using testosterone pellets rather than switching back to testosterone ester injections (Handelsman et al. 1990), an observations confirmed by other investigators (Jockenhövel et al. 1996; Zacharin and Warne 1997). The most frequently cited desirable features of testosterone implants were the lack of wide swings in androgen effects (including mood) and the long inter-treatment interval.

12.7.2 LH and FSH suppression

Suppression of elevated LH and FSH levels was studied in the 22 men with hypergonadotropic hypogonadism (Fig. 12.5). Plasma LH and FSH were markedly suppressed in a dose-dependent fashion by all three regimens. This reciprocal relationship reflects the strong inverse correlations of total and free testosterone with LH ($r=0.47$, 0.46 respectively) and FSH ($r=0.45$, 0.47). The parallel suppression of LH and FSH was also consistent with their high correlation ($r=0.87$) with each other. The 1200 mg (6×200 mg) regimen produced significantly greater and more sustained suppression of LH and FSH (both $p<0.001$) than the two 600 mg regimens while the two 600 mg regimens had very similar time-courses for plasma LH and FSH levels (both $p>0.30$).

The 600 mg dose regimens produced nadir LH levels between one and three months with a significant increase commencing by four months and return to baseline at five months. In contrast, the 1200 mg dose produced nadir LH levels between one and four months with return to baseline only at six months. Nadir LH levels achieved were comparable with eugonadal controls on the 1200 mg dose but remained elevated with both 600 mg dose combinations. The time-course for suppression of FSH levels was similar. Both 600 mg dose regimens produced nadir FSH levels between one and two months. At three months the 6×100 mg combination maintained suppres-

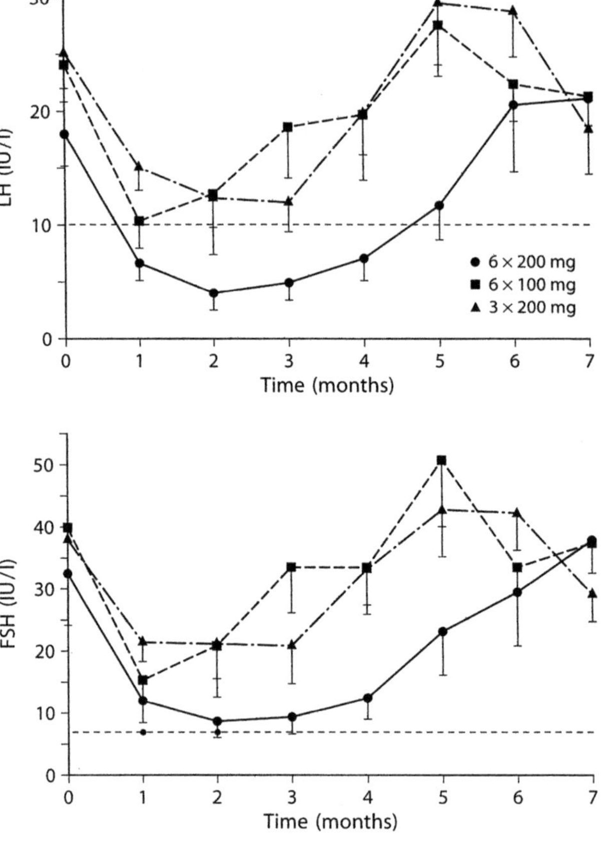

Fig 12.5. LH (upper panel) and FSH (lower panel) levels over 6 months in 22 men with primary (hypergonadotropic) hypogonadism on one of the 3 pellet regimens 6 × 200 mg (filled circles), 6 × 100 mg (filled squares) and 3 × 200 mg (filled triangle). Dashed line indicates the upper limit of normal eugonadal male range. Mean LH levels are suppressed into or just above this range whereas mean FSH levels remain consistently elevated. The fall and rise of LH and FSH mirrors closely the reciprocal changes in testosterone levels. Data is plotted as mean and standard error of mean

sion of FSH returning to baseline levels at four months whereas the 3 × 200 mg dose produced shorter duration of FSH levels which had returned to baseline by three months. In contrast the 1200 mg dose induced sustained FSH suppression with return to baseline levels only after six months. Nadir FSH levels remained elevated in all three treatment regimens compared with eugonadal controls.

The suppression of elevated gonadotropins in men with primary hypogonadism mirrored closely both the time-course of clinical androgenic effects and the maintenance of physiological testosterone levels. Thus the clinical monitoring of androgen replacement therapy by observing efficacy of androgen replacement can be augmented by measuring blood testosterone concentrations and, in men with hypergonadotrophin hypogonadism, gonadotropin suppression. The ability of steady-state testosterone supply to suppress nadir LH levels to eugonadal levels while FSH levels remained supra-normal is consistent with the role of non-steroidal testicular factors such as inhibin in regulation of FSH.

12.7.3 SHBG

Plasma SHBG levels are not altered by implantation of 400–1200 mg testosterone pellets in either hypogonadal (Conway et al. 1988) or eugonadal (Handelsman et al. 1992, 1996) men. This differs from a smaller study using 1200 mg which showed a modest decrease in plasma SHBG (Jockenhövel et al. 1996). The minimal changes in blood SHBG concentrations after implantation contrasts markedly with parenteral testosterone esters and oral testosterone undecanoate, which consistently and markedly lower SHBG levels (Conway et al. 1988). These observations that physiological testosterone replacement has minimal if any effect on blood SHBG levels, whereas more unphysiological routes and/or supraphysiological doses lower SHBG, substantially supports the suggestion (Conway et al. 1988) that reduced SHBG levels are a manifestation of hepatic androgen over-dosage (whether via portal or systemic routes). This interpretation corresponds with current understanding of the route- and dose-dependent effects of estrogens on SHBG (von Schoultz and Carlstrom 1989) but supersedes the older interpretation that lowered SHBG is a physiological effect of androgen therapy (Anderson 1974).

12.7.4 Biochemistry and hematology

During the first four months after implantation of six 100 mg testosterone implants in androgen deficient men (Conway et al. 1988), hemoglobin levels rose while plasma iron and urea fell (data not shown) reflecting the anabolic effects of androgens on erythrocyte (Gardner and Besa 1983) and total body protein (Kochakian 1976; Mooradian et al. 1987). There are no significant changes in other biochemical (electrolytes, glucose, calcium, lipids) or hematological variables or in kidney or liver function tests following pellet implantation in androgen-deficient or eugonadal men.

12.8 Adverse effects and continuation rates

Pellet implantation has few side-effects and is generally very well tolerated. In a review of 973 consecutive implantations studied prospectively in 221 men over 13 years, the continuation rates were 93% overall, rising successively from 88% after the first implantation to 91% after the second, 95% after the third to the ninth and 99% after the tenth or subsequent implantation (Handelsman et al. 1997). One or more adverse events was reported after 11% of implantations consisting of extrusions (8.5%), bleeding (2.3%) and infections (0.6%).

Immediate side-effects of the implantation procedure include minor discomfort at the puncture site or along the implant track for a few hours in some men. Bleeding from the incision site is uncommon and is readily con-

trolled by topical pressure. After only 3/973 (0.3%) implantations was clinically significant bleeding observed. In two instances this required topical
pressure for 30–45 min with <50 ml blood loss and in the other topical
pressure for 2 hours with ~100 ml blood loss.

The most frequent adverse effect is pellet extrusion which characteristically occurs about two months after implantation. The extrusion rate depends on operator skill and can be maintained at ~5% with experience. Extrusions typically involve a single pellet exiting from the incision site or occasionally through the skin above its final location. Multiple extrusions are
more common than expected, suggesting common patient or procedure-related causative factors but the only known predictor of extrusion is higher
level of physical activity at work. Extrusion does not require any specific
management but multiple extrusions may shorten the effective duration of
that treatment cycle. Incipient extrusion often produces a low grade, foreign-
body inflammatory reaction which may be confused with dermal infection.
Culture rarely identifies known pathogens and the infection usually subsides
quickly with or without administration of a broad-spectrum antibiotic.

Palpable subdermal fibrosis at the sites of past implantations is uncommon and, when present, does not indicate residual unabsorbed steroid nor
influences subsequent pellet implantations or absorption. Fibrosis is less frequent with fused pellets than with older style compressed, cholesterol-containing implants (Bishop and Folley 1951). In over 1000 implantation procedures, no instances of allergy to local anaesthetic or keloid formation have
been recorded.

12.9 Clinical use of testosterone pellet implants

12.9.1 Indications, contra-indications and limitations

Androgen deficiency sufficient to justify androgen replacement therapy is an
indication for use of testosterone implants. There is no evidence of any differential responses according to the type or cause of the hypogonadism, the
clinical features or the patient's age. Pellet implants are particularly suitable
for androgen-deficient men who dislike or are unable to have regular injections (e.g. adolescents, frequent travellers). Due to the long-lasting effects
and the inconvenience of removal, pellets preferably should be used by men
in whom the beneficial effects and tolerance for androgen replacement therapy have already been established by treatment with shorter-acting testosterone preparations. Conversely, in the rare event that rapid interruption of testosterone therapy is necessary (e.g. diagnosis of prostate cancer), it is a limitation of implants that immediate cessation of androgen action requires minor surgery to remove pellets. The only contra-indications are those relating
to testosterone itself (e.g. prostate or breast cancer) and those relating to the
minor surgery of implantation (e.g. bleeding disorders, allergy to local anaes-

thetics). Caution is advised for keloid-prone individuals and in men with little subdermal fat in whom implantation may be more difficult.

12.9.2 Dose and monitoring

Since the pellet testosterone release rate is known (1.3 mg/day per 200 mg pellet), it is possible to replicate the daily testosterone production rate of 3–9 mg in eugonadal men (Gandy 1977; Southren et al. 1968) by a single implant of three to six 200 mg pellets (600–1200 mg) which will last for between 4 and 6 months. Indeed pellets constitute a highly flexible dosage form since by using various combinations of 100 mg and 200 mg pellets it is possible to administer testosterone at release rates of 0.65 to 7.8 mg per day in increments of 0.65 mg/day. Individual monitoring of androgenic effects can be readily performed by the observation of clinical effects, testosterone levels and, in men with primary hypogonadism, suppression of gonadotropin levels as for other testosterone preparations. Based on clinical pharmacology and experience the routine dose is 4×200 mg implants. Due to the predictability of the time-course, it is usually sufficient after an uncomplicated implant to review the patients after the third month, when clinical symptoms and blood testosterone concentrations require a further implantation procedure.

12.9.3 Comparison with other testosterone formulations

In a randomized, cross-over comparative study of the three most widely used testosterone formulations (oral, intramuscular injections, pellets), pellets were clearly superior in durability and stability of clinical effects (Conway et al. 1988). These findings were confirmed by other investigators (Jockenhövel et al. 1996; Zacharin and Warne 1997). The long duration of effect permits infrequent applications which, being far more convenient, in turn minimizes compliance problems and facilitates effective long-term androgen replacement therapy. Immediately after completion of the three phase cross-over study, patients expressed a preference for remaining on pellets (43%), returning to testosterone ester injections (43%) and very few wished to use oral medication (14%). After a further year nearly all (86%) had switched to pellet implants and none remained on the oral androgen. The principal reasons for choosing implants were the dislike of fluctuating androgen levels and the frequency of medications with other preparations.

12.9.4 Costs

The cost of androgen replacement therapy consist of payments for the drugs and for the associated medical care (visits to doctor and/or nurse for prescription, monitoring and treatment). Although exact cost and relativities depend on specific features of local health care systems, the annual drug costs for pellet implants is comparable with testosterone ester injections but substantially lower than oral testosterone undecanoate or transdermal testosterone patches, a fact which has to be balanced over time against the higher unit medical care costs for the implantation procedure and injections. Overall, however, within the UK (Cantrill et al. 1984) and Australian national health schemes, the patient out-of-pocket costs as well as the overall annual costs are lowest for testosterone implants, followed by testosterone ester injections with oral testosterone undecanoate and transdermal testosterone being most expensive. Thus testosterone pellets can represent an economical as well as convenient formulation compared with testosterone esters, which are comparable in cost but require much more frequent administration, and oral and transdermal formulations which are more expensive and require daily application.

12.10 Key messages

- The ideal androgen for long-term replacement therapy would be safe, effective, inexpensive, already marketed, long-acting due to depot, zero-order release properties.
- Although there is no perfect androgen formulation, testosterone pellet implants fulfil many of these criteria.
- Testosterone pellets are highly effective, economical, already marketed, have very long-acting properties and stability of effects with zero-order release pattern. In our opinion, this modality is superior to any presently available testosterone preparations.
- A single implantation procedure delivering four 200 mg pellets provides stable, effective and well-tolerated androgen replacement for four to six months.
- Total and free testosterone levels peak in the first month declining gradually over several months to baseline. SHBG levels are unaffected and, in men with hypergonadotropic hypogonadism, LH and FSH levels are markedly suppressed in a mirror-image of the testosterone levels.
- Drawbacks include the need for minor surgical skill for the implantation procedure and the cumbersome delivery system using unwieldly instruments which should be refined.
- The only significant side-effect is pellet extrusion which occurs after ~8.5% of implant procedures but which can be reduced with operator experience.

Acknowledgements

The author is particularly grateful to his colleagues at the Andrology Unit, Dr Ann Conway, Leo Turner, Chris Howe, Lyn Boylan, Mary-Anne Mackey, Susan Wishart and Dianne Quinn who conducted these studies with great dedication and expertise.

12.11 References

Anderson DC (1974) Sex hormone binding globulin. Clin Endocrinol 3:69–96

Behre HM, Nieschlag E (1998) Comparative pharmacokinetics of androgen preparations: application of computer analysis and simulation. In: Nieschlag E, Behre HM (eds) Testosterone: Action Deficiency Substitution. Springer-Verlag, Berlin, pp 329–348

Bishop PMF, Folley SJ (1951) Absorption of hormone implants. Lancet ii:229–232

Burris AS, Ewing LL, Sherins RJ (1988) Initial trial of slow-release testosterone microspheres in hypogonadal men. Fertil Steril 50:493–497

Butenandt A, Hanisch G (1935) Über die Umwandlung des Dehydroandrosterons in Androstenol-(17)-one-(3) (Testosterone); Umweg zur Darstellung des Testosterons aus Cholsterin (vorläufige Mitteilung). Z Physiol Chem 237:89–97

Cantrill JA, Dewis P, Large DM, Newman M, Anderson DC (1984) Which testosterone replacement therapy? Clin Endocrinol 24:97–107

Conway AJ, Boylan LM, Howe C, Ross G, Handelsman DJ (1988) A randomised clinical trial of testosterone replacement therapy in hypogonadal men. Int J Androl 11:247–264

David K, Dingmanse E, Freud J, Lacqueur E (1935) Über krystallinisches männliches Hormon aus Hoden (Testosteron), wirksamer als aus Harn oder aus Cholestrin bereitetes Androsteron. Z Physiol Chem 233:281–282

Deansley R, Parkes AS (1937) Factors influencing effectiveness of administered hormones. Proc Royal Soc Lond 124:279–98

Deansley R, Parkes AS (1938) Further experiments on the administration of hormones by the subcutaneous implantation of tablets. Lancet ii:606–608

Dewis P, Newman M, Ratcliffe WA, Anderson DC (1986) Does testosterone affect the normal menstrual cycle? Clin Endocrinol 24:515–521

Diaz-Sanchez V, Garza-Flores J, Larrea F, Richards E, Ulloa-Aguirre A, Veayra F (1989) Absorption of dihydrotestosterone (DHT) after its intramuscular administration. Fert Steril 51:493–497

Emmens W (1941) Rate of absorption of androgens and estrogens in free and esterified form from subcutaneously implanted pellets. Endocrinology 28:633–642

Forbes TR (1941) Absorption of pellets of crystalline testosterone, testosterone propionate, methyl testosterone, progesterone, desoxycorticosterone and stilbestrol implanted in the rat. Endocrinology 32:70–76

Foss GL (1939) Clinical administration of androgens. Lancet i:502–504

Frey H, Aakvag A, Saanum D, Falch J (1979) Bioavailability of testosterone in males. Eur J Clin Pharmacol 16:345–349

Gandy HM (1977) Androgens. In: Fuchs F, Klopper A (eds) Endocrinology of Pregnancy. Harper & Row, Hagerstown, Maryland, pp 123–156

Gardner FH, Besa EC (1983) Physiologic mechanisms and the hematopoietic effects of the androstanes and their derivatives. Curr Top Hematol 4:123–195

Hamilton JB (1937) Treatment of sexual underdevelopment with synthetic male hormone substance. Endocrinology 21:649–654

Hamilton JB, Dorfman RI (1939) Influence of the vehicle upon the length and strength of the action of male hormone substance, testosterone propionate. Endocrinology 24:711–719

Handelsman DJ, Conway AJ, Boylan LM (1990) Pharmacokinetics and pharmacodynamics of testosterone pellets in man. J Clin Endocrinol Metab 71:216–222

Handelsman DJ, Conway AJ, Boylan LM (1992) Suppression of human spermatogenesis by testosterone implants in man. J Clin Endocrinol Metab 75:1326–1332

Handelsman DJ, Conway AJ, Howe CJ, Turner L, Mackey MA (1996) Establishing the minimum effective dose and additive effects of depot progestin in suppression of human spermatogenesis by a testosterone depot. J Clin Endocrinol Metab 81:4113–4121

Handelsman DJ, Mackey MA, Howe C, Turner L, Conway AJ (1997) Analysis of testosterone implants for androgen replacement therapy. Clin Endocrinol 47:311–316

Hellman L, Bradlow HL, Frazell EL, Gallagher TF (1956) Tracer studies of the absorption and fate of steroid hormones in man. J Clin Invest 35:1033–1044

Howard JE, Vest SA (1939) Clinical experiments with male sex hormones. II Further observations on testosterone propionate in adult hypogonadism, and preliminary report on the implantation of testosterone. Am J Med Sci 198:823–837

Jockenhövel F, Vogel E, Kreutzer M, Reinhardt W, Lederbogen S, Reinwein D (1996) Pharmacokinetics and pharmacodynamics of subcutaneous testosterone implants in hypogonadal men. Clin Endocrinol 45:61–71

Kochakian CD (ed) (1976). Anabolic-Androgenic Steroids. Handbook of Experimental Pharmacology. Berlin, Springer-Verlag

Loeser AA (1940) Subcutaneous implantation of female and male hormone in tablet form in women. Br J Med: 479–482

Mooradian AD, Morley JE, Korenman SG (1987) Biological actions of androgens. Endo Rev 8:1–28

Nieschlag E (1996) Testosterone replacement therapy – something old, something new… Eur J Endocrinol 45:261–262

Nieschlag E, Cuppers HJ, Wickings EJ (1977) Influence of sex, testicular development and liver function on the bioavailability of oral testosterone. Eur J Clin Invest 7:145–147

Nieschlag E, Mauss J, Coert A, Kicovic P (1975) Plasma androgen levels in men after oral administration of testosterone or testosterone undecanoate. Acta Endocrinol 79:366–374

Parkes AS (1938) Effective absorption of hormones. Br J Med: 371–373

Petra P, Stanczyk FZ, Namkung PC, Fritz MA, Novy ML (1985) Direct effect of sex-steroid binding protein (SBP) of plasma on the metabolic clearance rate of testosterone in the rhesus macaque. J Steroid Biochem Molec Biol 22:739–746

Reiter T (1963) Testosterone implantation: a clinical study of 240 implantations in ageing males. J Am Geriat Soc 11:54–50

Ruzicka L, Wettstein A (1935) Über die krystallische Herstellung des Testikelhormons, Testosteron (androsten-3-on-17-ol). Helv Chim Acta 18:1264–1275

Southren AL, Gordon GG, Tochimoto S (1968) Further studies of factors affecting metabolic clearance rate of testosterone in man. J Clin Endocrinol Metab 28:1105–1112

Swyer GIM (1953) Effects of testosterone implants in men with defective spermatogenesis. Br J Med: 1080–1081

Thom MH, Collins WP, Studd JWW (1981) Hormonal profiles in postmenopausal women after therapy with subcutaneous implants. Br J Obstet Gynaecol 88:426–433

Thorn GW, Firor WM (1940) Desoxycorticosterone acetate therapy in Addisons disease. J Am Med Assoc 114:2517–2525

Vest SA, Howard JE (1939) Clinical experiments with androgens IV a method of implantation of crystalline testosterone. J Am Med Assoc 113:1869–1872

von Schoultz B, Carlstrom K (1989) On the regulation of sex-hormone-binding globulin. A challenge of an old dogma and outlines of an alternative mechanism. J Steroid Biochem Molec Biol 32:327–334

Wilson JD (1980) The use and misuse of androgens. Metabolism 29:1278–1295

Zacharin MR, Warne GL (1997) Treatment of hypogonadal adolescent boys with long acting subcutaneous testosterone pellets. Arch Dis Child (in press)

13 Long-term experience with testosterone replacement through scrotal skin

Linda E. Atkinson, Yu-Lin Chang, and Peter J. Snyder

Contents

13.1 Introduction

Testosterone is the primary endogenous androgenic hormone. Endogenous androgens, including testosterone and dihydrotestosterone (DHT), are responsible for the normal growth and development of the male sex organs and for the maintenance of secondary sex characteristics. The goals of treating male hypogonadism are the development or restoration of secondary sex characteristics, sexual function, and normal metabolic processes, and prevention of chronic bone loss (Bhasin 1992; Bhasin and Bremner 1997; Ghusn and Cunningham 1991). This chapter will summarize the clinical experience with Testoderm, a transdermal delivery system for testosterone, and the evidence of meeting treatment goals while providing a therapy for chronic use that is generally safe and does not produce untoward side effects with chronic treatment.

13.1.1 Androgen replacement therapy

Total daily testosterone secretion in normal young men has been variously reported to range from 4 to 9 mg/day (Wilson 1996) or more recently as 3.7±2.2 mg/day (Vierhapper et al. 1997). Circulating testosterone concentrations generally fall between 7–35 nmol/l (with the exact range depending on the assay employed). The production and secretion of testosterone by testicular Leydig cells is stimulated by pituitary luteinizing hormone (LH). The secretion of pituitary LH into the bloodstream is pulsatile, and serum testosterone levels fluctuate in a pulsatile manner during the course of the day (Bridges et al. 1993). When measured every few hours, testosterone concentrations show a circadian pattern, with levels peaking in the early morning and falling to a nadir in the evening. The diurnal pattern is less prominent and serum testosterone concentrations may be lower in older men (Tenover 1992). Intersubject variability in serum testosterone concentrations is large, approximately 30–40%. Because of the variability of testosterone concentrations in the normal male during the day, from person to person, and among assays, there is no accepted testosterone value used as a cutoff to define testosterone deficiency. Symptoms, etiology, clinical impression, and a very low or low-normal testosterone value aid the diagnosis of hypogonadism.

Testosterone, its long-acting esters, and 17α-alkylated androgens are currently approved for androgen replacement therapy in hypogonadal men in the USA. In order to be efficacious, orally administered testosterone must be modified to reduce the highly efficient hepatic first-pass metabolism (Bhasin 1992). Despite some success with modification (i.e., testosterone undecanoate, methyltestosterone, and fluoxymesterone), doses must be administered two to three times a day to obtain effective peak drug levels. In addition, use of the alkylated products risks hepatotoxicity with chronic administration (see also Chapter 10 in this volume).

The most common route of testosterone administration is intramuscular injection, although subcutaneous implants, oral tablets, buccal matrices, and transdermal dosage forms are also available. The disadvantages of injection therapy include the inconvenience of frequent office visits, the discomfort of injections, and most important, the extreme fluctuation in blood testosterone concentrations during the treatment interval. In order to obtain adequate testosterone replacement, 200–400 mg testosterone enanthate must be injected intramuscularly every two to four weeks. In hypogonadal men, injections produce supraphysiologic testosterone concentrations (35–70 nmol/l) during the first few days after an injection, which decline to subnormal values thereafter (Nieschlag et al. 1976; Snyder and Lawrence 1980). With this regimen, serum testosterone levels rise rapidly into the high normal or supraphysiologic range in the first two days after injection, and then gradually decline into the hypogonadal range at the end of the injection interval. These fluctuations in serum testosterone levels can be attended by extremes in the patient's mood, libido, sexual activity, and energy levels (see also Chapter 10 in this volume).

13.1.2 Role of transdermal products in androgen replacement therapy

Transdermal delivery of testosterone offers several benefits compared to oral or intramuscular routes of testosterone administration. Transdermal systems deliver daily, physiologic doses of unmodified testosterone that yield consistent therapeutic serum testosterone levels and mimic the circadian rhythm of testosterone production in eugonadal males. The unmodified hormone delivered through the skin avoids the hepatic first-pass metabolism and negates the potential problem of liver toxicity. Transdermal patches can be easily self-administered, and dosing can be discontinued immediately by removing the system. Unlike gel-based androgen solutions, which are administered percutaneously, transdermal delivery systems are limited in area and do not transfer drug to sexual partners. Also, overdosing and substance abuse are less likely to occur with transdermal administration than with oral or intramuscular testosterone replacement therapies.

Disadvantages of transdermal medication are the potential for skin irritation, contact sensitization to the drug or other components, and requirements of self-motivation and compliance for daily administration over many years. Individual differences in drug absorption through the skin require that doses are monitored a few weeks after initiation of therapy. Patients may want to wear patches in concealed body areas, to avoid embarrassing questions. Because transdermal systems are a newer therapeutic modality, they are more expensive. Nevertheless, transdermal products have attracted users and account for about one-third of the testosterone prescriptions written in the USA.

13.2 Testoderm Testosterone Transdermal Systems

13.2.1 Description

The first transdermal testosterone system, Testoderm Testosterone Transdermal System (Testoderm), provides for the controlled delivery of testosterone through skin. It delivers testosterone to the circulation at a total daily dose of approximately 4 or 6 mg, less drug per day than obtained in injection regimens. Testoderm systems have been evaluated in clinical trials for the past 13 years, and available in the USA since 1994. Testoderm is also approved in several European and Asian countries, and it is indicated for replacement therapy in males for conditions associated with a deficiency or absence of endogenous testosterone.

Testoderm systems are thin (0.14 mm), rectangular films with rounded corners; when applied to genital skin, they cling without the need for an aggressive adhesive. Following placement of the system, testosterone is absorbed continuously through the genital skin into the bloodstream. Each Testoderm system is designed to deliver testosterone for one day and is worn for 22 to 24 hours after application. The system delivers approximately one-third of its total drug content, with the remaining two-thirds providing thermodynamic energy for drug delivery. The dose (rate of drug delivery) depends on the size (area) of the system.

After the protective release liner is peeled away from the system, the Testoderm is applied firmly to a clean, intact area of skin on the scrotum. Hair on the scrotum should be dry-shaved before applying the system to ensure good skin contact. Warming the system with hands or a hair dryer before placement ensures good adherence. Narrow, thin adhesive strips have recently been added to the contact surface in order to facilitate application and adherence. This modified product has been shown to be bioequivalent to original Testoderm.

13.2.2 Special characteristics of Testoderm

13.2.2.1 Scrotal application

Testosterone does not readily pass through body skin in quantities that are sufficient for therapy, because of a thick stratum corneum barrier. Percutaneous gels of testosterone and dihydrotestosterone, available in France, provide normal levels of testosterone or DHT if applied to a large area of the thigh or chest. Scrotal skin, in contrast to other areas of skin, has a thin stratum corneum and a rich superficial vascular supply. The contrast of testosterone transport through scrotal skin and chest skin is illustrated in Fig. 13.1. The increase in testosterone area-under-the-curve (AUC) per unit area is almost 40 times greater with Testoderm applied to the scrotum (60 cm^2) than with Testoderm applied to the chest (200 cm^2). However, the barrier to testosterone transport in trunk or appendage skin is partially over-

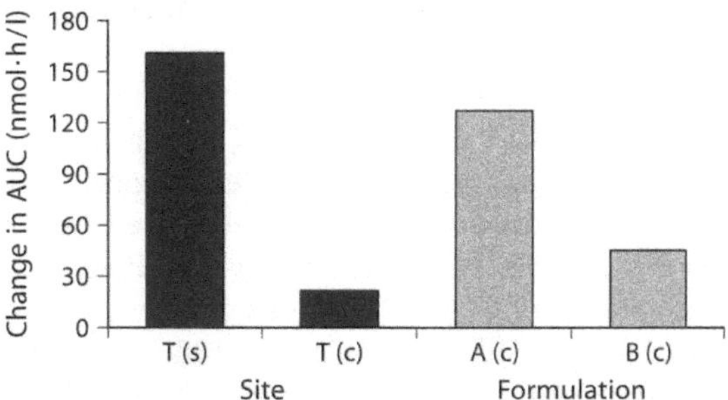

Fig. 13.1. Skin site and formulation affects testosterone concentratons. AUC, mean Area-Under-the-Curve; T(s), Baseline-corrected AUC when Testoderm is applied to scrotal skin; T(c), Baseline-corrected AUC when Testoderm is applied to the chest; A(c), Baseline-corrected AUC when Formulation A containing skin permeation enhancers is applied to the chest; B(c) Baseline-corrected AUC when Formulation B containing skin permeation enhancers is applied to the chest. Data from 12 normal men

come by the incorporation of permeation enhancers into the transdermal formulation, as demonstrated in the higher testosterone levels recorded after application of formulations A and B to chest skin.

Application to scrotal skin is well-tolerated. Of 72 patients responding to daily questionnaires in clinical trials of Testoderm, 7% reported itching, 4% discomfort, and 2% irritation at the application site. In a comparative study, the moderately intense irritation recorded immediately after system removal was significantly less than the topical irritation observed after removal of a nonscrotal testosterone transdermal system. Furthermore, allergic contact dermatitis occurred in 12% of subjects using the nonscrotal patches but was not observed in subjects using Testoderm. The absence of allergic reaction in subjects who used scrotal patches after having a reaction with nonscrotal systems indicates that testosterone is not the allergen (Jordan 1997).

13.2.2.2 Elevated serum dihydrotestosterone (DHT) concentrations

Because of high 5α-reductase activity in scrotal skin (Kuttenn et al. 1980), a portion of the testosterone delivered by Testoderm is metabolized to DHT, increasing the average serum DHT concentration two- or three-fold to 5.5 nmol/l (normal range = 1.0–3.0 nmol/l) when the average peak serum testosterone concentration is 20 nmol/l. Clinical studies have shown that total serum androgen concentrations (testosterone + DHT) are maintained within the normal range throughout Testoderm treatment. Moreover, the range observed in the serum testosterone/DHT ratio during Testoderm therapy (0.7–12.5) shows considerable overlap with the range in normal serum testosterone/DHT ratios (3.6–15.2) reported in the literature. DHT is not secreted by the testes in significant amounts (Ito and Horton 1971), but rather is generated by the metabolic conversion of testosterone in peripheral tissues such as the skin and prostate, where DHT acts as a paracrine hormone. Testosterone passively enters cells across a concentration gradient at the membrane and in target cells is converted to DHT by 5α-reductase. Thus, the androgen concentration gradient in the prostate favors influx of testosterone but not DHT. Since DHT serum concentrations are normally low, the clinical significance of elevated DHT concentrations is not known and there has been concern that the prostate, a DHT-dependent organ, would be adversely affected. However, percutaneous DHT has been used for androgen replacement therapy (see Chapter 15 by Schaison and Couzinet, this volume) and testosterone undecanoate tablets which are used extensively outside the USA, produce elevated serum DHT levels (Gooren 1986). No reports of adverse prostatic events have been associated with either of these therapies. de Lignieres (1995) has observed that levels of serum DHT between 3 and 40 nmol/l are not correlated with any clinical, sonographic, or biochemical evidence of prostate growth stimulation when men are treated with percutaneous DHT gel.

Hirsutism and acne – side effects that might be associated with excess androgen activity – were observed infrequently in Testoderm clinical trials. The low frequency of these events, coupled with the observation that LH levels in men with primary hypogonadism were not suppressed to subnormal values (Findlay et al. 1989), suggest that total androgen activity was not supraphysiologic.

13.2.3 Summary of early clinical studies

The results of the initial Testoderm clinical trials have been reported by Ahmed et al. (1988), Bals-Pratsch et al. (1986, 1988), Carey et al. (1988), Cunningham et al. (1989) and Findlay et al. (1987, 1989) and summarized by Place et al. (1990) These trials established that transdermal delivery of testosterone could significantly improve mood, energy and sexual function (libido, spontaneous erections, etc.) in hypogonadal men in comparison to their pretreatment condition. These improvements were accomplished by physiologic doses of testosterone that restored serum testosterone concentrations to normal levels (≥ 12 nmg/l). Furthermore, Testoderm therapy proved to be safe over a two-year evaluation period. These were no significant adverse changes in hemotology or hepatic function testes. Both serum cholesterol and high density lipoprotein cholesterol decreased significantly ($p < 0.01$) in the first three months of Testoderm administration but stabilized thereafter.

13.3 Clinical measures of efficacy

13.3.1 Pharmacokinetics

The biological responses to androgens are well known, but are not easily translated into clinical endpoints that can be readily evaluated (Bhasin 1992). The most frequently used endpoints are libido and number of spontaneous erections, but they generally take two to eight weeks to stabilize (Place et al. 1990). Suppression of high gonadotropins in response to testosterone administration is also prolonged (Findlay et al. 1989), and despite levels of testosterone that provide positive effects on mood, energy, and sexual function, the gonadotropins may not be normalized (Place et al. 1990). Other measurements, such as body composition, have awaited specialized technological methods, (e.g. dual energy x-ray absorptiometry [DEXA]) that are now available but are expensive and not used routinely in practice to evaluate therapeutic success. Thus, the pharmacokinetics of testosterone preparations may act as a surrogate in determining the efficacy of testosterone replacement. Clinical studies have demonstrated that testosterone doses that restore serum concentrations to the low- to mid-range values (approximately 8–20 nmol/l) observed in normal men will support improvements in sexual function, mood, energy, muscle mass, and bone mineral density (Cofrancesco and Dobs 1996; Place et al. 1990). Some evidence exists that a threshold for improvements in sexual function is found with fairly low testosterone concentrations (8 nmol/l), but clinical experience has shown that this is highly variable. Evidence seems to be lacking that serum testosterone concentrations in the upper end of the normal range provide more benefit than those in the lower half of the normal values (Buena et al. 1993).

13.3.1.1 Pharmacokinetics of Testoderm

Steady-state serum testosterone and DHT concentrations in 30 hypogonadal men before and during treatment with Testoderm are shown in Fig. 13.2. The temporal pattern of endogenous testosterone concentrations over 24 hours in hypogonadal men is without a circadian rhythm. During a 22-hour period when Testoderm is applied, the daily circadian pattern of testosterone is mimicked. Testosterone levels rise to a maximum two to four hours after application and are maintained in the normal range (12–27 nmol/l in the assay used) until the systems are removed. Testosterone concentrations then return rapidly toward baseline. Dihydrotestosterone concentrations are below normal (1–3 nmol/l) before application of Testoderm, but after Testoderm is applied, DHT levels rise above normal within a few hours and remain unfluctuating during the treatment period (ALZA Corp, unpublished).

13.3.1.2 Pharmacokinetics of transdermal delivery and intramuscular injections

Testoderm treatment produces relatively stable and physiologic concentrations of testosterone in serum that may circumvent the mood swings and variations in libido associated with injections of testosterone ester. The comparative pharmacokinetic parameters of testosterone after Testoderm (testosterone transdermal system) (Testoderm) therapy and Depo-Testosterone (testosterone cypionate solution) treatment were evaluated in an open-label, randomized, crossover study with 11 patients (Mooradian and Atkinson, unpublished). Testosterone pharmacokinetics were estimated over the 21-day

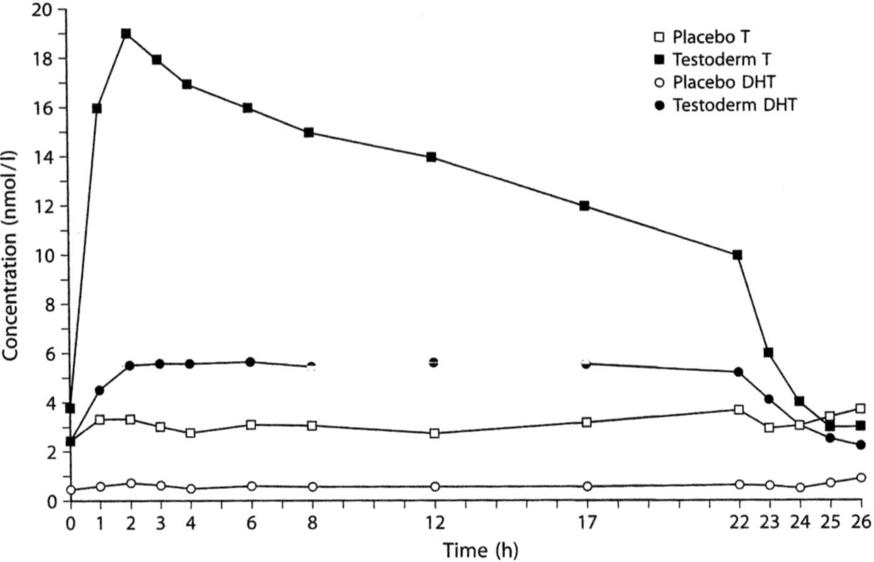

Fig. 13.2. Testosterone replacement therapy with Testoderm. 32 hypogonadal men wore a placebo transdermal system or Testoderm applied to scrotal skin at hour 0 and removed at hour 22. Serum samples were analyzed for testosterone (T) and dihydrotestosterone (DHT). The normal range for T is 12–27 nmol/l. The normal range for DHT is 1-3 nmol/l

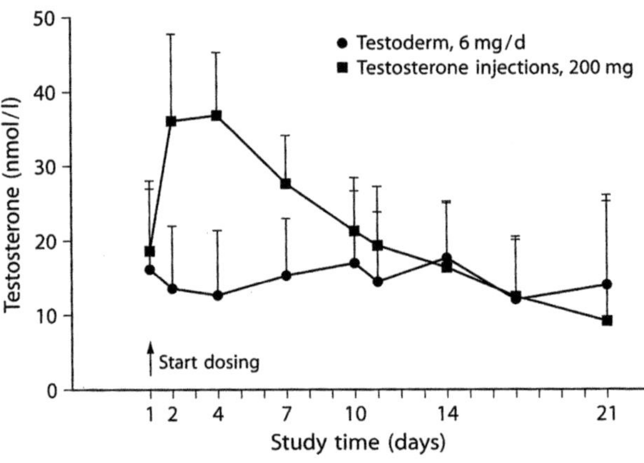

Fig. 13.3. Comparison of transdermal and injection therapy testosterone levels. Crossover study design of 11 hypogonadal men injected with 200 mg testosterone cypionate and Testoderm, 6 mg/day. Testosterone concentrations were monitored over a 21 day interval (one injection and 21 daily applications of Testoderm, applied to the scrotum

period after a 200-mg intramuscular injection of Depo-Testosterone and during a 21-day period of daily scrotal applications of Testoderm. Blood samples for testosterone, free (unbound) testosterone, estradiol (E_2), and dihydrotestosterone were obtained immediately pre-dose and on days 1, 2, 4, 7, 10, 14, and 21 of the treatment period. Depo-testosterone treatment was associated with marked changes in mean serum total testosterone concentrations from beginning to end of the 21 days, as illustrated in Fig. 13.3. The mean estimated testosterone AUC over the 21 days was 10361±2593 nmol h/l, with the mean C_{max} of 39.7±8.6 nmol/l occurring on days 2 to 4. In contrast, Testoderm treatment resulted in physiologic concentrations of testosterone in serum with minimal variation over 21 days. Values for the 21-day AUC were 6855±3449, 66% of the comparable injection AUC. Mean serum concentration of testosterone at C_{max} was 22.7±11.9 nmol/l, with little change in maximum values from day to day. Testoderm treatment resulted in more variable testosterone levels (CV = 50%) compared to injection (CV = 25%) during the 21-day treatment period.

The concentration-time profile of free testosterone serum concentrations generally followed that of testosterone. During both the injection and the Testoderm treatments, free testosterone concentrations were in the normal range. Testoderm treatment was associated with approximately two-fold higher concentrations of dihydrotestosterone compared with Depo-Testosterone throughout the treatment, due to the first pass metabolism of testosterone in scrotal skin. Concentrations of estradiol in serum were significantly higher during Depo-testosterone injections than during Testodermtreatment and were above the normal range for the first week after the testosterone injection.

13.3.2 Body composition studies

13.3.2.1 Bone

Hypogonadism was identified in 5 to 33% of adult men treated for vertebral fractures and osteoporosis. Men with iatrogenic hypogonadism were found to have reduced bone mineral density (BMD), with more pronounced loss from the spine. The role of testosterone and/or dihydrotestosterone in maintenance of bone is not well understood. However, androgen receptors are found in osteoblastic cells and in quantities that are physiologically important and have been shown to affect a variety of processes that increase bone formation (Orwoll and Klein 1995, see also Chapter 6 by Finkelstein in this volume).

Behre et al. (1997) measured BMD in lumbar spine by quantitative computer tomography (QCT) longitudinally in 72 men who were treated for hypogonadism. BMD was either normalized from levels below the fracture threshold or maintained without diminishment for up to 16 years of treatment. No difference in treatment outcome was observed between men using testosterone injections (250 mg/3 weeks) and 11 men who wore scrotal patches.

Snyder (unpublished) is investigating the effect of treatment of hypogonadism with Testoderm, 6 mg/day, on BMD in 11 hypogonadal men who were previously untreated for one year. Bone mineral density in four areas (lumbar spine, femoral neck, Ward's triangle, and femur trochanter) was estimated by DEXA before treatment initiation and every 6 months during treatment (Fig. 13.4). In the first year of hormone replacement, BMD had in-

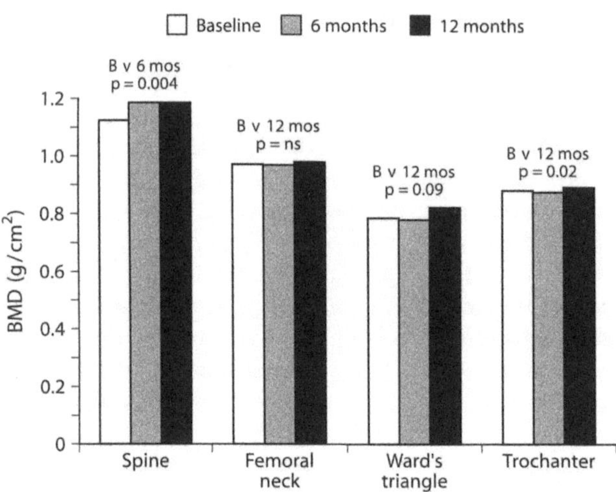

Fig. 13.4. Bone mineral density during Testoderm treatment. 11 previously untreated hypogonadal men were administered Testoderm, 6 mg/day. The bars represent the mean concentrations of bone mineral density (BMD), g/cm² obtained by dual energy x-ray absorptiometry (DEXA) for the lumbar spine (spine), the femoral neck, Ward's triangle and the femur trochanter. P values were obtained by the Wilcoxon test, and compared month 6 or month 12 to baseline (B), the measurement before treatment was started

creased significantly from pretreatment in the lumbar area after 6 and 12 months and in the femur trochanter after 12 months. BMD in Ward's triangle and the femur neck was maintained at pretreatment levels, but did not increase. This study confirms that testosterone replacement with Testoderm maintained BMD in cortical bone and increased BMD rapidly in vertebral bone.

13.3.2.2 Muscle mass and strength

13.3.2.2.1 Hypogonadal men

Unlike the well-documented effects of testosterone replacement therapy on the restoration of sexual function, mood, and energy levels for hypogonadal men, the effects of physiologic testosterone doses on fat, muscle mass, and muscle strength have not been studied thoroughly. Such studies are confounded by duration of hypogonadism, prepubertal or postpubertal onset, diet, exercise, and possibly the dose of testosterone. Bhasin et al. (1997) have demonstrated that testosterone administration (100 mg/week) to untreated hypogonadal men for ten weeks will produce an increase in lean tissue mass (primarily skeletal muscle) and muscle strength (see also Chapter 7 by Bhasin et al. in this volume).

A three-year study of changes in lean muscle mass and strength of hypogonadal men treated with Testoderm, 6 mg/day, is in progress (Snyder, unpublished). Eleven males who had testosterone levels of < 8 nmol/l for approximately one year and who had no diseases that would affect bone metabolism participated in the study. Their fat and lean tissue mass was estimated by DEXA before treatment and every six months thereafter. Muscle strength of the knee on the dominant side was measured by Biodex Dyna-

Table 13.1 Change in body composition and muscle strength in hypogonadal men administered Testoderm

	Months of treatment (mean (SD) of response)				
	0	3	6	12	p-values
Total tissue (kg)	85.2 (10.0)	nd*	86.4 (10.7)	86.7 (12.3)	all >0.2
Fat tissue (kg)	30.1 (7.4)	nd*	28.7 (7.2)	29.5 (8.3)	0.02[a]
Peak hand grip (kg)	45 (12)	48 (16)	52 (15)	42 (7)	0.005[a]
Knee flexion, (torque/lbs)	55 (20)	57 (17)	58 (19)	56 (13)	all >0.06

*nd = not done, [a] = month 0 vs month 6

mometer before treatment and every three months during treatment. Hand grip strength was also documented. The response to the first year of treatment showed an increase in nonfat tissue and a decrease in fat tissue (Table 13.1). However, a slight initial increase in muscle strength was not maintained during the first year. Maximum change in all parameters occurred during the first six months of treatment and did not increase further at the 12 month assessment.

13.3.2.2.2 Testosterone-deficient men with AIDS

Testosterone deficiency in HIV-infected men is well documented and may be as prevalent as 50% of male AIDS patients (Cofrancesco et al. 1997). Testosterone concentrations decline as the disease progresses, and low concentrations are correlated with lymphocyte depletion and weight loss. Grinspoon and colleagues (1996) noted that the degree of loss of lean body and muscle mass correlated with lower total and bioavailable testosterone concentrations in these patients. Unlike the improvement in lean body mass that Testoderm administration produces in hypogonadal patients who are not infected, Testoderm administration was not able to reverse the loss of body cell mass in men who had AIDS, a 5 to 20% weight loss, low normal ($\leq$14 nmol/l) testosterone concentrations, and below normal free testosterone ($\leq$55 pmol/l). Testoderm was administered in a double-blind, placebo-controlled, parallel group study that included 133 patients. There were statistically significant increases in testosterone and free testosterone in the active treatment arm at weeks 4, 8, and 12 of the 12-week study (Table 13.2). Despite the adequate serum levels of testosterone, testosterone replacement did not increase body cell mass more significantly than placebo (Table 13.3), thus failing to correct pretreatment weight loss. In addition, average weight loss did not continue to decline in either the testosterone or placebo groups. These results suggest that testosterone replacement by itself is inadequate to improve body cell mass in HIV-related weight loss (Dobs et al. 1998).

Table 13.2. Changes in serum free testosterone (pmol/L) and total testosterone (nmol/L) after Testoderm administration in patients with AIDS

	Serum free testosterone		Serum total testosterone	
	Testoderm (n = 62)	Placebo (n = 53)	Testoderm (n = 62)	Placebo (n = 53)
Baseline	47.0	43.0	15.0	15.4
Week 4	76.0*	46.0	22.9	15.7
Week 8	77.0*	48.0	23.1	15.0
Week 12	83.0*	45.0	24.1	15.9

*$p \leq 0.005$ as compared to baseline

Table 13.3. Change in body cell mass (Kg) from baseline to week 12 after Testoderm administration in patients with AIDS

	Testoderm	Placebo	Difference	p-value
	Study 1 (single center)			
	(n = 17)	(n = 19)		
Total Body Potassium	+0.80	+0.22	+0.58	0.40
Bioimpedance Analysis	+0.69	+0.39	+0.30	0.57
	Study 2 (multicenter)			
	(n = 58)	(n = 52)		
Bioimpedance Analysis	+0.23	+0.32	−0.09	0.812

13.4 Evaluation of the safety of chronic therapy with Testoderm

13.4.1 Eight-year follow-up

The systemic tolerance and topical safety of Testoderm systems with chronic administration was evaluated in patients who participated in the first efficacy and safety clinical trials of Testoderm. Eleven centers continued with patients throughout eight years, with yearly reevaluation and continuation at the discretion of the investigator. Patients continued on the same Testoderm dose they had been using at the end of the efficacy study; however, dosage changes were permitted at the discretion of the investigator and the patient. During each year of treatment, patients made three study visits at four-month intervals, with visits at six-month intervals during years 6 through 8. Baseline data for all analyses originated in the first study the patients entered. Other than excluded medications (glucocorticoids [except for hypopituitarism] and major tranquilizers), use of concomitant medications was not restricted during this study. At

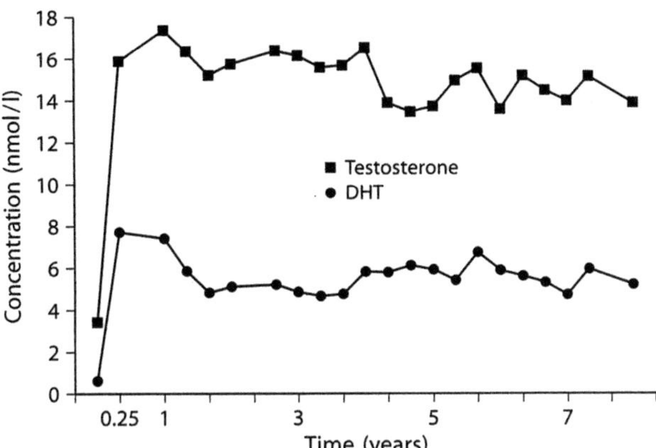

Fig. 13.5. Androgen concentrations during 8 years of treatment. A cohort of 25 hypogonadal men were observed for 8 years of Testoderm, 4 or 6 mg/day, therapy. Testosterone and dihydrotestosterone (DHT) were measured at each clinic visit

each clinic visit, blood samples were drawn for assessment of serum concentrations of sex hormones (testosterone, DHT, and estradiol).

Results from a cohort of 25 men who were treated for the entire eight-year study duration are summarized in the following discussions. Their serum concentrations of testosterone and DHT over the eight years are illustrated in Fig. 13.5. Testosterone concentrations are maintained in the normal range, although a slight downward trend can be observed, which may be due to reduced compliance over the long run or temporal changes in the radioimmunoassay. Dihydrotestosterone serum concentrations remain elevated but stable.

13.4.2 Observed changes in prostate

13.4.2.1 Benign prostatic hyperplasia

Benign prostatic hyperplasia (BPH) is the most common neoplastic growth in aging men, with 75% of men over age 50 having some symptoms arising from BPH. Histopathology studies show that BPH is a very slow growing tumor with low prevalence in the fourth decade of life, increasing to about 70% prevalence in the seventh decade of life. There are a few anecdotal reports in the literature of prostate complications during androgen replacement therapy (Jackson et al. 1989). In one study, the incidence of prostate abnormalities in 100 men over age 45 using testosterone propionate was similar to that in 100 age-matched men who had not received exogenous testosterone (Lesser 1946).

Prostate surveillance was included in long-term trials of Testoderm treatment. The following section comprises case reports and general findings regarding normal and abnormal observations. Case reports of BPH are summarized in Table 13.4. The size and design of the studies do not permit an estimate of incidence of prostate disease with Testoderm use nor an estimate of prevalence of prostate disease in treated hypogonadal men.

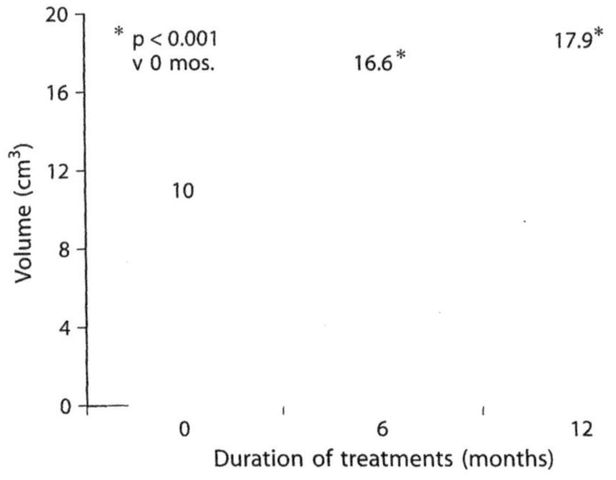

Fig. 13.6. Change in prostate volume during Testoderm treatment. 11 previously untreated hypogonadal men were administered Testoderm, 6 mg/day for 12 months. The bars represent the mean (*) prostate volume in cubic centimeters (cm^3) determined by transrectal ultrasound before treatment (0 months) and at 6 and 12 months of treatment. P values were obtained from the Wilcoxon test, comparing mean values at 0 months to 6 and 12 months

The first trial enrolled 11 patients who had not been given androgen replacement for at least one year (Snyder, unpublished). Prostates were small prior to treatment (10.1 ± 0.1 g) and enlarged significantly ($p \leq 0.001$) to 16.6 ± 3.2 g by 6 months of treatment. At 12 months the prostates were about the same size as at 6 months (17.9 ± 3.3 g), as shown in Fig 13.6. Prostate-specific antigen (PSA) concentrations were normal during the 12-month period, with the exception of one patient who was diagnosed with BPH at month 6 (Table 13.4, Case 8).

The second surveillance followed 63 patients who were using Testoderm for up to three years. In addition to serum testosterone concentration measurements, digital rectal examinations of the prostate, PSA levels, and ultrasound of the prostate were done. The majority of patients in this trial had been previously treated with testosterone before study entry. All patients had normal (< 2.0 ng/ml) PSA values upon entry into the study. At the three-month ultrasound examination and biopsy, one patient was found to have adenocarcinoma of the prostate, with a PSA of 4.0 ng/ml. Ultrasonography revealed that three patients entered the study with enlarged prostates, although they had no obstructive symptoms. Two of the patients (Table 13.4, Case # 9, 10) had normal PSA values during the trial, and one (Table 13.4, Case # 11) had occasionally elevated PSA concentrations (4.5–6.9 ng/ml) which returned to normal again. During the course of the study three other patients (Table 13.4, Case #5, 6, 7) were found to have mild-to-moderate BPH, as diagnosed by ultrasound and biopsy. Their PSA values were normal at all times, and they were without obstructive symptoms.

Growth of the prostate with transdermal testosterone therapy was evaluated in a subset of 34 patients who had a prostate volume recorded at study entry and at least one other time during treatment. The results shown in Table 13.5 indicate that on average, prostate volume increased 15 to 17% over three years of treatment. Of 34 patients, 11 had prostate volume increases of $> 15\%$ between start of study and 18 months of treatment and 13 of 24 patients had a prostate enlargement that was greater than 15% between baseline and 36

Table 13.4. Individual cases of benign prostatic hyperplasia noted in trials with Testoderm

Case #	Age	BPH Severity	Symptoms or Treatment	Prostate Enlargement, cm^3	Prostate-Specific Antigen, ng/ml	Latency, Years*
1	53	mild	yes	12, biopsy: hyperplasia	–	1
2	77	severe	TURP**	–	–	7
3	64	moderate	TURP**	35	normal: 0.1	4+
4	73	mild	yes	–	–	6
5	57	mild	none	13 → 25	normal: <2 ,3, <2	9+
6	53	mild	none	21 → 28	normal: 2.7, <2, 3.2, <2	1
7	68	mild	none	na, biopsy: hyperplasia	normal: 2, 2.1, <2	1
8	74	mild	none	30 → 39	elevated: <2, 5.4, <2	0.5
9	60	mild	none	–	normal: <2, 2, 2.3	2
10	51	mild	none	–	normal: <2	1
11	57	mild	none	39 → 38	elevated: <2, 4.5, 6.6, <2, 6.9, <2	5
12	67	mild	yes	33	normal: <2	0
13	55	mild	yes	36 → 44 g	elevated: 2.0–4.0	na
14	41	mild	none	40 → 51 g	normal: <2	6

* time from first testosterone replacement therapy
** TURP = transurethral resection of the prostate

Table 13.5. Change in prostate volume during chronic Testoderm treatment

Duration of treatment	Mean (SD) volume, cm^3	Mean percent change from baseline (SD)
Baseline (n=34)	19.6 (9.5)	–
<18 months (n=25)	21.1 (9.5)	15.5 (31)
18–36 months (n=24)	23.7 (12.0)	16.9 (16.9)

months of treatment. Meikle et al. (1997) and Behre et al. (1994) have also observed that some increase in prostate volume does occur with testosterone replacement. This would not be unexpected in an androgen-sensitive organ.

In summary, reports of prostate disease (carcinoma or hyperplasia) related to androgen replacement therapy have been infrequent. In our overall clinical experience of 681 patients and normal subjects treated with Testoderm, 14 cases of BPH were identified. The ages of the patients ranged from 41 to 77 years, and the cumulative androgen treatment duration was 6 months to more than nine years.

13.4.2.2 Prostate cancer

Preclinical cancer (defined as the presence of microscopic foci) is found in about 50% of 60- to 70-year-old men worldwide. Progression to clinically observable cancer is more prevalent in some countries (USA, Western Europe) than others (Asia). Risk factors for prostate cancer include genetics, environment, and amount of animal fat consumed. There are no data indicating that androgens stimulate the progression of cancer from a preclinical to clinical stage. No causative association between androgens and prostate disease has been established, although prostate disease is not believed to occur in eunuchs or other men with androgen deficiency (see also Chapter 9 by Frick et al. in this volume).

Four cases of carcinoma of the prostate have been observed in clinical trial experience that included 681 men exposed to Testoderm. The patients were 51 to 86 years of age, with one to four years of androgen treatment. One of the cases was identified because of a routine ultrasound examination. One case was identified by digital rectal examination at the post-study physical of a four-month comparison of Testoderm and IM injections. Two of the cases occurred in a blinded study and were identified because of elevated PSA levels. Their actual treatment – placebo or Testoderm – is unknown at this time. All androgen replacement therapies are contraindicated in patients with prostate cancer, because of the supportive effect of androgens on androgen-responsive prostate cancer cells. There are sufficient case reports of carcinoma of the prostate in hypogonadal men who are receiving testosterone replacement therapy in the literature and in this study to conclude that patients should be screened and evaluated for disease in the same manner as if they were eugonadal men.

13.4.3 Lipids

Changes in lipid concentration, particularly in total cholesterol and high density lipoprotein (HDL) cholesterol, have been reported in connection with testosterone treatment of hypogonadism (for review see Chapter 8 by von Eckardstein in this volume). The impact of restoring testosterone to normal male levels in hypogonadal men has been associated with an increase, a decrease, or no changes in HDL. Total cholesterol has also been reported to increase or decrease (Zgliczynski et al. 1996). Early experience with Testoderm showed that HDL and cholesterol declined significantly 15% and 11%, respectively by the first three months of treatment in a cohort of 48 subjects, but had not changed further at one year and two years of treatment (Place et al. 1990). Observation of a cohort of 25 subjects over eight years revealed the same decrease in cholesterol and HDL occurred by eight weeks, remained unchanged thereafter through the first year of therapy, with cholesterol declining further at the end of year 2. However, by the end of the third year, HDL had decreased by 15% from pretreatment values, and it continued at

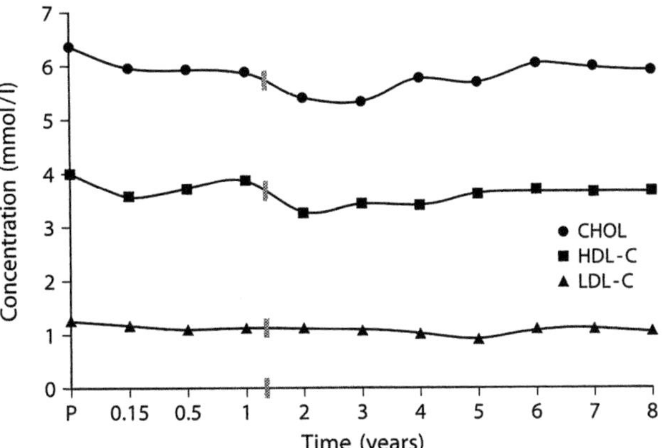

Fig. 13.7. Lipids during 8 years of Testoderm treatment. Total cholesterol (CHOL), high density lipoprotein cholesterol (HDL), and low density lipoprotein cholesterol (LDL) serum concentrations were measured frequently over 8 years of Testoderm therapy

this lower level through the end of the study (Fig 13.7). The decline may be accounted for by the age and health status of the patients. Although the lipids were measured in the nonfasting state, variability was reasonable and values were consistent over the duration of the study. In a second Testoderm study, HDL concentrations also declined significantly ($p = 0.01$) from untreated values – 17% at 3 months and 14% 12 months – while cholesterol values did not change in this time period in a group of 11 previously untreated adult hypogonadal men (Snyder, unpublished).

In order to evaluate the impact of Testoderm treatment on hypogonadal men in a controlled and comparative fashion, a study was conducted with hypogonadal men who had normal lipid concentrations for age, were nondiabetic, and who were currently using testosterone injection therapy. Sex hormones (testosterone, DHT, bioavailable testosterone [Bio-T], 3α-androstanediol glucuronide [3α-Adiol G], estradiol, sex hormone binding globulin [SHBG] and lipids and lipoproteins (total cholesterol, HDL and HDL subfractions, low density lipoproteins [LDL], triglycerides, apolipoprotein A_1 [Apo A], apolipoprotein B [Apo B]) were measured under fasting conditions during the third month of injection therapy. Randomization of patients to a 3-month placebo transdermal treatment or Testoderm followed the injection period with a crossover to the opposite treatment for 3 months. Blood samples were obtained at 12 weeks of placebo and Testoderm treatment. Mean serum testosterone concentrations in the placebo, transdermal, and injection treatments were 7.2, 14.1, and 18.9 nmol/l, respectively. The mean lipid concentrations after three months on each treatment for 16 patients who had complete data are shown in Fig 13.8. Compared to concentrations during placebo treatment, cholesterol, HDL, LDL, and Apo B showed nonsignificant decreases ($p > 0.10$) during Testoderm use. However, triglycerides and Apo A declined 26% ($p = 0.01$) and 7% ($p = 0.03$), respectively during Testoderm treatment. Compared to concentrations during the placebo phase, testosterone injection therapy produced a significant ($p < 0.05$) lowering of cholesterol (10%) and a highly significant ($p = 0.003$) decrease in Apo A concentrations

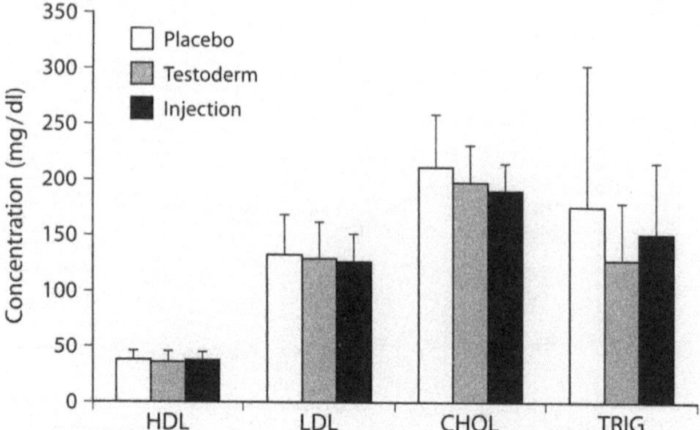

Fig. 13.8. Lipids during testosterone therapy. Sixteen hypogonadal men participated in a crossover design study utilizing testosterone injection (200 mg), (Injection), Testoderm, 6 mg/day (Testoderm) and placebo each administered for a 12 week period. Values shown are from Week 12 of each treatment. The bars represent the mean (SD) serum lipid concentrations: total cholesterol (CHOL), high density lipoprotein cholesterol (HDL), low density lipoprotein cholesterol (LDL), and triglycerides (TRIG). P values were obtained from a one-sample t test

of 9.5%. Significant differences in lipids or lipoprotein concentrations were not observed when Testoderm treatment was compared to testosterone injection treatment. It is of interest that a concomitant decrease in HDL and HDL_2 was not observed with the decline in Apo A, as reported by von Eckardstein et al. (1997).

We investigated the association of hormone concentrations and the corresponding lipid and lipoprotein concentrations in all treatment periods. The mean of each patient's slope from regression was determined with a specific hormone (testosterone, DHT, estradiol, Bio-T, SHBG, and 3α-Adiol G) as independent variable and a lipid or lipoprotein as dependent variable during the entire study. Of all the variables, only Apo A_1 was correlated with changes in hormone concentrations (Table 13.6), most significantly with testosterone and less so with Bio-T, 3α-Adiol G, and estradiol.

Table 13.6. Correlation of apoprotein A_1 and serum sex hormone levels

Sex Hormone	Mean regression slope	
	(SD)	p-value
Testosterone	−0.016 (0.02)	0.008
Bio-T	−0.023 (0.04)	0.028
DHT	−0.056 (0.41)	0.607
Estradiol	−2.443 (4.14)	0.038
SHBG	32.097 (72.29)	0.108
3α-Adiol G	−0.020 (0.03)	0.037

13.4.4 Hematocrit and hemoglobin

Androgens are known to stimulate erythropoesis, but polycythemia may be an unwanted outcome of testosterone replacement therapy as Krauss and colleagues (1991) have reported. Testoderm therapy over eight years was not associated with rising hemoglobin or hematocrit values, although they remained significantly lower than pretreatment levels throughout the treatment period (Table 13.7). In the one-year study conducted by Snyder (personal communication), hemoglobin and hematocrit mean values increased significantly from the untreated condition with testosterone (Testoderm) replacement (Table 13.7). Hemoglobin and hematocrit changes were compared in a crossover study design during three-month treatment periods with testosterone injection, placebo, and Testoderm. The differences between placebo and Testoderm hemoglobin and hematocrit values were not significant. However, treatment with testosterone injections significantly ($p < 0.001$) raised both parameters, although within the normal range (Table 13.7). From these studies we can conclude that Testoderm does not generally raise hemoglobin or hematocrit concentrations above the normal range. In addition, Testoderm treatment may be beneficial for hypogonadal men who have low or normal hemoglobin and hematocrit, but its effect on subjects with elevated hemoglobin is not known.

Table 13.7. Hemoglobin and hematocrit levels before and after chronic Testoderm treatment

Clinical Trial	Hemoglobin, g/dl mean (SD)			Hematocrit, % mean (SD)		
	Prestudy	Poststudy	p-value	Prestudy	Poststudy	p-value
1-year (n=11)	13.5 (1.0)	15.3 (1.0)	0.001	38 (3.7)	44 (3.6)	0.005
8-year (n=25)	15.5 (1.5)	14.4 (2.0)	0.002	46.2 (4.4)	43.3 (6.1)	0.02
3-month (n=16)						
Placebo		14.3 (1.4)	–		40.9 (3.4)	–
Testoderm		14.7 (1.5)	a, b		42.7 (4.3)	a, c
Injection		15.5 (1.2)	d		44.6 (4.6)	d

a: $p \geq 0.1$ for Testoderm vs placebo
b: $p = 0.001$ for testosterone injection vs Testoderm
c: $p = 0.01$ for testosterone injection vs Testoderm
d: $p \leq 0.001$ for testosterone injection vs placebo

13.5 Overview of safety of extended testosterone replacement therapy

Adverse events are routinely collected in all Testoderm clinical trials, including those specific studies discussed above that evaluated the effect of chronic Testoderm use in organs or systems (bone, muscle, prostate, lipids, hematology) that have potential for benefits as well as risks. The current database holds adverse events for 681 patients and normal men who have been treated with Testoderm between one week and eight years. The most frequently recorded ($\geq 1\%$) adverse events, thought by the investigator to be related or of unknown relationship to study drug, are named in the following section.

13.5.1 Summary of adverse drug reactions in clinical trials

As expected from earlier clinical trials, the most often reported adverse events considered related in some way to Testoderm treatment were application site reactions (4.6%), itching (2.9%), and hemorrhage (shaving cut) (0.3%). Headache (2.1%), flu syndrome (1.5%), asthenia (1.0%), and diarrhea (1.0%) were systemic effects of unknown relationship to Testoderm treatment. Side effects that are generally thought to be androgen-related but were not always attributed to study drug, were as follows: prostatic disorder (2.2%), acne (1.3%), abnormal liver function tests (0.4%), gynecomastia (0.4%), seborrhea (0.3%), and 0.1% each for polycythemia, libido decreased, libido increased, hirsutism, breast pain, impotence, and priapism. The prostatic disorder category contained cases of benign prostatic hyperplasia, prostate cancer, and prostititis.

13.5.2 Summary of spontaneously reported adverse drug reactions (pharmaceutical sales)

Of the postmarketing adverse drug experiences recorded for Testoderm during the first three years in the USA market, there were no serious adverse effects judged to be caused by the product. All reactions for Testoderm were reported with a frequency of 0.001% or less based on the number of systems shipped to suppliers. The most frequently reported adverse events probably or possibly related to Testoderm treatment were application site reactions including erythema, itching, and pain. The spontaneous nature of the postmarketing adverse event reporting system precludes any assurance that all events have been reported. However, it is reassuring that the low incidence of side effects observed in clinical trials is upheld in the marketplace.

13.6 Next generation of Testoderm

Existing transdermal testosterone replacement therapies present certain disadvantages to some users. The currently available Testoderm system can be worn only on the scrotum and requires shaving of the scrotum for proper adherence. While Androderm can be applied to the thighs or upper body, it has been reported to be associated with significant skin irritation that requires rotation of skin application sites. Testoderm TTS was developed to provide alternative skin site locations for the scrotal Testoderm and to improve upon the skin site irritation profile of the marketed nonscrotal patch.

Two studies described the testosterone pharmacokinetics of Testoderm TTS. This nonscrotal patch restored serum testosterone concentrations of hypogonadal men to normal, while approximating the endogenous pattern of serum testosterone found in normal males (Fig. 13.9) In one of the studies, three sites of application were determined to be bioequivalent – the upper arm, back, and upper buttocks (Yu et al. 1997a). Results of the second study showed that one and two systems produced dose-proportional increases in serum testosterone concentrations (Yu et al. 1997b). Both studies found that Testoderm TTS provided a nominal testosterone dose of 5 mg/day. It is expected that this physiologic replacement of testosterone will be efficacious in the treatment of male hypogondism. Application site reactions are acceptable with an incidence in 457 Testoderm TTS study participants of 12% itching, 3% moderate or severe erytherma and 1% burning sensations (ALZA Corp., unpublished).

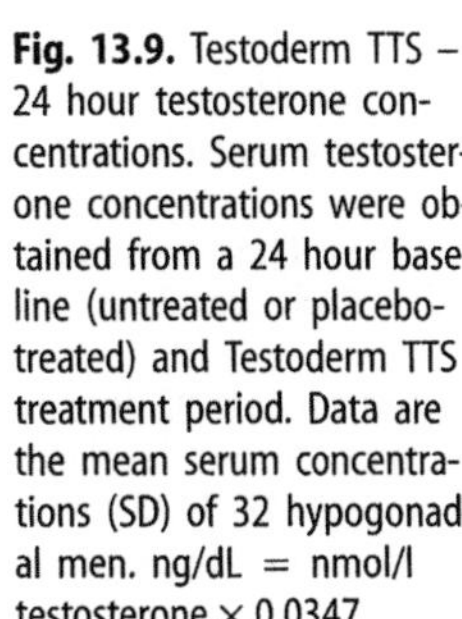

Fig. 13.9. Testoderm TTS – 24 hour testosterone concentrations. Serum testosterone concentrations were obtained from a 24 hour baseline (untreated or placebo-treated) and Testoderm TTS treatment period. Data are the mean serum concentrations (SD) of 32 hypogonadal men. ng/dL = nmol/l testosterone × 0.0347

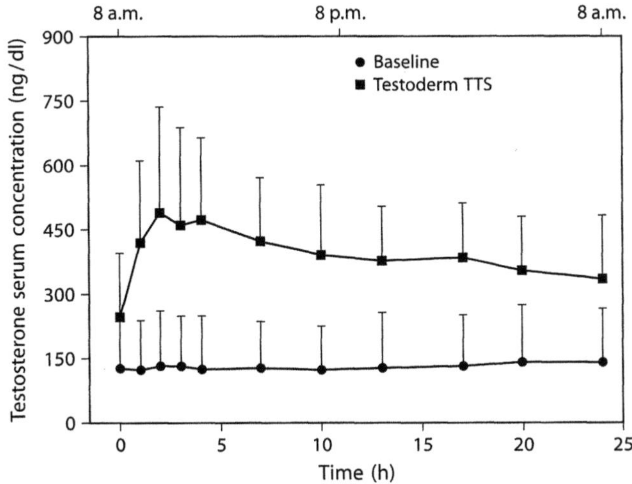

13.7 Key messages

- Serum testosterone, DHT, and estradiol concentrations remain stable during chronic Testoderm use.
- Long-term surveillance of Testoderm patients in clinical trials reveals no increasing rates or unexpected adverse changes regarding serum lipids, hematocrit, and hepatic function.
- Testoderm use has the expected positive effects on bone mineral density and body composition during one year.
- Cases of severe prostate disease are infrequent, although prostate volume may increase from pretreatment as with other testosterone replacement therapies.
- Testoderm-related reports of acne, gynecomastia, hirsutism, and edema are few.

13.8 References

Ahmed SR, Boucher AE, Manni A, Santen RJ, Bartholomew M, Demers LM (1988) Transdermal testosterone therapy in the treatment of male hypogonadism. J Clin Endocrinol Metab 66:546–551

Bals-Pratsch M, Knuth UA, Yoon Y-D, Nieschlag E (1986) Transdermal testosterone substitution therapy for male hypogonadism. Lancet 2:943–946

Bals-Pratsch M, Langer K, Place VA, Nieschlag E (1988) Substitution therapy of hypogonadal men with transdermal testosterone over one year. Acta Endocrinol 118:7–13

Behre HM, Bohmeyer J, Nieschlag E (1994) Prostate volume in testosterone-treated and untreated hypogonadal men in comparison to age-matched normal controls. Clin Endocrinol 40:341–349

Behre HM, Kliesch S, Leifke E, Link TM, Nieschlag E (1997) Long-term effect of testosterone therapy on bone mineral density in hypogonadal men. J Clin Endocrinol Metab 82:2386–2390

Bhasin S (1992) Clinical review 34: Androgen treatment of hypogonadal men. J Clin Endocrinol Metab 74:1221–1225

Bhasin S, Bremner WJ (1997) Emerging issues in androgen replacement therapy. J Clin Endocrinol Metabol, 82:3–7

Bhasin S, Storer TW, Berman N, Yarasheski KE, Clevenger B, Phillips J, Lee WP, Bunnell TJ, Casaburi R (1997) Testosterone replacement increases fat-free mass and muscle size in hypogonadal men. J Clin Endocrinol Metabol 82:407–413

Bridges NA, Hindmarsh PC, Pringle PJ, Matthews DR, Brook CGD (1993) The relationship between endogenous testosterone and gonadotrophin secretion. Clin Endocrinol 38:373–378

Buena F, Swerdloff RS, Steiner BS, Lutchmansingh P, Peterson MA, Pandian MR, Galmarini M, Bhasin S (1993) Sexual function does not change when serum testosterone levels are pharmacologically varied within the normal male range. Fertil Steril 59:1118–1123

Carey PO, Howards SS, Vance ML (1988) Transdermal testosterone treatment of hypogonadal men. J Urol 140:76–79

Cofrancesco Jr J, Dobs AS (1996) Transdermal testosterone delivery systems. Endocrinologist 6:207–213

Cofrancesco Jr J, Walen III JJ, Dobs AS (1997) Testosterone replacement treatment options for HIV-infected men. J Acquired Immune Deficiency Syndrome and Human Retrovirology 16:254–265

Cunningham GR, Cordero E, Thomby JI (1989) Testosterone replacement with transdermal therapeutic systems; physiological serum testosterone and elevated dihydrotestosterone levels. JAMA 261:2525–2530

de Lignieres B (1995) Effect of high dihydrotestosterone plasma levels on prostate of aged men. Abstract 2nd International Androgen Workshop. February 17–20 1995. Long Beach CA

Dobs A et al. (1998) submitted for publication

Findlay JC, Place VA, Snyder PJ (1987) Transdermal delivery of testosterone. J Clin Endocrinol Metab 64:266–268

Findlay JC, Place V, Snyder PJ (1989) Treatment of primary hypogonadism in men by the transdermal administration of testosterone. J Clinical Endocrinol Metab 68:369–373

Ghusn HF, Cunningham GR (1991) Evaluation and treatment of androgen deficiency in males. Endocrinologist 399–408

Gooren LJG (1986) Long-term safety of the oral androgen testosterone undecanoate. Inter J Androl 9:21–26

Grinspoon S, Corcoran C, Lee K, et al. (1996) Loss of lean body and muscle mass correlates with androgen levels in hypogonadal men with acquired immunodeficiency syndrome and wasting. J Clin Endocrinol Metab 81:4051–4058

Ito, T, Horton R (1971) The source of plasma dihydrotestosterone in man. J Clin Invest 50:1621–1627

Jackson J, Waxman J, Spiekerman A (1989) Prostatic complications of testosterone replacement therapy. Arch Intern Med 149:2365–2366

Jordan WP Jr (1997) Allergy and topical irritation associated with transdermal testosterone administration; a comparison of scrotal and nonscrotal transdermal systems. Amer J Contact Dermatitis 8:108–113

Krauss DJ, Taub HA, Lantinga LJ, Dunsky MH, Kelly CM (1991) Risks of blood volume changes in hypogonadal men treated with testosterone enanthate for erectile impotence. J Urol 146:1566–1570

Kuttenn F, Mowszowicz I, Mauvais-Jarvis P (1980) Androgen metabolism in human skin. In: Mauvais-Jarvis P, Vickers CFH, Wepierre J (eds) Percutaneous absorption of steroids. Academic Press, London

Lesser MA (1946) Testosterone propionate therapy in one hundred cases of angina pectoris. J Clin Endo 6:549–557

Meikle AW, Arver S, Dobs AS, Adolfsson J, Sanders SW, Middleton RG, Stephenson RA, Hoover DR, Rajaram L, Mazer NA (1997) Prostate size in hypogonadal men treated with a nonscrotal permeation-enhanced testosterone transdermal system. Urology 49:191–196

Nieschlag E, Cuppers HJ, Wieglmann W, Wickings FJ (1976) Bioavailability and LH suppressing effects of different testosterone preparations in normal and hypogonadal men. Horm Res 7:138–145

Orwoll ES, Klein RF (1995) Osteoporosis in men. Endocr Rev 16:87–116

Place VA, Atkinson LE, Prather DA, Trunnell N, Yates FE (1990) Transdermal testosterone replacement through genital skin. In: Nieschlag E and Behre H (eds). Testosterone: action, deficiency, substitution. Springer-Verlag, Heidelberg Berlin, pp 165–181

Snyder PJ, Lawrence DA (1980) Treatment of male hypogonadism with testosterone enanthate. J Clin Endocrinol Metab. 51:1335–1339

Tenover JS (1992) Effects of testosterone supplementation in the aging male. J Clin Endocrinol Metab 75:1092–1098

Vierhapper H, Nowotny P, Waldhausl W (1997) Determination of testosterone production rates in men and women using stable isotope/dilution and mass spectrometry. J Clin Endocrinol Metab 82:1492–1496

von Eckardstein A, Kliesch S, Nieschlag E, Chirazi A, Assmann G, Behre HM (1997) Suppression of endogenous testosterone in young men increases serum levels of high density lipoprotein subclass lipoprotein A-1 and lipoprotein (a). J Clin Endocrinol Metab 82:3367–3372

Wilson JD (1996) Androgens In: Hardman JG, Limbird LE, Molinoff PB, Ruddon RW, Gilman AG (eds) Goodman and Gilman's *The Pharmaceutical Basis of Therapeutics, Ninth Edition.* McGraw-Hill, New York, pp 1441–1457

Yu Z, Gupta SK, Hwang SS, Cook DM, Duckett MJ, Atkinson LE (1997a) Transdermal testosterone administration in hypogonadal men: comparison of pharmacokinetics at different sites of application and at the first and fifth days of application. J Clin Pharmacol 37:1129–1138

Yu Z, Gupta SK, Hwang SS, Kipnes MS, Mooradian AD, Snyder PJ, Atkinson LE (1997b) Testosterone pharmacokinetics after application of an investigational transdermal system in hypogonadal men. J Clin Pharmacol 37:1139–1145

Zgliczynski S, Ossowski M, Slowinska-Szrednicka J, Brzezinska A, Zgliczynski W, Soszynski P, Chotkowska E, Srzednicki M, Sadowski Z (1996) Effect of testosterone replacement therapy on lipids and lipoproteins in hypogonadal and elderly men. Atherosclerosis 121:35–43

14   A permeation-enhanced non-scrotal testosterone transdermal system for the treatment of male hypogonadism

A. Wayne Meikle

Contents

14.1 Introduction

Male hypogonadism (testosterone deficiency) is a relatively common disorder in clinical practice and may present with a variety of clinical symptoms, including sexual dysfunction, fatigue, depressed mood, and the absence or regression of secondary sexual characteristics. Testosterone deficiency may occur as the result of Leydig cell dysfunction from primary disease of the testes or inadequate luteinizing hormone secretion from diseases of the pituitary or impaired gonadotropin-releasing hormone (Gn-RH) secretion by the hypothalamus (Griffin and Wilson 1992; Santen 1991). Infertility may be observed in men with testosterone deficiency, seminiferous tubule disease, or hypogonadotropic hypogonadism.

Some causes of hypogonadism are relatively common, while others are rare. Klinefelter's syndrome is a primary testicular disorder occurring in about 1 in 500 men and resulting in both androgen deficiency and infertility (Luciani and Guichaoua 1985; Nielsen and Wohlert 1991; Tunte and Niermann 1968). Infertility in men with primary testicular disease is irreversible, but those with gonadotropin deficiency and infertility can often be treated successfully (Huang and Huang 1994; McClure 1987; Nachtigall et al. 1997). Men with primary gonadal failure usually have azoospermia or oligospermia and testosterone deficiency. When successful fertility is improbable or not desired, testosterone replacement therapy is offered to men with androgen deficiency.

When selecting testosterone replacement therapy, a safe general principle is to mimic the normal concentrations of testosterone (350–1050 ng/dl) and its active metabolites (Behre et al. 1990; Cantrill et al. 1984; Meikle et al. 1992, 1996b; Matsumoto 1994, 1995). This will minimize unphysiologically high testosterone serum concentrations to prevent possible side-effects or low concentrations to prevent androgen deficiency. While it is known that patients experience symptoms with androgen deficiency, it is unknown if un-physiologically high concentrations carry health risks. Thus, a safe course would be to duplicate normal physiology as closely as possible. If this goal is met, physiological responses to androgen replacement therapy can be expected when therapy is used to virilize prepubertal males and restore or preserve virilization in postpuberal men. The therapy should be safe without untoward effects on the prostate, serum lipids, and cardiovascular, liver and lung function. Self-administration, convenience, minimal discomfort and reproducible pharmacokinetics are important considerations as well as cost. A permeation enhanced non-scrotal testosterone transdermal system (TTD, Androderm and other names in various countries) was designed for treatment of male hypogonadism (Mazer et al. 1992; Meikle et al. 1992) and achieves many of these goals.

Contraindications and precautions for prescribing androgen replacement therapy include men with known prostate or breast cancer. In those with polycythemia, water retention, sleep apnea and symptomatic benign prostatic hyperplasia, androgen therapy for hypogonadism should be monitored for potential worsening of symptoms.

In this review, the safety and efficacy of TTD will be presented and compared with testosterone enanthate injection therapy. Dose response relationships, pharmacokinetics of 2.5 mg and 5 mg patch systems will be covered. The biological effects of the system in normalization of hormone concentrations, improvement of libido and mood, restoration of sexual function and virilization will be reviewed. In addition, safety issues related to lipid profiles, prostate growth, hematocrit and skin irritation and associated management will be discussed.

14.2 Methods

14.2.1 Design of the TTD (Androderm) system

The TTD system has six components as shown in Fig. 14.1. The drug reservoir (component 2) contains 12.2 mg of testosterone in a proprietary, permeation enhancing vehicle composed of water, ethyl alcohol, glycerin, glycer-

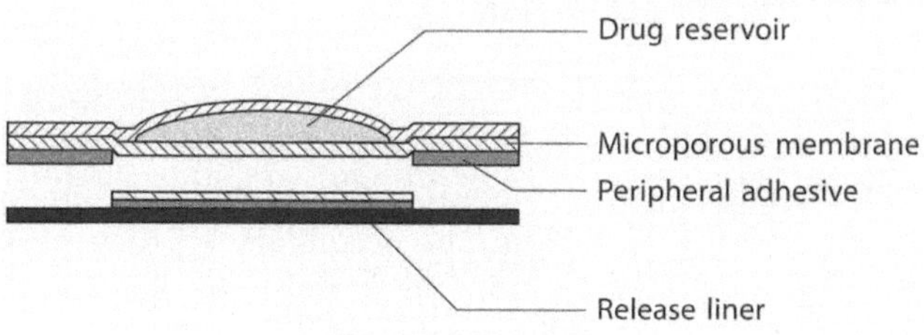

Fig. 14.1. Cross-sectional schematic diagram of the TTD, a testosterone transdermal system (TTD). The disc/release liners are removed prior to application of the system. (Meikle et al. 1996 a)

ol monooleate, methyl laurate and pharmaceutical gelling agents (Mazer et al. 1992; Meikle et al. 1992, 1996 a).

When applied to non-scrotal skin, the 2.5 and 5.0 mg systems are designed to deliver approximately 2.5 mg or 5 mg of testosterone, respectively, over a 24 hour period, with approximately 60% of this amount absorbed during the first 12 hours. For adult hypogonadal men, the standard TTD dosing regimen is two 2.5 mg systems or one 5.0 mg system (patch) applied nightly to the abdomen, back, thighs or upper arms. These regimens deliver an average of 5 mg testosterone per 24 hours, comparable to the daily testosterone production rate of eugonadal men (Meikle et al. 1988; Southren et al. 1968; Vermeulen et al. 1972).

14.2.2 Selection of subjects

This presentation will summarize the key results of four Phase III clinical trials involving TTD therapy of hypogonadal men. The overall objectives were to evaluate the pharmacokinetics, metabolism, efficacy and safety during TTD treatment.

A total of 116 hypogonadal men (ranging in age from 15 to 65 years) participated in the four studies, and about one-third had primary hypogonadism and two-thirds pituitary or hypothalamic dysfunction. Twenty-three had received no previous androgen replacement therapy. The study population included men with a variety of testicular, pituitary and hypothalamic causes of hypogonadism. Hypogonadism was defined by morning serum testosterone of less than 306 ng/dL. All but four patients had developed secondary sexual characteristics before entering the studies.

14.2.3 Clinical trials

Information on the design and results of the individual clinical trials has appeared in papers or abstract form (Arver et al. 1996; Meikle et al. 1996a, 1996b, 1997b). In brief, all four studies were multi-center phase III trials in hypogonadal men. Three of the studies had open-label TTD treatment periods of 6 to 12 months duration. The fourth study was an open-label extension protocol.

14.2.4 Efficacy parameters

The primary efficacy parameters were morning hormone concentrations and 24-hour profiles of: testosterone, bioavailable testosterone, dihydrotestosterone, estradiol, sex hormone binding globulin, luteinizing hormone, and follicle-stimulating hormone. The secondary efficacy parameters included hypogonadal symptoms, mood, and sexual function. The sex hormone parameters were measured by radioimmunoassays.

14.2.5 Safety parameters

The safety parameters included a blood chemistry panel, complete blood counts, urinalysis, prostate size, and appearance measured by transrectal ultrasound, prostate specific antigen, coronary lipid risk profile and local skin tolerability.

14.2.6 Influence of dose

The pharmacokinetics of testosterone were evaluated after application of one, two, or three testosterone transdermal delivery systems (TTD) to 12 hypogonadal men (mean age 46.6±10.5 years) enrolled in an open-label, randomized, crossover study (Brocks et al. 1996). The application periods included an initial two-day washout period with no testosterone therapy followed by two days of therapy with one, two, or three TTD applied daily to the patient's back. On day 2 of active patch therapy, serial blood samples were collected for determination of total and non-sex hormone binding globulin bound serum testosterone concentrations. Serum concentrations of testosterone were determined using radioimmunoassay, and patch testosterone depletion analysis of used transdermal systems permitted estimation of testosterone delivery through the skin (Mazer et al. 1992; Meikle et al. 1996b, 1997b).

14.2.7 Bioequivalence of two 2.5 mg patches and one 5 mg patch

A randomized, steady-state, replicate-design, crossover study compared the bioequivalence of two, TTD 2.5 mg/day TTD with a single TTD 5 mg/day TTD in 21 postpubertal hypogonadal men (Meikle et al. 1997d).

For this study four treatment periods were used, and each subject received each TTD regimen (two 2.5 mg/day and one 5 mg/day) during two separate treatment periods. TTD systems were applied nightly to the upper arm on day one, the thigh on day two and the back on day three. On day three using the back, blood samples were obtained for analysis of total serum testosterone over a 24-hour period. The two formulations were considered bioequivalent if the 90% confidence intervals for the ratios TTD one 5 mg/day: TTD 2 2.5 mg/day for both area under the serum concentration-time curve {AUC (0–24 hours)} and maximum observed serum concentration were completely contained in the interval (0.80, 1.25).

14.2.8 Pretreatment with topical glucocorticoids to assess effects on skin irritation

14.2.8.1 Pretreatment in normal men

Three regimens were applied daily (Monday through Friday) to the upper back (sites rotated) of healthy men for 6 weeks: TTD 2.5 mg/day patch with no pretreatment (A); TTD 2.5 mg/day patch with pretreatment of the application site with triamcinolone acetonide cream 0.1% (A+TA); inactive occlusive control (C). A five-point skin irritation system was used (0: no erythema (B), 1: minimum B, 2: moderate B, 3: intense B+edema, 4: intense B+edema+vesicles) at the time of patch removal and at the next two consecutive clinic visits. If a subject had a grade 3 skin reaction, he was withdrawn from the study (Wilson et al. 1997) .

14.2.8.2 Pretreatment in hypogonadal men

In 16 hypogonadal men, the effect of pretreatment of the skin with triamcinolone acetonide 0.1% cream on the severity and frequency of skin irritation associated with the TTD treatment and on the pharmacokinetics was investigated. For this study sixteen hypogonadal men (ages 28 to 63 years) applied two TTD 2.5 mg/day patches nightly during each treatment period. In the control period, no pretreatment was used. In the triamcinolone acetonide period, 100 mg of triamcinolone acetonide 0.1% cream, U.S.P., was spread over a 7.5 cm^2 area of skin of the back before applying the active absorptive area of each system. Serial serum testosterone profiles were measured by radioimmunoassay on day 2 of each period, with systems applied to the back. Erythema was assessed with a 4 point scale (0=none, 1=mild, 2=moderate and 3=severe) at 1 and 24 hr after removing the day 1 systems (Meikle et al. 1997a).

14.3 Results

14.3.1 Initial pharmacokinetics and metabolism study
(Meikle et al. 1996b)

14.3.1.1 Hormone concentration profiles

Nightly application of two TTD at approximately 22h to the back resulted in mean serum concentrations of testosterone, bioavailable testosterone, dihydrotestosterone and estradiol that rose from a hypogonadal baseline to mimic the normal ranges and circadian variation observed in healthy young men (Fig. 14.2). Maximum concentrations of testosterone and bioavailable testosterone showed parallel time courses and reached peak levels during the early morning hours (eight hours after application) and then slowly declined with minimum concentrations in the evening. Dihydrotestosterone and estradiol were parallel, but reached a peak at about 13 hours after application, then declined slightly before system removal and reached the baseline more slowly than either testosterone or bioavailable testosterone.

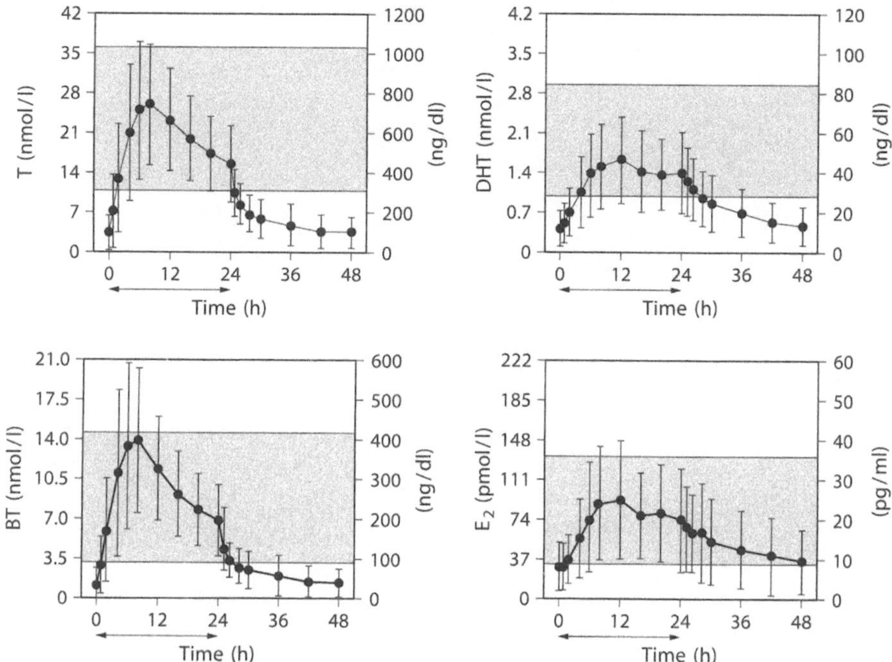

Fig. 14.2. Serum concentration profiles of testosterone, bioavailable testosterone, dihydrotestosterone, and estradiol during and after the nighttime application of two TTD systems to the backs of 34 hypogonadal men (mean ±SD). The shaded areas represent 95% confidence intervals for morning hormone levels in normal men between the ages of 20–65 yr. The arrow denotes the 24-h duration of TTD system application. (Reproduced with permission J Clin Endocrinol Metab 81:1832–40, 1996) (Meikle et al. 1996 b)

14.3.1.2 Pharmacokinetic limits

Table 14.1 summarizes the pharmacokinetic data derived from the profiles of testosterone, bioavailable testosterone, dihydrotestosterone and estradiol. The maximum serum concentration for testosterone and bioavailable testosterone were 24.2±10.4 nmol/L and 13.5±6.45 nmol/L, respectively; and the time-averaged steady state levels were 18.1±7.49 nmol/L and 9.09±3.99 nmol/L, respectively. Using the steady-state values for each hormone, the ratios between the hormonal measurements were as follows: bioavailable testosterone/ testosterone, 0.53±0.19; dihydrotestosterone/testosterone, 0.063±0.018; and estradiol/testosterone, 0.0033±0.0018. The estimated half-life for testosterone and bioavailable testosterone were slightly over one hour and about 3–4 hours for dihydrotestosterone and estradiol. Of the various preparations available for androgen replacement therapy, only transdermal systems achieve circadian variations of serum testosterone concentrations (Behre and Nieschlag 1992; Behre et al. 1990; Bhasin et al. 1992; De Lignieres 1993; Handelsman et al. 1990; Nieschlag and Behre 1990; Place et al. 1990; Schaison et al. 1990; Snyder and Lawrence 1980). Injectable esters of testosterone produce wide fluctuations with peaks and troughs during the dosing interval.

Table 14.1. Pharmacokinetic parameters from the initial study with two TTD systems applied to the back for 24 h. (Reproduced with permission J Clin Endocrinol Metab 81:1832–40, 1996) (Meikle et al. 1996b)

Hormone (unit)	BL^a	$C'max^b$	Testosterone max(h)[c]	$C'ss^d$	R'^e	$t_{1/2}(h)^f$
T (nmol/L)	3.22±2.81	24.1±10.4	8.1±3.2	18.1±7.49	–	1.29±0.71
BT (nmol/L)	1.35±1.25	13.5±6.45	8.4±3.6	9.08±3.99	0.53±0.19	1.21±0.75
DHT (nmol/L)	0.44±0.32	1.27±0.62	12.9±5.3	1.15±0.60	0.063±0.018	2.83±0.97
E2 (pmol/L)	33±26	74±40	13.5±7.3	59±33	0.0033±0.0018	3.53±1.93

Values are the mean ±SD (n=34)

[a] Baseline level defined as the average of 0 and 48 h hormone concentrations.
[b] Maximum hormone concentration of the baseline-subtracted profile.
[c] Time of the maximum hormone concentration.
[d] Time-average steady state hormone concentration (C'ss) derived from baseline-subtracted profile.
[e] Ratio of C'ss level of hormone to C'ss level of testosterone. Units correspond to ng/dL concentrations of each hormone.
[f] Apparent elimination of half-life.

T=testosterone; BT=bioavailable testosterone; DHT=dihydrotestosterone; E2=estradiol

Oral and sublingual androgen therapy results in short-lived peaks at 1–6 hours and troughs at 2–6 hours.

14.3.1.3 Delivery kinetics

The steady cumulative input of testosterone during the 24 hours of patch application was 5.48±2.48 mg. Approximately 60% was delivered during the first 12 hours and 40% during the next 12 hours. The estimates from the kinetic calculations were in excellent agreement with the patch depletion measurements (5.48±2.13 mg). These results indicate that the systemic dose and depletion dose were about equal. When the TTD or the scrotal testosterone patches are applied to the skin in the evening, the higher delivery of testosterone during the first 12 h than during the second 12 h produces a pattern of serum testosterone that mimics the circadian variation observed in normal men (Bremner et al. 1983). Other formulations of testosterone have not as yet been prepared that have a circadian pattern of testosterone delivery, and estimates of testosterone delivery with other regimens may exceed normal production rates of testosterone.

14.3.2 Sex hormone binding globulin, luteinizing hormone, follicle-stimulating hormone concentrations

Luteinizing hormone and follicle-stimulating hormone concentrations remained stable during the 48 h interval, even in the nine patients with primary hypogonadism (time-averaged luteinizing hormone, 22.4±8.7; and follicle-stimulating hormone, 29.7±19.6 IU/L, respectively). Sex hormone binding globulin also showed no significant variation (time-averaged level, 32.6±22.2 nmol/L.

These results indicate that short-term testosterone therapy resulting in normal physiologic concentrations of testosterone do not significantly influence gonadotropins or sex hormone binding globulin, suggesting that longer-term administration is required to suppress gonadotropins.

14.3.3 Application site influence

14.3.3.1 Performance of application sites, testosterone and bioavailable testosterone data

Most skin except for the scrotum is relatively impermeable to steroid hormones. Enhancers in the TTD permit the delivery of testosterone across relatively impermeable skin. Comparative studies were limited, but hydrocortisone applied to the skin in an acetone solution was shown to be about 40 times more permeable through scrotal skin than the ventral forearm (1.0), back (1.7), lateral ankle (0.42) or soles of the feet (0.14) (Feldmann and Mai-

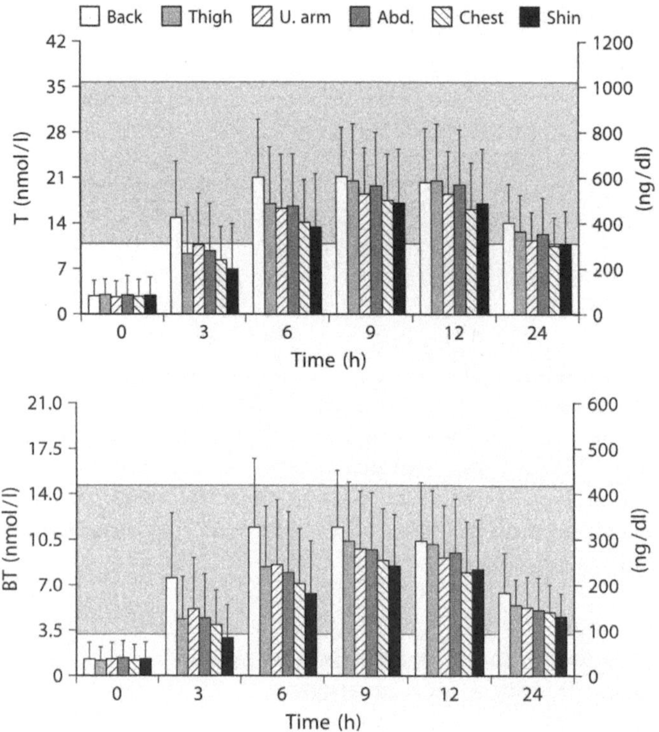

Fig. 14.3. Serum concentrations of testosterone and bioavailable testosterone during the 24-h application of two TTD systems to the back, thigh, upper arm, abdomen, chest, and shin (mean ±SD). Individually marked columns (shown between upper and lower panels) correspond to the respective application sites at each measured time point. The shaded areas represent 95% confidence intervals for morning hormone levels in normal men between the ages of 20–65 yrs. Time zero values represent nighttime baseline levels, obtained just before the TTD application. (Reproduced with permission J Clin Endocrinol Metab. 81:1832–40, 1996). (Meikle et al. 1996b)

bach 1967). It was important to determine the performance of the TTD to deliver testosterone through various skin sites so that satisfactory sites could be discovered for testosterone replacement therapy. Although sequential single dose applications of two TTD to the abdomen, back, chest, shin, thigh and upper arms resulted in similar profiles for testosterone and bioavailable testosterone, the different steady-state concentrations during the 24 hours of observation were quantitatively different (Fig. 14.3). The highest concentration of bioavailable testosterone in descending order was the back> thigh>upper arm>abdomen>chest>shin. Despite a slight reordering of the effectiveness of each site for testosterone, the chest and shin had the lowest concentrations of either bioavailable testosterone or testosterone.

For all application sites, the normal range for testosterone was achieved in about 80–90% and the percentage below the normal range for testosterone averaged 17.6% for both the chest and shin with the other sites being more efficient. Bioavailable testosterone values below the normal range were also higher for the chest and shins than other sites. The TTD was not designed for scrotal use and should not be worn on the scrotum. Although regional differences were observed for delivery of testosterone with the TTD, they were qualitatively similar to those reported for hydrocortisone. The magnitude of differences between skin sites was much less for the TTD than for topical hydrocortisone. This may imply that the TTD enhancers overcome differences in skin permeability to testosterone through non-scrotal skin. Much larger quantities of dihydrotestosterone gel applied to skin are required to produce therapeutic levels of dihydrotestosterone than needed with the TTD or transcrotal system, confirming that non-scrotal skin is relatively impermeable to androgens without the use of enhancers and possibly occlusion. The range for bioavailable testosterone/testosterone was similar for all sites and averaged 0.51–0.54.

14.3.3.2 Application site influence on testosterone delivery

The testosterone profiles and patch depletion data were in good agreement for all sites but the input parameters between various sites were highly significantly different (p < 0.0001). The basis for the preferred sites for applying the TTD depended on the influence of skin sites on delivery of testosterone, as estimated from patch depletion and pharmacokinetic results. Two patches applied to the back, upper arms, abdomen and thigh delivered about 4.5 mg/day and the chest and shin about 3–4 mg/day. Intersubject variability of testosterone delivery was greater (33–43%) for the four best sites and even higher for the shin (48%) and chest (56%) than the intrasubject variation (average 26% for the back, abdomen, thigh and upper arm). The explanation for the poorer performance by the shin and chest compared to the other sites is unknown but could be related to differences in skin permeability affected by cutaneous blood flow and degree of adhesion.

14.3.3.3 Calculation of testosterone and bioavailable testosterone clearance rates and concentration of testosterone dependence on clearance and input

The clearance rate of testosterone and bioavailable testosterone can be calculated from the delivery of testosterone divided by their respective steady-state serum concentrations. Based on 235 observations during the initial and site to site studies, the correlation between the steady-state testosterone concentration and the 12 h input of testosterone (r = 0.564) was lower than the correlation between the steady-state concentration of bioavailable testosterone and the testosterone input (r = 0.754). Individuals with high clearance rates and high input had relatively low steady-state testosterone and bioavailable testosterone values. The mean clearance rate of testosterone was 1284 liters/day, which is in good agreement with values determined by isotope dilution methods in eugonadal men. Estimation of the clearance rate of bioavailable testosterone was about twice that of testosterone. The high clearance rate of bioavailable testosterone and the high correlation between the ratio of bioavailable testosterone/testosterone and the testosterone clearance rate support the results of Vermuelen and Ando (1979) that the non-sex hormone bound testosterone fraction of testosterone is metabolized by splanchnic and non-splanchnic tissues.

14.3.3.4 Correlations between the clearance rates of testosterone and bioavailable testosterone and bioavailable testosterone/ testosterone ratios

Clearance rates varied in proportion to the bioavailable testosterone/testosterone ratio ($r=0.742$; $p<0.001$; $n=34$). In contrast, no significant relationship ($r=-0.395$; $p>0.05$) existed between the bioavailable testosterone clearance rate and the bioavailable testosterone/testosterone ratio. Morning serum T concentrations reached the normal range during the first day of dosing (Mazer et al. 1992; Meikle et al. 1996b). There was no accumulation of testosterone during continuous treatment. Upon removal of the TTD systems, serum testosterone concentrations decreased with an apparent half-life of approximately 70 minutes. Hypogonadal baseline concentrations were reached within 24 hours after removal of the system. During chronic treatment, these baseline concentrations decreased via feedback suppression of the pituitary-gonadal axis.

In both single dose and steady-state pharmacokinetic studies dihydrotestosterone and estradiol levels remained within the normal reference ranges, and the dihydrotestosterone/testosterone and estradiol/testosterone ratios were comparable to those in eugonadal men, approximately 1:10 and 1:200, respectively (Meikle et al. 1986). These findings indicate that the 5α-reductase and aromatase activities of non-scrotal skin do not produce any significant pre-systemic metabolism of testosterone during TTD treatment (Meikle et al. 1992, 1996a), and contrast with the high dihydrotestosterone concentrations observed with the scrotal patch (Place et al. 1990; Schaison et al. 1990). These results indicate that local metabolism in the skin can affect the pattern of hormones measured in the circulation. The disproportionately high concentrations of estradiol observed with injections of testosterone enanthate may imply local conversion of testosterone to estradiol. In contrast, dihydrotestosterone values were proportional to the testosterone following both the TTD and IM testosterone preparations.

14.3.4 Hormone levels

14.3.4.1 Normalization of hormone levels

The hormonal effects of TTD were demonstrated in 94 hypogonadal men (Arver et al. 1996; Meikle et al. 1996a, 1996b). In this population, 93% of patients were treated with two TTD daily, 6% used three systems daily, and 1%

used one system daily. On these dosing regimens TTD produced mean morning serum concentrations of testosterone within the normal reference range in 92% of patients, bioavailable testosterone in 88%, dihydrotestosterone in 85% and estradiol in 77%.

14.3.4.2 Comparison with intramuscular testosterone

In another of the trials, sixty-six patients, previously treated with testosterone injections, received TTD or intramuscular testosterone enanthate (200 mg every 2 weeks) treatment for 6 months (Arver et al. 1996; Meikle et al. 1996a, 1996b). Using the TTD the percentage of time that serum concentrations measured throughout the dosing interval remained within the normal range was 82%, 87%, 76% and 81% for testosterone, bioavailable testosterone, dihydrotestosterone and estradiol, respectively. At day seven after IM testosterone enanthate, the percentages were 72%, 39%, 70% and 35%, respectively. For both TTD and IM testosterone replacement, the dihydrotestosterone/testosterone ratio using the time-average levels of each hormone was about 1:13, within the normal range seen in healthy males. However, disproportionately elevated mean levels of estradiol (approximately 4 ng/dL) were observed at one and seven days after IM testosterone compared to TTD (2 ng/dL), suggesting that local conversion of testosterone to estradiol by aromatase of muscle may occur at the injection site of testosterone.

14.3.5 Dose proportionality

14.3.5.1 Pharmacokinetics of testosterone in hypogonadal men after transdermal delivery: influence of dose

The pharmacokinetics of testosterone were evaluated after application of one, two, or three TTD to 12 hypogonadal men (mean age 46.6±10.5 years) enrolled in an open-label, randomized, crossover study (Brocks et al. 1996). Serum concentrations of testosterone were determined using RIA, and patch testosterone depletion analysis of used TTD permitted estimation of testosterone delivery through the skin. In general, serum concentrations of testosterone correlated well with a dose increase (Fig. 14.4). Based on a strict bioequivalence approach to dose proportionality, the increases in area under the concentration-time curve and morning concentrations were somewhat less than proportional with an increase from one to two TTD, but were proportional to the increase in dose from two to three TTD. Non-sex hormone binding globulin-bound serum testosterone concentrations closely paralleled those of total serum testosterone, and the unbound testosterone fraction was inversely but non-linearly related to the serum sex hormone binding globulin. Applying three TTD yielded serum concentrations of testosterone that more frequently exceeded the upper limit of the normal range than after one or two patches. The area under the curve and cumulative release of testosterone correlated well with the number

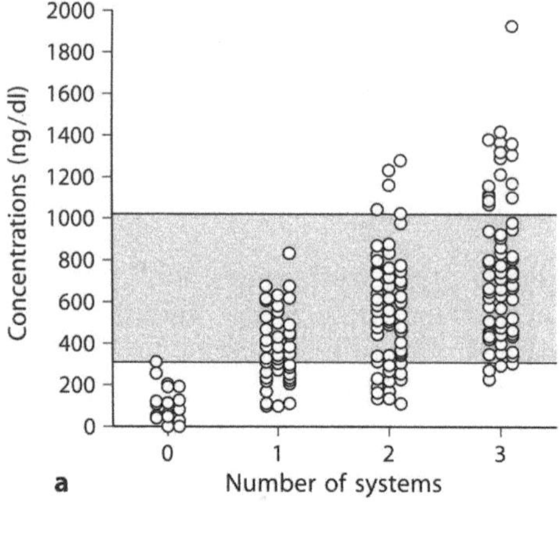

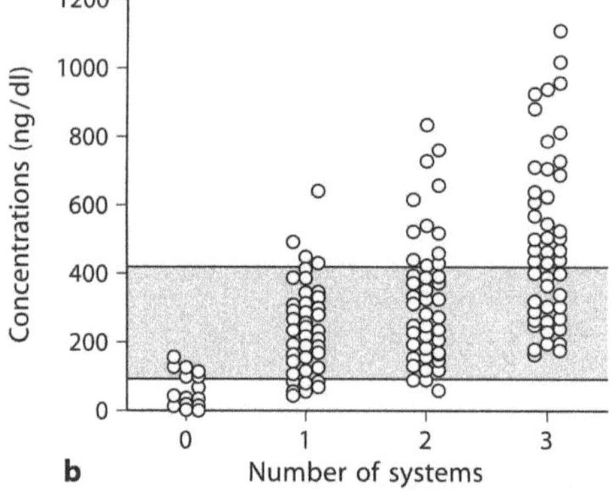

Fig. 14.4. Individual serum concentrations (not baseline-corrected) of testosterone from patients (n = 12) before and after application of 1, 2, or 3 testosterone transdermal delivery systems: (**a**) total testosterone; (**b**) non-sex hormone binding globulin {SHBG}-bound testosterone. Horizontal lines represent upper and lower limits of the normal reference range. (Reproduced with permission J Clin Pharmacol 36:732–9, 1996). (Brocks et al. 1996)

of applied systems. Because of interindividual differences, one dose does not fit all; dosing should be based on serum testosterone concentrations to assess the treatment requirements for each hypogonadal patient. These observations suggest that in hypogonadal men treated with physiologic replacement doses of testosterone, the serum unbound fraction of testosterone is determined by the serum concentration of sex hormone binding globulin. It also suggests that men with low sex hormone binding globulin concentrations, as observed in obese men, will have higher clearance rates of testosterone and require higher doses than men with higher sex hormone binding globulin levels, as observed in older men (Baker and Hudson 1983; Glass et al. 1977; Gray et al. 1991).

Serum testosterone concentrations obtained about 12h after skin application can be used to assess adequacy of replacement therapy following transdermal therapy. However, with testosterone injections, topical dihydrotestosterone, oral or sublingual androgen preparations, the timing of collection of samples for testosterone and dihydrotestosterone assays is more problematic and standards to assess adequacy of replacement therapy have not been established with them.

14.3.5.2 Comparison of the 2.5 mg patches and the 5 mg patch pharmacokinetics

Once it was observed that most hypogonadal men using TTD require two 2.5 mg patches daily or 5 mg/day to achieve physiologically normal serum testosterone levels (Meikle et al. 1997d), a study was conducted to assess the bioequivalence of TTD (two 2.5 mg patches) and a newly formulated testosterone transdermal system (one 5 mg patch). A single patch would have advantages in convenience of use over a two-patch system. A randomized, steady-state, replicate-design, crossover study compared the bioequivalence of two 2.5 mg/day TTD and a single 5 mg/day TTD in 21 postpubertal hypogonadal men. For this study four treatment periods were used, and each subject received each TTD regimen twice (two 2.5 mg/day and one 5 mg/day) during four separate treatment periods. On study day three and while using the back, blood samples were obtained for analysis of total serum testosterone over a 24-hour period. Twenty hypogonadal patients completed all four phases of the study. The two formulations resulted in similar concentrations maximum values with arithmetic mean of 925±340 ng/dL (mean±SD) for TTD two-2.5 mg/day and 905±254 ng/dL for TTD one 5 mg/day (point estimate =0.99, 90% confidence interval 0.92–1.07). The time to achieve peak serum concentration was also comparable and ranged from 3.49 to 9.52 hours for TTD 2′ 2.5 mg/24 hr and from 3.50 to 10.08 hours for TTD one 5 mg/day, with a median time of 5.99 hours for both formulations. Values for the area under the curve (0–24 hours) were also comparable 14710±4050 ng h/dL for TTD two 2.5 mg/24 h and 14890±4230 ng.h/dL for TTD one 5 mg/24 hr (point estimate =1.02, 90% confidence interval =0.96–1.08). In conclusion, the results indicated the TTD one 5 mg/day patch was bioequivalent to TTD two 2.5 mg/day patches.

14.4 Biologic effects

14.4.1 Long-term hormonal patterns

Studies comparing various testosterone replacement regimens for treatment of male hypogonadism are limited. This study permitted comparison of the commonly used regimen, testosterone enanthate 200 mg every two weeks with TTD. Thirty-four patients entered the long-term TTD treatment period, and 29 (nine with primary and 20 with secondary hypogonadism) completed the study and were included in the overall evaluation of efficacy and safety (Arver et al. 1996). Five patients withdrew from the study during TTD systems treatment because of adverse events (three), noncompliance (one), and personal reasons (one).

Morning testosterone levels (mean±SD) measured after 200 mg of IM testosterone enanthate, androgen withdrawal for 8 weeks and one year of thera-

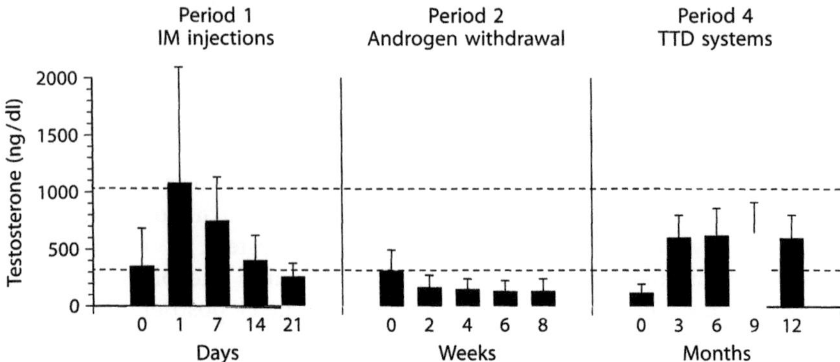

Fig. 14.5. Mean plus standard deviation (error bars) morning serum testosterone levels during intramuscular (IM) injection, period of androgen withdrawal and treatment with testosterone transdermal (TTD) systems. Evaluation interval 0 in period 4 shows baseline level before system application. Physiological levels are achieved within 24 hours. Dashed lines indicate normal reference range. (Reproduced with permission J Urol 155:1604–8, 1996) (Arver et al. 1996).

py with TTD are shown in Fig. 14.5. During IM testosterone, a mean±SD serum testosterone level of 1086.4±1619.0 ng/dL occurred one day after injection, and by 21 days afterwards the mean level was 248.0±121.4 ng/dL. Despite the mean maximum and minimum levels outside of the reference, the mean (586.8±262.5 ng/dL) for the entire 21-day dosing interval was within the normal range (306 to 1031 ng/dL). During the period of androgen withdrawal, all patients had testosterone values below the limit of normal.

During the 12 months of TTD use, morning serum testosterone levels were measured monthly at approximately 12 hours after daily application. The mean monthly concentrations ranged between 522.6 ng/dL and 642.1 ng/dL and were all within the normal limits. The mean morning testosterone concentrations averaged 599.8±199.6 ng/dL for months 3, 6, 9, and 12 and are displayed in Fig. 14.5. The 24-hour pharmacokinetic profile for testosterone (measured at months 3, 6, and 12) mimicked the normal circadian pattern, with a morning peak (740.9±278.2 ng/dL) and nighttime trough.

14.4.2 Objective nocturnal erectile response

Many studies have shown that androgen deficiency reduces erectile function, libido and frequency of nocturnal erections (McClure 1988; Morales et al. 1997; Noldus and Huland 1994). However, few studies have compared various therapies on erectile function. To be included in data analysis for comparisons between the study periods, patients had to have at least one acceptable RigiScan 7 tracing in each study period. Twenty-two of the 29 patients who completed the study met this criterion, including the two patients previously treated with the TTD systems during IM testosterone enathate period; but since they were excluded from the analysis, 20 patients participated in this aspect of the study.

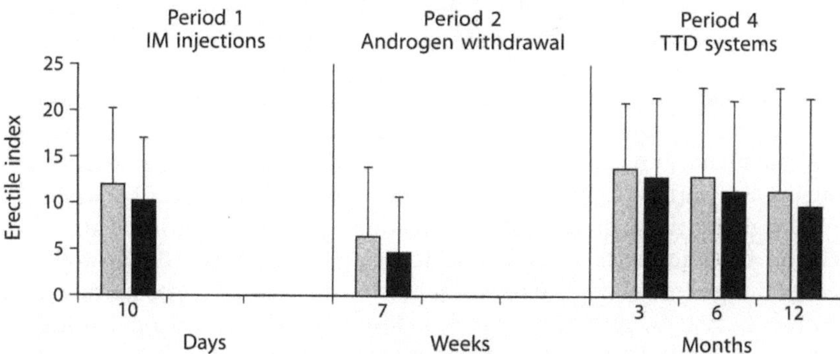

Fig. 14.6. Study period comparison of mean erectile index plus standard deviation (error bars) at penile base (■) and tip (▨) in 20 evaluable patients. (Erectile index=total events per hour × mean event duration × mean rigidity.) For base and tip differences were significant between periods I and II (p < 0.001), and periods II and IV (p = 0.002). Difference between periods I and IV was not significant (p = 0.341). IM, intramuscular. TTD, testosterone transdermal system. (Reproduced with permission J Urol 155:1604–8, 1996) (Arver et al. 1996)

During androgen withdrawal compared to IM testosterone, significant reductions were observed in the number of nocturnal erectile events per hour (p = 0.021), the mean duration of events (p = 0.007), and mean rigidity of erectile events at both the base and tip (p < 0.001). Treatment with TTD after the withdrawal significantly increased the number of erectile events per hour, mean duration of events, and mean rigidity (p = 0.047, p = 0.014, and p = 0.001, respectively), and responses of these measurements were similar during TTD and IM testosterone treatment. Further, the erectile index also showed a marked reduction (p < 0.001) during the androgen withdrawal period compared with the IM treatment period and a sustained and significant rise following treatment with the TTD systems (p = 0.002) (Fig. 14.6). The mean erectile index during treatment intervals with TTD or IM testosterone injections treatment was not significantly different. These findings indicate that the TTD is as effective as IM testosterone in correcting objective erectile dysfunction and are in agreement with other studies showing that androgen replacement therapy improves erectile function (McClure 1988; Morales et al. 1997; Noldus and Huland 1994).

14.4.3 Subjective assessment of sexual function and libido

This study also was designed to compare the efficacy of two forms of androgen replacement therapy on subjective aspects of sexual function. Overall sexual function assessed with the total Watts score was significantly lower during the androgen withdrawal period than during the IM treatment period (p ≤ 0.001) and compared with the withdrawal period, it significantly increased during the TTD systems treatment period (p = 0.003). Overall sexual function scores did not differ significantly when the treatment periods with the IM injections and TTD were compared (p=0.395). The individual variables of the Watts scoring system, which included sexual desire (libido), arousal, frequency of sexual activity, orgasm, and satisfaction, showed similar patterns to the overall scores during the three periods of observation. Overall sexual function as well as the individual parameters were maintained during the entire 12 months of treatment period with the TTD systems.

The mean number (±SEM) of erections per week recorded using the Davidson Questionnaire results significantly decreased (p < 0.001) following androgen withdrawal (2.3±0.6) compared with the IM testosterone treatment period (7.5±0.8), and from the low levels during androgen withdrawal, the erections significantly (p < 0.001) increased during the TTD systems treatment period (7.8±1.1) to values comparable to those observed with the initial IM injections.

The incidence of impotence, decreased libido, hot flushes, and depression increased during eight weeks of androgen withdrawal compared with IM testosterone replacement and TTD treatment, which were comparable to each other. During the androgen withdrawal period, the mean incidence of episodes of impotence was reported by 69% of the men and declined significantly to 17% of patients during long-term TTD treatment (p < 0.001). Median frequency of episodes of impotence was also reduced significantly between treatment and withdrawal periods (p < 0.0001). The mean incidence and median frequency of impotence were comparable between periods of treatment with IM testosterone treatment and TTD. Diminished libido was reported by 17% of patients during IM testosterone treatment, rose significantly (p < 0.001) to 79% of patients during androgen withdrawal, and declined to 3%–21% of patients during long-term TTD treatment. The differences between testosterone replacement by IM injection or TTD were not significant. The severity of impotence and decreased libido rose during the androgen withdrawal compared to the testosterone treatment periods, but the changes were not statistically significant. Thus, our studies confirmed that androgen replacement therapy with either IM testosterone and TTD compared to the hypogonadal state improved both unconscious and conscious aspects of sexual function and arousal and are consistent with other reports (McClure 1988; Morales et al. 1997; Noldus and Huland 1994).

14.4.4 Fatigue and mood

Androgens are known to influence subjective symptoms of fatigue and mood but this study allowed comparison of two testosterone replacement therapies on these symptoms. Fatigue was reported by 18.5% of patients seven days after IM testosterone by 79% of patients during androgen withdrawal, and from 7–31% of patients during long-term TTD treatment. The differences for fatigue, severity of fatigue, and median frequency of fatigue between testosterone replacement by IM injection or TTD and androgen withdrawal were significant (P < 0.001). No differences were noted in incidence, frequency, or severity of fatigue between two active testosterone treatment periods.

If patients reported hot flushes at more than half of their visits, the comparison of the incidence of hot flushes between study periods was analyzed. The incidence of hot flashes increased from 14% of patients during IM testosterone treatment to 28% of patients during androgen withdrawal, and then declined significantly to only 3% of patients with TTD treatment (P = 0.016).

Depression, part of the psychological state evaluation, showed the most significant worsening during androgen withdrawal. Mean scores for depression (Beck Depression Inventory) rose from 5.1 during IM treatment to 6.9 during the withdrawal period II (P < 0.05), and decreased to 3.9 during transdermal treatment (P < 0.05), indicative of improvement. Other psychological parameters, such as measures of aggression and hostility, showed no significant changes during the treatment periods of the study. These observations are in agreement with those reported by Wang et al. (1996) but contrast with observations made in older men by Sih et al. (1997).

14.4.5 Gonadotropin suppression and sex hormone binding globulin

The control of gonadotropin secretion and sex hormone binding globulin levels are complex and are in part regulated by hormones other than sex steroids (Anderson et al. 1997a, 1997b; Cantrill et al. 1984; Griffin and Wilson 1992; MacIndoe et al. 1997). The relationship between serum testosterone and estradiol and gonadotropins is imperfect. Sex hormone binding globulin is in part regulated by insulin, growth hormone and insulin-like growth factor (Crave et al. 1995; Ibanez et al. 1995; Vermeulen 1996). Luteinizing hormone levels generally decreased with TTD therapy and normalized in five of the nine patients with primary hypogonadism. However, follicle-stimulating hormone levels decreased slightly and became normal in only two of the nine patients with primary hypogonadism. Follicle-stimulating hormone concentrations in patients with secondary hypogonadism tended to be lower in parallel with the luteinizing hormone levels. Sex hormone binding globulin levels showed a slight but significant decrease during TTD treatment (0.84 vs 1.07 pg/dL; p < 0.001) when compared to the androgen withdrawal period.

In combined trials, using continuous TTD treatment, serum luteinizing hormone concentrations were reduced and normalized in 10 of 21 (48%) men with primary (hypergonadotropic) hypogonadism within six to 12 months of treatment. Although the luteinizing hormone concentrations declined substantially in many patients treated long-term, luteinizing hormone levels remained elevated in some patients despite serum testosterone concentrations within the reference range. Many studies conducted in normal men for suppression of spermatogenesis have established that supraphysiologic doses of androgens are needed to suppress gonadotropin secretion. Thus, it is not unexpected that normal or elevated concentrations of testosterone produced by TTD or IM testosterone do not routinely normalize gonadotropins in hypogonadal men (Behre and Nieschlag 1992; Behre et al. 1990; Bhasin et al. 1992; De Lignieres 1993; Handelsman et al. 1990; Nieschlag and Behre 1990; Place et al. 1990; Salehian et al. 1995; Schaison et al. 1990; Snyder and Lawrence 1980).

14.4.6 Adolescence bone parameters and virilization

Androgen therapy is known to affect virilization and increase bone density in hypogonadal boys and men (Behre et al. 1997; Finkelstein et al. 1992; Stepan et al. 1989), but experience with TTD in adolescents is limited. β-Thalassemia major often causes a severe form of hypogonadotropic hypogonadism affecting adolescents and young men (De Sanctis et al. 1997). Testosterone pharmacokinetics of the TTD in these adolescents and young men mimicked the nocturnal testosterone secretion and circadian patterns of puberty and young adulthood. Clinical experience of treating nine hypogonadal adolescents and young men (age 16.8–28.9 years) has been reported by De Sanctis et al. (1997).

Three dose regimens were used: two patients initially applied 1 TTD nightly for 12 h (regimen 1); after three months they applied 1 TTD nightly for 24 h (regimen 2); seven patients applied 2 TTD nightly for 24 h (regimen 3). Meaningful data were available in eight patients treated for 9 to 12 months. The effects of 9–12 months of TTD treatment increased morning hormone levels (P = 0.003), weight (P = 0.003), Tanner genital stage (P = 0.0005), height (P = 0.001), penile length (P = 0.0004) and upper lip hair (P = 0.02). Although these are preliminary observations with TTD in adolescents, they are consistent with other satisfactory androgen replacement therapies.

14.4.7 Safety assessment

No clinically significant changes were observed in any of the routine clinical laboratory parameters monitored during the study, including hematology and serum chemistries. No cases of sleep apnea were observed during our studies.

14.4.7.1 Hematocrit

Androgens are known to stimulate erythropoiesis, which accounts in part for the higher hematocrits in men than women. Comparisons of various androgen therapies on producing polycythemia, which may result in an elevated hematocrit (HCT) causing high blood viscosity and predispose to thromboembolic events are needed. In our study (Arver et al. 1997), the incidence of elevated HCTs was assessed in 66 hypogonadal men (mean age = 44.6 yr;

body mass index=27.7) treated in parallel for 24 weeks with either TTD or testosterone enanthate in a multicenter study.

Before entry into the study, all patients had been treated with testosterone enanthate injections at doses ranging from 100 to 400 mg given every one to four weeks. Thirty-three were randomized to treatment with two 2.5 mg TTD systems applied nightly (delivered dose of 5 mg testosterone/day), and thirty-three to treatment with testosterone enanthate, 200 mg administered by IM injection every two weeks. Twenty-six patients in the TTD group completed the 24-week treatment period, and 32 in the testosterone enanthate group. During their initial testosterone enanthate injection therapy, the HCT exceeded the upper limit of normal, 52%, in 4 of 26 patients in the TTD group and 3 of 32 patients in the testosterone enanthate group (2-tail Z-test; p>0.05). After 24 weeks of treatment, only 2 of 26 (8%) patients in the TTD group had an elevated HCT compared to 8 of 32 (25%) patients in the testosterone enanthate group (1-tail Z-test; p<0.05). The range of HCT was also higher in the testosterone enanthate group (42.0 to 61.6) than in the TTD group (39.6 to 53.6). The differences in HCT response between the two groups could not be accounted for by differences in ages and body mass index. For the two patients treated with TTD with an elevated HCT, morning levels of testosterone, bioavailable testosterone, and dihydrotestosterone were within normal limits; estradiol concentrations, however, were elevated. For the eight patients treated with testosterone enanthate and with elevated HCT, bioavailable testosterone values (measured seven days post injection) were elevated in six, and estradiol levels in all eight patients. For all 58 patients, HCT correlated significantly with bioavailable testosterone (r=0.46), estradiol (r=0.58), and age (r=0.36). Thus, hypogonadal men treated for 24 weeks with TTD appear less likely to have an elevated HCT than those treated with testosterone enanthate injections. Sih et al. (1997) also reported a high incidence (24%) of hematocrits of 52% or higher in older men treated with 200 mg of testosterone cypionate every two weeks, which is comparable to using testosterone enanthate. Whether the supraphysiologic levels of bioavailable testosterone and estradiol produced by the IM testosterone ester regimens account for the elevated HCT is possible but unknown. These findings should be taken into consideration for men prone to polycythemia and at risk of thromboembolic events associated with testosterone treatment for hypogonadism.

14.4.7.2 Effect on plasma lipids

An elevated serum concentration of low-density lipoprotein cholesterol and a low concentration of high-density lipoprotein cholesterol are known risk factors for cardiovascular disease. Although many studies have shown that parenteral androgen replacement therapy causes small alterations in serum lipids, dose-dependent reductions in serum high-density lipoprotein have been reported (Bagatell and Bremner 1995; Bagatell et al. 1992, 1994). This contrasts with methyltestosterone, which is known to produce substantial reductions in serum high-density lipoprotein cholesterol (Friedl et al. 1990). Since sex steroid concentrations are different in men treated with TTD and IM testosterone enanthate, it was of interest to determine if these differences in serum concentrations of sex steroids produced differences in lipid profiles. In patients treated for 6 to 12 months with TTD ($n = 67$), the average (SE) serum total cholesterol and HDL concentrations were 199±7.6 mg/dL and 46±2.3 mg/dL (Arver et al. 1996; Meikle et al. 1996a). Compared with values following eight weeks of androgen withdrawal in 29 patients (hypogonadal), the following alterations in lipids were observed during one year of TTD treatment: cholesterol decreased 1.2%; HDL decreased 8%; cholesterol/HDL ratios increased 9%. From week 4 to week 24 of TTD treatment, HDL increased by 5% from 47 to 50 mg/dL (upper limit of normal 69 mg/dL) and the cholesterol/HDL ratio decreased by 1% (from 4.15 to 4.12). In the IM parallel group, HDL and the cholesterol/HDL ratio were unchanged (41 mg/dL and 4.8 respectively) during the treatment period. No consistent differences in lipid profiles were observed between the two androgen regimens despite the differences sex steroid patterns.

14.4.7.3 Effects on the prostate

Androgens are known to result in prostate enlargement and a rise in prostate specific antigen at puberty and in aging eugonadal men (Meikle et al. 1995, 1997b, 1997c; Wilson 1987; Wilson et al. 1975). Serial observations of prostate volume and prostate-specific antigen during androgen replacement therapy in hypogonadal men are of interest related to the androgen hypothesis of prostate enlargement. Prostate size and serum prostate specific antigen concentrations during treatments with IM injections of testosterone enathate and TTD were comparable to values reported for eugonadal men (Meikle et al. 1996a, 1997b). One case of a prostate carcinoma was diagnosed during TTD treatment, and two cases occurred during IM treatment. An androgen withdrawal interval longer than eight weeks in hypogonadal men may be needed to result in a dissociation between age and prostate volume as observed by Behre and Nieschlag (Behre et al. 1994).

Compared to the initial testosterone enanthate therapy (mean 17 g, range 9–33 g), prostate volume measured by transrectal ultrasound (Fig. 14.7) declined significantly (about 18%, $p < 0.001$) after an eight-week interval of androgen withdrawal (mean 14 g; range 7–33 g). Following three months of TTD therapy the prostate volume (mean 18 g; range 9–32 g) had increased significantly (about 18%, $p < 0.001$). Despite continuous TTD therapy, prostate volume showed no significant change at subsequent three-month intervals (3, 6 and 12 months). Thus, TTD treatment caused regrowth of the prostate to its previous size during the initial three months of therapy but did not result in continuous growth over the next nine months of therapy.

Prostate-specific antigen (Table 14.2) also decreased significantly ($p < 0.001$) after androgen withdrawal and increased ($p < 0.006$) following TTD therapy for one year, but it did not become elevated. It was significantly

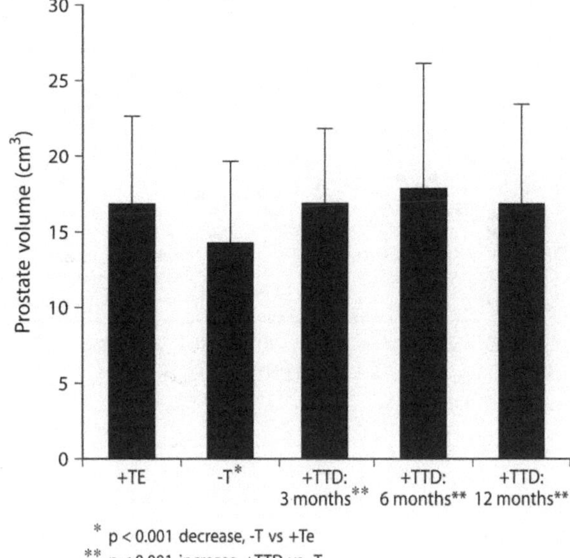

Fig. 14.7. Prostate volume measured by transrectal ultrasound. (T = testosterone; TE = testosterone enanthate; TTD = testosterone transdermal system.) (Reproduced with permission Urol. 49:191–6, 1997) (Meikle et al. 1997 b)

Table 14.2. Prostate specific antigen levels in ng/mL (Reproduced with permission Urol 49:191–6, 1997) (Meikle et al. 1997 b)

	Baseline (+ TE)	8 Week	1 Year
		Withdrawal (-Testosterone)	(+TTD)
Mean	1.0	0.51	0.66
Range	0–3.2	0–2.15	0–2.95
SEM	0.16	0.1	0.11
P Value	1/11 < 0.001	11/111 < 0.006	1/111 < 0.001

Normal range = 0–4 ng/mL

lower on TTD than on initial IM testosterone therapy. Other comparison studies are needed to confirm any differences in serum prostate-specific antigen concentrations between various androgen replacement regimens. Sih et al. (1997) reported that injections of testosterone to hypogonadal elderly men did not elevate prostate-specific antigen above values expected for controls of comparable age.

Behre (1994) reported that untreated hypogonadal men had prostate volumes below their age-matched control group, whereas androgen-treated hypogonadal men had prostate volumes comparable to their age-group. Neither dihydrotestosterone nor transcrotal testosterone therapy appears to enlarge the prostate above that expected for an age-matched group. In our study, prostate volume correlated significantly with age at all three periods of obser-

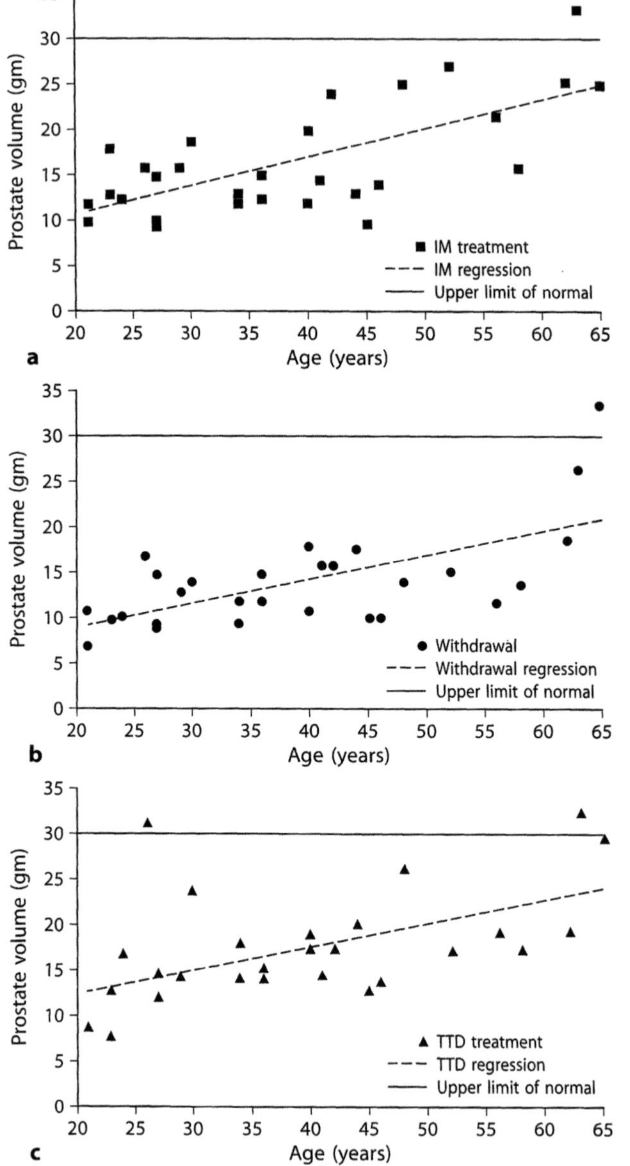

Fig. 14.8. Linear regression of prostate volume with age. (**a**) Prostate volume correlated significantly with age during therapy with testosterone enanthate (r = 0.69, P < 0.01). (**b**) Prostate volume correlated significantly with age after an 8-week testosterone withdrawal interval (r = 0.64, P < 0.01). (**c**) Prostate volume correlated significantly with age during 1 year of TTD treatment (r = 0.55, P < 0.01). (Reproduced with permission Urol. 49:191–6, 1997) (Meikle et al. 1997b)

vation: during initial therapy with testosterone enanthate (r = 0.69, p < 0.01, Fig. 14.8), following an eight-week withdrawal interval (r = 0.64, p < 0.01) and after one year of TTD treatment (r = 0.55, p < 0.01).

These results indicate the strong influence of age on prostate volume, with a smaller effect of androgen replacement therapy. The regression line for the therapy with testosterone enanthate was slightly greater than for the androgen withdrawal and TTD periods (b = 0.31, 0.26 and 0.25, respectively). However, pairwise comparisons showed no significant differences, suggesting that short-term androgen withdrawal and TTD therapy had no significant influence on the relationship between age and prostate volume as noted in patients on long-term testosterone enanthate. Available observations in aging hypogonadal men treated with androgen replacement suggest no acceleration of prostate growth beyond what would be expected in their eugonadal peers (Behre et al. 1994).

To determine if the cause of hypogonadism influenced prostate volume, we compared prostate volumes in men with primary hypogonadism (16.0±1.9 ml; SEM; n = 9) versus those with secondary or tertiary gonadal failure (16.9±1.4 ml; SEM; n = 20). During therapy with testosterone enanthate, no significant difference in prostate volume was observed between the two groups, suggesting that circulating gonadotropins do not affect prostate volume growth, as suggested by some investigators. To evaluate whether the hormone concentrations achieved during androgen replacement therapy were associated with prostate volume, correlations were made between prostate volume and hormone concentrations at steady state. In addition, none of the steady-state hormonal concentrations measured during TTD correlated significantly with prostate volume.

Waist/hip ratio correlated significantly with prostate volume (r = 0.39, testosterone enanthate; r = 0.57, androgen withdrawal; r = 0.56, TTD; all p < 0.05), but body weight, height, and body mass index showed insignificant correlations with prostate volume. Age also correlated significantly with waist/hip ratio (r = 0.57, p < 0.0001). When waist/hip ratio and prostate volume were controlled for age, the relationship between waist/hip ratio and prostate volume was insignificant (p > 0.05).

14.4.7.4 Gynecomastia as side-effect

Testosterone therapy is known to cause gynecomastia, presumably from the conversion of testosterone to estradiol. Since estradiol concentrations were higher following IM testosterone than after TTD, it was of interest to compare breast growth responses between the two regimens. In the randomized groups treated with TTD or IM testosterone enanthate, gynecomastia was present in ten patients randomized to the TTD group and nine randomized to the IM group. At the start of the study one patient in the TTD group also developed gynecomastia during treatment with TTD; it resolved in three of ten TTD patients, improved in one patient, and was unchanged in four patients (the outcome in the two remaining patients is unknown). Gynecomastia resolved in only one of nine men in the IM group, and the remaining eight patients showed no change. Although additional comparison studies of androgen replacement therapies are needed, these results are consistent with the postulate that higher serum estradiol concentrations produced by IM testosterone compared to TTD also contribute to the higher incidence of gynecomastia in the IM treated group.

14.4.7.5 Skin tolerability

Transdermal systems using alcohol or enhancer are known to have a relatively high incidence of skin irritation (Hogan and Maibach 1990). The skin of most men tolerated TTD well except for commonly observed mild local erythema, but local symptoms requiring management of the application sites were also observed. Of all patients exposed to TTD in the clinical trials (Mazer et al. 1992), only 5% stopped treatment because of chronic irritation and 4% from allergic contact dermatitis, which developed after 3 to 8 weeks of use (Arver et al. 1996; Meikle et al. 1996a). The frequency of local tolerability of TTD appears to be comparable to other transdermal products on the market (Hogan and Maibach 1990), but much higher than the incidence reported with the transcrotal patch, which employs no alcohol or enhancers.

Since glucocorticoids are known to reduce contact allergic responses of the skin, studies were conducted to test the hypothesis that topical application of a glucocorticoid would ameliorate the contact dermatitis associated with use of TTD without reducing testosterone delivery significantly. In 16 hypogonadal men, the effect of pretreatment of the skin with triamcinolone acetonide 0.1% cream on the severity and frequency of skin irritation associated with the TTD treatment and on pharmacokinetics was investigated (Meikle et al. 1997a).

Based on the ANOVA results, triamcinolone acetonide pretreatment with TTD was bioequivalent to the control regimen (confidence intervals for the concentrations' average and concentration maximum ratios were between 0.8 and 1.25). Erythema scores indicated significant improvement with triamcinolone acetonide pretreatment compared to control ($P < 0.0005$, Fig. 14.9).

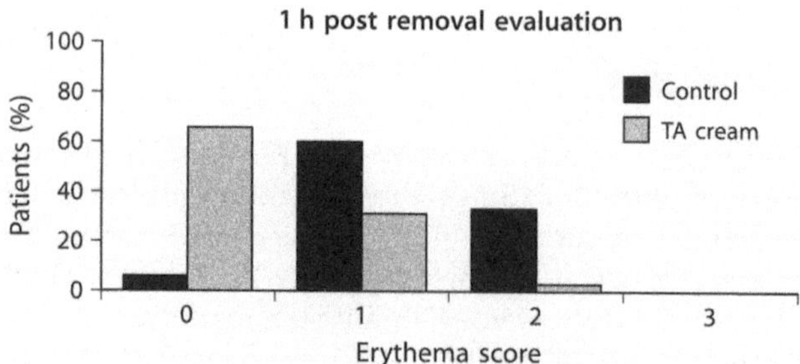

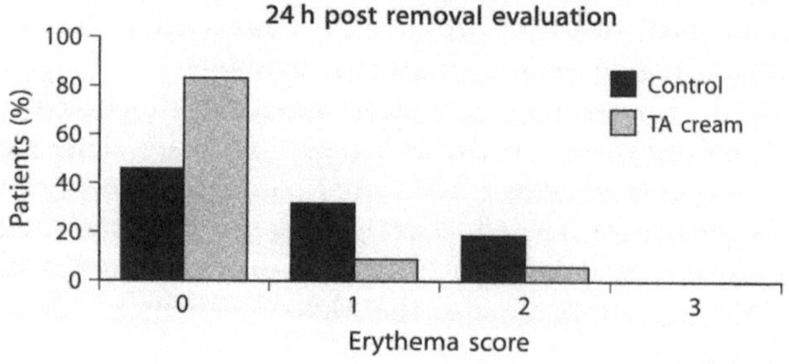

Fig. 14.9. Hypogonadal men and pretreatment effects on skin irritation (Meikle et al. 1997a)
Scoring system for erythema of application site (0 = none; 1 = mild, faint; 2 = moderate, bright pink; 3 = severe, beet red. n = 16. TA = 0.1% triamcinolone acetonide cream)

Pretreatment with triamcinolone acetonide 0.1% cream not only improved local tolerability but did not significantly alter the pharmacokinetics of the TTD treatment in hypogonadal men. In contrast, a hydrocortisone ointment resulted in approximately a 50% reduction in delivery of testosterone, indicating that ointments are not to be used for pretreatment. Approximately 0.2 mg of triamcinolone acetonide is applied with each 2.5 mg patch or 0.4 mg with two 2.5 mg patches or one 5 mg patch, which is below the calculated threshold of about 0.8 mg of the steroid that has been associated with suppression of the hypothalamic-pituitary-adrenal axis (McCubbin et al. 1995) .

In another study of eugonadal men the objective was to determine the effect of pretreatment with topical triamcinolone cream on the incidence and severity of skin irritations in healthy male volunteers receiving TTD applications (Wilson et al. 1997). Preliminary data for the first 62 subjects who completed the study comparing TTD to TTD plus 0.1% triamcinolone acetonide cream indicated that mild to moderate erythema scores were reduced about 60–70% by triamcinolone acetonide pretreatment. This study also confirmed that 0.1% triamcinolone acetonide cream pretreatment reduced the severity and incidence of skin reactions associated with use of the testosterone transdermal system.

14.5 Clinical application of the studies of the TTD system

Much of this review has dealt with technical aspects and clinical trials of new drug development. These studies have provided information that has application for the physician managing hypogonadal men who would benefit from androgen replacement therapy (Table 14.3). Practical uses of the TTD system will be summarized for the clinician. Many satisfactory forms of androgen replacement therapy are available for the clinician to use. Of the therapies available, the scrotal and non-scrotal transdermal testosterone formulations deliver testosterone in a more physiologic way than any other available. However, the scrotal system results in disproportionately high ratios of dihydrotestosterone/testosterone. If testosterone enanthate or cypionate were given in smaller doses at more frequent intervals (50–100 mg/ week), the testosterone fluctuations would be less. Whether the propensity of 200 mg of testosterone enanthate injected every two weeks to elevate hematocrit and prostate-specific antigen and induce gynecomastia would also be less common with smaller more frequent doses of IM therapy requires further study. The TTD would appear to be less prone to cause polycythemia and gynecomastia than IM testosterone because it produces normal rather than elevated estradiol concentrations. Thus, until it is known whether attempts to duplicate normal physiology are important, mimicking normal physiology seems a reasonable goal to achieve. The non-scrotal TTD has two dose preparations, which provide versatility to titrate the serum testosterone concentration to the desired level.

Table 14.3. Treatment of male hypogonadism

Group	Goal of therapy	Plasma testosterone	Preparation	Usual dose
Delayed puberty	Short term maintenance, initial	100–300 ng/dL	hCG	500 IU IM 1–2 x = s/wk
			Androderm	2.5 mg patch 12 h at night
			TE or TC	50–100 mg q 3–4 wks
	Subsequent	300–400 ng/dL	Androderm TE or TC	2.5 mg/daily 100 q 2 weeks
Adult	Long-term maintenance	400–1000 ng/dL		
Hypogonadotropic hypogonadism			GnRH*	5–30: g SC q 2 h
Hypogonadism			hCG	1000–4000 IU IM 1–3 × = s/wk
			Androderm	5 mg/day
			Testoderm	4 or 6 mg/day

* Experimental, requires programmed pump

hCG, human chorionic gonadotropin; GnRH, gonadotropin-releasing hormone; TE, testosterone enanthate; TC, testosterone cypionate

The pharmacokinetic studies indicate that when the TTD is applied in the evening it will produce patterns of testosterone and its metabolites that mimic normal physiology. By measuring concentrations of testosterone about 12 hours after patch application, proper assessment of dosing can be made. Some men have normal testosterone levels with one 2.5 mg patch but most require a 5 mg dose (which is now easiest to produce with one 5 mg patch daily). Most men use the thigh and abdomen as an application site because these sites are readily available for self application. Experience with the TTD in hypogonadal adolescent boys is limited. However, until more studies are conducted, dosing of prepubertal hypogonadal boys could be accomplished by administering a 2.5 mg for 12 hours for several months which would approximate early puberty (Griffin and Wilson 1992; Santen 1991). During Tanner stage II–III, a single patch worn for 24 hours should deliver mid-pubertal concentrations of testosterone. As Tanner stage IV approaches and during adult life, a 5 mg dose could be used indefinitely.

Hypogonadal men may have impotence and erectile dysfunction that are not correctable by androgen replacement therapy. Before studies of erectile dysfunction are performed in men with hypogonadism, androgen replacement therapy should be continued for 4–6 months. If no improvement in erectile function occurs, then urologic evaluation should be done. Even if hypogonadal men have causes of erectile dysfunction other than testosterone deficiency, androgen therapy provides other benefits (such as improved libido, mood, lean body mass, bone density and less fatigue).

Androgen replacement therapy may have adverse effects on prostate diseases, lipids, polycythemia, edema and liver. Since the TTD system contains testosterone, hepatic dysfunction would be unexpected. The TTD appears to have a lower incidence of elevated hematocrit than injectable testosterone esters; therefore, if a man has a tendency to polycythemia, the TTD may be a better choice than injectable forms of testosterone esters.

The relationship between prostate volume, including the transition zone where benign prostatic hyperplasia develops, and symptoms of urinary voiding is weak (Lepor et al. 1997; Meikle et al. 1997c). However, before beginning androgen replacement therapy in hypogonadal men age 45 years or older, a good practice is to perform screening for prostate disease by administering a prostate symptom questionnaire, performing a digital rectal examination and measuring serum prostate-specific antigen. If no evidence of prostate cancer is evident, androgen therapy may be initiated. Thereafter, surveillance for prostate cancer should be done at intervals recommended for men of comparable age. Evidence that androgen therapy will cause symptomatic benign prostatic hyperplasia is weak. While it appears safe to administer androgen therapy to hypogonadal men with asymptomatic benign prostatic hyperplasia, men with substantial symptomatic disease should have appropriate medical or surgical management of benign prostatic hyperplasia before beginning TTD or other androgen replacement therapy.

TTD-induced symptoms of contact dermatitis, which occurs in about 10% of men, can be reduced substantially in many men by pretreatment with a topical glucocorticoid creams only a few men ever develop an allergic reaction, mainly in response to the alcohol component of the TTD, but it precludes continued use. Redness and itching are symptoms that require glucocorticoid pretreatment to permit continued use. If these symptoms occur, the patient should be advised to contact the physician. Pretreatment with 0.1% triamcinolone acetonide has been shown to reduce substantially skin irritation so that continued use is possible. This topical glucocorticoid pretreatment does not significantly affect testosterone pharmacokinetics and delivery of the TTD. Estimates of the quantity of triamcinolone acetonide absorbed under the patch appears to be insufficient to cause any suppression of the hypothalamic-pituitary-adrenal axis.

14.6 Conclusions and future goals

The TTD is patient-friendly and delivers physiological amounts of testosterone, produces normal range concentrations of testosterone and its metabolites which mimic the circadian rhythms of healthy young men, and improves hypogonadal symptoms, mood and sexual function. No currently available delivery system comes closer to replicating normal physiologic testosterone replacement than TTD.

Despite the simplicity of use of the TTD system and its favorable pharmacokinetic profile compared to other treatment modalities, improvements could be made. The use of a single patch rather than two per day is a major advantage to the patient. A system designed to deliver testosterone for several days in a physiological way would also be an improvement. For some men, the visibility of the patch presents social embarrassment, which could be overcome by development of a less visible patch or an invisible gel or paste. Skin irritation with use of the TTD presents a limitation in about 10% of men and for some precludes continued use. While pretreatment with topical glucocorticoids does not alter the performance of the patch, it is an added nuisance for the physician and the patient. Incorporating the glucocorticoid into the patch for such men would overcome part of the problem. A better solution would be to prepare a less irritating formulation that still has favorable performance. Cost reduction is always welcome.

14.7 Key messages

- A permeation enhanced non-scrotal testosterone transdermal system (TTD) was designed to treat male hypogonadism.
- In the clinical trials, 93% of patients used the standard two 2.5 mg patches applied nightly to the back, abdomen, upper arms or thighs. This regimen delivered (now also available as a 5 mg patch) an average of 5 mg/24 h of testosterone, which is comparable to normal production rates.
- During chronic treatment (n = 94), morning testosterone values were within the normal reference range (306–1031 ng/dL) in 92% of patients.
- The 24 hour testosterone profiles mimicked the morning peak and nighttime nadir observed in healthy young men.
- In comparison to the untreated hypogonadal state (or androgen withdrawal period), TTD was comparable to 200 mg of injectable testosterone enanthate and resulted in significant reductions in hypogonadal symptoms, improved mood and improved sexual function.
- Safety evaluations also showed comparable results to injectable testosterone-enanthate treatment with regard to prostate, lipid profiles and systemic laboratory parameters.
- A total of 9% of patients stopped treatment due to chronic skin irritation (5%) or local allergic contact dermatitis to the system (4%). Pretreatment with 0.1% of triamcinolone acetonide did not significantly affect the pharmacokinetics of TTD, but substantially reduced the degree of skin irritation in hypogonadal men.
- Thus, a 5 mg TTD or two 2.5 mg patch(es) applied nightly to the skin of hypogonadal men is a physiological, convenient and "patient-friendly" modality for testosterone replacement therapy.

Acknowledgements

Support for these studies was provided by USPHS grants MO1 RR-00064, DK-45760, DK-43344 and TheraTech, Inc. AWM is independent from TheraTech and contributed to the experimental design and data analysis of the investigation.

The clinical trials were funded by TheraTech, Inc and SmithKline Beecham. Thanks are due to the investigators and their associates at each center.

Stephan Arver, MD, Ph.D., Karolinska Institute, Stockholm, Sweden
Adrian Dobs, MD, MPH, Johns Hopkins University, Baltimore, Maryland
Norman Mazer, MD, Ph.D., TheraTech, Inc, Salt Lake City, Utah
Douglas K. Wilson, PharmD, SmithKline Beecham, King of Prussia, Pennsylvania
Dion R. Brocks, Ph.D., SmithKline Beecham, King of Prussia, Pennsylvania
Richard G. Middleton, MD, University of Utah, Salt Lake City, Utah
Stephen W. Sanders, Pharma, TheraTech Inc., Salt Lake City, Utah
Robert A. Stephenson, MD, University of Utah, Salt Lake City, Utah

14.8 References

Anderson RA, Wallace AM, Kicman AT, Wu FC (1997a) Comparison between testosterone oenanthate-induced azoospermia and oligozoospermia in a male contraceptive study. IV. Suppression of endogenous testicular and adrenal androgens. Hum Reprod 12:1657–1662

Anderson RA, Wallace EM, Groome NP, Bellis AJ, Wu FC (1997b) Physiological relationships between inhibin B, follicle-stimulating hormone secretion and spermatogenesis in normal men and response to gonadotrophin suppression by exogenous testosterone. Hum Reprod 12:746–751

Arver S, Dobs AS, Meikle AW, Allen RP, Sanders SW, Mazer NA (1996) Improvement of sexual function in testosterone deficient men treated for 1 year with a permeation enhanced testosterone transdermal system. J Urol 155:1604–1608

Arver S, Meikle AW, Dobs AS, Sanders S, Mazer NA (1997) Hypogonadal men treated with the Androderm testosterone transdermal system had fewer abnormal hematocrit elevations than those treated with testosterone enanthate injections. Endocrine Soc. 79th Annual Meeting:P01–327

Bagatell CJ, Bremner WJ (1995) Androgen and progestagen effects on plasma lipids. Prog Cardiovasc Dis 38:255–271

Bagatell CJ, Knopp RH, Vale WW, Rivier JE, Bremner WJ (1992) Physiologic testosterone levels in normal men suppress high-density lipoprotein cholesterol levels [see comments]. Ann Intern Med 116:967–973

Bagatell CJ, Knopp RH, Rivier JE, Bremner WJ (1994) Physiological levels of estradiol stimulate plasma high density lipoprotein 2 cholesterol levels in normal men. J Clin Endocrinol Metab 78:855–861

Baker HW, Hudson B (1983) Changes in the pituitary-testicular axis with age. Monogr Endocrinol 25:71–83

Behre HM, Nieschlag E (1992) Testosterone buciclate (20 Aet-1) in hypogonadal men: pharmacokinetics and pharmacodynamics of the new long-acting androgen ester. J Clin Endocrinol Metab 75:1204–1210

Behre HM, Oberpenning F, Nieschlag E (1990) Comparative pharmacokinetics of testosterone preparations: application of computer analysis and simulation. In: Nieschlag E, Behre HM (eds) Testosterone: action, deficiency, substitution. Springer-Verlag, Berlin, Heidelberg, New York pp 115–134

Behre HM, Bohmeyer J, Nieschlag E (1994) Prostate volume in testosterone-treated and untreated hypogonadal men in comparison to age-matched normal controls. Clin Endocrinol (Oxf) 40:341–349

Behre HM, Kliesch S, Leifke E, Link TM, Nieschlag E (1997) Long-term effect of testosterone therapy on bone mineral density in hypogonadal men. J Clin Endocrinol Metab 82:2386–2390

Bhasin S, Swerdloff RS, Steiner B, Peterson MA, Meridores T, Galmirini M, Pandian MR, Goldberg R, Berman N (1992) A biodegradable testosterone microcapsule formulation provides uniform eugonadal levels of testosterone for 10–11 weeks in hypogonadal men. J Clin Endocrinol Metab 74:75–83

Bremner WJ, Vitiello MV, Prinz PN (1983) Loss of circadian rhythmicity in blood testosterone levels with aging in normal men. J Clin Endocrinol Metab 56:1278–1281

Brocks DR, Meikle AW, Boike SC, Mazer NA, Zariffa N, Audet PR, Jorkasky DK (1996) Pharmacokinetics of testosterone in hypogonadal men after transdermal delivery: influence of dose. J Clin Pharmacol 36:732–739

Cantrill J, Dewis P, Large D, Newman M, Anderson D (1984) Which testosterone replacement therapy? Clin Endocrinol (Oxf) 21:97–107

Crave JC, Lejeune H, Brebant C, Baret C, Pugeat M (1995) Differential effects of insulin and insulin-like growth factor I on the production of plasma steroid-binding globulins by human hepatoblastoma-derived (Hep G2) cells. J Clin Endocrinol Metab 80:1283–1289

De Lignieres B (1993) Transdermal dihydrotestosterone treatment of andropause. Ann Med 25:235–241

De Sanctis V, Vullo C, Urso L, Rigolin F, Cavallini A, Carmelli K, Daugherty C, Mazer N (1997) Clinical experience using the Androderm testosterone transdermal system (TTD) in hypogonadal adolescents and young men with β-thalassemia major. Endocrine Soc. 79th Annual Meeting:413, P412–515A

Feldmann R, Maibach H (1967) Regional variation in percutaneous penetration of 14C cortisol in man. J Invest Dermatol 48:181–183

Finkelstein JS, Neer RM, Biller BM, Crawford JD, Klibanski A (1992) Osteopenia in men with a history of delayed puberty. N Engl J Med 326:600–604

Friedl K, Hannan CJ J, Jones R, Plymate S (1990) High-density lipoprotein cholesterol is not decreased if an aromatizable androgen is administered. Metabolism 39:69–74

Glass AR, Swerdloff RS, Bray GA, Dahms WT, Atkinson RL (1977) Low serum testosterone and SHBG in massively obese men. J Clin Endocrinol Metab 45:1211–1219

Gray A, Feldman HA, McKinlay JB, Longcope C (1991) Age, disease, and changing sex hormone levels in middle-aged men: results of the Massachusetts Male Aging Study. J Clin Endocrinol Metab 73:1016–1025

Griffin JE, Wilson JD (1992) Disorders of the Testes and Male Reproductive Tract. In: Wilson JD, Foster DW (eds) William's Textbook of Endocrinology. 8th ed. Saunders, Philadelphia, pp 799–852

Handelsman DJ, Conway AJ, Boylan LM (1990) Pharmacokinetics and pharmacodynamics of testosterone pellets in man. J Clin Endocrinol Metab 71:216–222

Hogan DJ, Maibach HI (1990) Adverse dermatologic reactions to transdermal drug delivery systems. J Am Acad Dermatol 22:811–814

Huang CC, Huang HS (1994) Successful treatment of male infertility due to hypogonadotropic hypogonadism – report of three cases. Chang Keng I Hsueh 17:78–84

Ibanez L, Potau N, Georgopoulos N, Prat N, Gussinye M, Carrascosa A (1995) Growth hormone, insulin-like growth factor-I axis, and insulin secretion in hyperandrogenic adolescents. Fertil Steril 64:1113–1119

Lepor H, Nieder A, Feser JCOC, Dixon C (1997) Total prostate and transition zone volumes, and transition zone index are poorly correlated with objective measures of clinical benign prostatic hyperplasia. J Urol 158:85–88

Luciani J, Guichaoua M (1985) Chromosome abnormalities in male infertility. Ann Biol Clin (Paris) 43:71–74

MacIndoe JH, Perry PJ, Yates WR, Holman TL, Ellingrod VL, Scott SD (1997) Testosterone suppression of the HPT axis [In Process Citation]. J Investig Med 45:441–447

Matsumoto AM (1994) Hormonal therapy of male hypogonadism. Endocrinol Metab Clin North Am 23:857–875

Matsumoto AM (1995) Clinical use and abuse of androgens and antiandrogens. In: Becker KL (eds) Priniciples and Practice of Endocrinology and Metabolism. 2nd ed. J.B. Lippincott, Philadelphia, pp 1110–1122

Mazer NA, Heiber WE, Moellmer JF, Meikle AW, Stringham JD, Sanders SW, Tolman KG, Odell WD (1992) Enhanced transdermal delivery of testosterone: a new physiological approach for androgen replacement in hypogonadal men. J contr Rel. 19:347–362

McClure RD (1987) Endocrine investigation and therapy. Urol Clin North Am 14:471–488.

McClure RD (1988) Endocrine evaluation and therapy of erectile dysfunction. Urol Clin North Am 15:53–64

McCubbin MM, Milavetz G, Grandgeorge S, Weinberger M, Ahrens R, Sargent C, Vaughan LM (1995) A bioassay for topical and systemic effect of three inhaled corticosteroids. Clin Pharmacol Ther 57:455–460

Meikle AW, Bishop DT, Stringham JD, West DW (1986) Quantitating genetic and nongenetic factors that determine plasma sex steroid variation in normal male twins. Metabolism 35:1090–1095

Meikle AW, Stringham JD, Bishop DT, West DW (1988) Quantitating genetic and nongenetic factors influencing androgen production and clearance in men. J Clin Endocrinol Metab 67:104–109

Meikle AW, Mazer NA, Moellmer JF, Stringham JD, Tolman KG, Sanders SW, Odell WD (1992) Enhanced transdermal delivery of testosterone across nonscrotal skin produces physiological concentrations of testosterone and its metabolites in hypogonadal men. J Clin Endocrinol Metab 74:623–628

Meikle AW, Stephenson RA, McWhorter WP, Skolnick MH, Middleton RG (1995) Effects of age, sex steroids, and family relationships on volumes of prostate zones in men with and without prostate cancer. Prostate 26:253–259

Meikle AW, Arver S, Dobs AS, Sanders SW, Mazer NA (1996a) Androderm: a permeation enhanced non-scrotal testosterone transdermal system for the treatment of male hypogonadism. In: Bhasin S, HL Gabelnick, JM Spieler, RS Swerdloff, C Wang (eds) Pharmacology, biology and clinical applications of androgens. Wiley-Liss, Inc, New York, pp 449–458

Meikle AW, Arver S, Dobs AS, Sanders SW, Rajaram L, Mazer NA (1996b) Pharmacokinetics and metabolism of a permeation-enhanced testosterone transdermal system in hypogonadal men: influence of application site – a clinical research center study. J Clin Endocrinol Metab 81:1832–1840

Meikle AW, Annand D, Hunter C, Caramelli KE, Daugherty CA, John VA, Mazer NA (1997a) Pre-treatment with a topical corticosteroid cream improves local tolerability and does not significantly alter the pharmacokinetics of the androderm testosterone transdermal system in hypogonadal men. Endocrine Soc. 79th Annual Meeting:P01–322

Meikle AW, Arver S, Dobs AS, Adolfsson J, Sanders SW, Middleton RG, Stephenson RA, Hoover DR, Rajaram L, Mazer NA (1997b) Prostate size in hypogonadal men treated with a non-scrotal permeation-enhanced testosterone transdermal system. Urology 49:191–196

Meikle AW, Stephenson RA, Lewis CM, Middleton RG (1997c) Effects of age and sex hormones on transition and peripheral zone volumes of prostate and benign prostatic hyperplasia in twins. J Clin Endocrinol Metab 82:571–575

Meikle AW, Wilson DE, Boike SC, Fairless AJ, Etheredge RC, Jorkasky DK (1997d) A study to assess the bioequivalence of Androderm (2 X 2.5 mg patches) and a newly formulated testosterone transdermal system (1 × 5 mg patch). Endocrine Soc. 79th Annual Meeting:215, P211–321A

Morales A, Johnston B, Heaton JP, Lundie M (1997) Testosterone supplementation for hypogonadal impotence: assessment of biochemical measures and therapeutic outcomes. J Urol 157:849–854

Nachtigall LB, Boepple PA, Pralong FP, Crowley WF, Jr (1997) Adult-onset idiopathic hypogonadotropic hypogonadism – a treatable form of male infertility. N Engl J Med 336:410–415

Nielsen J, Wohlert M (1991) Chromosome abnormalities found among 34,910 newborn children: results from a 13-year incidence study in Arhus, Denmark. Hum Genet 87:81–83

Nieschlag E, Behre HM (1990) Pharmacology and clinical uses of testosterone. In: Nieschlag E, Behre HM (eds) Testosterone: action, deficiency, substitution. Springer-Verlag, Berlin, Heidelberg, New York pp 92–108

Noldus J, Huland H (1994) Erectile dysfunction and hypogonadism. Is routine endocrine screening necessary? Urologe A 33:73–75

Place VA, Atkinson L, Prather DA, Trunell N, Yates FE (1990) Transdermal testosterone replacement through genital skin. In: Nieschlag E, Behre HM (eds) Testosterone: action, deficiency, substitution. Springer-Verlag, Berlin, Heidelberg, New York pp 165–180

Salehian B, Wang C, Alexander G, Davidson T, McDonald V, Berman N, Dudley R, Ziel F, Swerdloff R (1995) Pharmacokinetics, bioefficacy, and safety of sublingual testosterone cyclodextrin in hypogonadal men: comparison to testosterone enanthate – a clinical research center study. J Clin Endocrinol Metab 80:3567–3575

Santen RJ (1991) Male Hypogonadism. In: Yen SSC, Jaffe RB (eds) Reproductive Endocrinology. 3rd ed Saunders, Philadelphia, pp 739–794

Schaison G, Nahoul K, Couzinet B (1990) Percutaneous dihydrotestosterone (DHT) treatment. In: Nieschlag E, Behre HM (eds) Testosterone: Action, deficiency, substitution. Springer-Verlag, Berlin, Heidelberg, New York pp 155–164

Sih R, Morley J, Kaiser F, Perry HM, Patrick P, Ross C (1997) Testosterone replacement in older hypogonadal men: a 12-month randomized controlled trial. J Clin Endocrinol Metab 82:1661–1667

Snyder PJ, Lawrence DA (1980) Treatment of male hypogonadism with testosterone enanthate. J Clin Endocrinol Metab 51:1335–1339

Southren A, Gordon G, Tochimoto S (1968) Further study of factors affecting the metabolic clearance rate of testosterone in man. J Clin Endocrinol Metab 28:1105–1112

Stepan JJ, Lachman M, Zverina J, Pacovsky V, Baylink DJ (1989) Castrated men exhibit bone loss: effect of calcitonin treatment on biochemical indices of bone remodeling. J Clin Endocrinol Metab 69:523–527

Tunte W, Niermann H (1968) Incidence of Klinefelter's syndrome and seasonal variation. Lancet 1:641.

Vermeulen A (1996) Decreased androgen levels and obesity in men. Ann Med 28:13–15

Vermeulen A, Ando S (1979) Metabolic clearance rate and interconversion of androgens and the influence of the free androgen fraction. J Clin Endocrinol Metab 48:320–326

Vermeulen A, Rubens R, Verdonck L (1972) Testosterone secretion and metabolism in male senescence. J Clin Endocrinol Metab 34:730–735

Wang C, Alexander G, Berman N, Salehian B, Davidson T, McDonald V, Steiner B, Hull L, Callegari C, Swerdloff RS (1996) Testosterone replacement therapy improves mood in hypogonadal men – a clinical research center study. J Clin Endocrinol Metab 81:3578–3583

Wilson DE, Kaidbey K, Boike SC, Jorkasky DK (1997) Use of topical corticosteroid cream in the pretreatment of skin reactions associated with Androderm testosterone transdermal system. Endocrine Soc. 79th Annual Meeting:P01–323

Wilson JD (1987) The testes and the prostate. N Engl J Med 317:628–629

Wilson JD, Gloyna RE, Siiteri PK (1975) Androgen metabolism in the hypertrophic prostate. J Steroid Biochem 6:443–445

15 Percutaneous dihydrotestosterone treatment

Gilbert Schaison and Béatrice Couzinet

Contents

15.1 Introduction

The aim of hormone replacement therapy is to obtain normal physiologic levels of sex steroids.

Various forms of androgen therapy are readily available for the treatment of hypogonadism in men (Bhasin et al. 1997; Cantril et al. 1984). The commonly used long-acting, injected, testosterone esters produce wide variations of serum androgen concentrations. High levels in the supraphysiological range are obtained within 24 hours after the injection and subnormal levels in the hypogonadal range after 15 days (Snyder and Lawrence 1980; Sokol et al. 1982). In addition, an elevation in the estradiol-testosterone ratio can cause gynecomastia. A long-acting testosterone ester, testosterone buciclate, which has a 12 to 16 week duration of action, is under investigation (Behre and Nieschlag 1992). Oral therapy requires the administration of multiple daily doses of testosterone. Preparations which are 17α-alkylated androgens, such as fluoxymesterone and methyltestosterone, are potentially hepatotoxic and should not be used on a long-term basis. Testosterone undecanoate, administered orally, produces only short-lived testosterone peaks and thus requires repeated doses three times a day for clinical effects. Testosterone incorporated into microspheres may be slowly and steadily released from the intramuscular site over extended periods of time. The main advantage of this microsphere formulation is to maintain estradiol and dihydrotestosterone (DHT) levels in the normal male range (Bhasin et al. 1992).

15.1.1 Percutaneous drug delivery

In France, as early as 1942, M.F. Jayle was interested in the transcutaneous absorption of molecules and initiated research on steroid absorption through the skin (Jayle 1980). Since then, the ability of the skin to absorb steroids has been widely recognized and extensively studied. The skin of an average adult covers a surface area of approximately 2 m^2 and receives one third of the blood circulation. The phenomenon of percutaneous absorption embraces more than just diffusion through the stratum corneum (Sitruk-Ware 1989). Several different processes occur in sequence. First, absorption of the molecules within the stratum corneum, secondly, retention in the stratum corneum (reservoir effect) and thirdly, diffusion through the epidermis until the molecules reach the capillary plexus and are transferred to the circulation. Percutaneous or transdermal medications offer advantages over conventional delivery. They bypass hepatic "first pass" and prevent gastrointestinal side-effects. They optimize the blood concentration/time profile, avoiding the peaks and troughs associated with intermittent dosage forms.

15.1.2 Percutaneous androgen application

When applied percutaneously to human skin in an alcohol solution, steroids rapidly penetrate the stratum corneum (within 10 minutes after application). The diffusion of the steroid through the epidermis and dermis occurs over a period of several hours. The rate of absorption depends on the dose that is applied topically and the surface of application. About 10% of the total dose passes through the cutaneous barrier, is transferred to the vascular system and is excreted in urine over the following 72 hours.

Regional differences of transdermal absorption of drugs exist. The scrotal skin has a unique superficial vascularity. The stratum corneum has the highest rate of steroid absorption, about 30 to 40 fold higher than the forearm. This remarkable absorption rate is sufficient to bring the blood level of testosterone into the normal range via a skin surface of only 40 to 60 cm^2.

Thus, a scrotal transdermal testosterone delivery system has been shown to be an effective modality for the treatment of male hypogonadism. This scrotal testosterone patch when applied daily produces normal testosterone levels in hypogonadal men 4–8 hours after application followed by gradual decrease in serum testosterone levels over the next 24 hours. Plasma DHT levels increase to the supra-normal range while plasma estradiol levels remain low (Ahmed et al. 1988; Bals-Pratsch et al. 1986; Findlay et al. 1989; Korenman et al. 1987).

A permeation-enhanced non-genital patch can be applied on the nonscrotal skin. After nightly application of two patches, serum testosterone levels are in the normal range 8–12 hours after application. Unlike the scrotal delivery system, the non-genital patch produces physiological levels of serum dihydrotestosterone levels and a normal dihydrotestosterone to testosterone ratio (Meikle et al. 1992).

DHT can also be used percutaneously. In a hydroalcoholic gel, it has been shown to be readily absorbed through the skin (Delignieres and Morville 1980; Fiet et al. 1982). The effects of daily percutaneous administration of DHT have been studied in hypogonadal patients, normal men and subjects with so-called idiopathic gynecomastia (Schaison et al. 1980). The hydroalcoholic gel contained a 2.5% solution of DHT and was packed in 80 g tubes. Graduations indicated doses of 5 g of gel or 125 mg DHT (Andractim, Besins Iscovesco, Paris, France). The rapid penetration of the gel obviates the need for an occlusive dressing. Men were advised to apply DHT to a large area of the skin in the morning and to take a shower, washing the area of application 10 minutes after gel administration.

15.2 Effects of percutaneous DHT treatment in hypogonadal men

15.2.1 Pharmacokinetics of DHT gel

The topical administration of a single daily dose of 125 mg of DHT to the abdominal skin of 10 hypogonadal patients immediately increases the plasma levels of this steroid from 0.12±0.05 ng/ml before treatment to 2.51±0.32 ng/ml on day 4. The application of a dose of 250 mg of DHT (125 mg twice daily at 0800 h and 2000 h) increases the plasma level of DHT to 4.43±0.56 ng/ml on day 4 of treatment (Fig. 15.1). Repeated evaluation of plasma DHT by hourly sampling from 0800 to 2000 h demonstrated the stability of diurnal DHT levels and its regular distribution from a presumably cutaneous reservoir (Fig. 15.2). Indeed, DHT levels plateaued after four days. Long-term percutaneous DHT therapy for up to three months maintained stable serum DHT concentrations.

When the gel was applied to an area of the abdomen less than 20×20 cm, the levels observed varied from one subject to another. Individual variability of percutaneous absorption may explain this phenomenon. However, when

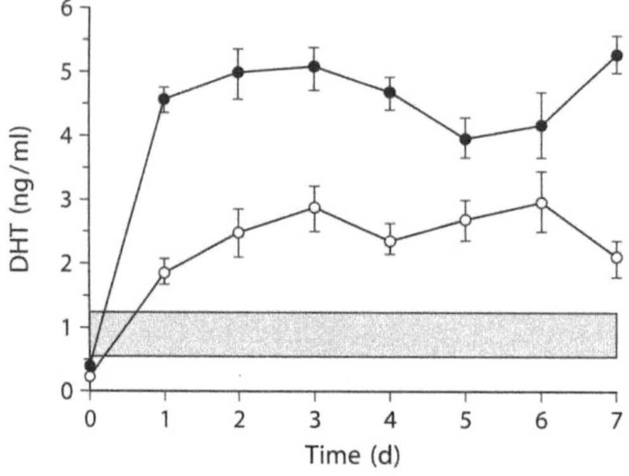

Fig. 15.1. Plasma dihydrotestosterone (DHT) levels after topical administration of DHT 125 mg per day (○-○) or twice daily (●-●) for seven days in hypogonadal men (The shaded area represents the range for normal men)

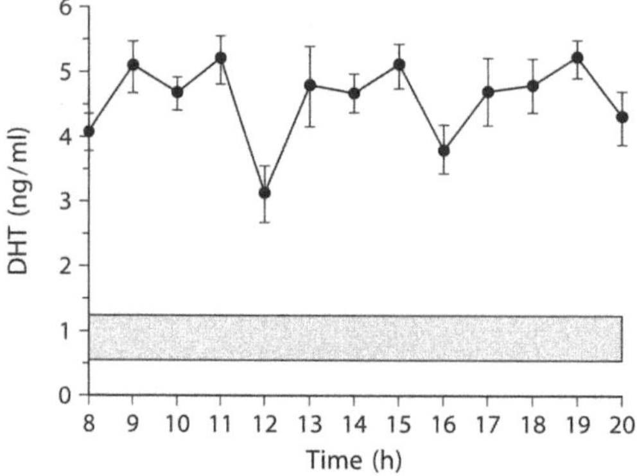

Fig. 15.2. Plasma DHT levels from 0800 h to 2000 h on day 4 of DHT administration 125 mg twice daily. (The shaded area represents the range for normal men)

the gel was applied to a large area (thorax, shoulders and arms) plasma DHT levels were similar in all subjects. Plasma levels of DHT was also studied after scrotal application of the gel. Treatment with the same dose (125 mg twice daily) of transscrotal DHT induced similar plasma levels. Although the scrotal application site of steroids results in a higher absorption rate, the relatively small amount of scrotal skin available makes it difficult to absorb 5 g hydroalcoholic gel applied twice daily.

Plasma testosterone levels measured after ether extraction and celite chromatography and plasma estradiol levels were low before treatment and, as expected, were not modified by DHT administration (Fig. 15.3). SHBG capacity, measured by binding, did not decrease significantly and the bioavailable testosterone, measured after precipitation of SHBG with amonium sulfate, remained low (Fig. 15.3). In contrast, plasma 3α-androstanediol glucuronide (3α-Adiol G), which is a main metabolite of DHT in peripheral target tissues, followed the increase of plasma DHT treatment (Fig. 15.4). Finally, in patients with primary disorders of testicular function there was no modification of plasma LH and FSH levels.

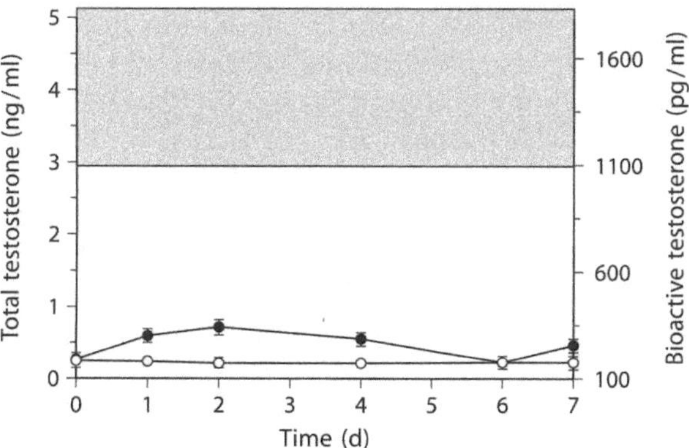

Fig. 15.3. Plasma total (○–○) and bioactive (●–●) testosterone levels are not modified by DHT administration. (The shaded area represents the range for normal men)

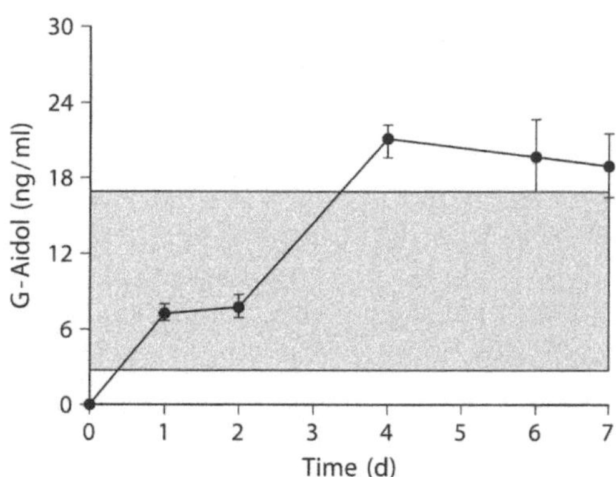

Fig. 15.4. Plasma 3α-Adiol G levels (G-Adiol) after topical administration of DHT 125 mg twice daily in hypogonadal men. (The shaded area represents the range for normal men)

15.2.2 Clinical effects

Clinically, all men treated with percutaneous DHT reported improvement in sexual functions. In patients with previously untreated hypogonadotropic hypogonadism, virilization developed within six months of repeated daily topical application of 250 mg of DHT. Plasma testosterone levels were not modified but the treatment induced male puberty and increased muscular mass. Patients with primary hypogonadism reported improvement in libido, erection, sexual intercourse or masturbatory activity. Thus, DHT was able to induce the same effects as testosterone in hypogonadal men.

15.2.3 Effects on plasma lipids

Testosterone effects on plasma lipids remain controversial. It has been shown that suppression of endogenous testosterone in young men increased serum levels of HDL cholesterol and lipoprotein (a) (von Eckardstein et al. 1997).

Administration of testosterone to orchidectomized patients as well as administration of supraphysiological doses of testosterone to healthy men are associated with significant decrease in LPa. This antiatherogenic effect also observed with estrogens is antagonized by the atherogenic risk induced by the concomitant decrease of HDL cholesterol.

Nonaromatizable androgens produce a greater decrease in plasma HDL levels than aromatizable androgens. Vermeulen and Deslypere (1985) reported that long-term transdermal DHT therapy in elderly males resulted in a moderate decrease in plasma HDL cholesterol levels. In our patients treated with DHT, the percutaneous route explained that plasma LDL, HDL and LPa remained unchanged after three months of treatment.

15.2.4 Effects on bone mineral density (BMD)

The effects of testosterone on BMD in men are well documented. Behre et al. (1997) studied BMD in 72 hypogonadal patients under testosterone substitution therapy that continued for up to 16 years. The most significant increase in BMD was seen during the first year of testosterone in previously untreated patients. Long-term testosterone treatment maintained BMD in the age-dependant reference range in all 72 hypogonadal men. The exact mechanism by which androgens affect BMD is not fully elucidated. However, in male patients with estrogen receptor mutation (Smith et al. 1994) or aromatase deficiency (Morishima et al. 1995), there is no doubt that estrogens play a major role in bone mineralization in men. BMD measurements and biochemical markers of bone formation and resorption have not been studied so far during prolonged treatment with DHT. However, non-aromatizable androgens have certainly little or much less effect than testosterone itself in preventing bone loss in hypogonadal men.

15.2.5 Long-term effects of DHT

The main therapeutic issue is the potential long-term effect of abnormally elevated serum DHT concentrations. Indeed, percutaneous DHT treatment increases DHT/T ratio to around 5, whereas in normal men the DHT/T ratio ranges between 0.1 to 0.2. Transscrotal testosterone therapy produces also an increase in serum DHT concentrations higher than the normal range. The high serum DHT levels are presumably due to increased metabolism of testosterone to DHT by the 5α-reductase activity in the scrotal skin. However, the DHT/T ratio under transdermal testosterone therapy remains below 1.

DHT is considered the hormonal mediator for hyperplastic prostatic growth. Short-term DHT elevations have not been reported to have any adverse clinical effects. In our patients treated daily for several months, topical administration of DHT did not lead to any increase in prostate size as showed by ultrasonography. However, the effects of chronically elevated serum DHT concentration are unknown. We are particularly concerned about the possible effects on the prostate in young men who will need continuous treatment for many years. Percutaneous DHT treatment is associated with abnormally high serum DHT levels without any increase of plasma estradiol levels. Thus, the potential long-term effect of DHT treatment on the prostate and other tissues needs to be further studied.

15.3 Effects of percutaneous DHT treatment in normal men

In ten normal adult men, DHT was applied on a large area of the skin at a dose of 125 mg twice daily. The mean plasma DHT level before treatment was 0.76±0.09 ng/ml. A significant increase occurred on day 2 of treatment and DHT plateaued thereafter (4.26±0.66 ng/ml) throughout the study (Fig. 15.5). This rise in plasma DHT concentration was accompanied by a concomitant increase in plasma 3α-androstanediol glucuronide and a fall in plasma testosterone and estradiol levels (Figs. 15.6 and 15.7) (Korenman et

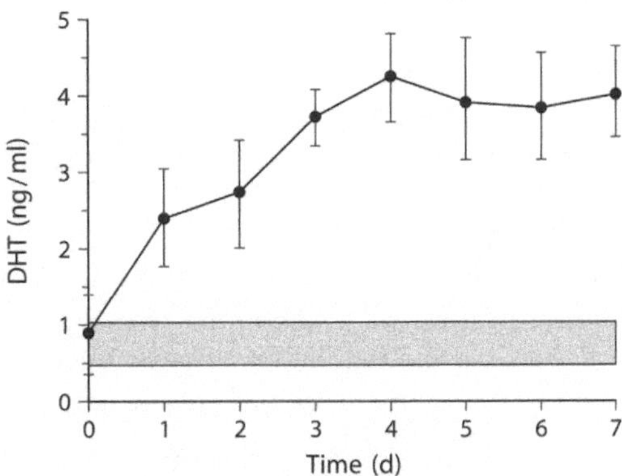

Fig. 15.5. Plasma DHT levels after topical administration of DHT 125 mg twice daily in normal men. (The shaded area represents the range for normal men)

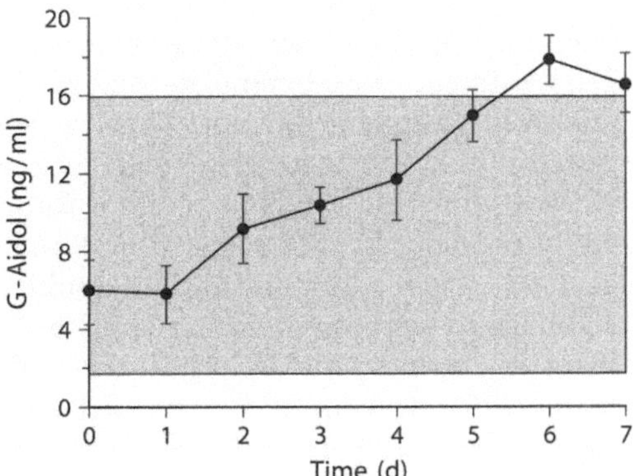

Fig. 15.6. Plasma 3α-Adiol G levels (G-Adiol) after topical administration of DHT 125 mg twice daily in normal men. (The shaded area represents the range for normal men)

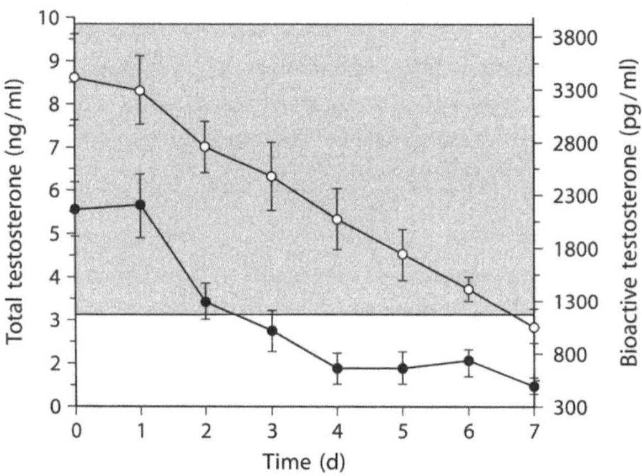

Fig. 15.7. Plasma total and bioactive testosterone levels after DHT administration in normal men. (The shaded area represents the range for normal men)

al. 1987). There was a significant negative correlation between changes in plasma DHT and testosterone levels during DHT therapy. Moreover, there was a highly significant fall in plasma bioavailable testosterone (Fig. 15.7). SHBG binding capacity was not significantly reduced by ten days of DHT treatment. DHT affinity for SHBG is three times higher than that of testosterone and can displace testosterone from its binding site on SHBG. In the present study, plasma bioavailable testosterone decreased following the fall in plasma total testosterone levels. The correlation between changes in total and unbound testosterone levels cannot be explained by the increased occupancy of the SHBG binding site by DHT which would have increased the percentage of bioavailable testosterone (Kuhn et al. 1984). Thus, the decrease of plasma concentration of testosterone was the consequence of the inhibitory activity of DHT on the regulation of gonadotropin secretion.

The mechanism of sex steroid-gonadotropin feedback in male patients is mainly mediated by estrogens. The concomitant administration of testosterone and of an aromatase inhibitor prevents testosterone-induced suppression of gonadotropins. Male patients with estrogen receptor mutation (Smith et al. 1994) have elevated gonadotropin concentrations despite normal testosterone secretion. In a man with aromatase deficiency (Carani et al. 1997), estrogen treatment caused complete suppression of serum gonadotropins, whereas androgen treatment did not. Thus, testosterone must be converted to estrogens to induce the negative feed-back on gonadotropin secretion in men. However, in male estrogen receptor knockout (ERKO) mice, testosterone directly regulates hypothalamic GnRH and circulating levels of LH (Lindzey et al. 1997). DHT treatment has no significant effect on wild type castrates but significantly depressed plasma levels of LH and restores hypothalamic GnRH levels in castrated ERKOs. Thus, the availability of DHT to regulate hypothalamic GnRH content and serum LH levels in ERKO male mice provide evidence that androgens can directly regulate GnRH and LH synthesis and secretion without activation of the estradiol receptor.

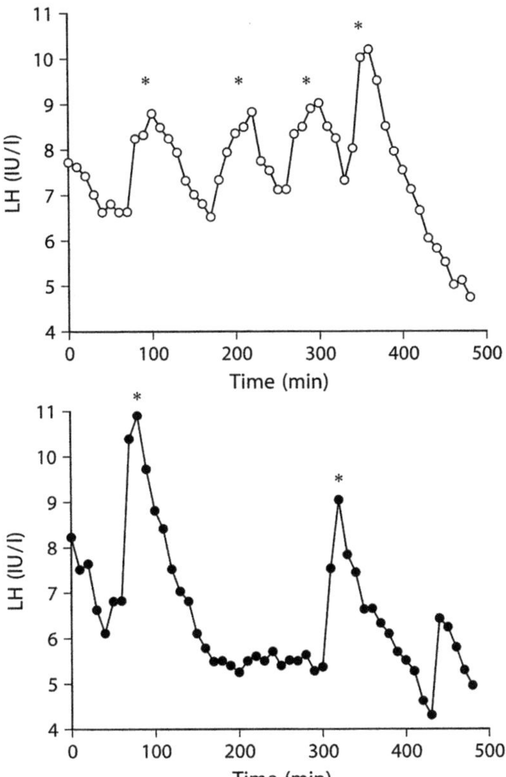

Fig. 15.8. LH pulse frequency and amplitude before and after DHT administration in normal men. The * indicate significant pulses

To further study the role of DHT on gonadotropin secretion, we have assessed the gonadotropin response to GnRH before and after percutaneous DHT administration in normal adult men. In addition, the baseline study of pulsatile LH secretion was performed with sampling every 10 minutes for 8 hours. Sampling was then repeated on day 4 of DHT treatment. The LH and FSH response to GnRH was similar before and after DHT administration. In contrast, on day 4 of DHT treatment, there was an abrupt and significant decrease of LH pulse frequency from 4.1±0.4 to 2.5±0.6 pulses/8 hrs (Fig. 15.8). The mean plasma LH level decreased slightly. The LH pulse amplitude was not modified. Plasma FSH levels did not change during the study. These data suggest a suppressive effect of DHT on GnRH secretion and confirm previous studies showing that DHT may act at the hypothalamic level (Veldhuis et al. 1984).

15.4 Other indications for percutaneous DHT treatment

15.4.1 Idiopathic gynecomastia

Idiopathic gynecomastia is a common disorder. Various medical treatments have been proposed but their effectiveness is variable or even doubtful. In view of the possible abnormality of the androgen/estrogen balance in gynecomastia, the effects of treatment by percutaneous DHT have been studied.

The use of DHT appears to have several advantages over other available treatments. DHT is non-aromatizable and it is well known that aromatizable androgens may by themselves induce gynecomastia. In addition, the fall in plasma estradiol induced by the treatment may be beneficial. The effectiveness seems to be due principally to systemic effects. Indeed, the results are the same when DHT is applied by the patients either on the abdominal skin or directly on the gynecomastia.

In one study conducted for 18 months in 40 men with idiopathic gynecomastia, twice daily application of 125 mg of percutaneous DHT resulted in complete disappearance of gynecomastia in ten patients, partial regression in 19 and no change in 11 patients after 4 to 20 weeks (Kuhn et al. 1983). Thus, the conclusion was that chronic percutaneous DHT administration was an efficient treatment for persistent idiopathic gynecomastia since it caused a reduction in 29 out of 40 cases.

Using the same dose of DHT (125 mg twice daily) and a radiological examination of the breast before and after treatment, we observed less than 50% of success in patients with persistent pubertal gynecomastia. This result was not significantly different when compared to an untreated control group. In some cases it was not possible to exclude a spontaneous regression. In contrast, in many cases, the fact that the gynecomastia did not disappear might be related to a lack of effect of the treatment on the fibrous structure of a longstanding gynecomastia.

As gynecomastia is the sign of an estrogen-androgen imbalance, we have also studied aromatase activity in pubic skin fibroblasts from patients with isolated gynecomastia (Bulard et al. 1987). We found increased aromatase activity in peripheral tissue which could represent one element leading to the appearance of gynecomastia. Thus, therapeutic use of aromatase inhibitors seem much more promising than the use of percutaneous DHT to reverse gynecomastia.

15.4.2 5α-reductase deficiency

5α-reductase type 2 deficiency leading to male pseudohermaphroditism is a rare autosomal recessive disorder. Patients with 5α-reductase 2 deficiency are characterized by an external female phenotype at birth which can range from almost normal female structure to a male phenotype with hypospadias, microphallus, bilateral testes and normally virilized wolffian structures. Affected boys undergo various degrees of virilization during puberty and develop male behavior in 70% of cases.

It is well known that in children with micropenis, judicious use of testosterone for a short period can enhance penile growth. If the diagnosis of 5α-reductase deficiency is made in infancy with the use of biochemical and molecular genetics, early correction of cryptorchidism and hypospadias is mandatory (Katz et al. 1997). Dihydrotestosterone gel may be used to enlarge the phallus before reconstructive surgery and to facilitate the correction of hypospadias. In men who were not treated during childhood, topical dihydrotestosterone therapy may also stimulate penile growth after puberty.

15.4.3 Acquired immunodeficiency syndrome and wasting

The acquired immunodeficiency syndrome (AIDS) wasting syndrome (AWS) is characterized by a disproportionate decrease in lean body mass. The pathogenesis of the AWS remains unknown but it has been reported that endogenous secretion of testosterone is decreased in 30–50% of men with AIDS. This decrease is likely the consequence of the nutritional insufficiency observed in these patients.

Grinspoon et al. (1996) demonstrated that change in body composition including loss of lean body mass and muscle mass, and deterioration in exercise function capacity are highly correlated with androgen levels in hypogonadal men with AWS. The use of testosterone in immunodeficiency virus infected men is growing. Spark et al. (1997) recently reported that men with AWS converted T to DHT inefficiently. They studied the effects of percutaneous DHT gel on weight and lean body mass. After ten weeks, mean weight gain was 2.2 Kg and lean body mass 3.4%. Percutaneous DHT gel is well tolerated and may benefit men with AWS if androgens are needed to supplement endogenous testosterone secretion.

15.5 Side-effects of percutaneous DHT treatment

The lack of local or systemic side-effects in all cases confirmed that percutaneous DHT administration is a relatively safe modality of androgen therapy. However, androgenization of female partners of men treated with testosterone in an alcoholic solvant applied daily on the abdomen has been reported. Among 17 couples, there were 11 women with increased plasma testosterone levels and four with hirsutism (Delanoe et al. 1984). In only one case the woman had applied testosterone to her partner. In the other cases, even after evaporation of the alcoholic solvant, the testosterone may be transferred to the skin of the female partner during sexual intercourse (Moore et al. 1988). We did not observe such an androgen side-effect with the percutaneous DHT treatment when the men applied the gel in the morning and took a shower ten minutes later.

15.6 Key messages

- Short-term percutaneous DHT treatment is an effective and well-tolerated modality for the treatment of male hypogonadism.
- The 24 hour stability of DHT plasma level is an interesting pharmocokinetic characteristic of percutaneous absorption.
- The administration of non-aromatizable DHT has the advantage over testosterone that it avoids estrogenization.
- In contrast, this treatment leads to high plasma levels of DHT and an unacceptable DHT/T ratio. We have to be concerned about the potential long-term effect of high DHT levels on the prostate.
- This steroid is less efficient than testosterone to prevent bone loss in hypogonadal men.
- The aim of hormone replacement therapy is to obtain normal, physiologic levels of sex steroids. Thus, at the present time and in all cases, transdermal preparations of testosterone are more advisable than percutaneous DHT in the treatment of male hypogonadism.

15.7 References

Ahmed R, Boucher A, Manni A, Santen R, Bartholomew M, Demers L (1988) Transdermal testosterone therapy in the treatment of male hypogonadism. J Clin Endocrinol Metabol 66:546–551

Bals-Pratsch M, Knuth UA, Yoon YD, Nieschlag E (1986) Transdermal testosterone substitution therapy for male hypogonadism. Lancet 25:943–945

Behre HM, Nieschlag E (1992) Testosterone buciclate (20 Act-1) in hypogonadal men: pharmacokinetics and pharmacodynamics of the new long acting androgen ester. J Clin Endocrinol Metabol 75:1204–1210

Behre HM, Kliesch S, Leifke E, Link TM, Nieschlag E (1997) Long term effect of testosterone therapy on bone mineral density in hypogonadal men. J Clin Endocrinol Metabol 82:2386–2390

Bhasin S, Swerdloff RS, Steiner B, Peterson MA, Meridores T, Galmirini M, Pandian MR, Goldberg R, Berman N (1992) A biodegradable testosterone microcapsule formulation provides uniform eugonadal levels of testosterone for 10–11 weeks in hypogonadal men. J Clin Endocrinol Metabol 74:75–83

Bhasin S, Bremner WJ (1997) Emerging issues in androgen replacement therapy. J Clin Endocrinol Metabol 82:3–8

Bulard J, Mowszowicz I, Schaison G (1987) Increased aromatase activity in pubic skin fibroblasts from patients with isolated gynecomastia. J Clin Endocrinol Metabol 64:618–623

Cantril JA, Dewis P, Large DM, Newman M, Anderson DC (1984) Which testosterone replacement therapy? Clin Endocrinol (Oxf), 21:97–107

Carani C, Qin K, Simoni M, Faustini-Fustini M, Serpente S, Boyd J, Korach KS, Simpson ER (1997) Effect of testosterone and estradiol in a man with aromatase deficiency. N Engl J Med 337:91–95

Delanoe D, Fougeyrollas B, Meyer L, Thonneau P (1984) Androgenization of female partners of men on medroxyprogesterone acetate/percutaneous testosterone contraception. Lancet 4:276–277

Delignieres B, Morville R (1980) Treatment of male hypogonadism by topical administration of androgens. Percutaneous absorption of steroids. P Mauvais-Jarvis, CFH/Vickers, J Wepierre (eds) Academic Press, pp 273–283

Fiet J, Morville R, Chemama D, Vilette JM, Gourmel B, Brerault JL, Dreux C (1982) Percutaneous absorption of 5α dihydrotestosterone in man. I. Plasma androgen and gonadotropin levels in normal adult men after percutaneous administration of 5α didydrotestosterone. Int J Androl 5:586–594

Findlay JC, Place VA, Snyder PJ (1989) Treatment of primary hypogonadism in men by the transdermal administration of testosterone. J Clin Endocrinol Metab 68:369–373

Grinspoon S, Corcoran C, Lee K, Burrows B, Hubbard J, Katznelson L, Walsh M, Guccione A, Cannan J, Heller H, Basgoz N, Klibanski A (1996) Loss of lean body and muscle mass correlates with androgen levels in hypogonadal men with acquired immunodeficiency syndrome and wasting. J Clin Endocrinol Metab 81:4051–4058

Jayle MF (1980) In memoriam. Percutaneous absorption of steroids. P Mauvais-Jarvis, CFH Vickers, J Wepierre (eds) Academic Press, 273–283

Katz MD, Kligman I, Cai LQ, Zhu YS, Fratianni CM, Zervoudakis I, Rosenwaks Z, Imperato-McGinley J (1997) Paternity by intrauterine insemination with sperm from a man with 5α-reductase 2 deficiency. N Engl J Med 336:994–997

Korenman S, Viosca S, Garza D, Guralnik M, Place V, Campbell P, Davis S (1987) Androgen therapy of hypogonadal men with transscrotal testosterone systems. Am J Med 83:471–478

Kuhn JM, Roca R, Laudat MH, Rieu M, Luton JP, Bricaire H (1983) Studies on the treatment of idiopathic gynaecomastia with percutaneous dihydrotestosterone. Clin Endocrinol 19:513–520

Kuhn JM, Rieu M, Laudat MH, Forest M, Pugeat M, Bricaire H, Luton JP (1984) Effects of 10 days administration of percutaneous dihydrotestosterone on the pituitary-testicular axis in normal men. J Clin Endocrinol Metab 58:231–235

Lindzey J, Wetsel WC, Stoker T, Cooper R, Couse JF, Korach KS (1997) Steroid regulation of hypothalamic GnRH and serum LH levels in male estrogen receptor knock out mice. The Endocrine Society 79th meeting. Minneapolis, MN: Abstract OR 16–3

Meikle AW, Mazer NA, Moellmer JF, Stringham JD, Tolman KG, Sanders SW, Odell WD (1992) Enhanced transdermal delivery of testosterone across the nonscrotal skin produces physiological concentrations of testosterone and its metabolites in hypogonadal men. J Clin Endocrinol Metab 74:623–628

Moore N, Paux G, Noblet C, Andrejak M (1988) Spouse-related drug side-effects. Lancet 27:468–469

Moore RJ, Gazak JM, Quebbeman JF, Wilson JD (1979) Concentration of dihydrotestosterone and 3α-androstanediol in naturally occurring and androgen-induced prostatic hyperplasia in the dog. J Clin Invest 64:1003–1009

Morishima A, Grumbach MM, Simpson ER, Fisher C, Qin K (1995) Aromatase deficiency in male and female siblings caused by a novel mutation and the physiological role of estrogens. J Clin Endocrinol Metab 80:3689–3698

Schaison G, Renoir M, Lagoguey M, Mowszowicz I (1980) On the role of dihydrostestosterone in regulating luteinizing hormone secretion in man. J Clin Endocrinol Metab 51:1133–1137

Sitruk-Ware R (1989) Transdermal delivery of steroids. Contraception 39:1–191

Smith EP, Boyd J, Frank GR, Takahashi H, Cohen RM, Specker B, Williams TC, Lubahn DB, Korach KS (1994) Estrogen resistance caused by a mutation in the estrogen-receptor gene in a man. N Engl J Med 331:1056–1061

Snyder PJ, Lawrence DA (1980) Treatment of male hypogonadism with testosterone enanthate. J Clin Endocrinol Metab 51:1335–1339

Sokol RZ, Palacios A, Campfield LA, Saul C, Swerdloff RS (1982) Comparison of the kinetics of injectable testosterone in eugonadal and hypogonadal men. Fertil Steril 37:425–430

Spark R, Brewster SE, Libman H, Dezube BJ (1997) Dihydrotestosterone gel: a novel androgen for AIDS wasting syndrome. The Endocrine Society 79th meeting. Minneapolis, MN: Abstract P1–319

Veldhuis JD, Rogol AD, Samojlik E, Ertel MH (1984) Role of endogenous opiates in the expression of negative feed back actions of androgens and estrogen on pulsatile properties of luteinizing-hormone secretion in man. J Clin Invest 74:47–55

Vermeulen A, Deslypere JP (1985) Long-term transdermal dihydrotestosterone therapy: effects on pituitary gonadal axis and plasma lipoproteins. Maturitas 7:281–287

von Eckardstein A, Kliesch S, Nieschlag E, Chirazi A, Assmann G, Behre HM (1997) Suppression of endogenous testosterone in young men increases serum levels of high density lipoprotein subclass lipoprotein 11 and lipoprotein (a). J Clin Endocrinol Metabol 82:3367–3372

16 Androgens in male senescence

Jean Marc Kaufman and Alex Vermeulen

16.1 Introduction

An "andropause", defined as the male equivalent of the menopause, which in women signals the end of reproductive life and a near total cessation of gonadal sex steroid production, does not exist. Indeed, aging in healthy men is normally not accompanied by abrupt or drastic alterations of gonadal function, and androgen production as well as fertility can be largely preserved until very old age.

The available data, although still limited, suggest that aging has only limited influence on sperm quality and fertilizing capacity (Nieschlag et al. 1982; Rolf et al. 1996), changes in semen parameters being essentially limited to a decrease of ejaculate volume and sperm motility (Rolf et al. 1996). Moreover, a decreased ejaculatory frequency, as observed in elderly men (Rolf et al. 1996), might account for at least part of these age-related changes, but may also mask more subtle changes in spermatogenetic activity (Cooper et al. 1993).

As to the endocrine testicular function, it is now well established that mean serum testosterone levels decrease progressively in healthy elderly men, even though there is a considerable interindividual variability in the extent of the changes (Vermeulen 1991). More than 20% of otherwise apparently healthy men over 60 years of age present with subnormal testosterone levels as compared to the serum levels in young adults. Moreover, this age-dependent decline in androgen production can be accentuated by intercurrent diseases with transient or more permanent adverse effects on Leydig cell function.

The extent to which a relative hypoandrogenism in the elderly contributes to clinical signs and symptoms of aging remains a largely underexplored issue that certainly deserves further attention as many clinical features of aging in men are reminiscent of those of hypogonadism in younger subjects.

16.2 Declining endocrine testicular function in senescence

16.2.1 Testosterone production and serum levels

Early reports of decreased spermatic vein testosterone blood concentrations (Hollander and Hollander 1958) and decreased testosterone blood production rates (Kent and Acone 1966) in elderly men as compared to younger individuals, have subsequently been confirmed by several studies performed in the seventies (Baker et al. 1977; Giusti et al. 1975; Vermeulen et al. 1972). However, the observation of a reduced testosterone blood production rate does not necessarily imply lower bioavailable testosterone plasma levels. Indeed, the blood production rate is the product of the mean plasma levels and the metabolic clearance rate, the latter having been shown to be also reduced in elderly men (Kent and Acone 1966; Vermeulen et al. 1972).

Whether aging in healthy men is also associated with decreased serum testosterone concentrations has long been a highly controversial issue addressed in a rather large series of cross-sectional studies (Vermeulen 1990, 1991 for review). The early reports of decreased mean serum testosterone levels in elderly men, dating from the late sixties and early seventies, were followed by several studies that failed to confirm an age-related decline of testosterone levels.

These discrepancies may be explained at least in part by biased selection of the study populations (Gray et al. 1991) with some of the early studies having included subjects who did not qualify as healthy ambulatory elderly, while some of the later studies may have selected exceptionally healthy elderly men, no longer representative of an ambulant elderly male population in general good health, by applying unreasonably stringent inclusion criteria. Furthermore, in some of these studies reported testosterone serum concentrations in young men were surprisingly low. Finally, in some studies blood sampling was performed in the late afternoon when, due to the diurnal rhythmicity of testosterone secretion, serum levels are lower and the effect of age is minimal.

In any case, more recent studies which included exclusively healthy ambulatory young and elderly men have again confirmed the existence of an age-associated decline of serum testosterone levels, albeit of lesser magnitude than reported in the early studies (Vermeulen 1991 for review), mean values at age 75 years being about two thirds of those at age 25 years (Fig 16.1). The occurrence of an age-related decrease of serum testosterone levels has recently also been documented in longitudinal studies (Morley et al. 1996; Pearson et al. 1995).

Whereas mean serum testosterone levels in the adult male population decrease with age, there is at any age a remarkable interindividual variability of serum testosterone levels, which is rather striking in the elderly, with some men having frankly low serum testosterone levels while many others have a perfectly preserved testosterone secretion with serum levels well within the normal range for young adults. With advancing age, a progressively larger proportion of men present with subnormal values relative to those in young adults; in a group of 300 healthy men aged 20 to 100 years (Vermeulen et al. 1996), we observed a subnormal testosterone level in less than 1% of men below age 40 years but in more than 20% of men older than 60 years (Fig 16.2).

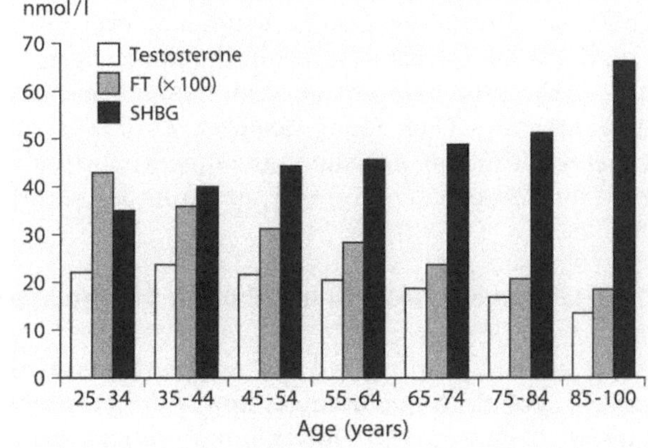

Fig. 16.1. Mean serum levels of testosterone, free testosterone (FT) and sex hormone binding globulin (SHBG) according to age in a cross-sectional study of 300 healthy men (according to data in Vermeulen et al., 1996)

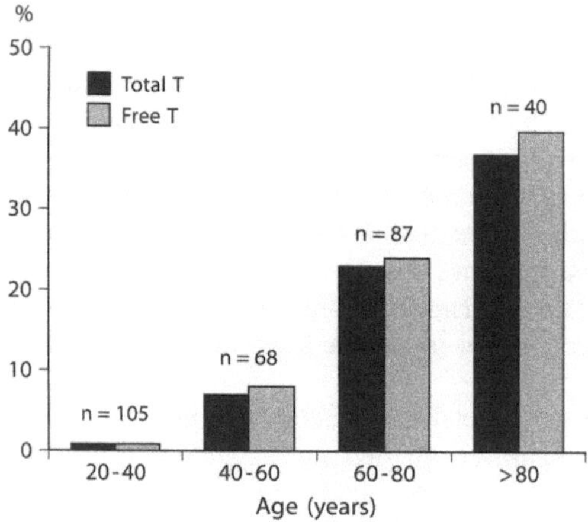

Fig. 16.2. Proportion of healthy men presenting with subnormal serum levels for total testosterone ($<$ 11 nmol/l) or free testosterone ($<$ 0.18 nmol/l) (from Kaufman and Vermeulen, 1997)

16.2.2 Sex hormone binding globulin and free testosterone serum levels

Whereas some authors may still argue that total testosterone concentrations are not reduced in perfectly healthy elderly men, there is virtually unanimity amongst authors that the free and non specifically bound serum testosterone, which are generally considered to represent the biologically available serum testosterone fractions, does indeed decrease with age (Vermeulen 1991 for review). In healthy ambulatory men mean free serum testosterone decreases by as much as 50% between age 25 and 75 years, the sharper decline of free testosterone in comparison with total testosterone being explained by an age-dependent increase of sex hormone binding globulin (SHBG) concentrations. In 300 healthy men aged 25–100 years we observed an approximately log linear decrease of free testosterone levels at a rate of 1.2% per year (Fig 16.1), while total serum testosterone remained relatively stable up to the age of 55 years and declined thereafter at a rate of 0.85% per year (Vermeulen et al. 1996).

As is the case for total testosterone, there is a great interindividual variability in prevailing free testosterone levels in elderly men, ranging from markedly low levels to levels in the upper normal range for young adults (Fig 16.3), the proportion of men with subnormal free testosterone levels increasing with age (Fig 16.2). However, as discussed in further sections of this chapter, limits of normality are somewhat arbitrary as the sensitivity threshold for androgen action may vary from tissue to tissue and according to age.

16.2.3 Tissue levels and metabolism of androgens

It is generally considered that free testosterone diffuses passively within the cell. It should be mentioned, however, that during transit through the capillaries, a large part of the albumin-bound testosterone dissociates and be-

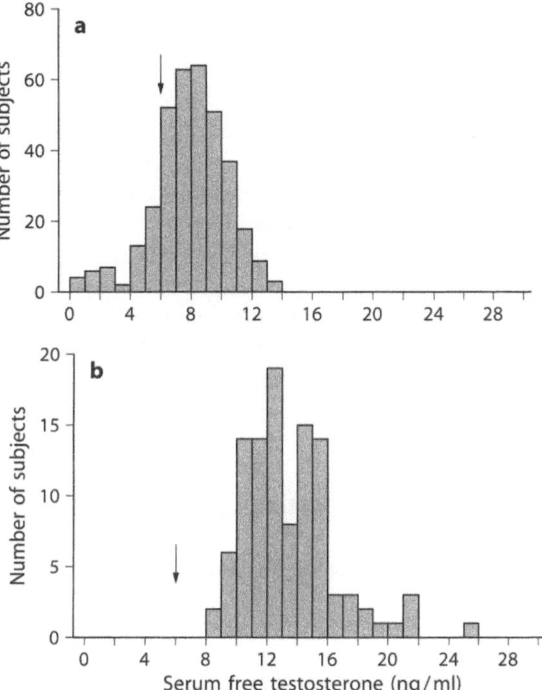

Fig. 16.3. Histogram for the distribution of serum free testosterone in **a)** 353 community-dwelling elderly men without major health problems, aged 70 to 85 years and in **b)** a younger control population of healthy men aged 23 to 58 years; the lower limit for the laboratory normal range is indicated by the arrows

comes bioavailable. SHBG-bound testosterone is not directly available to the cells and the physiopathology of SHBG is still incompletely understood. It has been shown that in some tissues (e.g. in the prostate and epididymis) cellular membranes carry SHBG receptors (Porto et al. 1992; Rosner et al. 1991), the activation of which stimulates intracellular cAMP production. This binding protein has strong similarities with two membrane proteins, laminine and merosine (Joseph and Baker 1992). Moreover, SHBG may be internalized by several cell types.

Androgen receptors are present in the highest concentration in the accessory sex organs whereas the concentrations in skeletal muscle tissue and the heart are much lower. This concentration of receptors is affected by the concentration of androgens and by age (Blondeau et al. 1982; Rajfer et al. 1989). As both circulating steroids and tissular receptors concentration decrease with age, it is not surprising that tissular androgen concentration also decreases with age (Deslypere and Vermeulen 1981, 1985).

The influence of the aging process on the metabolism of testosterone manifests itself essentially by a decrease of the metabolic clearance rate which results from an age-associated decrease of cardiac output and of hepatic as well as tissular perfusion, and from increased binding of plasma testosterone to SHBG.

No data are available concerning the possible influence of aging on the blood conversion rates of testosterone to androstenedione and dihydrotestosterone (DHT). There is no apparent age-related decrease of DHT levels, but serum DHT is not considered a reliable parameter of tissular DHT formation (Labrie et al. 1995). The conversion of testosterone to estradiol increases with age (Siiteri and MacDonald 1975), which results in an increase of the estradiol over testosterone ratio in plasma (Vermeulen 1976). It is not evident, however, whether this is the consequence of the aging process itself or of the increase in fat mass in the elderly. Serum levels of 5α-androstane-$3\alpha17\beta$ diol-glucuronide decrease significantly with age (Deslypere et al. 1982), a consequence of the decreased serum concentrations of the precursors (70% testosterone and 30% dehydroepiandrosterone-sulfate), as well as an age-associated decrease of type 2 5α-reductase activity (activity enhanced by androgens) (Kuttenn et al. 1977), which is also reflected in the decreased ratio of $5\alpha/5\beta$ metabolites in urine (androsterone/etiocholanolone glucuronides and sulfates, respectively) (Vermeulen et al. 1972; Zumoff et al. 1976).

16.2.4 Factors affecting serum testosterone levels in elderly men

16.2.4.1 Influence of physiological factors and lifestyle

The physiological basis which underlies the large interindividual variation in serum testosterone levels seen at any age is not yet fully elucidated, but several physiological variables and factors related to lifestyle have been identified that account for part of the wide range of normal values observed in healthy men. In any case, the apparent interindividual variability of testosterone levels is not merely artefactual as a result of the cross-sectional design of the clinical studies, as single-point plasma testosterone estimates reflect longer term androgen status in healthy men fairly well (Vermeulen and Verdonck 1992), as illustrated in Fig. 16.4 for 298 men aged 70 to 85 years studied on two separate occasions at one year intervals.

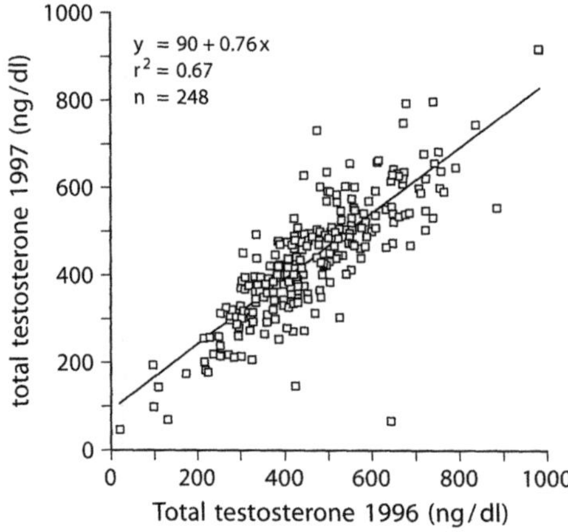

Fig. 16.4. Total testosterone assessed at one year interval in 248 men aged 70 to 85 years (part of the same population as in Fig. 16.3)

Heredity appears to play an important role, with Meikle et al. (1986, 1988) concluding from the study of monozygotic and dizygotic twins that up to 60% of the variability of serum testosterone and up to 30% of the variability of SHBG, after normalisation for body surface area, may be attributable to genetic factors. Nevertheless, also according to the authors of the latter studies, non-genetic factors still have a substantial impact on testosterone serum levels.

The ultradian pattern of episodic testosterone secretion undoubtedly contributes to the variability of testosterone levels (Naftolin et al. 1973; Spratt et al. 1988; Veldhuis et al. 1987). The circadian variation of serum testosterone, with highest levels in the early morning and lowest levels in the late afternoon, should not play an important role in the wide range of normal testosterone levels as they are, as a rule, evaluated in the early morning. The occurrence of circannual variations with amplitude of about 25% have been reported with the highest testosterone levels usually observed around October (Smals et al. 1976) or somewhat later, in December according to Dabbs (1990).

Several metabolic and hormonal factors influence SHBG serum levels. Insulin and insulin like factor-I (IGF-I) inhibit SHBG production by hepatoma cells in vitro and in clinical studies insulin has been found to be inversely correlated with serum SHBG and testosterone levels (Haffner et al. 1988, 1994; Simon et al. 1996; Vermeulen et al. 1996).

In clinical studies, the body mass index (BMI) emerges as an important determinant of SHBG levels (Demoor and Goossens 1970) with, for the whole range of clinically encountered BMI values, a highly significant negative correlation with SHBG and testosterone serum levels, which is explained at least in part by increased insulin levels (Giagulli et al. 1994; Khaw and Barrett-Connor 1992; Plymate et al. 1988). In elderly men this inverse relationship between BMI and SHBG levels can be clearly demonstrated notwithstanding the background of an age-related rise of SHBG levels (Vermeulen et al. 1996). Whereas moderate obesity affects only total serum testosterone by lowering SHBG binding capacity, in morbid obesity (BMI > 35–40) free testosterone levels are decreased as well as a result of neuroendocrine disturbances (Giagulli et al. 1994).

Serum SHBG and testosterone levels have been reported to be inversely correlated to 24 hour growth hormone and IGF-I levels (Erfurth et al. 1996; Pfeilschifter et al. 1996; Vermeulen et al. 1996) and it has been proposed that decreased activity of the somatotropic axis may play a role in the age-associated increase of SHBG levels and decrease of free testosterone levels (Vermeulen et al. 1996).

It has long been known that alterations in thyroid hormone levels can have a marked effect on SHBG levels, with thyreotoxicosis resulting in a severalfold increase of SHBG levels and marked increase of total serum testosterone (Vermeulen et al. 1971). In this regard, it is interesting to note that more subtle changes in thyroid hormone levels within the "normal range" can also affect SHBG and testosterone levels, with subclinical hyperthyroidism, characterized by suppressed serum levels of thyroid-stimulating hormone without clinical symptoms of hyperthyroidism or elevation of thyroid hormone levels above normal, resulting in a significant increase of SHBG and testosterone levels (Faber et al. 1990).

Besides specific high-affinity binding of testosterone to SHBG, 20 to 40% of testosterone in the circulation is albumin-bound, so that any substantial variation in albumin concentration will be reflected by changes of total testosterone levels (Vermeulen et al. 1996).

As to factors related to lifestyle, reports on the effects of diet on serum testosterone levels have not always been concordant, but from the whole of the available data it seems possible to conclude that diet influences testosterone levels mainly indirectly through changes in SHBG levels, fiber-rich, vegetarian diets being associated with higher SHBG and testosterone levels than western-type diets and more particularly those diets with high fat content (Adlercreutz 1990; Belanger et al. 1989; Key et al. 1990; Meikle et al. 1990; Reed et al. 1987). From the available data, it is not clear to what extent changes in SHBG related to diet may be mediated through changes in insulin secretion, vegetarians having generally lower fasting insulin levels than omnivores.

Prolonged fasting may transiently affect testosterone production through diminished gonadotropic testicular drive (Cameron et al. 1991). Smoking has been associated with higher testosterone levels than in non-smokers (Dai et al. 1988; Deslypere and Vermeulen 1984; Field et al. 1994). This effect is observed in both young and elderly men, the difference amounting to 5–15% of the levels in non-smokers for both total and free testosterone levels (Vermeulen et al. 1996). Alcohol abuse, also in absence of cirrhosis of the liver, may accentuate the age-associated decrease of testosterone levels, estradiol serum levels being increased (Cicero 1982; Ida et al. 1992; Irwin et al. 1988).

Both physical and psychological stress, as well as strenuous physical activity have been shown to result in depressed testosterone levels (Nilsson et al. 1995; Opstad 1992; Theorell et al. 1990).

16.2.4.2 Testosterone serum levels in disease

It is now generally accepted that the aging process *per se* adversely affects testosterone production, it is nevertheless also evident that the age-associated decline in testosterone levels may often be accentuated by intercurrent disease (Handelsman 1994; Turner and Wass 1997; see also Chapter 17 by Handelsman "Androgen therapy in non-gonadal disease" in this volume).

Acute critical illness (Dong et al. 1992; Impallomeni et al. 1994; Spratt et al. 1993; Woolf et al. 1985), acute myocardial infarction (Swartz and Young 1987; Wang et al. 1978a) and surgical injury (Wang et al. 1978b) have been reported to cause profound, but generally transient decreases of (free) testosterone levels.

A series of chronic diseases can induce more longstanding decreases in testosterone levels. Both testosterone and SHBG levels tend to be decreased in elderly men with diabetes mellitus (Barrett-Connor et al. 1990). Impaired glucose tolerance and noninsulin-dependent diabetes mellitus (NIDDM), with high prevalence in elderly persons, are associated with decreased testosterone levels (Andersson et al. 1994; Chang et al. 1994), in agreement with the observations of a negative correlation between insulin levels and testosterone levels.

Coronary atherosclerosis has been reported to be accompanied by lower testosterone levels (Phillips et al. 1994) and there have also been several reports of decreased testosterone levels in survivors of myocardial infarction as compared to controls (Lichtenstein et al. 1987; Mendoza et al. 1983; Poggi et al. 1976; Sewdarsen et al. 1990; Swartz and Young 1987), although it is not clear whether the decreased testosterone levels represented a consequence of atherosclerosis or rather a pre-existent risk factor for cardiovascular disease.

Chronic renal failure is often accompanied by a hypogonadotropic hypogonadism with impaired pulsatile release of pituitary luteinizing hormone (LH) (Handelsman and Dong 1993). In chronic liver disease, decreased (free) testosterone levels are accompanied by increased SHBG, androstenedione and estrone levels (Baker et al. 1979; Elewaut et al. 1979). Hypogonadism in hemochromatosis is multifactorial with possible contribution of pituitary insufficiency, primary testicular defects, cirrhosis of the liver and diabetes mellitus (Kelly et al. 1984).

In the elderly, as in the young, Leydig cell function may be adversely affected by endocrine diseases such as Cushing's syndrome (Luton et al. 1977; Mc Kenna et al. 1979) and pituitary tumors, in particular prolactinomas.

Sleep apnoea syndrome is accompanied by a relative hypogonadotropic hypogonadism (Worstman et al. 1987), with massive obesity often being a contributing factor to the hypogonadism in these patients (Grunstein et al. 1989).

Moderate impairment of testicular function has been observed in periarteritis nodosa, during acute flares of rheumatoid arthritis and in active ankylosing spondylitis (Gordon et al. 1988; Tapia-Serrano et al. 1991).

Finally, among drugs not uncommonly used in the elderly and that may impair Leydig cell function, special mention should certainly be made of chronic use of glucocorticoids, which often induces a marked suppression of (free) testosterone levels by combined actions at the testicular and hypothalamo-pituitary level, and by a decrease of SHBG serum levels (MacAdams et al. 1986). LH secretion and Leydig cell function may be adversely affected by hyperprolactinemia during chronic use of neuroleptic drugs and related compounds (Bixler et al. 1977).

16.3 Physiopathology of declining Leydig cell function in senescence

16.3.1 Primary testicular changes

Primary testicular factors undoubtedly play an important role in the age-associated decline of Leydig cell function as indicated by a reduced absolute secretory response to stimulation with human chorionic gonadotropin (hCG), albeit the relative testosterone response may be normal (Harman and Tsitouras 1980; Longcope 1973; Nankin et al. 1981; Nieschlag et al. 1973, 1982; Rubens et al. 1974). This decrease in testicular reserve for testosterone secretion appears to be underlied by a reduced number of Leydig cells (Harbitz 1973; Neaves et al. 1984; Sniffen 1950). There is, moreover, evidence for involvement of vascular changes (Sasano and Ichijo 1969; Suoranta 1971), the deficient oxygen supply inducing changes in testicular steroid metabolism (Pirke et al. 1980; Vermeulen and Deslypere 1986).

In apparent agreement with the view of a primary testicular defect are observations of moderate increases of basal gonadotropin levels in elderly men as observed in several, but not all, studies on the age-related decline of testicular function (Tsitouras and Bulat 1995; Vermeulen 1991 for review). Although there have been some discordant findings (Mitchell et al. 1995; Urban et al. 1988), this increase is not limited to immunoreactive forms of the gonadotropins, as it is also demonstrated for gonadotropin levels measured by bioassay (Kaufman et al. 1991; Matzkin et al. 1991; Tenover et al. 1988).

16.3.2 Altered neuroendocrine regulation

Although the combined observations of a diminished testicular reserve for testosterone secretion and increased basal gonadotropin levels seem consistent with the view that the age-related decline of Leydig cell function results from primary testicular dysfunction, closer examination of the data suggests that other mechanisms must also be involved. Indeed, the observed responses to hCG challenges in elderly men indicate that the secretory reserve of the Leydig cells, albeit diminished, should still be amply sufficient to allow for a normalisation of plasma testosterone levels, provided the endogenous drive by pituitary LH is adequate. In fact, in the face of a persistent status of relative hypoandrogenism, the only modestly increased basal levels of pituitary luteinizing hormone (LH) should be regarded as inappropriately low. Furthermore, in contrast to previous reports of a delayed or diminished LH response upon stimulation with pharmacological doses of gonadotropin-releasing hormone (GnRH) (Nieschlag et al. 1982; Rubens et al. 1974; Winters and Troen 1982), assessment of pituitary secretory capacity for (immunoreactive as well as bioactive) LH by challenges with small physiological doses of GnRH, clearly indicates a well-preserved or even increased pituitary secretory reserve in elderly men (Kaufman et al. 1991).

From the whole of these observations it can be concluded that elderly men present not only with a primary testicular defect but also with alterations of the neuroendocrine control of Leydig cell function, with failure of the feedback regulatory mechanisms to normalize the testosterone levels notwithstanding the existence of an adequate testosterone and LH secretory capacity (Kaufman et al. 1990; Vermeulen and Kaufman 1992; Winters and Troen 1982). The relative inadequacy of the gonadotropin response to hypoandrogenism in elderly men is further underlined by the observation that the limited increase of basal LH levels might to some extent reflect reduced plasma clearance rather than only increased secretion of LH (Kaufman et al. 1991).

Several changes in the neuroendocrine control of Leydig cell function have been documented in elderly men. Circadian rhythmicity of LH and testosterone secretion is clearly blunted in elderly men (Bremner et al. 1983; Deslypere and Vermeulen 1984; Plymate et al. 1989; Tenover et al. 1988). Furthermore, the pulsatile pattern of LH secretion is altered with unchanged LH pulse frequency (Tenover et al. 1987, 1988; Urban et al. 1988; Vermeulen et al. 1989; Winters et al. 1984) but decreased mean LH pulse amplitude as a consequence of a reduced number of LH pulses with larger amplitude (Veldhuis et al. 1992; Vermeulen et al. 1989).

Indirect evidence suggests that the main changes occur at the level of the hypothalamic GnRH-secreting neuronal system. Indeed, as the responsiveness of the pituitary gonadotrophs to "physiological" doses of GnRH is preserved (Kaufman et al. 1991), the decreased LH pulse amplitude is most likely due to decreased stimulation by endogenous GnRH, with reduced size of the bolus of the neuropeptide intermittently released into the pituitary portal circulation. Moreover, the LH pulse frequency, governed by the hypothalamic GnRH pulse generator and expected to increase in a state of hypoandrogenism (Plant 1986), has been found by most authors to remain unchanged and thus inappropriately low, an increased LH pulse frequency in elderly men having been reported only by one group (Mulligan et al. 1995; Veldhuis et al. 1992). Additional evidence of altered hypothalamic regulation of gonadal function in elderly men is provided by the observation of a clearly increased sensitivity to the negative feedback effects of sex steroids in comparison to the situation in young adults (Deslypere et al. 1987; Winters et al. 1984, 1997). Furthermore, the LH response to opioid receptor blockade in elderly men is blunted in comparison to that in young individuals (Mikuma et al. 1994; Vermeulen et al. 1989), the receptor blockade failing to produce the expected increase in LH pulse frequency and amplitude observed in the young (Vermeulen et al. 1989).

From the latter studies it can be concluded that alterations in LH secretion in elderly men is not due to increased endogenous opiod tone and at the present time the mechanisms underlying the apparent deficiency of GnRH secretion in elderly men remain to be fully elucidated. In any case, the observed changes in LH secretion with decreased mean LH pulse amplitude can be expected to have a significant impact on testosterone secretion as there exists a linear correlation between LH pulse amplitude and plasma testosterone levels (Veldhuis et al. 1992; Vermeulen et al. 1993).

16.3.3 Increase of serum SHBG

The progressive increase of plasma SHBG binding capacity with age should be regarded as a third important aspect of the physiopathological mechanisms responsible for the age-related changes in circulating testosterone levels. Indeed, against the background of a relative inability of elderly men to respond to hypoandrogenism by increased testosterone secretion, as a consequence of primary testicular changes and altered neuroendocrine regulation, a progressive increase of SHBG binding capacity will result in an even steeper decline of free and not specifically bound (i.e. albumin-bound), bioavailable testosterone levels.

The increase of SHBG concentrations in elderly men is remarkable as it occurs in the face of an increase of fat mass and insulin levels, factors known to be inversely correlated to SHBG levels, but the cause of this increase of SHBG levels remains unclear. It is unlikely that the decreased testosterone levels *per se* are responsible, as the increase in SHBG levels is observed at an earlier age than the decrease of testosterone levels; the estradiol serum levels are rather similar in young and elderly men (Vermeulen et al. 1996). Indirect evidence suggests that the increase of SHBG concentrations in the elderly may be related to the age-associated decline in circulating growth hormone and IGF-I levels, this hypothesis remaining to be confirmed (Vermeulen et al. 1996).

16.4 Clinical relevance of hypoandrogenism of senescence

16.4.1 General background

Many of the clinical features of aging in men are reminiscent of the clinical changes seen in hypogonadism in young men, with a decrease in general well-being, mood changes, decrease of energy and virility, decrease in sexual pilosity and skin thickness, decrease in muscular mass and strength, increase in upper and central body fat, decrease in bone density and increased prevalence of osteoporotic fractures, decrease in libido and sexual drive, and a markedly increased prevalence of impotence. It certainly seems a reasonable working hypothesis that at least part of these clinical changes observed in aging men are causally related to declining Leydig cell function. Unfortunately, this possibility has not been systematically explored and it is equally plausible that many of the observed clinical changes and the age-related decrease in serum testosterone levels are coincident and independent consequences of aging.

The bulk of available information is derived from cross-sectional studies, which revealed mostly only weak correlations between androgen status and clinical parameters. This is probably not surprising as clinical changes are often multifactorally determined and cross-sectional study designs are certainly not most appropriate to unravel the complex clinical picture arising from concurrent effects of the aging process *per se*, a relative hypoandrogenism and intercurrent diseases. In this regard, the few studies which examined the effects of androgen treatment in elderly men provide precious information but should also be interpreted with caution as the demonstration of beneficial effects of pharmacological intervention does not necessarily imply that the corrected clinical sign or symptom was due to pre-existent androgen deficiency.

In this context, it should be reminded that in elderly men there is a striking interindividual variability not only of the prevailing testosterone levels but also of the pace of "clinical aging". If the age-associated decline of testosterone levels does play a significant role in the clinical aging process, carefully performed longitudinal studies should allow such a causal relationship to be established and might provide us with more definitive answers in the future. In any case, as discussed in further sections of this chapter, the decreased androgen levels can at best be responsible for part of the clinical changes observed in aging men.

16.4.2 Hypoandrogenism of senescence and sexual activity

Aging in men is accompanied by a decrease in libido and sexual activity. Indeed, whereas mean coital frequency at age 20–25 years is reported to be about four times a week, it would decrease to once a week at age 55–60 years, three times a month at age 70 and less than two times a month between age 75 and 80 years (Masters 1986; Tsitouras-Bulat 1995). Nevertheless, only 15% of men over 60 years old deny any sexual interest (Verwoerdt et al. 1969) and 80% of men over 60 years old remain sexually active (Kaiser 1992).

Whereas normal libido requires adequate testosterone levels, as shown by the effect of testosterone replacement therapy in hypogonadal males, the correlation of libido with plasma testosterone levels is rather poor (Bagatell et al. 1994a; Davidson et al. 1983; Schiavi 1996) and the testosterone concentration required to sustain sexual activity and maintain libido appears to be rather low (Gooren 1987; Schiavi 1996). Indeed, androgen replacement therapy after chemical castration with GnRH analogues, at a dose maintaining testosterone levels at approximately half baseline levels, sustained normal sexual activity (Bagatell et al. 1994b) and there is good evidence that healthy adults have substantially higher androgen levels than required for normal sexual behaviour (Udry et al. 1985). Nevertheless, Schiavi et al. (1990) in a study involving 77 healthy men, aged 45–74 years, observed that bioavailable testosterone was higher in men desiring sexual activity with their wife more frequently than once a week, than in men desiring a lower frequency. The same authors (Schiavi et al. 1988) also observed that men with primary hypoactive sexual desire disease had significantly lower plasma testosterone levels than controls. Udry et al. (1985) reported similar results, whereas Tsitouras et al. (1982) reported that as a group, older men with higher testosterone levels were more likely to exhibit higher levels of sexual activity than men with low testosterone levels; however, the overlap of testosterone levels between groups was important and only a small fraction of men with low sexual activity had abnormally low testosterone levels.

Although potency and nocturnal penile tumescence (NPT) require adequate testosterone levels and although several studies show that hormonal alterations might play a role in 6–45% of cases (Morley 1986), most frequently the cause of impotence in elderly males is non-hormonal.

There exists, nevertheless, a clearly significant correlation between bio-available testosterone levels and the frequency of NPT (Morley 1986; Schiavi et al. 1988). Erections in response to visual stimuli, on the other hand, are largely testosterone-independent in normal subjects (Bancroft 1984; Kwan et al. 1983), suggesting that whereas testosterone is required to sustain NPT, it is much less so to maintain response to external stimuli. Still, testosterone levels have been reported to correlate inversely with the latency of erection in response to visual stimuli (Lange et al. 1980; Rubin et al. 1979).

Schiavi (1996), observing a significant association between the levels of bioavailable testosterone and the duration of NPT in man aged 55–64 years, but not in younger or older age groups, suggested that circulating androgen levels might be well above the threshold of activation of NPT in young men, fall within the threshold in middle-aged men and become no longer sufficient to sustain NPT in elderly men. It is not evident whether testosterone itself or its metabolites, DHT or estradiol, are required for this action. The mechanism may be a receptor-mediated effect, androgen receptors having been described in the corpora cavernosa (Horwitz and Horwitz 1982), with highest levels at puberty followed by a progressive, irreversible decline with age (Gonzales-Cadavid et al. 1991). It is generally believed that testosterone stimulates the activity of the nitric oxide synthase enzyme (Lugg et al. 1995), the concentration of which is restored by testosterone in castrated animals (Mills et al. 1996). Davidson et al. (1982) suggested that the effects of testosterone may be peripheral, via changes in genital sensitivity.

Nocturnal penile tumescence is clearly androgen-dependent, but Schiavi et al. (1990) did not observe any correlation between NPT and erectile problems in the elderly, suggesting that their erectile problems are largely non-hormonal in origin. Among non-hormonal factors that may influence the frequency of impotence in elderly men one can mention:
- the overall health status of both partners, diabetes mellitus being a common cause of impotence at any age;
- boredom with, or loss of attractiveness of the (same) sexual partner, as well as monotony of sexual life;
- low level of sexual activity in young age, the activity of the aging male being strongly correlated with the activity in younger age (Martin 1975; Pfeiffer 1974);
- medications (Tsitouras and Bulat 1995) such as psychotropic drugs (tricyclics; MAO inhibitors; phenothiazines; hypnotics), antihypertensive compounds (β-blockers; guanethidine; prazozine; angiotensine-converting enzyme inhibitors), H_2-antihistaminics, drug abuse (alcohol; heroine; marijuana) (Kligman 1991; Tsitouras and Bulat 1995);
- psychopathology such as stress and depressive states;
- atherosclerosis and cardiovascular disease, being the most frequent causes of erectile dysfunction in the elderly and accounting for about 50% of cases of impotence (Kaiser 1992; Virag et al. 1985)

– neurological factors with decreased sensory, neural and autonomic functioning, an important cause of impotence in the elderly and together with atherosclerosis the most frequent cause in diabetics.

In conclusion, whereas adequate testosterone levels are required for libido and NPT, it appears that testosterone levels required for normal sexual activity are rather low, and although testosterone co-determines potency, the factors most commonly involved in impotence in elderly men are not hormonal.

16.4.3 Testosterone and cardiovascular risk profile

Cardiovascular disease (CVD) and cardiovascular mortality are much more frequent in men than in women of reproductive age, whereas hyperandrogenic women have an atherogenic lipid profile and are at increased risk of CVD (see also Chapter 8 by von Eckardstein "Androgens, lipids and the cardiovascular system" in this volume). It is generally considered, therefore, that androgens increase the risk of CVD, by decreasing HDL-C and increasing LDL-C. Indeed, before puberty there is no sex difference in lipid levels, but at puberty, in parallel with the increase of testosterone levels in boys, there is a dissociation of lipid profiles between sexes with a decrease of HDL-C levels and an increase of LDL-C in boys but not in girls (Kirkland et al. 1987). Suppression of plasma testosterone levels in men, by GnRH analogues, increases HDL-C (Bagatell et al. 1992; Goldberg et al. 1985; Moorjani et al. 1987), whereas co-administration of testosterone enanthate 100 mg/2 weeks (Moorjani et al. 1987) for 20 weeks, abolishes this effect. Treatment of female-to-male transsexuals with testosterone undecanoate, aiming at physiological male testosterone levels (Asscheman et al. 1994), causes a decline of HDL-C and HDL2-C levels. All these data are in accordance with the generally held view that testosterone is atherogenic.

Most studies on the relationship between testosterone and plasma lipid levels in healthy males lead, however, to a different conclusion. Indeed, in a review article Bagatell and Bremner (1995) observed that out of 20 studies involving healthy males, 15 reported a positive correlation between testosterone and HDL-C levels, two reported a negative correlation (Handa et al. 1997; Semmens et al. 1983) and three found no correlation at all. The observations of positive correlations are not limited to total testosterone, as a similar correlation is found between free testosterone and HDL-C (Haffner et al. 1993; Heller et al. 1983; Hamäläinen et al. 1986) and this as well in Eastern (Ooi et al 1996) as in Western countries.

These results are, at first sight, difficult to reconcile with the hypothesis that the endogenous androgen levels are responsible for the more atherogenic lipid profile found in adult males as compared to females. As it is well known that the generally modest increase of testosterone levels in hyperandrogenic women (hirsutism; polycystic ovary syndrome) is accompanied by an evident atherogenic lipid profile (Graf et al. 1990; Lithell et al. 1987;

Mahabeer et al. 1990; Rojanasakul et al. 1988; Talbott et al. 1995; Wild et al. 1985; Wild and Bartholomew 1988), and that increasing the testosterone levels from the normal female range to the physiologic male levels, in female to male transsexuals or in males rendered hypogonadal with GnRH analogues, is also accompanied by an atherogenic lipid profile, it might be hypothesized that testosterone levels in normal men or (elderly) men with partial androgen deficiency are above the threshold causing deleterious effects on plasma lipids, whereas in women or castrated men, levels are below this threshold. As to the mechanism of the inverse correlation between plasma testosterone and lipid levels, there is evidence that insulin resistance is involved (see Chapter 8 by von Eckardstein in this volume).

Besides effects on the lipid profile, testosterone decreases also lipoprotein(a) levels, an independent risk factor for coronary artery disease (Zmuda et al. 1996). Not only is there a negative correlation between endogenous testosterone levels and the atherogenic lipid profile, but Phillips et al. (1994) in a study involving 55 males, also observed an inverse correlation between (free) testosterone levels and the degree of coronary artery disease. Moreover, most authors (Jaross et al. 1989; Lichtenstein et al. 1987; Luria et al. 1982; Phillips 1976, 1978; Poggi et al. 1976; Swartz and Young 1987) observed that male survivors of myocardial infarction have lower testosterone levels, both in the short term as well as months to years later. It is, however, not evident, whether the decline in testosterone levels is antecedent to, or a consequence of the infarction, the more so that several authors (Barrett-Connor and Khaw 1988; Cauley et al. 1987) observed that both levels of testosterone and estradiol were similar in subjects who later developed cardiac infarction and in controls.

Surprisingly, no correlation has been found between testosterone levels and cardiovascular mortality, neither by Barrett-Connor and Khaw (1988) in a study involving 1000 Caucasian men, nor by Cauley et al. (1987) or by Haffner (1996). The apparent discrepancy between the association of high testosterone levels with a favorable lipid profile and the absence of any effect on cardiovascular mortality suggests that testosterone might have direct vascular effects. Such effects have been reported by Herman et al. (1997), who observed that low testosterone levels were significantly associated with increased endothelium-dependent vasodilation after bilateral orchidectomy for prostatic carcinoma.

In conclusion, high testosterone levels (within the physiological range) do not cause an atherogenic lipid profile; on the contrary low testosterone levels are accompanied by low HDL-C and high LDL-C levels. Nevertheless, a correlation between testosterone levels and cardiovascular mortality has, so far, not been observed.

16.4.4 Sarcopenia

Aging in men is associated with a decrease of lean body mass and an increase of fat mass, especially in the upper body and central body regions (Forbes and

Reina 1970; Swerdloff and Wang 1993; Tenover 1994). The loss of muscle mass is accompanied by diminished muscle strength, which occurs regardless of the level of physical activity (Rogers and Evans 1993). Muscle weakness is an important clinical problem in old age, contributing to functional limitation for daily activities and related problems such as an increased risk for falls (Bhasin and Tenover 1997; Dutta and Hadley 1995; Guralnik et al. 1995; Rubenstein et al. 1994; see also Chapter 7 by Bhasin "Androgens and muscles" in this volume). The limited data available suggests the existence of a correlation between serum testosterone levels and muscle function (Abbasi et al. 1993), testosterone levels being also correlated with training-induced gain of strength (Häkkinen and Pakarinen 1994). Short-term administration (4 weeks to 3 months) of testosterone to elderly men, aimed at increasing initially low testosterone serum levels to values within the normal range for young men, has been reported to increase lean body mass (Tenover 1992), muscle strength (Morley et al. 1993; Urban et al. 1995) and skeletal muscle protein synthesis (Urban et al. 1995). However, the latter studies, besides being of short duration, also included only limited number of subjects. Recently, Sih et al. (1997) reported on the effects of testosterone in men over the age of 50 years (mean age 68 years) with low bioavailable serum testosterone, in a prospective, randomized, placebo-controlled trial of 12 months duration. They observed a significant increase in grip strength in the testosterone-treated men; lower extremity muscle strength was not evaluated in this study.

Testosterone replacement in elderly men with initially low testosterone serum levels appears to result in a rather modest decrease of fat mass (Tenover 1994); in their longer term study Sih et al. (1997) observed no significant change in fat mass as assessed by bioelectrical impedence, but the treatment did result in a significant decrease of serum leptin levels.

The cause of sarcopenia in elderly men is most probably multifactorial (Tenover 1994), possible contributing factors including the decreased activity of the somatotropic axis, the decrease in physical activity level and hypoandrogenism. The data discussed in this section seem to indicate that the decrease of testosterone production may indeed be involved in the sarcopenia of elderly men, at least in those men with more pronounced decrease of serum testosterone, and that testosterone replacement may improve muscle strength in the latter group. However, these initial reports need confirmation, with more particular attention for the relationship between androgen status and muscle strength in the lower extremities (Bhasin and Tenover 1997).

16.4.5 Senile osteoporosis

Aging of men is accompanied by progressive bone loss, which persists and may even be accelerated in old age. Recently, osteoporosis in men is increasingly being recognized as a significant problem of public health (Orwoll and Klein 1995). As many as one out of four patients suffering hip fractures are males (Cooper et al. 1992). Moreover, the prognosis of hip fracture appears to be worse in men as compared to women in terms of both morbidity and mortality (Poor et al. 1995).

The role of androgens in bone metabolism is complex and still not fully understood (see Kaufman 1996 for review; see also Chapter 6 by Finkelstein "Androgens and bone metabolism" in this volume). In any case, profound hypogonadism in younger men has been shown to result in accelerated bone loss (Stepan et al. 1989) and there is indication that testosterone replacement therapy in men with acquired hypogonadism may result in partial recovery of bone density (Behre et al. 1997; Devogelaer et al. 1992; Finkelstein et al. 1989; Katznelson et al. 1996).

Whether the relative hypoandrogenism in elderly men is instrumental in the development of senile osteopenia is an intriguing, but not fully explored possibility. Indices of bone mass have been reported to be positively correlated to androgen levels in elderly men (Foresta et al. 1984). However, both androgen levels and bone mass are known to decline with age, so that a causal relationship is not easily established.

In a relatively small group of men, Kelly et al. (1990) observed a positive correlation of a free testosterone index with the bone mineral density at the forearm, but not at the spine and the proximal femur, after correction for age. Murphy et al. (1993) found a weak but significant positive correlation between a free androgen index and bone mineral density at the proximal femur after adjustment for the effect of BMI and age in a group of 134 men with mean age of 69.5 years. Although Rudman et al. (1994) found total testosterone to be correlated with femoral neck bone mineral density in a group of men age 58–95 years, no association, independent of age was found between total or free androgen levels and bone mineral density at multiple skeletal sites in elderly men studied by Meier et al. (1987) and Drinka et al. (1993). In a subgroup of 57 healthy elderly men participating in an ongoing population-based study, we observed a significant correlation between testosterone levels and bone mineral density at the forearm, but not at the hip or the spine, after correction for the effect of age and BMI (Kaufman 1996). More recently, Greendale et al. (1997) observed in a group of 534 men with mean age of 68.6 years a positive correlation between bioavailable, but not total, testosterone levels and bone mineral density at the radius, the spine and the hip, bone mineral density at the different skeletal sites being also positively correlated to serum estradiol levels.

Hypogonadism has been reported to be a risk factor for hip fracture in elderly men (Boonen et al. 1997; Jackson et al. 1992; Stanley et al. 1991), but besides low bone mass, other factors related to testosterone, such as muscle weakness may be involved in the occurrence of hip fractures.

Presently, there is no controlled data available on the effects of testosterone replacement therapy on bone mineral density in elderly men. Limited data available on the effects of androgen treatment in elderly men on biochemical indices of bone turnover have been rather inconsistent. Morley et al. (1993) observed an increase of serum levels of osteocalcin, a marker of osteoblastic activity, during androgen treatment. Tenover (1992) reported a reduced hydroxyprolinuria in a small group of elderly men treated with androgen. Orwoll and Oviatt (1992) found no significant effect of androgen treatment on biochemical indices of bone turnover in a larger group of elderly men. No change in serum osteocalcin or alkaline phosphatase was observed in the longer term study by Sih et al. (1997).

In conclusion, evidence for the involvement of (relative) hypoandrogenism in senile osteoporosis in men remains limited and additional studies are certainly needed to clarify this clinically important point.

16.4.6 Conclusions

The extent of the clinical consequences of the relative hypoandrogenism of elderly men remains unclear. It seems likely that hypoandrogenism may play a role in several aspects of the clinical changes in elderly men, but it is also evident that the androgen status is only one of many factors that influence the aging process. Although association between plasma androgen levels and clinical signs or symptoms suggestive of hypoandrogenism have been generally weak or even absent, this must not necessarily mean that the decline in testosterone levels in elderly men is not clinically relevant. Indeed, the rather inconclusive character of some studies may be due to problems such as a lack of representativeness of the androgen measurements for the androgen status in preceding years, a lack of statistical power in small studies and, mainly, the cross-sectional rather than longitudinal design of most studies. A crucial question that remains to be elucidated is that of the possible existence of threshold concentrations for the action of androgens in target tissues and their level in the elderly.

16.5 Androgen substitution in the elderly men

16.5.1 Who should be considered for treatment?

The age-associated decrease in serum testosterone levels raises the problem of androgen substitution in elderly males: who should be treated, how and for how long?

As to the first question, it seems evident that elderly men with androgen levels within the normal range do not a priori require androgen supplementation. What, however, are the criteria for normalcy of testosterone levels in elderly men? In the absence of a reliable parameter of androgen activity, it remains difficult to determine the normal range of serum testosterone. The generally accepted lower limit of normal testosterone serum concentration, 11 nmol/l (320 ng/dl) is based on values obtained in young adults, levels below this limit being observed in only 1% of healthy men aged 20–40 years old, but in over 20% of men over 65 years (Fig 16.2). Parameters of the biologically active fraction of serum testosterone, i.e. serum free testosterone or non-specifically bound (i.e. albumin-bound) testosterone, is in principle more appropriate for the evaluation of the androgen status. Their use will result in classification of an even larger proportion of elderly men as hypoandrogenic, as the age-dependent decrease is steeper than for total testosterone.

It has already been mentioned that, as far as stimulation of sexual activity is concerned, androgen levels within the normal range for young adults may be insufficient to sustain erection in elderly men! In the absence of clinical signs of androgen deficiency, and taking into account the paucity of data on the consequences of relative hypoandrogenism and the risks and benefits of androgen supplementation, it is certainly not advisable to propose systematic androgen supplementation on the basis of testosterone levels at or just below the lower limit for young adults.

As to the objective signs of relative androgen deficiency, a decrease of muscle mass and strength and a concomitant increase in central body fat and osteoporosis can most easily be objectified. Decreased libido and sexual desire, loss of memory, difficulty in concentration, forgetfulness, insomnia, irritability, depressed mood as well as decreased sense of well-being, are rather subjective feelings or impressions, not easily objectified and certainly difficult to differentiate from hormone-independent aging. Complaints of excessive sweating and hot flashes, although occurring in severe hypogonadism are relatively rare (but do occur) in elderly men with partial androgen deficiency. The recently reported data on the possible influence of decreased dehydroepiandrosterone sulfate levels on the symptoms of aging in men add to the difficulty of evaluating the role of subnormal testosterone levels in the symptomatology of elderly men.

In the present state of the art, androgen supplementation should probably only be considered in the presence of androgen serum levels below the lower normal limit (11 nmol/l total testosterone or 0.25 nmol/l free testosterone) for young men, together with unequivocal signs and symptoms of androgen deficiency, in the absence of other reversible causes of decreased androgen levels and after careful screening for contra-indications. Indeed, it should be realized that many other hormonal and non-hormonal factors, related to the aging process may play a role in the symptomatology. The decision to treat will finally depend upon the balance between possible benefits and risks. Unfortunately, few controlled studies, generally of rather short duration and involving only small number of subjects have assessed this balance.

16.5.2 Potential benefits

As to the benefits several studies (Hajjar et al. 1997; Marin et al. 1993; Morley et al. 1993; Tenover 1992, 1996) report an improvement of the sense of general well-being, of libido, or muscle strength with an increase in lean body mass and a decrease in body fat. Tenover (1992, 1996) reports, moreover, an improvement in spatial cognition, but Sih et al. (1997) observed no

effects on memory, recall or verbal fluency tests. There is presently no conclusive data about possible beneficial effects on bone mineral density, reliable
evaluation of skeletal effects requiring controlled trials of long duration.

16.5.3 Potential risks

As to the risks of androgen replacement therapy in elderly men, we consider
here only effects of "physiological" doses of testosterone and not the effects
of massive pharmacological doses, as used by body builders.

The possible effects on plasma lipids, and the possible atherogenic effects
of androgens have been the subject of major concern. We have already discussed that there appears to exist a relatively low threshold value of serum
testosterone above which testosterone would not be atherogenic and hence,
does not increase cardiovascular morbidity and mortality. In accordance
with this view is the observation that testosterone treatment of female-to-
male transsexuals (Asscheman et al. 1994; see also Chapter 19 by Gooren
"Testosterone treatment of transsexuals" in this volume) or of chemical or
surgical castrates (Bagatell et al. 1992; Goldberg et al. 1985; Moorjani et al.
1987) with the aim of achieving male testosterone levels, is clearly atherogenic, whereas most studies concerning treatment with physiological supplementation doses of testosterone of older males with partial androgen deficiency (Conway et al. 1988; Ellyin 1995; Marin et al. 1992, 1996; Morley et al.
1993; Sih et al. 1997; Tenover et al. 1992) generally show no effect or even favorable effects on the plasma lipids, and more specifically on HDL-C. As already mentioned, favorable effects are probably mediated by an improvement
of insulin sensitivity (Friedl et al. 1989) and are only seen if the androgen
can be aromatized to estrogens, treatment with non-aromatizable androgens
such as methyltestostrone or simultaneous administration of an aromatase
inhibitor such as spironolactone, decreasing significantly HDL-C levels
(Friedl et al. 1990). A similar atherogenic effect is seen when supraphysiological doses of androgen are administered, for example for contraceptive purposes (Andersson et al. 1995; Bagatell et al. 1994; Bhasin et al. 1996; Wu et
al. 1996). However, it should be added that, whereas several studies show
that physiological doses of testosterone do not induce an atherogenic lipid
profile, none addresses the effects on cadiovascular morbidity or mortality.

As to other cardiovascular effects of androgens, both favorable and unfa-
vorable effects have been reported. Occasionally thrombotic complications
have been reported after androgen treatment (Nagelberg et al. 1986; Shiozo-
wa et al. 1982) but, on the other hand, also low testosterone levels have been
associated with increased thrombotic risk (Bonithon-Kopp et al. 1988). Testosterone, administered to female-to-male transsexuals increases endothelin
(a potent vasoconstrictor) levels, while the antiandrogen cyproterone-acetate
together with ethinylestradiol decreases endothelin levels in male-to-female
transsexuals (Polderman et al. 1993). In monkeys, testosterone significantly
inhibits prostacyclin and increases tromboxane B 2 (Ajayi 1995), which stim-

ulates platelet aggregation. Adams et al. (1995) observed that testosterone administration to female cynomolgus monkeys increased atherosclerotic plaque formation, but, on the other hand, reversed the atherosclerosis-related impairment of endothelium-dependent vasodilatatory respone to acetylcholine.

On the favorable side, one can mention the fibrinolytic effects (Fearnley and Chakrabarti 1962) and the inverse correlation between testosterone and plasminogen activation inhibitor type I (PAI-1) (Beer et al. 1996). However, it remains unclear whether replacement therapy with physiological doses of testosterone affects these different factors.

Another possible side-effect of androgen substitution, even in physiologic doses, is polycythemia. Whereas a moderate increase in hematocrit in elderly males is probably beneficial, Hajjar et al. (1997) observed that out of 47 elderly, hypogonadal males, receiving 200 mg of testosterone enanthate or cypionate every two weeks, 11 (24%) developed polycythemia sufficient to require phlebotomy or temporary withholding of testosterone, one third of which occurred less than one year after starting treatment. Sih et al. (1997) reported a similarly frequent development of polycythemia.

Androgens may exacerbate obstructive sleep apnea (Matsumoto et al. 1985; Sandblom et al. 1983). Therefore, patients should be specifically questionned for symptoms of sleep apnea, and chronic obstructive pulmonary disease, especially in overweight subjects or heavy smokers, is a relative contra-indication to androgen therapy.

Gynecomastia, related to the conversion of testosterone to estradiol in peripheral tissues, mainly fat tissue, which is relatively increased in elderly men, is a not uncommon but benign side-effect in elderly men, especially in the obese. Although testosterone causes some sodium and water retention (Wilson 1988), this effect does usually not cause a problem, except in patients with congestive heart failure, hypertension or renal insufficiency. Hepatoxicity is very rare when non-oral galenic forms of testosterone are used.

Of greater concern are the possible effects on the prostate, which is an androgen-dependent organ (see also Chapter 9 by Frick "Androgens and the prostate" in this volume). As far as benign prostatic hyperplasia (BHP) is concerned, studies to date failed to observe an important growth of the prostate (Behre et al. 1994; Wallace et al. 1993) and all studies have failed to find any relationship between plasma and BPH tissue levels of testosterone, DHT or estradiol. It appears that tissue levels are determined by the enzyme activity in the tissue itself, rather than by surrounding plasma androgen levels. Prostate specific antigen (PSA) levels, a parameter of androgen stimulation of the prostatic tissue, increases moderately during treatment but usually within the normal range and, after stopping treatment the values return to pretreatment levels (Behre et al. 1994; Hajjar et al. 1997; Tenover 1994, 1996).

Clinical prostatic carcinoma undoubtedly is an androgen-sensitive tumor (Goldenberg et al. 1995): hence presence of a clinical prostatic carcinoma is an absolute contraindication to testosterone suplementation. Subclinical carcinoma, only detectable by prostate biopsy but undetectable by biochemical or clinical procedures, is found in more than 50% of males over 70 years old.

Only a small minority of these subclinical carcinomas will further develop to a clinical carcinoma. It is not known whether testosterone treatment would stimulate the progression of subclinical carcinoma and so far no data are available indicating that testosterone substitution will activate subclinical carcinoma (Schröder 1996; Jackson et al. 1989). However, all studies so far concern only small numbers of carefully selected elderly males treated for relatively short periods of time. In any case, before starting testosterone supplementation careful exclusion of the presence of a prostatic carcinoma by rectal examination and PSA, when required supplemented by echography, with subsequent 6-monthly controls is mandatory.

16.5.4 Modalities of androgen substitution

In young healthy men, testosterone levels vary between 11 and 35 nmol/l and show a circadian variation with an amplitude of ± 35%, highest levels being reached in the early morning and nadir values in the evening around 6–8 p.m. Therefore, when supplementing testosterone, the aim should be to alleviate the symptoms related to the relative androgen deficiency, if possible by achieving plasma testosterone levels that mimic the levels and their nycthemeral variations as found in young adults.

As the hypothalamo-pituitary-testicular axis is very sensitive to negative feedback, and even more so in elderly males (Deslypere et al. 1987; Winters et al. 1984, 1997) it is important to ascertain that the dose administered increases the testosterone levels up to the physiological range and does not merely suppress LH secretion with only replacement of the deficient testosterone production by an inadequate dose of exogenous testosterone. In practical terms, full replacement doses are usually required. The pharmacology and practical aspects of testosterone replacement are discussed in detail in Chapter 10 by Nieschlag and Behre "Pharmacology and clinical use of testosterone" and several other chapters of this volume.

16.6 Key messages

- Mean total serum testosterone decreases progressively in healthy men over the age of 55 years (30% decrease between age 25 and 75 years). The age-associated decrease of the bioavailable fractions of serum testosterone is steeper as a consequence of an age-related increase of serum SHBG-binding capacity (50% decrease of free testosterone between age 25 and 75 years).
- There is great interindividual variability of prevailing androgen levels in the elderly, ranging from perfectly preserved to frankly hypogonadal. Part of the interindividual variability in serum testosterone levels is explained by heredity, physiological factors and lifestyle.

- The proportion of men with "subnormal" testosterone increases with age (>20% after age 60 years).
- The age-related decline in Leydig cell function can transiently or more permanently be accentuated by intercurrent disease.
- Many of the clinical features of aging in men are reminiscent of the clinical changes seen in hypogonadism in younger men; relative hypoandrogenism may be involved in some, but certainly not all clinical changes.
- Testosterone levels required for normal sexual activity are rather low and although testosterone levels codetermine potency, the factors most commonly involved in impotence in elderly men are not hormonal. Hypoandrogenism may be involved in the sarcopenia of elderly men; its role in male senile osteoporosis remains to be confirmed.
- In the present state of the art, androgen supplementation should only be considered in the presence of androgen serum levels below the lower normal limit for younger men, together with unequivocal signs and symptoms of androgen deficiency, in the absence of other reversible causes of decreased androgen levels and after screening for contra-indications.
- Possible benefits of the treatment include an improved sense of general well-being, of libido and of muscle strength, with increase of lean body mass and limited decrease of fat mass.
- So far, the still relatively limited data on safety of testosterone replacement therapy in the elderly has been rather reassuring: larger scale studies of longer duration are still needed to reassure completely on the issue of safety at the prostate level; development of polycythemia seems to emerge as one of the most troublesome side-effects, which may not unfrequently necessitate (temporary) interruption of treatment.
- Androgen replacement therapy in the elderly requires careful monitoring by an experienced physician.

Acknowledgement

Part of this work was supported by the Fonds voor Wetenschappelijk Onderzoek Vlaanderen, Grants G.3062.92 and G.0058.97.

16.7 References

Abbasi A, Drinka PJ, Mattson DE, Rudman D (1993) Low circulating levels of insulin-like growth factors and testosterone in chronically institutionalized elderly men. J Am Geriatr Soc 41:975–982

Adams MR, Williams JK, Kaplan JR (1995) Effects of androgens on coronary artery atherosclerosis and atherosclerosis-related impairment of vascular responsiveness. Arterioscl Thromb Vasc Biol 15:562–570

Adlercreutz H (1990) Western diet and Western diseases: some hormonal and biochemical mechanisms and associations. Scan J Clin Lab Inv 201 (Suppl):S3–S23

Ajayi AA (1995) Testosterone increases human platelet thromboxane A2 receptor density and aggregation responses. Circulation 91:2742–2747

Andersson B, Björntorp P, Marin P, Lissner L, Vermeulen A, (1994) Testosterone concentration in women and men with NIDDM. Diabetes Care 17:405–411

Andersson RA, Wallace BM, Wu FCW (1995) Effects of testosterone enanthate on serum lipoproteins in man. Contraception 52:115–119

Asscheman H, Gooren LJG, Megens JAJ, Nauta J, Kloosterboer HJ, Eikelboom F (1994) Serum testosterone is the major determinant of male-female differences in serum levels of high density lipoprotein cholesterol and HDL2 cholesterol. Metabolism 43:935–939

Bagatell CJ, Bremner WJ (1995) Androgens and progestogen effects on plasma lipids. Prog Cardiovasc Dis 38:255–271

Bagatell CJ, Knopp RH, Vale WW, Rivier JE, Bremner WJ (1992) Physiologic testosterone levels in normal men suppress high-density lipoprotein cholesterol levels. Ann Int Med 116:967–973

Bagatell CJ, Heiman JR, Rivier JE, Bremner WJ (1994a) Effects of endogenous testosterone and estradiol on sexual behavior in normal young men. J Clin Endocrinol Metab 78:711–716

Bagatell CJ, Heiman JR, Matsumoto AM, Rivier JE, Bremner WJ (1994b) Metabolic and behavioral effects of high dose exogenous testosterone in healthy men. J Clin Endocrinol Metab 79:561–567

Baker HWG, Burger HG, de Kretser DM, Dulmanis A, Hudson B, O'Connor S, Paulson CA, Purcell N, Rennie GC, Seah CS, Taft HP, Wang C (1979) A study of the endocrine manifestations of hepatic cirrhosis. Q J Med 177:145–178

Baker HWG, Burger HG, de Kretser DM, Hudson B (1977) Endocrinology of aging: pituitary testicular axis. In: James VHT (ed) Proc of the 5th International Congress of Endocrinology, Excerpta Medica Foundation, Amsterdam, pp 179–183

Bancroft J (1984) Androgens, sexuality and the aging male. In: Labrie F, Proulx L (eds) Endocrinology, Elsevier, Amsterdam, pp 913–916

Barrett-Connor EL, Khaw KS (1988) Endogenous sex hormones and cardiovascular disease in men: a prospective population based study. Circulation 78:539–545

Barrett-Connor EL, Khaw KT, Yen SCC (1990) Endogenous sex hormone levels in older adult men with diabetes mellitus. Am J Epidemiol 132:895–901

Beer NA, Jakubowicz Ds, Malt DW, Beer RM, Nestler JE (1996) Oral dehydroepiandrosterone administration reduces plasma levels of plasminogen activator type I and tissue plasminogen activator antigen in men. Am J Med Sc 311:205–210

Behre HM, Bohmeyer J, Nieschlag E (1994) Prostate volume in testosterone-treated and untreated hypogonadal men in comparison to age-matched normal controls. Clin Endocrinol 40:341–349

Behre HM, Kliesch S, Leifke E, Link TM, Nieschlag E (1997) Long term effect of testosterone therapy on bone mineral density in hypogonadal men. J Clin Endocrinol Metab 82:2386–2390

Belanger A, Locong A, Noel C, Cusan N, Dupont A, Prevost J, Caron S, Sevigny J (1989) Influence of diet on plasma steroids and plasma binding globulin levels in adult man. J Steroid Biochem 32:829–833

Bhasin S, Tenover JS (1997) Age-associated sarcopenia – Issues in the use of testosterone as an anabolic agent in older men. J Clin Endocrinol Metab 82:1659–1660

Bhasin S, Storer TW, Berman N, Callegari C, Clevenger B, Phillips J, Bunnell TJ, Tricker R, Shirazi A, Casaburi R (1996) The effects of supraphysiological doses of testosterone on muscle size and strength in normal men. N Engl J Med 335:1–7

Bixler EO, Santen RJ, Kales A (1977) Inverse effects of thioridazine (Melleril) on serum prolactin and testosterone concentrations in normal men. In: Troen P, Nankin HR (eds) The testis in normal and infertile men. Raven, New York, pp 405–409

Blondeau JP, Baulieu EE, Robel P (1982) Androgen dependent regulation of androgen receptor in the rat ventral prostate. Endocrinology 110:1926–1932

Bonithon-Kopp, Scarabin PY, Bara I, Castanier M, Jacqueson A, Roger M (1988) Relationship between sex hormones and haemostatic factors in healthy middle aged men. Atherosclerosis 71:71–76

Boonen S, Vanderschueren D, Xiao GC, Verbeke G, Dequeker J, Geusens P, Broos P, Bouillon R (1997) Age related (Type II) femoral neck osteoporosis in men: biochemical evidence for both hypovitaminosis D- and androgen deficiency induced bone resorption. J Bone Miner Res 12:2119–2126

Bremner WJ, Vitiello MV, Prinz PN (1983) Loss of circadian rhythmicity in blood testosterone levels with aging in normal men. J Clin Endocrinol Metab 56:1278–1281

Cameron JM, Weltzin TE, McConaha C, Helmreich DL, Kaye WH (1991) Slowing of pulsatile luteinizing hormone secretion in men after forty-eight hours of fasting. J Clin Endocrinol Metab 73:35–41

Cauley JA, Gutai JP, Kuller LH, Dai WS (1987) Usefulness of sex steroid hormone levels in predicting coronary artery disease in men. Am J Cardiol 60:771–777

Chang TC, Tung CC, Hsiao YL (1994) Hormonal changes in elderly men with non insulin dependent diabetes mellitus and the hormonal relationships to abdominal adiposity. Gerontology 40:260–267

Cicero TJ (1982) Alcohol induced defects in the hypothalamo-pituitary luteinizing hormone action in the male. Alcoholism 6:207–215

Conway AJ, Boylan LM, Howe C, Ross G, Handelsman DJ (1988) Randomized clinical trial of testosterone replacement therapy in hypogonadal men. Int J Androl 112:47–26

Cooper C, Campion G, Melton LJ III (1992) Hip fractures in the elderly: a world wide projection. Osteoporosis Int 2:285–289

Cooper TG, Keck C, Oberdieck U, Nieschlag E (1993) Effects of multiple ejaculations after extended periods of sexual abstinence on total motile and normal sperm numbers as well as on accessory gland secretions from healthy normal and oligospermic men. Hum Reprod 8:1251–1258

Dabbs JM (1990) Age and seasonal variation in serum testosterone concentration among men. Chronobiol Int 7:245–248

Dai WS, Gutai JP, Kuller LH, Cauley JA (1988) Cigarette smoking and serum sex hormones in men. Am J Epidemiol 128:796–805

Davidson JM, Kwan M, Greenleaf WJ (1982) Hormonal replacement and sexuality in men. Clin Endocrinol Metab 11:594–623

Davidson JM, Chen JJ, Crapo L, Gray GD, Greenleaf WJ, Catania JA (1983) Hormonal changes and sexual function in aging men. J Clin Endocrinol Metab 57:71–77

Demoor P, Goossens JV (1970) An inverse coorelation between body weight and the activity of the steroid binding globulin in human plasma. Steroidologia 1:129 136

Deslypere JP, Vermeulen A (1981) Aging and tissue androgens. J Clin Endocrinol Metab 53:430–434

Deslypere JP, Vermeulen A (1984) Leydig cell function in normal men: effect of age, lifestyle, residence, diet and activity. J Clin Endocrinol Metab 59:955–962

Deslypere JP, Vermeulen A (1985) Influence of age on steroid concentration in skin and striated muscle in women and in cardiac muscle and lung tissue in men. J Clin Endocrinol Metab 60:648–653

Deslypere JP, Sayed A, Punjabi U, Verdonck L, Vermeulen A (1982) Plasma 5α-androstane-3α,17-diol and urinary 5α-androstane, 3α,17-diol glucuronide, parameters of peripheral androgen action: a comparative study. J Clin Endocrinol Metab 54:386–391

Deslypere JP, Kaufman JM, Vermeulen T, Vogelaers D, Vandalem JL, Vermeulen A (1987) Influence of age on pulsatile luteinizing hormone release and responsiveness of the gonadotrophs to sex hormone feedback in men. J Clin Endocrinol Metab 64:68–73

Devogelaer JP, De Cooman S, Nagant de Deux Chaisnes C (1992) Low bone mass in hypogonadal males. Effect of testosterone substitution therapy, a densitometric study. Maturitas 15:17–23

Dong Q, Hawker F, Mc William D, Bangah M, Burger H, Handelsman DJ (1992) Circulating immunoreactive inhibin and testosterone levels in men with critical illness. Clin Endocrinol 36:399–404

Drinka PJ, Olson J, Bauwens S, Voeks SK, Carlson I, Wilson M (1993) Lack of association between free testosterone and bone density separate from age in elderly males. Calcif Tissue Int 52:67–69

Dutta C, Hadley EC (1995) The significance of sarcopenia in old age. J Gerontol 50A:1–4

Elewaut A, Barbier F, Vermeulen A (1979) Testosterone metabolism in normal males and male cirrhotics. Z Gastroenterol 17:402–405

Ellyin FM (1995) The long-term beneficial effects of low dose testosterone in the aging male. Program and Abstracts of the 77th Annual Meeting of the Endocrine Society (USA), The Endocrine Society, Washington DC, p 322 (Abstract P2-127)

Erfurth EM, Hagmar LE, Sääf M, Hall K (1996) Serum levels of insulin-like growth factor I and insulin-like growth factor binding protein 1 correlate with serum free testosterone and sex hormone binding globulin in healthy young and middle-aged men. Clin Endocrinol 44:659–664

Faber J, Perrild H, Johansen JS (1990) Bone Gla protein and sex hormone-binding globulin in nontoxic goiter: parameters for metabolic status at the tissue levels. J Clin Endocrinol Metabol 70:49–55

Fearnley CR, Chakrabarti R (1962) Increase of blood fibrinolytic activity by testosterone. Lancet 2:128–132

Field AE, Colditz GA, Willett WC, Longcope C, McKinlay JB (1994) The relation of smoking, age, relative weight and dietary intake to serum adrenal steroids, sex hormones and sex hormone binding globulin in middle age men. J Clin Endocrinol Metab 79:1310–1316

Finkelstein JS, Klibanski A, Neer RM, Doppelt SH, Rosenthal DI, Segre RV, Crowley WF (1989) Increase in bone density during treatment of men with idiopathic hypogonadotropic hypogonadism. J Clin Endocrinol Metab 69:776–783

Forbes GB, Reina JC (1970) Adult lean body mass declines with age: some longitudinal observations. Metabolism 19:653–663

Foresta G, Ruzza G, Mioni R, Guarnieri G, Gribaldo R, Meneghello A, Mastrogiacomo I (1984) Osteoprosis and decline of gonadal function in the elderly male. Horm Res 19:18–22

Friedl KE, Jones RE, Hannan CJ Jr, Plymate SR (1989) The administration of pharmacological doses of testosterone or 19-nortestosterone to normal men is not associated with increased insulin secretion or impaired glucose tolerance. J Clin Endocrinol Metab 68:971–975

Friedl KE, Hannan CJ Jr, Jones RE, Plymate SR (1990) High density lipoprotein cholesterol is not decreased if an aromatizable androgen is administered. Metabolism 39:69–74

Giagulli VA, Kaufman JM, Vermeulen A (1994) Pathogenesis of decreased androgen levels in obese men. J Clin Endocrinol Metab 79:997–1000

Giusti G, Gonnelli P, Borreli D, Fiorelli G, Forti G, Pazzagli M, Serio M (1975) Age related secretion of androstenedione, testosterone and dihydrotestosterone by the human testes. Exp Gerontol 10:241–245

Goldberg RB, Rabin AN, Alexander AN, Coelle GC, Getz GE (1985) Suppression of plasma testosterone leads to an increase in serum total and high density lipoprotein cholesterol and apoproteins A-1 and B. J Clin Endocrinol Metab 60:203–207

Goldenberg SL, Bruchowsy N, Gleave ME, Sullivan LD, Akakura K (1995) Intermittent androgen suppression in the treatment of prostate cancer. Urology 45:839–845

Gonzales-Cadavid NF, Swerdloff RS, Lemmi CAE, Rajfer J (1991) Expresssion of androgen receptor gene in rat penile tissue and cells during sexual maturation. Endocrinology 129:1671–1678

Gooren LJG (1987) Androgen levels and sex functions in testosterone treated hypogonadal men. Arch Sex Beh 16:463–476

Gordon D, Beastall GH, Thomson JA, Sturrock RD (1988) Prolonged hypogonadism in male patients with rheumatoid arthritis during flares in disease activity. Brit J Rheumatol 27:440–444

Graf MJ, Richards CJ, Brown V, Meissner L, Dunaif A (1990) The independent effects of hyperandrogenemia, hyperinsulinaemia and obesity on lipid and lipoprotein profiles in women. Clin Endocrinol 33:119–131

Gray A, Berlin JA, McKinlay JB, Longcope C (1991) An examination of research design effects on the association of testosterone and male aging. Results of a meta-analysis. J Clin Epidemiol 44:671–684

Greendale G, Edelstein S, Barrett-Connor E (1997) Endogenous sex steroids and bone mineral density in older women and men: the Rancho Bernardo study. J Bone Miner Res 12:1833–1843

Grunstein RR, Handelsman DJ, Lawrence SJ, Blackwell C, Caterson ID, Sullivan CE (1989) Neuroendocrine dysfunction in sleep apnea: reversal by nasal continuous positive airway pressure. J Clin Endocrinol Metab 68:352–358

Guralnik JM, Ferrucci L, Simonsick EM, Salive ME, Wallace RB (1995) Lower extremity function in persons over 70 years as a predictor of subsequent disability. N Engl J Med 322:556–561

Haffner SM (1996) Androgens in relation to cardiovascular disease and insulin resistance in aging men. In: Oddens B, Vermeulen A (eds) Androgens and the aging Male. Parthenon Publishing Group, New York, pp 65–93

Haffner SM, Katz MS, Stern MP, Dunn JF (1988) The relationship of sex hormones to hyperinsulinemia and hyperglycemia. Metabolism 37:683–688

Haffner SM, Mykkänen L, Valdez RA, Stern MP, Katz MS (1993) Relationship of sex hormones to lipids and lipoproteins in non-diabetic men. J Clin Endocrinol Metab 77:1610–1615

Haffner SM, Valdez RA, Mykkänen L, Stern MP, Katz MS (1994) Decreased testosterone and dehydroepiandrosterone sulfate concentrations are associated with increased glucose and insulin concentrations in non-diabetic men. Metabolism 43:599–603

Hajjar RR, Kaiser FE, Morley JE (1997) Outcomes of long-term testosterone replacement in older hypogonadal males: a retrospective analysis. J Clin Endocrinol Metab 82:3793–3796

Häkkinen K, Pakarinen A (1994) Serum hormones and strength development during strength training in middle-aged and elderly males and females. Acta Physiol Scand 150:211–219

Hamäläinen E, Adlercreutz H, Ehnholm C, Puska P (1986) Relationships of serum lipoproteins and apoproteins to the binding capacity of sex hormone binding globulin (SHBG) in healthy Finnish men. Metabolism 35:535–541

Handa K, Ishii H, Kono S, Shinchi K, Imanishi K, Mehara H, Tanaka K (1997) Behavioral correlates of plasma sex hormones and their relationships with plasma lipids and lipoproteins in Japanese men. Atherosclerosis 130:37–44

Handelsman DJ (1994) Testicular dysfunction in systemic disease. Endocrin Metab Clin 23:839–852

Handelsman DJ, Dong Q (1993) Hypothalamo-pituitary-gonadal axis in chronic renal failure. Endocrin Metab Clin 22:145–161

Harbitz TB (1973) Morphometric studies of Leydig cells in elderly men, with special reference to the histology of the prostate. Acta Pathol Microbiol Scand 81:301–314

Harman SM, Tsitouras PD (1980) Reproductive hormones in aging men I. Measurement of sex steroids, basal luteinizing hormone and Leydig cell response to human chorionic gonadotropin. J Clin Endocrinol Metab 51:35–40

Herman SM, Robinson JTC, McCredie J, Adams MR, Boyer MJ, Celermajer DS (1997) Androgen deprivation is associated with enhanced endothelium-dependent dilatation in adult men. Arterioscler Thromb Vasc Biol 17:2004–2009

Heller RF, Wheeler MJ, Micallef J, Miller NE, Lewis B (1983) Relationship of high density lipoprotein cholesterol with total and free testosterone and sex hormone binding globulin. Acta Endocrinol 104:253–256

Hollander N, Hollander VP (1958) The microdetermination of testosterone in human spermatic vein blood. J Clin Endocrinol Metab 18:966–970

Horwitz KB, Horwitz LD (1982) Canine vascular tissues are targets for androgens, estrogens, progestogens and glucocorticoids. J Clin Invest 69:750–758

Ida Y, Tsyjimaru S, Nakamura K, Shirao I, Mukasa H, Egarni H, Nakazawa H (1992) Effect of acute and repeated alcohol ingestion on hypothalamic-pituitary gonadal and hypothalamic-pituitary-adrenal functioning in normal males. Drug Alcohol Dep 31:57–64

Impallomeni M, Kaufman BM, Palmer AJ (1994) Do acute diseases transiently impair anterior pituitary function in patients over the age of 75? A longitudinal study of the TRH test and basal gonadotropin levels. Postgrad Med J 70:86–91

Irwin M, Dreyfus E, Baird D S, Smith TL, Schuckit M (1988) Testosterone in chronic alcoholic men. Br J Addic 83:949–953

Jackson JA, Waxman J, Spiekerman M (1989) Prostatic implications of testosterone replacement therapy. Arch Intern Med 149:2364–2366

Jackson JA, Riggs MW, Spiekerman M (1992) Testosterone deficiency as a risk factor for hip fracture in men: A case control study. Am J Med Sci 304:4–8

Jaross VW, Schollberg K, Seiler E, Wilke W, Wilke B, Schirmer G (1989) Besonderheiten im hypophysären und Geschlechts-Hormonhaushalt bei Männern nach überstandenem Herzinfarkt. Z Med Lab Diagn 30:32–34

Joseph DR, Baker MF (1992) Sex homone binding globulin, androgen binding protein and vitamin K dependent protein S are homologous to laminin, merosin and Drosophila crumbs protein. FASEB 6:2477–2481

Kaiser FE (1992) Impotence in the elderly. In: Morley J, Krenman P (eds) Endocrinology and Metabolism in the elderly. Cambridge-Blackwell, Cambridge, pp 262–271

Katznelson L, Finkelstein JS, Schoenfeld DA, Rosenthal DI, Anderson EJ, Klibanski A (1996) Increase in bone density and lean body mass during testosterone administration in men with acquired hypogonadism. J Clin Endocrinol Metab 81:4358–4365

Kaufman JM (1996) Androgens, bone metabolism and osteoporosis. In Oddens B, Vermeulen A (eds) Androgens and the aging male. Parthenon Publish Group, New York, pp 39–60

Kaufman JM, Vermeulen A (1997) Declining gonadal function in elderly men. Baillière's Endocrinol Metab 11:289–309

Kaufman JM, Deslypere JP, Giri M, Vermeulen A (1990) Neuroendocrine regulation of pulsatile luteinizing hormone secretion in elderly men. J Steroid Biochem 37:421–430

Kaufman JM, Giri M, Deslypere JP, Thomas G, Vermeulen A (1991) Influence of age on the responsiveness of the gonadotrophs to luteinizing hormone-releasing hormone in males. J Clin Endocrinol Metab 72:1255–1260

Kelly TM, Edwards CQ, Meikle AW, Kushner JP (1984) Hypogonadism in haemochromatosis – Reversal with iron depletion. Ann Intern Med 101:629–632

Kelly PJ, Pocock NA, Sambrook PN, Eisman JA (1990) Dietary calcium, sex hormones, and bone mineral density in men. Br Med J 300:1361–1364

Kent JZ, Acone AB (1966) Plasma androgens and aging. In: Vermeulen A, Exley D (eds) Androgens in normal and pathological conditions. Excerpta Medica Foundation, Amsterdam, pp 31–35

Key T, Roe L, Thorogood M, More JW, Clark MC, Wang DY (1990) Testosterone, sex hormone binding globulin, calculated free testosterone and oestradiol in male vegetarians and omnivores. Br J Nutr 64:111–119

Khaw KT, Barrett-Connor E (1992) Lower endogenous androgens predict central obesity in men. Ann Epidemiol 2:675–682

Kirkland RT, Keenan BS, Probstfield JL, Patsch W, Tsai-Lien L, Clayton GW, Insull W (1987) Decrease in plasma high density lipoprotein-cholesterol levels at puberty in boys with delayed adolescence: Correlation with plasma testosterone levels. JAMA 257:502–507

Kligman EW (1991) Office evaluation of sexual function and complaints. Clin Ger Med 7:25–39

Kuttenn F, Mowszowicz I, Schaison G, Mauvais-Jarvis P (1977) Androgen production and skin metabolism in hirsutism. J Endrocrinol 75:83–91

Kwan M, Greenleaf WJ, Mann J, Crapo L, Davidson J (1983) The nature of androgen action on male sexuality: a combined laboratory/self report study on hypogonadal men. J Clin Endocrinol Metab 57:557–562

Labrie F, Bélanger A, Simard J, Van Luu-The, Labrie C (1995) DHEA and peripheral androgen and estrogen formation: intracrinology. Ann N.Y. Acad Sc 774:16–28

Lange JD, Brown WA, Wincze J, Zwick W (1980) Serum testosterone concentration and penile tumescence changes in men. Horm and Behav 14:267–270

Lichtenstein MJ, Yarnell JWG, Elwood PC, Beswick AD, Sweetnam PM, Marks V, Teale D, Riad-Fahmy D (1987) Sex hormones, insulin, lipid and prevalent ischemic heart disease. Am J Epidemiol 126:647–657

Lithell H, Nillus SJ, Bergh T, Selinus I (1987) Metabolic profile in obese women with the polycystic ovary syndrome. Int J Obesity 11:1–9

Longcope C (1973) The effect of human chorionic gonadotropin on plasma steroid levels in young and old men. Steroids 21:583–592

Lugg J, Rajfer J, Gonzales-Cadavit NF (1995) Dihydrotestosterone is the active androgen in the maintenance of nitric-oxide mediated penile erection in the rat. Endocrinology 136:1495–1501

Luria MH, Johnson MW, Pego R, Seuc CA, Manubens SJ, Wieland MR, Wieland RG (1982) Relationship between sex hormones, myocardial infarction and occlusive coronary disease. Arch Intern 142:42–44

Luton JP, Thieblot P, Valcke JC, Mahoudeau JA, Bricaire H (1977) Reversible gonadotropin deficiency in male Cushing's disease. J Clin Endocrinol Metab 45:488–495

MacAdams MR, White RH, Chipps BE (1986) Reduction of serum testosterone levels during chronic glucocorticoid therapy. Ann Intern Med 104:648–651

Mahabeer S, Naidoo C, Norman RJ, Jiala I, Riddi K, Joubert JM (1990) Metabolic profiles and lipoprotein lipid concentration in non-obese and obese patients with polycystic ovarian disease. Horm Metab Res 22:401–406

Marin P, Holmäng S, Jönsson L, Sjöström L, Kvist H, Holm G, Lindstedt G, Björntorp P (1992) The effects of testosterone treatment on body composition and metabolism in middle aged men. Int J Obesity 16:991–997

Marin P, Holmäng S, Gustafsson C, Jönsson L, Kvist H, Elander A, Eldh J, Sjöström L, Holm G, Bjöntorp P (1993) Androgen treatment of abdominally obese men. Obesity Res 1:245–251

Marin P, Lönn L, Anderson B, Oden B, Olbe L, Bengtsson BA, Björntorp P (1996) Assimilation of triglycerides in subcutaneous and intraabdominal adipose tissues in vivo in men: effects of testosterone. J Clin Endocrinol Metab 81:1018–1022

Martin CE (1975) Marital and sexual factors in relation to age, disease and longevity. In: Wirdt RD, Winokur G, Ruff M (eds) Life history research in psychopathology, University of Minnesota Press, Minneapolis, pp 326–347

Masters WH (1986) Sex and aging – Expectations and reality. Hosp Pract 15:175–198

Matsumoto AM, Sandblom RE, Schoene RB, Lee KA, Giblin EC, Pierson DJ, Bremner WJ (1985) Testosterone replacement in hypogonadal men: effects on obstructive sleep apnoea, respiratory drives, and sleep. Clin Endocrinol 22:713 721

Matzkin H, Braf Z, Nava D (1991) Does age influence the bioactivity of follicle stimulating hormone in men. Age and Ageing 20:199–205

McKenna TJ, Lorber D, Lacroix A, Rabin D (1979) Testicular activity in Cushing's disease. Acta Endocrinol 91:501–510

Meier DE, Orwoll ES, Keenan EJ, Fagerstrom RM (1987) Marked decline in trabecular bone mineral content in healthy men with age: lack of association with sex steroid levels. J Am Geriatr Soc 35:189–97

Meikle AW, Bishop DT, Stringham JD, West DW (1986) Quantitating genetic and non genetic factors to determine plasma sex steroid variation in normal male twins. Metabolism 35:1090–1095

Meikle AW, Stringham JD, Bishop T, West DW (1988) Quantitation of genetic and non genetic factors influencing androgen productions and clearance rates in men? J Clin Endocrinol Metab 67:104–109

Meikle AW, Stringham JD, Woodward MG, McMurry MP (1990) Effects of fat containing meal on sex hormones in men. Metabolism 39:943–946

Mendoza SG, Zerpa A, Carrasco H, Colmenares O, Rangel A, Gardside PS, Kashyap ML (1983) Estradiol, testosterone, apolipoproteins, apolipoprotein cholesterol and lipolytic enzymes in men with premature myocardial infarction and angiographically assessed coronary occlusion. Artery 2:1–23

Mills TM, Reilly CM, Lewis RW (1996) Androgens and penile erection – A review. J Andrology 17:633–638

Mikuma N, Kumamoto Y, Maruta H, Nitta T (1994) Role of the hypothalamic opioidergic system in the control of gonadotropin secretion in elderly men. Andrologia 26:39–45

Mitchell R, Hollis S, Rothwell C, Robertsons WR (1995) Age related changes in the pituitary testicular axis in normal men; lower serum testosterone results from decreased bioactive LH drive. Clin Endocrinol 42:501–507

Moorjani S, Dupont A, Labrie F, Lupien PJ, Brun D, Gagné C, Giguère M, Bélanger A (1987) Increase in plasma high density lipoprotein concentration following complete androgen blockade in men with prostatic carcinoma. Metabolism 36:244–250

Morley JE (1986) Impotence. Am J Med 80:897–906

Morley JE, Kaiser FE, Perry HM, Patrick P, Morley PMK (1996) Longitudinal changes in testosterone, SHBG, LH, and FSH in healthy older males. Abstracts of the 10th International Congress of Endocrinology, San Francisco, p 174 (abstract P1–158)

Morley JE, Perry HM, Kaiser FE, Kraenzle D, Jensen J, Houston K, Mattammel M, Perry HM (1993) Effects of testosterone replacement therapy in old hypogonadal males: a preliminary study. J Am Geriatr Soc 41:149–152

Mulligan T, Iranmanesh A, Gheorghiu S, Godschalk M, Veldhuis JD (1995) Amplified nocturnal luteinizing hormone (LH) secretory burst frequency with selective attenuation of pulsatile (but not basal) testosterone secretion in healthy aging men: possible Leydig cell desensitization to endogenous LH signaling—a clinical research center study. J Clin Endocrinol Metab 80:3025–3031

Murphy S, Khaw KT, Cassidy A, Compston JE (1993) Sex hormones and bone mineral density in elderly men. J Bone Miner Res 20:133–40

Naftolin E, Judd JH, Yen SSC (1973) Pulsatile pattern of gonadotropin and testosterone in men. Effects of clomiphene with and without testosterone. J Clin Endocrinol Metab 36:285–288

Nagelberg SB, Lane L, Loriaux DL, Liu L, Sherin RJ (1986) Cardiovascular accident associated with testosterone therapy in a 20 year old hypogonadal man (letter). N Engl J Med 314:649–6450

Nankin HR, Lin T, Murono EP, Osterman J (1981) The aging Leydig cell III. Gonadotropin stimulation in men. J Androl 2:181–189

Neaves WB, Johnson L, Porter JC, Parker CR, Petty CS (1984) Leydig cell numbers, daily sperm production and gonadotropin levels in aging men. J Clin Endocrinol Metab 59:756–763

Nieschlag E, Kley KH, Wiegelmann W, Solback HG, Krüskemper HL (1973) Lebensalter und endokrine Funktion der Testes des erwachsenen Mannes. Deut Med Wochenschr 98:1281–1284

Nieschlag E, Lammers U, Freischem CW, Langer K, Wickings EJ (1982) Reproductive functions in young fathers and grandfathers. J Clin Endocrinol Metab 55:676–681

Nilsson P, Møller L, Solstad K (1995) Adverse effects of psychosocial stress on gonadal function and insulin levels in middle-aged males. J Intern Med 237:479–486

Ooi LS, Panesar NS, Masarei JR (1996) Urinary excretion of testosterone and estradiol in Chinese men and relationships with serum lipoprotein concentrations. Metabolism 45:279–284

Opstad PR (1992) The hypothalamo-pituitary regulation of androgen secretion in young men, after prolonged physical stress combined with energy and sleep deprivation. Acta Endocrinol 127:231–236

Orwoll ES, Klein RF (1995) Osteoporosis in men. Endocr Rev 16:87–115

Orwoll ES, Oviatt S (1992) Transdermal testosterone supplementation in normal older men. Program of the 74th Annual Meeting of the Endocrine Society. The Endocrine Society, Bethesda, Abstract 1071

Pearson UJD, Blackman MR, Metter EJ, Waclawiw Z, Carter HB, Harman SM (1995) Effect of age and cigarette smoking on longitudinal changes in androgens and SHBG in healthy males. Abstracts of the 77th annual Meeting of the Endocrine Society, The Endocrine Society, Bethesda p 322 (Abstract P2-129)

Pfeiffer E (1974) Sexuality in the aging individual. Arch Sex Behaviour 22:481

Pfeilschifter J, Scheidt-Nave C, Leidig-Bruckner G, Woitge HW, Blum WF, Wüster C, Haack D, Ziegler R (1996) Relationship between circulating insulin-like growth factor components and sex hormones in a population based sample of 50 80 year old men and women. J Clin Endocrinol Metab 81:2534-2540

Phillips GB (1976) Evidence for hyperoestrogenemia as a risk factor for myocardial infarction in men. Lancet 2:14-18

Phillips GB (1978) Sex hormones, risk factors and cardiovascular disease. Am J Med 65:7-11

Phillips GB, Pinkernell BJ, Jing TY (1994) The association of hypotestosteronemia with coronary heart disease. Arterioscler Thromb 14:701-706

Pirke KM, Sintermann R, Vogt HJ (1980) Testosterone and testosterone precursors in the spermatic vein and in the testicular tissue of old men. Gerontology 26:221-230

Plant M (1986) Gonadal regulation of hypothalamic gonadotropin-releasing hormone release in primates. Endocr Rev 7:75-88

Plymate JR, Marej LA, Jones RE, Friedl KE (1988) Inhibition of sex hormone binding globulin production in human hepatoma (Hep G2) cell-line by insulin and prolactin. J Clin Edocrinol Metab 67:460-467

Plymate SR, Tenover JS, Bremner WJ (1989) Circadian variation in testosterone, sex hormone-binding globulin, and calculated non sex hormone-binding globulin bound testosterone in healthy young and elderly men. J Androl 10:366-371

Poggi UI, Arguelles AE, Rosner J, de Laborde NP, Cassini JH, Volmer MC (1976) Plasma testosterone and serum lipids in male survivors of myocardial infarction. J Steroid Biochem 7:229-231

Polderman KH, Stehouwer CDA, Van de Kamp GJ, Dekker GA, Verheugt FWA, Gooren LJG (1993) Influence of sex hormones on plasma on plasma endothelin levels. Ann Int Med 118:429-432

Poor G, Atkinson EJ, Lewallen DG, O'Fallon WM, Melton LJIII (1995) Age related hip fractures in men: clinical spectrum and short-term outcomes. Osteoporosis Int 5:419-426

Porto CS, Abreu LC, Gunsalus GL, Bardin CW (1992) Binding of sex hormone binding globulin (SHBG) to testicular membranes and solubilized receptor. Mol Cell Endocrinol 89:33-38

Rajfer JK, Namkun PC, Petra P (1989) Identification, partial characterization and age associated changes of a cytoplasmatic androgen receptor in the rat penis. J Steroid Biochem 33:1489-1492

Reed MJ, Cheng RW, Simmonds M, Richmond W, James VHT (1987) Dietary lipids: an additional regulator of plasma levels of sex hormone binding globulin. J Clin Endocrinol Metab 64:1083-1085

Rogers MA, Evans WJ (1993) Changes in skeletal muscle with aging: effects of exercise training. Exercise Sports Sci Rev 21:65-102

Rojanasakul A, Chailurkit L, Sirimongkolkasem R, Chaturachinda K (1988) Serum lipid and lipoprotein levels in women with polycystic ovarian disease with different body mass index. Int J Obstet Gynecol 27:401-406

Rolf C, Behre M, Nieschlag E (1996) Reproductive parameters of older compared to younger men of infertile couples. Int J Androl 19:135-142

Rosner W, Hryb DJ, Kahn MS, Nakhla HM, Romas NA (1991) Sex hormone binding globulin: anatomy and physiology of a new regulating system. J Steroid Bioch Molec Biol 40:813-820

Rubens R, Dhont M, Vermeulen A (1974) Further studies on Leydig cell function in old age. J Clin Endocrinol Metab 39:40-45

Rubenstein LZ, Josephson KR, Robbins AS (1994) Falls in the nursing home. Ann Intern Med 121:442-451

Rubin HB, Henson DE, Falvo RE, High RW (1979) The relationship between men's endogenous levels of testosterone and their penile response to erotic stimuli. Behav Res Ther 17:305–312

Rudman D, Drinka PJ, Wilson CR, Mattson DE, Scherman F, Cuisinier MC, Schultz S (1994) Relations of endogenous anabolic hormones and physical activity to bone mineral density and lean body mass in elderly men. Clin Endocrinol 40:653–661

Sandblom RE, Matsumoto AM, Scoene RB, Lee KA, Giblin EC, Bremner WJ, Pierson DJ (1983) Obstructive sleep apnea induced by testosterone administration N Engl J Med 308:508–510

Sasano N, Ichijo S (1969) Vascular patterns of the human testes with special reference to its senile changes. Tohoku J Exp Med 99:269–280

Schiavi RC (1996) Androgens and sexual function in men. In: Oddens B, Vermeulen A (eds). Androgens and the aging male. Parthenon Publishing Group, New York, pp 111–128

Schiavi RC, Schreiner-Engel P, White D, Mandeli J (1988) Pituitary-gonadal function during sleep in men with hypoactive sexual desire and in normal controls. Psychosom Med 50:304–318

Schiavi RC, Schreiner-Engel P, Mandeli J, Schanzer H, Cohen E (1990) Healthy aging and male sexual function. Am J Psychiatry 147:776–771

Schröder FH (1996) The prostate and androgens: the risk of supplementation. In: Oddens B, Vermeulen A (eds) Androgens and the aging male. Parthenon Publishing Group, New York, pp 223–26

Semmens J, Rouse I, Beilin LJ, Masarei JR (1983) Relationship of plasma HDL cholesterol to testosterone, estradiol and sex hormone binding globulin levels in men and women. Metabolism 32:428–432

Sewdarsen M, Vythilingum S, Jiadal I, Desai RK, Becker P (1990) Abnormalities in sex hormones are a risk factor for premature manifestation of coronary artery disease in South African Indian men. Atherosclerosis 83:111–117

Shiozawa Z, Yamada H, Mabuchi C, Saito H, Sobue I, Huang YP (1982) Superior sagittal sinus thrombosis associated with androgen therapy for hypoplastic anemia. Ann Neurol 12:57–88

Siiteri PK, MacDonald PC (1975) Role of extraglandular estrogen in human endocrinology. In: Greep RD, Astwood B (eds) Handbook of Physiology. Vol II. American Physiological Society, pp 491–508

Sih R, Morley JE, Kaiser FE, Perry HM III, Patrick P, Ross C (1997) Testosterone replacement in older hypogonadal men: a 12 month randomized controlled trial. J Clin Endocrinol Metab 82:1661–1667

Simon D, Nahoul K, Charles MA (1996) Sex hormones, ageing, ethnicity and insulin sensitivity in men: an overview of the TELECOM study. In: Oddens B, Vermeulen A (eds) Androgens and the aging male. Parthenon Publishing Group, New York, pp 95–101

Smals AGH, Kloppenburg PWC, Benraad TJ (1976) Circannual cycle in plasma testosterone levels in man. J Clin Endocrinol Metab 42:979–982

Sniffen RC (1950) The testes. I. The normal testis. Arch Pathol 50:259–284

Snyder PJ, Lawrence DA (1980) Treatment of hypogonadism with testosterone enanthate. J Clin Endocrinol Metab 51:1335–1339

Spratt DI, O'Dea L, Schoenfeld D, Butler J, Narashimha H, Rao P, Crowley WF (1988) Neuroendocrine-gonadal axis in men: frequent sampling of LH, FSH and testosterone. Am J Physiol 254:E658–E666

Spratt DI, Cox P, Orav J, Moloney J, Bigos T (1993) Reproductive axis suppression in acute illness is related to disease severity. J Clin Endocrinol Metab 76:1548–1554

Stanley HL, Schmitt BP, Poses RM, Deiss WP (1991) Does hypogonadism contribute to the occurrence of a minimal trauma hip fracture in elderly men? JAGS 39:766–771

Stepan JJ, Lachman M, Zverina J, Pacovsky V, Baylink DJ (1989) Castrated men exhibit bone loss: effect of calcitonin treatment on biochemical indices of bone remodeling. J Clin Endocrinol Metab 69:523–527

Suoranta H (1971) Changes in small vessels of the adult testes in relation to age and some pathological conditions. Virchows Archiv A: Pathological Anatomy and Histopathology 352:765–781

Swartz CM, Young MA (1987) Low serum testosterone and myocardial infarction in geriatric male patiens. J Am Geriatr Soc 35:39–44

Swerdloff RS, Wang C (1993) Androgens and aging in men. Exp Gerontol 28, 435 446

Talbott E, Guzick D, Clerici A, Berga S, Detre K, Weimer K, Kuller L (1995) Coronary heart disease risk factors in women with polycystic ovary syndrome. Atheroscl Thromb Vasc Biol, 15:821–826

Tapia-Serrano R, Jimenez-Baldera FJ, Murrieta S, Bravo-Gatica C, Guerra R, Mintz G (1991) Testicular function in active ankylosing spondylitis – Therapeutic response to human chorionic gonadotrophin. J Rheumatol 18:841–848

Tenover JS (1992) Effect of testosterone supplementation in the aging male. J Clin Endocrinol Metab 75:1092–1098

Tenover JS (1994) Androgen administration to aging men. Endocrinol Metab Clin 23:877–889

Tenover JS (1996) Effect of androgen supplementation in the aging male in: Oddens B & Vermeulen A (eds) Androgens and the aging male. Parthenon Publishing Group, New York pp 191–221

Tenover JS, Matsumoto AM, Plymate SR, Bremner WJ (1987) The effects of aging in normal men on bioavailable testosterone and luteinizing hormone secretion: response to clomiphene citrate; J Clin Endocrinol Metab 65:1118–1126

Tenover JS, Matsumoto AM, Clifton DK, Bremmer WJ (1988) Age related alterations in the circadian rhythms of pulsatile luteinizing hormone and testosterone secretion in healthy men. J Gerontol 43:M163–M169

Theorell T, Karasek RA, Eneroth P (1990) Job strain variation in relation to plasma testosterone in working men – a longitudinal study. J Intern Med 227:31–36

Tsitouras PD, Bulat T (1995) The aging male reproductive system. Endocrinol Metab Clin 24:297–315

Tsitouras PD, Martin CE, Harman SM (1982) Relation of serum testosterone to sexual activity in healthy elderly men. J Gerontol 37:288–293

Turner HE, Wass JAH (1997) Gonadal function in men with chronic illness. Clin Endocrinol 47:379–403

Udry JR, Billy JO, Morris NM, Groff TR, Raj MH (1985) Serum androgenic hormones motivate normal behavior in adolescent boys. Fertil Steril 43:90–94

Urban RJ, Veldhuis JD, Blizzard R, Dufau ML (1988) Attenuated release of biologically active luteinizing hormone in healthy aging men. J Clin Invest 81:1020–1029

Urban RJ, Bodenburg Y, Gilkison C, Foxworth J, Coggan AR, Wolfe RR, Ferrando A (1995) Testosterone administration to elderly men increases skeletal muscle strength and protein synthesis. Am J Physiol 269:E820–E826

Veldhuis JD, King JC, Urban RJ, Rogol LD, Evans WS, Kolp LA, Johnson ML (1987) Operating characteristics of the male hypothalamo-pituitary-gonadal axis. Pulsatile release of testosterone and follicle stimulating hormone and their temporal coupling with luteinizing hormone. J Clin Endocrinol Metab 68:929–947

Veldhuis JD, Urban RJ, Lizarralde G, Johnson ML, Iranmanesh A (1992) Attenuation of luteinizing hormone secretory burst amplitude as a proximate basis for the hypoandrogenism of healthy aging men. J Clin Endocrinol Metab 75:707–713

Vermeulen A (1976) Testicular hormonal secretion and aging in males. In: Grayhack St, Wilson J, Scherbenski MJ (eds) Benign prostatic hyperplasia. DHEW Publications (NIH), Bethesda, pp 177–182

Vermeulen A (1990) Androgens and male senescence. In: Nieschlag E, Behre HM (eds) Testosterone: Action, deficiency, substitution. Springer Verlag, Berlin, pp 261–276

Vermeulen A (1991) Androgens in the aging male – Clinical review 24. J Clin Endocrinol Metab 73:221–224

Vermeulen A, Deslypere JP (1986) Intratesticular unconjugated steroids in elderly men. J Steroid Biochem 24:1079–1083

Vermeulen A, Kaufman JM (1992) Role of the hypothalamo-pituitary function in the hypo-androgenism of healthy aging. J Clin Endocrinol Metab 74:1226A–1226C

Vermeulen A, Verdonck G (1992) Representativeness of a single point plasma testosterone level for the long term hormonal milieu in men. J Clin Endocrinol Metab 74:939–942

Vermeulen A, Stoica T, Verdonck L (1971) The apparent free testosterone concentration, an index of androgenicity. J Clin Endocrinol Metab 33:759–767

Vermeulen A, Rubens R, Verdonck L (1972) Testosterone secretion and Metabolism in male senescense. J Clin Endocrinol Metab 34:730–735

Vermeulen A, Deslypere JP, Kaufman JM (1989) Influence of antiopioids on luteinizing hor-mone pulsatility in aging men. J Clin Endocrinol Metab 68:68–72

Vermeulen A, Kaufman JM, Deslypere JP, Thomas G (1993) Attenuated LH pulse frequency and its relation to plasma androgens in hypogonadism of obese men. J Clin Endocrinol Metab 76:1140–1146

Vermeulen A, Kaufman JM, Giagulli VA (1996) Influence of some biological indices on sex hormone binding globulin and androgen levels in aging and obese males. J Clin Endo-crinol Metab 81:1821–1827

Verwoerdt A, Pfeiffer E, Wangh AS (1969) Sexual behaviour in senescence. Geriatrics 24:137–154

Virag R, Bocully P, Frydman D (1985) Is impotence an arterial disorder? A study of arterial risk factors in 440 impotent men. Lancet i:181–183

Wallace EM, Pye SD, Wild ST, Wu FCW (1993) Prostate specific antigen and prostate gland size in men receiving exogenous testosterone for male contraception. Int J Androl 16:35–40

Wang C, Chan V, Tse TF, Yeung RTT (1978 a) Effect of acute myocardial infarction on pitui-tary-testicular function. Clin Endocrinol 9:249–253

Wang C, Chan V, Yeung RTT (1978 b) Effect of surgical stress on pituitary testicular func-tion. Clin Endocrinol 9:255–266

Wild RA, Painter PC, Coulson PB, Carruth BK, Ranney GB (1985) Lipoprotein lipid concen-trations and cardiovascular risk in women with polycystic ovary syndrome. J Clin Endo-crinol Metab, 61:946–951

Wild RA, Bartholomew MJ (1988) The influence of body weight on lipoprotein lipids in pa-tients with polycystic ovary syndrome. Am J Obst Gynecol 1529:423–427

Wilson JD (1988) Androgen abuse by athletes. Endoc Rev 9:203–208

Winter SJ, Troen P (1982) Episodic luteinizing hormone (LH) secretion and the response of LH and follicle-stimulating hormone to LH-releasing hormone in aged men: evidence for coexistent primary testicular insufficiency and an impairment in gonadotropin secre-tion. J Clin Endocrinol Metab 55:560–565

Winters SJ, Sherins RJ, Troen P (1984) The gonadotropin suppressive activity of androgen is increased in elderly men. Metabolism 33:1052–1059

Winters SJ, Atkinson L for the Testoderm Study Group (1997) Serum LH concentrations in hypogonadal men during transdermal testosterone replacement through scrotal skin: further evidence that ageing enhances testosterone negative feedback. Clin Endocrinol 47:317–322

Woolf PD, Hamill RW, McDonald JV, Lee LA, Kelly M (1985) Transient hypogonadotropic hypogonadism caused by critical illness. J Clin Endocrinol Metab 60:444–450

Worstman J, Eagleton LE, Rosner W, Dufau ML (1987) Mechanism for the hypotestosterone-mia of the sleep apnea syndrome. Am J Med Sci 293:221–225

Wu FCW, Farley TMM, Peregourdov A, Waites GMH and the World Health Organisation task force on methods for the regulation of male fertility (1996). Effects of exogenous testosterone in normal men – Experience from a multicenter efficacy study using testos-terone enanthate. Fert Steril 65:626–636

Zmuda JN, Thompson PD, Dickenson R, Bausserman LL (1996). Testosterone decreases lipoprotein (a) in men. Am J Cardiol 77:1244–1247

Zumoff B, Bradlow L, Finkelstein J, Boyar RM, Hellman L (1976) The influence of age and sex on the metabolism of testosterone. J Clin Endocrinol Metab 42:703–706

17 Androgen therapy in non-gonadal disease

Peter Y. Liu and David J. Handelsman

Contents

17.1 Introduction

Systemic non-gonadal disease strongly influences all four dimensions of male reproductive health – sexuality, virilisation, fertility and ageing – although these effects are not always immediately apparent. Even during ill-health, men value their reproductive health and this includes a deeply-held but often unarticulated belief that their sexual function and fertility will be preserved. Recognising such unstated but important expectations is necessary for providing satisfying medical care. Consequently the impact of systemic disease and its management on male reproductive health needs to be much better understood.

This review will focus on the potential role for androgen therapy as an adjunct to standard medical care. Androgen therapy must be considered as either physiological androgen replacement or pharmacological androgen therapy. The key distinction is that androgen replacement therapy aims to replicate (but not exceed) tissue androgen exposure of healthy eugonadal men, whereas pharmacological androgen therapy aims to utilise androgens to their maximal efficacy within adequate safety limits. This distinction imposes tight dosage restrictions for androgen replacement therapy; by not exceeding endogenous androgen exposure, the expectation for safety may be regarded as comparable with the benchmark of the life-long health experience of eugonadal men. Thus androgen replacement therapy is judged by how well it replicates endogenous testosterone concentrations, thereby limiting it to using testosterone in physiological doses. In contrast, pharmacological androgen therapy is no different from pharmacotherapy with any xenobiotic drug used to achieve a therapeutic goal. This requires establishing rigorously the empirical safety profile that is circumscribed by the scope of the controlled follow-up studies completed. Pharmacological androgen therapy is judged by the efficacy, safety and cost-effectiveness standards applicable to any drug without any *a priori* limitations on dose or class of androgen. Most applications of androgen therapy have utilised synthetic androgens rather than testosterone itself. This has been based either on the oral activity of 17α-alkylated androgens or on the physiologically misguided distinction between anabolic and androgenic effects which is now considered obsolete. The usefulness of testosterone itself in more convenient formulations or of selective androgens (based on tissue-specific activation to aromatised and/or 5α-reduced metabolites) warrants further consideration.

The goals of androgen therapy for non-gonadal disease must be considered in relation to the natural history of the underlying disease. Androgen deficiency is a common accompaniment of systemic disease and its management may contribute to morbidity arising from the underlying disease. Since complete androgen deficiency due to congenital androgen resistance (Quigley et al. 1995) or early-life castration (Nieschlag et al. 1993) does not reduce life expectancy, androgen replacement therapy in itself is unlikely to influence mortality. Hence most studies of androgen therapy in systemic disease utilise pharmacological androgen therapy aiming to modify the natural history of the underlying disease (in either mortality or morbidity) and must be judged by their efficacy, safety and cost-effectiveness usually in comparison with placebo. Similarly in reconciling the potential risks and benefits of androgen therapy for non-gonadal disease, the major potential risks comprise negative effects on the natural history of the disease. Depending on the prognosis of the underlying disease, the long-term hazards of androgen therapy applicable to otherwise healthy men considering androgen-based treatment for ageing or hormonal male contraception, such as acceleration of prostate or cardiovascular disease, are less significant. Conversely, however, the benefits – which are achieved much more quickly – remain similar in comprising effects on bone, muscle, cognitive function and quality of life.

In order to promote evidence-based andrology, this review will focus on controlled clinical studies reported over the last three decades, rather than simply recounting the plethora of studies performed in the six decades since testosterone became available for clinical treatment (Foss 1939; Hamilton 1937). A comprehensive account of early, mostly uncontrolled studies of androgen therapy up to the mid-1970's is contained in two classical textbooks (Kopera 1976; Kruskemper 1968). Recognising there are few if any well-established indications for pharmacological androgen therapy, placebo controls are an essential requirement for high-quality studies. In addition such studies should have adequate power and duration and utilise objective end-points in order to define the role of androgen therapy. Unfortunately, very few studies in the biomedical literature fulfil these basic requirements, with the remainder of anecdotal, observational or poorly controlled studies contributing mainly to confusion. This review will exclude disorders of the reproductive system (including puberty) or ageing, laboratory models and/or observational studies of systemic disease effects on male reproductive health, which are reviewed elsewhere (Handelsman 1997).

17.2 Liver disease

Androgen therapy for acute or chronic liver disease has been studied extensively in attempts to modify the natural history of the underlying liver disease, mostly via nutritional improvement and/or promotion of liver regeneration. Virtually all major studies have examined alcoholic liver disease and there are no controlled therapeutic studies of androgen therapy in non-alcoholic liver disease or after liver transplantation. Furthermore, the utility of androgen replacement therapy for its effects on bone, muscle, cognition, psychosexual function or quality of life have not been examined.

17.2.1 Cirrhosis

The earliest controlled studies of androgen therapy to ameliorate the natural history of alcoholic cirrhosis claimed a survival benefit in 26 men treated with testosterone propionate (100 mg alternate days for ~4 weeks) compared with 27 placebo-treated men (Wells 1960). This study had defective randomisation and could not be replicated in another study of 17 men treated with either 100 mg of testosterone propionate or methenolone acetate every second day for one month. They had no survival advantage after 6 months compared with 10 placebo-treated controls (Fenster 1966). Neither study was large nor long enough to be definitive. The best evidence is derived from the Copenhagen Study Group for Liver Disease, which enrolled 221 men with alcoholic cirrhosis in a three-year prospective double-blinded, randomised, placebo-controlled study testing oral micronised testosterone (600 mg daily). This study showed convincingly no benefit in mortality (Copenhagen Study Group for Liver Diseases 1986), hepatic histology (Gluud et al. 1987a), liver hemodynamics and biochemical function (Gluud et al. 1987c) or improvement in sexual dysfunction (Gluud et al. 1988b). The comprehensively negative outcome with sufficient power to exclude a 35% decrease in mortality was at variance with many enthusiastic but poorly controlled previous reports (Kopera 1976). The observation of portal vein thrombosis in three men treated with testosterone may be related to the extreme portal testosterone levels created by oral administration of very high androgen dosage. Characteristic of testosterone pharmacokinetics in chronic liver disease (Nieschlag et al. 1977), this regimen produced markedly supraphysiological peripheral blood testosterone concentrations (Gluud et al. 1987b) which suggests even more extreme portal testosterone concentrations.

17.2.2 Hepatitis

A prospective randomised multi-centre Veterans Administration study claimed a mortality benefit after 30 days of oxandrolone treatment (80 mg daily), compared with placebo in 263 men presenting with alcoholic hepatitis (Mendenhall et al. 1984). The poorly defined entry and end-point definitions have been criticised (Maddrey 1986) and the benefits, if any, appeared to be short-term. The same authors reported a further study of 271 poorly nourished men with alcoholic hepatitis randomised to treatment with oxandrolone plus high calorie food supplements compared with a group receiving placebo without dietary supplementation (Mendenhall et al. 1993). This study showed no overall survival benefit on an intention-to-treat analysis; however, subgroup analysis demonstrated a significant doubling of survival at one month which persisted for six months in those with "moderate" but not severe malnutrition at entry. Due to the study design, the benefit of androgen therapy relative to enhanced nutrition could not be resolved. Another randomised controlled study of 19 men and 20 women with alcoholic hepatitis treated with 80 mg oxandrolone, parenteral nutrition, both or neither for 21 days demonstrated modest improvement in hepatic biochemical function but did not report other clinical end-points (Bonkovsky et al. 1991 a; 1991 b).

Overall androgen therapy appears unlikely to provide a significant reduction in mortality from acute or chronic alcoholic liver disease. The effects of androgen therapy on non-alcoholic liver disease or on the morbidity of chronic liver disease (including secondary androgen deficiency) have been little studied and warrant further investigation. In balancing the risks and benefits of androgen therapy, the hepatotoxicity of 17α-alkylated androgens suggests that, where possible, other safer oral and parenteral androgens should be preferred.

17.3 Hematological Disease

17.3.1 Anemia due to marrow failure

In severe aplastic anemia, a major study of 110 patients compared HLA-identical marrow transplantation with oral, intramuscular or no androgen therapy (Camitta et al. 1979). This showed a major survival advantage (70% vs 35% six month survival) for 47 patients having HLA-identical bone marrow transplantation compared with 63 patients in whom no donor was available who were randomised to oral (oxymetholone 3–5 mg/kg/day), intramuscular (nandrolone decanoate 3–5 mg/kg/wk) or no androgen therapy (Camitta et al. 1979). The latter three groups did not differ in survival, a finding consistent with another small randomised study that showed no survival benefit due to androgen therapy (50–100 mg nandrolone phenylpropionate weekly) compared with placebo vehicle injections (Branda et al. 1977).

In standard non-transplantation treatment for aplastic anemia, a randomised cross-over study of 44 patients which concluded that anti-thymocyte globulin (ATG) was superior to androgen therapy (nandrolone decanoate 5 mg/kg/wk) was, however, flawed as half the patients had failed prior androgen therapy, thus constituting an entry bias against androgen therapy (Young et al. 1988). Coupled with ATG, androgen therapy appears to offer morbidity but not mortality benefit in aplastic anemia. A randomised, controlled multi-centre study of the European Bone Marrow Transplantation in Severe Aplastic Anemia Study Group of 134 patients with newly diagnosed severe aplastic anemia receiving standard therapy (including ATG and methylprednisolone) demonstrated an improvement in transfusion independence due to treatment with oxymetholone (2 mg/kg/day) compared with placebo (Bacigalupo et al. 1993). However, there was no overall benefit in survival that was determined principally by the severity of disease based on leucocyte count. These findings confirmed the benefit of androgen therapy on transfusion independence but not survival from two smaller randomised placebo-controlled studies involving 61 patients using oral methenolone acetate (2–3 mg/kg/day) (Kaltwasser et al. 1988; Li Bock et al. 1976) but contradict another randomised placebo-controlled study which found no benefit from androgen therapy (fluoxymesterone 25 mg/m^2/day or oxymetholone 4 mg/kg/day) over placebo in 53 patients (Champlin et al. 1985). All studies observed female virilisation frequently but no formal evaluation of quality of life was reported.

An important pair of studies attempted to define the optimal dosage and type of androgen therapy for aplastic anemia (French Cooperative Group for the Study of Aplastic and Refractory Anemieas 1986). In the first study, 110 patients were randomised into four groups according to androgen (norethandrolone, fluoxymesterone) and dose (high 1 mg/kg/day, low 0.2 mg/kg/day). Survival was mainly influenced by disease severity but in less severe cases high-dose androgen therapy significantly improved survival over low-dose androgen therapy. Despite randomisation, there were imbalances between treatment groups with respect to disease severity and age that undermine the interpretability of the findings. In the second study, 125 patients were randomised to four different androgens – norethandrolone, stanozolol, fluoxymesterone (all at 1 mg/kg/day) or testosterone undecanoate (1.7 mg/kg/day). The fluoxymesterone treatment group had the best and stanozolol the worst survival with norethandrolone and testosterone undecanoate being equivalent and intermediate in efficacy. Once again, however, the treatment groups were unbalanced with respect to disease severity and age and the apparent benefit was restricted to the less severe and older (>30 yr) cases. The superiority of any specific androgen remains to be unequivocally demonstrated and, although in one study higher androgen doses appear to have advantages, it remains unclear whether differences between androgens are related to class (oral vs injectable, 17α-alkylated or not) or simply dose.

The French Cooperative Study Group has also performed a series of cohort studies examining the efficacy of androgen therapy in patients with aplastic anemia. Their initial cohort randomised 352 men and women to treatment with methandrostenolone (1 mg/kg/day), oxymetholone (2.5 mg/ kg/day), methenolone acetate (2.5 mg/kg/day) or norethandrolone (1 mg/kg/ day). The methandrostenolone group had the best, whereas oxymetholone and methenolone groups exhibited equally the worst two-year survival from randomisation (Cooperative Group for the Study of Aplastic and Refractory Anemias 1979). Unfortunately treatment groups were unbalanced for disease severity which is the principal determinant of survival. Despite post-hoc stratified analyses, ultimately in the absence of *a priori* stratified randomisation it remains difficult to conclude whether underlying disease prognosis or drug effects explained the differences in group survival. In a follow-up study from the same cohort who survived at least two years from initial randomisation, 137 patients were re-randomised to rapid (3 month) or slow (20 month) withdrawal of their original androgen therapy. The slow withdrawal group had a higher rate of maintained remission consistent with androgen therapy, having maintained a clinical benefit, presumably via maintenance of hemoglobin levels, although survival was not specifically reported (Najean and Joint Group for the Study of Aplastic and Refractory Anemias 1981).

Overall, androgen therapy does not improve survival in aplastic anemia but provides a morbidity benefit by maintaining hemoglobin and transfusion independence, but the degree of improvement in quality of life has not been quantified. In severe aplastic anemia bone marrow transplantation from an HLA-identical sibling (if feasible) is the preferred treatment and superior to androgen therapy. Androgen therapy may be useful in less severe aplastic anemia for which bone marrow transplantation is not available or justified. The relative merits of androgen therapy compared with HLA non-identical bone marrow transplantation have not been clearly defined. Similarly a role for androgen therapy with failing or failed bone marrow transplantation remains to be clarified. The preponderant use of oral 17α-alkylated androgens in aplastic anemia appears unjustified. Although it is unquestionably prudent to avoid injectable androgens in a population that may be thrombocytopenic, non-hepatotoxic oral androgens such as 1-methyl androgens (methenolone, mesterolone) and testosterone undecanoate appear to be equally effective as 17α-alkylated androgens without the risk of hepatotoxicity.

17.3.2 Myeloproliferative disorders

The use of androgen therapy in other causes of bone marrow failure has been less extensively studied. One controlled study of 29 patients with myeloproliferative disorders randomised patients to treatment with fluoxymesterone (30 mg daily) compared with transfusions alone but was terminated prematurely due to slow recruitment and poor hemoglobin response with only 4/14 achieving an increase of > 10 g/l (Brubaker et al. 1982). These findings are supported by another randomised study of 56 patients with myelodysplasia which found oral methenolone acetate (2.5 mg/kg/day) no better than intravenous cytosine arabinoside or symptomatic maintenance therapy (Najean and Pecking 1979).

17.3.3 Thrombocytopenia

A beneficial effect of androgen therapy in thrombocytopenia due to marrow failure has been suggested by a study in myelodysplasia associated with thrombocytopenia in which 20 patients were randomised to receive either danazol (600 mg daily) or fluoxymesterone (1 mg/kg/day). Although both groups had an impressive response in termination of clinical bleeding (6/6) and increasing platelet count (11/20), the lack of a placebo group means that the contribution of natural remission could not be evaluated (Wattel et al. 1994).

The role of androgen therapy in immune thrombocytopenic purpura (ITP) remains poorly defined in the absence of controlled clinical trials. Two short-term observational studies have reported that danazol increases platelet counts in ITP as well as decreasing prednisone requirement (Ambriz et al. 1986) and reducing platelet-reactive IgG (Ahn et al. 1983). Danazol is regarded as especially beneficial in premenopausal women with menorrhagia in whom the associated oligo/amenorrhea adds to its efficacy (Ambriz et al. 1986).

17.4 Renal disease

Although gonadal dysfunction is a consistent feature of end-stage renal disease (Handelsman 1985; Handelsman and Dong 1993), few studies have evaluated androgen replacement therapy in patients with end-stage renal disease, during dialysis or after renal transplantation.

17.4.1 Anemia of end-stage renal failure

Androgen therapy has been shown to increase hemoglobin in patients with end-stage renal failure in two well-controlled studies. One randomised 21 men to nandrolone (100 mg weekly) or placebo vehicle injections for 5 months in a cross-over design (Hendler et al. 1974), while another randomised 18 patients to nandrolone decanoate (200 mg weekly) for three months (Williams et al. 1974). Both found significant increases in mean hemoglobin (15 g/L and 10 g/l, respectively) and one reported a clinically significant decreased transfusion requirement (Hendler et al. 1974). A further study confirmed the beneficial effects of nandrolone decanoate (200 mg weekly) compared with placebo vehicle injections for four months (Buchwald et al. 1977), whereas three smaller and less well-conducted studies failed to show an increase in hemoglobin (Li Bock et al. 1976; Naik et al. 1978; van Coevorden et al. 1986). A further randomised, controlled clinical study compared four androgen regimens in dialysed patients, finding that testosterone enanthate (4 mg/kg/wk) and nandrolone decanoate (3 mg/kg/wk) were more effective in

increasing hematocrit than oxymetholone (1 mg/kg/day) and fluoxymesterone (0.4 mg/kg/day). However, whether these differences reflected different
effective androgen doses, the androgen class (17α-alkylated or not) or route
of administration (including pharmacokinetics) remains unclear (Neff et al.
1981).

The effect of androgen therapy on hemoglobin involves both increased circulating erythropoietin (Epo) concentration (Buchwald et al. 1977) and augmentation of Epo action (Ballal et al. 1991). The role of androgen therapy
relative to Epo in patients with end-stage renal disease is complex, particularly since the advent of recombinant human Epo. Direct comparisons between
androgen therapy and Epo in well-controlled clinical trials are not available.
One prospective study found very similar hemoglobin responses and safety
profiles for 18 men over 50 yrs of age treated with androgen (nandrolone decanoate 200 mg/wk) compared with six men under 50 yrs and 16 women receiving Epo (6000 U/wk); however, the lack of randomisation and non-com-
parability of groups by age and gender limits the interpretation of these findings (Teruel et al. 1996b). Other studies have examined the effects of androgen therapy in combination with Epo. One small non-randomised study reported significantly greater hematocrit responses among eight men choosing
to receive nandrolone decanoate (100 mg weekly) plus Epo (6000 U/wk) compared with Epo alone (Ballal et al. 1991). Another small but randomised
study employing a higher Epo dose (120 U/kg/wk), however, was unable to
detect any benefit of nandrolone decanoate (2 mg/kg/wk) for 16 weeks plus
Epo compared with the same dose of Epo alone in 12 dialysed patients
(Berns et al. 1992). Whether these discrepancies are due to study design or
Epo dose remains to be clarified.

In general, although androgen therapy is considerably cheaper, its efficacy
is modest but consistent and adverse effects, notably virilisation of women
and children and hepatotoxicity of 17α-alkylated androgens are significant
limitations. Restricting use to non-17α-alkylated androgens in men, particularly at advanced ages (Teruel et al. 1996a), may avoid most of the hazards.
Present practice appears to suggest the role of androgen therapy in end-stage
renal failure is limited to patients in whom Epo is contraindicated or unavailable. Whether androgen therapy is more effective according to the degree of
androgen deficiency induced by renal failure remains to be established, as do
any potential benefits of androgen replacement therapy in end-stage renal
failure.

17.4.2 Growth

One small double-blind, placebo-controlled cross-over study examined the
effects of testosterone on short-term growth in boys with short stature on
hemodialysis (Kassmann et al. 1991). After an 8 week run-in, 8 boys (mean
3.9 SD below mean height for age) on regular hemodialysis were randomised
to start on one of two 4-week treatment periods separated by a 6-week wash-

out period before crossing over to the other treatment. Treatment consisted of 2 gm/m^2/day of a transdermal gel corresponding to a topical daily dose of 50 mg/m^2/day testosterone or placebo. Although a significant increase in short-term growth velocity (using knemometry) was reported overall, the randomisation was unbalanced, with the four boys with subnormal pretreatment growth velocity all allocated (by chance) to start on active treatment and showing a therapeutic benefit over placebo. The other four boys with normal pretreatment growth velocity, however, were all randomised to start with placebo and showed no significant acceleration of growth velocity. No acceleration of bone maturation was evident, presumably due to the short-term study using a testosterone dose that maintained blood testosterone concentrations appropriate for age and pubertal stage. The small sample size and unbalanced randomisation makes it difficult to interpret the short-term growth results of this study. Since it is now clear that accelerating short-term growth of non-growth hormone deficient children does not predict any gain of final height, the clinical significance of these findings remains unclear. Further larger and longer studies would be needed before even low-dose androgen therapy could be considered effective or safe.

17.4.3 Enuresis

Following suggestions from the 1940's that androgen therapy might improve childhood enuresis, a recent controlled clinical trial involving 30 boys aged 6–10 yrs has claimed a benefit for oral mesterolone treatment compared with placebo (El-Sadr et al. 1990). This study may have been flawed as the method of randomisation leading to 20 being treated with mesterolone (20 mg daily for 2 weeks) compared with 10 on placebo (vitamin C) was not explained. The statistically significant increase in cystometric bladder capacity in mesterolone-treated group was attributable to six boys who had dramatic increases, whereas the remainder did not differ from the ten placebo-treated boys. Although no adverse effects were reported, the well-known potential hazards of androgen therapy in prepubertal children, including premature closure of epiphyses and short stature, precocious sexual maturation and psychological sequelae would require detailed safety evaluation before androgen therapy could be considered acceptable for a benign functional disorder with favourable natural history in otherwise healthy children.

17.5 Bone disease

17.5.1 Idiopathic osteoporosis

Clinical trials in osteoporosis have the unique limitation that fracture endpoint studies are frequently very prolonged due to the low rate of fractures even among high-risk groups. This leads to the need for very large, long-term studies to estimate reliably anti-fracture therapies. For example, a randomised, placebo-controlled, triple-dummy, double-blind study involving 327 patients for nine months treatment and 1 year of follow-up proved to have very low power to detect effects of androgen therapy (methandienone 2.5 mg daily) on fracture rates (Inkovaara et al. 1983). Even studies of accepted surrogate endpoints such as bone mineral density usually require two years to record significant therapeutic effects given the precision of the densitometric equipment, together with the slow metabolic turnover of bone. Nevertheless, androgen therapy has a long established role in treatment of osteoporosis. This may include therapeutic effects beyond increases in bone density, as the muscular and motivational effects of androgens may reduce risk of falls due to frailty which are a major contributory factor to osteoporotic fractures. Recent studies demonstrating a congenital osteodystrophy with low mature bone density in men (Smith et al. 1994) and mice (Lubahn et al. 1993) with mutated, non-functional estrogen receptor have been interpreted to show that aromatisation is necessary for androgen action on mature male bone. These animals still have intact, functional estrogen receptor and may therefore not be estrogen resistant. Furthermore, the blood estradiol concentrations in healthy young men which are extremely low compared with age-matched women as well as the strikingly positive effects of non-aromatisable androgens on mature bone density are inconsistent with androgen effects on mature bone requiring aromatisation.

The clearest evidence for the efficacy of testosterone in osteoporosis is derived from a study of 34 postmenopausal women randomised to estradiol (50 mg) or estradiol (50 mg) plus testosterone (50 mg) implants each six months for two years which showed a clear increase in hip and lumbar spine bone mineral density (Davis et al. 1995). This is supported by a study of 20 postmenopausal women treated with oral estrogens of whom 10 chose to switch to implanted estradiol (75 mg) plus testosterone (100 mg) which one year later had increased hip bone mineral density (Savvas et al. 1992). This non-randomised study, however, did not optimise estrogen therapy, making its findings difficult to interpret. As virilisation is the dose-limiting toxicity in treatment of postmenopausal osteoporosis, androgen doses well below androgen replacement levels for men have usually been utilised. For example, one randomised study treated 45 postmenopausal women with six months of androgen therapy (nandrolone decanoate 50 mg monthly) or six months of placebo injections in a cross-over design (Need et al. 1993). Even this short-term study showed a significant increase in forearm (but not lumbar spine) bone mineral density, although with virilisation reported by ~50% of patients. Supportive findings are reported in another randomised study of 60 patients with symptomatic osteoporosis in whom nandrolone decanoate (50 mg each 3 weeks) significantly increased radial bone mineral density more than 1-hydroxyvitamin D_3 or calcium infusion (Geusens and Dequeker 1986). Although the beneficial effect of low-dose androgen therapy is well established in women, its place among the expanding repertoire of anti-resorptive therapies including estrogen replacement therapy, bisphosphonates, calcitriol and calcium remains uncertain. These findings do, however, raise the question of how much more effective full androgen replacement doses of nandrolone would be in eugonadal men with osteoporosis.

The prevalence and public health significance of osteoporosis in ageing men has been underestimated. In some population-based studies the age-specific prevalence of osteoporotic fractures may be higher for men (Santavirta et al. 1992) although for osteoporosis as well as many other medical conditions women utilise medical services more frequently. Nevertheless, randomised controlled studies of adequate design and size to evaluate any therapies for male osteoporosis have not been reported and are needed.

17.5.2 Steroid-induced osteoporosis

One secondary cause of osteoporosis amenable to preventative androgen therapy is bone loss due to high dose glucocorticoid therapy used for anti-inflammatory purposes. A recent study of glucocorticoid-dependent men reported that testosterone may reverse the bone loss due to high-dose glucocorticoid therapy (Reid et al. 1996). In this study 15 glucocorticoid-dependent men with severe asthma were randomly allocated to monthly testosterone injections (250 mg mixed testosterone esters) or no treatment for 12 months, with the control group crossing over to testosterone treatment for the second 12 month period. In this study, 12 months of testosterone treatment increased lumber spine bone mineral density by 5% compared with no change on placebo. Although unblinded, this study of a low (sub-replacement) dose of testosterone still demonstrated a striking and clinical significant increase in bone mass within the relatively short period of one year. Further studies of more appropriate doses of testosterone or other androgens would be of great interest.

17.6 Neuromuscular disease

A recent well-designed and conducted study demonstrated that supraphysiological androgen therapy (testosterone enanthate 600 mg weekly for 10 weeks) increases muscle mass and strength in healthy eugonadal men compared with placebo injections after stratifying for exercise training and controlling for nutrition (Bhasin et al. 1996). The durability of these responses, their dose-dependency and their applicability to men with systemic disease remain, however, to be determined as does whether these findings have any useful application to men with neuromuscular or wasting disorders.

17.6.1 Muscular dystrophies

The effects of androgen therapy on neuromuscular disorders have been best studied by Griggs et al. in a series of careful studies of myotonic dystrophy (MD), a genetic myopathy recently identified as being due to a trinucleotide (CTG) repeat mutation in the gene for myotonin, a protein kinase. MD is associated with testicular atrophy and disproportionate hypogonadism. This was illustrated by a study of 22 men with myotonic dystrophy who had lower testosterone and higher gonadotropin concentrations than age-matched groups of 36 healthy men and 16 men with muscle wasting due to other neuromuscular disorders (Griggs et al. 1985). Serum testosterone concentrations did not correlate with degree of muscle wasting, however. Since life expectancy in MD is determined by respiratory muscular weakness leading to terminal pneumonia, androgen therapy might reverse hypogonadism and/or improve muscular strength and hence prolong life in this condition. In order to determine whether androgen therapy could provide any clinical benefit, a randomised placebo-controlled study was undertaken in 40 men with MD who were treated with either testosterone enanthate (3 mg/kg) or placebo injections each week for 12 months (Griggs et al. 1989). In a well-designed two-site study, muscle mass was increased as indicated by creatinine excretion and total body potassium, but there was no difference in quantitative measures of manual or respiratory muscle strength. Crucially the lack of improvement in pulmonary function implies that mortality benefits would be unlikely.

These findings make it seem improbable that the non-specific increase in muscle mass due to androgen therapy would have any real impact on the natural history or morbidity from this disorder. Androgen therapy may simply increase the mass of dysfunctional muscle. Whether the same can be extrapolated to other forms of genetic or degenerative myopathies or neuromuscular disorders would require further evaluation.

17.6.2 Headache

The role of androgen withdrawal and therapy in men with cluster headache, an almost exclusively male disorder, has been examined in two controlled studies. In one 60 men with chronic cluster headache were randomised single-blind to treatment with a single dose of a GnRH analog (3.75 mg leuprolide depot) or vehicle injection (Nicolodi et al. 1993a). Self-reported frequency, intensity and duration of headache as well as sexual activity declined progressively during three successive ten-day periods after injection compared with pre-injection baseline in those treated with leuprolide, whereas there was no change in placebo-treated men. The therapeutic response was delayed in onset corresponding temporally to the onset of castrate testosterone concentrations and the benefit persisted in most men for the one-month post-treatment follow-up period, while no changes were noted at any stage in the placebo group. As headache is a remitting illness with subjective study endpoints, the unmasking of active drug by the regular occurrence of sexual dysfunction in the treated group undermines the validity of the placebo control group. The surprising absence of a placebo effect in the intended control group reinforces the possibility of an observer bias. Subsequently, another study of 12 men with chronic cluster headache and 12 non-headache controls who underwent treatment with very high dose androgen therapy (testosterone propionate 100 mg daily) for 14 days (Nicolodi et al. 1993b). Remarkably, this produced a dramatic increase in self-reported sexual activity in the cluster headache, but not the control, group. These curious findings warrant more rigorous study with a double-blind study design utilising more objective end-points.

17.6.3 Depression

The role of androgens in mood has also long been debated. A double-blind laboratory study showing that a single dose of mesterolone (1–25 mg) mimics the effects of tricyclic antidepressants on the electroencephalogram led to a patent predicting that androgens might have beneficial effects on clinical depression (Itil et al. 1974). This was, however, refuted in a double-blind clinical trial which randomised 52 depressed men to treatment with mesterolone (150–450 mg daily) or placebo for 6 weeks (Itil et al. 1984). Both groups improved equally in scores for global clinical impression, physician's checklist for depression, self-rating and Hamilton depression rating. There were no differences in electroencephalogram measures or plasma monoamine oxidase levels. The abundant development of modern pharmacotherapy for depression leaves little scope for further studies of androgen therapy for depression in eugonadal men.

17.7 Rheumatological diseases

17.7.1 Hereditary angioedema

The efficacy of oral 17α-alkylated androgens in hereditary angioedema was established by a small, double-blind, placebo-controlled randomised crossover study (Spaulding 1960) in which six members of a single family received multiple periods of treatment or placebo. This study clearly demonstrated the efficacy of oral methyltestosterone in reducing the frequency of attacks well before the disease pathogenesis was understood. Subsequent studies have confirmed these observations and showed that androgen therapy increases C1-esterase inhibitor concentration partially rectifying the underlying biochemical deficiency responsible for the disorder (Sheffer et al. 1977). Although other 17α-alkylated oral androgens such as fluoxymesterone, oxymetholone and stanozolol have been used, danazol has become standard prophylactic therapy. This followed a randomised double-blind cross-over study which showed increased blood C1-esterase inhibitor concentration together with a dramatic decrease (94% vs 2%) in attack-free 28 day periods using 600 mg danazol daily compared with placebo in 93 courses among nine patients (Gelfand et al. 1976). Danazol doses are tapered to minimal levels that maintain adequate control of attack and this dose minimisation may explain the anecdotal impression that such danazol therapy has minimal effects on male fertility although quantitative studies have not been reported. Recent studies suggest that stanozolol (1–2 mg daily) is about as effective as danazol (50–200 mg daily) but, despite their efficacy, hepatotoxicity and female virilisation remain problems (Cicardi et al. 1997, Hosea et al. 1980). While it is assumed that the beneficial effects of androgen therapy for angioedema are only exhibited by 17α-alkylated androgens, only very limited studies of non-17α-alkylated androgens such as nandrolone, 1-methyl androgens or testosterone (Spaulding 1960) have been reported. Since angioedema requires lifelong prophylaxis, studies of non-hepatotoxic androgens should be undertaken.

17.7.2 Rheumatoid arthritis (RA)

The rationale for androgen therapy in RA is that
- the lower prevalence in men suggests a protective role for androgens,
- active disease is associated with reduction in endogenous testosterone production,
- androgen effects on muscle and bone may improve morbidity in RA and
- androgen effects (e.g. fibrinolysis) may reduce disease activity.

The best designed and conducted study of androgen therapy involved 107 women with active RA (ACR criteria) on stable standard (steroid, NSAID) treatment for at least 3 months who were randomised to treatment with fortnightly injections of either androgen therapy (testosterone propionate 50 mg plus progesterone 2.5 mg) or placebo for one year (Booij et al. 1996). The inclusion of a very low dose of progesterone, which the authors claim was biologically ineffective, was based on an old clinical practice aiming to reduce virilisation from testosterone. Evaluated on a double-blinded, intention-to-treat basis this study demonstrated significant improvement in the ESR, pain and disability scores and ACR improvement criteria, but not in the numbers of tender or swollen joint or joints requiring intra-articular steroid injections. There was a high dropout rate (39/107), mostly (28/39) due to inefficacy defined as any mid-study increase in anti-rheumatic medication; however, these were evenly distributed between treatment groups. As expected, virilisation was the major adverse effect reported but there were few other side-effects and tolerability was good as most androgen-treated patients (67% vs 37% on placebo) wished to continue their allocated medication at the end of the study. The significant benefits of androgen therapy over placebo were predominantly in subjective measures rather than objective signs of disease activity. This raises the possibility that androgen therapy may preferentially improve mood or tolerance of disability rather than actually modifying disease. This thorough study is a model for future investigation of the role of androgen therapy in systemic disease.

Other studies of androgen therapy in RA are small, poorly designed and inconclusive. One uncontrolled study of seven men with RA treated with six months of androgen therapy (oral testosterone undecanoate 120 mg daily) observed a decline in disease activity (reduced numbers of tender joints and analgesic usage) together with minor immunological changes uncorrelated with disease activity during androgen therapy and some patients subsequently relapsed after completion of the study. However, the lack of a placebo group in a disease with a remitting natural history renders such observations unconvincing (Cutolo et al. 1991). A larger study of 35 men with definite RA randomised them to injections of testosterone enanthate (250 mg monthly) or placebo for nine months (Hall et al. 1996). This study noted that overall disease activity (defined by biochemical variables and clinical scales) was not improved by androgen therapy and indeed, significantly more men on testosterone therapy experienced disease 'flare' during the study. This study, however, was poorly designed to include men with inactive RA as well as utilising an inadequate testosterone dose that had to be doubled after six months so that only three months observation at effective androgen dosage was possible. An older double-blind study randomised 40 patients with definite RA on stable NSAID to treatment with stanozolol 10 mg daily or placebo for 6 months on the basis that androgen therapy might increase fibrinolysis (Belch et al. 1986). This study found a significant improvement in the composite Mallaya disease activity index combining objective (ESR, hemoglobin, articular scores) and subjective (pain, morning stiffness) dimensions, despite the failure to influence measurable fibrinolysis. Adverse effects appeared under-reported although the authors warned about hepatotoxicity and female virilisation.

The role of androgen therapy in men with RA requires further investigation with better designed and powered studies to determine whether androgens can modify the natural history of the underlying disease and/or whether they improve perception and toleration of disease. No controlled studies examining whether improvement in muscle strength, bone density and other androgen-sensitive variables improve morbidity or quality of life in RA have been reported.

17.7.3 Other rheumatological disorders
(SLE, Raynauds, systemic sclerosis, and Sjogrens disease)

Few well-controlled studies of androgen therapy have been reported in men with other rheumatological disorders. This is not just due to paucity of cases in the female-preponderant autoimmune diseases as there are no controlled studies of androgen therapy even in ankylosing spondylitis or gout, the male preponderant rheumatological diseases.

In systemic lupus erythematosus, only two small uncontrolled studies (including together five men among 17 patients) using androgen therapy (nandrolone decanoate) have been reported (Hazelton et al. 1983; Lahita et al. 1992). This information is so limited that no conclusions can be drawn without larger and better-designed studies. Another double-blind study randomised 28 women with mild to moderate SLE to treatment with DHEA (200mg daily) or placebo for 3 months. Treatment with this weak androgen precursor did not improve SLE disease activity index, number of flares, prednisone usage or physician overall assessment, although there was an improvement in the patients' overall assessment of well-being (Van Vollenhoven et al. 1995).

One study has examined the effects of treatment with stanozolol (10 mg daily) or placebo for 24 weeks in primary Raynaud's phenomenon and systemic sclerosis (Jayson et al. 1991). Although 43 patients (19 Raynaud's, 24 systemic sclerosis; including only 4 men) entered, only 28 patients (11 Raynauds, 17 systemic sclerosis) completed the study. Compared with placebo, stanozolol significantly improved ultrasonic Doppler index as well as finger pulp and nail bed temperatures but there was no difference in reported frequency or severity of vasospastic attacks, scleroderma skin score or grip strength. The clinical significance of the changes in digital small vessel function recorded in the absence of vasospasm and without reduction in attack rates is unclear.

A more convincing double-blind study randomised 20 women with primary Sjogren's syndrome to treatment with androgen (nandrolone decanoate 100 mg fortnightly) or placebo for six months (Drosos et al. 1988). Androgen therapy did not produce any significant improvement over placebo in objective validated measures of xerostomia (stimulated parotid flow rate measurements, labial salivary gland histology), xerophthalmia (Schirmer's I test, slit lamp eye examination after rose Bengal staining) or systemic disease (ESR) although the subjective assessment of xerostomia by patients and physicians as well as overall patient's well-being assessment were significantly better on nandrolone. Virilisation was reported in nearly all nandrolone-treated women with this relatively high androgen dose but none discontinued for this reason. Again these studies reinforce the observations that androgen therapy may significantly improve feelings of well-being regardless of the underlying disease activity.

17.8 Critical illness, trauma and surgery

Any catabolic state – whether after surgery, trauma, illness or malnutrition – inevitably leads to muscle breakdown which is then rectified during recovery. Such catabolism is accompanied by a hypothalamic response creating a hypogonadotrophic, androgen-deficient state that may endure as long as the underlying catabolic state persists. This has long led to the hope that androgen therapy, whether in rectifying the conditional androgen deficiency or as pharmacological therapy to enhance the effects of nutritional supplementation or by other means, might improve mortality or morbidity associated with the catabolic state. For evaluation of androgen therapy, the objectives might be to improve mortality if the natural history can be modified or to enhance recovery and/or reduced complications. Such benefits would most likely arise from improved muscle and/or bone structure or function so that key outcome variables in evaluating the efficacy of androgen therapy in catabolic state would be muscle mass and strength. The only available information, however, comes from older studies restricted to surrogate endpoints for muscle such as improved nitrogen balance, the clinical significance of which is nebulous.

A number of studies have examined the effects of androgen therapy as an adjunct to elective surgery using improved nitrogen balance as their endpoint. The most comprehensive and best designed study randomised 60 patients after colorectal cancer surgery to receive either a single injection of stanozolol (50 mg) or no extra treatment. Participants were also randomised among three types of post-operative, peripheral-vein nutrition (standard dextrose-saline, amino acid supplementation or glucose-amino acid-fat mixture) and stratified by gender (Hansell et al. 1989). The primary endpoint was cumulative nitrogen balance for the first four post-operative days and this was consistently and significantly influenced only by nutritional supplementation. Stanozolol augmented nitrogen balance only on the third post-operative day in the group receiving amino acid supplements. This was largely attributable to its effects in women and was no improvement over standard post-operative care on other post-operative days, with other nutritional supplements or had any influence on a wide range of other metabolic variables. In addition neither convalescence nor complication rates were influenced by androgen or nutritional therapy.

These findings were largely confirmed by four other studies. The first randomised 44 men with tuberculosis requiring pulmonary resection to treatment with either high-dose norethandrolone (50 mg daily) or no extra treatment within strata of different intensity of postoperative hyperalimentation (Webb et al. 1960). This showed a modest, transient effect of androgen therapy on positive nitrogen balance restricted to the first three postoperative days which was absent during the second three postoperative days. The second study randomised 36 patients to one injection of stanozolol (50 mg) or placebo one day before surgery with similar outcomes (Blamey et al. 1984). A third study randomised 30 men after gastric surgery for duodenal ulcer (vagotomy/pyloroplasty) to a single post-operative injection of nandrolone decanoate (50 or 100 mg), parenteral nutrition, both, or to standard treatment (Tweedle et al. 1973). This study reported that the eight day post-operative nitrogen balance was best with the combination of nandrolone plus parenteral nutrition and that each alone was superior to standard treatment but no clinical outcome measures were reported. Finally, the fourth study randomised 20 patients recovering from multiple trauma to receive either nandrolone decanoate injections (50 mg on day 3 plus 25 mg on day 6) or no extra treatment. It found that nandrolone plus standard enteral or parenteral nutrition was superior to no extra treatment in nitrogen balance, urinary 3-methyl histidine excretion and amino acid retention for the first 10 days of hospitalisation (Hausmann et al. 1990). The only clinical outcome measure, however, was six-month survival, which did not differ according to androgen therapy.

Some studies have been unable to detect any clinical benefits. One well-designed study randomised 48 patients requiring hyperalimentation to supplemental treatment with either nandrolone decanoate (50 mg) or placebo injections biweekly aiming to determine whether nitrogen balance could be improved within the first 21 days postoperatively (Lewis et al. 1981). No benefit was observed in nitrogen balance, weight gain, creatinine output, and serum albumin or immune function. These negative findings were supported by another study that examined a higher nandrolone dose. This study randomised 24 patients requiring intravenous alimentation to nandrolone decanoate (100 mg before starting and repeated one week later) or no extra treatment and found increased fluid but not nitrogen balance and did not find any clinical benefits (Young et al. 1983).

Although controlled, few of these short-term studies of one or two doses of androgen therapy examined any important morbidity measures in clinical outcomes such as convalescence or complication rates. Longer studies with muscular strength and/or function as well as clinical outcome measures as primary end-points and clarifying the relationship with nutritional supplementation are needed before androgen therapy can be considered a useful adjunct to the care of critically ill men.

17.9 Immune disease – HIV/AIDS

Androgen therapy for HIV/AIDS has been mainly investigated for its effects on disease-associated morbidity (weight loss, weakness, quality of life) rather than to influence the underlying disease natural history. One rationale for androgen therapy stems from the observation that body weight loss is an important terminal determinant of survival in AIDS and other fatal diseases (Grunfeld and Feingold 1992). It has been estimated that death occurs when lean body mass reaches 66% of ideal (Kotler et al. 1989), leading to the proposition that if androgens (or other agents including megestrol or growth hormone) increased appetite and/or body weight, death may be delayed.

The only prospective, randomised placebo-controlled study of androgen therapy in HIV-positive men reported improved body weight and well-being in 63 HIV seropositive men suffering from wasting and weakness (Berger et al. 1996). This study randomised men to either 15 mg or 5 mg oxandrolone daily or placebo for 16 weeks. Both oxandrolone (but not control) groups demonstrated transient weight gain within the first month, peaking at the first week. Subsequently while the high-dose group maintained mean weight gain and the other groups less so, the within-group variance increased, suggesting major within-group heterogeneity in time-course. There was also no clear dose-response relationship. As body composition was not studied, the effects on lean body mass remain unknown, as does whether these findings would be replicated among HIV seropositive men without wasting. Another prospective non-randomised study of 60 AIDS patients who chose either 150 mg oxandrolone daily or no treatment for 30 weeks observed that the peak (but not average) body weight was increased (Hengge et al. 1996) but lean body mass was not studied. Most studies of androgen therapy for HIV/AIDS have, however, been uncontrolled and unconvincing (Rabkin et al. 1995; Gold et al. 1996). A double-blind, randomised placebo-controlled study showing the efficacy of recombinant growth hormone and involving 172 men with HIV/AIDS for 12 months demonstrates the feasibility of well-designed studies in this population (Schambelan et al. 1996). Despite this demonstration of efficacy, growth hormone is much more expensive than androgens and convincing studies to test the relative efficacy of androgen therapy are awaited.

Two large double-blind randomised, placebo-controlled clinical trials involving a total of 367 men and four women have demonstrated the efficacy of high-dose (600 mg daily) oral megestrol acetate as an appetite stimulant to cause weight gain and enhanced well-being (Oster et al. 1994; van Roenn et al. 1994). Both studies demonstrated an increase predominantly in fat rather than lean body mass, a finding not unexpected as a result of non-specific appetite stimulation. Megestrol also produces a major suppression in endogenous testosterone levels (Engleson et al. 1995), indicating these effects may be partly due to, or confounded by, androgen deficiency.

One well-controlled study has demonstrated convincingly that intralesional injection of commercial hCG caused macroscopic regression of 10/12 lesions in 6 men with AIDS-related Kaposi's sarcoma (Gill et al. 1996). The mechanism of action of intralesional hCG injection is unclear. One possibility, that it is due to bioactive peptide contaminants of commercial hCG preparations co-purified from pregnancy urine, is supported by differences between different commercial preparations and could be tested by examining the efficacy of recombinant hCG that is free of urinary contaminants. Despite the indirect evidence that the hCG injections increased endogenous testosterone production, it seems unlikely that the hCG effect is due to increased endogenous androgen concentrations.

17.10 Malignant disease

Androgen therapy in malignant disease can be considered either as adjunctive therapy to improve mortality or morbidity (maintenance of weight, hemoglobin, muscle or bone) by modifying the underlying disease process or as a cytoprotective regimen to reduce impact of oncological therapy on spermatogenesis and fertility.

17.10.1 Effects on morbidity and mortality

Androgen therapy could influence mortality from malignant disease via direct antitumour effects or improve morbidity by maintaining weight, hemoglobin, neutrophil count, muscle mass and bone mass through its known actions. Reduced morbidity may also augment treatment by creating greater tolerance for more aggressive cytotoxic therapy. Despite encouraging results from animal models and uncontrolled clinical reports, human studies are so far less convincing. Although older studies demonstrate a consistent but modest effect of androgen therapy in reducing the magnitude, duration and/ or complications from chemotherapy-induced neutropenia, few well-controlled clinical studies have shown unequivocal benefits of androgen therapy. The recent availability of recombinant human G-CSF/GM-CSF with its greater efficacy and better tolerability (albeit at greater cost) reduces the benefits from androgen-induced prevention of neutropenia to second-line status.

One open controlled study randomised 33 patients with lung or other non-hormone responsive solid cancers to standard chemotherapy plus nandrolone decanoate (200 mg weekly) or no additional treatment. In this study androgen therapy produced better maintenance of body weight, hemoglobin and less transfusion requirement, but no improved survival or physical performance (Spiers et al. 1981). Similarly, a cohort of 23 patients with inoperable lung cancer requiring palliative chest radiotherapy were randomised to receive or not to receive additional treatment with nandrolone phenylpropionate (loading dose 100 mg followed by 50 mg weekly during hospitalisation). During radiotherapy (4500 cGy), androgen therapy maintained higher hemoglobin and lower transfusion requirements (Evans and Elias 1972). In contrast, however, a third study failed to demonstrate a definite benefit of androgen therapy. In this study 37 patients with unresectable non-small cell lung cancer requiring standard combination chemotherapy were randomised to receive or not additional treatment with nandrolone decanoate (200mg weekly for four weeks). Androgen therapy was associated with only a non-significant statistical trend towards improved survival (median 8.2 vs 5.5 months) and less weight loss but no improvement in marrow function (Chlebowski et al. 1986). It is possible that the greater myelosuppression associated with the more aggressive modern combination chemotherapy negates any morbidity benefits.

Androgen therapy has also been trialed in maintenance therapy for acute non-lymphocytic leukemia (ANLL) on the basis that enhanced proliferation of residual normal hematopoietic precursors would suppress competitively the growth of the leukemic clones. Among 114/212 patients with newly diagnosed ANLL who obtained complete remission after standard induction chemotherapy, 82 agreed to be randomised to undergo standard maintenance chemotherapy alone or in combination with BCG vaccination, stanozolol (0.1 mg/kg/day) or BCG vaccination plus stanozolol. After three years follow-up, all four arms had similar rates of remission and adverse events (Mandelli et al. 1981).

For malignant disease, the relative contributions of direct antitumour effects on mortality compared with indirect effects on morbidity may be difficult to determine. On balance, the role of androgen therapy in malignant disease requires further evaluation by randomised controlled studies using morbidity rather than mortality endpoints.

17.10.2 Cytoprotection

Spermatogenic damage is an inevitable consequence of modern combination chemotherapy and radiotherapy for the malignant diseases of men in the reproductive age group. These include mainly testicular and hematological malignancies all usually treated with curative intent as well as soft tissue sarcoma and bone marrow transplantation where prognosis for long-term survival is more variable. While initial damage is invariably severe, recovery depends on the regimen utilised with the prognosis for recovery of spermatogenesis and fertility varying from predictable within a few years to essentially irreversible damage. Men wishing to father children prior to recovery of spermatogenesis may utilise sperm cryopreservation as an useful form of fertility insurance. In addition, it has been proposed that the adjunctive hormonal treatment may protect the testes from cytotoxic damage. This stems from an experimental observation that pretreatment with a GnRH analog protected the mouse testis from spermatogenic damage from the alkylating drug cyclophosphamide (Glode et al. 1981). This, together with the clinical observation that the gonads of prepubertal leukemia patients were relatively less damaged compared with their post-pubertal counterparts suggested that quiescent gonads were protected against cytotoxic damage. However subsequently these experimental findings were not reproducible (da Cunha et al. 1987) and the clinical prepubertal protection was also recognised to be illusory (Shalet et al. 1978).

Most clinical studies of adjuvant cytoprotective therapy have used GnRH superactive agonists but have shown little promise so far. These include only one randomised (Waxman et al. 1987), two non-randomised controlled (Brennemann et al. 1994; Kreusser et al. 1990) and one uncontrolled (Johnson et al. 1985) studies. These studies have not been well-designed (lacking randomisation, adequate length of follow-up) and have been unable to test definitively the underlying hypothesis of the potential protective role of gonadal quiescence as the degree and duration of gonadal regression has been limited. Given the time constraints between the start of adjuvant cytoprotective therapy and the imperative to commence chemotherapy, GnRH superactive agonists with their inevitable flare of gonadotrophins and testosterone are a poor choice to test this concept. Indeed, the negative clinical findings so far may be interpreted as reinforcing the hypothesis and demonstrating the need to test a pure GnRH antagonist which eliminates the initial flare reaction.

Based on animal models, suppression of spermatogenesis by androgens with or without other sex steroids prior to and during cytotoxic therapy has been proposed to protect the germinal epithelium against cytotoxic or irradiation damage (Morris 1993). In practice, however, androgen-based cytoprotection therapy has been little studied clinically, largely because of the slowness of androgen-induced spermatogenic suppression when life-saving cancer treatment has to be delayed. A recent pilot study randomised ten men about to commence monthly cyclophosphamide bolus injections for nephrotic syndrome to testosterone injections (mixed esters 100mg every 15 days) or no extra treatment for 30 days prior to and during the six months of cyclophosphamide therapy (Masala et al. 1997). During the six months of cyclophosphamide treatment, sperm concentrations were severely reduced to near-azoospermia in all ten men. Subsequently by the third and sixth month after cessation of cyclophosphamide all five androgen-treated men returned to normal sperm concentrations. In contrast, even at six months after cessation of cyclophosphamide, four out of five controls remained azoospermic with the other having a subnormal (11 million/ml) sperm concentration. Despite the small numbers, these results were impressive and unprecedented and warrant further study.

It remains unclear whether such an androgen-based cytoprotective regimen is feasible or effective for the more intensive combination chemotherapy used for the treatment of malignancy. These findings also contrast favourably with the consistently negative findings from the use of GnRH agonists in potential cytoprotective regimens. Ultimately long-term studies comparing cytoprotection regimens for efficacy, safety and cost-effectiveness compared with sperm cryopreservation plus artificial reproductive technologies would be of interest.

17.11 Respiratory disease

17.11.1 Chronic obstructive lung disease

Advanced chronic airflow limitation is associated with weight loss and muscle depletion, possibly due to the increased energy requirements required for breathing. Interventions aimed at improving muscle bulk such as nutrition, exercise or androgens may therefore have an impact on the morbidity and/or mortality of the underlying respiratory disease. One large well-conducted prospective study demonstrated that short-term low-dose androgen therapy (nandrolone decanoate) augmented the effects of nutritional supplementation in patients with moderate to severe chronic airways disease (Schols et al.

1995). From 233 consecutive patients with stable, moderate to severe and bronchodilator-unresponsive pulmonary disease admitted to an intensive pulmonary rehabilitation program, 217 were randomised into three groups.

These were to receive eight weeks of treatment with
- placebo injections,
- a nutritional supplement (one high fat, high calorie drink daily) plus placebo injections or
- a nutritional supplement plus androgen injections (nandrolone decanoate [50 mg men, 25 mg women]) with intramuscular injections given fortnightly.

Participants were also stratified according to the degree of baseline muscle depletion (body weight <90% and/or lean mass <67% ideal or not) at entry. During the study all patients underwent a standardised exercise program. Both nutrition and androgen therapy increased body weight over placebo, with androgen therapy having more prominent effects on lean body mass and respiratory muscle strength although there was no measurable improvement in submaximal exercise tolerance (12 minute walking distance) nor any major adverse effects. Although the lack of an androgen-alone arm and blinding with respect to nutritional supplementation made it difficult to evaluate the impact of androgen therapy relative to improved nutrition, these encouraging short-term results warrant further study including the long-term follow-up reportedly underway (Schols et al. 1995).

Improvement in the pathogenesis of the underlying pulmonary disease may itself ameliorate the gonadal dysfunction of systemic disease. In one study of men with chronic obstructive pulmonary disease with severe hypoxia and impotence, long-term oxygen therapy improved total and free testosterone and lowered SHBG (without changes in LH or FSH) in five men who had improved sexual function. The remaining seven who had unimproved sexual function had no changes in circulating hormone concentrations (Aasebo et al. 1993).

17.11.2 Cystic fibrosis and lung transplantation

Cystic fibrosis is now among the leading diseases leading to lung transplantation. Following lung transplantation, respiratory muscle and general debility are important factors predicting poor outcome. For this reason, short-term androgen therapy prior to lung transplantation is worthy of evaluation by a controlled clinical trial. Although there are no controlled studies of androgen therapy for cystic fibrosis, this disorder is unusual among illnesses of young men in that obstructive azoospermia and infertility are virtually invariable so that any deleterious effects of androgen therapy on spermatogenesis and natural fertility would be of little relevance. After transplantation iatrogenic factors including steroids (Reid et al. 1985) and ketoconazole (Keogh et al. 1995) therapy may further exacerbate the androgen deficiency associated with end-stage cardiorespiratory disease.

17.11.3 Obstructive sleep apnea

The adverse effects of sleep apnea on reproductive function (Grunstein et al. 1989) and its precipitation by androgen therapy (Sandblom et al. 1983) have only recently been recognised. Overt sleep apnea in an obese but otherwise healthy man was first reported as an idiosyncratic reaction to androgen therapy (Sandblom et al. 1983). A subsequent observational study (including this original case) claims a general effect of androgen therapy on sleep breathing (Matsumoto et al. 1985). The non-independence of the study sample and the marked skewing of results by including the original case are more consistent with an idiosyncratic rather than general effect of androgen therapy on regulation of sleep breathing. Whether this involves central chemoreceptor-mediated regulation or increased obstruction of the upper airways remains unclear.

A randomised crossover study of 11 hypogonadal men receiving testosterone enanthate (200–400 mg per fortnight) or no therapy reported that androgen therapy increased sleep arousals. This study compared somnography during androgen therapy (3–7 days after a testosterone injection) with a no-treatment group consisting of patients after withdrawal (mean 53 days post-injection) of androgen therapy (Schneider et al. 1986). Anatomical and functional evaluation of the upper airway patency in 4 patients showed no treatment-related difference but this finding is inconclusive due to the small sample size. These findings, although lacking placebo or blinding, might suggest an effect of very high peak circulating testosterone concentrations following testosterone ester injection, although the effects of more physiological androgen replacement remain to be studied. Another observational study examined the prevalence of obstructive sleep apnea in hemodialysed men and the potential role of testosterone ester injections in its causation (Millman et al. 1985). Obstructive sleep apnea symptoms were common (12/29, 41%), particularly in those receiving regular testosterone enanthate injections (250 mg weekly) to stimulate erythropoiesis (9/12, 75%) compared with those not receiving testosterone (6/17, 35%). Withdrawal of testosterone, however, did not alter the signs or symptoms of sleep apnea in the five men studied both during and two months after cessation of testosterone treatment. This suggests that testosterone ester injections are not a regular precipitant of obstructive sleep apnea.

A low frequency of obstructive sleep apnoea complicating androgen therapy as an idiosyncratic effect cannot be excluded. Whether this idiosyncratic reaction is related to the pharmacokinetics of the testosterone formulation used, such as the extreme peak serum testosterone following intramuscular injections, has yet to be determined. Whether similar effects would occur with more physiological testosterone formulations remains to be established in properly controlled clinical trials, although the low frequency of such reactions would require very large studies.

17.11.4 Asthma

One small double-blind study of 15 steroid-dependent asthmatic boys randomised to ethylestrenol (0.1 mg/kg/day) or placebo for 12 months reported a significant improvement in peak expiratory flow rate in the androgen group compared with the placebo group (Kerrebijn and Delver 1969). Despite the claim of no acceleration of bone maturation (according to the ratio of bone-age/height velocity) in this older study, the safety of such androgen therapy in boys prior to completion of puberty is very doubtful and androgen therapy has no place in the modern treatment of adolescent asthma.

17.11.5 Tuberculosis

A variety of controlled clinical studies have suggested a benefit of androgen therapy in pulmonary tuberculosis, predominantly among patients with wasting and weight loss. The original studies reported some decades ago (Kopera 1976) are now of historical interest only. Improved general nutrition and anti-tuberculosis chemotherapy have largely precluded any further role for androgen therapy in the management of tuberculosis.

17.12 Vascular disease

17.12.1 Cardiovascular disease

The effects of androgen therapy in cardiovascular disease have been well reviewed in detail elsewhere (Alexandersen et al. 1996; Barrett-Connor 1996). Very few controlled clinical studies using clinical endpoints such as cardiovascular events have been reported. A single placebo-controlled, cross-over study examined 62 elderly men with established ischemic heart disease who were randomised to commence treatment with either testosterone undecanoate (120 mg per day for 2 weeks followed by a maintenance dose of 40 mg per day for another 2 weeks) or placebo following which, after a 2 week washout period, subjects then crossed-over to the other treatment (Wu and Wenig 1993). In this study dramatic improvement in cardiac ischemia by both subjective (77% vs 7% with angina symptoms) and objective criteria (ECG [69% vs 8%], Holter [75% vs 8%]) although no change in cardiac function (echocardiography) was observed. This short-term study was, however, apparently unblinded and the objective scales were not reported.

Epidemiological studies show a consistent inverse correlation between blood testosterone concentrations and cardiovascular events as well as a positive correlation between blood testosterone and cholesterol (total & HDL) fractions (Barrett-Connor 1996). These findings predict that androgen therapy would have a protective or neutral effect on cardiovascular disease in men but empirical evaluation by surveillance in the course of longer-term studies would be required. A large body of observational studies has been accumulated relating endogenous and exogenous androgens to cardiovascular risk factors as a surrogate for cardiovascular disease or events. These correlational studies have focussed almost exclusively on lipids which represent only one of a large constellation of correlated cardiovascular risk factors. This preoccupation coexists with a neglect of numerous other hormonally-sensitive vasoactive factors (e.g. endothelin, nitric oxide, ANP, prostaglandins; endothelial cell and smooth muscle function; thrombosis, fibrinolysis, hyperviscosity and blood flow; insulin resistance, body fat and its distribution; sodium and fluid retention) which, together with lifestyle differences, may be relevant to explaining the gender disparity in cardiovascular disease rates. Androgen effects on lipids are related to dose, type of androgen (Friedl et al. 1990) and route of administration (Thompson et al. 1989) but, given recent large coronary prevention studies using lipid-lowering agents geared towards lowering total cholesterol, it is salient to note that androgens consistently lower total as well as subfractions of cholesterol. Supraphysiological androgen doses lower HDL cholesterol (Bagatell and Bremner 1995); however, the clinical significance of such pharmacological effects when total cholesterol is concurrently lowered remain to be clarified. In summary, these considerations raise the possibilities of using androgen therapy to reduce progression in coronary heart disease and that androgen supplementation in ageing men or systemic disease may be accompanied by an amelioration of coronary heart disease.

17.12.2 Arterial-peripheral vascular disease

The rationale for androgen therapy in arterial peripheral vascular disease is obscure and the results of controlled studies unconvincing of any benefit. Four randomised-controlled studies of short duration have examined the effects of androgen therapy in peripheral vascular disease. The first placebo-controlled, double-blind study involved 44 non-diabetic men with claudication randomised to start treatment with either testosterone isobutyrate (300 mg) or control (meprobamate as a placebo) every fortnight over a 12-week study period (Dohn et al. 1968). Subsequently after an undefined wash-out period subjects crossed over to the other treatment. There were no improvement in foot pulses, metronome walking distance (a measure of claudication), laboratory plethysmographic parameters or symptoms. As this study's use of a crossover might have led to a null bias, a subsequent study was performed without a crossover design and using a higher dose of testosterone for a longer duration. In this study 39 men with claudication, many previously treated and some continuing to use anticoagulants and/or smoke, were randomised to either testosterone enanthate (200 mg weekly for 3 weeks then fortnightly) or placebo oil vehicle injections for six months (Hentzer and Madsen 1967). Again no difference in metronome walking distance, claudication symptoms, palpable pulses, venous filling time or grip strength were recorded. An objective measures of muscle blood flow (Xenon clearance) was reported to be improved by 19% whereas the control decreased by 7%. The third double-blind study randomised 22 untreated, non-diabetic men with claudication to treatment with either testosterone (100 mg daily plus various other hormones six days per week) or placebo vehicle injections for six weeks (Albrechtsen et al. 1972). No difference in pain or standardised walking test was reported. Finally a fourth study randomised 26 patients to injections of 100 mg testosterone propionate or placebo oil vehicle three times weekly for 2–6 months without any subjective or objective benefit (Genster and Oram 1971). On balance, androgen therapy appears to offer little promise of benefit in arterial peripheral vascular disease to justify the much longer studies that would be required to evaluate its role definitively.

17.12.3 Venous disease

The use of androgen therapy in acute or chronic venous disease arises from their fibrinolytic effect, which may reduce venous fibrin plugging. One study of chronic venous insufficiency aiming to test whether androgen therapy would reduce the rate of venous ulceration involved 60 patients with venous skin changes but no ulceration being treated with below-knee compression stockings as standard therapy (McMullin et al. 1991). They were randomised to receive either stanozolol (10 mg daily) or placebo tablets for six months and androgen therapy produced a significant but modest reduction in the area of venous skin changes but no change in prospective rate of new ulcers or skin oxygenation. The side-effects comprised mostly virilisation presumably due to stanozolol treatment of women.

Another prospective two-centre study examined the role of androgen therapy in prevention of post-operative deep venous thrombosis (DVT). In this study 200 patients scheduled for elective major abdominal surgery were randomised into three groups (Zawilska et al. 1990). The first received inhaled heparin (800 units/kg) one day prior to surgery alone, a second group received the same dose of inhaled heparin plus a single injection of nandrolone phenylpropionate (50 mg) and the third group received standard heparin prophylaxis (5000 units twice daily sc). Treatments were from the day before surgery until the fifth post-operative day. Using daily ^{125}I-fibrinogen scanning to detect DVT in 183 evaluable patients, there was no significant difference in post-operative DVT or clinically significant bleeding episodes among the three groups. Unfortunately the study had major between-centre differences and used a suboptimal detection method. It was also underpowered to reliably evaluate the claim that addition of nandrolone to nebulised heparin was as effective as standard heparin but with much lower bleeding risk. Larger and better designed studies of the effects of androgen therapy on venous disease in men seem warranted.

17.13 Women

17.13.1 Menopausal symptoms

The role of androgen therapy for menopausal symptoms is controversial (Carson and Carson 1996; Davis and Burger 1996). Androgen therapy has been proposed for treatment of some menopausal symptoms noting that it has undoubted corollary benefits for bone density. As there is no change in circulating testosterone after natural menopause and only mild decrease after oophorectomy, such a minor androgen deficiency would justify only ultra-low dose testosterone replacement therapy. No studies using such low doses aiming to replace physiological testosterone concentrations in postmenopausal women have, however, been reported. Thus androgen therapy for post-menopausal women must be evaluated as a pharmacological therapy with regard to safety, efficacy and cost-effectiveness as for other drug therapies. With subjective symptomatic end-points, carefully designed placebo-controlled trials of adequate duration and size are essential to evaluate efficacy. The first controlled studies used very high androgen doses equivalent to those used for androgen replacement in men (Greenblatt et al. 1950; Sherwin and Gelfand 1985, 1987) who have ~30 times the testosterone production rate of women. Not surprisingly these produced quite significant virilisation (Urman et al. 1991), although this was not studied quantitatively or objectively in the original studies. In addition, these studies of unselected post-menopausal women utilising primarily subjective endpoints lacked adequate design features such as randomisation (Greenblatt et al. 1950; Sherwin and Gelfand 1987), objective and validated end-points and quantitative analysis methodology (Greenblatt et al. 1950).

Recent studies have used more appropriate testosterone doses and aimed to treat menopausal symptoms refractory to an adequate trial of estrogen therapy (Burger et al. 1987). One controlled trial randomised 20 postmenopausal women with severe loss of libido refractory to estrogen therapy to treatment with either an estrogen implant (40 mg) alone or together with a testosterone implant (50 mg). While the addition of testosterone improved subjective reports of sexuality, the study did not involve a placebo and all women in the control arm switched to the active treatment arm at the first visit so efficacy data are only evaluable for the initial six weeks of treatment (Burger et al. 1987). These findings are inconsistent with another study of 40 postmenopausal women with symptoms including loss of libido who were randomly allocated to treatment with a 50 mg estradiol implant with or without a 100 mg testosterone implant (Dow and Hart 1983). In that study a higher testosterone dose produced no measurable benefit in symptoms including sexuality but no safety evaluation was reported.

While androgen therapy has clear benefits for bone density in unselected menopausal women (Davis et al. 1995), its efficacy for estrogen-resistant menopausal symptoms requires further clarification. In addition, the safety of such androgen therapy warrants more objective, quantitative evaluation of virilisation as well as long-term surveillance of hormone-dependent cancers and the cardiovascular system in placebo-controlled studies of adequate power and duration.

17.13.2 Vaginal skin atrophy

Vaginal skin atrophy (known by many synonyms including kraurosis vulvae, senile atrophy, vulvar lichen sclerosis, atrophic pruritus vulvae) causes erosions and fissuring resulting in sharp pain, soreness and dyspareunia. Androgen therapy as a topical cream containing testosterone or other androgens is a traditional therapy although no large placebo-controlled clinical studies have been reported. One small open study suggests that testosterone (1 mg 2% testosterone propionate petrolatum ointment daily for 4 weeks) is effective at reducing visible skin lesions, vulval pain and itching (Joura et al. 1997). Another study reported that 2% DHT is as effective as 2% testosterone propionate in white petrolatum ointment (Paslin 1996). Unfortunately the effects of placebo petrolatum were not studied and virilisation due to systemic absorption was common (Joura et al. 1997) so the mechanism of topical androgen therapy remains unclear.

17.13.3 Breast cancer

Androgen therapy continues to have an established role in late-stage advanced breast cancer usually as a third or fourth line therapy after failure of other hormonal therapies and when the virilising side-effects are less unacceptable (Ingle 1984). Few recent studies have included androgen therapy and it now has a residual but diminishing role relative to modern hormonal and cytotoxic chemotherapy for breast cancer.

17.14 Body weight

17.14.1 Wasting

Many older studies examined the role of androgen therapy to augment body weight in patients with wasting or cachexia from a variety of underlying medical diseases as well as for cosmetic reasons in otherwise healthy people. For example, one double-blind study treated 28 healthy men and women and 26 male patients with wasting associated with chronic diseases (e.g. tuberculosis, chronic degenerative disorders) with placebo or one of two doses (25 mg or 50 mg daily) of norethandrolone for 12 weeks (Watson et al. 1959). The placebo group subsequently also crossed over to active treatment for another 12 weeks. Compared with placebo, both androgen groups had significantly improved body weight gain and reported improved appetite and well being but there was no dose-response relationship. Most patients had abnormal BSP retention and nearly all women experienced some virilisation. Very few other studies, however, were well-controlled and the end-point of weight gain has little validity in isolation outside the context of the overall objectives of medical management for specific illnesses (see HIV/AIDS).

17.14.2 Obesity

Few controlled clinical trials of androgen therapy in obesity have been reported. Although massive obesity is associated with lowering of total testosterone, there have been no controlled studies aiming to rectify any consequent androgen deficiency. A series of studies by Marin has raised interesting questions about the role of pharmacological androgen therapy in obesity. A pilot study reported reduced waist/hip circumference and improved insulin sensitivity following three months transdermal treatment with testosterone (250 mg in 10 gm gel daily) in eight men but not with dihydrotestosterone (250 mg in 10 gm gel daily) in nine men (Marin et al. 1992). The study design, lacking placebo controls or any dose finding, did not allow any conclusion as to whether this difference arose from differences in skin bioavailability or androgen type (aromatisable or not) or potency. The same investigators then reported a double-blind study in which 27 middle-aged men with abdominal obesity were randomised to placebo, testosterone or dihydrotestosterone treatment by daily topical application of a transdermal gel (125 mg in 5 mg gel daily) for 9 months (Marin et al. 1995). Testosterone treatment inhibited lipid uptake into adipose tissue triglycerides, decreased lipoprotein lipase activity, reduced visceral fat stores (CT scan) and increased euglycemic clamp insulin sensitivity compared with dihydrotestosterone and placebo groups (Marin 1995). The results of Marin et al. were not confirmed by another study which randomised 30 obese middle-aged men into 3 groups to receive oral oxandrolone (10 mg/day), testosterone enanthate (150 mg) injections fortnightly or placebo treatments for nine months using a double-dummy, double-blinded design (Lovejoy et al. 1995). Due to lowering of HDL cholesterol by oral oxandrolone, a monitoring committee required the oxandrolone arm be switched to injections of nandrolone decanoate (30 mg) fortnightly. None of the androgens (oxandrolone, nandrolone, testosterone) had any consistent overall effect on muscle or fat mass but the interim change in study design reduced its power. The discrepancies between these studies require clarification with large sample size, longer duration and more clinically meaningful endpoints.

17.15 Dermatological disease

The frequency and cosmetic impact of male pattern balding has, over millennia, led to innumerable attempted "cures", even to attempts to use testosterone topically. Prompted by a paradoxical claim that topical testosterone could cause hair regrowth, a double-blind, randomised study of 51 balding men showed that topical application of 1% testosterone propionate cream daily to one side of the scalp for a median of 4–5 months was no more effective than placebo applied to the other half of the scalp (Savin 1968). Given the dependence of male pattern balding on masculine levels of androgen exposure after puberty, acceleration of hair loss might have been expected but the study endpoints (investigator and patient subjective global grading of regrowth) were not designed to detect this. More recently controlled studies of a 5α-reductase inhibitor have added a selective anti-androgen to the already vast list of baldness cures (Rittmaster 1994).

17.16 Gastrointestinal disease

There are no controlled studies of androgen therapy reported in gastroenterological disorders although observational studies indicate impaired reproductive function in inflammatory bowel disease. Older uncontrolled studies in patients with inflammatory bowel disease, malabsorption and ulcer disease aiming to provide palliative improvement in weight loss and/or protein deficiency (Kopera 1976) are now considered clinically obsolete. The development of definitive anti-infective therapy for peptic ulcer disease, superseding already highly effective anti-acid treatments, leave no place for consideration of adjunctive androgen therapy. Despite the lack of definitive treatment for inflammatory bowel disease, there seems little clear rationale for androgen therapy for the underlying disease, although it may be considered for amelioration of the effects of anti-inflammatory glucocorticoid treatment on bone or muscle.

17.17 Key messages

- Androgen replacement therapy aims to replicate (but not exceed) tissue androgen exposure of healthy eugonadal men and hence is limited to testosterone in physiological dosage. Although androgen deficiency is a common accompaniment of systemic disease, androgen replacement therapy may influence morbidity but is unlikely to influence mortality.
- Pharmacological androgen therapy aims to utilise androgens to maximal efficacy within adequate safety limits. Such treatment must be judged by the efficacy, safety and cost-effectiveness standards applicable to any

drug. Given the few well-established indications for pharmacological androgen therapy, this requires placebo controls, studies of adequate power and duration and objective endpoints. Unfortunately, very few studies of pharmacological androgen therapy fulfil these basic requirements.

- Androgen therapy appears unlikely to provide a significant reduction in mortality from acute or chronic alcoholic liver disease but the effects on non-alcoholic liver disease have not been studied.
- Hepatotoxicity of 17α-alkylated androgens suggests that, where possible, other safer oral and parenteral androgens should be preferred.
- Androgen therapy does not improve survival in aplastic anemia but provides a morbidity benefit by maintaining hemoglobin and transfusion independence.
- In anemia of end-stage renal failure, androgen therapy is cheaper but less effective than erythropoietin and its use may be restricted to older men where erythropoietin is contraindicated or unavailable.
- Many important questions and opportunities remain for androgen therapy in non-gonadal disease.
- Traditional indications for androgen therapy (e.g. osteoporosis, anemia due to marrow or renal failure, advanced breast cancer) remain until more specific therapeutic agents are available when the role for androgen therapy diminishes unless new clinical data justifies their retention based on efficacy, safety or cost-effectiveness criteria.
- The best opportunities for future evaluations of androgen therapy include osteoporosis (especially steroid-induced), chronic respiratory and rheumatological disorders. Such future studies require randomisation, placebo controls, stratification for *a priori* factors, adequate power and duration, detailed dose-finding and ideally real rather than surrogate endpoints.
- Future pharmaceutical research is also needed to develop better products including improved formulations for more convenient and effective delivery of testosterone as well as more potent, selective androgens.

17.18 References

Aasebo U, Gyltnes A, Bremnes RM, Aakvaag A, Slordal L (1993) Reversal of sexual impotence in male patients with chronic obstructive pulmonary disease and hypoxemia with long term oxygen therapy. J Steroid Biochem Molec Biol 46:799–803

Ahn YS, Harrington WJ, Simon SR, Mylvaganam R, Pall LM, So AG (1983) Danazol for the treatment of idiopathic thrombocytopenic purpura. N Engl J Med 308:1396–1399

Albrechtsen O, Barfod B, Barfod E, Laursen NP, Nordentoft B, Yde H (1972) Hormonal treatment of arterial insufficiency in the lower extremities. Dan Med Bull 19:157–159

Alexandersen P, Haarbo J, Christiansen C (1996) The relationship of natural androgens to coronary heart disease in male: a review. Atherosclerosis 125:1–13

Ambriz R, Pizzuto J, Morales M, Chavez G, Guillen C, Aviles A (1986) Therapeutic effect of danazol on metrorrhagia in patients with idiopathic thrombocytopenic purpura (ITP). Nouv Rev Fr Hematol 28:275–279

Bacigalupo A, et al. (1993) Treatment of aplastic anaemia (AA) with antilymphocyte globulin (ALG) and methylprednisolone (MPred) with or without androgens: a randomized trial from the EBMT SAA working party. Br J Haematol 83:145–151

Bagatell CJ, Bremner WJ (1995) Androgen and progestagen effects on plasma lipids. Prog Cardiovas Dis 38:255–271

Ballal SH, Domoto DT, Polack DC, Marciulonis P, Martin KJ (1991) Androgens potentiate the effects of erythropoietin in the treatment of anemia of end-stage renal disease. Am J Kidney Dis 17:29–33

Barrett-Connor E (1996) Pharmacology, biology, and clinical applications of androgens: Current stuatus and future prospects. In: Bhasin S, Gabelnick HL, Spieler JM, Swerdloff RS, Wang C, Kelly C (eds) New York: Wiley-Liss, pp 215–223

Belch JJ, Madhok R, McArdle B, McLaughlin K, Kluft C, Forbes CD, Sturrock R (1986) The effect of increasing fibrinolysis in patients with rheumatoid arthritis: a double blind study of stanozolol. Q J Med 58:19–27

Berger JR, Pall L, Hall CD, Simpson DM (1996) Oxandrolone in AIDS-wasting myopathy. AIDS 10:1657–1662

Berns JS, Rudnick MR, Cohen RM (1992) A controlled trial of recombinant human erythropoietin and nandrolone decanoate in the treatment of anemia in patients on chronic hemodialysis. Clin Nephrol 37:264–267

Bhasin S, Storer TW, Berman N, Callegari C, Clevenger B, Phillips J, Bunnell TJ, Tricker R, Shirazi A, Casaburi R (1996) The effects of supraphysiologic doses of testosterone on muscle size and strength in normal men. N Engl J Med 335:1–7

Blamey SL, Garden OJ, Shenkin A, Carter DC (1984) Modification of postoperative nitrogen balance with preoperative anabolic steroid. Clin Nutr 2:187–192

Bonkovsky HL, Fiellin DA, Smith GS, Slaker DP, Simon D, Galambos JT (1991a) A randomized, controlled trial of treatment of alcoholic hepatitis with parenteral nutrition and oxandrolone. I. Short-term effects on liver function. Am J Gastroenterol 86:1200–1208

Bonkovsky HL, Singh RH, Jafri IH, Fiellin DA, Smith GS, Simon D, Cotsonis GA, Slaker DP (1991b) A randomized, controlled trial of treatment of alcoholic hepatitis with parenteral nutrition and oxandrolone. II. Short-term effects on nitrogen metabolism, metabolic balance, and nutrition. Am J Gastroenterol 86:1209–1218

Booji A, Biewenga-Booji CM, Huber-Bruning O, Cornelis C, Jacobs JW, Bijlsma JW (1996) Androgens as adjuvant treatment in postmenopausal female patients with rheumatoid arthritis. Ann Rheum Dis 55:811–815

Branda RF, Amsden TW, Jacob HS (1977) Randomized study of nandrolone therapy for anemia due to bone marrow failure. Arch Intern Med 137:65–69

Brennemann W, Brensing KA, Leipner N, Boldt I, Klingmuller D (1994) Attempted protection of spermatogenesis from irradiation in patients with seminoma by D-Tryptophan-6 luteinizing hormone releasing hormone. Clin Investigator 72:838–842

Brubaker LH, Briere J, Laszlo J, Kraut E, Landaw SA, Peterson P, Goldberg J, Donovan P (1982) Treatment of anemia in myeloproliferative disorders: a randomized study of fluoxymesterone v transfusions only. Arch Intern Med 142:1533–1537

Buchwald D, Argyres S, Easterling RE, Jr FJO, Brewer GJ, Schoomaker EB, Abbrecht PH, Williams GW, Weller JM (1977) Effect of nandrolone decanoate on the anemia of chronic hemodialysis patients. Nephron 18:232–238

Burger HG, Hailes J, Nelson J (1987) Effect of combined implants of oesteradiol and testosterone on libido in postmenopausal women. Br J Med i:936–937

Camitta BM, Thomas ED, Nathan DG, Gale RP, Kopecky KJ, Rappeport JM, Santos G, Gordon-Smith EC, Storb R (1979) A prospective study of androgens and bone marrow transplantation for treatment of severe aplastic anemia. Blood 53:504–514

Carson PR, Carson SA (1996) Androgen replacement therapy in women: myths and realities. Int J Fertil 41:412–422

Champlin RE, Ho WG, Feig SA, Winston DJ, Lenarsky C, Gale RP (1985) Do androgens enhance the response to antithymocyte globulin in patients with aplastic anemia? A prospective randomized trial. Blood 66:184–188

Chlebowski RT, Herrold J, Ali I, Oktay E, Chlebowski JS, Ponce AT, Heber D, Block JB (1986) Influence of nandrolone decanoate on weight loss in advanced non-small cell lung cancer. Cancer 58:183–186

Cicardi M, Castelli R, Zingale LC, Agostoni A (1997) Side effects of long-term prophylaxis with attenuated androgens in hereditary angioedema: comparison of treated and untreated patients. J Allergy Clin Immunol 99:194–196

Cooperative Group for the Study of Aplastic and Refractory Anaemias (1979) Androgen therapy of aplastic anaemia: a prospective study of 352 cases. Scand J Haematol 22:343–356

Copenhagen Study Group for Liver Diseases (1986) Testosterone treatment of men with alcoholic cirrhosis: a double-blind study. Hepatology 6:807–813

Cutolo M, Balleari E, Giusti M, Intra E, Accardo S (1991) Androgen replacement therapy in male patients with rheumatoid arthritis. Arthritis Rheum 34:1–5

da Cunha MF, Meistrich ML, Nader S (1987) Absence of protection by a GnRH analogue against cyclophosphamide induced testicular cytotoxicity in the mouse. Cancer Res 47:1093–1097

Davis SR, Burger HG (1996) Clinical review 82: androgens and the postmenopausal woman. J Clin Endocrinol Metab 81:2759–2763

Davis SR, McCloud P, Strauss BJG, Burger H (1995) Testosterone enhances estradiol's effects on postmenopausal bone density and sexuality. Maturitas 21:227–236

Dohn K, Hvidt V, Nielsen J, Palm L (1968) Testosterone therapy in obliterating arterial lesions in the lower limbs. Angiology 19:342–350

Dow MGT, Hart DM (1983) Hormonal treatments of sexual unresponsiveness in postmenopausal women: a comparative study. Br J Obstet Gynaecol 90:361–366

Drosos AA, van Vliet-Dascalopoulos E, Andonopoulos AP, Galanopoulou V, Skopouli FN, Moutsopoulos HM (1988) Nandrolone decanoate (deac-durabolin) in primary Sjogren's syndrome: a double blind study. Clin Exp Rheumatol 6:53–57

El-Sadr A, Sabry AA, Abdel-Rahman M, El-Barnachawy R, Koraitim M (1990) Treatment of primary nocturnal enuresis by oral androgen mesterolone. A clinical and cystometric study. Urology 36:331–335

Engleson ES, Pi-Sunyer FX, Kotler DP (1995) Effects of megestrol acetate therapy on body composition and circulating testosterone concentrations in ptients with AIDS. AIDS 9:1107–1108.

Evans JT, Elias EG (1972) The erythropoietic response to anabolic therapy in patients receiving radiotherapy. J Clin Pharmacol New Drugs 12:101–104

Fenster LF (1966) The nonefficacy of short-term anabolic steroid therapy in alcoholic liver disease. Ann Intern Med 65:738–744

Foss GL (1939) Clinical administration of androgens. Lancet i:502–504

French Cooperative Group for the Study of Aplastic and Refractory Anaemias (1986) Androgen therapy in aplastic anaemia: a comparative study of high and low-doses and of 4 different androgens. Scand J Haematol 36:346–352

Friedl KE, Hannan CJ, Jones RE, Plymate SR (1990) High-density lipoprotein cholesterol is not decreased if an aromatisable androgen is administered. Metab 39:69–74

Gelfand JA, Sherins RJ, Alling DW, Frank MM (1976) Treatment of hereditary angioedema with danazol: reversal of clinical and biochemical abnormalities. N Engl J Me. 295:1444–1448

Genster HG, Oram V (1971) Medical treatment of arterial insufficiency in the lower limbs. Ugeskr Laeg 133:244–246

Geusens P, Dequeker J (1986) Long-term effect of nandrolone decanoate, 1a-hydroxyvitamin D_3 or intermittent calcium infusion therapy on bone mineral content, bone remodeling and fracture rate in symptomatic osteoporosis: a double-blind controlled study. Bone and Mineral 1:347–357

Gill PS, Lunardi-Ishkandar Y, Louie S, Tulpule A, Zheng T, Espina BM, Besnier JM, Hermans P, Levine AM, Bryant JL, Gallo RC (1996) The effects of preparations of human chorionic gonadotropin on AIDS-related Kaposi's sarcoma. N Engl J Med 335:1261–1269

Glode LM, Robinson J, Gould SF (1981) Protection from cyclophosphamide-induced testicular damage with an analogue of gonadotrophin-releasing hormone. Lancet 1:1132–1134

Gluud C, Bennett P, Dietrichson O, Johnsen SG, Ranek L, Svendsen LB, Juhl E (1981) Short-term parenteral and peroral testosterone administration in men with alcoholic cirrhosis. Scand J Gastro 16:749–755

Gluud C, Bennett P, Svenstrup B, Micic S, Copenhagen Study Group for Liver Diseases (1988a) Effect of oral testosterone treatment on serum concentrations of sex steroids, gonadotrophins and prolactin in alcoholic cirrhotic men. Aliment Pharmacol Therap 2:119–128

Gluud C, Wantzin P, Eriksen J, Copenhagen Study Group for Liver Diseases (1988b) No effect of oral testosterone treatment on sexual dysfunction in alcoholic cirrhotic men. Gastroenterology 95:1582–1587

Gluud C, Christoffersen P, Eriksen J, Wantzin P, Knudsen BB, Copenhagen Study Group for Liver Diseases (1987a) No effect of long-term oral testosterone treatment on liver morphology in men with alcoholic cirrhosis. Am J Gastroenterol 82:660–664

Gluud C, Dejgard A, Bennett P, Svenstrup B (1987b) Androgens and oestrogens before and following oral testosterone administraton in male patients with and without alcoholic cirrhosis. Acta Endocrinol 115:385–391

Gluud C, Henricksen JH, Copenhagen Study Group for Liver Diseases (1987c) Liver haemodynamics and function in alcoholic cirrhosis. J Hepatol 4:168–173

Gold J, High HA, Li Y, Michelmore H, Bodsworth NJ, Finlayson R, Furner VL, Allen BJ, Oliver CJ (1996) Safety and efficacy of nandrolone decanoate for treatment of wasting in patients with HIV infection. AIDS 10:745–752

Greenblatt RB, Barfield WE, Garner JF, Calk GL, Harrod JP (1950) Evaluation of an estrogen, androgen, estrogen-androgen combination, and a placebo in the treatment of the menopause. J Clin Endocrinol Metab 10:1547–1558

Griggs RC, Kingston W, Herr BE, Forbes G, Moxley RT (1985) Lack of relationship of hypogonadism to muscle wasting in myotonic dystrophy. Arch Neurol 42:881–885

Griggs RC, Pandya S, Florence JM, Brooke MH, Kingston W, Miller JP, Chutkow J, Herr BE, Moxley RT (1989) Randomized controlled trial of testosterone in myotonic dystrophy. Neurology 39:219–222

Grunfeld C, Feingold KR (1992) Metabolic disturbances and wasting in the Acquired Immunodeficiency Syndrome. N Engl J Med 327:329–37

Grunstein RR, Handelsman DJ, Lawrence SJ, Blackwell C, Caterson ID, Sullivan CE (1989) Hypothalamic dysfunction in sleep apnea: reversal by nasal continuous positive airways pressure. J Clin Endocrinol Metab 68:352–358

Hall GM, Larbre JP, Spector TD, Perry LA, Silva JAD (1996) A randomized trial of testosterone therapy in males with rheumatoid arthritis. Br J Rheumatol 35:568–573

Hamilton JB (1937) Treatment of sexual underdevelopment with synthetic male hormone substance. Endocrinology 21:649–654

Handelsman DJ (1985) Hypothalamic-pituitary gonadal dysfunction in chronic renal failure, dialysis, and renal transplantation. Endo Rev 6:151–182

Handelsman DJ, Dong Q (1993) Hypothalamo-pituitary gonadal axis in chronic renal failure. Endocrin Metab Clin 22:145–161

Handelsman DJ (1997) Testicular dysfunction in systemic diseases. In: Nieschlag E, Behre HM (eds) Andrology: Male reproductive health and dysfunction. Springer-Verlag, Berlin, pp 227–237

Hansell DT, Davies JW, Shenkin A, Garden OJ, Burns HJ, Carter DC (1989) The effects of an anabolic steroid and peripherally administered intravenous nutrition in the early postoperative period. J Parent Ent Nutr 13:349–358

Hausmann DF, Nutz V, Rommelsheim K, Caspari R, Mosebach KO (1990) Anabolic steroids in polytrauma patients. Influence on renal nitrogen and amino acid losses: a double-blind study. J Parent Ent Nutr 14:111–114

Hazelton RA, McCruden AB, Sturrock RD, Stimson WH (1983) Hormonal manipulation of the immune response in systemic lupus erythematosus: a drug trial of an anabolic steroid, 19-nortestosterone. Ann Rheum Dis 42:155–157

Hendler ED, Goffinet JA, Ross S, Longnecker RE, Bakovic V (1974) Controlled study of androgen therapy in anemia of patients on maintenance hemodialysis. N Engl J Med 291:1046–1051

Hengge UR, Baumann M, Maleba R, Brockmeyer NH, Goos M (1996) Oxymetholone promotes weight gain in patients with advanced human immunodeficiency virus (HIV-1) infection. Br J Nutr 75:129–138

Hentzer E, Madsen PC (1967) Testosterone in the treatment of arterial insufficiency of the lower limbs. Scand J Clin Lab Invest 99:198–206

Hosea SW, Santaella ML, Brown EJ, Berger M, Katusha K, Frank MM (1980) Long-term therapy of hereditary angioedema with danazol. Ann Intern Med 93:809–812

Ingle JN (1984) Additive hormonal therapy in women with advanced breast cancer. Cancer 53: 766–777

Inkovaara J, Gothoni G, Halttula R, Heikinheimo R, Tokola O (1983) Calcium, vitamin D and anabolic steroid in treatment of aged bones: double-blind placebo-controlled long-term clinical trial. Age and Ageing 12:124–130

Itil TM, Cora R, Akpinar S, Herrmann WM, Patterson CJ (1974) "Psychotropic" action of sex hormones: computerized EEG in establishing the immediate CNS effects of steroid hormones. Curr Therap Res 16:1147–1170

Itil TM, Michael ST, Shapiro DM, Itil KZ (1984) The effects of mesterolone, a male sex hormone in depressed patients (a double blind controlled study). Meth and Find Exptl Clin Pharmacol 6:331–337

Jayson MI, Holland CD, Keegan A, Illingworth K, Taylor L (1991) A controlled study of stanozolol in primary Raynaud's phenomenon and systemic sclerosis. Ann Rheum Dis 50:41–47

Johnson DH, Linde R, Hainsworth JD, Vale W, Rivier J, Stein R, Flexner J, R van Welch, Greco FA (1985) Effect of a luteinizing hormone releasing hormone agonist given during combination chemotherapy on posttherapy fertility in male patients wth lymphoma: preliminary observations. Blood 65:832–836

Joura EA, Zeisler H, Bancher-Todesca D, Sator MO, Schneider B, Gitsch G (1997) Short-term effects of topical testosterone in vulvar lichen sclerosis. Obstet Gynecol 89:297–299

Kaltwasser JP, Dix U, Schalk KP, Vogt H (1988) Effect of androgens on the response to anti-thymocyte globulin in patients with aplastic anaemia. Eur J Haematol 40:111–118

Kassmann K, Rappaport R, Broyer M (1992) The short-term effect of testosterone on growth in boys on hemodialysis. Clin Nephrol 37:148–154

Keogh A, Spratt P, McCosher C, McDonald P, Mundy J, Kaan A (1995) Ketoconazole to reduce the need for cyclosporine after cadiac transplantation. N Engl J Med 333:628–633

Kerrebijn KF, Delver A (1969) Ethylestrenol (Orgabolin): effects on asthmatic children during corticosteroid treatment. Scand J Resp Dis 68:70–77

Kopera H (1976) Miscellaneous uses of anabolic steroids. In: Kochakian CD (ed) Anabolic-androgenic steroids. Berlin: Springer-Verlag, pp 535–625

Kotler DP, Tierney AR, Wang J, Pierson RN (1989) Magnitude of body-cell-mass depletion and the timing of death from wasting in AIDS. Am J Clin Nutr 50:444–447

Kreusser ED, Hetzel WD, Hautmann R, Pfeiffer EF (1990) Reproductive toxicity with and without LHRHa administration during adjuvant chemotherapy in patients with germ cell tumors. Hormon Metab Res 22:494–498

Kruskemper HL (1968) Anabolic steroids. Academic Press, New York

Lahita RG, Cheng CY, Monder C, Bardin CW (1992) Experience with 19-nortestosterone in the therapy of systemic lupus erythematosus: worsened disease after treatment with 19-nortestosterone in men and lack of improvement in women. J Rheumatol 19:547–555

Lewis L, Dahn M, Kirkpatrick JR (1981) Anabolic steroid administration during nutritional support: a therapeutic controvery. J Parent Ent Nutr 5:64–66

Li Bock E, Fulle HH, Heimpel H, Pribilla W (1976) Die Wirkung von Mesterolon bei Panmyelopathien und renalen Anaemien. Med Klin 71:539–547

Lovejoy JC, Bray GA, Greeson CS, Klemperer M, Morris J, Partington C, Tulley R (1995) Oral anabolic steroid treatment, but not parenteral androgen treatment, decreases abdominal fat in obese, older men. Int J Obesity 19:614–624

Lubahn DB, Moyer JS, Golding TS, Couse JF, Korach KS, Smithies O (1993) Alteration of reproductive function but not prenatal sexual development after insertional disruption of the mouse estrogen receptor gene. Proc Natl Acad Sci 90:11162–11166

Maddrey WC (1986) Is therapy with testosterone or anabolic-androgenic steroids useful in the treatment of alcoholic liver disease? Hepatology 6:1033–1035

Mandelli F, Amadori S, Dini E, Grignani F, Leoni P, Liso V, Martelli M, Neri A, Petti MC, Ferrini PR (1981) Randomized clinical trial of immunotherapy and androgenotherapy for remission maintenance in acute non-lymphocytic leukemia. Leuk Res 5:447–452

Marin P (1995) Testosterone and regional fat distribution. Obesity Res 3 (Suppl 4): 609S–612S

Marin P, Holmang S, Jonsson L, Sjostrom L, Kvist H, Holm G, Lindstedt G, Bjorntorp P (1992) The effects of testosterone treatment on body composition and metabolism in middle-aged obese men. Int J Obesity 16:991–997

Marin P, Oden B, Bjorntorp P (1995) Assimilation and mobilization of triglycerides in subcutaneous abdominal and femoral adipose tissue in vivo in men: effects of androgens. J Clin Endocrinol Metab 80:239–243

Masala A, Faedda R, Alagna A, Satta A, Chiarelli G, Rovasio PP, Ivaldi R, Taras MS, Lai E, Bartoli E (1997) Use of testosterone to prevent cyclophosphamide-induced azoospermia. Ann Intern Med 126:292–295

Matsumoto A, Sandblom RE, Schoene RB, Lee KA, Giblin EC, Pierson DJ, Bremner WJ (1985) Testosterone replacement in hypogonadal men: effects on obstructive sleep apnea, respiratory drives and sleep. Clin Endocrinol 22:713–721

McMullin GM, Watkin GT, Coleridge Smith PD, Scurr JH (1991) Efficacy of fibrinolytic enhancement with stanozolol in the treatment of venous insufficiency. Aust NZ J Surg 61:306–309

Mendenhall CL, Anderson S, Garcia-Pont P, Goldberg S, Kierman T, Seeff L, Sorrell M, Tamburro C, Weesner R, Zetterman R, Chedid A, Chen T, Rabin L, Veterans Administration Cooperative Study on Alcoholic Hepatitis (1984) Short-term and long-term survival in patients with alcoholic hepatitis treated with oxandrolone and prenisolone. N Engl J Med 311:1464–1470

Mendenhall CL, Moritz TE, Roselle GA, Morgan TR, Nemchausky BA, Tamburro CH, Schiff ER, McClain CJ, Marsano LS, Allen JI (1993) A study of oral nutritional support with oxandrolone in malnourished patients with alcoholic hepatitis. Hepatology 17:564–576

Millman RP, Kimmel PL, Shore ET, Wasserstein AG (1985) Sleep apnea in hemodialysis patients: the lack of testosterone effect on its pathogenesis. Nephron 40:407–410·

Morris ID (1993) Protection against cytotoxic-induced testis damage – experimental approaches. Eur Urol 23:143–147

Naik RB, Gibbons AR, Gyde OH, Harris BR, Robinson BH (1978) Androgen trial in renal anaemia. Proc Eur Dial Trans Assoc 15:136–143

Najean Y, Joint Group for the Study of Aplastic and Refractory Anaemias (1981) Long-term follow-up in patients with aplastic anemia. A study of 137 androgen-treated patients surviving more than two years. Am J Med 71:543–551

Najean Y, Pecking A (1979) Refractory anemia with excess of blast cells: prognostic factors and effect of treatment with androgens or cytosine arabinoside. Results of a prospective trial in 58 patients. Cooperative Group for the Study of Aplastic and Refractory Anemias. Cancer 44:1976–1982

Need AG, Nordin BEC, Chatterton BE (1993) Double-blind placebo-controlled trial of treatment of osteoporosis with the anabolic steroid nandrolone decanoate. Osteoporosis Int (Suppl 1): S218–S222

Neff MS, Goldberg J, Slifkin RF, Eiser AR, Calamia V, Kaplan M, Baez A, Gupta S, Mattoo N (1981) A comparison of androgens for anemia in patients on hemodialysis. N Engl J Med 304:871–875

Nicolodi M, Sicuteri F, Poggioni M (1993 a) Hypothalamic modulation of nociception and reproduction in cluster headache. I. Therapeutic trials of leuprolide. Cephalgia 13:253–257

Nicolodi M, Sicuteri F, Poggioni M (1993b) Hypothalamic modulation of nociception and reproduction in cluster headache. II. Testosterone-induced increase of sexual activity in males with cluster headache. Cephalgia 13:258–260

Nieschlag E, Cüppers HJ, Wickings EJ (1977) Influence of sex, testicular development and liver function on the bioavailability of oral testosterone. Eur J Clin Invest 7:145–147

Nieschlag E, Nieschlag S, Behre HM (1993) Lifespan and testosterone. Nature 366:215

Oster MH, Enders SR, Samuels SJ, Cone LA, Hooton TM, Browder HP, Flynn NM (1994) Megestrol acetate in patients with AIDS and cachexia. Ann Intern Med 121:400–408

Paslin D (1996) Androgens in the topical treatment of lichen sclerosis. Int J Dermatol 35:298–301

Quigley CA, DeBellis A, Marschke KB, El-Awady MK, Wilson EM, French FF (1995) Androgen receptor defects: historical, clinical and molecular perspectives. Endo Rev 16:271–321

Rabkin JG, Rabkin R, Wagner G (1995) Testosterone replacement therapy in HIV illness. Gen Hosp Psychiat 17:37–42

Reid IR, Wattie DJ, Evans MC, Stapleton JP (1996) Testosterone therapy in glucocorticoid-treated men. Arch Intern Med 156:1173–1177

Reid IR, Ibbertson HK, France JT, Pybus J (1985) Plasma testosterone concentrations in asthmatic men treated with glucocorticoids. Br J Med 291:574

Rittmaster RS (1994) Finasteride. N Engl J Med 330:120–125

Sandblom RE, Matsumoto AM, Scoene RB, Lee KA, Giblin EC, Bremner WJ, Pierson DJ (1983) Obstructive sleep apnea induced by testosterone administration. N Engl J Med 308:508–510

Santavirta S, Konttinen YT, Heliovaara M, Knekt P, Luthje P, Aromaa A (1992) Determinants of osteoporotic thoracic vertebral fracture. Screening of 57,000 Finnish women and men. Acta Orthoped Scand 63:198–202

Savin RC (1968) The ineffectiveness of testosterone in male pattern baldness. Arch Dermatol 98:512–514

Savvas M, Studd JWW, Norman S, Leather AT, Garnett TJ (1992) Increase in bone mass after one year of percutaenous oestradiol and testosterone implants in post-menopausal women who have previously received long-term oral ostrogens. Br J Obstet Gynaecol 99:757–760

Schambelan M, Mulligan K, Grunfeld C, Daar ES, LaMarca A, Kotler DP, Wang J, Bozette SA, Breitmeyer JB, Serostim Study Group (1996) Recombinant human growth hormone in patients with HIV-associated wasting. Ann Intern Med 125:873–882

Schneider BK, Pickett CK, Zwillich CW, Weil JV, McDermott MT, Santen RJ, Varano LA, White DP (1986) Influence of testosterone on breathing during sleep. J Appl Physiol 61:618–623

Schols AM, Soeters PB, Mostert R, Pluymers RJ, Wouters EF (1995) Physiologic effects of nutritional support and anabolic steroids in patients with chronic obstructive pulmonary disease. A placebo-controlled randomized trial. Am J Respir Crit Care Med 152:1268–1274

Shalet SM, Beardwell CG, Jacobs HS, Pearson D (1978) Testicular function following irradiation of the human prepubertal testis. Clin Endocrinol 9:483–490

Sheffer AL, Fearon DT, Austen KF (1977) Methyltestosterone therapy in hereditary angioedema. Ann Intern Med 86:306–308

Sherwin BB, Gelfand MM (1985) Differential symptom response to parenteral estrogen and/or androgen administration in the surgical menopause. Am J Obstet Gynecol 151:153–160

Sherwin BB, Gelfand MM (1987) The role of androgens in the maintenance of sexual functioning in oophorectomized women. Psychosom Med 49:397–409

Smith EP, Boyd J, Frank GR, Takahashi H, Cohen RM, Specker B, Williams TC, Lubahn DB, Korach KS (1994) Estrogen resistance caused by a mutation in the estrogen-receptor gene in a man. N Engl J Med 331:1056–1061

Spaulding WB (1960) Methyltestosterone therapy for hereditary episodic edema (hereditary angioneurotic edema). Ann Intern Med 53:739–745

Spiers ASD, DeVita SF, Allar MJ, Richards S, Sedranak N (1981) Beneficial effects of an anabolic steroid during cytotoxic chemotherapy for metastatic cancer. J Med 12:433–446

Teruel JL, Aguilera A, Marcen R, Antolin JN, Otero GG, Ortuno J (1996a) Androgen therapy for anaemia of chronic renal failure. Scand J Urol Nephrol 30:403–408

Teruel JL, Marcen R, Navarro-Antolin J, Aguilera A, Fernandez-Juarez G, Ortuno J (1996b) Androgen versus erythropoietin for the treatment of anemia in hemodialyzed patients: a prospective study. J Am Soc Nephrol 7:140–144

Thompson PD, Cullinane EM, Sady SP, Chenevert C, Saritelli AL, Sady MA, Herbert PN (1989) Contrasting effects of testosterone and stanozolol on serum lipoprotein levels. J Am Med Ass 261:1165–1168

Tricker R, Casaburi R, Storer TW, Clevenger B, Berman N, Shirazi A, Bhasin S (1996) The effects of supraphysiological doses of testosterone on angry behavior in healthy eugonadal men – a clinical research center study. J Clin Endocrinol Metab 81:3754–3758

Tweedle D, Walton C, Johnston IDA (1973) The effect of an anabolic steroid on postoperative nitrogen balance. Br J Clin Practice 27:130–132

Urman B, Pride SM, Yuen BH (1991) Elevated serum testosterone, hirsutism, and virilism associated with combined androgen-estrogen hormone replacement therapy. Obstet Gynecol 77:1124–1131

van Coevorden A, Stolear JC, Dhaene M, Herweghem JLV, Mockel J (1986) Effect of chronic oral testosterone undecanoate administration on the pituitary-testicular axes of hemodialyzed male patients. Clin Nephrol 26:48–54

van Roenn JH, Armstrong D, Kotler DP, Cohn DL, Klimas NG, Tchekmedyian NS, Cone L, Brennan PJ, Weitzman SA (1994) Megesterol acetate in patients with AIDS-related cachexia. Ann Intern Med 121:393–399

Van Vollenhoven RF, Engleman EG, McGuire JL (1995) Dehydroepiandrosterone in systemic lupus erythematosus. Arthritis Rheum 38:1826–1831

Watson RN, Bradley MH, Callahan R, Peters BJ, Kory RC (1959) A six-month evaluation of an anabolic drug, norethandrolone, in underweight persons. Am J Med 26:238–248

Wattel E, Cambier N, Caulier MT, Sautiere D, Bauters F, Fenaux P (1994) Androgen therapy in myelodysplastic syndromes with thrombocytopenia: a report on 20 cases. Br J Haematol 87:205–208

Waxman JH, Ahmed R, Smith D, Wrigley PFM, Gregory W, Shalet S, Crowther D, Rees LH, Besser GM, Malpas JS, Lister TA (1987) Failure to preserve fertility in patients with Hodgkin's disease. Cancer Chemother Pharmacol 19:159–162

Webb WR, Doyle RS, Howard HS (1960) Relative metabolic effects of calories, protein, and an anabolic steroid (19-nortestosterone) in early postoperative period. Metabolism 9:1047–1057

Wells R (1960) Prednisolone and testosterone propionate in cirrhosis of the liver: a controlled trial. Lancet ii:1416–1419

Williams JS, Stein JH, Ferris TF (1974) Nandrolone decanoate therapy for patients receiving hemodialysis. A controlled study. Arch Intern Med 134:289–292

Wu S, Weng X, (1993) Therapeutic effects of an androgenic preparation on myocardial ischemia and cardiac function in 62 elderly male coronary heart disease patients. Chin Med J 106:415–418

Young GA, Yule AG, Hill GL (1983) Effects of an anabolic steroid on plasma amino acids, proteins, and body composition in patients receiving intravenous hyperalimentation. J Parent Ent Nutr 7:221–225

Young N, Griffith P, Brittain E, Elfenbein G, Gardner F, Huang A, Harmon D, Hewlett J, Fay J, Mangan K (1988) A multicenter trial of antithymocyte globulin in aplastic anemia and related diseases. Blood 72:1861–1869

Zawilska K, et al. (1990) Nebulised heparin and anabolic steroid in the prevention of postoperative deep venous thrombosis following elective abdominal surgery. Folia Haematologica 117:699–707

18 Testosterone in male contraception

Eberhard Nieschlag and Hermann M. Behre

Contents

The Weimar Manifesto on Male Contraception

Leading researchers from all over the world met in Weimar (Germany) on June 29, 1997 for a Summit Meeting to discuss the current status and prospects for male hormonal contraception. The researchers represented academic institutions as well as international organisations such as the Population Council (New York) and the World Health Organisation (Geneva).

All agreed that contraception is an essential component of reproductive health for men and women. Without contraception, all socioeconomic progress and the future of this planet are endangered. Apart from vasectomy and condoms, no contraceptives are available for men. New forms of male contraception are required to involve men more actively in family planning and to share the benefits and burdens of contraception more equitably.

The researchers agreed that clinical research had shown that a reversible male hormonal contraceptive is feasible and effective. However, it has not yet been possible to show wide acceptability due to the inadequate funding for largescale clinical trials. There is a pressing need for product development and clinical trials of new, long-acting testosterone preparations and combinations with other agents. With adequate funding, a male hormonal contraceptive could be available in the near future.

Since any contraceptive method can only be brought to practical usage by pharmaceutical companies, the researchers urged the pharmaceutical industry to become actively involved in the development of novel male contraceptives. They also appealed in the spirit of the UN Cairo Declaration to politicians and research foundations to commit themselves to the development of male contraception for the sake of future generations.

Prof. E. Nieschlag (Chair)
University of Münster, Germany

Dr. H. M. Behre
University of Münster, Germany

Prof. W. Bremner
University of Washington, Seattle, USA

Prof. E. Diczfalusy
Karolinska Hospital, Stockholm, Sweden

Prof. D. J. Handelsman
University of Sydney, Australia

Prof. I. Huhtaniemi
University of Turku, Finland

Dr. E. Johansson
Population Council, New York, USA

Prof. M. T. Mbizvo
WHO, Geneva, Switzerland

Dr. C. Meriggiola
University of Bologna, Italy

Dr. K. Sundaram
Population Council, New York, USA

Prof. R. Swerdloff
University of California, Los Angeles, USA

Prof. C. Wang
University of California, Los Angeles, USA

Dr. F. Wu
University of Manchester, UK

18.1 General prospect

The first of a series of topics that Kate Noble lists in an article in a special TIME-Magazine issue (Winter 1997/1998) on "Shape of things to come" in the years 1999 to 2500 is: "Male birth control pill or contraceptive injection becomes commonly available". This prediction does not come as a surprise to the researcher active in the field. The surprising element is the year designated to witness the "common availability": 1999! From all we know today, this male contraceptive will be hormone-based and while scientists have believed for a long time that the principle of hormonal male contraception is ready to be converted into a viable consumable drug, the pharmaceutical industry, for whom drug development is an inherent task, has refused to fulfill this task for hormonal male contraception. Confronted with this reluctance, leading scientists active in the field of male contraception felt compelled to draft the "Weimar Manifesto on Male Contraception" in June 1997 (see insert on opposite page) and to urge industry to assume an active role in male contraceptive development. Since then, at least three companies have publicly declared their new commitment to male contraception development. An enthusiastic article in the Guardian on October 6, 1997 ("Trials may put men on the pill by 2000") predicts general availability of a hormonal male contraceptive by 2000. This statement probably gave rise to TIME's optimism.

All hormonal male contraceptives clinically tested to date are based on testosterone, either on testosterone alone or on a combination of testosterone with other hormones, in particular with either gestagens or GnRH analogues. Therefore it is appropriate to include an overview on current hormonal approaches to male contraception in this volume.

18.2 Principle of hormonal male contraception

The testes have an endocrine and an exocrine function: the production of androgens and of male gametes. Suppression of gamete production or interference with gamete function without affecting the endocrine function is the goal of endocrine approaches to male fertility regulation. However, since the two functions of the testes are closely interdependent, it has remained impossible so far to suppress spermatogenesis exclusively and reversibly without significantly affecting androgen synthesis.

FSH and LH/testosterone are responsible for the maintenance of fully normal spermatogenesis (for review see Chapter 4 by Weinbauer and Nieschlag in this volume and Weinbauer and Nieschlag 1996). If only one of the two is eliminated, spermatogenesis will be reduced, but only in quantitative terms, i.e. fewer but normal sperm will be produced and azoospermia will not be achieved. This has been demonstrated in monkeys by the elimination of FSH by immunoneutralization, resulting in reduced sperm numbers but not in complete azoospermia (Srinath et al. 1983), which – at least until quite recently – was considered to be required for an effective male method. Therefore, even if new modalities for the selective suppression of FSH or FSH action should become available, it remains doubtful whether they would lead to a method for male contraception (Nieschlag 1986). However, in bonnet monkeys immunization against FSH or the FSH receptor led to an impairment of the fertilizing capacity of sperm (Moudgal et al. 1992, 1997a). Ongoing trials will show whether similar effects can be obtained in men (Moudgal et al. 1997b, 1997c).

Until such results become available, the concept of azoospermia remains valid as a prerequisite for effective hormonal male contraception. To achieve this goal not only FSH must be suppressed, but also the testes must be depleted of testosterone. Since testosterone alone can maintain spermatogenesis and much lower testosterone concentrations appear to be necessary for maintenance of spermatogenesis than previously considered, the depletion of intratesticular testosterone must be quite complete (Weinbauer and Nieschlag 1996). In order to maintain androgenicity, including libido, potency, male sex characteristics, psychotropic effects, protein anabolism, bone structure and hematopoesis, testosterone levels in the general circulation have to be replaced while the testes themselves are depleted of testosterone.

This leads to the general principle of endocrine regulation of male fertility, namely the suppression of FSH and LH, resulting in a depletion of intratesticular testosterone and cessation of spermatogenesis, while at the same time, peripheral testosterone is substituted with an androgen preparation. This can be achieved by testosterone alone, or by the combination of testosterone with other substances suppressing pituitary gonadotropin secretion. As in female hormonal contraceptives, gestagens as such pituitary-suppressing agents have been tested in men in combination with androgens. GnRH agonists as well as antagonists are being considered as further possible combinations with androgens.

18.3 Testosterone alone

18.3.1 Testosterone enanthate

According to the principle outlined above, testosterone should be the first choice for hormonal male contraception since it not only suppresses pituitary LH and FSH secretion, but also replaces testosterone. Indeed, since the 1970s various investigations have been undertaken to suppress spermatogenesis with testosterone. Not until 1990 was an initial study testing this form of male contraception published by the WHO, the first study ever performed on the efficacy of hormonal male contraception. Volunteers in ten centers on four continents participated and received 200 mg testosterone enanthate intramuscularly per week. Those volunteers developing azoospermia within the first six months continued to receive injections for a further year. In this period (efficacy phase) couples refrained from using any further contraceptive methods. A total of 137 men reached the efficacy phase. During this period only one pregnancy occurred. This high rate of efficacy is well comparable to that of established female methods. This was a very encouraging result. However, only about two-thirds of all participants developed azoospermia. The other volunteers showed strong suppression of spermatogenesis, as evidenced by oligozoospermia in the ejaculate (WHO 1990).

In order to answer the question of whether men developing oligozoospermia can be considered infertile, a second worldwide multicenter study followed (WHO 1996). In this study azoospermia, again, proved to be a most effective prerequisite for contraception. If sperm concentrations, however, failed to drop below 3×10^6/ml, resulting pregnancy rates were higher than when using condoms. When sperm concentrations decreased below 3×10^6/ml, which was the case in 98% of the participants, protection was not as effective as for azoospermic men but was better than that offered by condoms.

The WHO multicenter studies revealed an interesting phenomenon: the rate of azoospermia was greater in East-Asian than in Caucasian men (WHO 1995). This finding was also confirmed by independent studies using testosterone enenthate injections in men in Indonesia (Arysad 1993), Thailand (Aribarg et al. 1996) and China (Cao et al. 1996).

Even if these WHO studies represented a breakthrough by confirming a principle of action, they did not offer a practicable method. For a method requiring weekly intramuscular injections is not acceptable for broad use. Moreover, several months (on average four months) were required before sperm production reached significant suppression. For this reason current research is concentrating on the development of long-acting testosterone preparations and on action to improve overall effectiveness.

18.3.2 Testosterone buciclate

Under the auspices of WHO a synthesis program identified testosterone buciclate as a testosterone ester with long-lasting effectiveness. First tested in monkeys and then in hypogonadal patients, it showed a long effective phase of 3–4 months after a single injection (see Chapter 11 by Behre and Nieschlag in this volume). A single injection of 1200 mg in a contraceptive study resulted in suppression of spermatogenesis comparable to that of weekly enanthate injections (Behre et al. 1995a). However, in this early study only eight volunteers were included and results of extended trials are eagerly awaited. So far, testosterone buciclate has proved to be the longest-acting injectable testosterone preparation and thus most suited for contraceptive purposes.

18.3.3 Testosterone undecanoate

If testosterone is suited for contraception, the orally effective testosterone undecanoate should provide the male contraceptive "pill". This possibility was tested in the early phase of development. However, even when high doses of 3×80 mg were taken daily for 12 weeks only one of seven volunteers developed azoospermia (Nieschlag et al. 1978). Although this result was disappointing, the study demonstrated that constant levels of testosterone in serum are important to suppress pituitary gonadotropins.

Currently, testosterone undecanoate is under development as an injectable testosterone preparation (see Chapter 10 by Nieschlag and Behre in this volume). Although the half-life is not quite as long as that of testosterone buciclate, it is considerably longer than that of testosterone enanthate and therefore appears to be well suited for contraceptive purposes. Clinical trials have started and results are anticipated with great interest.

18.3.4 Testosterone implants

Implants consisting of pure testosterone are used for substitution in hypogonadism in some countries (see Chapter 12 by Handelsman in this volume). In male contraceptive studies one-time application showed efficacy comparable to weekly testosterone enanthate injections (Handelsman et al. 1992). The disadvantage of minor surgery required for implantation under the abdominal skin is compensated for by their low price and long duration of action. Studies using repeated implant applications of extended periods are awaited.

18.3.5 19-Nortestosterone

When searching for preparations with longer-lasting effectiveness, 19-nortestosterone-hexoxyphenylpropionate was tested whose spectrum of effects is very similar to that of testosterone, and which has been used as an anabolic since the 1960s (molecular structure in Fig. 10.1). The 19-nortestosterone ester injected every three weeks enabled azoospermia to be reached by as many men as by testosterone enanthate. Thus the 19-nortestosterone ester is as effective as testosterone enanthate but allows a longer injection interval (Behre et al. 1992; Knuth et al. 1985; Schürmeyer et al. 1984). However, 19-nortestosterone is not fully equivalent to testosterone as it is converted to estrogens to a lesser degree than testosterone. Although we could not detect any side-effects in the trials using 19-nortestosterone, long-term untoward effects e.g. on bones cannot be excluded. In the light of emerging long-acting testosterone preparations 19-nortestosterone therefore appears less attractive for contraception.

18.4 Testosterone combined with gestagens

In order to accelerate the onset of testosterone effectiveness and to increase azoospermia rates, it was attempted to combine testosterone with various gestagens.

18.4.1 "Early" combinations

Gestagens are potent inhibitors of LH and FSH secretion and therefore have been tested, in combination with testosterone, for male fertility regulation. Norethindrone, medroxyprogesterone acetate (MPA), depot-MPA (DMPA), 17-hydroxyprogesterone capronate and megestrol acetate have been used in clinical trials initiated by the WHO (1972 – 1983) and the Population Council (Schearer et al. 1978). The most favorable combination was the monthly intramuscular injection of 200 mg DMPA plus 200 mg testosterone enanthate or testosterone cypionate; this combination gave the best results in suppressing spermatogenesis and the incidence of untoward side-effects was low. No negative effects on libido and potency were observed. However, this combination did not produce azoospermia uniformly; therefore its possible efficacy remained uncertain.

To develop a method that could be self-administered, 20 mg of MPA was given orally, together with 50 to 100 mg of testosterone percutaneously via a cream for one year (Soufir et al. 1983). Sperm counts in the six volunteers were suppressed to less than 5 million/ml for most of the study time, but azoospermia was not achieved. Guerin and Roullet (1988) tried various combinations of oral gestagens and transdermal or oral testosterone preparations. In all 12 volunteers treated with 5 or 10 mg/day 19-norethisterone orally, plus 250 mg (!) percutaneous testosterone daily, azoospermia was achieved within eight weeks. These favorable results were mitigated by the elevated testosterone levels in the blood of female partners. In a similar study, the wives of five of the twelve men treated developed hirsutism secondary to testosterone being transferred to their skin from their husband's abdomen or buttocks (Delanoe et al. 1984). These untoward side-effects made this approach, based on topical androgen application, unattractive, although it demonstrated the potential of a combined regimen.

Danazol, a derivative of ethinyltestosterone, strongly inhibits LH and FSH secretion when given orally, but is devoid of other significant estrogenic or gestagenic activities. Various combinations of daily and oral doses of danazol with monthly injections of testosterone enanthate have been tested (Leonard and Paulsen 1978). A daily oral dose of 600 – 800 mg danazol, in combination with 200 mg of testosterone enanthate monthly, induced azoospermia, or severe oligozoospermia. However, since the rate of azoospermia was not significantly higher than that achieved with testosterone alone, and since the daily steroid load was quite high and liver toxicity had been reported by female users (for treatment of endometriosis), these studies were not continued.

18.4.2 Testosterone or 19-nortestosterone plus DMPA

Since monotherapy with the long-acting androgen ester 19-nortestosterone injected every three weeks resulted in effective suppression of spermatogenesis to azoospermia in about 70% of the volunteers (Schürmeyer et al. 1984) we tested the possibility of even more complete suppression of spermatogenesis (Knuth et al. 1989). Twelve volunteers were injected weekly with 200 mg 19-nortestosterone hexoxyphenylpropionate, followed by injections with the same dose every three weeks up to week 15. In addition, the volunteers were injected with 250 mg DMPA in weeks 0, 6 and 9. Azoospermia was achieved in nine of twelve volunteers during the study course, while in three of the remaining four volunteers, spermatogenesis was suppressed to single sperm, and in one volunteer to a sperm concentration of 1,3 mill/ml.

The promising results prompted the WHO Task Force on the Regulation of Male Fertility to launch a large-scale multicenter trial in five centers in Indonesia, comparing the effectiveness of testosterone enanthate, or 19-nortestosterone hexyoxyphenylpropionate in combination with DMPA (WHO 1993). Surprisingly, 43/45 and 44/45 subjects in the testosterone and the 19-nortestosterone groups respectively suppressed to azoospermia. Unfortunately, this study had failed to include groups treated with the androgens alone, so that it remained unclear whether the azoospermia rates of 97% and 98% were due to the combination treatment or could also be achieved by the androgens alone. The answer to this question was indirectly provided by Arysad (1993) in an independent study who achieved azoospermia in 7/7 men from Palembang (Indonesia) by weekly injections of 100 mg testosterone enanthate.

The results from Germany and Indonesia quoted above highlight the ethnic differences in response to the same contraceptive agent which is also seen in Chinese men compared to Caucasians. The results reveal that testosterone alone may be a much more effective contraceptive in East-Asian than in Caucasian men. Notwithstanding these promising results, it should be kept in mind that in the Indonesian, as well as in European studies, it took twenty weeks to achieve azoospermia. Therefore, even in the good responders, i.e. East-Asian men, another agent may be required to suppress spermatogenesis more rapidly. Perhaps such an additional agent may only be required at the beginning of the contraceptive phase. From all published evidence, however, it appears that DMPA may not be the appropriate candidate for faster suppression.

18.4.3 Testosterone and levonorgestrel

Levonorgestrel has been widely used for contraception in females either orally or as an implant and has proved safe and effective. Although early studies combining levonorgestrel with testosterone enanthate were not very encouraging (Fogh et al. 1980), a recent controlled trial comparing testosterone enanthate (100 mg/week) with testosterone enanthate in combination with 0,5 mg levonorgestrel orally showed that the combination resulted in a more pronounced suppression of spermatogenesis than testosterone enanthate alone (Bebb et al. 1996).

Encouraged by the renewed interest in levornorgestrel we recently concluded a trial combining oral levonorgestrel with transdermal testosterone applied to non-genital skin. The advantage of such a combination would be that it is completely self-administered and thus independent of medical personnel. Unfortunately, the results were disappointing as only insufficient suppression of spermatogenesis occurred. We presume that the testosterone dose absorbed from the transdermal systems was too low and often impeded by inadequate adhesiveness to the skin of the systems (Büchter et al. unpublished). The study again emphasizes the need for constant serum testosterone levels to suppress gonadotropins, even when co-administered with a potent gestagen.

18.4.4 Testosterone and cyproterone acetate

Animal studies and studies in sexual delinquents have shown that the antiandrogen cyproterone acetate, which can be considered a potent gestagen, suppresses spermatogenesis, an effect exerted through the suppression of pituitary gonadotropin secretion. In clinical trials using 5 to 20 mg cyproterone acetate per day for up to 16 weeks, sperm counts and motility were reduced markedly (Fogh et al. 1979; Moltz et al. 1980; Wang and Yeung 1980). Thus, cyproterone acetate appeared to be a possibility for male fertility control. However, decreases in serum testosterone levels to below normal were also observed. Some of the volunteers complained of fatigue, lassitude and decrease in libido and potency attributable to the diminished testosterone levels. Thus, although cyproterone acetate produces infertility, clinical trials have not been continued as the lowered testosterone levels were considered intolerable. The consequence would have been to combine it with testosterone. The combination of an androgen with an antiandrogen, however, appeared somewhat illogical and it therefore took some time until this lead was taken up again.

In a recent study cyproterone acetate (50 or 100 mg/day) was combined with testosterone enanthate injections (100 mg per week for 4 months). All ten subjects receiving either dose of cyproterone acetate reached azoospermia, on average 7 weeks after initiation of treatment. Thus, a very good result in terms of sperm suppression was achieved. However, these volunteers showed a decrease in red blood cells, hematocrit and body weight, probably due to the high dose of antiandrogenic gestagen used in the study (Meriggiola et al. 1996).

In an attempt to create a complete oral male contraceptive, Meriggiola et al. (1997) administered cyproterone acetate (12.5 mg twice daily) and testosterone undecanoate (80 mg twice daily) to eight volunteers. Unfortunately, only one became azoospermic and five had sperm counts suppressed below 3 mill/ml. Whether this low suppression of spermatogenesis was due to the oral testosterone undecanoate (instead of the injected testosterone enanthate) or due to the lower cyproterone acetate dose cannot be decided since both medications were changed simultaneously. Although the cyproterone acetate dose was reduced, a decrease in hemoglobin and hematocrit occurred, but no other sign of decreased androgenicity could be detected. Essentially, this result is not much better than when oral testosterone undecanoate was given alone (Nieschlag et al. 1977) and once again, the final formula for male contraception remained elusive.

18.5 Testosterone and GnRH analogues

GnRH analogues are promising hormone preparations for enhancing suppression of gonadotropins and spermatogenesis. These peptide hormones act specifically on the GnRH receptor in the anterior pituitary gland. In contrast to naturally-occurring GnRH, GnRH agonists – after producing an initial stimulation of gonadotropin release for approximately two weeks – lead to GnRH receptor down-regulation and thereby to suppression of LH and FSH synthesis and secretion. GnRH antagonists cause a competitive GnRH receptor blockage and effect an immediate and effective suppression of gonadotropins.

Although GnRH antagonists, with their immediate suppressive effect on gonadotropins, appear more attractive for contraceptive purposes, it took much longer to develop them than GnRH agonists, because of their more complex chemical structure and synthesis. Initial contraceptive trials were performed using testosterone plus GnRH agonists and GnRH antagonists were tested only recently.

18.5.1 Testosterone and GnRH agonists

Between 1979 and 1992, 12 trials using GnRH agonists, mostly in combination with testosterone for hormonal male contraception, were published (for review Nieschlag et al. 1992). Altogether 106 volunteers participated in these trials. The GnRH agonists decapeptyl, buserelin and nafarelin were administered at daily doses of 5–500 µg/volunteer for periods of 10–30 weeks. In about 30% of men, sperm production could be suppressed below 5x106/ml and azoospermia occurred in 21 men, while in the remaining volunteers, sperm numbers were only slightly reduced or remained unaffected. One explanation for the ineffectiveness of GnRH agonist plus androgen is the escape of FSH suppression after several weeks of GnRH agonist treatment (Behre et al. 1992; Bhasin et al. 1994).

Altogether, GnRH agonists in combination with testosterone did not prove useful in male contraception. At times it has been suggested that higher doses of the GnRH agonists should be used, but currently no further clinical studies appear to be under way.

18.5.2 Testosterone and GnRH antagonists

In contrast to GnRH agonists, GnRH antagonists produce a precipitous and prolonged fall of LH and FSH serum levels in men (e.g. Behre et al. 1994; Pavlou et al. 1989). As mentioned above, it took much longer to develop GnRH antagonists that were suitable for clinical application than it did for GnRH agonists, and clinical trials using GnRH antagonists for male contraception started some 12 years later than those using agonists.

Initial studies carried out in cynomolgus monkeys showed that GnRH antagonists and testosterone can effectively suppress spermatogenesis in primates (Weinbauer et al. 1987). However, in further studies it became evident that the concomitant administration of GnRH antagonists and testosterone prevented the rapid and complete suppression of spermatogenesis (Weinbauer et al. 1988), whereas delaying the testosterone substitution following GnRH administration led to complete suppression, as evidenced by testicular histology and azoospermia (Weinbauer et al. 1989). The interval between the administration of the GnRH antagonist and testosterone could be as short as two weeks (Weinbauer et al. 1992). At variance with these results were findings by other investigators that in cynomolgus monkeys azoospermia could be achieved with simultaneous antagonist and testosterone application (Bremner et al. 1991). Despite this inconsistency, monkey studies have played an important part in paving the way for subsequent clinical trials. To date, results have become available from five clinical trials using GnRH antagonists for male contraception (for review Nieschlag and Behre 1996).

Overall, 35 of the 40 volunteers (88%) who took part in these studies became azoospermic, most within three months. This is a much better rate of complete suppression than that produced by the administration of testosterone enanthate alone. Although these phase I clinical investigations with GnRH antagonist involved only a few subjects, it seems that – in the human – GnRH antagonist administration followed by delayed testosterone administration (azoospermia in 20/22 men) offers little advantage over concomitant GnRH and testosterone administration (azoospermia in 15/18 men). It should also be noted that, in the two most recent studies, all 14 volunteers became azoospermic (Behre et al. 1995b; Pavlou et al. 1994). Although the number of subjects is still small, this is a promising result. In addition, the mean time to achieve azoospermia in these two studies was 6−8 weeks which is considerably shorter than the mean of 17 weeks that is required in Caucasian men when testosterone alone is used (WHO 1995).

These results are promising. However, the antagonists and regimen tested to date require daily injections which makes them unacceptable for contraceptive purposes. The development of depot formulations is therefore expected with great eagerness, but such development appears to be much more difficult than it had been for GnRH agonists. Furthermore, it could be argued that the high price of GnRH antagonists may preclude them from development as male contraceptives, which need to be affordable and in the same price range as comparable female methods. If prices for GnRH antagonists were to remain in the same range as those of GnRH agonists prescribed for the treatment of prostate carcinoma, they are unlikely ever to become part of a male contraceptive. However, prices are dictated by many variables and production costs will decrease with greater demand.

Finally, after orally effective GHRH secretagogues have been synthezised and proven useful in clinical trials, it may be hoped that similarly effective oral GnRH antagonists can be synthezised and become available for male contraception – at a reasonable price.

18.6 Acceptability and side-effects

Researchers active in the field recognize with satisfaction that the public demand for male contraception is growing. The media are eager to report about the current status and new developments and the public consumes such reports with great interest. It is felt that men should share the burden of contraception with women and there should be equal opportunities and duties for both sexes. Population conferences in Mexico (1993) and Cairo (1994) explicitly call for new male contraceptive methods. In particular, younger politicians (male and female!) take up the cause of male contraception.

The question remains whether this is all lip service or whether a male contraceptive would indeed be accepted if it should become available on the market. Although it is very difficult to predict acceptability before the final drug entity is really available, there are a number of opinion polls indicating that in many societies and parts of the world men and women would accept such a method (e.g. Anderson and Baird 1997). However, there are a number of prerequisites that must be fulfilled, most importantly, lack of side-effects and full reversibility.

Concerning the latter, it can be stated that in all volunteers who ever participated in clinical trials on hormonal male contraception, sperm parameters returned to pretreatment levels. In particular trials in recent years paid special attention to side-effects. If enough testosterone was provided to the volunteers there were no negative effects on libido and potency. In particular, when supraphysiological serum testosterone levels occurred, moderate weight gain, acne and rise in hematocrit could be observed in some trial participants. Recently attention has been focused on blood lipids and prostate parameters. While no serious effects were observed, it again appears difficult to make predictions without the final hormone or hormone combination being available for longer-term testing. The principle governing testosterone substitution in hypogonadism also ermerges as the guideline for hormonal contraception i.e. to aim for physiological or only moderately elevated serum testosterone levels. These should guarantee the lowest rate of side-effects or at least a rate not higher than maleness as such is connected with under normal conditions.

18.7 Future developments

The foregoing paragraphs have summarized the current status of hormonal male contraception. Although great progress has been made over the past decade, the optimism of TIME, as quoted in the introduction, that a male contraceptive will be generally available in 1999 appears to be unrealistic. The pace of development now rests with the pharmaceutical industry and its willingness to collaborate with research. Increasing competition between the companies may spur drug development.

While hormonal contraception has finally reached the stage of active involvement of all concerned, it is encouraging to see that another lead in male contraception promoted by WHO (see Hamilton and Waites 1989) has also been taken up by research foundations working jointly with industry (i.e. Rockefeller/Schering Foundations in 1997): the post-testicular approach to male contraception. However, as this approach is still deeply rooted in basic research, there is a long way to go before clinical application will emerge. Nevertheless, it is comforting to note that while a first generation of pharmacologic male contraceptives is being clinically developed, the search for the second generation is also taking shape.

18.8 Key messages

- Testosterone enanthate-induced azoospermia leads to effective, safe and reversible male contraception. Under severe oligozoospermia, however, pregnancies still occur.
- About two thirds of Caucasian and almost all East-Asian men reach azoospermia during weekly testosterone enanthate injections.
- In order to extend the injection intervals long-acting testosterone preparations are being tested (testosterone implants) or developed (testosterone buciclate and undecanoate).
- In order to speed up suppression of spermatogenesis and increase the rate of azoospermia testosterone is combined with either gestagens or GnRH antagonists.
- All effective approaches tested so far require injections or implantations. Self-administered modalities (oral or transdermal) did not yet prove to be effective.

18.9 References

Anderson RA, Baird DT (1997) Progress towards a male pill. IPPF Med Bull 31:3–4

Aribarg A, Sukcharoen N, Chanprasit Y, Ngeamvijawat J, Kriangsinyos R (1996) Suppression of spermatogenesis by testosterone enanthate in Thai men. J Med Assoc Thai 79:624–629

Arysad KM (1993) Sperm function in Indonesian men treated with testosterone enanthate. Int J Androl 16:355–361

Bebb RA, Anawalt BD, Christensen RB, Paulsen CA, Bremner WJ, Matsumoto AM (1996) Combined administration of levonorgestrel and testosterone induces more rapid and effective suppression of spermatogenesis than testosterone alone: a promising male contraceptive approach. J Clin Endocr Metab 81:757–762

Behre HM, Nashan D, Hubert W, Nieschlag E (1992) Depot gonadotropin-releasing hormone agonist blunts the androgen-induced suppression of spermatogenesis in a clinical trial of male contraception. J Clin Endocr Metab 74:84–90

Behre HM, Böckers A, Schlingheider A, Nieschlag E (1994) Sustained suppression of serum LH, FSH and testosterone and increase of high-density lipoprotein cholesterol by daily injections of the GnRH antagonist Cetrorelix over 8 days in normal men. Clin Endocr 40:241–248

Behre HM, Baus S, Kliesch S, Keck C, Simoni M, Nieschlag E (1995a) Potential of testosterone buciclate for male contraception: endocrine differences between responders and nonresponders. J Clin Endocri Metab 80:2394–2403

Behre HM, Kliesch S, Lemcke B, Nieschlag E (1995b) Suppression of spermatogenesis to azoospermia by combined administration of GnRH antagonist and 19-nortestosterone cannot be maintained by 19-nortestosterone alone in normal men. 77th Annual Meeting of the Endocrine Society. Washington, DC, USA, OR29–6

Bhasin S, Berman N, Swerdloff RS (1994) Follicle-stimulating hormone (FSH) escape during chronic gonadotropin-releasing hormone (GnRH) agonist and testosterone treatment. J Androl 15:386–391

Bremner WJ, Bagatell CJ, Steiner RA (1991) Gonadotropin-releasing hormone antagonist plus testosterone: a potential male contraceptive. J Clin Endocr Metab 73:465–469

Cao J, Yuan J, Jin W (1996) Clinical trial of an anti-fertility method with testosterone enanthate in normal men. Chung Hua I Hsueh Tsa Chih 76:335–337

Chen ZD, Cai SL, Chan YZ, Jian R, Zhao QW, Gu WZ, Fang RY, Zhang YP (1986) Clinical study of testosterone undecanoate compound on male contraception. J Clin Androl 1:18–20

Delanoe D, Fougevrollas B, Meyer L, Thonneau P (1984) Androgenisation of female partners of men on medroxyprogesterone acetate/percutaneous testosterone contraception. Lancet 1:276

Fogh M, Corker CS, Hunter WM, McLean H, Philip J, Schon G, Skakkebaek NE (1979) The effects of low doses of cyproterone acetate on some functions of the reproductive system in normal men. Acta Endocrinol 91:545

Fogh M, Corker CS, McLean H (1980) Clinical trial with levornorgestrel and testosterone oenanthate for male fertility control. Acta Endocrinol 95:251–257

Guerin JF, Roullet J (1988) Inhibition of spermatogenesis in men using various combinations of oral progestagens and percutaneous or oral androgens. Int J Androl 11:187–199

Hamilton DW, Waites GMH (1989) (eds) Scientific basis of fertility regulation: Cellular and molecular events in spermatogenesis. Cambridge University Press, Cambridge, pp 315–321

Handelsman DJ, Conway AJ, Boylan LM (1992) Suppression of human spermatogenesis by testosterone implants. J Clin Endocr Metab 75:1326–1332

Knuth UA, Behre HM, Belkien L, Bents H, Nieschlag E (1985) Clinical trial of 19-nortestosterone-hexoxyphenylpropionate (Anadur) for male fertility regulation. Fertil Steril 44:814–821

Knuth UA, Yeung CH, Nieschlag E (1989) Combination of 19-nortestosterone-hexyoxyphenylpropionate (Anadur) and depot-medroxyprogesterone-acetate (Clinovir) for male contraception. Fertil Steril 51:1011–1018

Leonard JM, Paulsen CA (1978) Contraceptive development studies for males: oral and parenteral steroid hormone administration. In: Pantanelli DJ (ed) Hormonal control of male fertility. Department of Health, Education and Welfare, National Institutes of Health, Bethesda, p 223

Meriggiola MC, Bremner WJ, Paulsen CA, Valdiserri A, Incorvala L, Motta R, Pavani A, Capelli M, Flamigni C (1996) A combined regimen of cyproterone acetate and testosterone enanthate as a potentially highly effective male contraceptive. J Clin Endocr Metab 81:3018–3023

Meriggiola MC, Bremner WJ, Costantino A, Pavani A, Capelli M, Flamigni C (1997) An oral regimen of cyproterone acetate and testosterone undecanoate for spermatogenic suppression in men. Fertil Steril 68:844–850

Moltz L, Römmler A, Post K, Schwartz U, Hammerstein J (1980) Medium dose cyproterone acetate (CPA): effects on hormone secrection and on spermatogenesis in man. Contraception 21:393

Moudgal NR, Ravindrath N, Murthy GS, Dighe RR, Aravindan GR, Martin F (1992) Long-term contraceptive efficacy of vaccine of ovine follicle-stimulating hormone in male bonnet monkeys (Macaca radiata). J Reprod Fertil 96:91–102

Moudgal NR, Sairam MR, Krishnamurthy HN (1997a) Immunization of male bonnet monkeys (Macaca radiata) with a recombinant FSH receptor preparation affects testicular function and fertility. Endocrinology 138:3065–3068.

Moudgal NR, Jeyakumar M, Kishnamurthy HN, Sridhar S, Krishnamurthy H, Martin F (1997b) Development of male contraceptive vaccine — a perspective. Hum Reprod Update 3:335–346

Moudgal NR, Murthy GS, Prasanna Kumur KM, Martin F, Suresh R, Medhamurthy R, Patil S, Sehgal S, Saxena B (1997c) Responsiveness of human male volunteers to immunization with ovine follicle stimulating hormone vaccine: results of a pilot study. Hum Reprod 12:457–463

Nieschlag E (1986) Reasons for abandoning immunization against FSH as an approach to male fertility regulation. In: Zatuchni GI, Goldsmith A, Spieler JM, Sciarra J (eds) Male contraception: Advances and Future Prospects. Harper & Row, Philadelphia pp 395–400

Nieschlag E, Behre HM (1996) Hormonal male contraception: Suppression of spermatogenesis with GnRH antagonists and testosterone. In: Filicori M, Flamigni C (eds) Treatment with GnRH analogs: Controversies and perspectives. Parthenon, London, 243–248

Nieschlag E, Hoogen H, Bölk M, Schuster H, Wickings EJ (1978) Clinical trial with testosterone undecanoate for male fertility control. Contraception 18:607–614

Nieschlag E, Behre HM, Weinbauer GF (1992) Hormonal male contraception: A real chance? In: Nieschlag E, Habenicht UF (eds) Spermatogenesis – fertilization – contraception. Molecular, cellular and endocrine events in male reproduction. Springer, Heidelberg, pp 477–501

Pavlou SN, Wakefield G, Schlechter NL, Lindner J, Souza KH, Kamilaris TC, Konidaris S, Rivier JE, Vale WW, Toglia M (1989) Mode of suppression of pituitary and gonadal function after acute or prolonged administration of a luteinizing hormone-releasing hormone antagonist in normal men. J Clin Endocrinol Metab 73:1360–1369

Pavlou SN, Herodotou D, Curtain M, Minaretzis D (1994) Complete suppression of spermatogenesis by co-administration of a GnRH antagonist plus a physiologic dose of testosterone. Proceedings of the 76th meeting of the Endocrine Society, Anaheim, CA, 1324

Schearer SB, Alvarez-Sanches F, Anselmo G, Brenner P, Coutinho E, Lathen-Faundes A, Frick J, Heinild B, Johansson EDB (1978) Hormonal contraception for men. Int J Androl Suppl 2:680–712

Schürmeyer T, Knuth UA, Freischem CW, Sandow J, Akhtar FB, Nieschlag E (1984) Suppression of pituitary and testicular function in normal men by constant gonadotropin-releasing hormone agonist infusion. J Clin Endocr Metab 59:1–6

Soufir JC, Jouannet P, Marson J, Soumah A (1983) Reversible inhibition of sperm production and gonadotropin secretion in men following combined oral medroxyprogesterone acetate and percutaneous testosterone treatment. Acta Endocrinol 102:625–632

Srinath BR, Wickings EJ, Witting CH, Nieschlag E (1983) Active immunization with follicle stimulating hormone for fertility control: a 4 1/2-year study in male rhesus monkeys. Fertil. Steril. 40:110–117

Wang C, Yeung KK (1980) Use of a low dosage oral cyproterone acetate as a male contraceptive. Contraception 21:245

Weinbauer GF, Nieschlag E (1996) The Leydig cell as a target for male contraception. In: Payne AH, Hardy MP, Russell LD (eds), The Leydig Cell. Cache River Press, Vienna, pp 629–662

Weinbauer GF, Surmann FJ, Nieschlag E (1987) Suppression of spermatogenesis in a nonhuman primate (*Macaca fascicularis*) by concomitant gonadotropin-releasing hormone (GnRH) antagonist and testosterone treatment. Acta Endocrinol 114:138–146

Weinbauer GF, Göckeler E, Nieschlag E (1988) Testosterone prevents complete suppression of spermatogenesis in the gonadotropin-releasing hormone (GnRH) antagonist-treated non-human primate (*Macaca fascicularis*). J Clin Endocr Metab 67:284–290

Weinbauer GF, Khurshid S, Fingscheidt U, Nieschlag E (1989) Sustained inhibition of sperm production and inhibin secretion induced by a gonadotropin-releasing hormone antagonist and delayed testosterone substitution in non-human primates (*Macaca fascicularis*). J Endocr 123:303–310

Weinbauer GF, Limberger A, Behre HM, Nieschlag E (1992) The GnRH antagonist cetrorelix inhibits spermatogenesis in the non-human primate: effects of delayed high dose testosterone substitution. Proceedings of the 9th International Congress of Endocrinology, Nice, France, 419

WHO (1972–1995) Special Programme of Research, Development and Research Training in Human Reproduction. Annual and Biannual Reports, WHO, Geneva

World Health Organization Task Force on Methods for the Regulation of Male Fertility (1990) Contraceptive efficacy of testosterone-induced azoospermia in normal men. Lancet 336:955–959

World Health Organization Task Force on Methods for the Regulation of Male Fertility (1993) Comparison of two androgens plus depot-medroxyprogesterone acetate for suppression to azoospermia in Indonesian men. Fertil Steril 60:1062–1068

World Health Organization Task Force on Methods for the Regulation of Male Fertility (1996) Contraceptive efficacy of testosterone-induced azoospermia and oligozoospermia in normal men. Fertil Steril 65:821–829

World Health Organization Task Force on Methods for the Regulation of Male Fertility (1995) Rates of testosterone-induced suppression to severe oligozoospermia or azoospermia in two multinational clinical studies. Int J Androl 18:157–165

19 Androgen treatment of female-to-male transsexuals

Louis J. Gooren

Contents

19.1 Female-to-male transsexuals

19.1.1 Description of transsexualism and of population studied

Transsexualism is the condition in which a person with apparently normal somatic sexual differentiation is convinced that he/she is actually a member of the opposite sex. It is associated with an irresistable urge to be hormonally and surgically adapted to that sex. Traditionally conceptualized as a psychological phenomenon, research on the brains of male-to-female transsexuals has found that the sexual differentiation of one brain area (the bed nucleus of the stria terminalis) follows a female pattern (Zhou et al. 1995). The latter finding may lead to a concept of transsexualism as a form of intersex, where the sexual differentiation of the brain is not consistent with the other variables of sex such as chromosomal pattern, nature of the gonad and nature of internal/external gonads. Basal serum levels of sex steroids and levels after endocrine stimulation/inhibition are not different between transsexuals and non-transsexuals (Gooren 1990). The prevalence of transsexualism, assessed in the Netherlands, is 1:11900 men and 1:30400 women (Van Kesteren et al. 1996a).

The authors' clinic, the Division of Andrology of the University Hospital of the Vrije Universiteit of Amsterdam, provides comprehensive care to transsexual patients; this includes cross-sex hormonal treatment. This chapter primarily addresses the effects of androgen administration to female-to-male transsexuals. To the extent that the effects of androgen deprivation in the opposite group, the male-to-female transsexuals, lead to a better understanding of androgen (patho)physiology, relevant research findings will also be described. Most female-to-male transsexuals start hormonal treatment between the ages of 20–25 years. Between 1975 and 1996 293 female-to-male transsexuals received treatment with androgens. Their ages ranged from 17 to 70 years with a mean age of 34 years. The total duration of exposure to androgens of this group amounted to 2481 patient years. After psychological evaluation of their gender problem, subjects receive androgen treatment. If they make a successful change to the male sex role, surgery is performed after 18–24 months of androgen treatment, including mastectomy, ovariectomy, hysterectomy and phalloplasty. Following ovariectomy androgen therapy is continued to maintain virilization and, importantly, to prevent osteoporosis.

19.1.2 Treatment regimen

In most cases subjects are treated with parenteral androgens/2 weeks (Sustanon-250 (Organon) or Testoviron-180 (Schering Pharma)) and sometimes, if injections pose a problem, with oral androgens (testosterone undecanoate (Andriol, Organon), 4 to 6 capsules of 40 mg per day). With the latter virilization tends to be sluggish and menstrual activity, subjectively experienced as undesirable, is often not suppressed; for this purpose a progestational compound has to be added. So far testosterone patches have not been available in the Netherlands.

19.1.3 Resulting endocrine profiles

As a consequence of treatment with parenteral androgens resulting serum testosterone levels vary strongly over time following administration (Behre et al. 1998). Serum testosterone levels, measured cross-sectionally, are 29±14 nmol/L (mean±SD). Serum 17β-estradiol values fall only modestly, a change which becomes significant only in studies with large numbers of subjects. For instance, in a study with 17 subjects serum levels fell from 163±53 to 131±33 pmol/L (p < 0.05) upon testosterone administration. These values remain biologically significant. Ovariectomy does not lead to a difference in serum levels of testosterone and 17β-estradiol in female-to-male transsexuals treated with androgens, but serum levels of FSH/LH are much higher following ovariectomy (LH respectively 19.3±6.4 U/L versus 2.9±1.1; FSH 32±10 versus 3.7±2.2 U/L) (Spinder et al. 1989a, 1989b) . Therefore we believe that serum 17β-estradiol in not yet ovariectomized female-to-male transsexuals is mainly derived from peripheral aromatization of testosterone rather than from ovarian production. On average plasma levels of sex hormone-binding globulin fall 50%. We have found that testosterone administration to female-to-male transsexuals increases both basal and adrenocorticotropin-stimulated secretion of adrenal androgens (Polderman et al. 1994a, 1995).

19.1.4 The female-to-male transsexual as model of the hypogonadal male

Studies into the biological effects of cross-sex hormones in transsexuals are sometimes regarded. as idiosyncratic. They may, however, yield important scientific information, particularly when the effects of androgen administration to female-to-male transsexuals are compared to the effects of androgen deprivation in male-to-female transsexuals, thus providing a mirror image of the induction of androgen effects in female-to-male transsexuals. Androgen effects are strikingly similar in female-to-male transsexuals and hypogonadal men. Throughout our experiments we have gained the impression that the female-to-male transsexual subject undergoing androgen administration can serve as a model of the hypogonadal male. There is as yet no information that in terms of androgen (receptor and post-receptor) physiology human males differ from females. In fetal rats there is an identical androgen receptor expression pattern in females and males (Bentvelsen et al. 1995). For the study of androgen effects female-to-male transsexuals as group offer an advantage over hypogonadal men since the former are all equally 'androgen-deficient', which is almost never the case with a group composed of hypogonadal men which are, in any event, difficult to recruit. Furthermore, female-to-male transsexuals are healthy and in an age range between 20–35 years when they start androgen treatment.

19.2 Androgen effects

19.2.1 Anabolic effects

Data were collected on the weight changes in 86 transsexual females treated with androgens in the first year of hormone administration. There was an increase of on average 5.2 kg (range 1–14 kg) of weight in 69 subjects, a decrease of on average 5.6 kg (range 4–12 kg) in 8 subjects, and in 9 subjects no weight changes were recorded. Part of the weight gain must be explained by an increase in muscle mass, on average 4.1 kg in this group of androgen-naive subjects. This compares well with a recent study of Bhasin et al. (1997) on the effects of androgens in hypogonadal men. They found an increase in body weight of 4.5 ± 0.6 kg with an increase in fat-free mass of 5.0 ± 0.7 kg after 10 weeks of administration of testosterone enanthate 100 mg/week. In our study muscle surface measured with MRI at the level of the thigh increased by 20%, slightly more than noted by Bhasin et al. (1997). No studies of muscle strength have been performed. While subcutaneous fat depots decreased in size as measured by MRI, an increase of visceral fat was observed, progressively with time (Elbers et al. 1997). Comparing the effects of anti-androgen+estrogen administration to male-to-female transsexuals, it became apparent that in an androgenic hormonal milieu fat storage occurs predominantly (but not exclusively) in the visceral fat depot, whereas in an estrogenic milieu this is predominantly subcutaneous.

19.2.2 Virilizing effects

In almost all subjects the pitch of the voice drops and a coarsening of the voice occurs within 6–12 weeks of androgen administration. The development of sexual hair follows essentially the pattern observed in pubertal boys: first on the upper lip, cheek and pubic area, and after longer treatment also on the chin, neck and legs and arms and chest. If there is a pre-existent degree of hirsuteness on legs or arms, for instance, androgen exposure increases hair growth first in these areas. This probably represents an expression of high local skin sensitivity to female levels of androgens. There is a strong genetic component, as judged from the family history, in both the rate and degree of development of hairiness, as is also the case with the development of alopecia androgenetica. The development of sexual hair and its hormonal correlates are presently the subject of scientific investigation. In male-to-female transsexuals sexual hair decreases upon androgen deprivation, but facial hair is resilient; its growth is slowed and diameter and pigmentation decline, but it remains present, even after many years of androgen deprivation. Severe acne upon androgen administration was observed in 80 of the 293 cases, predominantly on the upper back, as is the case in men starting androgen administration rather late in life. This problem could be acceptably treated with topical agents, such as benzoyl peroxide with or without miconazole.

19.2.3 Effects on breasts, uterus, clitoris and ovaries

Two studies have reported findings of androgen effects on the breasts (Futterweit and Schwartz 1988; Sapino et al. 1990). The effects varied strongly from one subject to the other: most subjects showed intralobular fibrous stroma and some extralobular fibrous stroma; approximately half showed lobular atrophy. Effects on both endometrium and breasts probably reflect the simultaneous action of androgens and estrogens generated by the high levels of androgens in these subjects. Anecdotally, a female-to-male transsexual (not belonging to this population and living in England) already having undergone mastectomy, reported the occurrence of a metastasized breast carcinoma probably originating from residual mammary glandular tissue. There is probably no relationship with previous androgen treatment. In women with the Stein-Leventhal syndrome the risk of developing postmenopausal breast cancer is not increased (Anderson et al. 1997). In men genetic factors are significant in the development of breast tumours. Most breast tumours in men are estrogen-receptor positive and respond to tamoxifen (Hecht and Winchester 1994). Since a part of the administered androgens are aromatized to estrogens, it seems advisable that candidates who have an own or family history of breast cancer be carefully monitored when they receive androgens.

The study of Futterweit and Deligdisch (1986) describes the effects of long-term androgen administration on the endometrium in 12 female-to-male transsexuals. Six had a proliferative endometrium, four a secretory and two an inactive endometrium. In women with the polycystic ovarian syndrome there is some concern that they may develop endometrial carcinoma (McDonald et al. 1976). They have high circulating estrogen levels which, in their effects on the endometrium, are not opposed by progesterone since they are generally anovulatory. A similar situation exists in androgen-treated female-to-male transsexuals whose gonadotropin secretion is suppressed by the high dose of exogenous androgens which generate biologically active estrogen levels. Therefore we recommend hysterectomy, which the vast majority wishes anyway. As with breast cancer, subjects with a history of endometrial cancer need to be monitored carefully since the exposure to estrogens continues when androgen treatment is given.

Several authors have studied the effects of androgen administration on ovarian morphology. The universal finding is that long-term androgen exposure (>18 months) induces the morphological features characteristic of the polycystic ovary as encountered in polycystic ovarian disease (PCOD). The ovaries are enlarged and show a two-fold increase in cystic and a three-fold increase in atretic follicles. The ovarian cortex is three times thicker than normal and is collagenized. The ovaries show theca interna hyperplasia, luteinization, and stromal hyperplasia. These characteristics can apparently be induced by exogenous androgens, without an intra-ovararian source of androgens (Futterweit and Deligdisch 1986; Pache et al. 1991; Spinder et al. 1989 a). Moreover, the hormonal contents of the ovarian follicles (17β-estradiol, androstenedione and inhibin) of androgen-treated female-to-male transsexuals were similar to those of women with PCOD (Pache et al. 1992). The accompanying serum LH levels in the transsexuals are suppressed by high-dose testosterone administration, unlike the elevated serum LH levels typical of PCOD (Spinder et al. 1989a). Since elevated serum LH values are a significant diagnostic criterium in the diagnosis of PCOD, the androgen excess of PCOD is probably a result and not the prime mover of PCOD.

We recommend removing the ovaries routinely in view of the increased chance of malignant degeneration of polycystic ovaries (Schildkraut et al. 1996).

The volume of the clitoris increases four to eight times upon testosterone administration. Like penile growth there is only a limited potential for growth. The younger the patient at the start of androgen administration, the more growth is encountered. In some the enlarged clitoris is sufficiently large to serve as a small phallus (Hage 1996).

19.2.4 Effects on bones

Sex steroids are important regulators of bone metabolism and bone mineral density (BMD) (Orwoll and Klein 1995). They play a substantial role in bone growth and the pubertal acquisition of peak bone mass and maintenance of bone mass in adulthood in both sexes. The question poses itself whether androgens can protect 'female bones' from estrogen deprivation and vice versa. The latter can be positively answered (Lips et al. 1989). In a number of studies we have addressed the effects of cross-sex hormone administation on bone. One year of administration of testosterone esters (250 mg/2 weeks) led to a significant increase in insulin-like growth factor-I (without a change in insulin-like growth factor-binding protein-3), and an increase in bone alkaline phosphatase. There was no significant increase in BMD after one year of testosterone administration (Van Kesteren et al. 1996b). In another study histomorphometry was performed in iliac crest biopsies of 15 female-to-male transsexuals (age 30±6.1 years) who had been treated for a median period of 39 months with a similar androgen treatment regimen. The findings were compared to those of eugonadal men and postmenopausal women. Biochemical variables of bone formation were lower than in the two comparison groups. The Z-score (an age-adjusted measure of bone mineral density, expressed as standard deviations from the mean of that age) was as expected for age. Cortical thickness was higher in this group than in the comparison subjects (probably pointing to an androgenic-anabolic effect) and trabecular bone structure was similar to the two comparison groups. The eroded surface was lower than in postmenopausal women so there was probably a low bone turnover. The conclusion was that androgens protect the bone of female-to-male transsexuals from the usual effects of estrogen deprivation as are observed in women (Lips et al. 1996). However, in a longer-term follow-up study of 28–63 months in 19 female-to-male transsexuals (age 16–39 years; median age 25) (13–35 months after ovariectomy) there was a significant decrease in BMD in the group over this period (from 1.14±0.15 to 1.09±0.12 g/cm^2). The best predictor of bone loss was the elevation of serum LH levels. We have interpreted this correlation as indicative of a probable undersubstitution with androgens in the longer term (Van Kesteren et al. 1998). Clinically it is of note that our subjects take androgen dosages that compare to those in androgen-treated hypogonadal men (testosterone esters 250 mg per 3 weeks or oral testosterone undecanoate 40 mg 4 capsules per day).

19.2.5 Effects on prostate-specific antigen

Prostate-specific antigen (PSA) is a serine protease with chymoptrypsin-like enzymatic activity, produced in men by the prostate gland stimulated by androgens. PSA has been detected in some female tissues (breast, ovarian and endometrial tissues) and body fluids (amniotic fluid, milk and breast cyst fluid). A recent study showed that PSA levels in women are correlated with the degree of hirsutism and with levels of the androgen marker 3α-androstanediol glucuronide (Melegos et al. 1997). In our subjects we were able to confirm that administration of androgens leads to an increase of PSA levels (from median 28 (range 1–250) to median 68 (range 19–575 pg/ml) (p < 0.02). We are presently investigating the relative contributions of the breasts and of the uterus/ovaries to serum PSA levels which are surgically removed in female-to-male transsexuals.

19.2.6 Effects on psychological functioning

In adulthood men and women differ in a number of psychological variables: men are more aggression-prone, have a higher sexual arousability and outperform women in visuospatial and mathematical ability, while women perform better on tasks of verbal skill, perceptual speed and fine manual dexterity. We studied (Van Goozen et al. 1995) the effects of testosterone administration to female-to-male transsexuals. While no significant differences were encountered with control women before testosterone administration, androgen administration was clearly associated with an increase in aggression-proneness (the self-imagined preparedness to act), sexual arousability and spatial ability performance. Verbal fluency tasks deteriorated. Following androgens their scores were not very different from those of eugonadal men which led us to believe that early imprinting (so-called organizational effects of sex steroids) is not a prerequisite in the human for these effects to become apparent. We have seen no adverse effects on aggression or on sexual behaviour of testosterone administration in female-to-male transsexuals.

19.3 Observed side-effects of androgen administration

19.3.1 The liver

Elevation of liver enzymes was observed in 45 cases, of which eight were pre-existent, three related to alcohol, one to a cytomegaly virus infection. The elevations never exceeded 2.5 times reference laboratory values. In 33 androgen administration might have been the cause of the disturbance. In 13 cases this lasted for less than six months, in the remaining 20 longer but this did not require discontinuation of androgen administration and disappeared in all but two after 12–15 months. Liver tumours were not observed. In this context it should be pointed out that 17α-methyl testosterone preparations were not administered to our population (Van Kesteren et al. 1997).

19.3.2 Cardiovascular risk factors

We have not found a higher degree of mortality and morbidity in this group of androgen-treated subjects in comparison with age-matched subjects (Van Kesteren et al. 1997). As for the shift to a higher cardiovascular risk profile in female-to-male transsexuals following androgen administration: we have observed one case of myocardial infarction and one of angina pectoris. This is not higher than expected on the basis of disease prevalence data. In any case the relationship between androgens and cardiovascular disease is complicated. No relationship could be established between endogenous sex hormone levels and cardiovascular disease in men (Barrett-Connor and Khaw 1988), while men with cardiovascular disease may have lower-than-normal serum testosterone levels (Phillips et al. 1994).

19.3.2.1 Blood pressure and fluid retention

All androgens cause some degree of sodium retention and expansion of extracellular fluid volume. This effect is usually small; weight increases by about 3% in healthy individuals (Wilson 1987). In 12 (of 293) subjects elevated blood pressure was found, which is in accordance with the expected number on the basis of prevalence studies in healthy subjects (Van Kesteren et al. 1997). Sodium and water retention was observed in five cases but was corrected with a dose reduction of androgens.

19.3.2.2 Vasoactive substances, homocysteine, t-PA and PAI-1

Among the most significant vasoactive substances are the endothelins. Studies with endothelins and specific endothelin-receptor antagonists have suggested that these peptides are important in vascular physiology and disease. It is likely that endothelin-1 plays a role in atherosclerosis and cardiac hypertrophy (for review Levin 1995). In a study we found that testosterone administration to female-to-male transsexuals increased plasma endothelin levels, while androgen deprivation with cyproterone acetate with simultaneous administration of ethinyl estradiol decreased endothelin levels (Polderman et al. 1993).

An elevated plasma level of total homocysteine (tHcy) level is an independent risk factor for cardiovascular disease. Until now it is not known whether the higher total plasma homocysteine (tHcy) in healthy men versus premenopausal women can be explained by their differences in sex steroids. We determined the effects of administration of androgens in female-to-male and of estrogens+anti-androgens in male-to-female transsexuals on tHcy levels. Besides a direct effect of sex steroids, changes in plasma tHcy levels could be secondary to other biological effects (anabolic/catabolic) of these steroids. In male-to-female transsexuals the plasma tHcy level decreased from median 7.2 µmol/L to 6.0 µmol/L (p = 0.0003) upon administration of an anti-androgen+estrogens and in female-to-male transsexuals it increased from median 8.2 µmol/L to 8.4 µmol/L (p = 0.01) upon androgen administration. Changes in serum albumin levels correlated positively with changes in tHcy levels in male-to-female transsexuals. Changes in serum creatinine levels correlated positively with changes in tHcy levels in both male-to-female and female-to-male transsexuals. The sex difference in plasma tHcy levels between men and women seems to be associated with the differences in sex steroid milieu. Whether this is a direct or indirect effect of sex steroids is difficult to discern since a large number of factors also influenced by sex steroids affect plasma tHcy levels (Giltay et al. 1998).

The process of fibrinolysis plays a pivotal role in the etiology of arterial and venous thrombosis. Tissue-type plasminogen activator (t-PA) activates plasmin which acts as a circulating thrombolytic agent. Its action is counterbalanced by plasminogen activator inhibitor type 1 (PAI-1). Several cardiovascular risks appear to be associated with elevated levels of PAI-1, thus enhancing potential thrombosis formation. The effects of androgen administration on t-PA and PAI-1 appeared to be neutral. The von Willebrand factor (vWF) is the carrier protein of factor VIII in the cascade of coagulation factors. An elevation of factor VIII and of vWF are significant factors in platelet adhesion and coagulation. Androgen appeared to lower the concentrations of vWF in our populations (Van Kesteren et al. submitted), thus playing a favourable role in cardiovascular disease.

19.3.2.3 Serum lipids

There is no consensus on the effect of androgens on serum lipids and lipoproteins. This is (partially) due to differences in study designs and to whether endogenous or exogenous androgen effects are investigated. Evaluation is particularly complicated since part of both endogenous and exogenous testosterone is metabolized to estrogens.

Serum high density lipoprotein (HDL) cholesterol is lower in eugonadal men than in premenopausal women while these differences are not manifest before puberty. Highly convincing evidence that testosterone is involved in this sex difference comes from experiments wherein androgen levels were severely lowered, whereupon HDL cholesterol levels increased. When serum androgens levels subsequently returned to normal, serum levels of HDL cholesterol decreased to normal male levels (von Eckardstein et al. 1997).

In our own study the (aromatizable) oral androgen testosterone undecanoate was administered to both hypogonadal men and to previously non-treated female-to-male transsexuals. While serum estradiol levels in the females were 3 to 4 times higher during testosterone administration, in both sexes levels of HDL cholesterol and HDL2 cholesterol declined and were eventually of the same magnitude. This led us to conclude that testosterone is indeed the major determinant of the sex difference in HDL cholesterol levels (Asscheman et al. 1994). In line with this finding is that women with hyperandrogenism show "male-like" plasma lipoproteins profiles (Adams et al. 1995). Serum total cholesterol and triglycerides did not change and serum apolipoprotein A1 declined upon oral androgen administration. With parenteral testosterone administration likewise a decrease of HDL-cholesterol was noted and also a small decrease of total cholesterol (Elbers et al. 1997).

19.3.2.4 Insulin resistance

Evidence that insulin resistance induces an unfavourable lipid profile, a decrease in plasma high density lipoprotein cholesterol and an increase in plasma triglycerides, has become quite convincing (Laws and Reaven 1993). There is very likely a link between hyperinsulinemia (or insulin resistance, with which it is inextricably associated) and hypertension in otherwise healthy persons in whom high plasma insulin levels are the only abnormality (Zavaroni et al. 1985). Coexistence of insulin resistance and hyperandrogenism has been described frequently, mainly in women with the polycystic ovary syndrome. An improvement of insulin resistance could be observed when the hyperandrogenism was corrected in these women (Shoupe and Lobo 1984). Several reports describe induction of insulin resistance in men taking anabolic steroids (Cohen and Hickman 1987).

In our own experiments, using the hyperinsulinemic-euglycemic clamp method, we studied the effects of three months' administration of 250 mg testosterone esters/2 weeks on insulin resistance in a group of female-to-male transsexuals (Polderman et al. 1994b). In hyperinsulinemic-euglycemic clamp studies insulin is infused at preset standardized infusion rates and the amount of glucose required to keep blood glucose levels constant provides a reliable indication of insulin sensitivity or, alternatively, of insulin resistance. In our study glucose utilization decreased significantly, showing that less glucose entered the cells; in other words, insulin became biologically less effective in its effect to deliver glucose to the cell, i.e. the development of peripheral insulin resistance. (Polderman et al. 1994b). In a similar study design, but now after 12 months' testosterone administration, such an effect could no longer be demonstrated so clearly. Whether androgen-induced insulin resistance is a transient phenomenon or whether the two studies have a different outcome remains unresolved.

19.3.2.5 Body fat distribution

From puberty onwards males and females differ in their body distribution of fat over the subcutaneous and intra-abdominal (or visceral) fat depots. Subcutaneous fat largely disappears in normal weight pubertal boys and visceral fat stores are also small but start to become manifest with increasing age. From age 20–30 years on men have approximately twice as much visceral fat storage as women with a comparable body mass index. If men store fat, they preferentially accumulate it intra-abdominally, while in premenopausal women this occurs preferentially subcutaneously on hips and thighs. That androgens are involved in this sex-specific fat distribution pattern has become increasingly clear, though the precise mechanism remains unknown. Women with elevated levels of serum androgens have more visceral fat than control women. In our own experiments we studied the effects of androgen administration on fat distribution in female-to-male transsexuals undergoing sex reassignment. It could be shown that androgens lead to a quick depletion of subcutaneous fat, while it takes between 1–3 years for visceral fat accumulation to become manifest (Elbers et al. 1997).

We could further establish that sex steroids have an effect on fat cell size. Androgen administration to female-to-male transsexuals led to a decrease in fat cell surface, both in the gluteal and abdominal area, while estrogen administration to male-to-female transsexuals had an opposite effect. In the abdominal area androgen administration led to an increase in the basal rate of lipolysis (measured in vitro); no such an effect was observed in the gluteal adipocytes. Estrogen administration to male-to-female transsexuals inhibited basal lipolysis both in the gluteal and abdominal area.

Visceral fat accumulation has been found to be a risk factor for cardiovascular disease and for non-insulin-dependent diabetes mellitus. Its mechanism has not been fully clarified but it has been found that visceral fat is metabolically very active with a high turnover of triglycerides. The visceral adipocytes show an enhanced sensitivity to β-adrenergic lipolytic stimuli, resulting in an increased production of free fatty acids. The subsequent load of triglycerides and/or free fatty acids reaching the liver via the portal vein interferes with the hepatic handling of insulin, leading to hyperinsulinism and a deteriorated glucose tolerance and further to a lowering of HDL cholesterol and higher serum levels of triglycerides. This constitutes a risk for cardiovascular disease (Seidell et al. 1990). In one of our studies (Giltay et al. submitted) it could be shown that a factor affecting coagulation, PAI-1 is positively correlated with the amount of visceral fat in healthy young men and women.

Although the above described visceral fat accumulation seems to be androgen-induced, studies have found that ageing men with visceral accumulation of body fat often have relatively low testosterone (Seidell et al. 1990). The

Table 19.1. Effects of sex steroids on cardiovascular risk factors

	Androgens	Estrogens
Total cholesterol	neutral	neutral
HDL-cholesterol	–	++
Triglyceride	neutral	–
Endothelin-1	–	++
Homocysteine	–	++
Plasminogen activator inhibitor type 1	neutral	++
Tissue-type plasminogen activator	neutral	++
von Willebrand factor	++	–
Insulin sensitivity	neutral	–
Visceral fat accumulation	–	neutral

++ positive effects; – negative effects

mechanism of this association remains to be elucidated. It has been hypothesized that supplementation of testosterone in middle-aged viscerally obese men induces mobilization of fat from these visceral fat depots. The mechanism is supposedly a stimulation of the lipolytic activity and an inhibition of lipoprotein lipase resulting in a decrease in the amount of visceral fat. Table 19.1 provides an overview of our studies on the effects of androgens on risk factors of cardiovascular disease, as opposed to those of estrogens.

19.4 Conclusions

19.4.1 The scientific value of female-to-male transsexuals as a model

It is our belief that close monitoring of the effects of androgen administration to female-to-male transsexuals provides important scientific information for our understanding of androgen (patho)physiology both in men and in women. It is obviously unethical to administer androgens to non-transsexual women over a prolonged period. Monitoring this process in female-to-male transsexuals provides clues to the pathology of longstanding hyperandrogenism in women. And, as stated above, a cohort of uniformly hypogonadal patients in a similar age range is difficult to recruit by a single research centre. They might thus serve as a model for the effects of androgens in hypogonadal men. Most transsexuals are young and healthy when they start sex reassignment so there are fewer variables to consider; and they themselves are interested in the effects of long-term testosterone on their health, thus providing motivated research subjects.

19.4.2 The relative safety of androgen administration

Administration of androgens to 293 female-to-male transsexuals with a total exposure of 2481 patient years (individual exposure varied from 1 to 20 years) was a safe procedure. There were no major complications. An expected side effect, acne, could be remedied acceptably with topical treatment. Assessment of cardiovascular risk factors following androgen administration showed a shift to a more negative risk profile for a number of them but not for all. The cardiovascular mortality and morbidity actually observed was not higher than expected on the basis of Dutch health statistics. But it is of note that few female-to-male transsexuals in our population are yet over age 50. The observation that cardiovascular risk factors shift to a male-like risk profile has prompted us to encourage female-to-male transsexuals to adopt a healthy life style: avoidance of overweight, no smoking, a low fat diet and exercise.

19.5 Key messages

- The study of the effects of androgen administration to female-to-male transsexuals provides a better understanding of androgen pathophysiology in both women and men.
- Polycystic ovarian disease (PCOD) is not uncommon in women. Our studies have shown that the ovarian histopathology of PCOD probably can be explained as a resulting from hyperandrogenism. Further, the higher cardiovascular risk profile in women with PCOD is likely to be partially attributable to androgen effects. We found, however, no evidence that insulin resistance often associated with PCOD is the result of hyperandrogenism.
- Men have a higher risk of developing cardiovascular disease than women. This is traditionally ascribed to the negative effects androgens supposedly have on risk factors. But in fact, our studies show that this view is too negative and must be modified. Several effects of androgens on these risk factors are more or less neutral while the effects on clotting (the von Willebrand factor) are even positive (Table 19.1) but admittedly, some effects are negative. It is worth mentioning that not all estrogen effects are positive in this regard (Table 19.1).
- In both sexes sex steroids are essential for maintaining bone mass. There are indications that estrogens (derived from aromatization of androgens) are required to maintain bone mass in men. We found that the dosages of testosterone usually given for androgen replacement in men were not fully capable of maintaining bone mass in female-to-male transsexuals while estrogens more successfully maintained bone mass in male-to-female transsexuals. In both sexes the serum LH level was

the best predictor of loss of bone mass. This information compels us to monitor bone mineral density in (severely) hypogonadal men on androgen replacement more closely, while, similar to the significance of serum TSH levels in the treatment of hypothyroidism, serum LH possibly provides a marker of the adequacy of androgen replacement.

19.6 References

Adams MR, Williams JK, Kaplan JR (1995) Effects of androgens on coronary artery atherosclerosis and atherosclerosis-related impairment of vascular responsiveness. Arterioscler Thromb Vasc Biol 15:562–70

Anderson KE, Sellers TA, Chen PL, Rich SS, Hong CP, Folsom AR (1997) Association of Stein-Leventhal syndrome with the incidence of postmenopausal breast carcinoma in a large prospective study of women in Iowa. Cancer 79:494–499

Asscheman H, Gooren LJG, Megens JAJ, Nauta J, Kloosterboer HJ, Eikelboom F (1994) Serum testosterone level is the major determinant of the male-female differences in serum levels of high density lipoprotein cholesterol and HDL2 cholesterol. Metabolism 43:935–939

Barrett-Connor E, Khaw K-T (1988) Endogenous sex hormones and cardiovascular disease in men. Circulation 78:539–545

Behre HM, Nieschlag E (1998) Comparative pharmacokinetics of androgen preparations: application of computer analysis and simulation. In: Behre HM, Nieschlag E (eds) Testosterone: Action, Deficiency, Substitution. 2nd edn. Springer Verlag, Berlin, pp 349–364

Bentvelsen FM, Brinkmann AO, van der Schoot P, van der Linden JETM, van der Kwast ThH, Boersma WJ, Schröder FH, Nijman JM (1995) Developmental pattern and regulation by androgens of androgen receptor expression in the urogenital tract of the rat. Mol Cell Endocrinol 113:245–253

Bhasin S, Storer TW, Berman N, Yarasheski KE, Clevenger B, Phillips J, Lee WP, Bunell TJ, Casaburi R (1997) Testosterone replacement increases fat-free mass and muscle size in hypogonadal men. J Clin Endocrinol Metab 82:407–413

Cohen JC, Hickman R (1987) Insulin resistance and diminished glucose tolerance in power-lifters ingesting anabolic steroids. J Clin Endocrinol Metab 64:960–963

Elbers JMH, Asscheman H, Seidell JC, Megens JAJ, Gooren LJG (1997) Long-term testosterone administration increases visceral fat in female-to-male transsexuals. J Clin Endocrinol Metab 82:2044–2047

Futterweit W, Deligdisch L (1986) Histopathological effects of exogenously administered testosterone in 19 female-to-male transsexuals. J Clin Endocrinol Metab 62:16–21

Futterweit W, Schwartz IS (1988) Histopathology of the breasts of 12 women receiving long-term androgen therapy. Mt Sinai J Med 55:309–312

Giltay EJ, Hoogeveen EK, Elbers JMH, Gooren LJG, Asscheman H, Stehouwer CDA (1998) Effects of sex steroids on plasma total homocysteine levels; a study in male and female transsexuals. J Clin Endocrinol Metab (in press)

Giltay EJ, Elbers JMH, Gooren LJG, Emeis JJ, Kooistra T, Asscheman H, Stehouwer CDA. Visceral fat accumulation is an important determinant of plasminogen activator inhibitor-1 levels in young, non-obese men and women: administration of sex steroid hormones disrupts this association. (submitted)

Gooren LJG (1990) The endocrinology of transsexualism. Psychoneuroendocrinol 15:3–14

Hage JJ (1996) Metadoioplasty: an alternative phalloplasty technique in transsexuals. Plast Reconstr Surg 97:161–167

Hecht JR, Winchester DJ (1994) Male breast cancer. Am J Clin Pathol 102:S25–30

Laws A, Reaven GM (1993) Insulin resistance and risk factors for coronary heart disease. Bailliere's Clin Endocrinol Metab 7, 1063–78

Levin ER (1995) Endothelins. New Engl J Med 333:356–63

Lips P, Asscheman H, Uitewaal P, Netelenbos C, Gooren LJG (1989) The effect of cross-gender hormonal treatment on bone metabolism in male-to-female transsexuals. J Bone Miner Res 4:657–662

Lips P, Van Kesteren PJM, Asscheman H, Gooren LJG (1996) The effect of androgen treatment on bone metabolism in female-to-male transsexuals. J Bone Miner Res 11:1769–1773

McDonald TW, Malkasian GD, Gaffey TA (1976) Endometrial cancer associated with feminizing ovarian tumor and polycystic ovarian disease. Obstet Gynecol 49:654–658

Melegos DN, Yu H, Asok M, Wang C, Stanczyk F, Diamandis EP (1997) Prostate-specific antigen in female serum, a potential new marker of androgen excess. J Clin Endocrinol Metab 82:777– 780

Orwoll ES, Klein RF (1995) Osteoporosis in men. Endocr Rev 16:87–116

Pache TD, Chadha S, Gooren LJG, Hop WCJ, Jaarsma, Dommerholt HBR, Fauser BCJM (1991) Ovarian morphology in long-term androgen treated female-to-male transsexuals. A human model for the study of polycystic ovarian syndrome. Histopathology 19:445–452

Pache TD, Hop WCJ, De Jong FH, Leerentveld RA, Van Geldorp H, Van der Kamp TMM, Gooren LJG, Fauser BCJM (1992) 17β-estradiol, androstenedione and inhibin levels in fluid from indidvidual follicles of normal and polycystic ovaries and in ovaries from androgen-treated female-to-male transsexuals. Clin Endocrinol 36:565–571

Phillips GB, Pinkernell BH, Jing T-Y (1994) The association between hypotestosteronemia and coronary artery disease in men. Arterioscler Thromb 14:701–706

Polderman KH, Stehouwer CDA, Van de Kamp GJ, Dekker GA, Verheugt FWA, Gooren LJG (1993) Influence of sex hormones on plasma endothelin levels. Ann Int Med 118:429–32

Polderman KH, Gooren LJG, Van der Veen EA (1994a) Testosterone administration increases adrenal androgen response to adrenocorticotropin. Clin Endocrinol 40:595–601

Polderman KH, Gooren LJG, Asscheman H, Bakker A, Heine RJ (1994b) Induction of insulin resistance by androgens and estrogens. J Clin Endocrinol Metab 79:265–271

Polderman KH, Gooren LJG, Van der Veen EA (1995) Effects of gonadal androgens and estrogens on adrenal androgen levels. Clin Endocrinol 43:415–421

Sapino A, Pietribiasi F, Godano A, Bussolati G (1990) Effects of long-term administration of androgens on breast tissues of female-to-male transsexuals. Ann N Y Acad Sci 586:143–145

Schildkraut JM, Schwingl PJ, Bastos E, Evanoff A, Hughes C (1996) Epithelian ovarian cancer risk among women with polycystic ovary syndrome. Obstet Gynecol 88:554–559

Seidell J, Björntorp P, Sjöström L, Kvist H, Sannerstedt R (1990) Visceral fat accumulation in men is positively associated with insulin, glucose and C-peptide levels, but negatively with testosterone levels. Metabolism 39:897–901

Shoupe D, Lobo RA (1984) The influence of androgens on insulin resistance. Fertil Steril, 41:385–8

Spinder T, Spijkstra JJ, Van den Tweel JG, Burger CW, Van Kessel H, Hompes PGA, Gooren LJG (1989a) The effects of long-term testosterone administration on pulsatile luteinizing hormone secretion and on ovarian histology in eugonadal female-to-male transsexual subjects. J Clin Endocrinol Metab 69:151–157

Spinder T, Spijkstra, Gooren LJG, Hompes PGA, Van Kessel H (1989b) Effects of long-term testosterone administration on gonadotropin secretion in agonadal female-to-male transsexuals compared with hypogonadal and normal women. J Clin Endocrinol Metab 68:200–207

Van Goozen SHM, Cohen-Kettenis PT, Gooren LJG, Frijda NH, van de Poll NE (1995) Gender differences in behaviour: activating effects of cross-sex hormones. Psychoneuroendocrinology 20:343–363

Van Kesteren PJ, Gooren LJG, Megens JA (1996a) An epidemiological and demographic study of transsexuals in the Netherlands. Arch Sex Behav 25:589–600

Van Kesteren PJM, Lips P, Deville W, Popp-Snijders C, Asscheman H, Megens J, Gooren LJG (1996b) The effect of one year cross-sex hormonal treatment on bone metabolism and serum insulin-like growth factor-I in transsexuals. J Clin Endocrinol Metab 81:2227–2232

Van Kesteren PJM, Asscheman H, Megens J, Gooren LJG (1997) Mortality and morbidity in transsexuals treated with cross-sex hormones. Clin Endocrinol 47:337–342

Van Kesteren PJM, Lips P, Gooren LJG, Asscheman H, Megens J (1998) Long-term follow-up of bone mineral density and bone metabolism in transsexuals treated with cross-sex hormones. Clin Endocrinol (in press)

Van Kesteren PJM, Kooistra T, Lansink M, Van Kamp GJ, Asscheman H, Gooren LJG, Emeis JJ, Vischer UM, Stehouwer CDA. The effects of sex steroids on plasma levels of marker proteins of endothelial cell functioning. (submitted)

von Eckardstein A, Kliesch S, Nieschlag E, Chirazi A, Assmann G, Behre H (1997) Suppression of endogenous testosterone in young men increases serum levels of high density lipoprotein subclass lipoprotein A-1 and liporotein(a). J Clin Endocrinol Metab 82:3367–3372

Wilson, JD (1987) Androgen abuse by athletes. Endocr Rev 2, 181–99

Zavaroni I, Dall'Aglio E, Alpi O, Bruschi F, Bonora E, Pezzarossa A, Butturini U (1985) Evidence for an independent relationship between plasma insulin and concentrations of high density lipoprotein cholesterol and triglyceride. Atherosclerosis, 55:259–66

Zhou JN, Hofman MA, Gooren LJG, Swaab DF (1995) A sex difference in the human brain and its relation to transsexuality. Nature 378:68–70

20 Abuse of androgens and detection of illegal use

Wilhelm Schänzer

Contents

20.1 Introduction

Anabolic androgenic steroids (AAS) are known to be misused both in competitive and in non-competitive sports (Haupt and Rovere 1984; Wilson 1988; Yesalis et al. 1993). Moreover, it seems that AAS are becoming "social drugs", as even young people apply them as an expression of an improved "lifestyle".

The misuse of AAS in athletics has been observed for more than 40 years. The first rumours dated from 1954 and were attributed to weightlifters who seemed to have used testosterone (Wade 1972). By 1965 synthetic AAS had become widely popular among bodybuilders and weightlifters, but were also applied in other forms of sports.

By using these steroids athletes hoped to obtain increased muscle strength. Such improvements in muscle strength to increase physical performance in sport is naturally an essential effect of training. As AAS stimulate protein synthesis in muscle cells, athletes expect a performance-enhancing effect beyond that brought about training. At the end of the sixties the first anti-doping rules were established by international sport federations (1967 International Cycling Union and 1968 International Olympic Committee, IOC), but only stimulants and narcotics were banned (Clasing 1992). At that time the Medical Commission (MC) of the IOC was already aware that the misuse of AAS in sport was widespread. They were not banned because no reliable method was available to detect them (Beckett and Cowan 1979). Under these circumstances the first methods for AAS detection were developed (Brooks et al. 1975; Ward et al. 1975) and in 1974 the MC of the IOC and the International Amateur Athletic Federation (IAAF) banned the use of AAS for the first time. This prohibition referred only to synthetic steroids, such as metandienone, stanozolol etc. and the misuse of endogenous steroids, e.g. testosterone, was not restricted. At that time athletes misused only synthetic AAS. The reason was that AAS were used in human medicine to treat catabolic conditions; scientific data obtained from animal studies led to the conclusion that synthetic AAS are more anabolic and less androgenic than testosterone itself (Kochakian 1976).

Whether athletes experience a positive performance-enhancing effect or not when using AAS has been discussed controversially for many years. Nowadays it is known that androgens have muscle growth-promoting effects in boys, in women and in hypogonadal men. It has never been proven that androgens, when administered in pharmacological doses, have positive effects on muscle growth in adult men. The assumption that AAS have less effect on muscle growth in men is based on the fact that the androgen receptor in men is nearly completely saturated (Wilson 1996). However, muscle mass enhancing effects from the use of androgens have been reported (Bhasin et al. 1996). An unethical and secret program of hormonal doping of athletes in the former German Democratic Republic has been reported (Franke and Berendonk 1997) and performance-improving effects of AAS were elucidated. Nevertheless, the physiological mechanisms underlying these phenomena is still unclear. Furthermore the effect of training on muscle growth, which also remains without scientific explanation, should be considered when reasons for AAS misuse are discussed.

To control the (mis)use of synthetic AAS urine samples of athletes collected after competition events were tested. As AAS are not used directly during competition but rather during training to increase muscle strength, athletes stopped administration of AAS before competition, changing to those AAS (e.g. stanozolol) they believed could not be detected and endogenous androgens, such as testosterone, which were not banned. Investigations of test samples from the Olympic Summer Games in 1980 in Moscow showed that 2.1% of male and 7.1% of female athletes had elevated testosterone levels (Zimmermann 1986) in urine and the highest urinary testosterone levels were detected for women in swimming and track and field events (Donike 1982). These results could only be explained by exogenous application of testosterone. Donike developed a gas chromatographic/mass spectrometric (GC/MS) method for detection of urinary excreted testosterone and epitestosterone and proved that the ratio of testosterone to epitestosterone is significantly increased after application of testosterone (Donike et al. 1983). Based on these results sport federations also banned testosterone in 1984 and applied the T/E ratio measurements (cutoff level of six) to their rules. In addition to the ban of testosterone, other endogenous androgens such as dihydrotestosterone and dehydroepiandrosterone were added to the list of prohibited AAS during recent years.

20.2 Frequency of steroid hormone misuse

Considering this misuse of AAS, the question arises to what extent AAS are really misused? Is it restricted to a few high-level performing athletes in sports or is it so extensive as to be regarded as a social drug problem? Available data derive

- from positive findings resulting from checking AAS doping in competition sports and
- from results of questionnaires concerning the use of AAS in non-competitive sports.

20.2.1 Androgen misuse in controlled competition sports

Androgen misuse in competition sports is investigated by laboratories which are accredited by the IOC. Each year the accredited laboratories (25 laboratories in 1996, worldwide) report the positive findings from A-samples. The annual testing frequency is about 90,000 samples for all laboratories. From 1992–96 a total number of approximately 460,385 doping tests were analyzed by the IOC accredited laboratories for Olympic and non-Olympic sports and 2,657 AAS were confirmed. The number of 2,657 identified AAS does not represent the number of positive athletes. The use of more than one AAS by athletes diminished the number of positive athletes by approximately 10–15%. The statistics do not indicate whether sanctions against the athlete tested positively are applied by the sport federation or not.

Table 20.1 shows the type of sport in which the misuse of AAS represents a problem. For example, in bodybuilding AAS misuse continues to be a problem, whereas other sports federations have more effective control systems.

Table 20.1. AAS findings and number of worldwide doping tests in relation to type of sport 1992–96 (statistic of IOC accredited laboratories, results of A-samples without testosterone)

Type of sport	No. of tests	No. of detected AAS	%
Bodybuilding	3,429	900	26.25
Powerlifting	8,018	273	3.40
Weightlifting	21,255	456	2.15
Athletics	66,340	240	0.36
Cycling	52,020	138	0.27
Swimming	15,589	43	0.28
American football	90,717	215	0.24
Other sports	203,017	392	0.19
Total	460,385	2,657	0.58

20.2.2 Androgen misuse in non-controlled sports

In comparison to officially controlled competition sports, no analytical data from laboratories concerning the misuse of AAS are available for those areas of athletics and private life where no tests are performed. To overcome this lack of information scientists have performed surveys, however, only a few publications are available. Yesalis et al. (1993) published results of investigations in the United States and calculated that 1 million Americans had used AAS sometime in their lives including about 250,000 in the past year. In 1993 a Canadian study (Canadian Centre for Drug-Free Sport 1993) confirmed that in Canada 80,000 young people between the ages of 13 and 18 had used AAS. A self-report questionnaire about the misuse of AAS among 13,355 Australian high school students reported 3.2% male and 1.2% female users (Handelsman and Gupta 1997). Questionnaires from Switzerland were summarized by Kamber (1995), who concluded that AAS are a serious problem. A questionnaire from 16,000 recruits in Switzerland and 3,700 women of the same age showed that 1.8% of the recruits and 0.3% of the women had administered AAS in 1993.

One of the main unsolved problems in AAS misuse is the existence of a "black market" providing bodybuilders and athletes with AAS. As AAS are not available on the free market without medical prescription, their consumption is illegal and an offence against the law. Governmental control and penalties are only directed against those selling AAS and persons are only found guilty when a huge amount of AAS is confiscated.

20.3 Frequently misused steroid hormones

Among the most frequent steroid hormones misused in controlled competitive sports are the synthetic steroid nortestosterone, stanozolol, metandienone, methenolone etc. (Table 20.2) and the endogenous steroids, testosterone and dihydrotestosterone. Table 20.2 does not refer to testosterone misuse in sports as no reliable data are documented by the IOC statistics. Fig. 20.1 shows the chemical structure of these misused steroids. The data indicate that the misuse of AAS is limited to a number (17) of well-known AAS. The different types of AAS misused since 1976 show that no trend or systematic changes of the substances used was observed in the last 20 years. The first positive cases with AAS in 1976 were detected for metandienone which still remains one of the most frequently abused AAS. Moreover, the misuse of endogenous steroids since 1980 did not alter the clear preference for the "classic" AAS. These data allow speculations concerning the effectiveness of AAS, and the impression arises that these steroids are the most effective ones. Data concerning the most commonly misused AAS in non-controlled sports are only available via recommendations in magazines for bodybuilders, via

Table 20.2. Identified anabolic androgenic steroids misuse in controlled competition sports 1992 to 1996 worldwide (data of IOC accredited laboratories, results of A-analysis without testosterone)

Steroid	No.
Nandrolone	1,030
Metandienone	516
Stanozolol	441
Methenolone	247
Mesterolone	92
Methyltestosterone	71
Boldenone	58
Drostanolone	35
Dihydrotestosterone	31
Clostebol	29
Oxandrolone	29
Oxymetholone	29
Fluoxymesterone	23
Dehydrochloromethyl-testosterone	12
Formebolone	10
Trenbolone	4
Total	2,657

Total number of analyzed samples 460,385

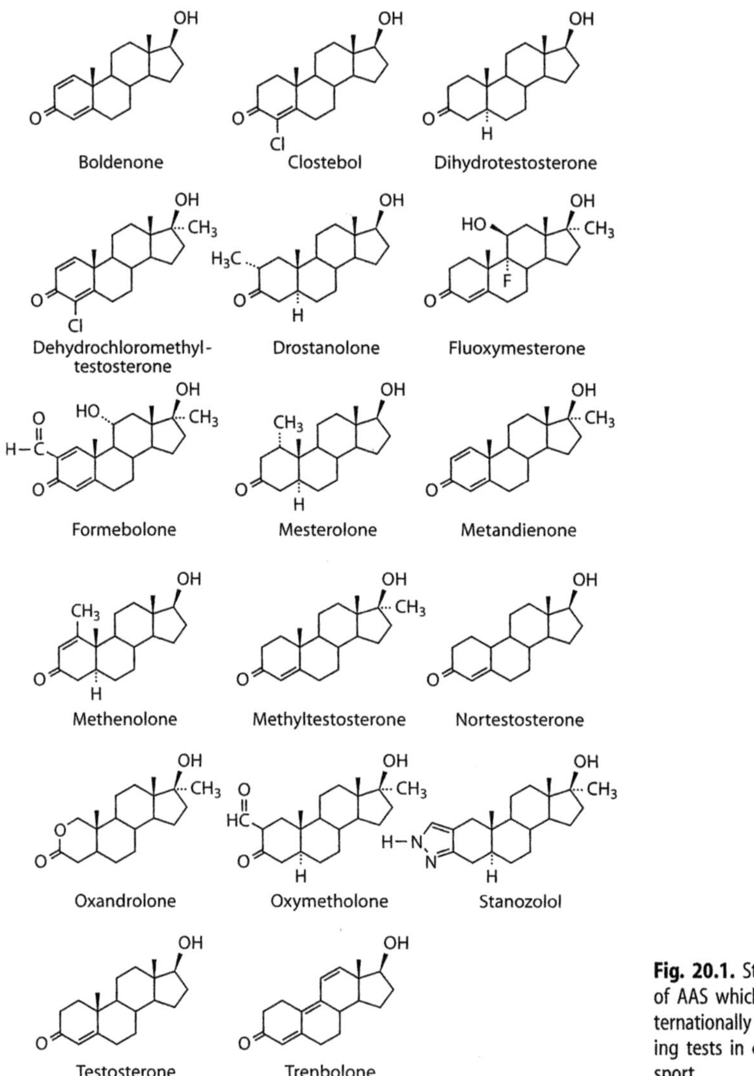

Fig. 20.1. Structure formula of AAS which have been internationally detected in doping tests in competetion sport

"underground" handbooks (Duchaine 1989; Grundig and Bachmann 1995; Taylor 1982), the Internet and via confiscated, smuggled substances and those obtained from black market sources. Frequently recommended AAS include boldenone undecylenate, drostanolone propionate, methenolone acetate, methenolone enanthate, metandienone, nandrolone decanoate, oxandrolone, oxymetholone, stanozolol, and testosterone in the form of different esters (cypionate, decanoate, enanthate, isocaproate, phenylpropionate, propionate and undecanoate).

These substances are largely indentical with those products which were confiscated and distributed by the "black market". Knuth et al. published a detailed questionnaire on the history of AAS use of 41 bodybuilders in 1989. Additionally to the misused AAS, which are in agreement with steroids in Table 20.2, the use of norethandrolone and oxabolone is reported.

Information on AAS for athletes and especially for bodybuilders is mainly provided by bodybuilder magazines and underground literature. The advantages and possible side-effects of nearly all AAS listed in Table 20.2 are discussed in such literature, with suggestions for the most effective AAS. Less effective steroids are also listed. Typical average athletic doses and frequence of intake are recommended. These recommendations are mainly based on reports by bodybuilders, their doctors and coaches experienced with AAS supplementation. Data about side-effects of AAS are not suppressed and, to the extent that they are recognized, extensively discussed. According to such information, one of the main side-effects of AAS use is gynecomastia, which is assumed to result from androgen conversion to estrogens and is associated with high doses of AAS which can be aromatized. To prevent this side-effect, the simultaneous administration of anti-estrogens such as tamoxifen and clomiphene is suggested. Moreover, for bodybuilders suffering from gynecomastia a few clinics worldwide which will perform surgical removal of increased breast tissue are recommended. There are also hints to athletes who may be tested in competition about when the use of the different AAS should be terminated in order to pass the test. All these sources of information give the impression that the use of AAS is quite normal and that toxic effects can be minimized and well controlled.

Finally, sources for legal (but mainly illegal) purchases all over the world are provided. "Black market" prices in different countries are listed and extensive news about fakes is included.

20.4 Detection of misuse of anabolic androgenic steroid hormones

20.4.1 Organization of doping tests

Doping control is organized by national and international sport federations for the different types of sports. In several countries (e.g. Norway, France and Germany) national anti-doping programs are administered by one overall organization. This strategy seems to be the most effective testing action as any possible intention by individual sport federations to hide positive cases and to protect their athletes can be excluded. The IOC only performed doping tests during the Olympic winter and summer games and it has no "out of competition testing program".

For a doping test athletes are selected according to the rules of the responsible sports federation. The doping test is carried out in two steps. The first step includes the sample-taking procedure and transportation of the urine specimens to an IOC accredited laboratory. In the following step the laboratory analyzes the sample for banned drugs. The sample-taking procedure is an important step. To avoid any manipulation athletes have to deliver a urine sample under visual inspection of an accredited supervisor. The urine is divided into an A- and a B-sample, both samples are sealed and then transported to the laboratory. All steps during this procedure are documented and the athlete has to sign a protocol of the sample-taking procedure and sealing of samples. All handling of a urine specimen (sample-taking, transportation containers and laboratory tests) must be documented and is designated as "chain of custody". The laboratory is not in possession of the athletes' name corresponding to the urine sample. For this reason all samples have code numbers. The reason for dividing the urine specimen into A- and B-samples is to guarantee the best "chain of custody": if the A-sample is tested positive, the B-sample will be analyzed in the presence of the athlete and his advisers. If the B-analysis confirms the A-result the sample is considered as positive. This result is the basis for the federation to impose sanctions on the athlete.

Doping test samples are analyzed by IOC accredited laboratories. Laboratories seeking accredition have to comply with the requirements for doping drug testing set by the IOC Medical Commission. The laboratory must show that it has the capability to analyze all banned substances below the specified concentration limits. Especially for the detection of synthetic AAS, which are misused mainly during the training period, the laboratory has to use highly sensitive methods. At the present time 25 laboratories all over the world (17 in Europe, 3 in North America, 3 in Asia, 1 in Australia and 1 in Africa) are accredited.

The prerequisite for this accreditation system is a standardization of analytical techniques and detection limits of banned substances among the different laboratories. Information concerning new doping drugs and doping techniques is rapidly distributed and in order to deal with new problems in a coordinated manner.

20.4.2 Detection and identification of misused anabolic androgenic steroids

Synthetic AAS were first banned in 1974. As no comprehensive analytical method for the detection of AAS in human urine was available in the beginning of the seventies, new methods had to be developed. The first methods were based on radioimmunoassay (RIA) techniques, e.g. in 1975 Brooks et al. developed an antiserum for metandienone with some cross reactivity to other 17α-methyl steroids. The RIA techniques were discouraging for several reasons: the method did not consider the high extent of metabolism of AAS (therefore the screening for the parent steroid was less successful), the antisera had only limited sensitivity for other steroids and the possibility of false positives, which was not acceptable for routine analysis. As early as 1975 Ward et al. presented a gas chromatography/mass spectrometry (GC/MS) method for the detection of the AAS metandienone, nortestosterone, norethandrolone and stanozolol. Nevertheless, the RIA technique was used as a

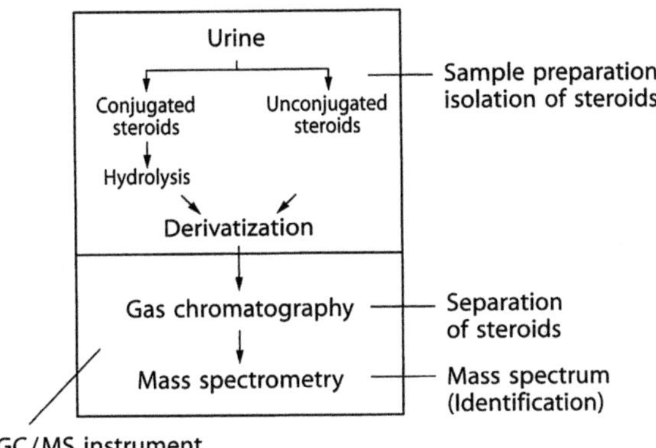

Fig. 20.2. Analysis of AAS:
a) Sample preparation and
b) GC/MS determination

screening method during the 1976 Olympic Games in Montreal and in Moscow in 1980, but confirmation of suspicious samples was performed by GC/MS. After 1981 all IOC accredited laboratories used GC/MS as the main analytical tool for AAS identification (Donike et al. 1984; Masse et al. 1989; Schänzer and Donike 1993).

Analysis of AAS can be divided and described in two steps (Fig. 20.2): first a sample preparation is performed with the aim to separate the banned substances from the biological matrix (urine) and to reduce biological interference (biological background). The sample preparation also includes a chemical modification (derivatization) of the isolated substances to improve their analytical detectability. The second step covers the analytical measurement, which is based on a physical principle, mainly on gas chromatography in combination with mass spectrometry (GC/MS). Additionally liquid chromatography and mass spectrometry (LC/MS) can be used for substances which show poor gas chromatographic properties or are sensitive to temperature.

The main advantage of chromatographic techniques such as GC/MS is the possibility of analyzing a high number of substances within one run. This minimizes costs and allows a high throughput of samples. In fact it is possible to run a maximum of 50 samples per day and per GC/MS instrument.

20.4.2.1 Metabolism

Anabolic androgenic steroids are metabolized to a large extent by phase I and phase II reactions and only a few AAS are excreted unchanged in urine for a short period of time after administration. To detect the misuse of AAS which are not excreted in urine or only to a small extent, the analytical method cannot rely on monitoring the parent steroid but on the identification of its metabolites. Detection of an AAS metabolite in urine is proof for the misuse of a banned anabolic androgenic doping substance. This presumes that the metabolite cannot be generated from endogenous steroids in

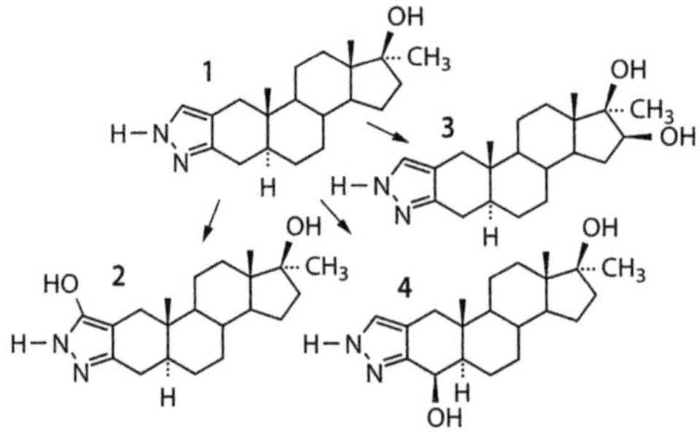

Fig. 20.3. The main metabolites in the metabolism of stanozolol (1), 3'-hydroxy-stanozolol (2), 16β-hydroxystanozolol (3), and 4β-hydroxystanozolol (4)

the body compartment. Metabolism of the most frequently misused anabolic steroids has been investigated by different working groups in recent years (Schänzer 1996). Basically the metabolism of AAS follows the metabolic pathways of the principal androgen testosterone. This includes reduction of the double bond at C4-C5 to form 5α- and 5β-isomers, the reduction of the 3α-keto group to a 3α-hydroxy function and, in case of 17β-hydroxy steroids with a secondary hydroxy group, the oxidation yielding a 17-keto function. Additionally many AAS are metabolized by cytochrom P-450 hydroxylation reactions, and steroids with hydroxy groups mainly at C-6β, C-16α and C-16β are produced. In the metabolism of stanozolol, a synthetic steroid with a condensed pyrazol ring on the steroid A-ring, further hydroxylation occurs at C-4β and C-3' of the heterocyclic ring (Fig. 20.3). In the course of phase I metabolism steroids are enzymatically transformed in general to more polar but pharmacologically inactive compounds. Phase I reactions are often followed by phase II processes, also known as phase II conjugation. In the case of AAS and their metabolites the reaction creates steroid conjugates with sulfate or glucuronic acid. These highly polar compounds are then rapidly eliminated in the urine.

Excretion studies in the last ten years performed with 17α-methyl steroids demonstrated that the metabolism of AAS is highly complex and the detection of more than 20 metabolites after administration of one single AAS is not unusual. Similar results regarding the high number of metabolites are already known for the metabolism of testosterone (Kochakian 1990).

20.4.2.2 Pharmacokinetics

A further important factor which has to be considered for detection of AAS is the pharmacokinetics of the parent compound and its excreted metabolites. As AAS are misused during training and the number of checks are limited, it is desirable to detect AAS as long as possible after their last administration. Analysis of the parent steroids and/or their metabolites, which are excreted very rapidly, are less effective for screening analysis than the detection of longterm excreted metabolites, which are steroids detectable for the longest period of time after administration (Schänzer et al. 1996). The main difference between the pharmacokinetics of AAS is caused by their pharmaceutical preparation and the kind of application. Depot preparations, e.g. 19-nortestosterone injected intramuscularly as its undecanoate ester (Deca-Duraboline), are detectable in urine for several weeks, whereas most oral preparations are completely eliminated within a few days after intake. Following these scientific data athletes changed their doping activities to AAS with short elimination times and to steroids which were believed to be not detectable. For example, stanozolol, an oral preparation, was assumed for several years to be nondetectable. One of the most famous doping cases was Ben Johnson, who was disqualified after he had won the 100 m final at the 1988 Olympic Games in Seoul. He tested positive for stanozolol, which was easily detected by the GC/MS method used by the testing laboratory. It was assumed that the steroid was administered within one week prior to competition (Donike 1989). An explanation for this foolish doping case was the assumption that stanozolol could not be detected.

20.4.2.3 Sample preparation

For sample preparation of anabolic steroids it has to be considered that most of the AAS and their metabolites are excreted in conjugated form. Following sample preparation unconjugated steroids can be separated by extracting an aliquot of urine (e.g. 2 ml) with an organic non water miscible solvent. Conjugated steroids, based on their polar and acidic character, are not extractable and remain in the aqueous layer. These conjugates (mainly glucuronides) can be liberated by enzymatic hydrolysis of the urine specimen. The enzyme used can be added directly to the urine or to an isolate obtained via an adsorber resin. Enzymatic hydrolysis is achieved completely using enzyme preparations with β-glucuronidase from E. coli or β-glucuronidase/arylsulfatase from Helix pomatia. The free steroids (conjugated fraction) are then extracted from the aqueous phase via a simple liquid extraction with tert-butyl methyl ether or in case of less polar steroids with an alkane (e.g. n-pentane).

The first analysis is a screening procedure by which all banned AAS are detected in one single analytical run. Suspicious samples are confirmed by a second aliquot of the same urine specimen, which is isolated using a substance-specific isolation technique. Identification of the misused steroids follows.

20.4.2.4 Derivatization

Based on the polar groups of AAS (hydroxy and keto groups) high interactions with polar functions of the GC-column phase reduce the detectability of AAS at low concentrations. Derivatization of polar functions of AAS can lead to a distinct improvement in peak intensity and detection limit of the analytical method. The most frequently used derivatization methods are acylation (e.g. trifluoroacetylation) and silylation (e.g. trimethylsilylation). For

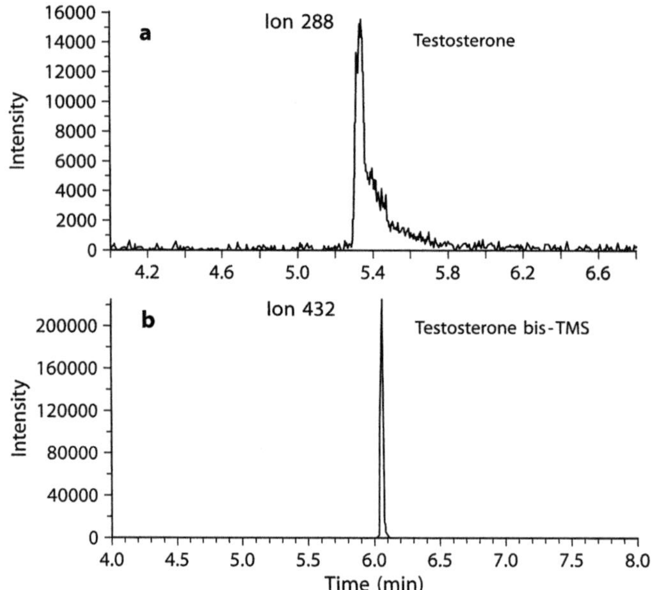

Fig. 20.4. GC/MS chromatogram of **a)** underivatized testosterone (50 ng on column) and **b)** bis-trimethylsilylated testosterone (10 ng on column)

doping analysis of AAS silylation is the method of choice and the introduction of a trimethylsilyl group to an AAS is the most common derivatization reaction, converting polar groups such as hydroxy and keto functions to less polar trimethylsilyl ethers with excellent GC behaviour. For this kind of derivatization a respectable reagent MSTFA (N-methyl-N-trimethylsilyltrifluoroacetamide) was developed (Donike 1969). The influence of derivatization with MSTFA on the gas chromatographic and mass spectrometric analysis is best demonstrated for testosterone in Fig. 20.4 and 20.5. Obviously the GC peak shape of the derivatized steroid is narrow and the peak height is increased. Additionally, the mass spectrum is generally changed to higher and more abundant molecular and fragment ions, which also improves the signal to noise ratio of the substance to be identified compared to the analytical and biological background. Therefore derivatization for GC/MS detection of substances isolated from biological fluids unequivocally yields a more accurate analytical result, which is an absolute requirement regarding the complex matrix and high number of possible interferences.

20.4.3 Detection of synthetic anabolic androgenic steroids

In some instances endogenous and exogenous AAS are distinguished according to their route of administration. The term "synthetic" should amplify the fact that these AAS are not produced in the body, they are chemically synthesized and can only enter the circulating blood system by exogenous application. AAS which are naturally synthesized in the glands of mammalian cells

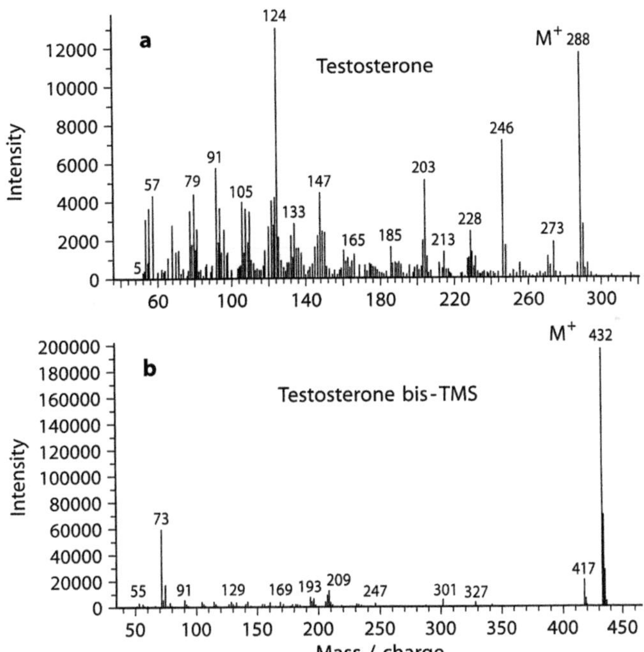

Fig. 20.5. Electron impact (EI) mass spectra of are distinguished **a)** underivatized testosterone and **b)** bis-trimethylsilylated testosterone

are called endogenous steroids, even though their application can be exogenous. As synthetic AAS and/or their metabolites are not present in the human organism, their identification in a urine sample of an athlete constitutes the misuse of a banned steroid. The criteria for identification of a substance are based on the analytical method applied.

In GC/MS identification of synthetic AAS obtained from a urine specimen it is mandatory to register a full mass spectrum or a selected ion monitoring (SIM) profile of the main abundant fragment ions. The mass spectrometrical data (MS spectrum or SIM profile) of the isolated substance should be in accordance with an authentic synthesized reference substance or, in the event that a synthesized reference metabolite is not available, with a well-characterized metabolite from an excretion study with the corresponding AAS. Additionally to the MS data, the GC retention time of the isolated steroid has to be in agreement with the GC retention time of the reference substance. For this purpose reference metabolites of AAS for frequently misused AAS which were not commercially available were synthesized (Schänzer and Donike 1993).

As an example Fig. 20.6 shows the criteria for a positive sample for stanozolol:

- registration of a full mass spectrum which can be compared with the reference spectrum or
- in case of low concentrations, a selected ion monitoring (SIM) profile with the main intense fragment ions of the metandienone metabolite 17,17-dimethyl-18-nor-5β-androst-1,3-dien-3α-ol.

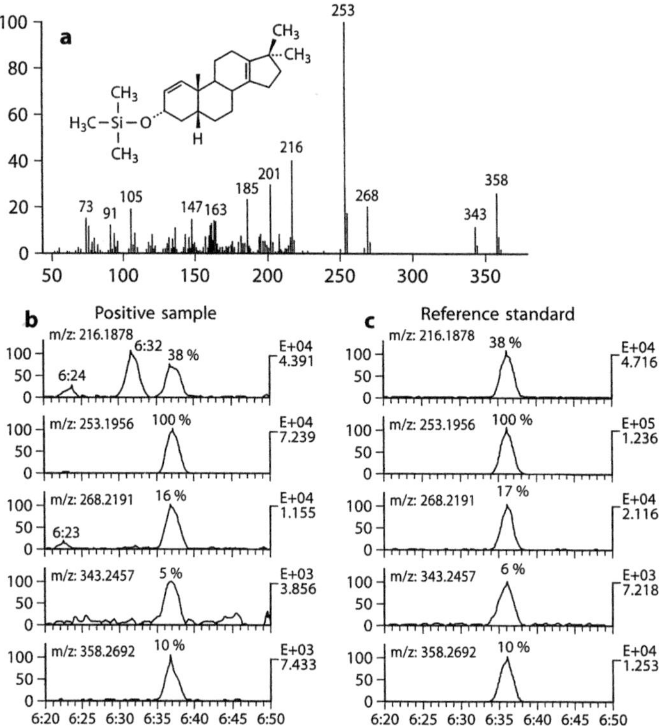

Fig. 20.6. Criteria for a positive confirmation: 1. The registrated EI mass spectrum (e.g. mass spectrum of an isolated metabolite of metandienone: 17,17-dimethyl-18-nor-5β-androst-1,13-dien-3α-ol TMS (**a**) has to be in accordance with the mass spectrum of an authentic reference substance, or 2. the main abundant fragment ions of the isolated substance show similar intensities (**b**) when selected ion monitoring (SIM) registration is applied in comparison to the intensities of the same fragments of the reference compound (**c**)

To increase the efficiency of AAS misuse testing and to detect AAS for a longer period of time after administration more selective and sensitive MS techniques were used during the last decade. The main improvements were first completed by installing more sensitive and selective mass spectrometers and second by substance-specific sample preparation (Schänzer et al. 1996). The use of high resolution mass spectrometry (HRMS) was announced to the public at the Olympic Games 1996 in Atlanta. This technique was established after 1992 in a few IOC accredited laboratories. The advantage of HRMS became apparent before Atlanta when, during doping testing by the International Weightlifting Federation, more than forty athletes were confirmed positive only by HRMS and not by the conventional MS technique. Following these results the IOC decided that it was neccessary that accredited laboratories use more sophisticated equipment, for instance, HRMS or MS/MS.

The basic principle of HRMS is based on the fact that elements do not have an integral number of atomic weight but a decimal form. Only carbon, as the reference element, has an integral number of 12 as its atomic weight. Thus hydrogen does not have an atomic weight of 1 but 1.00783, nitrogen the weight of 14.00307, and oxygen the weight of 15.99491. Molecular fragments with the same integral number of mass, e.g. the fragment ions $C_3H_6O^+$ and $C_3H_8N^+$ both have the rounded mass 58 but the exact calculated mass of 58.04186 for $C_3H_6O^+$ and 58.06567 for $C_3H_8N^+$. Neither mass fragments can be separated by conventional (low resolution) mass spectrometry but only by using high resolution MS with a resolution of 2500. Thus in practical terms, in this example the instrument (HRMS) can be set to detect only the signal of the mass of 58.04186 for $C_3H_6O^+$, and all masses differing by more than 0.0024 masses, such as 58.06567 for $C_3H_8N^+$, will be discriminated. Based on this fundamental physical principle HRMS analysis of AAS steroids and their metabolites isolated from urine reduces the biological background and increases the signal to noise ratio, yielding a much higher selectivity in screening and confirmation.

A further improvement in detection of synthetic AAS is achieved in sample preparation for confirmation analysis. If the screening procedure identifies a suspicious sample, it has to be reanalyzed. When the urinary concentration of the suspicious substance is appreciably low, a more specific isolation technique is used to reduce interfering compounds. Two methods mainly used are: 1. Immuno affinity chromatography (IAC) and 2. High performance liquid chromatography (HPLC) separation.

In IAC antibodies towards specific AAS are prepared and coupled to an agarose gel which is then filled in columns. The urine or a urinary extract is passed through the column, some steroids are specifically bound and separated from the complex matrix (Delahaut 1997; Schänzer et. al 1996).

HPLC separation is achieved using analytical or preparative reverse phase columns. The lipophilic urinary extract from 2–10 ml urine, containing the liberated steroids, is added to a LC column and the steroids are separated. The separated fractions are then analyzed by GC/MS or GC/HRMS, yielding a dramatic improvement in signal to noise ratios. This enables a longer retrospective assessment of whether AAS have been misused or not.

20.4.4 Detection of endogenous anabolic androgenic steroids: the testosterone problem

20.4.4.1 Indirect detection methods

The misuse of testosterone by athletes is also tested by GC/MS analysis of urinary extracts. However, the method reveals only the presence of testosterone and its ratio to epitestosterone. The mass spectrometrical data alone do not indicate whether testosterone originates exogenously (doping) or whether it was produced endogenously. In 1983 Donike et al. developed a method to calculate the urinary excreted testosterone by a ratio to 17-epitestosterone. Both isomeric steroid hormones are excreted as glucuronides which are enzy-

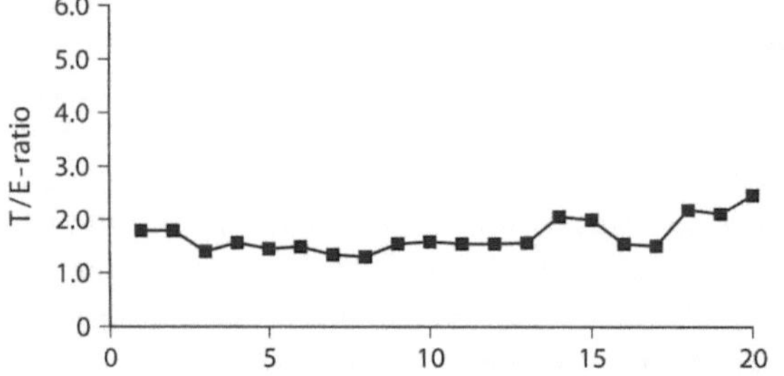

Fig. 20.7. Stability of the testosterone/epitestosterone ratio (T/E ratio) of a professional cyclist over a period of 20 days at the Tour de France 1993

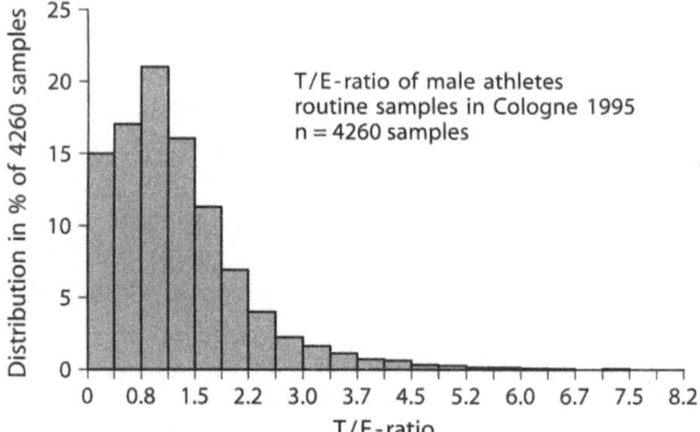

Fig. 20.8. Distribution of T/E ratios of male athletes analyzed 1995 in the IOC accredited Cologne laboratory

matically hydrolyzed before GC/MS analysis. The urinary testosterone/epitestosterone ratio (T/E ratio) represents a relatively constant factor in an individual person and alterations under physical excercise have not been noted (Fig. 20.7). Exogenous application of testosterone results in an increase in the urinary concentration of testosterone glucuronide, whereas epitestosterone glucuronide is not influenced. Based on measurements of large reference groups Donike proposed a T/E ratio of 6:1 as a marker to handle a urine specimen suspicious for testosterone misuse. As an example, Fig. 20.8 shows the distribution (population based reference ranges) of the T/E ratios of 4260 male athletes tested in Cologne in 1995. An increased T/E value (T/E > 6) is not immediately considered as a positive sample. Following the IOC rule the athlete has to be further investigated and it has to be determined that the increased value is not caused by physical or pathological conditions. In practical terms this requires an endocrinological profile or an evaluation of previous tests in order to establish the athlete's individual T/E reference values (subject-based reference values). The test sample is considered positive when the tested T/E ratio clearly exceeds the subject-based reference values (> mean + 3 standard deviations) of the athlete. In addition to the T/E ratio the testosterone and epitestosterone concentration as well as the concentrations of the main testosterone metabolites are assessed.

Doping with dihydrotestosterone (DHT) became public knowledge since the Asian Games in 1994 when 11 athletes were tested positive for DHT misuse. The criteria for DHT doping are also based on statistical methods and population-based reference values with limits for the ratios of DHT/epitestosterone, DHT/etiocholanolone, 5α-androstane-$3\alpha,17\beta$-diol/5β-androstane-$3\alpha,17\beta$-diol, and androsterone/etiocholanolone established (Donike et al. 1995; Kicman et al. 1995).

The main weakness of all the methods confirming doping with endogenous AAS is the application of statistical parameters. These methods are therefore indirect methods and they only confirm that an increased value varies from the normal values of the athlete. These methods do not identify any physical characteristics of the exogenous steroid differing from the endogenously produced steroid as direct proof.

20.4.4.2 Direct detection method: Gas chromatography/combustion/ isotope ratio mass spectrometry (GC/C/IRMS)

The disadvantage of the T/E ratio method can be overcome by a new method: gas chromatography/combustion/isotope ratio mass spectrometry (GC/C/ IRMS). This method was first introduced by Becchi et al. in 1994 and has been adopted by other research groups (Aguilera et al. 1996; Horning et al. 1997; Shackleton et al. 1997) with distinct modifications. The principle of IRMS is the precise measurement of the $^{13}C/^{12}C$ isotope ratio of organic compounds. This method became practical for trace analysis in doping control when instruments with the combination of gas chromatography and isotope ratio mass spectrometry were developed. Isotopes are elements with the same number of protons but different numbers of neutrons. Carbon occurs in three kinds of isotopes: ^{12}C (6 protons and 6 neutrons) with a frequency of approximately 98.9%, ^{13}C (6 protons and 7 neutrons) at a rate of 1.1% and ^{14}C (6 protons and 8 neutrons), a radioactive isotope with a half-life of 5760 years (used in determination of age), in traces. In the course of synthesizing organic compounds ^{12}C atoms react slightly faster than ^{13}C atoms. This effect results in a reduction of the ^{13}C amount compared to ^{12}C. The $^{13}C/^{12}C$ ratio is calculated in promill [$\delta^{13}C(\permil)$] relative to a reference gas with a standardized $^{13}C/^{12}C$ ratio. The value becomes more negative when the ^{13}C portion is reduced, as was explained during synthetic pathways. For isotope measurement steroids have to be isolated to high purity. Most research groups use HPLC separation of steroids or isolation of steroidal diols is performed via the Girard reagent (Shackleton et al. 1997). For gas chromatographic separation derivatization is applied using acetylation of steroids with the aim to improve GC peak shape or analysis refers to the underivatized steroids.

Steroids are separated by gas chromatography followed by complete oxidation to carbon dioxide in a combustion chamber. The carbon dioxide is then introduced to the mass spectrometer where the exact masses m/e 44 for $^{12}CO_2$ and m/e 45 for $^{13}CO_2$ are independently registered. For this kind of isotope ratio measurement a minimum of 20–30 ng of a steroid has to be

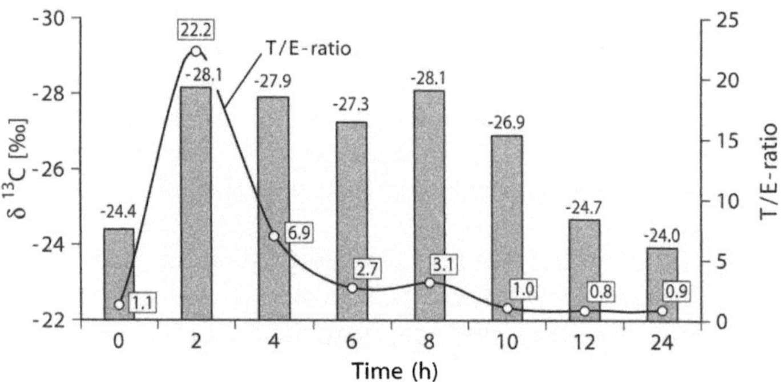

Fig. 20.9. Testosterone estimation in urine after oral application of 40 mg of Andriol (testosterone undecanoate): T/E ratio (line) and carbon isotope ratio mass spectrometry (column)

used to obtain precise data. The $^{13}C/^{12}C$ ratio can be estimated with a precision of $\pm 0.0002\%$ (± 0.2 permill to the $^{13}C/^{12}C$ ratio of the reference gas).

Fig. 20.9 presents data of GC/C/IRMS and T/E ratio analysis after oral administration of 40 mg of testosterone undecanoate (Andriol) to a single male volunteer. A direct proof of exogenous testosterone application is possible as the δ-values are decreased to –28 ppm after administration in comparison to –24 ppm before intake and at the end of the elimination curve. It is also obvious that the δ-values are still decreased when the T/E ratio drops below six and is close to the normal value. This method can therefore also be used when ethnic differences influence testosterone metabolism, e.g. in Asians who have low T/E ratios and when a testosterone application will not necessarily exceed the T/E ratio of six (de la Torre et al. 1997). Exogenous testosterone also influences the $^{13}C/^{12}C$ ratio of the metabolites of testosterone. Based on these data it was proved that precursors within the synthetic pathway of testosterone, such as pregnanediol, pregnanetriol (metabolites of progesterone and 17α-hydroxyprogesterone) and cholesterol are not influenced by exogenous testosterone, whereas testosterone and its metabolites have decreased δ-values indicating exogenous application. The results of a positve testosterone finding are presented in Fig. 20.10. The T/E-ratio of the positive urine sample was 14.7. Following the rules, the athlete was further investigated and 10 urine samples were collected over a period of two days and analyzed. The T/E ratio during this study was 1.0 ± 0.1 and confirmed that the sample with a T/E ratio of 14.7 was not in accordance with endogenous production of testosterone and was considered as an offence against the doping rules. The IRMS data of the corresponding positive urine sample (Fig. 20.10) show the decreased values for testosterone and the metabolites androsterone and etiocholanolone, whereas the higher δ-values of the precursors and the values obtained from the endocrinological study were in the same range.

It was also suggested to detect testosterone misuse by analysis of testosterone esters in blood (de la Torre 1995), but this method is limited to the ap-

		Positive sample	Endocrinological study
Indirect method	T/E-ratio	14.7	1.0[*]
	$^{13}C/^{12}C$ Isotope ratio in δ [‰]		
Direct method	Cholesterol	-24.0	-24.4
	Pregnanediol	-24.0	-24.2
	Testosterone	-29.1	-25.1
	Androsterone	-29.2	-25.0
	Etiocholanolone	-29.8	-25.9

[*] T/E-ratio 1.0 ± 0.10, n = 10 samples

Fig. 20.10. T/E ratio and IRMS results of a testosterone positive urine sample and a urine sample of the same athlete obtained during an endocrinological study

plication of testosterone esters in the form of injectable preparations and not applicable to the analysis of urine samples.

The isotope ratio mass spectrometry method can additionally be applied to detect and identify doping with other endogenous AAS such as dihydrotestosterone and dehydroepiandrosterone, where reliable methods are less efficient or not available.

20.5 Key messages

- Misuse of androgens in competitive sport has been banned since 1974 and tested by IOC accredited laboratories.
- Androgens are used by athletes during training to improve muscle strength. For this reason doping tests have been extended to out of competition tests.
- Questionnaires concerning the misuse in non-competitive sport show a high extent of anabolic androgenic steroids (AAS) misuse.
- Androgens are most frequently distributed illegally to athletes via a black market.
- Main misused AAS in controlled sports are testosterone, nandrolone, stanozolol and metandienone.
- Androgens are detected and identified by gas chromatographic/mass spectrometric analysis of urinary extracts.
- Derivatization methods for steroid analysis improve detection limits for anabolic steroids.
- Synthetic androgens are extensively metabolized and doping tests are focused on urinary excreted metabolites.
- Doping with endogenous steroids is controlled by indirect methods, e.g. testosterone misuse is tested by a ratio of testosterone to epitestosterone (normal < 6:1). Positive findings are followed by additional studies to exclude physiological and pathological influences.
- Direct methods, as gas chromatography/combustion/carbon isotope ratio mass spectrometry became recently available to identify unambiguously doping with endogenous steroids.

20.6 References

Aguilera R, Becchi M, Casabianca H, Hatton CK, Catlin DH, Starcevic B, Pope HG Jr (1996) Improved method of detection of testosterone abuse by gas chromatography/combustion/isotope ratio mass spectrometry analysis of urinary steroids. J Mass Spectrom 31:169–176

Becchi M, Aguilera R, Farizon Y, Flament MM, Casabianca H, James P (1994) Gas chromatography/combustion/isotope-ratio mass spectrometry analysis of urinary steroids to detect misuse of testosterone in sport. Rapid Commun Mass Spektrom 8:304–308

Beckett AH, Cowan DA (1979) Misuse of drugs in Sport. Brit J Sports Medicine 2:185–194

Bhasin S, Storer TW, Berman N, Callegari C, Clevenger B, Phillips J, Bunell TJ, Tricker R, Shirazi A, Casaburi R (1996) The effects of supraphysiologic doses of testosterone on muscle size and strength in normal men. J Med 335:1–7

Brand WA (1996) High precision isotope ratio monitoring techniques in mass spectrometry. J Mass Spectrom 31:225–235

Brooks RV, Firth R, Summer NA (1975) Detection of anabolic steroids by radio immunoassay. Br J Sports Med 9:89–92

Canadian Centre for Drug-Free Sport (1993) News Release – Over 80000 young Canadians using anabolic steroids

Clasing D (1992) Doping – verbotene Arzneistoffe im Sport. Gustav Fischer Verlag, Stuttgart – Jena – New York

Delahaut PH (1997) Immunoassay and immunoaffinity chromatography for the detection of drug residues. In: Schänzer W, Geyer H, Gotzmann A, Mareck-Engelke U (eds) Proceedings of the 14th Cologne Workshop on Dope Analysis 1996, Sport und Buch Strauß, Köln, pp 211–222

de la Torre X, Segura J, Polettini A, Montagna M (1995) Detection of testosterone esters in human plasma. J Mass Spectrom 30:1393–1404

de la Torre X, Segura J, Yang Z, Li Y, Wu M (1997) Testosterone detection in different ethnic groups. In: Schänzer W, Geyer H, Gotzmann A, Mareck-Engelke U (eds) Proceedings of the 14th Cologne Workshop on Dope Analysis 1996, Sport und Buch Strauß, Köln, pp 71-89

Donike M (1982 and 1989) Private communication.

Donike M (1969) N-Methyl-N-trimethylsilyl-trifluoracetamid, ein neues Silylierungsmittel aus der Reihe der silylierten Amide. J Chromatogr 42:103–104

Donike M, Bärwald K-R, Klostermann K, Schänzer W, Zimmermann J (1983) Nachweis von exogenem Testosteron. In: Heck H, Hollmann W, Liesen W, Rost R (eds) Sport: Leistung und Gesundheit, Deutscher Ärzte Verlag, Köln, pp 293–298

Donike M, Zimmermann J, Bärwald K.-R, Schänzer W, Christ V, Klostermann K, Opfermann (1984) Routinebestimmung von Anabolika im Harn. Deutsch Z Sportmed 1:14–23

Donike M, Ueki M, Kuroda Y, Geyer H, Nolteernsting E, Rauth S, Schänzer W, Schindler U, Völker E, Fujisaki M (1995) Detection of dihydrotestosterone (DHT) doping: alteration in the steroid profile and reference ranges for DHT and its 5α-metabolites. J Sports Med Phys Fitness 35:235–250

Duchaine D (1989) Underground steroid handbook II. Technical Books, Venice, USA

Franke WW, Berendonk B (1997) Hormonal doping and androgenization of athletes: a secret program of the German Democratic Republic government. Clin Chem 43:1262–1279

Grundig P, Bachmann M (1995) World anabolic review 1996. Sport Verlag Ingenohl, Heilbronn

Handelsman DJ, Gupta L (1997) Prevalence and risk factors for anabolic-androgenic steroid abuse in Australian high school students. Int J Androl 20:159–164

Haupt HA, Rovere GD (1984) Anabolic steroids: A review of the literature. Am J Sports Med 12:469–484

Horning S, Geyer H, Machnik M, Schänzer W, Hilkert A, Oeßelmann J (1997) Detection of exogenous testosterone by $^{13}C/^{12}C$ analysis. In: Schänzer W, Geyer H, Gotzmann A, Mareck-Engelke U (eds) Proceedings of the 14th Cologne Workshop on Dope Analysis 1996, Sport und Buch Strauß, Köln, pp 275–283

Kamber M (1995) Mitteilung in Doping – Information und Prävention. Maggelingen 7:4–7

Kicman AT, Coutts SB, Walker CJ, Cowan DA (1995) Proposed confirmatory procedure for detecting 5α-dihydrotestosterone doping in male athletes. Clin Chem 41:1617–1627.

Knuth UA, Maniera H, Nieschlag E (1989) Anabolic steroids and semen parameters in bodybuilders. Fertil Steril 52:1041–1047

Kochakian CD (ed) (1976) Anabolic-androgenic steroids. Springer Verlag, Berlin Heidelberg New York

Kochakian CD (1990) A steroid review: Metabolite of testosterone; significance in the vital economy. Steroids 55:92–97

Massé R, Ayotte C, Dugal R (1989) Integrated methodological approach to the gas chromatographic mass spectrometric analysis of anabolic steroid metabolites in urine. J Chromatogr 489:23–50.

Schänzer W (1996) Review – Metabolism of anabolic androgenic steroids. Clin Chem 42:1001–1020

Schänzer W, Donike M (1993) Metabolism of anabolic steroids in man: Synthesis and use of reference substances for identification of anabolic steroid metabolites. Anal Chim Acta 275:23–48

Schänzer W, Delahaut P, Geyer H, Machnik M, Horning S (1996) Longterm detection and identification of metandienone and stanozolol abuse in athletes by gas chromatography/high resolution mass spectrometry (GC/HRMS). J Chromatogr B 687:93–108

Shackleton CH, Phillips A, Chang T, Li Y (1997) Confirming testosterone administration by isotope ratio mass spectrometric analysis of urinary androstanediols. Steroids 62:379–387

Taylor WN (1982) Steroids and the athlete. McFarland & Company, London

Wade N (1972) Anabolic steroids: Doctors denounce them, but athletes aren't listening. Science 176:1399–1403

Ward RJ, Shackleton CHL, Lawson AM (1975) Gas chromatography – mass spectrometric methods for the detection and identification of anabolic steroid drugs. Br J Sports Med 9:93–97

Wilson JD (1988) Androgen abuse by athletes. Endocr Rev 9:191–199

Wilson JD (1996) Androgens. In: Hardmann JG, Limbird LE, Molinoff PB, Ruddon RW, Goodman Gilman A (eds) Goodman and Gilman's The Pharmacological Basis of Therapeutics, 9th edition, McGraw Hill, New York, pp 1441–1457

Yesalis CE, Kennedy NJ, Kopstein AN, Bahrke MS (1993) Anabolic-androgenic steroid use in the United States. Jama 270:1217–21

Zimmermann J (1986) Untersuchungen zum Nachweis von exogenen Gaben von Testosteron. Thesis, German Sports University, Cologne, Hartung-Gorre Verlag, Konstanz

E. Nieschlag, H. Behre (Hrsg.)

Andrologie

**Grundlagen und Klinik der
reproduktiven Gesundheit des Mannes**

1996. XXVIII, 465 Seiten, 158 Abbildungen
Gebunden DM 198,-
ISBN 3-540-60886-9

Die zentralen Themen der Andrologie sind
Infertilität, Hypogonadismus, Kontrazeption und
erektile Dysfunktion. Das Buch vermittelt einen
vollständigen Überblick über die klinische Praxis
dieses Gebiets und die relevanten naturwissen-
schaftlichen Grundlagen (einschließlich
Molekularbiologie und -genetik). Der Kliniker
wird umfassend informiert über die Aspekte
Paarbehandlung, andrologierelevante
Gynäkologie und moderne Verfahren der
assistierten Fertilisation, über psychologische
Faktoren der Infertilität, Endokrinologie und
reproduktive Funktionen des alternden Mannes
sowie über ein wichiges Teilgebiet der
Andrologie, den männlichen Beitrag zur
Kontrazeption. Juristische und ethische Fragen
kommen ebenfalls zur Sprache.

**Please order from
Springer-Verlag Berlin
Fax: + 49 / 30 / 8 27 87-301
e-mail: orders@springer.de
or through your bookseller**

Prices subject to change without notice.
In EU countries the local VAT is effective.
Errors and omissions excepted.

E. Nieschlag, H. Behre (Eds.)

Andrology

Male Reproductive Health and Dysfunction

1997. XXVIII, 437 pages, 158 figures
Hardcover DM 148,-
ISBN 3-540-61616-0

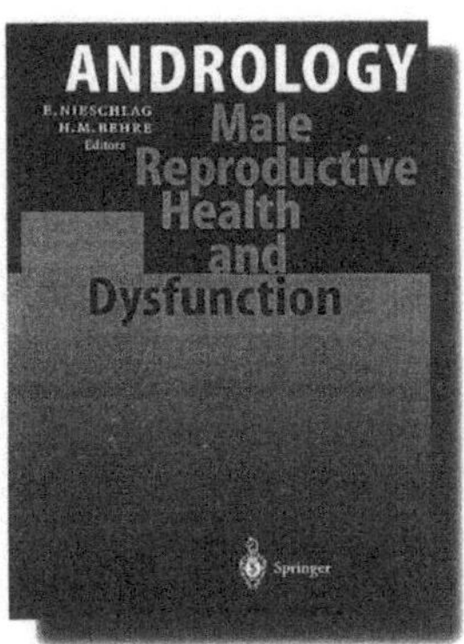

The book contains the scientific basis of andrology including molecular biology and molecular genetics and the clinical practice of andrology by dealing with the diagnosis and treatment of individual disorders classified according to the localisation of the cause. Management of the infertile couple and gynaecology relevant to andrology are described. Modern techniques of assisted fertilisation including IVF and ICSI are part of the book. Psychological factors relevant to infertility are discussed. One chapter is devoted to reproductive endocrinology of aging men. Male contraception e.g. vasectomy and approaches to hormonal male contraception are included. Ecological and toxic effects on the male reproductive system are summarized in one chapter. Legal and ethical aspects of andrology conclude the book.

Springer